Bildungswissenschaften und akademisches Selbstverständnis
in einer globalisierten Welt

Education and Academic Self-Concept
in the Globalized World

BALTISCHE STUDIEN ZUR ERZIEHUNGS- UND SOZIALWISSENSCHAFT

Herausgegeben von Gerd-Bodo Carlsburg,
Algirdas Gaižutis und Airi Liimets

BAND 28

Zu Qualitätssicherung und Peer Review der vorliegenden Publikation

Die Qualität der in dieser Reihe erscheinenden Arbeiten wird vor der Publikation durch Herausgeber der Reihe geprüft.

Notes on the quality assurance and peer review of this publication

Prior to publication, the quality of the work published in this series is reviewed by editors of the series.

Gerd-Bodo von Carlsburg
Thomas Vogel
(Hrsg./eds.)

Bildungswissenschaften und akademisches Selbstverständnis in einer globalisierten Welt

Education and Academic Self-Concept in the Globalized World

PETER LANG
EDITION

Bibliografische Information der Deutschen Nationalbibliothek
Die Deutsche Nationalbibliothek verzeichnet diese Publikation in der Deutschen Nationalbibliografie; detaillierte bibliografische Daten sind im Internet über http://dnb.d-nb.de abrufbar.

Library of Congress Cataloging-in-Publication Data
Bildungswissenschaften und akademisches Selbstverständnis in einer globalisierten Welt / Gerd-Bodo von Carlsburg, Thomas Vogel (Hrsg.) = Education and academic self-concept in the globalized world / Gerd-Bodo von Carlsburg, Thomas Vogel (eds.).
pages cm
Contributions in English and German.
ISBN 978-3-631-65610-5
1. Education and globalization–Cross-cultural studies. 2. International education–Cross-cultural studies. 3. Teachers–Professional relationships–Cross-cultural studies. 4. Group identity–Cross-cultural studies. I. Reinert, Gerd-Bodo, editor of compilation. II. Vogel, Thomas, 1959 May 27- editor of compilation. III. Title: Education and academic self-concept in the globalized world.
LC191.B555 2014
370.116–dc23
2014027551

Die Herausgeber danken dem Deutschen Akademischen Austauschdienst im Programm „Ostpartnerschaften“ und dem Programm LLP ERASMUS sowie der Pädagogischen Hochschule Heidelberg und der Litauischen Universität für Bildungswissenschaften in Vilnius für die finanzielle Förderung.

Gedruckt auf alterungsbeständigem, säurefreiem Papier.

ISSN 1434-8748
ISBN 978-3-631-65610-5 (Print)
E-ISBN 978-3-653-04836-0 (E-Book)
DOI 10.3726/978-3-653-04836-0

Peter Lang Edition ist ein Imprint der Peter Lang GmbH.

Peter Lang – Frankfurt am Main · Bern · Bruxelles · New York · Oxford · Warszawa · Wien

Diese Publikation wurde begutachtet.

www.peterlang.com

Giedre Riškutė, Assoc. Prof. Dr.
Geb. 1952, Litauische Universität für Bildungswissenschaften, Fakultät für Bildungswissenschaften, Lehrstuhl der Künste, Studium der Malerei und der Szenografie am Staatlichen Kunstinstitut, Mitglied der Litauischen Theater-Union und der Litauischen Künstlervereinigung.

Offizielle Webseiten:

www.riskute.lt / http://www.egalerija.net/main.php?id=51&lg=en&nr=1&artist=19

Born in 1952, Lithuanian University of Educational Sciences, Faculty of Education, Art Department, studies of painting and scenography at the State Institute of Art, Member of Lithuanian Theatre-Union and Lithuanian Artists' Association.

Prepared by www. riskute.lt and www.LEU-info.lt.

Assoziationen über diese Künstlerin
Der Name der Malerin Giedrė Riškutė steht in Litauen und in anderen europäischen Ländern, vor allem im Baltikum und seinen Nachbarstaaten, als Begriff für Innovation und künstlerische Kreativität. Diese Fähig- und Fertigkeiten versucht sie auch an ihre Studierenden der Litauischen Universität für Bildungswissenschaften weiterzuvermitteln, wo sie als Leiterin des Lehrstuhls für Kunst einen Ruf innehat.

Ihre künstlerischen Arbeiten haben eine Reihe von Museen im In- und Ausland erworben. Sie zählt als poetische Malerin lyrischen Weltbetrachtens, auf deren Schaffen die französischen sowie deutschen Impressionisten als auch die litauischen Maler einen bedeutenden Einfluss ausgeübt haben. Ihre Bilder atmen und strahlen den Rhythmus unwiederholbarer Farbnuancen, wundersame Augenblicke der Ruhe und Stille. Sie sind ein in die Realität geholtes Beispiel für Anspielungen auf die rätselhafte Schönheit unserer Natur, die Stille der Dinge und gedacht als Innehalten gegenüber unserer ständigen Unruhe.

Gerd-Bodo von Carlsburg, Algirdas Gaižutis

Inhalt

Welcoming Speech

Bernward Lange

A long period of cooperation, collegiality and student exchange unites the Lithuanian University of Educational Sciences (LEU) and the Heidelberg University of Education (PH HD). Our universities have achieved a remarkable level of steady international partnership to promote student exchange, academic cooperation, joint academic discussion, research and publication in order to mutually benefit from each other.

To look back on this engagement is very rewarding, but as our universities renewed their agreement of cooperation in 2013 our sight and will is as well adjusted to the future. From our side the academic partnership between LEU in Vilnius is the most intensive and constantly kept alive. Beside personal and institutional engagement it is owed to the facilitated conditions within bilateral state treaties, the German Academic Exchange Service DAAD, the Erasmus programme and European funding for academic exchange. But we can't neglect to face a decline of outgoing students in general. We have to concern ourselves with this issue.

It is an honour and a pleasure for me to give the foreword to the proceedings of the conference "Educational Science in the Search of Global Identity" 2013 in Vilnius. The topic was search for identity as educational researchers and teacher trainers inside a framework of globalization. I take the liberty to give some personal summary from presentations and personal communication on the conference.

Lithuania is undergoing a transition of a tremendous scale which is partly seen as an identity crisis. The membership of Lithuania in the EU is not only seen as a chance to preserve national identity but as an adaption to externally imposed standards, such as Bologna requirements. To some colleagues this transition process is even perceived as a threatening phenomenon, for example the individual being assigned to a degraded role of a consumer. We Germans tend to be reluctant when it comes to the question of national identity – for historical reasons. From different contributions we learned that national identity is valued very high in pedagogical thinking in Lithuania. This could be an impetus for us

Germans to re-reflect on this issue for ourselves in a more and more transcultural and globalized context.

Finally I want to draw the attention to an issue which to me personally is very remarkable, outlining the importance and the role of Lithuania within the European Union. At the very moment of the conference in the second half of the year 2013 the Republic of Lithuania presided the Consilium, i.e. the Council of the European Union and took major responsability within the EU. Europe looked and listened towards Lithuania. In May 2013 the President of the Republic of Lithuania, Dalia Grybauskaitė was awarded the International Charlemagne Price of the City of Aachen. Aachen was the major palace of emperor Carolus Magnus and the temporary capital of the Franconian Empire after the decline of the West Roman Empire. The high-ranking Charlemagne Price distinction is only reserved to laureates who contribute in an outstanding way to the integration of the European nations.

For us Germans your President's speech is worth reading. Perhaps we Germans are tending to take Europe as a given thing and a matter of course. President Dalia Grybauskaitė rather gave a striking and awful insight that the Lithuanian people had to fight for their right of independence, freedom and a future of their own. And that the Lithuanian people had to go through tough periods of austerity and financial discipline. I take the liberty to quote Dalia Grybauskaitė from her very remarkable speech at this occasion.

> *"Dear Europeans,*
>
> *the world has already recognized the role of Europe in advancing peace, reconciliation, democracy, and human rights. We must take the responsibility to transform these values that we all share into a vision of tomorrow and a daily reality for our neighbours and partners.* ***It is responsibility by example*** *which is expected from us. We Europeans are a role model of learning, listening in, negotiating, and finding solutions together."*

What we can contribute by our cooperation and by the 2013 conference is only kind of a small but visible part of "*responsibility by example*". Our cooperation is just a brick to the further erection of the European house. But every brick counts and contributes to the stability of the whole construction. So let us take it very seriously.

Heidelberg in May 2014
Bernward Lange
Vice-Rector for Research and International Relations
Heidelberg University of Education

Vorwort

Gerd-Bodo von Carlsburg

Die im Oktober 2013 an der Litauischen Universität für Bildungswissenschaften (Lithuanian University of Educational Sciences/Lietuvos edukologijos universitetas) in Vilnius stattgefundene zweitägige XIX. Internationale wissenschaftliche Konferenz (19th International Scientific Conference/XIX-oji Tarptautinė mokslinė konferencija) widmete sich der Themenstellung ***Bildungswissenschaft auf der Suche nach globaler Identität (Educational Sciences in Search of Global Identity)***, daher auch die Anlehnung des Titels dieses Bandes 28 der Reihe „Baltische Studien zur Erziehungs- und Sozialwissenschaft" an die Konferenzthematik, da die meisten Beiträge dieser Tagung gewidmet sind. Veranstaltet wurde selbige insbesondere vom Bildungswissenschaftlichen Institut (Department of Educational Sciences) der Bildungswissenschaftlichen Fakultät (Faculty of Educational Sciences/Ugdymo mokslµ fakultetas). Gemeinsamer Partner dieser Konferenz war 2013 die Pädagogische Hochschule Heidelberg – Heidelberg University of Education).

Alle jährlich stattfindenden Tagungen stehen unter der Prämisse *Bildungsreform und Lehrerbildung* (The Reform of Education and Teacher's Training/ Švietimo reforma ir mokytojų rengimas), ein Diskurs, der vom Ministerium für Bildung und Wissenschaft der Republik Litauen, der Litauischen Universität für Bildungswissenschaften (LEU) und der Pädagogischen Hochschule Heidelberg (Heidelberg University of Education) gemeinsam seit 2005 konzipiert wird. Fokus ist die stringentere Einbindung der pädagogischen Praxis (Trias: Handlungstheorie, -forschung, und -praxis), d.h. Bildsamkeit und Erziehbarkeit werden reflektiert unter der Prämisse einer *,reflexiv-modernen' Bildungswissenschaft*, will sagen: wie gestaltet sich eine inklusive, d.h. handlungsorientierte kooperative Schulbildung, im Hinblick auf die Herausforderungen des kommenden Jahrzehnts. Zu dieser Thematik diskutierten in Vilnius Wissenschaftler, Mitglieder (inter-)nationaler Bildungsorganisationen/-institutionen und praktizierende Pädagogen/ Philologen. Behandelt wurden aktuell anstehende Fragen, wie beispielsweise die vorangehende (Nach-)Reformierung des Studiums und die Qualitätsentwicklung (Festsetzung von *Bildungsstandards* im Hinblick auf eine Neuorientierung sowie Umstrukturierung hinsichtlich der Anforderungen internationaler Qualifikationen in Universität und Schule.

Die *Entwicklungsstrategie* der Litauischen Universität für Bildungswissenschaften bis zum Jahre 2014 sieht als *,Vision'* eine Vertiefung der wissenschaftlichen

Kompetenzen, die den internationalen Bildungs- und Qualitätsstandards entsprechen, vor, geplant ist die Gründung eines (Landes-)Instituts zur Weiterentwicklung und Normierung von Bildungsstandards nach dem Vorbild des IQB (Institut zur Qualitätsentwicklung im Bildungswesen an der Humboldt-Univer-sität zu Berlin). Dieses Qualitätsmanagement setzt somit den Erwerb interaktiver Kompetenzen im Bildungs-, Erziehungs-, Enkulturations- und Sozialisationsbereich voraus, die eine Befähigung zum *Identitätserwerb* ermöglichen sollen.

Dies impliziert primär einen *Perspektivenwechsel* im Bildungs- und Kulturbereich und eine Hinführung zur handlungsorientierten *‚reflexiv-modernen' Bildungswissenschaft*, die Katalysatorfunktion für den sozialen Kompetenzerwerb zeitigt und somit den Internalisierungsprozess zum Erwerb von Identität in Gang setzen kann.

Qualität von Bildung und Erwerb von nationaler sowie globaler Identität hängt primär von der Professionalisierung der Pädagogen/innen ab, von ihrer Befähigung, nicht nur fachliche Kompetenzen zu vermitteln, sondern auch kulturelle Werte und Normen, die der damit implizierten Persönlichkeitsbildung und somit evozierten Identitätsfindung dienen, der Soziabilisierung des Bürgers/der Bürgerin in einem modernen demokratischen Staatsgefüge.

Abb. 1: Saulius Ikamas: Vilnenser Türme (1992). Lithographie, Archiv: von Carlsburg

Folgende ***Wokshops*** (working groups) verdienen unter dem breiten Angebot von Vorträgen, Arbeitsgruppen... Erwähnung im Hinblick auf die Thematik dieses Bandes:

1. *Development of global identity.*
2. *National identity in the context of globalization: threats and perspectives.*
3. *Problems and opportunities of national identity promotion in the process of education.*

4. *Prerequisites of national identity education (spread, guidelines, contours, trends) in the process of pedagogue training.*
5. *New challenges of identity to Lithuanian education.*

Abb. 2: Ansicht von Vilnius (Wilna). Stahlstich (19. Jahrhundert). Archiv: von Carlsburg

Der Erziehungserfolg hängt zum großen Teil vom Engagement und Berufsethos der Pädagogen/-innen als auch von den Erfahrungen im Umgang mit Kindern, Jugendlichen und der Gesellschaft ab, wie TIMSS (Third International Mathematics and Science Study), PISA (Programme for International Student Assessment) und PIRLS/IGLU (Progress in International Reading Literacy Study/ Internatio-nale Grundschul-Lese-Untersuchung) eindeutig bestätigen. Auch Litauen hat die Ergebnisse verbessert.

Die ***Konferenzproblematik*** lässt sich somit auch auf folgende Schwerpunkte erweitern:

1. Faktoren, die die Bildungsqualität beeinflussen;
2. Probleme der Soziabilisierung;
3. Qualifizierung und Kompetenzen der Hochschullehrer;
4. Studieninhalte und -qualität;
5. Erziehungstheorie und -praxis im Hinblick auf Inklusion und *Identität*;
6. Zusammenwirken für eine zukünftige Bildungsplanung, die *globale Identität* als Bildungsziel projektiert.

Ein letzter, aber für Litauen ***prägender Exkurs*** dieser Tagung, der auch in den litauischen Beiträgen dieses Bandes diskursiven Widerhall findet, so insbesondere in dem Beitrag von A. Gaižutis, galt der litauischen Bildungs- und Literaturwissenschaftlerin, der Philologin sowie Philosophin Meilutė Julija Lukšienė-Matjošaitytė (geb. 1913 in Wien, gest. 2009 in Vilnius), die international eine hohe Reputation genoss: Dozentin, spätere Direktorin für Litauische Literatur an Universität Vilnius, letztlich Professorin an der Vytautas-Magnus-Universität in Kaunas. Einer ihrer Forschungsschwerpunkte beschäftigte sich mit dem Nachspüren und der Dissemination des Comenianischen Gedankenguts. Jan Amos Comenius (1592-1670), der theologische Pädagoge, Architekt eines neuzeitlichen christlich orientierten Schulsystems, Philosoph, Humanist, Enzyklopädist, Pansophist, der Visionär mit einer tiefen Sehnsucht nach Frieden, lebte in einer historischen Zeit gravierender gesellschaftlicher Umbrüche. Die traumatisierenden Erfahrungen des Dreißigjährigen Krieges schärften seine Imaginationskraft, die auch heute vielfältige Anregungen intendiert: *ein* Europa, *Diversivität* und *Heterogenität*, Frieden schaffen ohne Waffen. Sein nach Pantheismus suchender Geist – einerseits Gott und andererseits die Ratio als Orientierung, das ständige Abwägen, eins mit dem Makrokosmus, der Natur – bildet die Urkraft für menschliche Liebe und Erziehung zu derselben (Bindungspädagogik). J.A. Comenius firmiert für eine Neuordnung der Welt durch *Aufklärung.*

Abb. 3: Hauptgebäude (Rektorat) der Litauischen Universität für Bildungswissenschaften – Lietuvos edukologijos universitetas (LEU)

Auch Meilė Lukšienė sah das Comenianische Denken, Comenius' Auseinandersetzung mit René Descartes (1596-1650) – z.B. Les passions de l'âme, 1649 –, als Basis für Bildung und Lebenssinn: schon den Kindern von klein auf die Möglichkeit zu geben, sich selbst die Welt als Gottes Schöpfung und Ertrag seiner Weisheit, als Mikrokosmos im Makrokosmos, essentiell zu erschließen. Dafür erhielt sie 2004 die Jan-Amos-Comenius-Medaille der UNESCO. Die Reformierung der Lehrerbildung war ein weiteres Anliegen dieser engagierten Humanistin. Reform durch Evolution ist nicht nur in mittel- und osteuropäischen Ländern, sondern auch in Deutschland ein aktuelles Thema, hält man im Bundesland Baden-Württemberg mit seiner Grün-Roten-Landesregierung an der einzig in Europa bestehenden und nicht mehr zeitgemäßen, rückwärtsgewandten zweiteiligen Lehrerausbildung fest. Hinzu kommt eine zweiphasige Lehrerausbildung, zusammen mit Österreich eine europäische Spezies, die auch durch die angedachte School of Education keine Fortschritte bringen wird.

Abb. 4: Matthäus Merian d.Ä. – Das ChurFürstliche Pfältzische Schloß und Gartten, Zu Haydelberg. (V)on Mitternacht an zusehen (Theatrum Europaeum ca. 1643/45). Archiv: v. Carlsburg

In der Konferenz wurden u.a. Forschungsergebnisse vorgestellt, die verschiedene Lehr-/Lernetappen sowie soziologische als auch sozial-psychologische Kompetenzen tangieren. Die Auswahl der Beiträge in diesem Band kann nur einen Überblick vermitteln und motivieren, an der jährlich stattfindenden *Internationalen Konferenz* selbst aktiv teilzunehmen, um im Denken/Handeln aus *globaler*

Sichtweise den politischen und erziehungswissenschaftlich-kulturellen Paradigmenwechsel zu erörtern.

Abb. 5: Heidelberg um 1620, Kupferstich von Matthäus Merian d.Ä. Archiv: v. Carlsburg

Seit Mai 1994 besteht ein Austausch in der Lehrerbildung zwischen Heidelberg und Vilnius, wobei am 6. Oktober 1995 in Vilnius ein Kooperationsvertrag mit der Pädagogischen Universität Vilnius geschlossen wurde. Der Vertrag regelt den Austausch von Dozenten/innen und Studenten/innen und beinhaltet die gegenseitige Anerkennung von Studienleistungen bis hin zum Promotionsstudium. Durch den interkulturellen Transfer auf der Mikroebene sind Menschen in privater und dienstlicher Sphäre zusammengeführt worden; sehr wertvolle Freundschaften haben sich gebildet, die weiter vertieft wurden bzw. werden sollen. Somit entwickelte sich die Kooperation mit Heidelberg für litauische Lehrende und die Studierende zu einem sehr begehrten Studienort. Umgekehrt waren insbesondere die Praktika an den litauischen Schulen bei den Heidelberger Studierenden sehr begehrt; hier war die Nachfrage größer als das litauische Angebot.

Die XVII. Internationale Tagung wurde vorbereitet durch folgende Komitees:
Präsidium/Conference chairs and Deputy Chairs

Algirdas Gaižutis, Prof. Dr. Dr. h.c. mult.
Rektor der Litauischen Universität für Bildungswissenschaften
Mitglied der Litauischen Akademie der Wissenschaften

Gerd-Bodo von Carlsburg, Prof. Dr. Dr. h.c. mult.
Heidelberg University of Education, Institut für Erziehungswissenschaft
Senatsbeauftragter für Litauen (bis 31.12.2013)

Abb. 6: Vilnius, Mitte Hintergrund: Aušros vartai (1514), Tor der Morgenröte, weltbekanntes katholisches Heiligtum mit dem wundertätigen Bild der Barmherzigen Muttergottes (um 1600-1625) – die Madonna von Vilnius – in der Torkapelle, errichtet als Stadttor zusammen mit der Stadtmauer (1503-1522), Vordergrund links: Kirche St. Theresa (1633-1650) und das Kloster der Unbeschuhten Karmeliten (1621-1627), die Bierbrauer und Kerzengießer waren. Archiv: v. Carlsburg

Abb. 7: Heidelberg University of Education (Altbau). Archiv: von Carlsburg

Wissenschaftliches Komitee/Scientific Committee

Ona Monkevičienė, Prof. Dr.
Litauische Universität für Bildungswissenschaften
Dekanin der Fakultät für Bildungswissenschaften
Institut für Erziehungswissenschaft
Lehrstuhl für Frühe Kindheit

Thomas Vogel, Prof. Dr.
Heidelberg University of Education
Institut für Erziehungswissenschaft, Abteilung Schul-/Berufspädagogik
Senatsbeauftragter für Litauen (seit 01.01.2014)

Anmerkung: Alle Zitate in diesem Band wurden einheitlich kursiv gesetzt.

Literatur

Descartes, R. ([1641]/1992): Meditationes de Prima Philosophia. Meditationen über die Grundlagen der Philosophie. Philosophische Bibliothek, Band 250a. Lateinisch-Deutsch. Hamburg: Felix Meiner.

Descartes ([1649]/1989): The Passions of the Soul, translated by Voss, St.H. Indianapolis, Indiana (U.S.A.): Hackett Publishing Company.

Descartes, R. ([1649]/[2]1996]): Die Leidenschaften der Seele. Les passions de l'âme. Französisch-Deutsch. Vorwort, hrsg. u. übers. v. Hammacher, K. Hamburg: Felix Meiner.

Lukšienė, M. (1993a): Pedagogika ir kultūra. Lietuvos švietimo reformos gairės. Vilnius: valstybinis leidybos centras, p p. 25-45.

Lukšienė, M. (1993b): Atsinaujinimo (kaitos) nuostata ir poreikis. Lietuvos švietimo reformos gairės. Vilnius: valstybinis leidybos centras, pp. 102-112.

Lukšienė, M. (2000): Jungtys. Vilnius: Alma littera.

Heidelberg, im Juli 2014
Gerd-Bodo von Carlsburg

Educational Science in Search of a Global Identity: an Introduction

Thomas Vogel

Abstract
Educational science is a collective term covering a variety of scientific disciplines that deal with the matter of education in the broadest sense. The focus of the work carried out within these sciences is on analysing educational processes and systems, as well as on the importance of the various factors that influence these systems. Over the recent decades, the growing global networking of human activities – which we refer to as globalization – has become a very important factor influencing educational processes. As is so often the case throughout history, pedagogy and educational practice do not precede social developments in this process, rather they follow them. Their functions are often dictated by circumstances: they are intended to promote processes which are already taking place, or to limit any adverse effects. Although it is true that, in Germany at least, educational activities which are concerned with globalisation issues have been around for some time now, the search for a global identity for educational science is by no means complete and may even remain unfinished forever. The articles in this band discuss this endeavour and push forward with educational science's search for an identity. In order to set the scene for the task at hand, the author would like to clarify certain concepts, provide some background information and give a brief overview of the where the scientific debate currently stands.

Causes, dimensions and consequences of globalization

The term 'globalization' refers to the increasing economic and cultural connections between nations. Globalization is a complex process of international relations which manifests itself on different levels. The vast majority of the time, we view the process from an economic perspective. However, at the same time, globalization also has a significant impact on culture, national politics, ecology, the world of work, and last but not least, on educational processes. Given that the individual dimensions at play here are interdependent and mutually influential, it is difficult to develop a systematic description of the process of globalization that is both clearly and unambiguously oriented towards causes and consequences. The buzzword "global network" describes this difficulty. Sociologists and philosophers describe globalization as the way in which "everyday actions in the various dimensions of the economy, information, ecology, technology, transcultural conflicts and civil society are becoming boundless" (Beck 1997, 44), and as "the increasing scale and intensification of transport links, communication relationships and exchanges across national boundaries" (Habermas 1998, 101), as well as "the extension of the hold of a small number of dominant nations over the whole set of national financial markets" against the backdrop of "neoliberal market fundamentalism" (Bourdieu 1998, 47; Altvater/Mahnkopf 1996, 76), and "a growing separation of power and politics" (Baumann 2000, 174). Bourdieu refers to "globalization" as a "myth in the strong sense of the word" and an "an idée force, an idea which has social forces, which obtains belief." In view of the way in which a real belief in the constraints of globalization is steadily, subtly filtering through, Bourdieu believes that science is faced with an important task. First and foremost, it could follow up the creation and circulation of this discourse (Bourdieu 1998, 39 et seq.). In my view, Bourdieu has formulated a set of terms of reference here which could provide us with impetus over the course of this conference.

The causes of globalization are just as varied as its various incarnations and consequences. Historically speaking, the globalization processes are nothing new. International trade and labor cooperation between nations have existed for a long time. The situation is in fact quite the opposite, in that closed economies which have no trade relations with other nations are the exception in history. There is even evidence of a lively exchange of goods between different cultural areas in ancient times: for example, Chinese silk was worn in ancient Rome. Gold, silver, precious stones and spices were also traded alongside oriental fabrics during this period. Many have drawn a connection between the start of the current globalization process with Columbus' departure in search of a new sea route to India. This act symbolises Europe's thrust outward into the world and the linking of the world in accordance with the image of European culture. Over the coming centuries, hosts of missionaries, soldiers, settlers and merchants followed in the footsteps of Columbus and set about converting, conquering, civilizing or pleasing foreign nations. "The sailing ships eventually gave way to freighters and aircraft were used as vehicles of interconnection, and these in turn were compounded by telephones, TV programmes and electronic infrastructure. Now an arc stretches all the way from Columbus to CNN: the expansion of Euro-Atlantic civilization across all five continents. The Great Departure from Palos was the distant inauguration of a trans-epochal process which, by the end of the twentieth century, was being called 'globalization'." (Wuppertal Institute 2006, 13, translated into English in Wolfgang Sachs, Tilman Santarius (eds.), Fair Future: Resource Conflicts, Security & Global Justice, 1).

One of the important reasons why the process of globalization is of such great interest today is the particular scope and the large impact of development on almost all areas of life. The main causes of the major surge in globalization in terms of global history are as follows:

- the reduction of in the level of international conflict, particularly as a result of resolving the East-West conflict,
- a technological revolution caused by the development of information and communication technologies, in particular the internet,
- major liberalization in the general conditions for international trade,
- an increase in global ecological problems - in particular the greenhouse effect – which can only be solved at international level.

The rapid development of information and communication technologies and the liberalization of trade conditions in particular paved the way for an increased level of development in global business and production networks, as well as the accelerated internationalization of the financial markets. Liberalizing the trade conditions enabled particularly the financially powerful multi-national companies to reduce their production costs by setting up business in countries with lower labour costs, poorly developed social security systems and low environmental standards. In certain countries, the wages paid for simple jobs which require only a low level of qualification are so low that, despite the transport costs, the products manufactured there are cheaper than they would be had they been produced in developed countries.

The processes involved in globalization go hand in hand with risks and opportunities and simultaneously give rise to discrepancies and conflicts that must be resolved. Specifically, "separating the economic from the political sphere" (Wulf / Merkel 2002, 13), assuming independence for processes which are largely beyond the control of national politics. Generally speaking, a "regression of the State" (Bourdieu 1998, 42) has been generally evident throughout the course of globalization. It has led to numerous adverse developments in society, culture and the environment and has also been detected within education. People are forced into the developmental processes of globalization, with their many consequences. A "widespread, far-reaching view that freeing up the markets leads to emancipation and increased levels of autonomy for the people is incorrect" (ibid.). Quite on the contrary: the number of opportunities for citizens to be included in political decisions actually decreases throughout the course of globalization.

One picture can illustrate the many dimensions, problems and conflicts involved in globalization. This photo depicts a boy on a toxic waste dump near the capital of Ghana, Accra. Surrounded by highly toxic vapours and electronic waste from industrially developed countries, the young man holds up the remains of a screen. In order to get to the metal, he throws the screen onto the ground several times. On the one hand, the electrical scrap brings the hope of earning some money; however it also represents a sometimes deadly risk to public health and reaps devastating consequences for the environment. Children and young people need money to survive, but at the same time, their future is being destroyed. The picture illustrates the economic, social, cultural and environmental discrepancies between the different parts of the world:

- in order to save money and protect their economies, developed countries dispose of their waste as inexpensively as possible, in the so-called "underdeveloped" countries.[1]
- The social consequences of this are that people in the poorest countries of the world have become reliant on starvation wages for their survival and lose their health and future at an early age in the struggle for mere survival.
- Children and young people are denied the fundamental right to education and cultural participation, along with the chance of a better future.
- The environmental consequences of burning toxic materials and contaminating soil ultimately affect the world in its entirety.

The picture indicates the responsibility that follows from industrial production and consumption patterns in industrialized countries. It illustrates that fact that,

1 The United Nations estimates that around 100,000 tons of electronic waste is shipped to Africa each year from Germany alone.

in the context of economic globalization under the terms of neoliberal guidelines, poverty issues and discrepancies between income and opportunities have intensified throughout the world. Likewise, the globalization of the 'risks to civilization' has long since become a threat to the life of all human beings: cultural globalization is associated with conflicts which, for the time being, have not been successfully mitigated by applying global ethics which are tied to a collective commitment (cf. Steffens / White 2004, 16).

Given the complexity of the structures, disruptions and conflicts involved with globalization, however, there is no unified picture against which educational science might be able to develop an identity or find its direction. The dominance of the processes of globalization across all areas of life means that we need to address the question of how a social science should react to these open-ended and indeed largely contradictory processes. The role of science is to get to the bottom of complex relationships and, in the particular case of educational science, to open them up for practical use. We thus come closer to the question of what might be meant by educational science's search for a global identity.

Trends in developing a global identity for education

The term "identity" generally refers to the instance of complete conformity between two objects or situations, or to a phenomenon which remains completely unchanged despite the passing of time and the changing circumstances of the situation". (Brunner / Zeltner, 1980, 100). If we apply this definition to the problem at hand, the question of the identity of education in the process of globalization would firstly be; to what extent do globalization processes and the actions of educational science coincide, and do they pursue a common aim, for example, cosmopolitanism? The second part of the definition of "identity" given above does, however, allow us to ask: what might be a constant function of educational science that it retains despite the changes brought about by globalization and regardless of time and space? By viewing the matter from this perspective, the search for an identity for educational science would be guided more towards the question of a general essence of incarnation and the destiny of humankind, which remain valid irrespective of any cultural and social processes. Here, we need to ask what we mean when we speak of "general education" (in German: "Allgemeine Bildung") for all people regardless of their background.

The concept of identity is often used within the context of a person's psychological development. In psychology, individual identity is described as the "uniqueness of an individual, which is retained despite experiencing different situations throughout the course of life" and can also "be described as the sum of features and combinations of features that will always make an individual different from others (Brunner / Zeltner, 1980, 160 et seq.). If someone is searching for their identity, they are trying to find a genuine position in their social context. An identity develops and is asserted" against the pressure to adapt to social norms and whilst being overwhelmed by needs which press for gratification (Krappmann 1995, 434). If educational science (as a whole) is to set about the search for a global identity, the initial step must be to develop an "internal model" of itself, and not to adapt to social or possibly economic standards arising as a result of the globalization process. Although the process of creating an identity is always driven by the hope of being able bring it to its conclusion, the process remains eternally incomplete, both for the individual and for science. Just like a person, a social science will be continuously confronted with new situations which lack consistency and are partly contradictory, and it will be driven by expectations which are not fully understood and force them to reposition themselves.

Globalization brings with it a variety of exciting, often conflicting demands for the educational sciences; these demands must be clarified on the basis of this science's self-conception and integrated into the process of finding an identity. Tensions which are created by globalization and pose a particular challenge for educational science's search for an identity – and for our conference - were outlined in the UNESCO Report on Education for the 21st Century. It describes a total of seven areas of conflict that need to be addressed by education and the educational sciences: the tension between the 'global' and the 'local', the tension between the 'universal' and the 'individual', the tension between 'tradition' and 'modernity', the tension between long-term and short-term considerations, the tension between the need for competition and concern for equality of opportunity, the tension between the extraordinary expansion of knowledge and human beings' capacity to assimilate this knowledge and the tension between the spiritual and the material (cf. German Commission for UNESCO, 1997, 14 et seq.). The task of developing an identity proves extremely difficult in these areas of conflict. In almost all fields of practice, there is a wide range of challenges emerging for education within the conflict-ridden area of globalization. The phenomena of globalization are already represented in many areas within educational science. Three trends can be observed here:

1. The conditions in which human beings develop is changed and defined by the increasing level of globalization throughout the world of learning subjects, calling for a new discourse in educational theory.
2. There is an increasing level of globalization *in* education, which manifests itself in the global spread of schooling (a developmental obligation to offer

formal education has been introduced in state education systems in all of the countries throughout the world) and the way in which education is being brought in line with universal standards through global educational monitoring (e.g. through World Education Reports, Pisa studies, etc.).
3. Globalization is occurring *within* education, as can be seen in the growing level of deregulation in the provision of state education and its submission to market mechanisms. Another thing that must be taken into account here is the fact the the education market itself is being increasingly viewed as a service sector which has potential for growth (e.g. by offering the option of private schools and universities or by providing services on what is known as the further education market).

In the light of this trend, we must ask how educational science can position itself given the changed circumstances. There would appear to be three possible directions that the search for a global identity for educational science might take:

- The process could be pushed forward by educational science and practice according to neoliberal guidelines in the interests of economic globalization.
- Negative developments could be counteracted by the development and implementation of humane, socially just and ecologically sustainable principles, structures and practices with the support of educational science.
- Educational science and practice could be playthings in what would appear to be an inevitable progression in the liberalization of the global market.

Efforts to prepare people and curricula for global economic competition and to submit to the educational systems of neoliberal market mechanisms are currently the dominating force in national education debates on the issue of globalization. Reforms are enforced and education is adapted, as far as possible, in line with both real and speculative market expectations by appealing to international competition (cf. Steffens/Weiß 2004, 26). Here, it seems that the processes are being controlled by a kind of anonymous power and have been largely stripped of political structure. In this regard, Bourdieu says that the "social sciences are doomed either to abide by the unashamed wishes of companies or state bureaucracies or to perish by the censorship imposed by power and money." (Bourdieu 1998, 47). To explain this phenomenon, Koneffke states that globalization is actually a logical further development in the expansive nature of civil societies and their economies, which educational measures have had to follow since they first began. Globalisation and pedagogy therefore both have a common historical origin. They are however ultimately contradictory: According to Koneffke, the impetus of globalization has a desire for humankind to submit to its law of motion; on the other hand, education is geared towards ensuring that an educated, self-conscious humankind will define its own order (cf. Koneffke 2004, 241).

The question of the role of educational science in the globalization process is implicit in the subject of this conference, which is intended to call for a more detailed definition of the specific tasks with which educational science is faced in this changing situation, as well as to distinguish its self-image. Following from the discussion, however, we must not rule out the possibility of questioning the subject of the conference itself. By this I am referring to the question of whether educational science should develop a global identity at all or perhaps whether it might not also be capable of turning away from the demands imposed by globalization. In any case, it is important that we clarify the extent to which an identity that associates educational science with globalization in the objectives is possible and useful. Perhaps there is a question mark missing from the subject matter of this conference: does educational science actually need to search for a global identity, or can and should it turn away from the processes of globalization to a certain extent? On the other hand, one might also ask to what extent educational science is actually free to search for its own identity. Is it not the case that the conditions created as a result of globalization are so stringent that an independent search for an identity for educational science appears almost impossible?

The most important task for the global community will be the recovery of the political in the globalization process. In attempting to find a global identity for educational science, our aim at this conference is to analyse the difficulties involved in overcoming the tensions presented, to identify ways of achieving a balance between the individual poles and to develop suggestions as to how human beings can learn to contribute towards shaping peaceful, ecological and interdependent development on a global level.

References

Altvater, Elmar; Mahnkopf, Birgit (1996): Grenzen der Globalisierung. Ökonomie, Ökologie und Politik in der Weltgesellschaft. Münster.

Baumann, Zygmund (2000): Die Krise der Politik. Fluch und Chance einer neuen Öffentlichkeit. Hamburg.

Beck, Ulrich (1997): Was ist Globalisierung? Irrtümer des Globalismus - Antworten auf Globalisierung. 3. Aufl. Frankfurt am Main.

Blömeke, Sigrid; Gamm, Hans-Jochen; Gatzemann, Thomas u.a. (Hrsg.) (2004): Jahrbuch für Pädagogik 2004. Globalisierung und Bildung. Frankfurt am Main, Berlin, Bern u.a.

Bourdieu, Pierre (1998) Gegenfeuer. Wortmeldungen im Dienste des Widerstands gegen die neoliberale Invasion. Konstanz.

Brunner, Reinhard; Zeltner, Wolfgang (1980): Lexikon zur Pädagogischen Psychologie und Schulpädagogik. München u. Basel.

Deutsche UNESCO-Kommission (Hrsg.) (1997). Lernfähigkeit: Unser verborgener Reichtum. Unesco-Bericht zur Bildung für das 21. Jahrhundert. Neuwied, Kriftel, Berlin.

Habermas, Jürgen (1998): Jenseits des Nationalstaats? Bemerkungen zu Folgeproblemen der wirtschaftlichen Globalisierung. In Beck, Ulrich: Politik der Globalisierung. Frankfurt am Main. S. 67-84.

Koneffke, Gernot (2004): Globalisierung und Pädagogik - Bemerkungen zu einer alten, vertrackten Beziehung. In: Blömeke, Sigrid u.a. (2004). S. 237-254.

Krappmann, Lothar (1995): Identität. In: Lenzen, Dieter; Mollenhauer, Klaus (Hrsg.): Enzyklopädie Erziehungswissenschaft. Bd. 1 (Theorien und Grundbegriffe der Erziehung und Bildung). Stuttgart und Dresden. S. 434-437.

Steffens, Gerd; Weiß, Edgar (2004): Editorial. Zur Einführung - Globalisierung und Bildung - ein Problemaufriss. In: Blömeke u.a. 2004. S. 9-34.

Wulf, Christoph; Merkel, Christine (Hrsg.) (2002): Globalisierung als Herausforderung der Erziehung. Theorien, Grundlagen, Fallstudien. Münster, New York, München. European studies in education, Vol. 15).

Wuppertal Institut (Hrsg.) (2006): Fair Future. Begrenzte Ressourcen und Globale Gerechtigkeit. 3. Aufl. München.

The Globalizing World and Man's Education

Reflecting on Meilė Lukšienė's Insights – an introduction from Lithuanian perspective

Algirdas Gaižutis

Abstract

Not only in Europe, but also in all the other continents the national systems of education are being reformed aiming to discover the optimal perspectives and models for educating, teaching and developing the modern man. The routine and spirited discussions focus on new paradigms, the mission, the vision and strategies. The current debates inevitably embrace an intersection of traditionalism and postmodernism, the theory and practice of education, the school and other state and public organizations, the interests of various social groups and their frequent inability to go in one direction and pursue common goals. School or university teachers, working in both formal and informal education sectors, may sometimes fall into despair in the face of excessive emphasis on the significance of constant changes and reforms of the system of education without having any firm basis for that. It sometimes even seems that this sphere requires permanent alterations and bigger or smaller revolutions which, in their turn, have to be standardized, necessarily following identical norms, requirements and criteria.

I

Nevertheless, the undue passion for winds of change and continuous alterations, the constant development and its pace often makes one forget the issues which are no less fundamental – namely, the system of education as the basis of cultural and national identity which has to be preserved without withdrawing and avoiding provincial insularity in the open space of Europe and the world. This issue was and is especially significant for the countries that have liberated themselves from the ideological, economic and political grip of the Soviet Union and have become independent, new members of the European Union.

This paper will focus on how this issue was solved by Meilė Lukšienė, the well-known theorist of pedagogy and educational sciences. The article does not cover all her merits as a literaturologist, historian of the development of Lithuanian and not only Lithuanian – European education, culturologist and, finally, as a statesperson, the moving spirit of Sąjūdis, the Lithuanian Reform Movement. We will confine ourselves only to some major insights of hers.

Our considerations will begin with a short exposé of her biography which naturally can be considered to be the key to her prolific and highly topical research creation. Hence, who was and who is Meilė Lukšienė? We will start with a couple of eloquent details of her biography. The initiator of the education reform, literaturologist, education specialist, culturologist, an active member of the initiative group of Sąjūdis, the Lithuanian Reform Movement, Meilė Lukšienė (Matijošaitytė) was born on August 20, 1913 in Vienna. She came from a prominent family of Lithuanian cultural figures, educators and writers. In 1942 she completed her studies of Lithuanian Philology at Vytautas-Magnus-University (Kaunas); later she lectured at the same university and from 1949 till 1959 she was the Head of Department of Lithuanian Literature at Vilnius University.

Meilė Lukšienė, being a true patriot of Lithuania, but not of the Soviet Lithuania, a great scholar and educator of students, was continually watched by the Committee for State Security, the so-called KGB (Комитет государственной безопасности). As a courageous personality, enshrining the national values, she was accused of non-Marxist estimation of literature, inappropriate (non-communist) education of students. Thus, there appeared the famous "case" against Vilnius University in 1959-1961 which was supposed to bring university lecturers and intelligentsia to reason.

What was the case about? At the first sight, it was very simple. It was "proved" that the department headed by Lukšienė was ideologically unreliable: the students' (young poets' and prosaists') almanac included pessimistic poems, the topics of course and research papers lacked ideological and party-line focus, the Department published good reviews of bourgeois writers, etc. The work of the Department was highly criticized by the Lithuanian Communist Party Congress and the Congress of the Lithuanian Association for Writers. In 1959 in the Hall of Columns of Vilnius University there was a public, notorious trial-censure of the Department. The head of the Department, Meilė Lukšienė, did not repent, as it was typical in trials like that, but rather she expressed her content with the earnest, responsible work of the professors and conveyed her gratitude. After the "trial" Meilė Lukšienė was dismissed from the position of the head of the Department and was excluded from the University together with her closest friends – Vanda Zaborskaitė and Irena Kostkevičiūtė. The Rector of the University, the famous lawyer Juozas Bulavas, was discharged from the post of Rector and excluded from the Communist Party (Jungtys, p. 117).

However, Meilė Lukšienė was not discouraged. She concentrated all her energy and creative powers on the research on the Lithuanian education system and its history. She wrote her fundamental works "The Historic Features of Lithuanian Education in the First Half of the 20th Century" („Lietuvos švietimo istorijos bruožai XIX a. pirmojoje pusėje") (1970), "The Democratic Notion of Education in Lithuania, the Second Half of the 18th Century – the First Half of the 19th Century" („Demokratinė ugdymo mintis Lietuvoje, XVIII a. antroji–XIX a. pirmoji pusė") (1976), etc. Most of the scholar's papers were rejected by the censorship of the time and editorial offices and thus they were kept in her desk drawers for a long time. They did not decay, but rather they were humbly waiting for their time which eventually came. Symbolically, writer Mikhail Bulgakov in his well-known novel "The Master and Margarita" put the following memorable words into the lips of the protagonist Wolland: "Manuscripts don't burn..." There came time for all works written by Meilė Lukšienė. And that time came with Sąjūdis, the Lithuanian Reform Movement, and the beginning of the end of the Soviet occupation.

Then Meilė Lukšienė was as if reborn to a new life and became its inspiration and initiator of good, purposeful reforms, especially in education and other related spheres. Her whole life and the accumulated incredible experience evidenced that education is the fundamental factor for the development of a society, the essential basis for all sociological reforms. She became "the generator of ideas" in the Lithuanian Education Reform and modeled the principles of the future of the State. She is a co-author of the key strategic documents of Lithuanian education – *Conception of National School* (1988) (T*autinės mokyklos koncpecija*), *Conception of Lithuanian Education* (1992) (*Lietuvos švietimo koncepcija*), *Guidelines of Lithuanian Education Reform* (1993) (*Lietuvos švietimo reformos gairės*), *Programmes of General Education Schools* (1994) (*Bendrojo lavinimo mokyklos programos*), *Standards of General Education* (1997) (*Bendrojo išsilavinimo standartų*) and others. To implement the education reform, to reflect upon the strategy of its change, to observe and assess it, Meilė Lukšienė united and involved the communities of educational sciences and other spheres of science, theoreticians and practitioners. In 1999 the Lithuanian education system was assessed by Organization for Economic Cooperation and Development (OECD) and its experts acknowledged that in Lithuania there is a sound conceptual foundation for reforms.

In 2004 Lukšienė was awarded the UNESCO COMENIUS medal for her personal input into the scientific and practical consolidation of education as the basis for the establishment of the State. The UNESCO award says: "Mrs. Lukšienė is a scholar of European and global level and a personality who significantly contributed to the development of the democratic society and modern education."

In 2011 in Paris, during the UNESCO General Conference 36th session, the 100th anniversary of the birth of Meilė Lukšienė was included in the calendar of commemorative anniversaries and in 2013 it was commemorated in different countries of the world. The Seimas of the Republic of Lithuania proclaimed 2013 the year of Meilė Lukšienė, and during the year there were a lot of memorable events, exhibitions and international scientific conferences organized, there were books published to highlight the significance of this outstanding scholar and statesperson's merits both for the present and the future.

These scarce facts from Meilė Lukšienė's biography eloquently witness the fact that Lithuanian education has always (in times of wars, occupations, turmoil and challenges of globalization) been a vitally important battlefield and a way of struggle which helped to remember the native language and the nation's history, to preserve the cultural identity and to educate people in the trend of humanist ideals (even in the constricted conditions when Lithuania lost its statehood).

II

Meilė Lukšienė's scope of research interests and the universal character of her talent are also surprising. Her interests refer to diverse spheres of cognition and values – literaturology, pedagogy and, on a wider scale, educational sciences, philosophy and culturology. Such range of interests is a rare phenomenon. It is not approachable to everybody, it is not occasional. And of greatest importance is Meilė Lukšienė's underlying attentiveness to the singularity of these spheres, their historic tradition, the topicality of the problems and the junctures of these spheres. In her books and articles Meilė Lukšienė reveals herself as a great erudite. As time goes by, we are facing new challenges of the present day: demographic, financial and bank crises, massive emigration of young people to other countries and continents, unemployment, great numbers of grandchildren who are being raised by grandparents, drug dealers lurking at the school gates for our children, the banality of pop music and slick books, and other tricks, alleged novelties, contradictions and turmoil of the global world.

Meilė Lukšienė was interested in the issues of man's education from various perspectives: people's nature, their abilities and talent, rights and responsibilities, nationality and citizenship, morality and spirituality. Even in the Soviet period her articles and books had a different, bold, methodological background. We will venture to say that her approach was in some way similar to the French Annales School headed by Fernand Braudel, Jacques Le Goff, Georges Duby and others. They formulated the concept of a historic fact created by the historian himself, broadened the field of history research, employed the interdisciplinary method and paid special attention to the research on mentality.

When Meilė Lukšienė's articles and books about the development of our Lithuanian school and education system are read in the present day context, one feels that this issue is analyzed not as the development of structures or institutions, but rather her studies take into consideration the man, the everyday concerns and hopes, the changing rhythm of social life, the occupation of our country and the process of its liberation.

From the pagan times till the beginning of the present 21st century the scholar distinguished, using modern terminology, 5 paradigms, transformations, turning points in history. We will not elaborate on them. However, it is important to emphasize that a formal system of education appeared in Lithuania with the introduction of Christianity. The education system became an essential vehicle for the expression of the State culture. Thus, there appeared institutional education and schools of higher education in Lithuania.

Occasionally there occur queer situations when the value of the attempts and achievements of the educated people of the previous century are depreciated due to lack of specific knowledge or general ignorance. Meilė Lukšienė revealed that in those times the cultural development in our country was less regulated than in Central Europe, but it was based on the erudition of professorate as well as on the openness to ideas and experience. And, most importantly, it was based on a close relation with the culture of the country. The Education Commission (an analogue of the modern ministries of education) played a key role; it helped to develop civic education that covered morality, customs, legal backgrounds, religious beliefs, etc.

Certainly, each stage of the institutional education system had its own concerns, collisions, and later, approaching to the modern times, - political tensions and ideological conflicts. The system of education is inseparable from an attitude to man and community, the nation and state. In other words, it is related to a variety of mental images and beliefs which prosper in different layers of the society – among the aristocracy, nobility, townspeople, villagers, ordinary people and the authorities of different professions.

It would be especially interesting to investigate the highly significant analysis, conducted by Meilė Lukšienė, on the formation of worldview (which is a wider concept than world outlook). The analysis is presented in a unique book by a mother and a daughter – Julija Biliūnienė ir Meilė Lukšienė – "The Meanings of Time" ("Laiko prasmės"), published in 2004. It is a detailed investigation and consideration how a young learning person gets not only into a different material and cultural environment, but also into a much faster pace of change. The book tells how he, being not from the aristocracy or nobility, makes his way to European culture and eventually introduces original elements into the developing worldview. It is an original subject analysis based on the potentials of ethnography, social pedagogy and culturology. Such an approach gave an impetus to reveal and evaluate the complex genesis and structure of the Lithuanian intelligentsia – from its aristocratic, noble origin to peasantry. When through the profession of clergy and some other (townspeople's) occupations the peasants developed into quite a strong intelligentsia, it was strong both in the perception of its own value and spiritually. The intelligentsia was not numerous; nevertheless, it revealed its creative powers in different spheres of life in the interwar period.

This analysis was conducted on the basis of different biographies, memoirs and facts in the lives of Mykolas Riomeris, Jonas Biliūnas, Steponas Kairys, Povilas Višinskis, Feliksas Bugailiškis, Mykolas Biržiška, Liudvika Didžiulienė Žmona and others. And most importantly it was based on the recollections of her own, i.e. Meilė Lukšienė's, parents and relatives which makes those considerations not only profound, but also really personal and authentic. The scholar's contemplations fascinate by their multifaceted character and interpretations: the relationships of Lithuanians and Poles, search for God and his acknowledgment, ideological junctures, man's relations with the state and its authoritarian government, the values of the old and young generations, a way and scope of resistance to the Polish occupation of Vilnius region, Hitler's invasion and the Soviet occupation.

In the first half of the 20th century in Europe, and at the same time in Lithuania, the issues of the structure of the state as well as the interrelations of an individual and the state were actively discussed. Our intelligentsia, educated both abroad and in Lithuania, certainly played the lead as it diverse opinions concerning those issues. Nevertheless, as noted by the scholar, in the process of creating a higher level society, the majority were successful in separating their professional and personal affairs from political matters. There were fundamental goals to develop solidarity and to reduce constraints, violence and brutal fights in the society. Even though the intelligentsia, that had been trying to liberate itself from the agrarian and provincial insularity, was not numerous, it had a priceless virtue – idealism and belief in the magnanimity of the national state and national culture.

Certainly, all of it was undermined during the 50 soviet years of terror, losses, communist ideology, which restricted people's fundamental rights and liberties, and the tragic years of resistance. Meilė Lukšienė and her family suffered considerably from the Soviet regime; the scholar herself, as it has already been mentioned, was withdrawn from Vilnius University and was continuously distrusted. Nonetheless, being an outstanding personality guided by a creative hunger and strong creative will, she was disgusted with whimpering and animosity (which are still characteristic of some people even today). She was always loyal to the historical integrity of the Lithuanian intelligentsia and to herself.

Hence, when the USSR military power and ideological grip weakened, Meilė Lukšienė, being an erudite and well-aware of the events in Europe, was the first education specialist to turn to a different direction as she clearly saw the complexity and contradictions of the whole system of education. There appeared a necessity for a society of different mentality since the political and economic situation had changed dramatically and, moreover, there developed the crisis of the modern world as a result of which people's spiritual maturity was surpassed by technology and the accompanying mass culture and atmosphere as well as the tendency of relativizing the fundamental European values. In such a field of education system, full of diverse tensions, the solutions to narrow questions and attempts to be contented with minor "improvements", in Meilė Lukšienė's opinion, would not give tangible results and would drown (possibly good) initiatives in everyday trivia. The formation of the so-called information society mentality at the end of the 20th century required conceptuality, coherence and creativity of thinking. Hence, there appeared a new transformation, a new fundamental movement, a new paradigm.

Back then, UNESCO published its benchmark publications (Gaston Mialaret [1985], Jacques Delors [1996] and others). They expanded educational sciences on the global scale and deepened its interaction with the marginal disciplines – psychology, philosophy (educational philosophy), sociology, informatics, culturology. The interaction with the latter discipline, especially in the development of national identity, was viewed by Meilė Lukšienė as the distinguishing feature of the Lithuanian education reform. And later she was very glad about the introduction of philosophy of culture or cultural anthropology, culturology in the curriculum of schools of higher education. Culture was started to be conceived as a whole, as a vulnerable undivided organism rather than a combination of random facts.

In the 20th century crimes against humanity with a significant degree of fascism and communist ideologies, phenomena of totalitarianism and dehumanization oblige education to search for a reasonable, well-weighed balance, based on ethnic values, between man's material and spiritual interests. Moreover, it is important to remember to emphasize the supremacy of spiritual interests and spirituality as a continual goal, an ideal. The scholar agreed with Delors's idea about an increasing tendency in educational sciences to focus not only on "how", but also on "what" – the goals and objectives of education, their content and meaning. Such a tendency and trend with their four key supporting points – *learning to live together, learning to cognize, learning to act (work, create)* and *learning to be* – seemed to Meilė Lukšienė not only important, but also essential in the renewal of the education system. That is a real humanistic tendency and trend which enable a reasonable solution to diverse contradictions appearing in different segments of the education system as well as in the interaction of this system with the state and the objectives to preserve and improve its cultural identity (Jungtys, p. 59).

Hence, educology is a totality of specific disciplines which is much wider than pedagogy, its core. Education, teaching, training refer not only to an individual, but to diverse layers of the society, the whole of the society and constitute the basis of statehood. Unfortunately, even today there are people who do not comprehend this fact completely. Recently in the magazine "Veidas" and newspaper "Dialogas" there was published an interview with biochemistry Eugenijus Butkus in which he said that he knew nothing about such disciplines. Thus, it appears that he is an expert of all the other branches of science. Only another omniscient could debate such a position, but not the people who work and create in this sphere.

In the transition period from the Soviet ideologically marked school, Meilė Lukšienė thought it was essential to refuse the clichés of thinking, a priori schemes and to develop constructive, creative thinking both among learners and teachers. That would help to lay the foundations for analytical, critical and, simultaneously, global or holistic thinking. This does not refer to Descartes's "*Cogito ergo sum*" ("I think, therefore, I am") manifestation of rationalism. Nonetheless, the latter formula of the philosopher, this, in his opinion, "unassailable truth" emphasizes two fundamental matters – the significance of thinking and the significance of liberty. Essentially, nobody in the state or in the universe is able and can prohibit people from thinking freely even if they are in captivity, prison, exile or the underground.

It sometimes seems that nowadays, in times of challenges of globalization and abundance of information worldwide, it is especially difficult to "handle" young people's minds (not categorically, without lecturing or didactics, etc.). Now their minds are full of everything (other people's opinions, catchwords, affirmations and disaffirmations of the same matter, a variety of schemes, rumors, facts, which are said to be real, esoteric propositions, etc.). In brief, they possess a chaotic image of the world where one idea rejects or obscures another; they live in concord or strain one another. I believe there are two ways for a personality to prosper – systemic and mosaic education. Could both of them be productive? Let us try to explore and discuss this issue. There are educators who would wish a better pronounced order in young people's minds. Such order was already discussed by Eastern sages, Socrates, Plato and Aristotle in antiquity, Saint Augustine and St. Thomas Aquinas in the Middle Ages and others. Even modern philosophers and education specialists use a concept defined (more or less) by "order". Naturally, we are facing a change of generations. And every generation brings a lot of novelties: thinking and behavior patterns, ceremonials and rituals, communication and cooperation, value-related orientations, traditions and innovations in other spheres as well. But at the same time every generation brings not only its prominence, but also its risks. We, educators and teachers, are obliged to comprehend and be aware of both. We should not succumb to the confusion of novelties, postmodernities and lifestyles, to the colorful attractiveness of our own and foreign entities. The above mentioned risks have their own psychological and social nuances. Likewise, the individual's openness to the world and even the widely discussed tolerance are not very simple issues. Naïve people have always been and will always be handy for somebody since it is possible to exploit them for the achievement of either narrow egoistic or global goals.

Meilė Lukšienė was also concerned about this issue. She wrote about the dangers of relativism, about the modern relativization of values. Where does it lead to? What does it breed? Well, it results in ample individual needs to lead a comfortable life today as well as in benefits that carry no commitments to the society when the individual is perceived as an isolated value. The modern liberal market, which possesses some inclinations to ferity, is a good ground for the spread of pragmatism and hedonism. Meanwhile, it is essential that materiality does not upstage spirituality, self-development and self-creation, human individuality.

Therefore, Meilė Lukšienė's greatest concern and ache was how to turn school not into an object governed by the authorities, but rather into a subject agent. She believed that there is a lot to be fundamentally changed: there should appear a liberated process of teacher training, a respectful attitude of parents towards school, their interest and awareness, pupils' activation and self-government, school councils and mentorship, etc. Moreover, there should appear an atmosphere of work responsibility, a properly organized structure of school management as well as social partners' help. It is important to endeavor at every school's meaningful existence, their common goal and strategy for action.

Speaking about three types of school – fragmentary, project-oriented and problem-solving, with a special focus on the latter – a community that is renewed through school, the scholar expressed the following memorable ideas. The direction of education reform: a free personality, a free democratic society following shared moral principles, committed to the development of its native culture and eventually to the development of culture of the whole humankind. That is a susceptive, concise and well-polished formula.

The modernized school – just as the learning school – is facing a crucial objective, namely, the interaction between teachers and pupils as well as pupils' activity with all of its significant forms, self-expression and creativity. The development of a personality has to be consistently related to the vision of the society. Meilė Lukšienė wrote:

"When raising a man, it is essential to create a pluralistic, but, at the same time, solid, communicative democratic society and be able to live in it respectably and responsibly."

Unfortunately, things are not that simple and Meilė Lukšienė was well aware of this. She was concerned about the obstacles in the way of flexible, reasonable didactics at school. Primarily she was worried about the dispersal of school subjects, poor interdisciplinary relations and lack of integration. The study of language was impoverished by the abundance of grammar rules, psychologically inconsiderate abstractedness was introduced too early in the lower levels of schooling, e.g. mathematical terminology is used in physics, legal terms – in history, there are no deep associations between the native country study and geography and other disciplines, pupils' reading and writing skills have diminished, textbooks lack internal connections and there are many other problematic issues. In other words, it is crucial to alter the very structure of education.

In her formulation of the objectives of education reform and its modernization trends, Meilė Lukšienė laid special emphasis on the integration of moral education and training which should be applied in various spheres and on different levels. Integration is to be conceived as an antonym of disintegration which was a characteristic feature of the 50 years of the Soviet occupation. The disintegrative trend tried to build the education system as a whole despite various attempts and constraints to unify, collectivize and assimilate. However, in reality it was a mechanical fusion, based on and supported by ideology; professional groups withdrew into their professions, school subjects lost their essential connections, there developed a narrow professionalism, moral education was resolved into a variety of separate spheres (moral, patriotic, aesthetic, physical, military, etc.) as if they were parallel and independent of one another (Jungtys, p. 183).

The concept of integration was used by Meilė Lukšienė in its widest sense – the scholar applied it to both education and the preservation and development of cultural and national mentality. Integration facilitates a proper, promising juncture of education history, the accumulated European experience, the present and the future, and it does not obscure differentiated assignments at school. And in modern school classes have to be integrated on all levels, the interdisciplinary relations have to be enhanced, and, moreover, assignments have to be differentiated on the basis of learners' abilities.

Meilė Lukšienė spoke not only about differentiation of tasks according to learners' abilities and the integration of interdisciplinary relations in classes, but also about an "integrated nation wherever its members would live. The nation's spirit, existence and its very essence consist of its own culture. It is not a sum of random spheres, but rather a totality, and he who is aware of it, feels it and even perceives it, is a part of that, not of any other nation. Such a person is integrated and fused together with its culture, hence, with the nation itself" (Jungtys, p. 184).

Meilė Lukšienė was well aware of the ample routine in the Soviet school. It was related to ideology, administration, documentation, laws as well as the stereotypes of thinking and assessment. The scholar saw the Soviet school as an ocean with a variety of islands – different disciplines – without any connections and junctures; the freedom of teachers' creativity and initiative was diminished; pupils were reduced to a common denominator and their expression of individuality was belittled; the school lost its primary mission – being in the avant-garde of democratizing society. In her definition of the education reform as a long-term, coordinated process, Meilė Lukšienė wrote about a special role of the teacher-creator who has "a firm determination to develop himself, to reconsider his relation with the child, people, the society, and, especially, the state" (Jungtys, p. 265). Democracy cannot be created without changes at school irrespective of how well we think about ourselves. It is the teacher's and learner's mindset that has to change fundamentally, and primarily it is the teacher who should learn that.

It is namely Sąjūdis with Meilė Lukšienė at the fore who made a stand against the overthrow of the school, education and fostering. The scholar's ideas, interviews and previously unpublished texts presented in the book "*Jungtys*" ["Junctures"] (2000) are especially interesting and significant. They are numerous, written in the period from 1988 to 1989 and even later. They are sparkles of viable thinking. The education reform is, naturally, not a game with *Lego* bricks. The scholar was well aware of how difficult it is for most teachers to work innovatively since they were constrained by external control and their fear of choice even acknowledging that democracy is much more powerful than control. Moreover, the society had to disengage from chaos and the stiffening internal censor.

The teacher's role is the role of a certain creator envisaged by the scholar in her vision. The teachers' role is changing: they are no longer carriers and transmitters of knowledge, but rather they have become special filters of knowledge in the present-day world of excess information. Unfortunately, not all the information that enters the consciousness of children, adolescents and young people is authentic, correct, righteous, elevated and aimed at their internal harmony. Hence, in the modern world the teachers' role is becoming more complex: they are obliged to change themselves and alter their pupils' role, their abilities and creativity. The scholar wrote: "I would especially emphasize the significance of self-government. If pupils participate in the government of their school, they will be psychologically prone to communicate and cooperate. I believe that our schools should stop being so skeptical about pupils' self-government. To my mind, this is one of the widest fields of education – as children grow, they will learn to follow their beliefs in action. It is important that schools provide children with alternative knowledge, teach them to assess facts and develop their convictions. In the process of teaching teachers could not only state facts, but also present their possible evaluations and interpretations. This has not been taught yet. Let us take history teachers as an example. Choose a fact. Interpret it in the context of the English political tradition and then in respect to our political context. People will mature and learn to analyze the surrounding world." (Jungtys, 244). This is the only way to alter the school – to make it open-minded and democratic, civic, tolerant and responsible. And all of this is impossible without the teacher-creator, the leader of change and the architect of future generations. Hence, teacher training is a primary objective for the universities which are responsible for this mission. Meilė Lukšienė also concentrated on the issues of teacher training. Moreover, she was concerned about the "feminization" of the modern Lithuanian school. The scholar was convinced that rationality, a guarded sensitivity – the male element – and deep motherly love for people – the female element – should be balanced in the new school. Then spirituality – the possibility and significance of spiritual expression – would be recognized not only in religion or art, but also in our own system of education.

Back then and sometimes even today Meilė Lukšienė is sometimes reproved that the *Conception of the National School* is too declarative and lacks concreteness. However, Meilė Lukšienė knew very well where to start and what direction the reform should aim at. She believed that the reform is based on and is supposed to be based on two essential principles – democratization and humanization, the development of a free and responsible person. This primarily requires a substantial investment of spiritual and material capital into the system of education; and we can bravely claim that this investment is not sufficient today. No matter what we think and how we approach the issues, otherwise we will not strengthen the basis of our state and the worldview and values of the young generation. Moreover, Meilė Lukšienė was against the "sterile" national school. She herself attended a school which serenely tolerated diverse convictions.

The scholar was convinced about the following provision which should also be undeniably adopted by all of us. A citizen – a free man who is patriotically minded, who respects his national history and uniqueness and is able to function in a positive manner – can be developed only in a *depoliticized* school which embraces all democratic world outlooks. While the division into "black and white", which is so frequent in our present-day society, would definitely not help. Meilė Lukšienė believed that that intermediate layer of free and democratically mature personalities, who would not be wedged in the narrow frames of one or another pole, could only be developed in the new school. Pupils have to mature in the environment of pluralistic ideas.

Obviously, speaking about the *depoliticized* school Meilė Lukšienė did not mean that the school should exist beyond politics and its ongoing serious debates. Political ideas will always reach school in various other ways. However, the school cannot be one-sidedly oriented to the constantly changing programmes of political parties and their interest in school as a source of the future electorate.

"We should respect individuality", Meilė Lukšienė wrote.

"We should raise a man able to make choices and to assume responsibility for those choices. Responsibility is the main civic quality. If the school expands the spheres of responsibility, the citizen will mature" (Jungtys, p. 245).

The scholar was convinced that the school should be the environment for the development of free people – the disciplined ones, who do not easily succumb to the influence and appeal of random ideas and eventually to the turmoil of the conscious.

"One of the greatest tendencies nowadays is the pace of change, the increase of new spheres and aspects, and the fall of information. And the second is that we try to find solid ground under out feet, i.e. stability." (Jungtys, p. 253).

The quotation includes a nice metaphor used by the author – "solid ground under our feet". A nationally oriented school that respects and spares multiculturalism and is open to the world can guarantee that stability which is essential for individuals scarching for their true *Self*, when it sometimes seems that the ground is slipping from under our feet and you are trying to catch at a last straw.

These days there are discussions about leadership and management – both of schools and in them. A lot of respected professors and educational specialists speak about these issues since they are especially important for overcoming the undue rationality and routine of the school. The above mentioned leadership is viewed as an ongoing search for possibilities, the promotion of creativity, following the new paradigm of education and aiming at the systemic nature of school activity. There could be other possible pursuits, but these seem to be important for us. While the aforementioned management, administration are no more than a reduction of routine and irrelevant habits at school; a motivated implementation of innovations on the basis of fundamental supporting points: our nation, our state, our cultural identity and us, free citizens in our country and open to human, humanistic world, able to independently and self-sufficiently participate in dialogues and polilogues with other countries and cultures.

The opponents of Meilė Lukšienė reproved that she focused too much on the humanization and socialization of the school, that she was forgetting the significance of the natural sciences and opening the ways to the "sewage" of the West. They said that her reform was not national, but rather cosmopolitan. However, after the restoration of independence of Lithuania the scholar disfavored the so-called "globetrotters" who travelled widely and on return from different countries would spread ideas about how well the system of education worked in, let us say, Denmark or Germany, about an even better system in the USA, about a special system, let us say, in Japan, and the most wonderful system, let us say, in England, etc. – depending on what conferences, congresses or family reunions they had just returned from. Such universal imitation was not acceptable to Meilė Lukšienė; she even considered it to be rather rude. She said that one must learn from others, but not follow them blindly. This highly intellectual scholar was sure that the development of a free man in a free society requires not only some personal moral measure, but also a criterion of responsibility for one's nation – which is another great moral measurement. These matters were of special concern to Meilė Lukšienė. This is evidenced in her autobiographical writings and concise speeches.

In conclusion I would like to provide a couple of more impressions. In her book Marija Gimbutienė / "Laimos palytėta" [Marija Gimbutienė / "Touched by Laima"] (2002) Meilė Lukšienė wrote about her relative, a cousin and world-famous archeologist, an involuntary emigrant, and this is what made a great impression to me. The scholar said: Marija Gimbutienė took respect and love for work, strong will, a sense of and need for freedom, independence and beauty from her home. And later in her life she was very natural, her charm was artless, without a trace of artificiality, radiant and appealing to the people surrounding her. She was able to take pleasure not only in things, but also in other people's joy; she was a real scientist-creator, integrally associated with Lithuania, its land, nature, people and culture.

In my opinion, this is what Meilė Lukšienė herself was. Those qualities of a free man, who loves the history of his country and senses other people's longing for a better life, as well as the higher creative aspirations of a talented person were also characteristic of her friends – Vanda Zaborskaitė, Irena Kostkevičiūtė and others, and her family – her parents, husband, son, daughters and her granddaughter. Those women and men are the genuine spiritual power of Lithuania, the source of its vitality and future.

We will again refer to the great rationalist and intuitivist Descartes and his treatise "The Passions of the Soul", written in 1649, in which will and mind were transformed into spiritual qualities, able to judge, direct and, if the occasion arises, curb the bodily passions. The philosopher distinguished six primary passions: admiration, love, hatred, desire, joy and sorrow. Just like people's body, they can serve the soul or, conversely, they can enslave it. I believe in Meilė Lukšienė's case those passions faithfully served the high-minded soul. Hence, the road of her life is so luminous, revealing and magnetic.

References

Delors, J. (1996): L'éducation, un trésor est caché dedans. Paris: Unesco et Odile Jacob, pp. 91-105.

Descartes, R. ([1641]/²1992): Meditationes de Prima Philosophia. Meditationen über die Grundlagen der Philosophie. Philosophische Bibliothek, Band 250a. Lateinisch-Deutsch. Hamburg: Felix Meiner.

Descartes, R. ([1649]/1996]): Die Leidenschaften der Seele. Les passions de l'âme. Französisch-Deutsch. Vorwort, hrsg. u. übers. v. Hammacher, K. Hamburg: Felix Meiner.

Gimbutienė, M. (2002): Laimos palytėta: straipsniai, recenzijos, pokalbiai, pokalbiai, laiškai, vertinimai, prisiminimai /sudarytoja Austėja Ikamaitė. Vilnius: Scena, pp. 196-209.

Gimbutienė, M. (2002): Laimos palytėta: straipsniai, recenzijos, pokalbiai, pokalbiai, laiškai, vertinimai, prisiminimai /sudarytoja Austėja Ikamaitė. Vilnius: Scena, p. 231.

Lukšienė, M. (1993a): Pedagogika ir kultūra. Lietuvos švietimo reformos gairės. Vilnius: valstybinis leidybos centras, pp. 25-45.

Lukšienė, M. (1993b): Atsinaujinimo (kaitos) nuostata ir poreikis. Lietuvos švietimo reformos gairės. Vilnius: valstybinis leidybos centras, pp. 102-112.

Lukšienė, M. (2000): Jungtys. Sudarė Ramutė Bruzgelevičienė. Vilnius: Alma littera, p. 511 [(2013): Antrasis pataisytas leidimas, 2 egz. (2. durchges. Aufl.)]

Mialaret, G. (1985): Introduction to the educational sciences. Paris: Unesco, p. 106.

Picture certification

http://www.marsc.lt/files/marsclt/Meile%2520Luksiene.jpg

Is it Possible to Create a Global Identity of Educational Science?

Karl-Heinz Dammer

Abstract
Even though there are a lot of globally oriented large scale studies like e.g. PISA the creation of a global educational science is a task still to be done because this sort of studies is likely to produce a methodological "monoculture" which is additionally connected with a certain type of global governance mainly based on economic criteria. This "monoculture" is conflicting with the demands of methodological and theoretic pluralism we should strive for. Moreover, the method is indifferent concerning the contents of a global educational science: Is there a way to create a global common idea of education? This is an open question to be thoroughly discussed but one thing seems to be sure: There will be no globalization without diversity.

The subject of my contribution to this conference is directly derived from its title "Educational Science in Search of Global Identity" which I like to formulate as a question: Is it possible to create a *global* educational science and what would or what should it be like?

The title implies that the creation of a global educational science is a task still to be done and that it is worth to do so. I think it is not only worth but it is necessary to search a global attitude because educational problems, at least those concerning school, have to be discussed on a global level. Since TIMSS and PISA, two studies which aim at globalizing educational politics, educational science can no longer be treated as a national affair only. Thus the globalization itself can no longer be questioned but what we have to talk about is the way this globalization of educational science should go.

This seems to me very important as PISA and all the other large-scale-studies are more or less necessarily based on a certain type of research pursuing goals imperatively set by the OECD. As this is an economic setting, educational science is challenged to debate if or to what extent it wants to adapt.

Before dealing with this subject, it is necessary to clarify the meanings of the terms which are essential for our subject. First of all we have to note that the pedagogical tradition in Germany has established a distinction between "Erziehung" and "Bildung". The first, Erziehung, describes the process in which the elder generation makes the younger generation adapt common rules of the society of adults. "Bildung" then, refers to the personal development of an individual which is an autonomic activity, a sort of self-education. So the difference between these two terms is mainly the difference between heteronomy and autonomy even though these two processes are interdependent and should, from a normative point of view, be combined, "Bildung" being the aim of "Erziehung", "Erziehung" resulting in "Bildung".

Obviously, other languages, which are internationally used, do not care about this distinction: The German title of the annual OECD-report for example is "Bildung auf einen Blick" whereas the English version is entitled "Education at a glance", the French version "Regards sur l'éducation". So if the title of our conference in German is "Bildungswissenschaft auf der Suche nach globaler Identität", the translation "Educational Science in Search of Global Identity" might evolve misunderstandings as "Erziehungswissenschaft" and "Bildungswissenschaft" are less affiliated to each other than one might think.

"Erziehungswissenschaft", the literal translation of "educational science", has its origins in the 18th century, beginning with the first professorial chair occupied by Ernst Christian Trapp, who tried to develop a system of teaching based on observations of lessons. But it took far more than one century to establish "Erziehungswissenschaft" as an academic discipline on its own in the 1920ies. From the 1960ies on there is more and more theoretical diversification so that today we often use the plural "Erziehungswissenschaften".

"Bildungswissenschaft", in contrast, is a recent term which came up during the 1990ies and is used in two quite different ways. On the one hand it is just an administrative term within the academic curriculum of teacher training which includes all those disciplines dealing with "Bildung" in a more or less intensive way: These contain, besides educational science, sociology, psychology, theology, politology, media sciences and, of course, philosophy. The term has nothing to do with the identity of these disciplines; it just describes the subjects which the students have to deal with to a certain, often modest, amount apart from their teaching subjects. This idea of "Bildungswissenschaften", arising out of the specific German tradition of teachers' education, might be interesting for the teacher training in a global perspective but this would require a lot more thinking about the contents of the new discipline composed of more than half a dozen traditional academic domains: What can philosophy, sociology etc. contribute to a reasonable training of teachers in order to improve their professional competences? As even German universities are not homogenous as far as the disciplines belonging to "Bildungswissenschaften" are concerned, we are still quite far away from such a common national curriculum. So at the moment it might not be very useful to reflect about a global curriculum – quite apart from the fact that this concept of "Bildungswissenschaften" is unknown in other countries.

The second meaning of "Bildungswissenschaften" refers to a certain type of research in the context of international large-scale-studies as those mentioned above. As this type of research was global from the beginning, it might seem useful to pursue this semantic way looking for a global identity as we could perhaps more easily compromise here. This consensus, however, has first of all to be established, because these studies are linked to global politics of neoliberalism. The consensus means that we have to find an agreement about what we want to understand by "science" and by "education". This will be the subject of my further explanations as these two concepts are ambiguous. The way we define them in order to determine the identity of educational science has crucial influence on what we expect it to do.

If we use the singular of science we mostly refer to the quantifying model just mentioned. Even though this might be a simplification in a historic point of view, I will call it "positivism" in order to emphasize its political function. The term "positivism" has been introduced, as we know, by Auguste Comte in the first half of the 19th century. "Positive" in this sense means: related to observable and quantifiable facts scientifically treated with mathematical tools. The model of positivism, of course, was older: It traces back to the invention of modern science of nature by Bacon and Descartes. Comte's contribution to the history of science was the transposition of this model on social domains – a step prepared by Condorcet at the end of 18th century. Condorcet was convinced that the scientific methods providing the domination over nature could be applied as effectively to the control of social processes and thus could accelerate the social progress. In consequence, the main purpose of knowledge was to predict and thus to control social development, in Comte's statement: "savoir c'est prévoir".

As this paradigm has become more or less normative for international research it is well known, so that I don't have to explain it further, but there are two aspects to be pointed out in this context. The first one is that the paradigm is completely indifferent concerning its contents; it can be applied to any subject of any discipline and it is said to provide objective and reliable results guaranteed by a thorough application of mathematical devices. This use of objective devices also allows to abstract from the contexts determining the social meaning of the subject dealt with. The second aspect to be emphasized is the political nature historically inherent to this type of research: As Comte clearly pointed out it was always mingled with power hence becoming a sort of objective authority to justify political decisions.

So on the one hand, the question what should be the global identity of educational science can easily be answered: Educational science already has got a global identity which consists in the concept outlined above. On the other hand this answer has three consequences about which a discussion might be useful:

- If we opt for this kind of research as a base of scientific identity, we also opt for the political governance connected with it. Principally, positivism may serve any political purpose but nevertheless we have to consider that nowadays it is often linked to the system of New Governance pursuing neoliberal aims. As it is impossible to ignore this fact, we have to discuss whether global educational science shall serve these purposes or not.
- As positivism is applicable to any context, the global identity of educational science would be merely methodological. Therefore this science would be unable to elucidate the normative and practical problems resulting from different cultural contexts. We might call it a scientific identity but it has got nothing to do with education because here education is not explained from its content but from the method used. This method in turn tends to develop a sort of hidden normativity, so it defines the range of pedagogic aspects which science has to deal and not to deal with.
- The positivist paradigm is not as global as it might appear. Hattie's study for example, known and quoted worldwide, only refers to studies of anglophone countries, a selection implicating another tacit normativity which is not methodological but linguistic.
- Let me draw some first conclusions concerning the question of our conference: If we opt for positivism as the best solution to our problem, the identity of educational science would rather be determined by methodological constraints and political interests than by the specificities of education. Hence what would identity be like, if distinctive characteristics can be ignored? Is that the meaning of the term identity?
- Instead of speaking about educational science we should speak about educational sciences thus permitting a theoretical and methodological pluralism. This is no problem as long as we have a distinct idea of our subject which I will try to propose hereafter.
- In order to find a consensus, we must lead a free global discussion not guided by politics or languages, and of course we have to find a common base of communication. This might be English as it is a lingua franca. But using the English language should not lead to subdue cultural heterogeneity. Understanding a foreign culture and its pedagogical traditions is far more than a linguistic problem. We definitely need a comparative educational science which opens to us the historical, philosophical, political and social traditions of foreign cultures in order to really understand their ideas and practices of

education. Unfortunately this comparative branch of educational science has been, at least in Germany, marginalized during the last ten years, but I think that we depend on the results of comparative research if we are seriously interested in the differences and similarities of pedagogical traditions. Otherwise it will be very difficult to determine a common identity of educational science.

To sum up, it might be more promising to search global identity in contents rather than in methods or other external criteria. So what could an idea of education be like which is acceptable for scientists all over the world?

Let us begin with a rather skeptical consideration: I mentioned above that in the German terminological tradition there is a difference between "Bildung" and "Erziehung". It has been a common sense for a long time that the idea of "Bildung", which is associated to Wilhelm v. Humboldt, cannot be translated into other languages as it results from a special historic constellation in Germany, more precisely in Prussia, at the end of the 18th century. As it became a sort of cultural pattern which largely influenced social history up to the beginning of the 20th century, it is considered as a part of German identity. So, if it's true that "Bildung" is untranslatable and if we presume that other cultures would claim the same for their pedagogical concepts, it might be in vain to search for a global identity of just one educational science.

But actually I don't believe that this is a serious obstacle to our project for at least two reasons.

As far as "Bildung" is concerned, we have to admit that in consideration of its social effects – the separation of classes and creation of social privileges – the term has become quite ambiguous during the 19th century. Moreover, it has meanwhile become a sort of "container-term" appearing in different contexts with different implications. So even in German "Bildung" nowadays can no longer be considered as a concept which expresses a national identity.

The second reason is a more general one: "Bildung" and equivalent terms in other cultures may be untranslatable word by word but the idea behind it can be expressed and explained in other languages and thus be made comprehensible. So it is not important to use a common term of education but to develop a common idea of what we mean when using the term. In Saussure's words: It is not the "signifiant" that counts but the "signifié". So what could be a "signifié" of education acceptable by all or at least most of the cultures worldwide?

The answer to this question might be found in different ways but the most promising seems to be the historic way: Let's go back to the second half of the 18th century when Rousseau for the first time defined the subject of pedagogic and thus of educational science. The purpose of education is to help the growing person to become Himself, i. e. to develop his talents and capacities in order to become an intellectual and moral individuality being able to reflect and to decide on his own and to find his own way of understanding the conception of himself and the world. In a word, the aim of education is to help the growing individual to develop his autonomy.

Of course, on the other hand, an individual had always to be and has still to be adapted to society and its rules. So during the 18th century philosophers from Locke to Pestalozzi discussed a lot about how to harmonize individual and social demands and, as a consequence, their concepts are quite different some of them accentuating adaption (as for example Locke), some of them emphasizing autonomy (as for example Rousseau). But the nuclear idea is the same: An educated individual must be able to think, judge and act on his own and education has to form as well as to respect his nature.

The Swiss philosopher and novelist Peter Bieri, better known by his pseudonym Pascal Mercier, has tried to translate this nuclear idea to our times outlining the portrait of an ideal-typical educated person.

This person…

- is curious and has a large knowledge about the world, being able to tell important from less important things to be known
- is scrupulous concerning the subjects he is really interested in, knows that and how knowledge is produced and thus will less easily be victim of ideology or propaganda
- is conscious of the historic dimensions of his existence and its cultural relativity
- is self-reflected and thus able to evaluate the motives of his acts and thoughts, has moral sensitivity which enables him to understand and tolerate foreign cultures and different attitudes.

I think that this idea can form a nucleus valid for a global educational science, which allows us to draw two conclusions.

First, referring to its history, we are aware of the fact that education is more than the adaptation to current economic, political and social needs, even though there will always be a tension between adaptation and autonomy.

Second, we are interested in more than the scientific evaluation of the efficiency of school systems, which means we should also deal with the question how educational processes, here the forming of mind, are initiated and who is to what extent responsible for the initiation, the teacher or the individual himself?

Nevertheless, one argument remains against this idea of global educational science: It is rather occidental than global. In fact Eastern philosophies of education, those of China and Japan for example, are based on the idea of a harmonious community to which the individual person has to be morally adapted. Islamic cultures vary concerning their ideas of education, but – in contrast to western societies – these are all more or less based on religious obedience. So the educational and in the end cultural ideals of autonomy and individuality might not be accepted in these cultures. They might even be considered as a kind of modern colonialism, thus dictating a sort of monoculture.

So, will it be possible to find a consensus concerning the global identity of educational science including these cultures? And if we tried to find it: Who will be able to mediate? Here we should once more refer to the comparative educational science, a science which has unfortunately been weakened throughout the last decade.

So in the end, the search for identity will remain a project but nevertheless I am convinced that the way drafted above is the most promising one. The global identity of educational science has to be searched in respect of the methodological plurality and the heterogenity of cultural patterns. It has to be developed looking simultaneously for a common nuclear idea of education, common aims of research and common strategies of educational politics. This surely is an arduous way the results of which are difficult to be predicted. But at least it is a way traced by educational science itself and not by external criteria and it is a way in which we might learn a lot from each other. Thus diversity helps to live global.

References

Bieri, P. (2008): Wie wäre es, gebildet zu sein? In: Göppel, R./Lenhart, V./Rihm, Th./ Schön, B./Strittmatter-Haubold, V. (Hrsg.): Bildung ist mehr.... Potentiale über PISA hinaus. Heidelberg, S. 13-21.

Comte, A. ([1844]/1915): Abhandlung über den Geist des Positivismus. Leipzig.

Condorcet, M.-J.-A.-N. Caritat, Marquis de ([1795]/1976): Entwurf einer historischen Darstellung der Fortschritte des menschlichen Geistes, hrsg. v. Alff, W. Frankfurt.

Frost, U. (Hrsg.) (2006): Unternehmen Bildung. Die Frankfurter Einsprüche und kontroverse Positionen zur aktuellen Bildungsreform. Paderborn.

Heydorn, H.J. (1970): Über den Widerspruch von Bildung und Herrschaft. Frankfurt.

Humboldt, W. v. ([1798]/1903): Theorie der Bildung des Menschen. In: Humboldt, W. v.: Gesammelte Schriften, hrsg. v. d. Königlich Preußischen Akademie der Wissenschaften. Band I. Berlin, S. 282-87.

Jahnke, T./Meyerhöfer, W. (Hrsg.) (2006): Pisa & Co. Kritik eines Programms. Hildesheim/Berlin.

Koch, L. (2004): Normative Empirie. In: Heitger, M. u.a.: Kritik der Evaluation von Schulen und Universitäten. Würzburg, S. 39 – 55.

OECD (2013): Bildung auf einen Blick. Paris.

Ötsch, W. O. (2009): Mythos Markt. Marktradikale Propaganda und ökonomische Theorie. Marburg.

Pongratz, L.A. (2009): Pädagogische Gouvernementalität und neoliberale Regierungsweise – Gewinn und Grenze pädagogischer Gouvernementalitätsstudien. In: Kluge, S. (Hrsg.): Entdemokratisierung und Gegenaufklärung. Frankfurt, S. 187-199.

Terhart, E. (2012): „Bildungswissenschaften". Verlegenheitslösung, Sammelkategorie, Kampfbegriff? In: Zeitschrift für Pädagogik, Jg. 58, H. 1, S. 22-39.

Trapp, E.C. ([1780]/1977): Versuch einer Pädagogik. Paderborn.

About Nationality in the Globalized World

Libertas Klimka

Abstract
Historic time endows the concepts of nationality and citizenship as well as their relation with a different and shifting significance in the arena of world politics. The fall of totalitarian empires at the end of the 20th century conditioned a second wave of the establishment of national states which demonstrated the power and vitality of nationality. In the European Union similar processes are also happening today and they show that citizenship is just a derivative of the dimension of nationality.

This paper is a discourse on what could be the present strategy of reinforcing the fundamentals of the state. Has nationality, which is being instilled in public education as ethnic culture, lost its meaning for that purpose? And what is the future of national states? Will they evaporate in the realm of globalized world, or will they become colourful cultural blossoms in the economically and legally integrated community of European countries? In the context of this dilemma the paper discusses the goals of education in Lithuania. The conclusions state that the reinforcement of ethnic culture is the goal of national importance and it has to be efficiently implemented in the institutions of education and culture.

Keywords: national identity, citizenship, national state, globalization, EU politics, cultural politics

Introduction

The present-day complicated, constantly changing social, economic and cultural public environment is shaping the conception of national identity. And this is true not only of Lithuania, but also the whole of Europe. It is being argued that nowadays national identity is conceived in relation to otherness. Thus, it is indeed becoming an increasingly reflexive formation [1, 2]. In general, identity develops through public activity, it is reinforced by social activeness and creative work engaged for national community. On the other hand, identity gives meaning for the actions of separate individuals, especially if they specialize in culture or research. On the whole, the concept "national culture" is more relevant for the description of ethnic communities rather than political bodies. The nationality of modern societies is problematic since it requires forms of support which are based on the conscious decision and commitment of the citizens. Let us suppose that the Lithuanian folklore, which was a phenomenon of mass, hence, popular culture in the 19th century, at present is only a matter of elite stage expression. The conscious construction of identity presupposes the assimilation of spiritual values, world outlook and thinking [3]. Hence, it is a very significant goal of education as it encompasses the mission of national school [4].

Nationality in the Foundations of Statehood

The first and second periods of the independent Republic of Lithuania – in 1918 and in 1990 – were interrelated with the surge of national revival. What could be the strategy of reinforcing the foundations of the State today? Has nationality lost its significance for this purpose? A more substantial basis for the theoretical strategy requires a historical excursus. At the end of the 18th century and in the first half of the 19th century, when the nation became the basis for the legitimacy of political power, in Central and Western Europe there appeared the first national states. The philosophical grounding for these significant changes in political thought was provided by George Hegel. In his opinion, the political system of a nation depends on the structure of its cultural spirit since national law, economy, religion and political life are unified by the cultural foundation [5]. Hence, the state is a formation of spiritual nature which relies on morality, religion and philosophy. There is a close dialectical connection between nation and state; therefore, the moral and political responsibilities are closely interrelated. This Hegel's approach, which is called as national expressionism, rejected the sheer technological and bureaucratic conception of the nature of state mechanism. J. G. Herder became the apologist of the rights of a small nation within a state [6].

Generally speaking, in its strategy of development the national state follows two principal blocs of ideas – the universal principles of human rights, extensive democracy, and social welfare, as well as fostering the ethno-cultural identity. It is essential to retain the balance between the two blocs – the Lithuanian nation was taught this by the ideologist of National Rebirth, dr. Jonas Basanavičius (1851-1927), through his activity. In the project of the state, designed by him and other personalities of the time, political independence had to combine the human and national values, duties and responsibilities. They resembled two wings which could carry the state to its future… In the pre-war period the interrelations between nation and state were theoretically grounded by philosophers Stasys Šalkauskis (1886-1941) and Antanas Maceina (1908-1987) [7, 8]. Their insights are still relevant for the Lithuanian society even today [9].

Thus, there is no doubt that the power of national state and its potential to take care of the destiny of the nation and its people depend on the citizens' cultural awareness. It is an objective collective community which is shaped by the effort of generations [10]. This community should be aimed at by creating efficient images of "national pride": the media and artistic measures could be used to make the national history a topicality, unique ethno-cultural customs and traditions should be promoted, people's creative achievements in research and art could be emphasized in the society, sports achievements should be publicly announced.

However, at present the Lithuanian political expression is full of extremities of opinion. During the integration process to the EU structures everybody's attention was focused on political principles and the significance of national identity was nearly forgotten. There is also a possibility that it was deliberately ignored without a proper consideration of the EU regulations concerning national cultures. Nevertheless, as it is correctly summarized by philosopher Jokubaitis, "if there was no exclusivity of national culture, political independence would also lose its sense. A nation which is unable to preserve its cultural distinction cannot lay claim to political independence" [11]. Certainly, if one emphasizes the objectives of upholding the ethnic culture, it is not allowed to diminish most important political principles of the present day – human rights, expression of democracy, moral and cultural pluralism.

The attitudes to the further development of national states reveal two tendencies which can be called a quarrel between nationalists and liberals. The former critically assess the liberally conceived morality – excessive individualism, personal self-realization isolated from the society, depreciation of moral development ethics, and rejection of the idea of the generic good [12]. The residents of Central and Eastern Europe can still remember the times when there was no state, but they managed to preserve their identity by upholding their national culture [13]. Liberals, however, believe that the nation is a fictional, imaginary community or just a social construct, ideological mystification [14]. They raise human values above devotion and love for one's homeland. Nationalists, on the other hand, overemphasize multiculturalism and political egalitarianism. After all, the history of the tolerant politics in the Great Duchy of Lithuania is a great example for this. Nevertheless, most importantly citizenship stems from nationality, and this idea would not only eliminate a number of contradictions, but also would open new paths for the assembled work in order to reinforce the state.

The idea of a national state receives opposition from the dominant moral philosophies, for example, John Rawls' political liberalism [15]. In Lithuania there are active advocates of Jürgen Habermas's idea of constitutional patriotism [16]. It is likely that the trail of negative nationality assessment in Europe originates from the causes of the 2nd World War. There is no doubt that the oppressive impression was also created by the relatively recent events in the former Yugoslavia whose causes are still not clearly understood in the global community. In addition, misunderstandings appear because of the different terminology. For example, in some languages nationality is not distinguished from nationalism. Therefore, it is essential to discuss and define the meaning of the used terms.

On the other hand, the theoretical polemic is absolutely depreciated by the present-day political processes in Europe, for instance, the establishment of the Scottish Parliament in the United Kingdom or the preparations for a referendum on the status of Catalonia in Spain. Obviously, these processes are based on the idea of nationality, and thus prove its vitality.

The Practical Steps of Consolidating Identity

In Lithuania the positions of the national identity wing had to be reinforced in 1999 when the Law on the Principles of State Protection of Ethnic Culture was being passed. Nevertheless, one has to regret as most of the principles of this law, especially the ones connected with education, are not practically implemented. They are not implemented notwithstanding the General Programmes of Ethnic Culture for Schools of Basic and Secondary Education that were endorsed in 2012 [17]. The body that is responsible for implementation of the actions and is accountable to the Seimas of the Republic of Lithuania, - the Council for the Protection of Ethnic Culture – has turned into a routine bureaucratic body because of the defective principles of its formation.

The significance of scientific research of ethnic culture heritage and its exhibitions in museums is no longer questioned. Likewise, nobody doubts the importance of the volumes of Songbooks published by the Institute of Literature and Folklore. Research of ethnoculture and Lithuanistics is considered to be a priority research branch in the trends determined by the Research Council of Lithuania. However, when at the end of year the points given for research institutions for its publications are counted, these priorities mysteriously disappear. After all, these points determine the level of funding assigned for an institution for the following year… As if it is not evident that the state is created as a guarantee for the nation's survival and versatile evolution in the cultural context.

The society also has a controversial view of the attempts to revive and uphold the Lithuanian folklore, traditional customs and folk art. Even the opinions of the culture specialists diverge: they blame each other for ethnic sectarianism or cosmopolitanism. These extremities merely emphasize the sad situation our ethnic culture policy is finding itself in at the moment. The mass media broadcast only a couple of programmes representing the national culture. None of the television channels broadcasts a programme that would interpret the essence of Lithuanian folk art, folklore and customs in an ingenious, appealing way. Folk art is frequently equated to kitsch. Quality educational programmes are needed by both teachers and parents. The only tribute ethnic culture receives is sparse, superficial commentary from mass events.

The society's need for ethnic self-awareness is well reflected by the number of ethnographic ensembles. They were established during the period of the national rebirth, while getting rid of the soviet spiritual stagnation. Responding to the increased needs, the educational system of the Republic started training specialists who could lead the singing and dancing ensembles. Nowadays the specialty of ethnomusicology can be acquired at the Lithuanian Academy of Music and Theatre. However, there are hardly any students willing to study this programme, and the number of ensembles is also significantly decreasing. One of the reasons for these processes could be the absence of a solid system that would provide methodological support, would help to handle the issues concerning concert tours, the acquisition of concert costumes etc. In some parts of Lithuania this function is performed by centers of ethnic activity. Nevertheless, the situation is again problematic: the activity of such centers depends on the good will of municipality councils or on the way the heads of the regional culture divisions feel about ethnic culture.

The nationality priorities in the present-day cultural life as well as the urgency to pursue the national traditions can be grounded by the following arguments:

1. Ethnic culture is a continuous development of the all the nation – it happens in the family, in the circle of like-minded people, in the cell of the society, in the society as such. It provides the spheres of cultural expression with authenticity and viable traditions; and turns them into the spiritual heritage of the country;
2. During traditional holidays the commonality among members of the nation can be best felt; national self-awareness and the sense of citizenship develop and mature;
3. Ethnic culture is a certain way of acquiring universal spiritual values when people select the ones that correspond to the lifestyle and goals of the nation;
4. Ethnic culture is a basis for professionally created culture. It is open for various expression forms, therefore, it cannot be either compared to the Christian, professional European or global culture, or contrasted with them;
5. Ethnic culture unites the life of the nation, and by means of artistic forms it forwards the moral norms, ecological behavior and the principles of healthy lifestyle from generation to generation;
6. Traditions and national customs colour the nation's and state's way among the other European nations, make it interesting and attractive for the other members of the global community.

Nationality in the European Union

Today it is essential to determine the attitude of the European Union to nationality in the state politics. Moreover, it was not emphasized when Lithuania was joining the Union. A close analysis of the EU documents shows that the implicit position of the EU institutions is like this: the cultures of the member countries have to exist as individual and independent structures of those countries; and the function of the united Europe in the cultural sphere is just auxiliary – it has to guarantee the increasing contacts between the cultures, their cooperation, cognition and comprehension. Hence, the contribution of the EU institutions in the sphere of culture is focused on the enhancement of cultural cooperation. Cultural objects receive financial support, but it comes indirectly, through different programmes – promotion of tourism, countryside development, etc. (for example, support was given for the Lithuanian Museum of Ethnocosmology in Molėtai district, for the pottery craft center in Leliūnai, for the estate craft center in Zypliai). Such policy can be virtually acknowledged as an efficient factor of the vitality and development of every national culture since the cultural spread is implemented through interchange [18]. Cultural cooperation is based on the assumption that all national cultures have no difficulty in their individual development and cooperation with the other EU national cultures. Thus, the culture of the smaller nations should not face a greater danger of extinction than the cultures of the bigger European nations do. It has to be noted that this aspect is not included in the EU cultural policy at all. The upstaging threat is only seen in the context of American mass culture – i.e. globalization which comes with vigorous consumerism and pragmatism. Basically, the EU cultural policy encourages the state governments to take care of their national cultures so that every country would be really able to cooperate in cultural exchange and receive as much benefit as possible [19]. This is well-known for our specialists of culture who frequently express their concern "how interesting we can be for others"...

The EU institutions also have an implicit goal to uphold, create and develop the common European identity not only in the political, but also in the cultural sphere. The motif is clear and understandable: it will help to enhance the political identity of Europe and the EU unity [20]. It is interesting to notice such a metamorphosis: now it is becoming an efficient means against the leveling effect of globalization, and half a hundred years ago it was a weapon against the nationalism that flourished in the pre-war period. The orientation of the EU cultural policy to the common European cultural identity is not cosmopolitan [21]. Nevertheless, there is some concern since the concept of "European culture" has not been clearly defined yet in the EU documents.

The EU documents pay exceptionally great attention to regional culture. Such position is most probably useful for Lithuania: the historic events mercilessly diminished our ethnic lands, and such policy makes it possible to support the "Lithuanian islands" in Belarus, Poland and Latvia. In general, such attitude reflects a historically inherited and still living tradition of regional identity in Western Europe which has not been obliterated even by the establishment of national states.

The strategic EU documents emphasize the fact that in the future the role of culture will inevitably increase. It is being said that culture, its European dimension will become a way of the citizens' self-awareness and self-expression as the EU states will be more and more united due to economic integration [22].

What is the influence of people's free movement – student exchange, tourism, employment of specialists – on the development of national culture? It is ambiguous. Good news about Lithuania spread more and more widely – the culture promotes the country. But economically weaker EU member states, especially the new ones, will inevitably face bigger threats for their natural culture than the greater states. It becomes one more incentive to activate the country's attempts to preserve the national culture. This is what is surprising: having strong guarantees of citizens' freedom and rights, why don't we dare to return to the morality of the former idea of a national state and why don't we aim to exploit its strengths?

Conclusions

The historical time gives a different and changing significance for the concepts of nationality and citizenship in the arena of global politics. The collapse of totalitarian empires at the end of the 20th century caused a second wave of establishing national states which proves the vitality and power of nationality. In the European Union such processes are happening even today, which obviously demonstrates that citizenship results from the dimension of nationality. Hence, the reinforcement if ethnic culture is a task of national importance and it has to be efficiently solved in the institutions of education and culture.

References

1. Vabalaitė R.M. Modernioji tapatybės samprata // Lietuva globalėjančiame pasaulyje. Vilnius: Logos, 2006. – P. 129-148.

2. Jastrumskytė S. Reflektyvusis identitetas – lietuviškumas Europos akivaizdoje // Lietuviškojo europietiškumo raida: dabarties ir ateities iššūkiai. Vilnius: KFMI, 2006. – P. 101-115.

3. Roepstorff A., Simoniukštytė. Modernus lietuviškais identitetas – šimtmetį atkuriama tradicija // Lietuvos etnologija. 2001. Nr. 1 (10). – P. 145-166.

4. Klimka L. Pilietiškumas ir tautiškumas: laiko dimensijos ir aktualijos // Tautiškumas ir pilietiškumas: Atskirtis ar dermė? Vilnius: VPU leidykla, 20007. – P. 127-141.

5. Hėgelis G. Istorijos filosofija. Vilnius: Mintis, 1990.

6. Šešplaukis A. J.G. Herderis ir baltų tautos. Vilnius: Mokslo ir enciklopedijų leidykla, 1995.

7. Maceina A. Tauta ir valstybė // Naujoji Romuva. 1939, nr. 11. – P. 229-230.

8. Šalkauskis S. Dviejų pasaulių takoskyroje. Sintetinis esė apie Lietuvos tautinės globalizacijos problemą // Raštai. T.4. Vilnius: Mintis, 1995.

9. Asakavičiūtė V. Tarp Rytų ir Vakarų: tautinio tapatumo problema XIX a. pabaigos ir XX a. pradžios Lietuvos filosofijoje // Lietuvių tautos tapatybė. Tarp realybės ir utopijos. Vilnius: KFMI, 2007. – P. 194-217.

10. Girnius J. Tauta ir tautinė ištikimybė // Raštai. T.3. Vilnius: Mintis, 1995.

11. Jokubaitis A. Tautinė valstybė ir atsakomybė // Vertybių nykimas? Vilnius: Demokratinės politikos institutas, 2008. – P. 11-349.

12. Manent P. A World Beyond Politics? A Defence of the Nation State. Priceton, Oxford, 2006.

13. Rubavičius v. Neišgyvendinto sovietmečio patirtis: socialinė atmintis ir tapatumo politika // Lietuvių tautos tapatybė. Tarp realybės ir utopijos. Vilnius: KFMI, 2007. – P. 12-40.

14. Vyšniauskas G. „Lietuviškoji tapatybė“ – įsivaizduojamos bendruomenės ideologema // Lietuva globalėjančiame pasaulyje. Vilnius: Logos, 2006. – P. 162-181.

15. Rawls J. Lectures on the Moral Philosophy. Cambridge, London, 2000.

16. Habermas J. The Inclusion of the Other // Studies in Political Theory. Cambridge, 1999. – P. 105-127.

17. http://www.smm.lt/ugdymas/bendrasis/docs/EK%20programa%20Pagrindinis%20ug%20zin120412.pdf;http://www.smm.lt/ugdymas/bendrasis/docs/EK%20%20programa%20Vidurinis%20ug%20pask

18. Čičinskas J. Europos Sąjungos kultūros politika kaip nacionalinės kultūros veiksnys // Lietuva globalėjančiame pasaulyje. Vilnius: Logos, 2006. – P. 249-296.

19. Johler R. Europe, Identity Politics and the Production of Cultural Heritage // Lietuvos etnologija. 2002. Nr. 2 (11). – P. 9-22.

20. Rubavičius V. Naujas pilietinis europietiškumas ir racionalumo raiška // Lietuviškojo europietiškumo raida: dabarties ir ateities iššūkiai. Vilnius: KFMI, 2006. – P. 86-100.

21. EU directive 97/36/EC.

22. Tomlinson J. Globalisation and Cultural Identity. 2006.

Globalization and Education: Future Perspectives

Rimantas Želvys

Abstract

The paper investigates the influence of globalization on the development of education. Despite the different estimations of the extent and effect of globalization on education, it is obvious that globalization inevitably conditions the development of educational systems. Frequently the manifestations of globalization in education take the national states by surprise; hence, they are not always reacted to rationally and prudently. The paper presents three possible reactions: conservative ("we will continue following the same path"), fatalistic ("we will go with the flow") and prognostic ("it is essential to foresee the global changes as soon as possible and design an efficient policy of education"). The paper also reviews some future perspectives of education in this country on the basis of the current global tendencies.

Keywords: globalization, education policy, education change

Introduction

The term *globalization* is being increasingly used in the context of most varied aspects of social life. Initially the process of globalization appeared and developed as a consequence of economic integration, but later it started covering increasingly different spheres of people's social activity. Globalization is a term that is used in many ways, yet the definition provided by the World Commission on the Social Dimension of Globalization states that the principal underlying idea is the progressive integration of economies and societies (World Commission on the Social Dimension of Globalization, 2003). Frequently the effect of globalization on the social development is overrated, and sometimes it becomes the object of mythologization in the society (Rasmussen, 2003). On the other hand, the effect of globalization on the modern world cannot be overlooked. Despite the ambiguous estimations of the effect and consequences of globalization, there is one fact that causes no doubt: globalization cannot escape affecting the development of educational systems. According to Leišytė, who is researching the processes of globalization in higher education, in the face of globalization the state can be viewed in a variety of ways: as losing its power, as retaining its power or as transforming it. On the other hand, the state can also transform in the process of globalization (Leišytė, 2009). The transformation of the state conditions the changes of her social institutes, including education. Almost fifteen years ago, world famous education economist Martin Carnoy argued that globalization was making a huge influence on education and predicted that it would have even a greater influence in the future (Carnoy, 1999). The period of education development of the last decades visibly confirmed the correctness of this assumption. The fast rate of globalization frequently take the national states by surprise, hence the challenges of globalization are not always reacted to in the most appropriate ways. This paper aims to review the possible responses to the challenges of globalization.

The Extent of Globalization

In the present days of information technologies and fast communication, the processes of globalization manifest themselves even in the most distant parts of the world. In Asian, African, Latin American countries, in Australia and Oceania the systems of education are being reformed, and the extent and character of those reforms are close to the change of education happening in Northern America and Western Europe. Pasi Sahlberg, a famous researcher of education reforms, calls this phenomenon Global Education Reform Movement. According to the author, education reforms in different countries today share similar assumptions, values and characteristics due to the endless flow of information and harmonization of education policies through increased global educational borrowing and lending (Sahlberg, 2006). This idea can be illustrated by some eloquent examples. For instance, in Papua New Guinea the further plans of improving the system of education include prolonging the term of primary education since it is believed that this will help to guarantee a more efficient management of resources (Rena, 2011). In Nepal education experts prompt the modernization of the existing centralized system of school leaving examinations as it is insufficiently standardized. The achievements of pupils in this country can be hardly compared since every province gives different tasks in the test (Wagley, 2010). In Turkmenistan (Hasanov, 2013) and North Korea (Supreme Peoples' Assembly, 2012) the intention is to introduce the 12-year-long system of general education since the twelve-year-long learning at school is becoming a global aim of education.

Optimization of schools and classes, the centralized system of school leaving examinations, the twelve-year-long general education school – these are all undoubtedly tendencies of the global development of education. Why are these examples selected to illustrate the processes of globalization in education? Papua New Guinea is still viewed by the majority as one of the most distant and exotic places in the world. Nepal was closed for foreigners up to 1950. Turkmenistan and North Korea are considered to be the countries which are most secluded from the external world, which are hard to visit for tourists and which should be least affected by globalization because of their seclusion.

These and many other examples, which could also be presented, show the extent and rate of globalization. Most global phenomena take us by surprise in education. The representatives of the country's schools of higher education admit that they never expected such a fast and plentiful "drain" of the Lithuanian youth to foreign universities. A new challenge for the academic community is the open study programmes. Soon time will come when anybody online will be able to study the material presented by the best universities in the world. The minister of Science and Education of Lithuania recently expressed apprehensions that this might cause some universities close down. Some senior study programme assessment experts (including some experts from Western Europe) are still counting the computers in university classrooms when students have been carrying their iPads and using the service of wireless internet for a considerable time now. The world is quickly changing and our thinking is lagging behind the changes. Nevertheless, one way or another, the national states have to react to these challenges. Hence, what could be the national response to the challenges of globalization in education?

The Possible Reactions to the Challenges of Globalization

The first reaction is conservative: "we will continue following our path". Adherers of this approach prompt to continue forming the national system of education, initially taking into account the national mentality and historic context, partially returning to the roots of national education or searching for a distinctive, unique way. The critics of globalization quite frequently speak out in the public sphere criticizing, often justly, the flaws brought by globalization. The search for one's own way is by all means an interesting activity which can give significant moral satisfaction. This must be the reason why it is appealing to a certain part of the society. Unfortunately, most existing examples prove that this process does not answer the purpose in terms of both time and resources. All the time we encounter situations when, despite the wasted time and money, the countries which had been searching for "their own way" sooner or later have to return to the tracks of global development of education. The aforementioned Turkmenistan was searching for its distinct, unique way of development and in 1991 it switched to the system of 9-year-long secondary and 2-year-long higher education as the then president did not like "vague theories" (Turkmen Initiative for Human Rights, 2006). There were 22 years spent for searching for "their own way" and now the country is returning to the world practice. Eventually the government had to acknowledge the fact that the 12-year-long system of general education is a necessity of present-day modern education, and since September 1, 2013 Turkmenistan has taken over the 12-year-long system of secondary education.

Another example is the Bologna Process in higher education. During the period since 1999, when the Bologna Declaration was signed, this process has involved practically all European (and not only European) countries. One of the few exceptions is Belarus. At the beginning of 2013 the international expert assessment group received an application from the Belarusian non-governmental organizations to join the Bologna Process. This application reveals that the country still is not willing to reject the traditional five-year-long model of higher education and the centralized governing of higher education schools. Even if this application is going to be successful, the search for "its own way" will have cost Belarus at least fourteen years of integration into the international area of higher education. Similar examples can be found even nearer than Minsk. Just after the restoration of independence, in 1991 the then Supreme Council of Lithuania registered the Draft Law on Primary and Secondary Education of Republic of Lithuania. It included the following proposals: after the 6th form pupils attain a certificate of educational qualification. Afterwards children had to choose a type of school they want to attend further – a vocational school or a progymnasium; if they were unwilling to learn, they were allowed to drop out or to study several subjects at home. Now only a few education specialists still remember this fact, and the authors of the projects may not be willing to remember, but the aforementioned episode of the development of education in Lithuania is described in Lukšienė's book "Jungtys" (Lukšienė, 2000). Meilė Lukšienė, the major ideologist of the national school, whose 100th anniversary was recently solemnly commemorated, *en passant*, never rejected European and world tendencies of education development. On the contrary, in her works she projected the development of Lithuanian education in the wider context, and this fact can be confirmed by anyone who has read at least some of her works. Why, despite these and many other lessons of globalization, is searching for "one's own way" still fairly popular? Our response is simple: this way does not require a comprehensive cognition of educational science and the modern tendencies of education development, or, in other words, any self-taught thinker can consider himself an expert.

The second possible reaction is fatalistic: "we will go with the flow". If we are confident that globalization is such an invincible power, we can decide to passively accept whatever it brings us in the sphere of education. Moreover, there exist influential formal international organizations and structures which have been dictating education fashions to the world for several decades: Organization for Economic Co-operation and Development (OECD), United Nations

Educational, Scientific and Cultural Organization (UNESCO), the World Bank, the European Commission etc. Such an approach makes us passively implement the recommendations of these formal structures. It is frequently the case both in Lithuania and other countries of Central and Eastern Europe. When these countries joined the European Union, the local politicians often implement one or another education initiative arguing that those decisions were dictated by the European Commission. Some pessimistic theoreticians of education are speaking about the death of national ideas on education policy, while other educational scientists claim that comparative educology will disappear in a decade or two as there will be nothing to compare to. Quite frequently education (and not only education) officials even complain than earlier everything was directed from Moscow and now – from Brussels. And they want to say that we, the "tiny tots" of the European Union, cannot make any independent decisions. In the best case, this is the ignorance of the real situation, in the worst – malicious demagogy.

The absolute majority of educational documents, issued by the European Commission, are recommendatory. They present the possible trends of education development and certain guidelines, but do not impose any sanctions if those guidelines are not implemented or even not attempted to implement. Nobody has cancelled Chapter 3 Article 126 of the Treaty of Maastricht which states: "*The Community shall contribute to the development of quality education by encouraging co-operation between Member States and, if necessary, by supporting and supplementing their action, while fully respecting the responsibility of the Member States for the content of teaching and the organization of education systems and their cultural and linguistic diversity*" (European Communities, 1992). Thus, the European Union does not reject the independent policies of education of the member states, yet this situation is frequently a pretext to do nothing at all in the country's system of education. In other words, if the first type of reaction to globalization is conditioned by ignorance, the second type is the outcome of laziness and unwillingness to do something for the sake of the national system of education.

The third reaction is prognostic: theoreticians and practicians of education have to foresee the global tendencies of education development and, on their basis, form an active policy of education. The sooner those tendencies are anticipated, the more time we will have to prepare for their introduction and the greater freedom of action we will have. This is evidenced by numerous examples of successful and timely education reforms. When Lithuania started implementing the system of state school leaving examinations, in Central and Eastern Europe such a system existed only in Slovenia. Other countries of the region were intending to implement it or were having discussions on its necessity. Our early decision enabled us to create a system that seemed most relevant to us. The latecomers took our well-trodden path and implemented the models recommended by our consultants. The last on the list had to take over the established model even if it did not correspond to the context of the country – after all, the model had to match with the models of the other countries, i.e. it had to be comparable. Moreover, foreign consultants also recommended not trying to invent the wheel again and not to waste money on the creation of a new model since everything has already been designed.

Another example is the three-level system of studies of higher education. Lithuania switched to bachelor, master and doctor degree studies in 1994 – 5 years before the Treaty of Bologna. Nevertheless, the argumentation for the decision then was different from what we typically hear today – that under the influence of foreign experts, mainly Americans of Lithuanian origin, Lithuania expressed a preference for the model of higher education which dominated in Anglo-Saxon countries. However, as it appeared later, this decision was timely and successful. Back then this model was being implemented or was applied by very few countries of Central and Eastern Europe. Now the country has a twenty-year-long experience of the three-level system and representatives from the other post-communist countries come to learn from our experience. They study our study programmes attentively and rearrange their three-level systems according to the model we designed in due time. Moreover, back then we had a great freedom of choice, while their freedom of choice is limited to our experience.

This is the right time to speak about the practical benefit of the social sciences. As it is widely known, the public sphere is having a discussion, which was to some extent inspired by the officials of the Ministry of Education and Science, on which sciences should be a priority – the social sciences and humanities or the technological sciences. In summer of 2013 I read a critical article about the confused pupils who were still choosing the social sciences and humanities, while the popularity of the technological sciences was not increasing. Where did this article come from? In a South Korean newspaper which is published in English. It appears that South Korea, which is typically considered to be a leader in new technologies, is also having this problem. Thus, the target issue is neither national nor new. In about 1960 the Soviet Union was having a discussion between "physicists" and "lyrists". The circle of historic development has spun around and now we have returned to this discussion. Nowadays, in the times of efficient management of resources, one of the main arguments in the debate is that technological sciences create value for money, while the humanities and social sciences – do not. Without going deep into all the aspects of this discussion, it is possible to present one counterargument. Education is a big, complex, costly system; hence, the correct and timely projects of reforms can help to save not thousands or millions, but hundreds of millions. And conversely, useless reforms mean throwing hundreds of millions down the drain. It would be difficult to find a more successful value-for-money activity than the timely projects of global changes of education and appropriate reactions when implementing the necessary education reforms.

We have reviewed two fairly positive examples of education reforms in Lithuania. For the sake of objectivity, one example of a different character should be presented. When the system of vouchers in higher education was being considered in Lithuania, it was analyzed with one of the most authoritative experts of higher education from the USA - Aims McGuinnes. When he was asked for his opinion about this reform, he said that eventually the country would return to the pre-reform situation. Thus, the question if the system of vouchers in higher education is a step forward or a step back can be answered that it is neither the former nor the latter. We will stand about or walk around in circles, and sooner or later will return to the global tendency, i.e. tuition fees. Its size will depend on political decisions: two thousand or three thousand Litas, the same or different fee for all study programmes, compensated for the best students or not. It will be decided upon by the government of Lithuania. Nevertheless, we cannot remain the only country in Europe with a unique system of financing studies just as Belarus cannot remain a blank spot in the centre of the European map and not join the Bologna Process.

Conclusions

Hence, what are the future perspectives of education? According to the data of the Department of Statistics (Statistikos departamentas, 2013), the number of school-age children is going to decrease further, thus, fewer teachers and schools will be necessary. Respectively, the number of entrants to schools of higher education will also decrease. Consequently, we will have to reconsider the material and financial resources allocated to education and reassess the need for teacher training for all types of educational institutions. For one thing, the tendency of the decreasing number participants of the process of education does not put us in a positive frame of mind, yet, for another, the imperative to optimize the system creates favourable conditions to improve the quality of educators and the infrastructure of education. The standardization tendency is also going to be stronger revealed, thus there will appear more indicators and standards, tests and comparative studies. For one thing, standardization will increase the danger of bureaucracy, prompt the introduction of new mechanisms of education supervision and restrict the teachers' creativity. For another, intelligent orientation to the results will provide the conditions for rejecting a strict regulation of the education process and decentralization of the system of education. The emphasis is on "intelligent orientation" since the process and result cannot be equally strictly controlled: it would cost too much effort and money. If there is more orientation to the results of education, the process of education will have to be liberalized even more.

The decrease of students will make schools of higher education rearrange their internal structures and character of activity, with the increasing orientation to scientific research and students from non –EU countries. The schools of higher education which will not manage to re-orient themselves in time, will naturally disappear and there will be no need for "enforced" mergers. In our opinion, the Ministry of Education and Science of Republic of Lithuania has once wasted a lot of money on the preparation of several projects for the optimization of the system of schools of higher education. Even more money was spent by the ministry on the external assessment of universities which, as it is obvious now, is inevitably going to be improves and corrected. The only positive thing is that they did not manage to spend millions on the national system of rating universities as the European U-Multirank rating system is going to be introduced in 2014 (Van Vught, Ziegele 2011). All wiling universities of Lithuania will be able to participate in this rating system, prepared by some Western European centers of higher education. Thus, it is possible to assume that the rate and extent of globalization should encourage us not to passively accept the forthcoming innovations, but to actively and promptly respond to the changes in the global paradigm of education. The main advice for the developers and implementers of the education policy would be as follows: attentively follow the global tendencies, conceive what is to be done and what is not to be done in our system of education in time and globalization will not seem so frightening and determinative.

References

1. Carnoy, M. (1999) Globalization and Education Reform: What Plannners Need to Know. Paris: UNESCO.

2. European Communities (1992) Treaty on European Union. Access via the internet: http://eur-lex.europa.eu/lt/treaties/dat/11992M/htm/11992M.html#0001000001

3. Hasanov, H. (2013) Twelve-year secondary education to be introduced in Turkmenistan. Ashgabat: Trends. Access via the internet: http://en.trend.az/regions/casia/turkmenistan/2185059.html

4. Leišytė, L. Globalizacijos įtaka Lietuvos aukštojo mokslo kokybei ir institucijų autonomijai užtikrinti: teritorijų perskirstymas ir masto kitimas (2009). In.: Bulajeva, T. ir Duoblienė, L. Lietuvos švietimo politikos transformacijos. Vilnius: Vilniaus universiteto leidykla, p. 32-54.

5. Lukšienė, M. (2000) Jungtys. Vilnius: Alma littera.

6. Rasmussen, P. (2003) Global Change and Internationalization of Education. In: Rasmussen, P. (ed.) Education Policy and the Global Social Order. Aalborg: Aalborg University Printing Office, pp. 67-78.

7. Lithuanian Department of Statistics (2013) Education 2012. Vilnius: Lithuanian Department of Statistics.

8. Rena, R. (2011) Challenges for Quality Primary Education in Papua New Guinea – A Case Study. *Education Research International, vol. 11*. Access via the internet: http://www.hindawi.com/journals/edri/2011/485634/

9. Sahlberg, P. (2006) Education Reform for Raising Economic Competitiveness. *Journal of Educational Change, 2006*, pp. 1-29.

10. Supreme Peoples' Assembly (2012) 6th Session of the 12th Supreme Peoples' Assembly. Pyongyang: North Korean Economy Watch. Access via the internet: http://www.nkeconwatch.com/2012/09/25/6th-session-of-the-12th-supreme-peoples-assembly/

11. Turkmen Initiative for Human Rights (2006) Education in Turkmenistan. Report. Access via the internet: http://archive.chrono-tm.org/uploaded/2560683175898658.pdf

12. Van Vught, Frans & Ziegele, Frank (eds.). (2011) U-Multirank. Final Report. CHERPA-Network. Access via the internet: http://ec.europa.eu/education/higher-education/doc/multirank_en.pdf

13. Wagley, M. P. (2010) SLC syndrome: impact of the results. *The Himalayan Times Online, July 5, 2010*. Access via the internet: http://www.thehimalayantimes.com/

14. World Commission on the Social Dimension of Globalization (2003) The Social Dimension of Globalization. Geneva: International Labour Organization. Access via the internet: http://www.ilo.org/public/english/wcsdg/globali/globali.htm

Towards global identities: internationalization of teacher training at the Heidelberg University of Education

Henrike Schön, Anne Sliwka

Abstract

The article offers an insight into the complex problem area of strategic internationalization of teacher education as exemplified by the experience of the Heidelberg University of Education. This paper discusses three major issues: the state of the debate regarding internationalization, the obstacles and difficulties which counteract that implementation as well as the major forces driving it's implementation. This discussion highlights the difficult relationship between the demands of globalization and diversity in school life on the one hand and the yet to be established curricula for the internationalization of teacher training as a criteria of quality on the other hand. Finally, several aspects of the internationalization strategy of the Heidelberg University of Education are presented, from which the authors expect major advances in the training of teachers. The goal of teacher education must strive to sensitize teachers in the 21st century to adopt" global identities" and empower them to become cross-culturally competent to communicate and to act.

I.

The fundamental question of global identities in the 21st century asks what personality traits and skills are implied by this term. The definition of such traits and skills will have consequences in the context of university and school education and will require the internationalization of teacher education. In the second decade of the 21st century nothing less than training with an international focus can be organized for the training of young teachers. University faculty must address the implications of globalization and diversity and develop an education curriculum so that young people can seize the opportunities of globalization and overcome it's risks. In order for young teachers to be able to follow this path they must receive training which qualifies them professionally and personally to perform on the international stage or to deal with diverse cultures in their classrooms. Without intercultural communicative competence educator training in a heterogeneous world is hardly possible.

This paper will review the strategic internationalization efforts of the Heidelberg University of Education (Pädagogische Hochschule Heidelberg / PH HD). It deals with the current state of debate on the internationalization of teacher education and considers obstacles and difficulties in this process as well as the opportunities and driving factors. Finally, some aspects of the internationalization strategy of the University of Education Heidelberg are presented. The authors state general directions for the training of teachers which could lead to "Global Identities" which will sensitize and empower young teachers to become interculturally competent to communicate and act.

The Heidelberg University of Education trains teachers for the primary, secondary and special eduction schools. It also offers bachelor's degree programs in the field of education (Early Childhood Education and Health Promotion) as well as Master's degree programs with different profiles of Education (Educational Sciences, E-learning and Media Education, Foreign Languages). The majority of the 4,600 students enrolled in the teacher training courses, completes their studies with a state exam and strives for on appointment as teachers in civil service.

The university has supported the International Office for the past 15 years. The office currently coordinates 50 international university partnerships as well as the exchange and long term mobility of 250 students. Our university faculty and staff organize and execute study trips and excursions. The administration and faculty support the internationalization efforts. The International Office and the Vice President's Office for Research and International Affairs have provided major impetus to issues of internationalization of teacher education in Heidelberg.

During the past four years the discussion of internationalization efforts has been highly focused. Beginning with the establishment of a working group for an internationalization strategy (summer of 2010), the university progressed to the point that the University of Education was the only such institution in the state of Baden-Württemberg to complete the requirements of "Audit Internationalization of Higher Education" (Audit Internationalisierung der Hochschulen) of the German Rectors' Conference (Hochschulrektorenkonferenz / HRK). As a result, the university is in an excellent position with regard to the development of internationalization of teacher training – relative to other institutions in Germany.

Anne Sliwka, former Vice-Rector of the PH Heidelberg for Research, Internationality and Diversity and Henrike Schön, Head of the International Office and Erasmus Coordinator of the PH Heidelberg played leading roles in the discussion. This paper is the result of our focused discussions in many places and with many actors regarding this exciting, complex and multifaceted issue: Internationalization. This article attempts to capture the state of the debate from the perspective of an educational institution of higher education in the state of Baden-Württemberg, identify the factors that inhibit internationalization as well as those that drive it's development. Such efforts should provide an outlook on vital and essential factors of an internationalization strategy, which afford a custom fit internationalization program for teacher training.

Although significant progress was made during 2013, there is still no consensus or general social acceptance among the key players regarding the themes of internationalization of teacher education. Comparing the state of the debate regarding the international teacher training to a similar debate in the natural sciences or economics, huge differences can be discerned. Natural sciences and economics have a long history of internationalization and international networking. In teacher training one finds a strong nationally and regionally structured training, which, with the exception of foreign languages, has a difficulty in finding a rationale to identify and implement an internalization program. In addition, a socially accepted and practiced international networking at all levels of college and professional practice has yet to emerge. Such lack of consensus is due to the number of major players in teacher training such as: federal and state ministries, regional operating boards of education, the university administration, the rectors, deans and head of the institutes, the teachers and the students. Each group is responsible in their different functions for the quality of teacher education; however, each group assesses the need for internationalization in many very different ways.

I the past, the internationalization of teacher education have been promoted by convinced individuals: rectors and vice-rectors, some heads of the International Offices, professors with personal commitment and students desiring international experience. Despite a number of partners around the world, many universities lack a true commitment to internationalization. Many professors committed to international relations in their disciplines complain that their universities make little effort to recognize their efforts and few if any career advantages accrue to their commitment.

However, a paradigm shift seems to be emerging. Significant impetus toward internationalization has been achieved via international education studies and comparisons of academic achievement of pupils (TIMSS, PISA, IGLU) from countries across the globe. Research, conducted since 1998, resulted in the so-called „PISA shock" in which long held opinions of the quality of education of a given country were often found lacking. This shock has driven, the discussion regarding the quality of schools and teacher training. In addition, the public debate surrounding the reality in schools recognized a high degree of social and cultural diversity in the schools and has forced renewed thinking on the part of education officials and university administrations and faculty. The media in turn has recently focused on scandals and problems of the European education and mobility program ERASMUS. In addition to reports of scandal such as the fun and party generation, reputable reports of the low mobility of student teachers compared to participants from other disciplines were featured. The German Academic Exchange Service (DAAD) which acts as the independent National Agency for the Erasmus program in Germany confirmed on the basis of valid data that students of teaching positions in the exchange programs are significantly less internationally mobile than their fellow students in the sciences.

II.

This growing interest in teacher education is also reflected in the themes of meetings and conferences. For example the international Higher Education Fair and Conference "NAFSA" discussed in 2009, perhaps for the first time, the issue of "Internationalization Teacher Training Curricula". During the years 2012 and 2013 major organizations in Germany such as HRK, the DAAD and other intermediary organizations, have discussed the internationalization of teacher education in a level, manner and focus that would warrant the description of a "paradigm shift". Finally the recent conference in Vilnius on "Educational Science in Search of Global Identity" in October 2013 at the Lietuvos Edukologijos Universitet (Lithuanian University of Educational Sciences), internationalization received major attention. At the conference, a number of factors were identified which can either inhibit or promote internationalization. Some of these factors are described below.

The influence of ministries of culture on the framework and specifics of curricula in teacher education is considerable. Ministries send important signals regarding the design and execution of the programs of study for student teachers at the universities.

Reforms mandated by ministries can provide opportunities for the qualitative development of internationalization, but only under certain conditions: The reforms must meet the needs of the universities developmental requirements. The most recent reform 2011 in Baden-Württemberg correctly extended the period of study of primary school teachers from six to eight semesters. An international dimension however was not included during preliminary discussions. Thus there were no mandatory requirements to study abroad. Even foreign language teacher candidates are not required to study in the county of their language specialty. The provision of electives and options providing opportunities for internationalization and foreign study opportunities must be made available by the universities. An international dimension must be apparent in the curricula which also provides for "mobility windows" for faculty and students and includes the highest level of recognition, by decision-makers and stakeholders. A subsequent attempt to establish internationalization in developed curricula and degree programs which is already approved by the ministries is difficult and often impossible.

Other inhibiting factors are: requirements relating to the admission standards such as prerequisite degrees, the evaluation of language skills and various other qualifications of the students. These admission hurdles are particularly high in Baden-Württemberg.

Although the intention of the cultural and scientific ministries is to increase the numbers of foreign students, specific requirements often work against the academic success of such students. There is no question that a good command of the German language is essential to teaching in German schools. The poorly defined evaluation of foreign students German language skills often leads to an early termination of studies. In addition the universities fail to offer adequate support structures (German courses, tutorials, intercultural training, etc.). Frequently decisions about providing adequate help for foreign students are made by faculty and administrators based on narrow administrative policy and rules so that foreign students are not provided with chances to improve their language skills or are placed in teaching situations inappropriate to their language and cultural skill level.

In recent years, the relatively low mobility of prospective teachers was noticed. Studies by the DAAD on the ERASMUS program strongly suggest that aspiring teachers in ERASMUS significantly participate in fewer semesters abroad or in fewer foreign internships than students of other academic areas. This may have many reasons, which are currently under scientific investigation. One reason, however, is obvious: primary and secondary education and school life takes place in a countries official language. Thus teacher training is carried out in the local language. In fact, there are no good reasons why primary school teachers who will be teaching in Turkey or in Germany, should learn English, unless they qualify for a bilingual subject teaching. This fact presents a difficult and almost insurmountable obstacle to mobility.

Possible solutions to this problem may lie in models for joint multinational courses which are integrated into the teacher training in which foreign language requirements are significantly reduced but development and evaluation of intercultural skills are given high priority.

Another hindrance to mobility of teacher candidates is the lack of teaching positions available for students who have taught abroad. In Baden-Württemberg the assignment of teacher candidates to the few school openings is based solely on grades and family circumstances. No advantages accrue to an applicant who acquired a foreign language, studied abroad or who acquired intercultural competencies. As a consequence students in teacher training try to complete their studies in the shortest possible time in order to compete for the few openings available. It is hoped that the appointment to a teaching job will also be based on a candidates intercultural and language skills and sensitivity to diversity in schools.

If a critical analysis of the educational system is made, one could assume that the current system has favored those who are very immobile. However this conjecture must be examined empirically.

In contrast to hindrances there are, in fact, a number of factors which can promote mobility and internationalization. A major impetus is globalization. Globalization has already become a reality in the demographics of a school's pupil population. However, a huge discrepancy can be discerned between this very high diversity in the student body and the generally homogeneous structured teaching staff.

Here a circle is closed that begins with teacher training and ends in the cultural composition of school principals, school boards and ministries. The discussion of the diversity and heterogeneity of a school teaching staff and how the staff represents the demographics of the school's pupils and the society at large is extremely important, however this discussion should not lead to forms of positive discrimination in hiring teachers. The young Turkish woman is not a better educator or a language teacher simply because she comes from a bi- more aggressively cultural background.

There is no question, however, that educational institutions from the crib to the school in Germany have a serious need for multilingual and culturally competent educators. Thus teacher training universities will have to make corresponding changes in their curricula and hiring and recruitment practices. At this point, the recruitment of teachers must be pursued and in a more differentiated fashion in order to attract people with a migration background. It is also important to make the teaching profession more attractive just for this clientele. Unfortunately most of the very good graduates with non- homogeneous German background do not opt for the teaching profession, but study in the prestigious disciplines such as medicine or law.

Another factor that drives the internationalization of teacher education is the changed concept of the teaching profession. Professionalism of educators in the 21st century can no longer be based solely on a general culture-bound knowledge and personality traits. Curriculum content must be based on empirically testable, scientifically based, and thus globally valid knowledge. And this knowledge is not only the expertise in the secondary school teacher training in particular, but the professionally based knowledge of educational processes used in teaching students of all ages. Education as an academic discipline has developed a more international stance due to global surveys of academic achievement of students in various countries. However, this new research stance much result in a more international orientation in teacher training, based on research into educational processes.

Teacher training in Germany in this century stands unquestionably under pressure to innovate and such pressure is due to macro-societal challenges. The economic challenges confronting nation-states today, ultimately, impact the discussion of education, the quality of schools, the demands on content of teaching and on the issue of teacher training. In order for children and young people to obtain education and training they need for the world of the 21st century they will need teachers, who themselves are trained in an appropriate manner. Again, more internationalization of teacher training is necessary.

Corresponding to these trends and thereby generating synergies, internationalization has influenced universities in general and in particular teacher education. Over the past ten years strategic internationalization was demanded of the universities and funding was provided to promote and develop this stance. Almost all funding agencies require grant applications to provide highly specific statements regarding how the grant application fits into the internationalization strategy of the University. In particular, the German Academic Exchange Service (DAAD) and the German Rectors' Conference (HRK) require a university to adopt an internationalization stance and supports this stance by national and international workshops, seminars and conferences. This process has indeed reached teacher training albeit with a certain delay, but during the years 2012 and 2013 this process has increased dramatically. Both institutions have identified the internationalization of teacher education largely as a field with considerable pressure to act.

For the Heidelberg University of Education these pressures have produced recognizable consequences. The university has adopted many impulses, implemented them and developed them further. It has improved greatly in it's internationalization efforts and can enter into a dialogue with the University of Heidelberg. Following the summer term 2011 an internationalization strategy was approved by the Senate and by the University Council of the Heidelberg University of Education. The following autumn the PH HD entered into the "Audit Internationalization of Higher Education" with the HRK. During the one-year audit a "strengths and weaknesses" analysis, was completed and a recommendation report and a future workshop were proposed. Since all status groups of the university were included in the audit, the discussion of the necessity of internationalization was established on a much broader level. As a result, an awareness of the problem areas of internationalization at the university was identified. In addition, experiences and insights in the national and international discourse regarding the internationalization of teacher education could be introduced.

III.

Nevertheless, the PH Heidelberg faces major challenges to the internationalization of teacher education. Without a common internationalization cause, including the university management, the faculties and institutes internationalization may be only partially successful. This is a particularly sensitive issue since cooperation between The University of Education Heidelberg and the University of Heidelberg has been initiated. Thus new discussions of cooperation of specific scientific departments, subject didactics and educational sciences will take place, the results of which will shape or re-shape existing Bachelor's and Master's programs. At present, the following fields of activity can be identified, which can positively influence the internationalization effort. The creation of longer mobility windows is essential for students of all subjects not just students of foreign language. Such longer experiences in a foreign country are necessary in order to provide substantial experience with a gain in intercultural competence. Researchers, teachers and administrative staff must have the opportunity for professional development and to gain experience in a foreign country without this experience resulting in a detriment to his or her career. Such experience must serve the faculty member's occupational development. The much-quoted "internationalization at home", or internationally oriented measures at Heidelberg, such as guest lecturer programs or the establishment of English courses to improve mobility options, can be an enrichment at the university, but they are no substitute for personal experience in international contexts.

The creation of international mobility windows as well as a reorientation of the curricula must be mandatory and embedded in program of cooperation between grade schools and the university. A curricular anchoring of foreign mobility, for example, with the inclusion of intercultural competence into modules and specific course topics at the university must be mandatory. Such competencies must be an integral part of teacher certification and career promotion. The acquisition of language skills and the gain in intercultural communicative competence must be given the highest priority in teacher certification. Although such competencies have recently been expressed as a goal, students must identify themselves with these goals. Such measures cannot be emphasized enough in current descriptions of teacher competence. The competences must, in the future, be more precisely defined so that the gains in these competencies can be accurately expressed and credit points fairly assigned. Research and teaching can greatly benefit from the internationalization of education. The gain in knowledge of other teaching and educational cultures is essential in the international context of education. A science-based teacher training needs not only knowledge of the social processes of each region, but the accompaniment of an international online scientific discourse educational science. Finally, the educational landscape requires a new incentive system for the teaching profession: internationalization must be recognized for future teachers as a career advantage.

The Heidelberg University of Education has created with it's internationalization strategy, the justification for an internationally oriented educational institution of higher education and formulated a clear work program. With its implementation and further development at all levels of the university and the national and international dialogue, it will now be possible to make teacher education in Heidelberg sustainable and lasting.

References

Anderson, J.A. (2008): Driving change through diversity and globalization. Transformative leadership in the academy. Foreword by Ronald E. Crutcher. Sterling, Virgina.

Brewer, E./Cunningham, K. (Eds.) (2009): Integrating study abroad into the curriculum. Theory and practise across the disciplines. Foreword by Madeleine F. Green. Stearling, Virginia.

Deardorff, D. K./de Wit, H./Heyl, J. D./ Adams, T. (Eds.) (2012): The SAGE Handbook of International Higher Education. Singapore.

Deutscher Akademischer Austauschdienst (2013): Die Internationale Hochschule – Strategien anderer Länder. Bielefeldt.

Eriksson, B./Sieber, P. (Hrsg.) (2010): Internationale Zusammenarbeit in Bildungsfragen. Festschrift für Markus Diebold (=Internationale Perspektiven in Bildungsfragen, Bd.2) Wien.

Goetz, T./Jaritz, G. (Eds.) (2011): Pains and Gains of International Mobility in Teacher Education. Rotterdam, Boston, Taipei.

Hochschulrektorenkonferenz (Hrsg.) (2010): Kreative Vielfalt. Wie deutsche Hochschulen den Bologna-Prozess nutzen. Bonn.

Hochschulrektorenkonferenz (Hrsg.) (2012a): Bausteine für den Internationalen Erfolg. Stand und Perspektiven der Internationalisierung deutscher Hochschulen. Bonn.

Hochschulrektorenkonferenz (Hrsg.) (2012b): Die Deutschen Hochschulen internationalisieren! Internationale Strategie der HRK. Sprachenpolitik an deutschen Hochschulen. Bonn. (= Beiträge zur Hochschulpolitik 2/2012).

Hochschulrektorenkonferenz (2013): Empfehlungen zur Lehrerbildung. http://www.hrk.de/themen/lehre/arbeitsfelder/lehrerbildung/ (17.01.2014).

Kercher, J.(2013): Auslandsmobilität in der Lehrerausbildung. Vortrag im Rahmen der Tagung der IDA. (Unveröffentlichte Präsentation u.a. auf der Grundlage der DAAD/HIS-Mobilitätsstudie) Bonn.

NAFSA 2009: http://www.nafsa.org/Find_Resources/Internationalizing_Higher_Education/Network_Resources/Teaching_Learning_and_Scholarship/Internationalizing_Teacher_Education__Resources_from_NAFSA_s_2009_Colloquium/ (17.01.2014).

Nationale Agentur für EU-Hochschulzusammenarbeit im Deutschen Akademischen Austauschdienst (DAAD) (Hrsg.) (2013): ERASMUS Jahresbericht 2012. Bonn.

Pädagogische Hochschule Heidelberg – Heidelberg University of Education (2011): Internationalisierungsstrategie / Internationalization Strategy. http://www.ph-heidelberg.de/hochschule/ueber-uns/internationalisierung.html (17.01.2014).

Reil, T./Winter, M. (Hrsg.) (2002): Qualitätssicherung an Hochschulen: Theorie und Praxis. Bielefeldt. (= Forum der Hochschulpolitik, hrsg. v. d. Hochschulrektorenkonferenz)

Sterns, P.N. (2009): Education global citizens in colleges and universities. Challenges and Opportunities. London and New York.

Vbw – Vereinigung der Bayerischen Wirtschaft e.V. (Hrsg.) (2012): Internationalisierung der Hochschulen. Eine Institutionelle Gesamtstrategie. Münster.

Drawing a conclusion from Globalisation means rethinking our teaching and assessing our practice

Maren-Kristina Wilke

Abstract

This paper examines the links between globalisation and education with a primary focus on the importance of interdisciplinary competencies.

The leading questions are:

1. What does globalisation mean?
2. What challenges does globalisation set on education?
3. What consequences should educators and policy makers draw within these discussions?

The aim of this paper is to stimulate the rethinking of teaching practices in schools.

1. What does globalisation mean?

Globalisation seems to be an ever present element in our lives. Apparently, it effects our private life – as an example, through the insurmountable number of tablet PCs on the market. And it touches our professional life even more. We just need to realise the ever increasing speed of knowledge becoming obsolete so quickly.

The effects of globalisation on education may not be as obvious to us, but it will be for our future generations to come.

But what does globalisation actually mean? The following paragraph states what globalisation means in the present context of this paper:

Suárez-Orozco and Qin-Hillard (2004) point out that there is a lack of definition on globalisation. For some, globalisation just means free market, free travel or just capital flow. And the variety of the definitions of globalisation also varies with the number of disciplines. (Suárez- Orozco/ Qin- Hillard 2004: 13)

Coatsworth (2004) defines it as:

> "Globalization is what happens when the movement of people, goods, or ideas among countries and regions accelerates." *(cit. Coatsworth 2004: 38)*

This definition shows that globalisation is defined as something with a dynamic movement. But it still doesn't explain what globalisation exactly includes.

In addition to this, Suárez- Orozco and Qin- Hillard (2004) list a plethora of definitions from different disciplines and define globalisation in a summary of all perspectives, but with the focus on finding similarity in these definitions:

> "(...) we approach globalization as the process of change structured by four interrelated formations: (1) postnational forms of production and distribution of goods and service – fueled by growing levels of international trade, foreign direct investment, and capital flows; (2) information, communication, and media technologies that facilitate exchanges and instantaneously connect people across vast geographies and place a premium on knowledge-intensive work; (3) growing levels of worldwide migration; and (4) the resultant cultural transformation and exchanges that challenge traditional values and norms in both sending and receiving countries." *(cit. Suárez- Orozco/ Qin-Hillard 2004: 14)*

The citation shows, with the improvement in science and technology and the growing opportunities to travel, people have to face new challenges. This assumption is also confirmed by Menon (2007). (Menon 2007: 24)

For this paper the definition of globalisation follows the definition of Suárez-Orozco and Qin-Hillard, because it gives an understandable overview of the four areas which cause globalisation.

Besides this definition, Suárez-Orozco and Qin-Hillard (2004) give a nice example of the changes affected by globalisation. They describe that the world of our grandfathers began and ended at the beginning and the end of the village they lived in. (Suárez-Orozco and Qin-Hillard 2004: 2)

In contrast to the current world, the beginning and the end of the world is not as easily defined. Today, people can get all the information they want and need at any time. The more they have this access and the more they can travel, even more new perspectives spread out in the world and influence practices and thoughts. To carry the metaphor of Suárez-Orozco and Qin-Hillard further: now the world begins at the exit gate of peoples villages and leads to multiple ends which they have to design by themselves.

However, driven by progress of science and technology, globalisation is a process that challenges a series of domains.

The following paragraphs outline the challenges set on the domain of education, with the focus on interdisciplinary competencies needed to meet these challenges.

2. What challenges does globalisation set on education?

At the very least, the main effect of globalisation is that knowledge could become obsolete quickly. Hence any qualification one had aquired at any given time, will not last for an entire lifetime. More networking has to be done and international travel may follow. And what these small selection of the developments show is that the working environments and working tasks have become more complex.

The challenges set on education are enormous:

> *All experiences our youth and even adults make* "will be linked to economic realities, social processes, technological and media innovations, and cultural flows that traverse national boundaries with ever greater momentum." *(cit. Suárez-Orozco and Qin-Hillard 2004: 2)*

And as shown above, these links lead to huge complex areas. Thus, meeting these challenges require more than just having knowledge about a product or a process; it needs an education which helps to master these complexities.

Other than this professional knowledge, one needs personal competencies, like self-organisation and the ability to deal with conflicts. And one also needs social competencies, like communication skills and the ability to work in a team.

These developments will require a whole new teaching approach in our schools and universities. Teaching with the aim of just equipping pupils and students with knowledge will not be enough to master the new millennium world.

Suárez-Orozco and Qin-Hillard (2004) further describe that one of the main challenges in the new millennium world are the differences people have to face. They state that the possibility to migrate to other countries brings the challenge of having to deal with different cultural behaviors. (Suárez-Orozco / Qin-Hillard 2004: 4)

The constant increase of migration requires an educational change that gives the opportunity to learn and sense cultural differences, to understand other perspectives, accept these and to be tolerant of them. Within these perspectives pupils and students learn to take up other (cultural) perspectives and they also learn to consider multiple perspectives through other given situations. (ebd.)

Another change Suárez-Orozco and Qin- Hillard (2004) refer to is the increasing complexity of our lives we have reached and that there is still more to come.

> "(…)The skills needed [for this complexity] will require individuals who are cognitively flexible, culturally sophisticated, and able to work collaboratively in groups made up of diverse individuals." *(cit. Suárez- Orozco/ Qin Hillard 2004: 6)*

These skills can be taken together under the term interdisciplinary competencies.

What is needed to gain these competencies is an approach of teaching that gives the opportunity to gain such interdisciplinary competencies.

> "An education for globalization should therefore nurture the higher-order cognitive and interpersonal skills required for problem finding, problem solving, articulation ar-

guments, and deploying verifable facts or artefacts to substantiate claims." *(cit. Suárez-Orozco/ Qin- Hillard 2004: 8)*

These interdisciplinary competencies are as important as the basic knowledge in order to meet the challenges of globalisation.

Within the publication of the OECD "*The Nature of Learning*" Hanna Dumont and David Instance also follow the above arguments. Moreover, they describe these interdisciplinary competencies or in their own jargon "21st century skills", as follows:

> "higher-order thinking skills, learning to generate, process and sort complex situations, to think systematically and critically (…) to be adaptable and flexible to new information, to be creative, and to be able to identify and solve real-world problems. (…) Furthermore (…) students should develop into self-directed, lifelong learners, especially *(Dumont/Instance 2010: 23f) - and with it they refer to Darling-Hammond -* when education needs to prepare students (…) 'for jobs that do not yet exist, to use technologies that have not yet been invented, and to solve problems that we don´t even know are problems yet´" *(cit. Darling-Hammond 2008 in Dumont/Instance 2010: 24)*"

At this point, the perspective on an educational paradigm shift becomes very clear: Apparently we are living in the time of development which places the main responsibility on educators to rethink their teaching and assessing their practice in the direction which determines the approach they should apply. These skills cannot be taught by a teacher and it also cannot be taught without having self-determination and without being pro-active.

To support this paradigm shift the publication of the OECD "*The Nature of Learning*" gives a very interesting example. They summarize seven core principles in the design of learning environments, in which pupils and students can learn how to meet these challenges:

1. Learners should be seen as partners.
2. The learning environment should encourage learners towards co-operative learning.
3. Motivation and emotions must be considered by the teacher, because they are key players for and within the learning.
4. The teacher should be aware of the individual differences and prior knowledge of the students.
5. The seminars should be challenging but not overpowering the learners.
6. Expectations should be transparent at any time of the lesson and formative feedback should be applied.
7. Horizontal connectivity between subjects should be a primary concern. (Dumont/Instance 2010: 319-326)

Within these core principles, the paradigm shift becomes very clear and it may help us rethink our teaching.

3. What consequences should educators and policy makers draw within these discussions?

The above remarks show that teachers also have to face a series of new challenges in schools. This is a crucial point, because young teachers are usually in touch with new developments. But the ones who are already in school for a long time may not be used to that new perspective.

But it is important to ensure that these teachers can meet this paradigm shift as well, because this will influence their pupils' learning capacity.

In this context, Bloom (2004) said, policy makers need to invest in schools by increasing the possibility of inservice training to improve teachers personally and also to improve their quality of teaching. (Bloom 2004: 72)

Within the Hattie study (2009) it became obvious that a teacher's personality has a significant effect on the quality of teaching and learning.

> "Teachers are among the most powerful influence in learning." *(cit. Hattie 2009: 238)*

So the investment which Bloom (2004) stated was refering to the investment in the professional development of teachers.

Still further into the Hattie study, these remarks bring us to the point where it is not just about getting a new teaching kit but also a matter of changing old beliefs, habits and attitudes.

And Bloom (2004) also stated that a curriculum reform is badly needed.

> "Curricula....have failed to keep pace with the changes brought about by globalization." *(cit. Bloom 2004:73)*

It might be that some curricular reforms have failed, but in fact it´s not only the curricular reforms which failed but the implementation of it. And this again leads to the point where it is the teacher's personality which influences the levels of implementation. On one hand, this is because it is influenced by his or her own beliefs, habits and attitudes and on the other hand it is influenced by his or her professional development.

However, this could be summarized in the following to-do list:

Renewing curriculum

- Does the curriculum give the opportunity to apply competence oriented learning?

- Does the curriculum consider the request of the core principles?
- Can teachers use different forms of exams?
- Do teachers have enough flexibility to adapt teaching methods and exams in the class?

Improving the possibility to get inservice-training for teachers

- Do the teachers get time and space for inservice-training?
- Can the inservice-training be conducted as in-house training?
- Is the time allocated within their normal work hours or do they get real time off for these training opportunities?
- Will the teachers be used as a skill multiplier when they have been at an in-service training?

4. Prospects

Up to now it becomes clear that education needs a change from input orientation to outcome orientation, because this alone will enable students to become flexible human beings who can meet future challenges.

Additionally, it is really important to give current teachers and lecturers support for that change. This could be done by renewing curriculum design and increasing the possibility to get inservice training.

The fact remains that some of the above requests totally depend on politically-correct decisions. But the Hatti study also shows that the teacher's personality is the most influencing factor on learning. So the first step could be to support the paradigm shift by changing old beliefs, habits and attitudes.

But most of all, these can't be seen from a single perspective. So these "21st century competencies" should be an element to keep our eye on and keep working towards building a global identity for teacher training and also for educational science as a whole.

References

Bloom, D.E.: Globalization and Education. An Economic Perspective. In: Suárez-Orozco, M.M./Qin-Hillard, D. B. (Ed.): Globalization Culture and education in the new millennium.Berkley and Los Angeles, 2004.

Coatsworth, J.H.: Globalization, Growth and Welfare in History. In: Suárez-Orozco, M.M./Qin-Hillard, D. B. (Ed.): Globalization Culture and education in the new millennium. Berkley and Los Angeles, 2004.

Dumont, H./Istance, D./Benavides, F.: The nature of learning. Using research to inspire practice. Paris, 2010.

Hattie, J.: Visible learning. A synthesis of over 800 meta-analyses relating to achievement. Cornwall, 2009.

Menon, M.G.K. Globalisation and Education – An Overview. In: Sorondo, M.S./Malinvaud, E./Léna, P. (Ed.): Globalization and Education. Berlin and New York, 2007.

Ross-Holst, C.: Preface. In: Suárez-Orozco, M.M./Qin-Hillard, D.B. (eds.): Globalization Culture and education in the new millennium. Berkley and Los Angeles, 2004.

Suárez-Orozco, M.M./Qin-Hillard, D.B.: Globalization. Culture and Education in the New Millenium. In: Suárez-Orozco, M.M./Qin-Hillard, D.B. (eds.): Globalization Culture and education in the new millennium. Berkley and Los Angeles, 2004.

Lebensglück (ευτυχία στη ζωή) als Bildungsziel – als Katalysator für Identitätsfindung

Gerd-Bodo von Carlsburg, Martina Möller

Abstract

Happiness – a catalyst for identity determination?

Different to other languages there's only one term/word for "Glück" in the German language. In English you have different words: luck, bliss, happiness, fortune, felicity…

First you can make a difference between the objective side of happiness - you can have fortune - and the subjective side of happiness - to be happy.

An example for the objective side: You have a bad accident and by the will of god, you survive it without any kind of injury (blessure). Another example for the subjective side: You feel happy, because the weather is fine and the sun is shining on your body, you feel really good. It's a great day.

The old Greeks called happiness "**ευδαιμονία**". The idea of happiness has a long tradition and a very personal side.

The well-known Greek philosopher Aristotle claimed wisdom as the cardinal virtue of life. Wisdom or Sophia seemed for him as the unity of science and reason. Sophia is in his understanding an important virtue, because Sophia gives the community the greatest happiness. Bliss culminates in "well-being" by "well-doing". "Self-modesty" seems to be a key to bliss.

Bliss in the antique tradition of Aristotle is not the result of a hedonistic life.

The same perspective you can find by a serial of other philosophers, thinkers and psychologists. To struggle for more and more materials things, to struggle for more holiday adventures, for more fitness, for more beauty… it's not the way to became inwardly happy, feel blissed and a sense of life.

Therefor it's helpful to have a closer look in the philosophy of Viktor Emil Frankl, a man, who has survived as Jewish man a German "Konzentrationslager" (death camp) of the National Socialist regime. Also Erich Fromm has left us good advices, how to support human beings to live a biophilic life, to be a loving, creative and vital person.

The main questions are:

- How can teachers and school environment support young persons to become loving, creative and vital persons with happiness inside their hearts?
- Is there any chance to education youngsters in this way?
- And what are the professional requirements to do so?
- Is there a relationship between identity and happiness?

These will be the main subjects the following article will discuss.

Assoziationen zum ‚Glück' sind normativ determiniert und somit vielfältig ‚anders', d.h. nicht eineindeutig definierbar, da Normen einem ständigen Wandel und kulturethnologischen Evolutionen unterworfen sind, zugleich der Mikrokosmos von eigenen biografischen Erfahrungen und individuellem

Erleben geprägt, Glück rational und emotional anders erlebt wird. Doch zählt – fast erklärbar im Sinne eines angeborenen Auslösemechanismus' – das Streben nach Glück neben *Lebenskunst und Freiheitsdrang* zu den unveräußerlichen menschlichen Grundrechten, wie es insbesondere in der „Declaration of Independence" der Vereinten Staaten, verabschiedet im Kongress am 4. Juli 1776, verankert wurde:

Abb. 1: Lady Liberty

> *"We hold these truths to be self-evident, that all men are created equal, that they are endowed by their Creator with certain unalienable Rights, that among these are Life, Liberty and the pursuit of Happiness. -- That to secure these rights, Governments are instituted among Men, deriving their just powers from the consent of the governed, -- That whenever any Form of Government becomes destructive of these ends, it is the Right of the People to alter or to abolish it, and to institute new Government, laying its foundation on such principles and organizing its powers in such form, as to them shall seem most likely to effect their Safety and Happiness."*[1]

Im Museum im Sockel der Freiheitsstatue befindet sich auf einer Bronzetafel ein eingraviertes Sonett der amerikanischen Dichterin Emma Lazarus aus dem Jahre 1883, die 1903 vorerst am Sockel außen angebracht war und auf den Koloss von Rhodos verweist. Hier die freie Übersetzung des englischen Textes:

„Nicht wie der metallene Gigant von griechischem Ruhm,
Mit sieghaften Gliedern gespreizt von Land zu Land.
Hier an unserem meerumspülten hesperischen Tore soll stehen
Eine mächtige Frau mit Fackel, deren Flamme

Der eingefangene Blitzstrahl ist, und ihr Name
Mutter der Verbannten lautet. Von ihrer Leuchtfeuerhand
Glüht weltweites Willkommen, ihre milden Augen beherrschen
Den luftüberspannten Hafen, den Zwillingsstädte umrahmen.

‚Behaltet, o alte Lande, euren sagenumwobenen Prunk', ruft sie
Mit stummen Lippen. ‚Gebt mir eure Müden, eure Armen,
Eure geknechteten Massen, die frei zu atmen begehren,

Den elenden Unrat eurer gedrängten Küsten;
Schickt sie mir, die Heimatlosen, vom Sturme Getriebenen,

---

1 Quelle: http://www.archives.gov/exhibits/charters/declaration_transcript.html (18.12.2013). Abb. 1: Quelle: http://www.peterlang.com/index.cfm?cid=605 (Büro New York).

Hoch halt' ich mein Licht am gold'nen Tore!"[2]

Das Streben nach Glückseligkeit wurde als unveräußerliches Menschenrecht – gerade für jene Menschen, die geknechtet und Sklaven waren – festgehalten. Nicht zuletzt setzt hier auch der American Dream an, der nicht nur den sozialen und finanziellen Aufstieg (self-made-man) geschuldet ist, sondern essentiell auch die Entkopplung des Zusammenhangs von sozialer Herkunft mit ökonomischen Erfolg impliziert, das Versprechen, dass jeder Mensch ‚seines eigenen Glückes Schmied ist' und mit Mut, Fleiß, Engagement, Innovationen… sein Leben zum Besseren, zum Glücklichen zu wenden vermag – oder diesen Circulus vitiosus zumindest anstreben kann. Hierzu gehört der Wille und Freiheit, sein Leben so tatkräftig zu gestalten, sich als Subjekt so einbringen zu können, so autonom zu agieren, d.h. selbst zu entscheiden, dass letztlich das, was in der Aufklärung durch Immanuel Kant (1784) verdichtet wurde, zum Selbst-Credo stilisiert wird: „Habe Muth, dich deines *eigenen* Verstandes zu bedienen! …", d.h. „ohne Leitung eines andern".[3]

Das Streben nach Glückseligkeit – was immer das Individuum hierunter zu urteilen vermag – erscheint einerseits als anthropologisches Charakteristikum determiniert, andererseits definiert sich Glück durchaus als etwas Fragiles, einen epochalen Moment, der dem Menschen suggeriert, niemals dauerhaft zu sein, Glück als sicheren Besitzstand erscheinen zu lassen. Dies spiegelt sich in unserer Kultur insbesondere in Dichtung und Mundart wieder:

> *„Mit des Geschickes Mächten, ist kein ew'ger Bund zu flechten." (Friedrich v. Schiller, aus:* Das Lied von der Glocke *[1799; Z. 145])*
> *„Glück und Glas, wie leicht bricht das." (Sprichwort)*
> *„Glücklich ist, wer vergisst, was nicht mehr zu ändern ist."(Johann Strauß:* Die Fledermaus *[1874])*
> *„Jeder ist seines eigenen Glückes Schmied (Faber est suae quisque fortunae.)." (Sprichwort)*
> *„Das Glück ist mit dem Tüchtigen." (Volksmund)*
> *„Er/sie hatte mehr Glück (Schwein) als Verstand." (Volksmund)*
> *„Glück ist Liebe, nichts anderes. Wer lieben kann, ist glücklich." (Hermann Hesse)*
> *„Das höchste Glück im Leben ist, daß wir geliebt werden." (Victor Maria Hugo)*
> *„Das Glück gehört denen, die sich selbst genügen." (Arthur Schopenhauer)*
> *„Ein Augenblick des Glücks wiegt Jahrtausende des Nachruhms auf." (Friedrich II. der Große)*

---

2 Gedicht „The New Colossus" von Emma Lazarus: http://de.wikipedia.org/wiki/The_New_Colossus (18.12.2013).

3 Kant 1784, S. 481.

Glück und Unglück verbindet die Vergänglichkeit und damit den Lebensbezug, d.h.: Nutze jeden Augenblick. Das ist dein Weg zum Glück.

Das Glücksschwein steht für Freude und Feiern (z.B. Silvester, Kirmes…). Der Kranich als Glücksvogel symbolisiert die Sehnsucht nach Romantik, Ferne, Melodramatik… Der Scarabäus gilt in Ägypten als Glückskäfer (Symbol des Gottes Khephera [Sonnengottheit]) und wird als Amulett getragen, denn die in Wasser aufgelöste Puderform soll die Fruchtbarkeit fördern.

Abb. 2: Glück haben vs. Glücklich sein

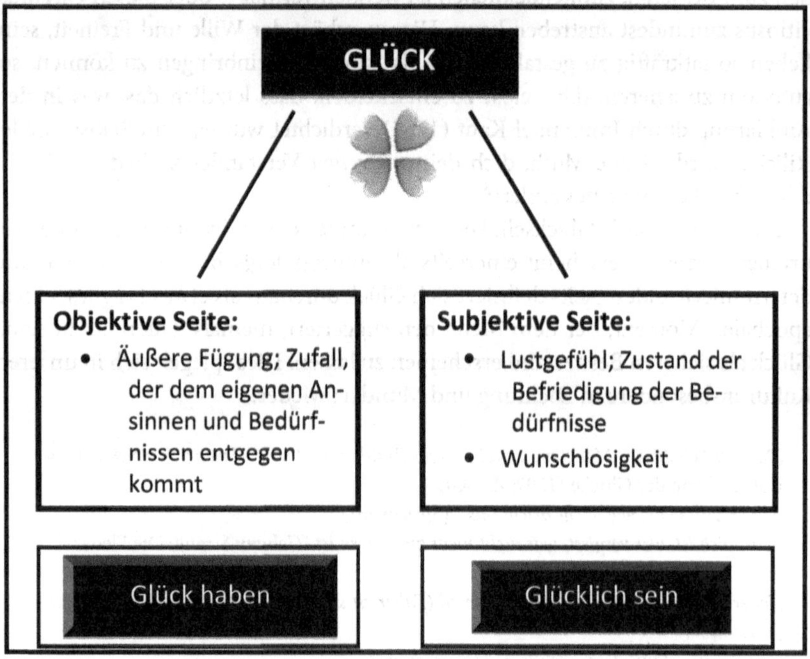

Bereits in den Sprichwörtern verdeutlicht sich somit folgende Assoziation: Jeder Mensch ist für sein Glück selbst *aktiv* verantwortlich, was letztlich auch *Innehalten* und Rückbesinnung im Sinne der Aristotelischen Maxime „Selbstgenügsamkeit" implizieren kann, somit auch die Schopenhauersche Deutung manifestiert.[4] Passivität führt zu Stillstand. Glück erscheint hier persönlich verantwortet, und der einzelne Mensch kann es nur subjektiv erfahren und leben.

---

4 Vgl. v. Carlsburg/Wehr 2013, S. 362-364; Schopenhauer 1902b, S. 368; Aristotelis de anima libri tres, 1833.

Auf der objektiven Seite gibt es Momente der glücklichen Fügung, Menschen berichten von Glück, wenn sie einen Unfall oder eine Naturkatastrophe wider alle Erwartung überlebt haben. Glück erscheint in diesem Falle als günstiger Zufall oder Schicksal, auf das der einzelne keinen Einfluss nehmen konnte, als Geschenk oder Gnade der ‚Vorsehung'.

Sowohl die objektive wie auch die subjektive Seite stellt für den Menschen etwas Erstrebenswertes und Positives dar. Trotz allem stellt sich die Frage, ob es auch ein Zuviel an Glück geben kann? Auch dieser Gedanke findet sich im Volksmund wieder: „Wenn dem Esel zu wohl ist, geht er auf's Glatteis." Hierin drückt sich die Annahme aus, dass ein zu glücklicher Mensch leichtsinnig und übermütig wird. Er fordert sozusagen sein Glück heraus und geht unvernünftig hohe Risiken ein, womit er schließlich seinen Glückszustand beendet:

- Glück als flüchtiger Zustand.
- Glück als vergängliches Moment.
- Ewiges Glück auf Erden nicht möglich. (?)

Permanentes Glück scheint dem sog. ‚normalen' Menschen nicht vergönnt zu sein, sehen wir einmal ab vom Einsiedler und dem Menschen, der den ‚Herausforderungen' entsagen kann! Aber empfindet dieser Mensch wirklich ‚Glück'? Niemand kann die Seele des Anderen schauen.

Abb. 3: Aristoteles (Αριστοτέλης) (384 v.Chr.- 322 v.Chr.)

Es sei dahingestellt, ob ein zu lange andauernder Glückszustand nicht langweilig wird, und der Mensch nach Affronts sucht bzw. sein Glück subjektiv gar nicht mehr spüren kann. Der Kontrast und die Balance zwischen Glück und Unglück oder dem Zustand des Strebens nach Glück scheint notwendig, um Glück als solches überhaupt *er*-leben zu können.

Bereits in der Antike galt ευδαιμονία – Glückseligkeit als Lebensziel. Aristoteles bringt in seiner Abhandlung „De anima libri tres (1833)/ Über die Seele (2011)" die Glückseligkeit in enge Verbindung mit der Tugendhaftigkeit, die als ein wesentliches Erziehungs- und Bildungsziel galt. Ein tugendhaftes Leben implizierte die Mäßigkeit und eine ausbalancierte Lebensführung jenseits der Extreme. Weisheit galt in der Antike als Kardinaltugend, die der Mensch erlangen sollte.[5]

> *„Weisheit (σοφία) stellt eine Einheit von beweisender Wissenschaft (επιστήμη) und Vernunft (νοῦς) dar; die Vernunft entdeckt die unbeweisbaren Grundprinzipien. Das Verhalten, welches den höchsten Wert besitzt, ist die Ausübung der theoretischen, anschauenden Lebensweise, die besagte Sophia (βίος θεωρητικός). Sie gewährt dem Menschen so das höchste Glück. Der Mensch als Gemeinschaftswesen soll aber auch die im gemeinschaftlichen Leben erforderlichen Tugenden ausüben."*[6]

Der Mensch wird bereits hier als ein vernunftbegabtes Wesen gesehen, das durch die theoretische Betrachtung, d.h. distanzierten und reflektierten Sicht auf die Welt, das Wahre und Gute erkennen kann. Einen Weg hierzu bieten sowohl die Vita contemplativa als auch die Philosophie. Innezuhalten, die Dinge zu betrachten und zu hinterfragen, ohne bereits Handlungsdruck zu haben, gehörte zu den Privilegien der oberen Klasse, die aus dem Arbeitsprozess weitgehend freigesetzt war und sich durch Bildung/Philosophie auf die Übernahme öffentlicher Ämter zum Gemeinwohl der Polis vorbereitete.

„Glückseligkeit (ευδαιμονία)" definiert sich für Aristoteles in der Tugendhaftigkeit der Seele.[7] Obgleich ein Mindestmaß an Gesundheit und materiellen Gütern notwendig für die ‚Glückseligkeit' erscheinen, erschöpft sich diese nicht in solchen Dingen. Vielmehr erscheint die ‚Selbstgenügsamkeit' als ein wesentlicher Schlüssel zur Glückseligkeit sowie die sinnliche Wahrnehmung und die Betrachtung (Erschauung). An dieser Stelle sei auf die Ideenlehre Platons verwiesen, nach der Mensch in der Lage ist, die Ideen hinter den konkreten empirischen Erkenntnisobjekten zu vernehmen und somit das Wesen/Urbild zu erkennen, da der Mensch alle Ideen bereits vorgeburtlich geschaut hat. Lernen erscheint in diesem Sinne ein Wiedererkennen der vorgeburtlich geschauten Ideen und bedarf des Innehaltens und der Vernunft.[8]

In seiner *Nikomachäischen Ethik*[9] stellt Aristoteles das Streben nach dem Guten, dem Tugendhaften als ein Charakteristikum der menschlichen Seele heraus. Tugend kann als Ergebnis der Natur, d.h. der menschlichen Seele, der Gewöhnung durch Erziehung und vernünftigen Handelns durch Bildung

---

5 Vgl. v. Carlsburg/Wehr, ebd., S. 360. Abb. 3: Aristoteles (Αριστοτέλης), Quelle: http://www.ulb.tu-darmstadt.de/media/ulb/fachinformation/hoelzinger/aristoteles_800_600.jpg.

6 Autorenkollektiv, ebd., S. 48; zit. ebenfalls in v. Carlsburg/Wehr, ebd., S. 360.

7 Vgl. Schwegler [1855] o.J. [ca. 1900], S. 158; v. Carlsburg/Wehr 2013, S. 361).

8 Vgl. z.B.: http://platon-heute.de/ideenlehre.html (4.9.2013).

9 Aristoteles 1985.

betrachtet werden. Der Mensch erscheint somit nicht per se tugendhaft von Geburt an, sondern bedarf der Erziehung und Bildung um seine menschliche Potenzialität zu entfalten. In einem tugendhaften Leben und Handeln findet der Mensch seine Glückseligkeit.[10]

A. Schopenhauer (1788-1860) rekurriert im V. Kap. – *Paränesen und Maximen* der *Aphorismen zur Lebensweisheit* – seines 1851 verfassten ersten Bandes *Parerga und Paralipomena* auf Aristoteles und dessen *Nikomachäische Ethik*.[11]

Lebensglück erscheint hier im Sinne einer Lebensweisheit und die Befähigung, sein Leben so vernünftig zu gestalten, dass Schmerz und negative Implikationen/Schäden möglichst vermieden werden. Arthur Schopenhauer formuliert es wie folgt:

> „A. Allgemeine.
>
> *1) Als die obere Regel aller Lebensweisheit sehe ich einen Satz an, den Aristoteles beiläufig ausgesprochen hat, in der Nikomachäischen Ethik (VII, 12): ὁ φρονιμος το αλυπον διωχνει ου το ἡδυ (quod dolore vacat, non quod suave est, perse quitur vir prudens. Besser noch deutsch ließe sich dieser Satz etwan so wiedergeben: ‚Nicht dem Vergnügen, der Schmerzlosigkeit geht der Vernünftige nach‘: oder ‚Der Vernünftige geht auf Schmerzlosigkeit, nicht auf Genuß aus.‘) Die Wahrheit desselben beruht darauf, daß aller Genuß und alles Glück negativer, hingegen der Schmerz positiver Natur ist. (...) Jeder Genuß besteht bloß in der Aufhebung dieser Hemmung, in der Befreiung davon, ist mithin von kurzer Dauer.*“[12]

Abb. 4: Arthur Schopenhauer (1788-1860)

Glück bedeutet in diesem Zusammenhang die Befähigung zum Zufriedensein mit sich selbst und mit dem Anderen, der Gemeinschaft, der Natur, mit

10 Vgl. Rolfes, E. (1911): Vorwort. Zu: Aristoteles: Nikomachische Ethik. Leipzig, Felix Meißner Verlag. http://gutenberg.spiegel.de/buch/2361/1 (4.9.2013); vgl. Schwegler, ebd., S. 159, v. Carlsburg/Wehr, ebd. S. 361. Abb. 4: Quelle: http://upload.wikimedia.org/wikipedia/commons/thumb/8/8a/Schopenhauer.jpg/220px-Schopenhauer.jpg.

11 Vgl. v. Carlsburg/Wehr, ebd., S. 363.

12 Schopenhauer 1902b, S. 367f.

Gott. ‚Selbstgenügsamkeit' attribuiert somit nicht das Streben nach Extremen – weder in materieller noch ideeller Hinsicht. Glück wird hier rein hedonistischen Bestrebungen enthoben und von dem Lusterleben distanziert.

> *„Der Thor lauft den Genüssen des Lebens nach und sieht sich betrogen: der Weise vermeidet die Uebel. (...)*
>
> *Das Verkennen dieser Wahrheit, durch den Optimismus begünstigt, ist die Quelle vielen Unglücks. Während wir nämlich frei von Leiden sind, spiegeln unruhige Wünsche uns die Chimären eines Glückes vor, das gar nicht existirt, und verleiten uns sie zu verfolgen: dadurch bringen wir den Schmerz, der unleugbar real ist, auf uns herab."*[13]

Parallelen erkennt man hier durchaus auch zur buddhistischen Lehre, denn Arthur Schopenhauer betrachtet das menschliche Dasein auf Erden tendenziell als Leidensgeschichte.[14] Im Buddhismus gelangt der Mensch erst durch den Eingang ins Nirwana aus dem Kreislauf der Leidensgeschichte von Geburt und Wiedergeburt heraus. Der Weg zum Glück erscheint ein Leben nach den Regeln Buddhas, das geprägt ist durch Meditation, eine asketischer Lebensführung und dem Streben nach der Wahrheit. Kontemplation ist hierbei ein wesentliches Moment auf diesem Wege.

Rekurriert man an dieser Stelle auf die Theodizee (θεία δίκη), die Gottesgerechtigkeit, so muss auf die Privationslehre verwiesen werden. Demnach ist im Ursprung alles wohl geschaffen, das Übel oder Böse in der Welt resultiert aus einem punktuellen Defizit – eben einem Mangel an Gutem. Dieser Gedanke zeigt Tradierungen von Augustinus (354 n. Chr.-430 n. Chr.) über Thomas von Aquin (um 1225-1274) bis zu Gottfried Wilhelm Leibniz (1646-1716) – dessen *Essais de Théodicée sur la bonté de Dieu la liberté de l'homme et l'origine du mal* 1710 erschienen.[15] Der Mensch müsste im Sinne des Jan Amos Comenius (1592-1670) Bildung erfahren, um danach Streben zu können, sich und die Welt im Sinne der Ideen Gottes zu vollenden. *Glück* kann hier *als Verheißung* betrachtet werden, als Gabe Gottes, wenn die Menschen nach seinen Geboten leben und Gottes Schöpfung zur Vollendung/Vollkommenheit verhelfen.

Der Glücksbegriff findet sich durchaus auch in der Bibel:

> *„Der Herr, dein Gott, wird dir Glück geben zu allen Werken deiner Hände." (5. Mose 30,9)*[16]

---

13 Schopenhauer, ebd., S. 369, vgl. auch Streminger 1992, S. 182.

14 Vgl. Streminger, ebd., S. 183.

15 Leibniz 1996, Bd. 2.1, S. 1-621; Bd. 2.2, S. 1-269, vgl. Leonhardt 2003, S. 75.

16 Weiterführend hierzu: https://www.ekd.de/aktuell/edi_2009_12_30_glueck.html (5.9.2013).

Obgleich bislang eine Zusammenschau unterschiedlicher Perspektiven erfolgte, zeigen diese jedoch eine entscheidende Gemeinsamkeit: *Wahres Glück bedarf der Bildung und Askese sowie der Ausrichtung eines transzendentalen Sinnhorizontes.*

Glück kann in diesem Zusammenhang somit durchaus auch als *Lebenskunst* betrachtet werden, die der Mensch sich erarbeiten muss. Dies betrifft vor allem die subjektive Seite von Glück. Lebenskunst beinhaltet in diesem Sinne nicht nur die Fähigkeit, Glück wahrzunehmen, sondern auch Glück zu teilen, es zu wahren und in die Welt hineinzutragen, anderen Menschen zugänglich zu machen. Insofern erscheint Glück durchaus mit der Persönlichkeit des Menschen verbunden, da es einen Zusammenhang gibt zwischen Glück, Identität und Persönlichkeitsentwicklung/-bildung.

Aus biografischer Perspektive heraus kann Glück als „ästhetisches Wohlbefinden beim narrativen Rückblick auf die eigene Lebensgeschichte“[17] umschrieben werden. Ein glücklicher Mensch muss demzufolge in der Lage sein, eine optimistisch-realistische, lebensgeschichtliche Narration zu entwerfen. Brüche, krisenhafte oder gar traumatische Erfahrungen müssen in die eigene Lebensgeschichte, das eigene Selbst integriert werden. Dem Individuum muss es gelingen, scheinbar zufälligen, auf den ersten Blick zumeist sinnlos erscheinenden (Krisen-)Erfahrungen retrospektiv Sinn zuzusprechen, um diese bewältigen und verarbeiten zu können.

Als wichtige Basis hierfür kann das *Urvertrauen* bezeichnet werden, dass der Mensch – folgt man Erik Homburger Erikson (1902-1994) – in der frühen Kindheit, exakter im Verlauf des ersten Lebensjahres, entwickelt, sofern es eine hinreichend gute primäre Bindung gibt, d.h. sensibel auf die Bedürfnisse des kleinen Kindes eingegangen wird.

Die moderne Bindungsforschung belegt den hohen Stellenwert der frühkindlichen Bindungserfahrung für die weitere Entwicklung und *Identitätsfindung* des Menschen.[18] Frühkindliche Bindungserfahrungen können als Grundlage für die Art und Weise, wie wir in Beziehung zur Welt treten, wie wir Identität entwickeln, betrachtet werden. Eine sichere Bindung bildet die Basis für das weitere Explorationsverhalten des Kindes und damit auch die Entfaltung der eigenen Potenziale. Einem sicher gebundenen Kind erscheint die Welt als positive Herausforderung, die es annehmen kann und der es offen und mit Neugier begegnet. Ein stabiles Selbstwertgefühl basiert auf Identität und erlaubt es dem Kind, mit Misserfolgen oder Rückschlägen produktiv umzugehen, ohne depressiv oder aggressiv zu werden.

---

17 Göppel 2010, S. 296, zit. in v. Carlsburg/Wehr, ebd., S. 366.

18 Vgl. Brisch 2013; Brisch/Hellbrügge 2008; 2012.

Mit umgekehrtem Reaktionen verhält sich das frühkindliche Wesen, wenn es in der frühen Mutter-Kind-Interaktion zu mangelnden Spiegelungsprozessen und fehlender Valorisierung kommt (vgl. Heinz Kohut [1913-1981])[19], kindliche Bedürfnisse traumatisierend missachtet wurden. Statt des gesunden primären Narzissmus', entwickelt sich der sekundäre, der sich später als narzisstische Persönlichkeits-/Identitätsstörung manifestieren kann. Der Mensch bedarf permanent der Bewunderung/Abstützung des fragilen Ichs von außen. Bekommt er diese nicht, so reagiert er mit narzisstischer Wut, eine ‚echte' störungsfreie Beziehungsfähigkeit ist nicht gegeben, die sozialen Strukturen sind belastet durch diese Persönlichkeitsstörung, die das eigene Ich stets grandios erleben muss, um sich seines Selbstwerts sicher zu sein. Der andere Mensch kann in seinem So-Sein nicht gesehen, anerkannt und wertgeschätzt werden, denn er wird instrumentalisiert für die eigene Bedürftigkeit.[20]

Nicht zu unrecht bringt Theodor Adorno (1903-1969) als Vertreter der Kritischen Theorie/Frankfurter Schule, den Stellenwert der frühkindlichen Beziehungserfahrungen in Verbindung mit dem Glücksbegriff:

> *„Glück ist nichts anderes als das Umfangensein, Nachbild der Geborgenheit in der Mutter.“*[21]

Bereits Johann Heinrich Pestalozzi (1746-1827) maß der Erziehung in der Wohnstube – und hier insbesondere der Mutter – in seinem anthropologischen Werk *Abendstunde eines Einsiedlers* wie auch in seinem berühmten Volksroman *Lienhard und Gertrud* zentrale Bedeutsamkeit für die Entwicklung des Kindes und Heranbildung einer die Sittlichkeit und den Gottesglauben vertiefenden ‚guten' Gesellschaft zu. Das Glück des Kindes, so könnte man es auch formulieren, liegt in den Händen der Mutter bzw. der primären Bezugspersonen als Geschöpf Gottes und somit in den Händen Gottes. Dem Kind bei seinem Start ins Leben Liebe, Wärme und Geborgenheit zu geben, seine primären Bedürfnisse zu stillen, stellt die Basis für eine gesunde – glückliche und Identität bildende – Entwicklung und somit den Grundstein für das Wohlbefinden des Kindes dar.[22] Letztlich bildet diese frühe Valorisierung das soziale Fundament zur Hinführung des Menschen als sittliches und autonom handelndes Wesens

---

19 Kohut 1993.

20 Vgl. Fally, M.: Ichhaftigkeit und Narzißmus. http://gestalttheory.net/download/narziss.pdf (5.11.2013); Kernberg/Hartmann 2010.

21 Adorno, Th.W. (1983): Minima Moralia. Frankfurt a.M., S. 143.

22 Vgl. Reinert [v. Carlsburg], G.-B./Arnhardt, G./Cornelius, P. 1996; Carlsburg, G.-B. v./Wehr, H. 2013a; Becker, M./Carlsburg, G.-B. v./Wehr, H. 2008.

im Kantischen Diktum, wie es auch Kerschensteiner als „Grundmerkmale der Erziehernatur" artikuliert:

> *„Das letzte Ziel aller Erziehung ist die sittlich autonome Persönlichkeit; der Weg zu ihr führt über die Güter Wert- und Lebensgemeinschaften (…)."*[23]

Rekurriert man die zeitgenössische Glücksforschung, so erfahren wir nach O.-A. Burow (2011, S. 102) vier zentrale Triebkräfte für „subjektives Wohlbefinden" (ebd.) und unser Glückserleben:

„→ *Glück als intensives, emotional positives Erlebnis, die ganze Person erfassend, überdauernd, sich im Lebenslauf entwickelnd;*
→ *Zufriedenheit als eher kognitive Einschätzung des eigenen Lebens als positiv, auf diversen Vergleichsprozessen (eigene Ansprüche, sozialer Vergleich) beruhend;*
→ *Freude als emotionaler Zustand des Sich-gut-Fühlens, eher kurzfristig, an konkrete Situationen gebunden;*
→ *Belastungsfreiheit als angenehmer Zustand der Unbeschwertheit und Entspannung, auf der Einschätzung der Abwesenheit von negativen Befindensfaktoren beruhend (Mayring 2007, S. 190)."*[24]

Die Reichweite als auch zeitlichen Dimensionen der genannten Burowschen „Faktoren" (ebd.) sind per definitionem Variablen, denn je nach biografischem Erleben und Identitätsentwicklung als auch Persönlichkeitsbildung sind sie unterschiedlich interpretierbar und veränderbar.

Der amerikanische Psychologe Martin Seligman (*1942) hat eine „Glücks-Formel" entwickelt, die konstante als auch variable Determinanten in sich trägt und somit in ihren Voraussagen entwicklungspsychologisch und erfahrungswissenschaftlich zu argumentieren versucht.[25]

Glück = Vererbung + Lebensumstände + Wille

Dabei erscheinen genetische Voraussetzungen und Lebenskonstellationen weniger leicht durch das Individuum beeinflussbar als der *eigene Wille*[26], der

---

23 Kerschensteiner 1930, S. 52.
24 Burow, O.-A. 2011, S. 102.
25 Seligman 2012, S. 85 (eigene Grafik); vgl. v. Carlsburg/Wehr, ebd., S. 366.
26 Vgl. v. Carlsburg/Wehr 2013, S. 367.

Abb. 5: Viktor Frankl (1905-1997)

somit zur ‚Stellschraube' des eigenen Glückserlebens wird. Der Mensch verfügt über einen freien Willen, womit sich der Bogen zur transzendentalen Betrachtung schließt. Der Mensch ist mehr als nur ein Reiz-Reaktions-Bündel, das Lust oder Unlust, Wohlbefinden oder Schmerz empfinden kann.

Der österreichische Neurologe und Psychiater Viktor Frankl (1905-1997)[27] setzt gerade an diesem Punkt in seiner *existenzanalytischen* Betrachtung an. Für ihn, der aufgrund seiner jüdischen Herkunft während der NS-Zeit in verschiedenen Vernichtungslagern interniert war und dort seine Familienangehörigen verloren hat, ist der Mensch vor allem ein *geistiges Wesen.*[28] Charakteristikum des menschlichen Daseins ist die ‚*Freiheit zur Stellungnahme*'[29], die innere Autonomie, der *Wille zum Sinn*[30] als entscheidender Faktor menschlicher Existenz, das freie Denken und Handeln, deren Voraussetzung letztlich „Selbstdistanzierung" und „*Selbsttranszendenz*"[31] impliziert.

> *„Existieren heißt, aus sich selbst heraus- und sich selbst gegenübertreten, wodurch der Mensch aus der Ebene des Leiblich-Seelischen heraustritt und durch den Raum des Geistigen hindurch zu sich selbst kommt. Existenz geschieht im Geist. Und sich selbst gegenüber tritt der Mensch insofern, als er qua geistige Person sich selbst qua psychophysischen Organismus gegenübertritt."*[32]

> *„In einem leer und sinnlos gewordenen Leben, in einem ‚existenziellen Vakuum' hypertrophiert der Hang und die Neigung, sich selbst zu bespiegeln, sich selbst zu beobachten, alles vor jedem auszubreiten, […] Ins existenzielle Vakuum wuchert anscheinend nicht nur die Neurose, sondern auch ihre Pseudotherapie".*[33]

---

27 Abb. 5: Quelle: http://upload.wikimedia.org/wikipedia/commons/f/fe/Viktor_Frankl2.jpg (19.8.2013).

28 Vgl. Waibel 2009, S. 70, Frankls (1991, S. 108ff.) Thesen zur Persönlichkeit rekurrierend, insb. Frankl 1990, S. 226).

29 Vgl. Waibel, ebd., S. 79.

30 Frankl 1991.

31 Waibel, ebd., S. 73, 79.

32 Frankl 1959, S. 665, zit. n. Riemeyer 2001, S. 66.

33 Frankl 1986, S. 42; zit. n. Waibel 2009, S. 73.

Gleich wie beelendend die gegenwärtige Situation des Menschen auch sein mag, so verfügt der Mensch doch über *'innere Freiheit'* im Sinne von Existenz.

> *„Dasein ist ein Sein, das immer noch entscheidet, was es ist."*[34]

Dies, was den Menschen letztlich zum Menschen werden lässt, ist die Fähigkeit zur *Selbstbestimmung*, zur Autonomie, sich selbst zur Chairperson seines eigenen Habitus' zu entwickeln, zu (re-)sozialisieren, und selbst ein Reflexionsrepertoire von der Innen- zur Außenperspektive zu konstruieren als sein eigenes anthropologisches Vermächtnis.

Über diese ‚innere Freiheit', die als *Katalysator für seine Identitätsfindung, als Lebensglück* definiert werden kann, als äußeres Vermächtnis, verfügt der Mensch, jedoch bedarf es der *Wertorientierung*, der Bildungsmächtigkeit, damit der Mensch – statt in ein existenzielles Vakuum zu verfallen oder zum Getriebenen seiner Bedürfnisse und Triebe zu werden – seiner Menschlichkeit, seinem *Humanitätssinn* und aus der Tradierung entwachsenem *Sein*, gerecht werden kann.

> *„Der Mensch ist nicht ein Ding unter Dingen, Dinge bedingen einander. Der Mensch jedoch ist insofern unbedingt, als er der Selbstbestimmung fähig ist. Er bestimmt jeweils, ob er sich von den Dingen bestimmen läßt oder nicht."*[35]

> *„Mensch-Sein ist immer auf etwas ausgerichtet, das nicht wieder es selbst ist – auf etwas oder auf jemanden, auf einen Sinn, den da ein Mensch erfüllt, oder auf ein anderes menschliches Sein, dem er da begegnet ist."*[36]

V.E. Frankl selbst gelang es, trotz Fleckfieber das Konzentrationslager Türkheim zu überstehen, da er versuchte, nachts sein Buch *Ärztliche Seelsorge*[37] zu rekonstruieren. Er konnte durch Selbsterfahrung erleben, wie es immer wieder Menschen gelang, die inhumanen, unwürdigen Lagerbedingungen zu überleben, wenn sie einen *Sinn im Leben*, eine lebensbejahende Identitätsförderung entwickelnd, sahen und als Ziel ihre Existenz vor Augen hatten.[38] „Ich hatte das Glück zu überleben". Diese Herausforderung bis an die Grenzen der Existenz lässt auch erahnen, warum Frankl den Terminus „Existenzanalyse" konstruierte.

Damit distanziert sich Viktor E. Frankl ein Stück weit von der Position Sigmund Freuds (1856-1939), der das Lust-Prinzip als zentrale Antriebsquelle

---

34 Frankl 1995, S. 225; zit. n. Riemeyer, ebd., S. 39.
35 Frankl 1980, S. 61.
36 Ebd., S. 183.
37 Frankl 2005.
38 Vgl. hierzu Frankl 1946.

im menschlichen Leben ansah, ebenso von Alfred Adler (1870-1937), dem Begründer der Individualpsychologie, der das Macht-Prinzip fokussierte, indem er diesen Prinzipien den *Willen zum Sinn*[39] als anthropologisches Streben entgegensetzte.

> *„Wenn nicht der im Menschen tief verankerte, unbewusste Willen zum Sinn (Frankl) gelebt werden kann, verfallen Menschen bei der Arbeit dem Willen zur Lust (Lustprinzip), wie es Siegmund Freud exzellent beschrieben hat, oder dem Willen zur Macht (Machtprinzip), wie es Alfred Adler schon richtig herausgearbeitet hat.“*[40]

Dabei leugnet V.E. Frankl nicht die Existenz der beiden anderen Prinzipien, sieht diese aber vor allem dann aktualisiert, wenn sich ein Mensch in einem ‚existenziellen Vakuum‘ befindet.

Somit stellt sich die Frage, wie ein Mensch sein *Lebens-Glück* finden will, wenn er stets mit sich selbst und der Frage konfrontiert ist, ob er wirklich glücklich ist. H. Ernst stellt die These auf, dass das Thema Glück als „heimliche(s) Leitmotiv des 21. Jahrhunderts“ die Diskussion kulminiert, d.h. dass wir in einer dem Glück ‚verfallenen' Sozialisation leben.[41]

Olaf-Axel Burow und Timo Hoyer sehen eine Verbindung von „Glückssuche“ und gesellschaftlichem Wandel.

> *„Das Bedürfnis nach Orientierung, nach Leitlinien eines guten, befriedigenden, erfüllten, eben glücklichen Lebens scheint in den Industrieländern unentwegt zu wachsen.“*[42]

Je mehr Optionen und Sinnhorizonte im menschlichen Dasein potenziell verfügbar sind, umso schwieriger erscheint der Umgang mit diesen Freiheiten. Ulrich Beck definiert diesbezüglich unser Gesellschaftssystem als „Risikogesellschaft“[43], da das Individuum zwar mehr Freiheiten besitzt, sein eigenes Leben autonomer gestalten kann als die Generationen davor, die Autonomie als Privileg ‚Auserwählter‘ *er*-lebten[44], auf der anderen Seite besteht durch mehr Freiheit, wenn sie beispielsweise als Anarchie missverstanden wird, auch die Gefahr des individuellen Scheiterns, d.h. die Entwicklung von Identität ist vergleichbar mit einem Pflänzchen, das einer professionellen Aufzucht eines/r Gärtner/in bedarf, der die Kunst des Lebens auch zu vermitteln in der Lage scheint. Der amerikanische Sozialkritiker Christopher Lasch (1932-1994) nennt seinen Band *Das*

---

39 Frankl 1991.

40 Graf 2005, S. 197.

41 Ernst 2012, S. 29, vgl. v. Carlsburg/Wehr, ebd., S. 367; Schmid 2012a, S. 20ff.

42 Burow/Hoyer 2012, S. 28; vgl. v. Carlsburg/Wehr, ebd., S. 369.

43 Beck 1986.

44 Vgl. Beck 2001; vgl. v. Carlsburg/Wehr, ebd., S. 369.

Zeitalter des Narzißmus, in dem das *eigentliche Ich* möglicherweise kleiner und fragiler denn je erscheint.[45]

Der Sozialpsychologe Heiner Keupp (*1943) proklamiert die „Patchwork-Identität“ als Identitätsform in der Gegenwart, die sich dadurch charakterisiert, dass jedes Individuum kreativ unterschiedliche Teilidentitäten – verkörpert in verschiedenen Rollen und Lebensbereichen – entwickeln und ausbalancieren muss. Es besteht die Notwendigkeit des „Boundary-Managements“, Grenzen müssen persönlich austariert werden, wobei das Individuum nicht frei von gesellschaftlichen und wirtschaftlichen Ansprüchen ist. [46]

Identität definiert sich nach Keupp nicht als Bildungsziel, das durch den junge Mensch bis zum frühen Erwachsenenalter einmal elaboriert worden ist und ein Stück Authentizität widerspiegelt, über die er dann sein Leben lang verfügt, sondern *Identität als Arbeits- und Erkenntnisprozess* wird zum permanenten, internalen aktiven *(Herstellungs-)*Prozess eines Aufbaus und Erhalts von *Kulturen.* Ja, sogar die Annahme, dass der erwachsene Mensch über eine annehmbare feste *Kern*-Identität verfügt, gewissermaßen ein AAM (angeborener Auslösemechanismus) – wie beispielsweise „Biophilie“ nach Fromm[47] – ist kritisch zu hinterfragen.

> *Erich Fromms (1900-1980) Biophiliebegriff als die „(...) >Liebe zum Lebendigen< beinhaltet Integration und strukturgemäßes individuelles als auch gesellschaftliches Wachstum. So manifestiert sich individuelles ‚Wohl-Sein' im strukturellen Zusammenhang. Eine humanistische Ethik der ‚Kunst des Lebens‘ wird deutlich, die sich (...) [dem humanistischen Gedankengut] verbunden fühlt. (...)“*[48]

Die Vision geht in Richtung einer ‚fluiden Identität' in einer „fluiden“ Gesellschaft[49], d.h. einer situativen und sequentiellen Konstruktion *adaptiver* Teilidentitäten. Der sog. ‚reflexiv-moderne' Mensch[50] bewegt sich in ganz verschiedenen, zum Teil durchaus auch widersprüchlichen Kontexten und muss seine Identität aktiv ausbalancieren. Dies impliziert allerdings fast ausschließlich die Gefahr einer essentiellen Hypertrophie, einer permanenten Selbstbespiegelung

---

45 Lasch 1980.

46 Vgl. Keupp, Vortrag vom 22.09.2003. http://www.ipp-muenchen.de/texte/identitaetskonstruktion.pdf (11.09.2013).

47 Vgl. insb. Fromm 1964a/GA II, S. 186; 1975c/GAV, S. 329f.

48 v. Carlsburg/Wehr, ebd., S. 385, Fußnote 27.

49 Keupp, Tagung vom 06.-07. April 2005. Folie 15. www.psz.co.at/img/30_jahresfeier/Festakt_ Vortrag_ Keupp_PPP.pdf (07.02.2011); vgl. v. Carlsburg/Möller 2011, S. 78.

50 Vgl. v. Carlsburg/Möller, ebd., S. 77-118.

als auch -suche, was letztlich Gegenteiliges bewirkt und selten zum *Glück* des Menschen reflexiv beiträgt.

Der Heidelberger Oberstudiendirektor Ernst Fritz-Schubert hat im Jahr 2007 das Fach Glück als Unterrichtsfach in der Willy-Hellpach Schule eingeführt. Zentrale Zielsetzungen seines Engagements sind:

- Vermittlung von Lebensfreude,
- Förderung der Persönlichkeitsentwicklung,
- Aufbau von Lebensbewältigungskompetenzen.

Obgleich diese Vision durchaus Plausibilität besitzt und wünschenswert erscheint, stellt sich die Frage, ob durch ein weiteres Unterrichtsfach die Kinder zur „Glückskompetenz" begleitend gelangen werden, oder ob nicht fächerübergreifend diese Konzeption durch alle zu vermittelnden Kompetenzen subkutan mit erlernt werden kann, wobei sich Fritz-Schubert gegen den Zweifel einer Erlernbarkeit von Glück, d.h. letztlich der sinngerichtete/-geleitete ‚richtige' Umgang mit Glück – wissenschaftlich, primär empirisch-analytisch –, ‚verteidigen' muss, insbesondere dann, wenn es um die Beantwortung der Frage geht, ob ‚Glück' auch als Katalysator zur Identitätsfindung determiniert werden kann.

O.-A. Burow (Gestaltpädagoge)/T. Hoyer bewerten das Unterfangen der Erlernbarkeit von Glück als Schuldisziplin äußerst kritisch und skeptisch:

> *„Das Glück, will man uns allerorten weismachen, wäre lernbar, herstellbar, beherrschbar. (...) Aber es gilt nicht für das Glück und auch nicht für das Wohlbefinden, nicht für Freude oder Zufriedenheit und schon gar nicht für die anspruchsvolle, schwer zu fassende Kategorie des glücklichen Lebens. (...). Es ist – zum Glück! – nicht einzustudieren, nicht trainierbar, nicht abrufbar, nicht anwendbar, kurzum: kein ‚learning outcome'".*[51]

Glück ist ebenso wenig wie Bildung von außen technologisch herstell- und formbar. Auch eine zu apodiktisch auf das Kind fokussierte, überbehütende und -fürsorgliche Erziehung bringt letztlich keine glücklichen Menschenkinder hervor, die gewappnet sind für die Herausforderungen des Lebens, die die Existenz auf dieser Erde schon als Glücksmoment definieren. Denn, um es mit den Worten Frankls zu artikulieren:

> *„Je mehr er sich selbst übersieht, je mehr er sich selbst vergißt, indem er sich hingibt einer Sache oder anderen Menschen, desto mehr ist er selbst Mensch, desto mehr verwirklicht er sich selbst. Erst Selbstvergessenheit führt zur Sensitivität und erst die Selbsthingabe zur Kreativität."*[52]

---

51 Burow/Hoyer 2012, S. 29; vgl. v. Carlsburg/Wehr, ebd., S. 368.

52 Frankl 1980, S. 184.

Die Erfahrung der eigenen „Selbstwirksamkeit“[53], das Aufbauen von Resilienzerfahrungen, Widerstand, das eigene Zutrauen zu mir selbst, das meine Zielsetzung sowie meine eigene produktive Tätigkeit bestimmt und nach Abschluss meiner Handlungskette vergleichend zu dem Ergebnis setzt, was man für ‚wert- und sinnvoll' befunden hat, wie groß ist die Übereinstimmung?, dies befördert *Glückserlebnisse*. Die subjektive Seite des Glücks setzt somit Bildungs- als Selbstbildungsprozesse voraus, die den Menschen auch in sein Verhältnis zur Welt setzen.

Wie bereits erörtert, setzt der Sozialpsychologe E. Fromm Glück in den Kontext einer „biophilen“ Lebensgestaltung, des „biophilen Charakters“, wie er es definiert.[54] Dabei grenzt er diese Orientierung vom sogenannten ‚Marketing-Charakter‘ ab, der sich in erster Linie daran orientiert, wie wirtschaftliche Märkte i.w.S., damit auch andere Menschen, das Individuum brauchen und handhaben können. Konsumorientierung, die Ausrichtung seines Lebens auf das Haben und nicht das Sein sind in diesem Zusammenhang dem tieferen Glück abträgliche Faktoren in der individuellen Lebensgestaltung. Glück bedeutet für E. Fromm, dass der Mensch *Antworten auf existenzielle Fragen* gefunden hat und mit sich und der Welt in Einklang, im Wohl-Sein, lebt.[55]

> *„Glück deutet darauf hin, daß der Mensch die Lösung des Problems der menschlichen Existenz gefunden hat: die produktive Verwirklichung seiner Möglichkeiten und somit das Einssein mit der Welt und das Bewahren der Integrität seines Selbst. Indem er seine Energie produktiv gebraucht, steigert er seine Kräfte: Er ‚brennt‘, ohne verzehrt zu werden. ... Glück ist das Kriterium der Tüchtigkeit in der Kunst des Lebens, also der Tugend im Sinne der humanistischen Ethik. (...) Das Gegenteil von Glück (...) ist die Depression, die aus innerer Sterilität und Unproduktivität entsteht.“*[56]

Erich Fromm geht davon aus, dass der Mensch nicht nur körperliche, sondern „existenzielle Bedürfnisse“[57] besitzt.

> *„Solche Bedürfnisse sind das Verlangen nach ‚Bezogenheit‘ – ‚Verwurzelung‘, – ‚Identitätserleben‘ – ‚Transzendenz‘, das Begehren nach einem ‚Orientierungsrahmen‘, Objekt der Hingabe und der Wunsch nach ‚Wirkmächtigkeit‘.“*[58]

53 Schwarzer/Jerusalem 2002, S. 28-53.

54 Fromm 1959b/GA IX, S. 340f.; v. Carlsburg/Wehr, ebd., S. 382f.

55 Fromm, E. [1974/75]/1989a/GA XII, S. 397; vgl. v. Carlsburg/Wehr, ebd., S. 375.

56 Fromm 1947a/GA II, S. 120.

57 Fromm 1955a/GA IV, S. 14, vgl. v. Carlsburg/Wehr, ebd., S. 371.

58 v. Carlsburg/Wehr, ebd., S. 372 mit Verweis auf Fromm 1973a/GA VII, S. 207-219 (Fromms Begriffsprägung der „Wirkmächtigkeit“).

Abb. 6: „Existenzielle Bedürfnisse" nach Fromm 1955a/GA IV, S. 14 (eigene Grafik)

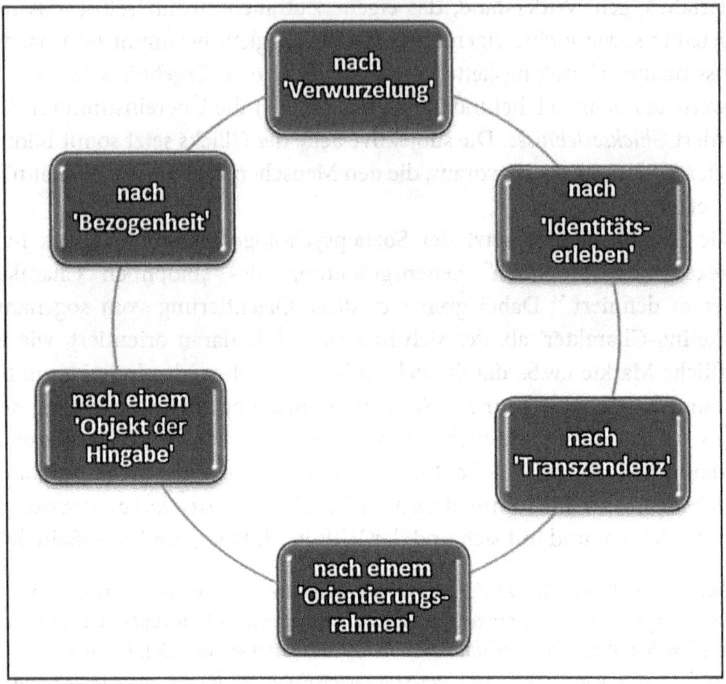

Der Mensch befindet sich somit in Spannungszuständen, die er zeitlebens niemals vollständig dauerhaft aufzulösen vermag. Diese Spannungszustände als essentielle und existenzielle Herausforderung an das Sein, die Identität evozieren, verlangen vielmehr nach einer dialektischen Lösung.

Damit stellt sich die Frage, inwiefern Unterricht einen realitätsbezogenen Beitrag leisten kann, damit junge Menschen unter Anleitung in vivo befähigt werden, sich zu ‚glücklichen Menschen' selbst heranzubilden?

Eine identitätsfördernde Bildungsarbeit erscheint hierbei als eine Möglichkeit, um Persönlichkeits-/Identitätsbildung zu evozieren, die sicherlich eine allseitige Kräfteentwicklung implizieren muss, statt zu einer kognitiven oder manuellen Verengung beizutragen. Nicht zuletzt muss auch der Unterstützung der moralischen Entwicklung eine zentrale Rolle zugesprochen werden. Für Frankl spielen das Gewissen und die Veredelung als ‚Verfeinerung' des Gewissens eine wichtige Rolle, geht es um die Ausrichtung auf den *Sinn.*[59]

59 Vgl. Frankl 1985/2011; 1991/2012; Waibel 2013.

„Auf der Suche nach Sinn leitet den Menschen das Gewissen. Mit einem Wort, das Gewissen ist ein Sinn-Organ. Es ließe sich definieren als die Fähigkeit, Sinngestalten in konkreten Lebenssituationen zu perzipieren.“[60]

Impulse hierfür lassen sich ebenfalls in der Themenzentrierten Interaktion (TZI) von Ruth C. Cohn (1912-2010) als auch bei Martin Buber (1878-1965) finden: ‚Dialogisches Prinzip', das ‚Personale Du', die ‚Echtheit' der ‚Ich-Du-Begegnung/Beziehung'. Es ließe sich ebenso auf die Pädagogik des brasilianischen Pädagogen Paolo Freire (1921-1997) verweisen, wie auch den polnischen Pestalozzi Janusz Korczak [Henryk Goldszmit] (1878-1942), der eine Pädagogik der gegenseitigen Valorisierung und Persönlichkeitsentwicklung, einer Autonomie des Kindes einforderte[61] und die „Rechte der Kinder“, insbesondere *Das Recht des Kindes auf Achtung*[62] proklamierte.

So war es Janusz Korczak ein Vermächtnis, die Not der ca. 200 jüdischen Waisenkinder lindern zu helfen und als ‚Vater', Pädagoge und Mediziner für diese Kinder zu leben, ihnen *Bindung* zu vermitteln und *Identität* aufzubauen, die ihnen auf grausame Weise genommen wurde. Und auch auf dem Wege in den Tod ließ er ‚seine' Waisenkinder nicht alleine, jedwede eigene Rettung ausschlagend – als diese zusammen mit ihm am 5. August 1942 vom Warschauer Ghetto in das Konzentrationslager Treblinka deportiert wurden.[63]

Paolo Freire (1921-1997) entwirft eine Befreiungspädagogik, fordert ebenso Autonomie des Denken und Handelns ein und fordert das Recht auf Freiheit und Bildung gerade für jene unterdrückten Menschen in Lateinamerika ein, die in Bildungsarmut verharren und ohne Chance sind, sich – weil ohne Lobby – Gehör zu verschaffen.[64] Freire fordert Bildung und Enkulturation als Chancen für Lebensglück und Identitätsentwicklung ein, definiert den Unterricht als einen dialogischen Prozess zwischen Lehrenden und Lernenden, aufgebaut auf Kongruenz, auf Handlungsaktivitäten, die durch unterstützendes, kooperatives Reflektieren zu neuen Erkenntnissen führen sollen, Unterricht als Katalysator für Identitätsfindung und -entwicklung, als Baustein für eine ‚glücklichere' Zukunft.

---

60 Frankl 2012, S. 24, Zitat aus Frankl 2012 (dt. Übersetzung eines Vortrags anlässlich der Verleihung der Ehrendoktorwürde an der Loyola University of Chicago [6. Januar 1970]). Engl. Originalfassung (1972). In: „American Journal of Psychoanalysis“, XXXII, No. 1, pp. 85-89.

61 Vgl. Korczak 2008.

62 Korczak 2007.

63 Vgl. Beiner 1982; 1984.

64 Freire 2007; 2007a; 2008.

All jene genannten Persönlichkeiten der Pädagogik stehen hier exemplarisch für Menschen, die Stellung bezogen und auf konkrete Lebenssituationen geantwortet haben. Letztlich sind noch zwei wichtige Wegweiser zu nennen: Ernst Christian Trapp und Wilhelm Frhr. v. Humboldt.

Obgleich dem Glück als Dimension im gegenwärtigen pädagogischen Diskurs nur eine sehr marginale Rolle eingeräumt wird – das Glück wird hauptsächlich als gesellschaftliche oder politische Dimension reflektiert, die personale, identitätsfördernde Dimension tritt hierbei in den Hintergrund – findet Glück als pädagogische Kategorie bereits im 18. Jahrhundert eine starke Berücksichtigung, s. u.a. Schopenhauer.

So umreißt der Philanthrop E.C. Trapp seinen Erziehungsbegriff in dem *Versuch einer Pädagogik*[65] in der *Erste(n) Abtheilung. Von der Erziehung überhaupt. § 11.* wie folgt:

> *„Erziehung* ***ist Bildung des Menschen zur Glückseligkeit.*** *(…) Daß der Zweck dieser Bildung Glückseligkeit sei, wissen zwar nicht alle, die sich damit befassen, wenn nemlich* ***wissen so*** *viel heissen soll, als sich Etwas deutlich denken. (…) Auch einer Kindermuhme kann man begreiflich machen, daß Glückseligkeit der letzte Zweck aller Erziehung, so wie alles menschlichen Bestrebens ist und sein muß."*[66]

Und im Erste(n) Abschnitt. Von der allgemeinen und besondern Glückseligkeit, als dem Zweck aller Erziehung. § 12.:

> *„Wenn es nicht bestritten werden soll, daß Glückseligkeit die letzte Absicht aller Erziehung sei und sein müsse: so wird man sich erst darüber zu vereinigen haben, worin sie bestehe. Sprachgebrauch und eignes Bewußtsein lehren uns, daß glückselig sein heisse, angenehme Empfindungen zu haben, und daß folglich die* ***Glückseligkeit ein Zustand angenehmer Empfindung sei.*** *In dieser Erklärung sind, meines Bedünkens, alle* ***Arten und Grade*** *der Glückseligkeit betroffen."*[67]

Abb. 7: Ernst Christian Trapp (1745-1818)

Schließlich im *Dritte(n) Abschnitt. Vom Unterricht. § 75.*:

---

65 Trapp 1780, unveränd. Nachdr. 1977; Abb. 7: http://e-ducation.net/pedagogues/trapp.jpg

66 Ebd., S. 25, Nachdr. S. 33, s. auch Burow/Hoyer (2011): http://www.uni-kassel.de/fb1/burow/downloads/Schule%20muss%20.pdf, S.3; Burow (o.J.): URL: http://www.win-future.de/downloads/zurrenaissancedesgluecksolafaxelburow.pdf, S. 3; Burow (o.J.a):URL:http://www.uni-kassel.de/fb1/burow/downloads/Gl%9Fck-G%9Attingen%C9Endfass..pdf, S. 2, ebenso Hoyer 2005, S. 175.

67 Ebd., S. 26, Nachdr. S. 34.

„(…) Allgemeinnützig (…) **ist theils** *Alles, was zur Erhaltung oder Wiederherstellung (…) des nothdürftigen Unterhalts …, unumgänglich nöthig ist;* **theils** *Alles, was uns zum … Umgang (…) mit Menschen geschickt zu machen unentbehrlich ist;* **theils** *Alles, was … den Anfang aller menschlichen Kenntnisse enthält, und (…) doch allen Menschen … nützlich sein kann, und den allermeisten unentbehrlich ist.*

(…) Indessen scheint mir diese letztere für die Pädagogik und Erziehung brauchbarer, wenn gleich jene in so fern von Kenntnissen und Unterricht die Rede ist, philosophisch richtiger sein sollte. (…) Erziehung ist Modifikation des Menschen in der Absicht ihn glücklich zu machen."[68]

„(…) Allgemeinnützig ist die Bildung alsdann, wann sie sich mit dem beschäftigt, was zum physischen und moralischen Leben des Menschen nöthig ist."[69]

Abb. 8: Wilhelm v. Humboldt (1767-1835)

Die Philanthropen betonen das aus Gottes Wurzeln sprießende Gute im Menschen, die von Gott gelegten sittlichen Vorgaben, auf die sie (Gottes-) Vertrauen aufbauen können, dass diese aus den Wurzeln sich bahnenden Säfte (Metapher Baum [J.A. Comenius]) bei entsprechend positiv begleitender Pädagogik – Einklang von Seele/Herz mit Vernunft/Ratio – ihren (vorbestimmten) Weg bahnen werden.[70]

Mit der Verdinglichung des Bildungsbegriffs in der deutschen pädagogisch intendierten Tradition und im 20. Jahrhundert der zunehmenden Verwissenschaftlichung und Rationalität wird das *Glück als pädagogische Kategorie* eliminiert. So bestimmt Wilhelm v. Humboldt die Aufgaben des Staates und die Zielsetzung von Bildung in seinem 1792 verfassten und 1851 in Breslau erschienenen Werk *Ideen zu einem Versuch, die Gränzen der Wirksamkeit des Staats zu bestimmen* wie folgt:

II

„Der wahre Zwek des Menschen – nicht der, welchen die wechselnde Neigung, sondern welchen die ewig unveränderliche Vernunft ihm vorschreibt – ist die höchste und proportionirlichste Bildung seiner Kräfte zu einem Ganzen. Zu dieser Bildung ist Freiheit die erste und unerlassliche Bedingung. Allein außer der Freiheit erfordert die Entwikkelung der menschlichen Kräfte noch etwas andres, obgleich mit der Freiheit eng

---

68 Ebd., S. 329f., Nachdr. S. 265f.; s. auch Burow/Hoyer (2011): http://www.uni-kassel.de/fb1/burow/downloads/Schule%20muss%20.pdf, S.3. Burow (o.J.): URL: http://www.win-future.de/downloads/zurrenaissancedesgluecksolafaxelburow.pdf, S. 3; Burow (o.J.a):URL:http://www.uni-kassel.de/fb1/burow/downloads/Gl%9Fck-G%9Attingen%C9Endfass..pdf, S. 2.

69 Ebd., S. 333, Nachdr. S. 268.

70 Vgl. Lausberg 2008: http://www.tabularasa-jena.de/artikel/artikel_3288/ (04.01.14).

verbundenes, Mannigfaltigkeit der Situationen. Auch der freieste und unabhängigste Mensch, in einförmige Lagen versezt, bildet sich minder aus. (...) Jeder Mensch vermag auf Einmal nur mit Einer Kraft zu wirken (...). Denn auch durch alle Perioden des Lebens erreicht jeder Mensch dennoch nur Eine der Vollkommenheiten, welche gleichsam den Charakter [die ***Identitätssuche****, d.V.] des ganzen Menschengeschlechts bilden. (...) Diese Kraft nun und diese mannigfaltige Verschiedenheit vereinen sich in der Originalität, und das also, worauf die ganze Grösse des Menschen zuletzt beruht, wonach der einzelne Mensch ewig ringen muss, und was der, welcher auf Menschen wirken will, nie aus den Augen verlieren darf, ist Eigenthümlichkeit der Kraft und der Bildung."*[71]

III

„(...) Der Zwek des Staats kann nemlich ein doppelter sein; ***er kann Glük befördern, oder nur Uebel verhindern wollen,*** *und im lezteren Fall Uebel der Natur oder Uebel der Menschen. Schränkt / er sich auf das leztere ein, so sucht er nur Sicherheit, und diese Sicherheit sei es mir erlaubt, einmal allen übrigen Zwekken, unter dem Namen des positiven Wohlstandes vereint, entgegenzusezen."*[72]

Dies definieren auch v. Carlsburg/Möller in ihren *Prolegomena zu einer ‚reflexiv-modernen' Pädagogik*:

„Natürlich hat Schule gemäß ihres staatlichen Auftrags immer auch mehr zu leisten als eine reine Persönlichkeitsentwicklung. Allerdings darf nicht der Anspruch der (ganzheitlich geprägten) Allgemeinbildung und das Humboldtsche Bildungsideal des „autonomen Individuums" und dessen Vernetzung in einem „Weltbürgertum" pragmatischen Nützlichkeitserwägungen geopfert werden. Wilhelm Frhr. v. Humboldt (1767-1835) definiert Bildung u.a. als exemplarische Kraft für ein mündiges und zukunftsorientiertes Handeln und entwickelt ein Modell, das ebenso einem dynamischen Gleichgewicht geschuldet ist, wie bei R.C. Cohn und den integrativen Therapeuten, ohne dies explizit als solches in seiner 1793 verfassten ‚Theorie der Bildung des Menschen' zu artikulieren...".[73]

D.h. grundlegend für die Ausbildung aller Kräfte zu einem ‚*Ganzen*', will sagen ganzheitlichen Ineinanderfließen (Denken, Handeln, Verhalten) sind für W. Frhr. v. Humboldt nicht nur das Zusammenwirken vielfältiger Lebens- und Handlungssituationen, sondern die klassisch-humanistisch geprägte Tradierung, die sich den alten Sprachen verpflichtet fühlt, das Zurück zu den sprachlichen Quellen im Melanchthonschen wie Humboldtschen Sinne, eine Klassik, die auch Empathie, Kompetenzförderung im personalen, sozialen und fachlichen Bereich, *Bildung und Enkulturation als Basis sozialer Kompetenz* nicht ausschließt.

71 v. Humboldt 1980, S.64f. Abb. 8: Humboldt 1985, zw. S. 4/5.

72 Ebd., S. 70 (Hervorhebung durch d. Verf.)

73 v. Carlsburg/Möller, ebd., S. 110.

Allerdings definieren Münch/Wyrobnik das Humboldtsche Bildungsideal als „kontemplativ", d.h. vorwiegend betrachtend, anschaulich. Sie sehen allerdings „Emotionalität und Sozialität" als „wichtige Quellen des Glücks".[74]

Benner bezieht sich in seiner problemgeschichtlichen Interpretation der *Humboldtschen Bildungstheorie* gerade auf den ersten Satz des Zitats von Kap. III und meint dazu:

> *„Richtig verstanden faßt Humboldts Formel von der höchsten und proportionierlichsten Bildung aller Kräfte zu einem Ganzen die auf den Zweck des Individuums bezogene Seite menschlicher Bildung als eine unabschließbare Aufgabe. (...) Höchste und proportionierlichste Bildung aller Kräfte zu einem Ganzen stellt somit eine Aufgabe dar, die wir nie ganz und nie endgültig erfüllen können, sondern mit Notwendigkeit immer auch wieder verfehlen. (...) An unserer Bestimmung aber können wir nur arbeiten, wenn wir eine* **Identität** *anstreben in der wir (...) unsere Bestimmung aus unserer nicht abschließbaren Arbeit an unserer Bestimmung gewinnen."*[75]

Burow interpretiert W. v. Humboldts Bildungsideal derart, dass Vernunft und umfassende Entfaltung aller menschlichen Potenziale, bei v. Humboldt die inneren (Bildungs-)*Kräfte* (Plural), im Dreh- und Angelpunkt stehen sollen,[76] sie allein führen zu Lebensglück als Lebenskunst. Glück lässt sich nicht einfach als ‚Leitkategorie pädagogischen Handelns' generieren.[77]

Münch/Wyrobnik betonen, dass in der Humboldtschen Bildungstheorie eine Engführung auf die intellektuelle Bildung erfolgt, schließen aber nicht aus, dass Menschen dank der humanistisch-klassischen Bildung auch zu einer höheren Lebenszufriedenheit gelangen mögen. Hierbei artikulieren beide Autoren den Transfer zu Johann Wolfgang v. Goethes „‚Glück der Persönlichkeit'" mit Verweis auf den 1819 verfassten *West-östliche(n) Divan* und kanalisieren diesen Ansatz in Richtung einer „‚glücksbetonten' Auffassung der Persönlichkeit".[78]

> *Suleika:*
>
> *„Volk und Knecht und Überwinder, / Sie gestehn zu jeder Zeit: /* **Höchstes Glück der Erdenkinder / Sei nur die Persönlichkeit**.

74 Münch/Wyrobnik 2011, S. 31; vgl. v. Carlsburg 2009.

75 Benner 1990, S. 52 (Hervorhebung durch d. Verf.).

76 Vgl. Burow 2011a: URL: http://www.uni-kassel.de/fb1/burow/downloads/Burow Kunst.pdf, S. 1ff. (04.01.14).

77 Ebd.

78 Münch/Wyrobnik, ebd., vgl. auch S. 30.

Jedes Leben sei zu führen, / Wenn man sich nicht selbst vermißt; / Alles könne man verlieren, / Wenn man bliebe, was man ist.“[79]

v. Goethe kehrt am 18. Juni 1788 aus Arkadien („*Et in Arcadia ego*“) nach Weimar zurück, unterwegs dichtet er weiter an seinem Schauspiel *Torquato Tasso*, um den Schmerz, Rom verlassen zu müssen, zu ertragen.[80] Am 30. Mai 1814, nach 26 Jahren, offenbart er seinem Kanzler Friedrich v. Müller, dass er seit Rom „keinen rein glücklicheren Tag mehr gehabt“ habe.[81]

„***Prinzessin.***
Eleonore! **Glücklich?**
Wer ist denn glücklich? – *Meinem Bruder zwar*
Möcht' ich so nennen, denn sein großes Herz
Trägt sein Geschick mit immer gleichem Muth;
Allein, was er verdient, das ward ihm nie.
Ist meine Schwester … glücklich?
Das schöne Weib, das edle große Herz!
Sie bringt dem jüngern Manne keine Kinder; (…)
Was half denn unsrer Mutter ihre Klugheit? (…)
Sie ließ uns Kindern nicht den Trost, daß sie
Mit ihrem Gott versöhnt gestorben sey.

Leonore.
O blicke nicht nach dem, was jedem fehlt;
Betrachte, was noch einem jeden bleibt!
Was bleibt nicht dir, Prinzessin?
(…)

Prinzessin.
Ich bin gesund, das heißt, ich bin nicht krank;
Und manche Freunde hab' ich, deren Treue
Mich glücklich macht. (…)“[82]

Burow – unter Nennung von Hoyer – konstatiert, dass im Hinblick auf v. Humboldt beide folgende Erkenntnis gewonnen haben:

---

79 v. Goethe, o.J., Kap. 9: http://gutenberg.spiegel.de/buch/3656/9 (Hervorhebung durch d. Verf.); Münch/Wyrobnik, ebd., S. 31 u. vgl. ebd., S. 30.

80 Vgl. v. Carlsburg 2011, S. 69; Heinemann o.J., Bd. 14, S. 5, 7 u. 430; vgl. auch Koenig 1881, S. 457.

81 Burckhardt 1898, S. 10, zit. in Heinemann, ebd., S. 430; v. Carlsburg, ebd.

82 v. Goethe 1866: Torquato Tasso, S. 149-151 (Hervorhebung fett/kursiv durch d. Verf.).

„Anknüpfend an diese Position Humboldts mündeten unsere Nachforschungen in die Erkenntnis: Mit dem Aufstieg und der Aufwertung des Bildungsbegriffs, dem Stolz und Stolperstein der deutschen Pädagogik, verlor die Pädagogik das Glück des Einzelnen immer mehr aus den Augen.[83]

Das ist umso merkwürdiger, als doch Glück Strebens- und Lebensziel Nummer eins ist."[84]

Die „Vereinseitigung" der Bildung im Sinne Humboldts („höchste und proportionirlichste Bildung seiner Kräfte zu einem Ganzen"[85]) führe zu einer Negierung des Glücksgefühls, obwohl dieser Terminus als Wohlbefinden und Zufriedenheit mit sich und den Anderen aus unserer Sozialisation nicht zu eliminieren ist. Hierzu verweist Burow unter Zitat Menzes[86] auf v. Humboldt:

„‚Die Entwicklung aller Keime (...), die in der individuellen Anlage eines Menschenlebens liegen, halte ich für den wahren Zweck des irdischen Daseyns, nicht gerade das Glück.'" [87]

„Die Entwicklung aller Keime aber, die in der individuellen Anlage eines Menschenlebens liegen, halte ich für den wahren Zweck des irdischen Daseyns, nicht gerade Glück."[88]

„‚Die Entwicklung aller Keime [...], die in der individuellen Anlage eines Menschenlebens liegen, halte ich für den wahren Zweck des irdischen Daseyns, nicht gerade Glück'" (...).
Seit dieser Zeit ist die Zielstellung einer Pädagogik, die in erster Linie dem Wohlbefinden von Schülern und Lehrern dienen sollte – jedenfalls vom Mainstream – nicht weiter verfolgt worden.
Die zweite Austreibung des Glücks erleben wir gerade im Gefolge der internationalen Schulleistungsvergleichsstudien, die die Aufgabe der Schule fast ausschließlich aus dem Blick ökonomischer Anforderungen betrachten, die sich aus der konkurrenzorientierten globalisierten Wissensgesellschaft ergeben."[89]

---

83 Diesen Satz weist Burow 2010, S. 1 auch Hoyer zu (Zitat Hoyer): URL: http://www.kultur-macht-schule.de/fileadmin/user_upload/Download_Kulturelle_Schulentwicklung/Burow_PaedagogikdesGluecks_2010_02_25.pdf, S. 1.

84 Burow 2011a: http://www.uni-kassel.de/fb1/burow/downloads/BurowKunst.pdf, S. 3.

85 Humboldt 1980, S. 64.

86 Menze 1965, S. 125.

87 Burow 2011a: http://www.uni-kassel.de/fb1/burow/downloads/BurowKunst.pdf, S. 3.

88 Brief v. Humboldts an Charlotte Diede am 26.6.1832, korrigiert durch d. Verf. URL: http://www.gutzitiert.de/zitat_autor_wilhelm_von_humboldt_thema_sinn_zitat_18945.html (05.01.14), so auch bei Hoyer, S. 11. URL: http://www.umweltbildung.at/fileadmin/umweltbildung/dokumente/foodforthought/Vortrag_Dr_Hoyer_Glueck_in_der_Bildung.pdf, S. 11. Diesen Brief zitiert ähnlich ebenfalls Obermann 2013, S. 32 mit Verweis auf Humboldt 1832, S. 171.

89 Burow 2010: http://www.kultur-macht-schule.de/fileadmin/user_upload/Download_Kulturelle_ Schulentwicklung/Burow_PaedagogikdesGluecks_2010_02_25.pdf, S. 1.

An dieser Stelle bietet sich der Verweis auf den amerikanischen Philosophen, Pädagogen und Begründer des amerikanischen Pragmatismus' John Dewey (1859-1952) an. Seine Definition von Glück formuliert er in seinem 23. Kap. der *Demokratie und Erziehung: Die berufliche Seite der Erziehung, 2. Die Stelle der beruflichen Ziele in der Erziehung.*

> *„(...) Eine Berufsarbeit ist das einzige, was die besondere Fähigkeit eines Individuums mit seinem Dienst an der Gesellschaft in Einklang bringt.* ***Herauszufinden, wozu einer geeignet ist, und ihm die Möglichkeit geben, dies zu tun, ist der Schlüssel zum Glück.*** *Nichts ist tragischer, als seinen wahren Lebensberuf zu verfehlen oder zu entdecken, daß man durch den Zwang der Umstände in einen unpassenden Beruf geraten ist."*[90]

Burow zitiert in seiner *Positiven Pädagogik* Deweys Maxime, ja Monitum, in einer ähnlichen Übersetzung, allerdings mit der klareren Differenz *Glück - Glücklichsein*, was natürlich gerade andere Schlussfolgerungen bezüglich der Persönlichkeitsbildung und Identität zeitigen könnte.

> *„»Herauszufinden, wozu man sich eignet, und eine Gelegenheit zu finden, dies zu tun, ist der Schlüssel zum Glücklichsein.« John Dewey (1930, S. 360)"*[91]

Selbstbildungs- und Selbsterkenntnisprozesse bilden nach Dewey somit das Fundament, das es dem Individuum ermöglicht, sich in Übereinstimmung mit seinen individuellen Potenzialen, seinem reflexiven Ganzen, in die normativ-gesellschaftlichen Lebenszusammenhänge einbringen zu können.

Burow stellt das Glück in Beziehung mit künstlerischen und kulturellen Erlebnissen, die es dem Einzelnen ermöglichen, sich selbst zu entdecken, sich selbst zu finden und kreativ weiterzuentwickeln.[92] Glück kann in diesem Zusammenhang – ähnlich wie es bereits J. Dewey sah – als Resultat einer *Passung* zwischen den individuellen Potenzialen und den Tätigkeitsfeldern, in denen er seelisch und verstandgeleitet wirkt, betrachtet werden.

Hinsichtlich dieses Hintergrunds muss nochmals auf Comenius verwiesen werden. Der Terminus Glück kommt in verschiedenen Bedeutungen, ja Vorsehungen, bei ihm vor. Zuerst einmal sah er darin die politische Verantwortung des Staates für das „Glück des Volkes" im Sinne einer „Hilfe zur Selbsthilfe für alle in allen Lebensaltern", d.h. auch die Ermöglichung von Enkulturation, (beruflicher) Bildung, Erziehung und ein lebensbejahendes

---

90 Dewey 2000, S. 399 (Hervorhebung durch d. Verf.).

91 Burow 2011, S. 9.

92 Vgl. Burow 2011.

soziales Fundament.[93] Im Jahre 1654 verfasst, veröffentlicht J.A. Comenius das Traktat *Das Glück des Volkes* (1659). Auch hier wiederum sein Petitum, Gottesglaube, Religion und Bildung in Einklang mit friedfertiger, dem Volke dienender Politik (*Der Engel des Friedens*[94]) zu verbinden. Glück wird hier bereits als Erziehungsziel und pädagogische Kategorie definiert und essentiell in die Verantwortung des Staates/Souveräns gelegt. Dazu insbesondere neben § 7 (18 Teile und Grade des Völkerglückes)[95] sein § 22:

> „*XV. Gute Jugenderziehung.*
>
> **(22)** *Sicherlich besteht auch das Glück des Volkes in* guter Jugenderziehung, *denn (...) weise Führer der Völker rechneten es zu den größten Obliegenheiten ..., auf die Erziehung zu achten und sie zu fördern. (...)*
>
> Glücklich ist das Volk, das an guten Schulen, guten Büchern, guten Einrichtungen und Methoden für die Jugenderziehung reich ist.“[96]

Heinrich v. Kleist, der in den *Berliner Abendblätter(n)*, gegründet mit seinen Freunden am 1. Oktober 1810 und eingestellt am 30. März 1811, vorwiegend Beiträge politischen und nationalistischen Inhalts schreibt – von Reinhold Steig als „Parerga“[97] (πάρεργον) bezeichnet–, die oft einen stark patriotisch gefärbten Hintergrund beinhalteteten, wie Aufstand gegen den Kaiser Napoléon I., allerdings auch „Befreiung des Volkes aus den Fesseln der Aufklärung“ und „Kampf gegen Ifflands Berliner Theaterleitung“ (französicher künstlerischer Einfluss), auch „gegen die auf der Berliner Akademie betriebene Nachahmung nach berühmten Meistern und gegen die Übertreibungen der neuen christlichen Malerschule“[98] als auch Aufstand „gegen die Neugestaltung des Erziehungswesens, weil man in seinen Kreisen von der Anwendung des [in Preußen äußerst hochgeschätzten] Pestalozzischen Systems eine [starke] Abschwächung der nationalen Erziehung befürchtete.“[99] v. Kleist verfasste auch eine Reihe von Briefen und Aufsätze, insbesondere an seine Braut Wilhelmine v. Zenge, und die beiden frühesten an seinen Kameraden aus der gemeinsamen Kadetten- und Fähnrichszeit (1795)

93 Arnhardt/Reinert [v. Carlsburg] 1996, S. 27.
94 Comenius 1996, S. 645-670; geschrieben „Im Monat Mai des Jahres MDCLXVII“, ebd., S. 645.
95 Comenius, ebd., S. 556.
96 Ebd., S. 561, vgl. auch Münch/Wyrobnik, ebd., S. 12.
97 Steig, o.J., S. 53f.
98 Ebd., S. 53.
99 Ebd., S. 55.

und späteren Generalleutnant Otto August Rühle von Lilienstern, so auch den *Aufsatz, den sichern Weg des Glücks zu finden (...)* in Briefform, der „sich stellenweise mit Kleists Brief an seinen Hauslehrer Martini, aus Potsdam den 18. März 1799, berührt und dadurch zeitlich bestimmt ist."[100] Kleists Empfindung und Auslegung:

„Aufsatz, den sichern Weg des Glücks zu finden, und ungestört, auch unter den größten Drangsalen des Lebens, ihn zu genießen!

An Rühle.

Abb. 9: Heinrich v. Kleist (1777-1811)

(...) Was die Menschen also Glück und Unglück nennen, das sehn Sie wohl, mein Freund, ist es nicht immer; denn bei allen Begünstigungen des äußern Glückes haben wir Tränen in den Augen des erstern, und bei allen Vernachlässigungen desselben, ein Lächeln auf dem Antlitz des andern gesehen.

Wenn also die Regel des Glückes sich nur so unsicher auf äußere Dinge gründet, wo wird es sich denn sicher und unwandelbar gründen? Ich glaube da, mein Freund, wo es auch nur einzig genossen und entbehrt wird, im Innern.

Irgendwo in der Schöpfung muß es sich gründen, der Inbegriff aller Dinge muß die Ursachen und Bestandteile des Glückes enthalten (...). Denn glücklich zu sein, das ist ja der erste aller unsrer Wünsche, (...) der uns durch den ganzen Lauf unsers Lebens begleitet, der schon dunkel in dem ersten kindischen Gedanken unsrer Seele lag und den wir deutlich als Greise mit in die Gruft nehmen werden. (...)

Lassen Sie uns also das Glück nicht an äußere Umstände knüpfen, wo es immer nur wandelbar sein würde, wie die Stütze a, auf welcher es ruht; lassen Sie es uns lieber als Belohnung und Ermunterung an die Tugend knüpfen, dann erscheint es in schönerer Gestalt und auf sicheren Boden. (...)

Wenn ich Ihnen so das Glück als Belohnung der Tugend aufstelle, so erscheint zunächst freilich das erste als Zweck und das andere als Mittel.

(...) Es ist kein besserer Sporn zur Tugend möglich, als die Aussicht auf ein nahes Glück, und kein schönerer und edlerer Weg zum Glücke denkbar, als der Weg der Tugend.

(...) Ja, mein Freund, die Tugend macht nur allein glücklich. (...) (D)ie Tugend ist die Mutter des Glücks, und der Beste ist der Glücklichste. (...)

100 Ebd., S. 51, Fußnote S. 57 von Steig: „Rühle von Lilienstern, Kleists Regimentskamerad und intimer Freund (1780-1847), nachmals preußischer Generalleutnant und Chef des Großen Generalstabes."; vgl. auch Münch/Wyrobnik, ebd., S. 8f.

Ich nenne nämlich Glück nur die vollen und überschwenglichen Genüsse, die (...) in dem erfreulichen Anschaun der moralischen Schönheit unseres eigenen Wesens liegen. Diese Genüsse, die Zufriedenheit unsrer selbst, das Bewußtsein guter Handlungen, das Gefühl unsrer durch alle Augenblicke unsers Lebens vielleicht gegen tausend Anfechtungen und Verführungen standhaft behaupteten Würde, sind fähig, unter den scheinbar traurigsten, ein sicheres tiefgefühltes und unzerstörbares Glück zu gründen."[101]

v. Kleists Überzeugung, „daß die Tugend den Tugendhaften selbst im Unglück glücklich mach(e)"[102] und mit Blick auf die von der *Seele* intendierte *Tugendhaftigkeit* „von den Wegen, die zwischen dem höchsten äußern Glück und Unglück liegen, grade nur auf der Mittelstraße zu wandern", erinnert stark an die Aristotelische Maxime (Αριστοτέλης [384 v. Chr.-322 v. Chr.]) in seiner Schrift *De anima libri tres/Über die Seele.*[103]

Abb. 10: Glücksdreieck nach Martin Seel (eigene Grafik)

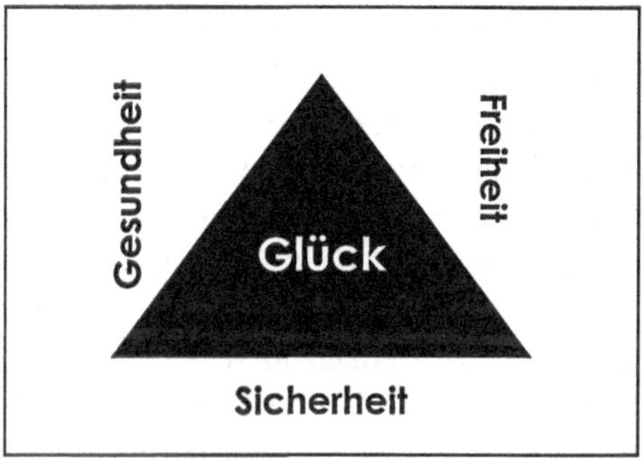

Burow verweist in seinem Beitrag *Durch Kunst und Kultur zu einer Pädagogik des Glücks* auf den an der Universität Frankfurt lehrenden bekannten Philosophen Martin Seel, der folgende drei Voraussetzungen zum Glücklichsein postuliert:[104]

Welche wichtigen Voraussetzungen müssen erfüllt sein, um Glück als Unterrichtsdisziplin erfolgreich zu lehren und mit allen Sinnen anschaulich zu internalisieren?

Lernen als multifaktorialer Prozess auf der Grundlage von (meta-)interaktiver pädagogischer Kompetenz entwickelndem Handeln zielt auf die

101 Kleist, o. J., S. 57-61.
102 Ebd., S. 63.
103 Vgl. v. Carlsburg/Wehr, ebd., S. 360 u. Schwegler [1855] o.J., S. 159.
104 Burow 2011a, URL: http://www.uni-kassel.de/fb1/burow/downloads/BurowKunst.pdf, S. 4 – mit Verweis auf Seel 1999.

Internalisierung von *Bildungswerten*, die der Vorbereitung auf das Leben in der Gemeinschaft dienen (***Sozialisation***), wie auf Handlungsfähigkeit (handelndes Subjekt), Erfahrungswissen und geistige Arbeit (***Bildung***), Bildung zur Selbstbildung, *Tugenden* – über die Urteilsfähigkeit, Willens-, Denk-, Verstandeskraft, Moral zu Charakterstärke, über die Identitätsfindung zu Persönlichkeitsbildung– (***Erziehung***), *Individualentwicklung* – Stärkung der Ich-Kräfte –, um durch dieses in- und external implizierte (Selbst-)Wachstum zu Emanzipation und Mündigkeit sowie zu Autonomieentfaltung und damit zu Sittlichkeit im Sinne (seiner) Moderne zu finden und diese als Kulturgut (***Enkulturation***) dem Nächsten i.w.S. als *Lebenskunst* zu vermitteln, wie es dereinst schon unsere genialen Leitbilder der Aufklärung expressiv und essentiell in ihren Wegweisungen als Maximen der Findung zu sich selbst (*personales Ich*), zu dem Anderen (*personales Du*), zur Welt (*personaler Globe*) generiert haben und wie es in der Humanistischen Pädagogik von Ruth C. Cohn als Konzept der Gruppenleitung proklamiert wird.[105] Es internalisiert planvolle Anleitung einer kompetenten Führung.

Auf dieser Grundlage als Fundament sozialer Kompetenzentwicklung ist Glück (*ευδαιμονία*) als die Disziplin zu verstehen, die als Fundament zur Erreichung von sozialer Kompetenz beginnt mit dem Denk-, Handelns und Verhaltensmodus der ‚*Selbstgenügsamkeit*' im Aristotelischem Sinne (*Nikomachische Ethik*), die auf die Tugendbildung der Jugend zielt und vor allem Arthur Schopenhauers metaphysisches Denken in seinen „Aphorismen zur Lebensweisheit" zeitigt, gewissermaßen Prolegomena in Form einer Spezies ‚Ratgeber über das Leben im weltlichen Paradies als glücklicher Daseinscharakter' unter artifiziellen Rahmenbedingungen verfasst.

Dieses (lebenslange) kooperative Lernen generiert transitive und intransitive *Erziehungswerte* (Ausbildung kommunikativer Kompetenzen) wie Mitgestaltung, -entscheidung und -verantwortung, Stärkung der ‚Verantwortung übernehmenden' geistigen Kräfte, Menschwerdung, Humanisierung, Vergeistigung ***(Enkulturation)***, die Glück entscheidende Spezies einer Gewinnung von *autonomer Freiheit* und Überwindung von „Furcht" vor besagter Freiheit im Frommschen Diktum.[106]

Diese Konzeption erfordert von den Lehrenden im Herbartschen Sinne ‚Denkkraft, Wissenschaft und pädagogischen Takt', eine Conditio sine qua non, die den Menschen zum Menschen als rational handelnde Gestalt definiert, die das Glück

105 Cohn 1975, S. 120ff.

106 Fromm, E. 1941a/GA I, S. 217-392.

des Selbst auch im Glück des Anderen und der ästhetischen Welterfassung durch Erfahrung seiner Kultur definiert, d.h. mehr als die alltägliche Weitergabe von Wissen an Jemanden durch einen Anderen mit dem Ziel hat, den weniger Wissenden zum Wissenden zu machen. Glück definiert sich immer als teleologisches Verfahren, als Prozess mit Finalstruktur!

Glück als ‚innere', ganzheitliche Vermittlungsdeterminante sollte somit folgende Prinzipien und Maximen berücksichtigen:

Abb.11: Prinzipien und Maximen für Glück als ‚innere', ganzheitliche Vermittlungsdeterminante

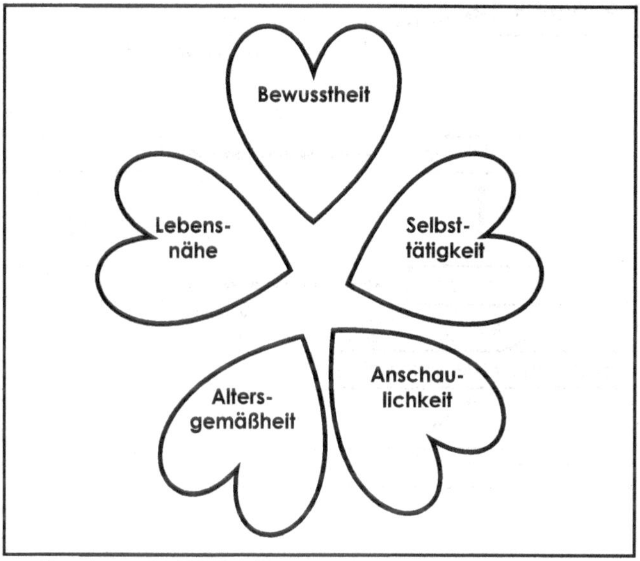

Das Glück per se erfordert letztlich eine motivierend-sozialisierte Balance zwischen beiden Begrenztheiten, letztlich wird aber immer das individuelle (seelische) Empfinden den Gestaltungscharakter von Glück implizieren.

Der Lehrende kann die Lebenskunst letztlich nicht didaktisieren, sondern stets nur als Vorbild/Wegweiser vorleben. Dies fordert beispielsweise auch der Sozialpsychologe/-philosoph E. Fromm:

> *„(D)er Lehrer … [hat] nicht in erster Linie die Aufgabe, Wissen zu vermitteln, sondern sollte bestimmte menschliche Haltungen lehren."*[107]

---

107 Fromm 1956a/GA IX, S. 509, zit. in: v. Carlsburg/Wehr, ebd., S. 388.

„Es ist die Haltung, die dem Kind jene Liebe zum Leben vermittelt, die ihm das Gefühl gibt: Es ist gut zu leben, es ist gut, ein kleiner Junge oder ein kleines Mädchen zu sein; es ist gut, auf dieser Welt zu sein!"[108]

Abb. 12: Mathetischer Erkenntnisweg (KdrV = Kritik der reinen Vernunft; Mathetik = Lernkunst [Comenius])

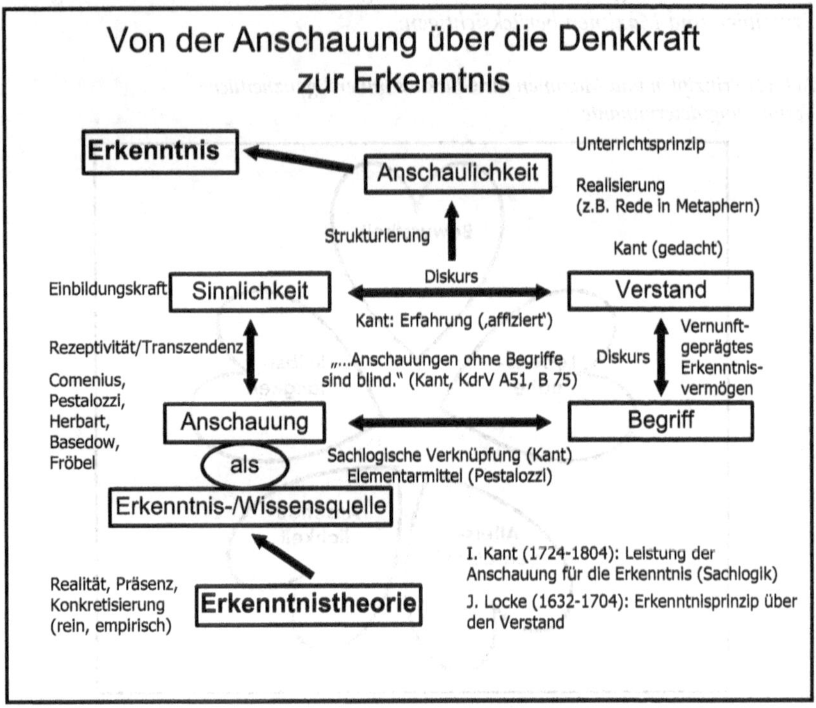

Glück erfordert – und dies haben alle unterschiedlichen Auffassungen und Dispute über die Zeiten nicht infrage gestellt – eine ***mathetische Methodologie*** (Comenius).[109]

Im Gegensatz zu Aristoteles, den Kritikern der Aufklärungsepoche, des Rationalismus', wie Johann Gottfried v. Herder (1744-1803), Johann Georg Hamann (1730-1788), Friedrich Heinrich Jacobi (1743-1819), auch im Großen und Ganzen Jean Paul [Friedrich Richter] (1763-1825) haben die vom Verstand geleiteten

108 Ebd., S. 469, zit. in v. Carlsburg/Wehr, ebd.; vgl. auch 1968a/GA IV, S. 321.
109 Grafik aus: v. Carlsburg 2013, S. 100.

Erkenntnistheoretiker so beispielsweise Immanuel Kant (1724-1804) und John Locke (1632-1804) die Urteilskraft vor die Seelenkraft in einem Feld von Aktivität und Passivität nach gesellschaftlich oktroyierten Normen postuliert.

„Jean Paul stellte sein Ideal und Freund Friedrich Heinrich Jacobi (1743-1819) über I. Kant. K. Fischer … meint in seiner Einführung über Jean Pauls Leben und Wirken, dass F.H. Jacobi die ‚Einschränkung der sinnlichen Erkenntnis' (**Denkkraft** *durch* **Anschaulichkeit**) *I. Kants bemängle, aber die ‚Kritik des Denkvermögens' in seiner ‚transzendentalen Logik' lobe, denn F.H. Jacobi zog in seiner Philosophie eine strikte Trennlinie zwischen ‚Denkvermögen und Gefühl'.*[110] *Was führt zur Klarheit, das emotionsgesteuerte Herz (*unklares Erkennen/Einsicht*) oder der von der Sachlogik, von einem rationalen Bewusstsein gesteuerte Verstand? Es blieb ein* Schwanken.*"* [111]

Wie essentiell das Individuum sein Lebensglück intendiert, hängt vor allem von seiner Ich-Stärke ab[112] ab. Dies postuliert schon Jean Paul in seiner *Levana.*

„Ich ist – Gott ausgenommen, dieses Ur=Ich und Ur=Du zugleich – das Höchste so wie Unbegreifliche, was die Sprache ausspricht und wir anschauen. (…)

Jedes Ich ist Persönlichkeit, folglich geistige Individualität (…)."[113]

Indes hat der Erzieher von der Individualität, die er wachsen läßt, eine andere zu trennen, die er beugen oder lenken muß; jene ist die des Kopfes, diese ist die des Herzens. (…)

Sollte man übrigens den Preis= und Idealmenschen in Worte übersetzen, so könnte man etwa sagen, er sei das harmonische Maximum aller individuellen Anlagen zusammengenommen (…)."[114]

„Sind Kopf und Herz zugleich zu heilen, so hat das Schicksal nur eine, aber eine lange Kur – das **Unglück**. *(…) Damit die Kinder durch diese trübe Erziehungszeit … hindurchkommen, sollen sie (…) mit drei Kräften ausgerüstet werden gegen die drei Entkräftungen des Willens, der Liebe und der Religion.*"[115]

Das Kap. V des *Vierte(n) Bruchstück(s)* wird nur paraphrasiert in dieser Ausgabe von Fischer (1894) wiedergegeben, wörtlich nur der folgende Satz, der als direkter Vorsatz im Original (1827) zu Jeans Pauls Meinung über das Glück hin-leitet. Daher erfolgt eine Übernahme aus der Werksausgabe im zweiten Zitat.

---

110 Konrad Fischer 1894, S. 45.
111 v. Carlsburg 2009, S. 58.
112 Vgl. Bürgermann/v. Carlsburg 2010.
113 Jean Paul 1894, *Zweites Bruchstück, Kap. II: Die Individualität des Idealmenschen,* § 29, S. 146.
114 Ebd., S. 148f.
115 Ebd., *Kap. III. Über den Geist der Zeit, § 35,* S. 153 (Hervorhebung fett durch d. Verf.).

„Wer keinen höhern und festern Himmel über seinem Auge hat, als den Thronhimmel aus Sam[m]t und Holz, ist sehr beengt[,] und hat über seinem Kopfe wenig Aussicht.“[116]

„Wer auf den blumigen Höhen der Menschheit doch kein Glück erreicht, der ist, wenn er ohne Gott im Innern ist, hülfloser, als der Niedrige, der wenigstens in der Anklage seiner tiefen Stellung die Hoffnung der Verbesserung sucht.“[117]

Doch auch die Auseinandersetzung mit *Bildungsgütern* kann als *Katalysator* menschlicher Glückserfahrung gelten, sofern ein real-reflexiver Bezug hergestellt werden kann. So betrachtet der an der Freien Universität Berlin lehrende Philosoph und Schriftsteller Peter Bieri *Bildung (als poetische Erfahrung)* als etwas, das das Selbst verändert und Wachstumsprozesse anregt. Bildung wird in diesem Sinne als Selbstbildung gedacht, ist nur durch das Individuum selbst zu erzielen.

„(E)s gibt Erfahrungen des Glücks, die aufs Engste mit (...) Bildung verknüpft sind: die Freude, an der Welt etwas besser zu verstehen; die befreiende Erfahrung, einen Aberglauben abschütteln zu können; das Glück beim Lesen eines Buchs, das einen historischen Korridor öffnet; die Faszination durch einen Film, der zeigt, wie ganz anders das Leben anderswo ist; die beglückende Erfahrung, eine neue Sprache für das eigene Erleben zu lernen; die freudige Überraschung, wenn man sich mit einem Mal besser versteht; die Erlösung, wenn es einem gelingt, eingefahrene Geleise des Erlebens zu verlassen und so mehr Selbstbestimmung zu erfahren; die überraschende Erfahrung, dass sich mit dem Anwachsen der moralischen Sensibilität der innere Radius vergrössert.“[118]

Lernen an sich, sich gemeinsam mit anderen oder durch die aktive Auseinandersetzung mit *kulturellen Objektivationen* weiterzuentwickeln, sich in eine vollkommen Sache zu vertiefen und vorübergehend die eigenen Grenzen zu verlieren, einen Zustand den Mihály Csíkszentmihályi als „Flow“[119] bezeichnet, Maria Montessori mit dem Begriff der ‚*Polarisation der Aufmerksamkeit*‘[120] belegt, die zur ‚Normalisierung‘ des Kindes beiträgt – bereits hier kann das subjektive Glück angebahnt werden und ***Identität*** gefördert oder verstärkt werden.

---

116 Ebd., *Viertes Bruchstück, Kap. V. Geheime Instruktion eines Fürsten an die Oberhofmeisterin seiner Tochter, § 99*, S. 312, unterschiedlicher Text gegenüber der Ausgabe 1827 ist durch eckige Klammer gekennzeichnet.

117 Jean Paul 1827, *Viertes Bruchstück, Kap. V. Geheime Instrukzion eines Fürsten an die Oberhofmeisterin seiner Tochter, § 101*, S. 115.

118 Bieri 2005, S. 6. http://www.hwr-berlin.de/fileadmin/downloads_internet/publikationen/Birie_Gebildet_sein.pdf (06.01.2014). Abb. 12: Archiv v. Carlsburg.

119 Csíkszentmihályi 2010.

120 Montessori 1996; Fuchs 203, S. 75-79.

Damit einher geht die Forderung, dass jeder Mensch nach seinem eigenen Tempo lernen *darf* und über gewisse Freiheitsgrade verfügt, damit Selbstbildungsprozesse möglich werden, die seine Persönlichkeitsbildung stärken. Hierzu gehört vielfach auch die ***Ent***schleunigung, anstatt das permanente Bestreben nach ***Be***schleunigung.

Dies setzt eine Pädagogik der Achtsamkeit, des Humanismus', voraus und die Anerkennung des Kindesrechts sowie sein So-Sein auf den aktuellen Tag – wie es J. Korczak[121] bereits formulierte – und um es mit einem Wort aus der Bibel zu belegen:

> *„Was nützt es dem Menschen, wenn er die ganze Welt gewinnt, aber Schaden leidet an seiner Seele?"*[122]

Abb.13: Daniel Nikolaus Chodowiecki (1726-1801):
Titelkupfer zu Zigenhagen:
Lehre vom richtigen Verhältnis von Gottes Schöpfung. (1791):
Setzt euch in das rechte Verhältnis unter einander
und mit der übrigen Schöpfung. **Weg zum wahren Glück**

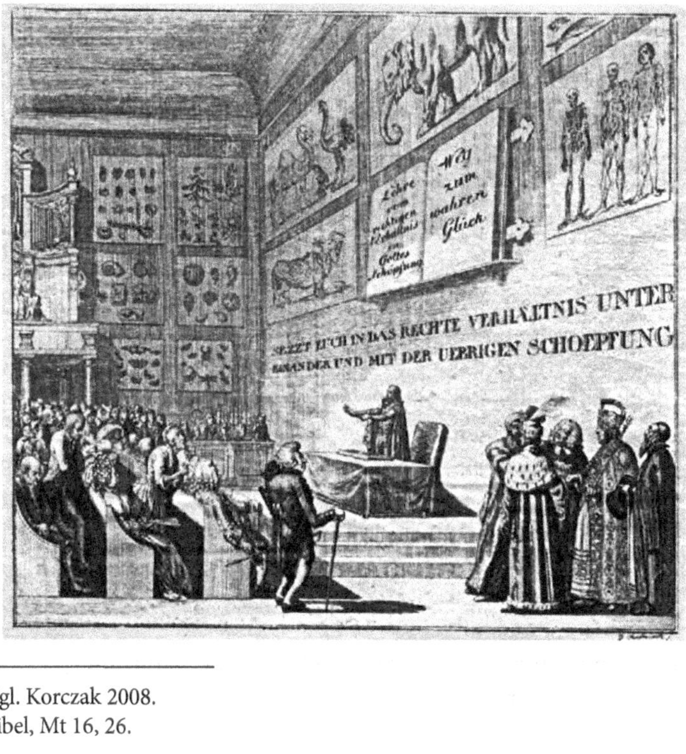

121 Vgl. Korczak 2008.
122 Bibel, Mt 16, 26.

Fazit

Bildung sollte sich stets am Individuum orientieren und es je nach seelischem und geistigem Potential in den Prozess integrieren. Jedoch muss der Bildungsprozess andererseits auch das Fundament legen, dass der Mensch über sich hinauswächst, seine Grenzen transzendiert und sich auf Werte/Sinnhorizonte außerhalb seines Ganzen einlassen.

Um Selbstbildungsprozesse zu intendieren, muss beim einzelnen Subjekt motivational begonnen werden, zugleich implizieren Bewährungsproben erst die Möglichkeit, über sich hinaus zu wachsen und Wege zum Glücklichsein selbst zu suchen und zu finden, Wege, die auch beschritten werden auf der Suche nach Identität. Reale Handlungsbezüge, die zur Stellungnahme herausfordern, mögen nicht nur der moralischen Entwicklung, den Tugenden, förderlich sein, sondern auch bei der Sinnorientierung im menschlichen Leben behilflich sein und somit indirekt auch zum *Lebensglück* beitragen.

Abb.14: Erkenntnisweg des Glücks (eigene Grafik)

Die ‚Vermittlungsdeterminante Glück' dient somit der Darbietung postpostmoderner Gedanken einer multikulturellen auf Inklusion zielenden Gesellschaft, d.h. das Individuum soll sich aktiv eingebunden fühlen in den Sozialisations-,

Bildungs-, Erziehungs- und Enkulturationsprozess. Dies impliziert ‚tatkräftiges' Lernen, um autonomes Denken zu lernen und wiederum zu lehren, nur so ist das Erlernen von *Glück als Katalysator* zur Einbindung in Sozialstrukturen zu verstehen, die Tugenden voraussetzen, die im Aristotelischen Sinne (*De anima libri tres*) *von der Seele* geleitet werden. Die menschliche Seele bildet den Kernpunkt des Lernens, mit der Außenwelt zurecht zu kommen, fähig zur Kommunikation mit seinem Glück und dem Glück des Anderen zu sein.

Sie sollte projekt-, handlungsorientiert, fachübergreifend und ganzheitlich (Kraft und Sinne) konzipiert sein und somit individualisiertes Lernen auf kooperativer Basis ermöglichen.

Teile das Glück mit einem Anderen, somit sind dann zwei Menschen glücklich, dies ist kein generalisierender Rat, denn jeder Mensch empfindet anders, bedingt durch sein biografisches Erleben, sein Wertesystem und sein gesellschaftliches Umfeld sowie insbesondere seine situative soziale Situation.

Vielleicht hat Leo N. Tolstoi (Graf Lew Nikolajewitsch Tolstoi [1828-1910]) mit seinem Roman *Anna Karenina* (1875/77, veröffentlicht 1877/78) auf seine Weise eine Lösung versucht, die allerdings in seiner Romanübersetzung (1. Satz) nicht so gut gelungen ist, wie in der Filmübermittlung zu Beginn des Abspanns.

> *„Alle glücklichen Familien sind einander ähnlich; aber jede unglückliche Familie ist auf ihre besondere Art unglücklich.“*[123]

> ***„Vielleicht sind deshalb alle glücklichen Familien einander ähnlich: Weil Glück tief in unserem Inneren für uns alle dasselbe bedeutet.“***[124]

Literatur

Adorno, Th.W. (1983): Minima Moralia. Frankfurt a.M.

Andresen, S./Hurrelmann, K. (2010): Was bedeutet heute „Glück“ für Kinder? In: Aus Politik und Zeitgeschichte, H. 38 (20. September), S. 3-8.

Antonovsky, A. (1997): Salutogenese. Zur Entmystifizierung der Gesundheit. Tübingen.

Aristotelis de anima libri tres [Aristoteleś Drei Bücher von der Seele] (1833), hrsg. u. kommentiert v. Trendelenburg, F.A. Jena.

123 Tolstoi 2010, S. 1.

124 Film Anna Karenina, ARD, 04.01.2014, Regisseur Christian Duguay, Drehbuch Francesco Arlanch, Italienisch-deutsche Produktion, gedreht in Litauen und Lettland, Text der Anmoderation in Modifikation zum Originaltext.

Aristoteles ([2]1877): De anima libri tres, interpret. u. kommentiert v. Trendelenburg, F.A. Berlin.

Aristoteles (1985): Nikomachische Ethik, übers. v. Rolfes, E., hrsg. v. Bien, G. Hamburg.

Aristoteles (1996): Von der Seele. München.

Aristoteles (2011): Über die Seele, übers. u. hrsg. v. Krapinger, G. Ditzingen/Stuttgart.

Arnhardt, G./Reinert [v. Carlsburg] (1996): Der Mensch als Wohlbegründetes, Ganzheitliches und Geschlossenes: Zum Pädagogischen im Weltbild des J.A. Comenius. In: Arnhardt, G./ Reinert [v. Carlsburg] (Hrsg.): Jan Amos Comenius.Über sich und die Erneuerung von Wissenschaft, Erziehung und christlicher Lebensordnung. Donauwörth, S. 12-32.

Autorenkollektiv (1982): Aristoteles. In: Autorenkollektiv. Philosophenlexikon, hrsg. v. Lange, E./Alexander, D. Berlin (Ost), S. 42-49.

Beck, U. (1986): Risikogesellschaft. Frankfurt a.M.

Becker, M./Carlsburg, G.-B. v./Wehr, H. (Hrsg.) (2008): Seelische Gesundheit und gelungenes Leben – Perspektiven der Humanistischen Psychologie und Humanistischen Pädagogik. Ein Handbuch. Frankfurt a.M.

Beiner, F. (1982): Janusz Korzak. Zeugnisse einer lebendigen Pädagogik. Korczak-Kolloquium. Vierzig Jahre nach seinem Tod. Heinsberg.

Beiner, F. (1984): Janusz Korzak. Zweites Wuppertaler Korczak-Kolloquium. Universitätsdruck. Wuppertal.

Benner, D. (1990): Wilhelm von Humboldts Bildungstheorie. Eine Problemgeschichtliche Studie zum Begründungszusammenhang neuzeitlicher Bildungsreform. Weinheim/München.

Bieri, P. (2005): Wie wäre es, gebildet zu sein? Festrede an der Pädagogischen Hochschule Bern. URL: http://www.hwr-berlin.de/fileadmin/downloads_internet/publikationen/Birie_ Gebildet_sein.pdf, S. 1-7.

Bieri, P. (2008): Wie wäre es, gebildet zu sein? In: Göppel, R./Lenhart, V./Rihm, T./Schön, B./Strittmatter-Haubold, V. (Hrsg.): Bildung ist mehr... Potentiale über PISA hinaus. Heidelberg, S. 13-21.

Brisch, K.H. (2013): Bindungsstörungen: Von der Bindungstheorie zur Therapie. Stuttgart.

Brisch, K.H./Hellbrügge, Th. (Hrsg.) (2008): Wege zu sicheren Bindungen in Familie und Gesellschaft: Prävention, Begleitung, Beratung und Psychotherapie. Stuttgart.

Brisch, K.H./Hellbrügge, Th. (Hrsg.) (2012): Bindung und Trauma: Risiken und Schutzfaktoren für die Entwicklung von Kindern. Stuttgart.

Bürgermann, S./Carlsburg, G.-B. v. (2010): Pädagogische Therapie: Identitätsprobleme erfolgreich behandeln. Ein integratives Konzept zur Selbstverwirklichung für Theorie und Praxis im Schulalltag. Augsburg.

Brumlik, M. (2002): Bildung und Glück. Versuch einer Theorie der Tugenden. Berlin/Wien.

Burkhardt, C.A.H. (Hrsg.) (1870/[2]1898): Goethes Unterhaltungen mit dem Kanzler Friedrich von Müller. 2. stark verm. Aufl. Stuttgart.

Burow, O.-A. (o.J.): Zur Renaissance des Glücks – Überlegungen zu einer vergessenen Dimension der Bildung. URL: http://www.win-future.de/downloads/zurrenaissancedesgluecksolafaxelburow.pdf, S. 1-23 (04.01.14).

Burow (o.J.a): Zur Renaissance des Glücks – Überlegungen zu einer vergessenen Dimension der Bildung. URL: http://www.uni-kassel.de/fb1/burow/downloads/Gl%9Fck-G%9Attingen%C9Endfass..pdf, S. 1-21 (04.01.14).

Burow, O.-A. (2010): Durch Kunst und Kultur zu einer Pädagogik des Glücks? URL: http://www.kultur-macht-schule.de/fileadmin/user_upload/Download_Kulturelle_Schulentwicklung/Burow_PaedagogikdesGluecks_2010_02_25.pdf, S. 1-2 (05.01.14).

Burow, O.-A. (2011): Positive Pädagogik. Sieben Wege zu Lernfreude und Schulglück. Weinheim/Basel.

Burow O.-A. (2011a): Durch Kunst und Kultur zu einer Pädagogik des Glücks. In: Kulturelle Bildung in der Schule. Durch Kooperation zum Erfolg. Werkbuch 03, hrsg. v. d. Arbeitsstelle Kulturelle Bildung in Schule und Jugendarbeit in NRW. Remscheid, S.10-27. URL: http://www.uni-kassel.de/fb1/burow/downloads/BurowKunst.pdf, S. 1-26. (04.01.14)

Burow, O.-A./Hoyer, T. (2011): Schule muss nicht bitter schmecken. Glück als unterschätzte Dimension der Ganztagsschule. In: Jahrbuch Ganztagsschule.

Schwalbach/Ts., S.48-60. URL: http://www.uni-kassel.de/fb1/burow/downloads/Schule%20muss%20.pdf, S. 1-11 (04.01.14).

Burow, O.-A./Hoyer, T. (2012): Positive Pädagogik oder: Wie das Glück aus der Schule verschwand – und wie es in die Ganztagsschule zurückkehren könnte. In: Lehren & Lernen, H. 6, S. 28-33.

Carlsburg, G.-B. v. (Hrsg.) (2009): Enkulturation und Bildung. Fundament sozialer Kompetenz. Frankfurt a.M.

Carlsburg, G.-B. v. (2009a): Bildung durch Erkenntnis. Das Problem der Verwendbarkeit von Erkenntnisquellen. In: Carlsburg, G.-B. v. (Hrsg.): Enkulturation und Bildung. Fundament sozialer Kompetenz. Frankfurt a.M., S. 61-85.

Carlsburg, G.-B. v. (2011): Der Einfluss der italienischen Kunst und Architektur des 18. und 19. Jahrhunderts auf die deutsche und baltische Malerei und Literatur. In: Carlsburg, G.-B. v. (Hrsg.): Enkulturation durch sozialen Kompetenzerwerb. Enculturation by Acquiring of Social Competences. Frankfurt a.M., S. 65-75.

Carlsburg, G.-B. v. (2013): Retrospektiven einer nach Identität suchenden ‚reflexiv-modernen' Bildungswissenschaft. In: Carlsburg, G.-B. v. (Hrsg.): Bildungswissenschaft auf der Suche nach globaler Identität. Educational Science in Search of Global Identity. Frankfurt a.M., S. 97-103.

Carlsburg, G.-B. v./Möller, M. (2011): Prolegomena zu einer ‚reflexiv-modernen' Pädagogik. In: Carlsburg, G.-B. v. (Hrsg.): Enkulturation durch sozialen Kompetenzerwerb. Enculturation by Acquiring of Social Competences. Frankfurt a.M., S. 77-118.

Carlsburg, G.-B. v./Wehr, H. (2010): Bildung zur Selbstbildung. Konzepte der Professionalisierung und Persönlichkeitsentwicklung im Lehrberuf. Ein Studienbuch. Augsburg.

Carlsburg, G.-B. v./Wehr, H. (2013): Glück (ευδαιμονία) – ein Katalysator für Identitätsfindung? In: Carlsburg, G.-B. v. (Hrsg.): Bildungswissenschaft auf der Suche nach globaler Identität. Educational Science in Search of Global Identity. Frankfurt a.M., S. 359-396.

Carlsburg, G.-B. v./Wehr, H. (2013a): Identität entfalten durch Valorisierung, Verständnis und Harmonie. In: Carlsburg, G.-B. v. (Hrsg.): Bildungswissenschaft auf der Suche nach globaler Identität. Educational Science in Search of Global Identity. Frankfurt a.M., S. 233-244.

Cohn, R.C. (1975/^{7}1986; ^{16}2009): Von der Psychoanalyse zur themenzentrierten Interaktion. Von der Behandlung einzelner zu einer Pädagogik für alle. Stuttgart.

Comenius, J.A. ([1659]/1996): Das Glück des Volkes. In: Arnhardt, G./Reinert [v. Carlsburg] (Hrsg.): Jan Amos Comenius. Über sich und die Erneuerung von Wissenschaft, Erziehung und christlicher Lebensordnung. Donauwörth, S. 554-603.

Comenius, J.A. ([1667]/1996): Der Engel des Friedens. In: Arnhardt, G./Reinert [v. Carlsburg] (Hrsg.): Jan Amos Comenius. Über sich und die Erneuerung von Wissenschaft, Erziehung und christlicher Lebensordnung. Donauwörth, S. 645-670.

Csíkszentmihályi, M. (2010): Flow – der Weg zum Glück. Der Entdecker des Flow-Prinzips erklärt seine Lebensphilosophie. Freiburg/Basel/Wien.

Dewey, J. (1916): Democracy and Education: An Introduction to the Philosophy of Education. New York.

Dewey, J. (1930): Democracy and Education: An Introduction to the Philosophy of Education. New York.

Dewey, J. (2000): Demokratie und Erziehung. Eine Einleitung in die philosophische Pädagogik, hrsg. u. mit einem Nachw. v. Oelkers, J. TB. Weinheim/Basel.

Erikson, E. (1971a): Identität und Lebenszyklus. Frankfurt a.M.

Ernst, H. (2012): Überglücklich? Die Schattenseiten des Glücks. In: Psychologie heute, H. 10, S. 28-33.

Fischer, K. (^{2}18949: Leben und Lehren Jean Pauls. In: Jean Paul: Levana oder Erziehlehre. 1. Abteilung, bearb. v. Fischer, K. Langensalza, S. 1-111.

Fleischer, M. (2004): Schopenhauer. Freiburg/Basel/Wien.

Frankl, V.E. (1946): Ein Psycholog erlebt das Konzentrationslager. Wien.

Frankl, V.E. (1959): Die Suche des Menschen nach Sinn. New York.

Frankl, V.E. (1985/^{24}2011): Der Mensch vor der Frage nach dem Sinn. München.

Frankl, V.E. (1990): Der leidende Mensch. München.

Frankl, V.E. (1991): Der Wille zum Sinn. Ausgewählte Vorträge über Logotherapie. München.

Frankl, V.E. (1995): … trotzdem Ja zum Leben sagen. Ein Psychologe erlebt das Konzentrationslager. München.

Frankl, V.E. ([11]2005): Ärztliche Seelsorge. Grundlagen der Logotherapie und Existenzanalyse. Überarb. Neuaufl., hrsg. v. Battyany, A. Wien.

Frankl, V.E. ([6]2012): Der Wille zum Sinn. Bern/Stuttgart/Wien.

Freire, P. (2007): Unterdrückung und Befreiung, hrsg. v. Schreiner, P./ Mette, N./ Oesselmann, D./Kinkelbur, D./in Kooperation mit Bernhard, A. Münster/ New York/München/Berlin.

Freire, P. (2007a): Bildung und Hoffnung, hrsg. v. Schreiner, P./ Mette, N./Oesselmann, D./Kinkelbur, D./in Kooperation mit Bernhard, A. Münster/New York/München/Berlin.

Freire, P. (2008): Pädagogik der Autonomie. Notwendiges Wissen für die Bildungspraxis, hrsg. v. Schreiner, P./Mette, N./Oesselmann, D./Kinkelbur, D. Münster/New York/München/Berlin.

Fritz-Schubert, E. ([3]2010): Glück kann man lernen. Was Kinder stark fürs Leben macht. Berlin.

Fritz-Schubert, E. ([2]2011): Schulfach Glück. Wie ein neues Fach die Schule verändert. Freiburg i.Br..

Fritz-Schubert, E. (2012): Dem Glück auf die Sprünge helfen. Das Geheimnis der Lebensfreude. Freiburg i.Br.

Fromm, E. (1999): Gesamtausgabe in zwölf Bänden (GA I-XII), hrsg. v. Funk, R. DVA/DTV. Stuttgart.

Fromm, E. (1941a/GA I): Die Furcht vor der Freiheit, S. 217-392.

Fromm, E. (1947a/GA II): Psychoanalyse und Ethik. Bausteine zu einer humanistischen Charakterologie, Kap. 4c: Lust und Glück, S. 110-124.

Fromm, E. (1955a/GA IV): Wege aus einer kranken Gesellschaft, S. 1-254.

Fromm, E. (1956a/GA IX): Die Kunst des Liebens, S. 439-518.

Fromm, E. (1959b/GA IX: Psychologie und Werte (Values, Psychology, and Human Existence, aus dem Amerik. v. Mickel, L./Mickel, E.), S. 331-341.

Fromm, E. (1964a/GA II): Die Seele des Menschen. Ihre Fähigkeit zum Guten und zum Bösen, Kap. 3: Die Liebe zum Toten und die Liebe zum Lebendigen, S. 179-198.

Fromm, E. (1968a/GA IV): Die Revolution der Hoffnung, S. 255-377.

Fromm, E. (1975c/GA V): Die Zwiespältigkeit des Fortschritts. Zum 100. Geburtstag von Albert Schweitzer, S. 329-333.

Fromm, E. ([1974/75] 1989a/GA XII): Vom Haben zum Sein. Wege und Irrwege der Selbsterfahrung, S. 393-483.

Fromm, E. ([1962] 1991f./GA XII): Zum Verständnis von seelischer Gesundheit, S. 143-160.

Fuchs, B. (2003): Maria Montessori. Ein pädagogisches Porträt. Weinheim/Basel.

Fuhr, T. (2002): Das Glück des Kindes. In: Zeitschrift für Pädagogik, Jg. 48, H. 4, S. 514-533.

Funk, R. (2006): Produktive Orientierung und seelische Gesundheit. In: Fromm Forum, 10. Tübingen, S. 4-8.

Göppel, R. (2007): Sind Kinder heute (noch) glücklich? Über den erstaunlichen Kontrast zwischen kulturkritischen Klagen und kindlichen Selbsteinschätzungen. In: Göppel, R.: Aufwachsen heute. Veränderungen der Kindheit – Probleme des Jugendalters. Stuttgart, S. 153-179.

Göppel, R. (2010): Von der „Leidabwehr“ zur „Lebenskunst“? Die Psychoanalytische Pädagogik und die Frage nach dem Glück. In: Göppel, R.: Pädagogik und Zeitgeist. Erziehungsmentalitäten und Erziehungsdiskurse im Wandel. Stuttgart, S. 284-300.

Goethe, J.W. v. (1817): Aus meinem Leben. Zweyte Abtheilung. Zweyter Theil. Auch ich in Arkadien! Stuttgart/Tübingen.

Goethe, J.W. v. (1819): West-östlicher Divan. Stuttgart/Tübingen.

Goethe, J.W. v. (1866): Torquato Tasso. Dritter Aufzug. Zweiter Auftritt. In: Goethes Werke. Sechster Band. Stuttgart, S. 90-204.

Goethe, J.W. v. (o.J. [um 1900ff.]): Goethes Werke. 14. Bd.: Italienische Reise. Auch ich in Arkadien, hrsg. v. K. Heinemann u. Mitwirk. mehrerer Fachgelehrter, bearb. v. R. Weber. Leipzig/Wien.

Goethe, J.W. v. (o.J. [um 1900ff.]): Goethes Werke. 15. Bd.: Zweiter Römischer Aufenthalt vom Juni 1787 bis April 1788, hrsg. v. K. Heinemann u. Mitwirk. mehrerer Fachgelehrter, bearb. v. R. Weber/K. Heinemann. Leipzig/Wien, S. 11-244.

Goethe, J.W. v. (2013): West-östlicher Divan. Berlin.

Goethe, J.W. v. (o.J.): Gesammelte Werke in sieben Bänden, hrsg. v. Heiseler, B. v. : West-östlicher Divan. Gütersloh. URL: http://gutenberg.spiegel.de/buch/3656/9 (05.01.14).

Graf, H. (2005): A Meaningful Occupation Assessment – MOA. Ein Testverfahren für Sinn- und Wertorientierte Personalentwicklung. In: Graf, H.: Mit Sinn und Werten führen: was Viktor E. Frankl Managern zu sagen hat; zum 100. Geburtstag von V.E. Frankl. Band 22 von Personal und Organisation. Münster.

Heinemann, K. (o.J. [um 1900ff.]): Anmerkungen des Herausgebers. In: Goethe, J.W. v.: Goethes Werke. 14. Bd.: Italienische Reise. Auch ich in Arkadien, hrsg. v. K. Heinemann u. Mitwirk. mehrerer Fachgelehrter, bearb. v. R. Weber. Leipzig/Wien o.J. (um 1900ff.), S. 5, 7-16, 428-484.

Heinemann, K. (o.J. [um 1900ff.]): Anmerkung des Herausgebers. In: Goethe, J. W. v.: Goethes Werke. 15. Bd.: Zweiter Römischer Aufenthalt vom Juni 1787 bis April 1788, hrsg. v. K. Heinemann u. Mitwirk. mehrerer Fachgelehrter, bearb. v. R. Weber/K. Heinemann. Leipzig/Wien o.J. (um 1900ff.) S. 7-10, 487-523.

Herbart, J.F. ([10]1903): Herbarts Erziehungs- und Unterrichtslehre (Pädagogik). 3. Abschnitt: Zucht. In: Vollständige Darstellung der Lehre Herbarts. (Psychologie, Ethik und Pädagogik.) Aus sämtlichen Werken und mit eigenen Worten des großen Denkers übersichtlich und systematisch geordnet und zusammengestellt. Zum Studium für Lehrer und Freunde der Pädagogik hrsg. v. Wagner, E. Langensalza, S. 285-333.

Herbart, J.F. ([7]1906): Über die ästhetische Darstellung der Welt, als das Hauptgeschäft der Erziehung [1804]. In: Herbart, J.F.: Pädagogische Schriften, mit Herbarts Biographie hrsg. v. Bartholomäi, neu bearb. u. m. erl. Anm. vers. v. Sallwürk, E. v. Langensalza, S. 201-219.

Herbart, J.F. (1968): Über die ästhetische Darstellung der Welt als das Hauptgeschäft der Erziehung [1804 – als Anh. zu Pestalozzis Idee eines ABC der Anschauung (²1802)].]. In: Herbart, J.F.: Kleine pädagogische Schriften. Bd. II, bes. v. Brückmann, A. Paderborn, S. 39-55.

Herbart, J.F. (⁶1983): Allgemeine Pädagogik aus dem Zweck der Erziehung abgeleitet, hrsg. v. Holstein, H. 6. durchges. u. verb. Aufl. Bochum.

Herbart, J.F. (1997): Über die ästhetische Darstellung der Welt als das Hauptgeschäft der Erziehung [1804]. In: Herbart, J.F.: Systematische Pädagogik. Bd. 1: Ausgewählte Texte, hrsg. v. Benner, D. Weinheim, S. 47-56.

Hoyer; T. (2005). Pädagogische Verantwortung für ein glückliches Leben oder: Vom „Glück" in der Pädagogik. In: Burkhart, H./Sikora, J./Hoyer, T.: Sphären der Verantwortung. Münster, S. 151-210.

Hoyer, T. (2007): Glück und Bildung. In: Hoyer, T. (Hrsg.): Vom Glück und glücklichen Leben. Sozial- und geisteswissenschaftliche Zugänge. Göttingen, S. 222-242.

Hoyer, T.: Bildung zum Glück? URL: http://www.umweltbildung.at/fileadmin/umweltbildung/dokumente/foodforthought/Vortrag_Dr_Hoyer_Glueck_in_der_Bildung.pdf, S. 1-15 (05.01.14).

Humboldt, W. v. (1956): Schriften zur Anthropologie und Bildungslehre, hrsg. v. A. Flitner. Düsseldorf/München.

Humboldt, W. v. (³1980): Ideen zu einem Versuch, die Gränzen der Wirksamkeit des Staats zu bestimmen [1792]. In: Humboldt, W. v.: Schriften zur Anthropologie und Geschichte. Werke in fünf Bänden, Bd. I, hrsg. v. Flitner, A./Giel, K. Stuttgart, S. 56-233.

Humboldt, W. v. (³1980a): Theorie der Bildung des Menschen [1793]. In: Humboldt, W. v.: Schriften zur Anthropologie und Geschichte. Werke in fünf Bänden, Bd. I, hrsg. v. Flitner, A./Giel, K. Stuttgart, S. 234-240.

Humboldt, W. v. (1980b): Über den Geist der Menschheit [1797]. In: Humboldt, W. v.: Schriften zur Anthropologie und Geschichte. Werke in fünf Bänden, Bd. I, hrsg. v. Flitner, A./Giel, K. Stuttgart, S. 506-518.

Humboldt, W. v. (1984): Theorie der Bildung des Menschen [1793]. In: Humboldt, W. v.: Schriften zur Anthropologie und Bildungslehre, hrsg. v. Flitner, A. Frankfurt a.M./Berlin/ Wien.

Humboldt, W. v. (41985): Theorie der Bildung des Menschen [1793]. In: Humboldt, W. v.: Bildung und Sprache, bes. v. Menze, C. Paderborn, S. 25-28.

Humboldt, W. v. ([1792]/1986): Ideen zu einem Versuch, die Grenzen der Wirksamkeit des Staats zu bestimmen. Ditzingen (Stuttgart).

Jean Paul (1827): Levana oder Erziehlehre. 2. Bändchen. In: Jean Paul's sämmtliche Werke XXXVII. Achte Lieferung. Zweiter Band. Berlin.

Jean Paul (21894): Levana oder Erziehlehre. 1. Abteilung, bearb. v. Fischer, K.. Langensalza, S. 112-313.

Kant, I. (1784): Beantwortung der Frage: Was ist Aufklärung? In: Berlinische Monatsschrift, H. 12, S. 481-494.

Kant, I. (1974): Kritik der Urteilskraft (1790), hrsg. v. Vorländer, K. Hamburg.

Kernberg, O.F./Hartmann, H.-P. (2010): Narzissmus. Grundlagen – Störungsbilder – Therapie. Stuttgart.

Kerschensteiner, G. (31930): Die Seele des Erziehers und das Problem der Lehrerbildung. Leipzig/Berlin.

Keupp, H. (o.J. [ca. 2002]): Sich selber finden – Identitätskonstruktionen heute und welche Ressourcen in Familie und Gesellschaft sie benötigen. München, S. 5. http://www.ipp-muenchen.de/texte/sich_selber_finden.pdf (07.02.11).

Keupp, H. (2003): Gesellschaftlicher Umbruch und seine Konsequenzen für die individuelle Lebensbewältigung. München. http://www.gestalt-institut-frankfurt.de/download/Keupp.pdf (16.11.2010).

Keupp, H. (2005): Förderung salutogenetischer Ressourcen durch Empowerment. Zusammenfassung und Ausblick. Salutogenese in der Psychiatrie. Tagung der Evangelischen Akademie Tutzing vom 06.-07. April 2005. Folie 15. URL: www.psz.co.at/img/30_jahresfeier/Festakt_ Vortrag_ Keupp_PPP.pdf (07.02.2011).

Keupp, H. (2006): Bildung zwischen Anpassung und Lebenskunst. Ressourcen der Lebensbewältigung, Vortrag auf dem „Tag der Weiterbildung“ (03.03.2006) in St. Johann im Pongau. In: http://www.eb.salzburg.at/download/Keupp-Referat.pdf (18.05.2009).

Keupp, H. (o.J. [ca. 2007]): Folie 1. http://www.pragdis.de/data/keupp_4053.pdf (07.02.11).

Keupp, H. (2008): Erziehungsberatung in bewegten Zeiten. Vortrag im Rahmen der Wissenschaftlichen Tagung und des Festaktes aus Anlass des fünfzigjährigen Bestehens der Landesarbeitsgemeinschaft für Erziehungs-, Jugend- und Familienberatung in Bayern e.V. am 27.10.2008 in Nürnberg, S. 5. http://www.dgvt.de/fileadmin/user_upload/Dokumente/Erziehungsberatung-Keupp.pdf (07.02.11).

Keupp, H: Identitätskonstruktion. Vortrag bei der 5. bundesweiten Fachtagung zur Erlebnispädagogik am 22.09.2003 in Magdeburg. http://www.ipp-muenchen.de/texte/identitaetskonstruktion.pdf (11.9.2013).

Klein, St. ([15]2008): Die Glücksformel oder Wie die guten Gefühle entstehen. Reinbek.

Kleist, H. v. ([1799]/o.J. [um 1900]): Aufsatz, den sichern Weg des Glücks zu finden, und ungestört, auch unter den größten Drangsalen des Lebens, ihn zu genießen! An Rühle. In: H. v. Kleists Werke. Kritisch durchges. u. erl. Gesamtausgabe. 5 Bde. Im Verein mit Minde-Pouet, G./Steig, R. hrsg. v. Schmidt, E. 4. Bd. Kleinere Schriften, hrsg. v. Steig, R. Leipzig/Wien, S. 57-73.

Kohut, H. (1993): Auf der Suche nach dem Selbst. Kohuts Seminare zur Selbstpsychologie und Psychotherapie. Leben Lernen 86, hrsg. v. Elson, M. Stuttgart.

Korczak, J. ([5]2007): Das Recht des Kindes auf Achtung. Fröhliche Pädagogik, hrsg. v. Beiner, F. Gütersloh.

Korczak, J. (2008): Wie man ein Kind lieben soll. Göttingen.

Lasch, Chr. (1980): Das Zeitalter des Narzißmus (The culture of narcissism, dt.). Aus d. Amerikan. v. Burmundt, G. Steinhausen.

Lausberg, M. (2008): Ernst Christian Trapp – Theoretiker der philanthropischen Erziehung. Tabula Rasa, Zeitschrift für Gesellschaft und Kultur, No. 33, 3. URL: http://www.tabularasa-jena.de/artikel/artikel_3288/)

Leibniz, G.W. ([1710]/1996): Philosophische Schriften. Bd. 2 (2.1/2.2): Die Theodizee, hrsg. u. übers. v. Herring, H. Frankfurt a.M.

Leonhardt, R. (2003): Skeptizismus und Protestantismus. Hermeneutische Untersuchungen zur Theologie 44. Tübingen.

Maslow, A.A.[H.] [1962](^{3}1990; ^{13}1997) Psychologie des Seins. Ein Entwurf. Fischer TB. Frankfurt a.M.

Maslow, A.H. [1954] (^{11}2008): Motivation und Persönlichkeit. Rowohlt TB. Reinbek.

Mayring, Ph. (2007): Glück – Wohlbefinden – Lebensqualität. Sozialwissenschaftliche und soziologische Konzepte. In: Hoyer, T. (Hrsg.): Vom Glück und glücklichen Leben. Sozial- und geisteswissenschaftliche Zugänge. Göttingen, S. 185-199.

Menze, C. (1965): Wilhelm von Humboldts Lehre und Bild vom Menschen. Ratingen.

Montessori, M. ([1916]/^{6}1996): Schule des Kindes. Montessori-Erziehung in der Grundschule, hrsg. u. eingel. v. Oswald, P./Schulz-Benesch, G. Freiburg.

Münch, J./Wyrobnik, I. (^{2}2011): Pädagogik des Glücks. Wann, wo und wie wir das Glück lernen. Baltmannsweiler.

Obermann, A. (2013): Im Beruf Leben finden. Allgemeine Bildung in der Berufsbildung – didaktische Leitlinien für einen integrativen Bildungsbegriff im Berufsschulreligionsunterricht. Göttingen.

Reinert [v. Carlsburg], G.-B./Arnhardt, G./Cornelius, P. (1996): Johann Heinrich Pestalozzi. Anthropologisches Denken und Handeln. Ein pädagogisches Konzept über die Zeiten. Studienbuch. Donauwörth.

Riemeyer, J. (2001): Die Logotherapie Viktor Frankls. Einführung in eine sinnzentrierte Psychotherapie. München/Wien.

Rolfes, E. (1911): Vorwort. Zu: Aristoteles: Nikomachische Ethik. Leipzig, Felix Meißner Verlag. http://gutenberg.spiegel.de/buch/2361/1 (04.09.2013).

Schmid, W. (^{10}2012): Glück. Alles, was Sie darüber wissen müssen, und warum es nicht das Wichtigste im Leben ist. Frankfurt a.M.

Schmid, W. (2012a): Die Menschen werden des Glücks müde. Wilhelm Schmid im Gespräch. In: Psychologie heute, 10, S. 20-27.

Schmidt, H. (Begr.) ([17]1965): Philosophisches Wörterbuch: Tugend, begründ. v., durchges., erg. u. hrsg. v. Schischkoff, G. Stuttgart, S. 611.

Schopenhauer, A. (o.J. [1902a]): Arthur Schopenhauers sämtliche Werke. In sechs Bänden. Mit einer biogr. Einl. v. Köhler, M.. Arthur Schopenhauer's sämmtliche Werke. Zweiter Band. Die Welt als Wille und Vorstellung. Erster Band. Vier Bücher, nebst einem Anhange, der die Kritik der Kantischen Philosophie enthält. Anhang. Kritik der Kantischen Philosophie. Berlin, S. 411-531.

Schopenhauer, A. (o.J. [1902b]): Arthur Schopenhauers sämtliche Werke. In sechs Bänden. Mit einer biogr. Einl. v. Köhler, M. Arthur Schopenhauer's sämmtliche Werke. Vierter Band. Parerga und Paralipomena. Kleine philosophische Schriften von Arthur Schopenhauer [1850]. Erster Band. Aphorismen zur Lebensweisheit. Berlin, S. 283-451.

Schwarzer, R./Jerusalem, M. (2002): Das Konzept der Selbstwirksamkeit. In: Zeitschrift für Pädagogik. Beiheft 44: Selbstwirksamkeit und Motivationsprozesse in Bildungsinstitutionen, hrsg. v. Jerusalem, M./ Hopf, D. Weinheim/ Basel, S. 28-53.

Schwegler, A. ([[2]1855] ca. 1900): Geschichte der Philosophie im Umriß. Ein Leitfaden zur Übersicht. Neue Ausg., durchges. u. erg. v. Stern, J. Leipzig.

Seel, M. (1999): Versuch über die Form des Glücks. Frankfurt a.M.

Seligman, M.E.P. ([9]2012): Der Glücks-Faktor. Warum Optimisten länger leben. Bergisch-Gladbach.

Steig, R. (o.J. [um 1900]): Einleitung des Herausgebers. In: H. v. Kleists Werke. Kritisch durchges. u. erl. Gesamtausgabe. 5 Bde. Im Verein mit Minde-Pouet, G./Steig, R. hrsg. v. Schmidt, E. 4. Bd. Kleinere Schriften, hrsg. v. Steig, R. Leipzig/Wien, S. 51-56.

Streminger, G. (1992): Gottes Güte und die Übel der Welt. Das Theodizeeproblem. Tübingen.

Tolstoi, L.N. (2010): Anna Karenina – Vollständige Ausgabe, übers. v. Röhl, H. Köln.

Trapp, E. C. (1780): Versuch einer Pädagogik. Berlin.

Trapp, E.C. (1977): Versuch einer Pädagogik. Unveränderter Nachdruck der 1. Ausgabe Berlin 1780. Mit Trapps hallischer Antrittsvorlesung, besorgt v. Herrmann, U. Paderborn.

Waibel, E.M. (2009): Erziehung zum Selbstwert. Persönlichkeitsförderung als zentrales pädagogisches Anliegen. Augsburg.

Waibel, E.M. (2013): Erziehung zum Sinn – Sinn der Erziehung. Grundlagen einer Existenziellen Pädagogik. Augsburg.

Wehr, H. (2011b): Wohl-Sein – der bedürfnis- und ressourcenorientierte Gesundheitsbegriff bei Erich Fromm. In: Knörzer, W./Rupp, R. (Hrsg.): Gesundheit ist nicht alles? – Was ist sie dann? Gesundheitspädagogische Antworten. Baltmannsweiler, S. 69-77.

Wilkinson, R./Pickett, K. (2009) Gleichheit ist Glück. Warum gerechte Gesellschaften für alle besser sind. Hamburg.

Wolf, U. ([2]2007): Aristoteles‘ >Nikomachische Ethik<. Darmstadt.

Über das Unvermögen der Sozialwissenschaften den Menschen zu bestimmen

Reet Liimets, Airi Liimets

Abstract
Authors argue that contemporary social sciences have transformed into a theoretically closed system, which does not permit to approach constructs such as 'self' and 'self-identity' interdisciplinarily. In order to substantiate these claims, authors address self-identity in the context of objective/subjective time and narrative theory. Consequently, it is found that social sciences have always struggled methodologically in defining the notions of subject, self, identity and individuality.

Keywords: identity, self, narrative, subjective and objective time

Einführung

In der gegenwärtigen qualitativen Humanwissenschaft ist bei der Deutung des Menschen als der Subjektivität oder der Ichidentität das Narrativ neben der Autobiographie und der Biographie zur wichtigsten Kategorie geworden. Dabei hat das Narrativ, im Vergleich zu den Begriffen Autobiografie und Biografie, eine verallgemeinernde Aufgabe zu erfüllen, denn sowohl Autobiographie als auch Biographie lassen sich ja auch durch die Begriffe Erzählung oder Narrativ bestimmen (Schulze 1993, 1993a; 1993b; Sarbin 1986; Aichinger 1998 usw.).

Donald E.Polkinghorne (1998) schreibt, dass die narrative Psychologie die Herangehensweise wieder auflebt, die vom Standpunkt das für menschliche Weltauffassung und Handlung wichtige Geschichtenerzählen wertet. Die qualitativ orientierte narrative Psychologie wird auf solche Weise als eine Gegenbewegung zur positivistischen Verhaltenspsychologie verstanden (Sarbin 1986; Schütze 1996; Nassehi & Weber 1990; Gergen 1985).

In der Psychologie wird der Anfang der psychoanalytisch orientierten narrativen Bewegung mit dem Namen von Theodore R. Sarbin und mit seinem Sammelband *Narrative Psychology: The Storied Nature of Human Conduct* verbunden. In der Einleitung dieses Werks schreibt Sarbin, dass das Narrativ (das er synonym mit einer Geschichte anwendet) eine Wurzelmetapher der menschlichen Tätigkeit ist, denn trotz sehr begrenzter Erkenntnisfähigkeit und linguistischer Fähigkeiten versucht jeder Mensch die Welt zu verstehen. Wenn die stattgefundenen Ereignisse sich miteinander nicht korrelieren, werden zwischen ihnen narrative Kohärenzverbindungen geschaffen (Sarbin 1986, S. 3–12).

Narrativ, objektive und subjektive Zeit

Im Zusammenhang mit der Narrativität in den Humanwissenschaften weist Donald E.Polkinghorne auf zwei Problemfelder hin, die a) mit dem Problem der Zeitwahrnehmung und der objektiven Zeit, b) mit dem Widerspruch zwischen der Bewusstseinszeit und der Tätigkeit verbunden sind (Röttgers 1988; Böhme 1990 usw.). Zusammenfassend ist das alles aber als eine Identitätsproblematik bestimmbar (Henrich 1979). Denn auch in der narrativen Herangehensweise wird immer wieder danach gefragt, wie die Zeit als ein objektives Phänomen mit dem subjektiven Selbstbewusstsein oder der subjektiven Zeit zu verbinden sei.

Polkinghorne (1998, S. 16) vertritt die Meinung, dass gerade die narrative Betrachtungsweise es vermöge, den Gegensatz zwischen dem Subjekt und Objekt oder der subjektiven und objektiven Zeit zu überwinden. Dem Standpunkt von Polkinghorne schließen sich zum Beispiel auch P.Ricœur (1991; 1996), W. Kraus (1996) und Z.Bauman (1994) an. Auch in diesem theoretischen Kontext ist Zeit also als die Verbindung der subjektiven oder der Zeit des Subjekts oder der Ich-Zeit und der sozialen oder der soziokulturellen Zeit bestimmbar. Dabei ähneln sich die eben erwähnten Betrachtungsweisen einander dadurch, dass seit der Entstehung der Moderne soziokulturelle Zeit als solche die Ich-Zeit oder die subjektive Identität gefährdet (Beck 1986; Bauman 1994; Giddens 1991). Das bedeutet, als ob das soziale System die Menschen mit deren Lebenszielen von sich abweist, denn durch die Vielheit verschiedener Lebensformen ist das soziale System zu seinem eigenen Gegensatz geworden, weil das Soziale in dem Maßen individualisiert worden ist, dass nur eine individualistische Lebensweise möglich geworden ist (Beck 1986; Hitzler & Honer 1994; Kraus 1996 usw.). Das Individuum als solches, so Martin Kohli (1988), sei zum Grundprinzip des sozialen Lebens geworden. Die individuelle oder die persönliche Identität wird also nicht mehr durch die Zugehörigkeit zu irgendwelcher sozialen Position oder sozialen Gruppe konstituiert, sondern das Konstruieren der persönlichen Identität wird solchen Institutionen übertragen, die eigentlich als keine Institutionen fungieren können – d.h. diese Aufgabe wird vom Subjekt als von einer Nicht-Institution übernommen.

Martin Kohli (1988) nennt eine solche sozialstrukturelle Konstituierung des Lebensprogramms für Institutionalisierung des Lebenslaufs, mit der die Erosion der gemeinschaftlichen Lebensformen einhergeht, was bedeutet, dass gesellschaftliche Moralnormen sich auflösen. Wegen der rasant steigenden Anzahl der Scheidungen büßt zum Beispiel die Familie als Institution immer mehr ihre Bedeutung ein. In dem Kontext entwickelt sich der individuelle Lebenslauf nach einem ganz neuen Muster: Der Selbstreflexion wird eine besondere Wichtigkeit beigemessen. Denn die Menschen müssen ständig über ihre Lebensorientierungen und Einstellungen nachdenken, weil es keine fortdauernden und allgemein akzeptierten Lebensmodelle mehr gibt. Die Autobiografie oder das Ich-Narrativ werden auf solche Weise zum Code der persönlichen Entwicklung, was bedeutet, dass die Chronologie der individuellen Zeit zu dominieren anfängt. Jeder muss somit dazu fähig sein, seine Lebensereignisse selber zu deuten und zu strukturieren, um die Stabilität des eigenen Individuums nicht zu verlieren (Kraus 1996). Das Leben des Menschen hängt davon ab, wie er mit seiner sog. biografischen Arbeit zurechtkommt. Die Ich-Kohärenz und die Ich-Kontinuität werden erreicht, indem man über das eigene Leben erzählt (Kraul & Marotzki 2002, S. 8–9). Vor dem Hintergrund des Gesagten erwirbt der Begriff Identität eine ganz andere Bedeutung (Klika 2000). Zentral wird hierbei die Frage, ob und auf welche Weise das Subjekt und seine Identität überhaupt möglich werden.

Historisch hat das genannte Problem seine Wurzeln im Prototyp des humanwissenschaftlichen Denkens des Abendlandes, nach dem die Entwicklung der westlichen Gesellschaft unvermeidlich in Richtung der Individualisierung verläuft (Elias 1990; 1992; Taylor 2000; Lotman 1999; 2001 usw.). Nach Paul Geyer (1997) entstammt die Auffassung, dass die Einzelperson als keine Repräsentation einer Ganzheit zu verstehen ist, aus der Neuzeit. Die sozialen Beziehungen verändern sich, weil das Subjekt diese Beziehungen vermittelnd sie privatisiert, sozusagen zu seinem eigenen persönlichen Eigentum macht. Die erwähnte Privatisierung fand in der westlichen Kulturgeschichte mutmaβlich zweimal statt – im antiken Rom und im Mittelalter. Nach den metaphysischen Auffassungen wird der Prozess als eine historische Tendenz der Entwicklung des Selbstbewusstseins verstanden, der die Bildung der abendländischen Persönlichkeit als solchen im Allgemeinen widerspiegelt. Die Auffassungen von Descartes, La Rochefoucauld, Montesquieu und Rousseau analysierend, zeigt Paul Geyer, dass die drei ersten Philosophen das Bewusstsein des Menschen noch nicht als einen Prozess, d.h. als zeitlich interpretierten. Sie untersuchten also, ob die soziale Identität als das Selbstbewusstscin gleichzeitig als Nicht-Identität fungieren kann (Frank 1991). Bei Rousseau dagegen entsteht zum ersten Mal das Problem, ob das Selbstbewusstsein als das Zeitliche mit sich selbst zugleich identisch und nichtidentisch sein kann. Rousseau kommt zur Schlussfolgerung, dass das Selbstbewusstsein selber die Zeit ist. Die Autonomiebestrebung wird zum wesentlichsten Merkmal der Subjektivität. Das Leben wird demnach als Selbstwerdungsprozess charakterisiert.

Die Bildung dieser neuen Selbstauffassung des Menschen wird sowohl in der sozial- als auch literaturwissenschaftlichen Tradition mit der Entstehung der modernen Autobiographie als der Gattung verbunden (Klika 2000; Holdenried 2000; Lejeune 1998 usw.), weil neue Kategorien zum Ausdrücken der

Subjekt-Objekt-Beziehungen erst dann entstehen können, wenn die sog. Einzelmonade (also das Subjekt bzw. das Ich oder die Persönlichkeit) die Fähigkeit der Selbstbeschreibung – d.h. sich als eine isolierte und einzigartige intellektuelle Entität zu modellieren – erreicht hat (Lotman 1999, S. 50).

Die in den „Bekenntnissen" von Rousseau beschriebene Struktur des Ichs unterschied sich radikal von der bisher dargestellten Selbstbeschreibungstradition. Da die Struktur des Ichs nach Lotman (2001) eines der Grundmerkmale der Kultur ist, dann ist es sowohl in verschiedenen humanwissenschaftlichen Disziplinen als auch in der Literaturwissenschaft zur Tradition geworden, Rousseau für den Begründer einer neuen Geisteshaltung in der westlichen Kultur zu halten (Holdenried 2000; Taylor 2000). Mit seinem Werk „Bekenntnisse" schuf Rousseau nämlich den Eigennamen *Ich* (Lotman 2001, S. 167). Lotman sagt (S.167–168): „Ab "Émile" bis zu dem Werk „Bekenntnisse" hatte Rousseau einen langen Weg hinter sich gelegt. In "Émile" deutet „ich" auf den Sprecher hin, der in der ersten Person spricht. In "Émile" wird über das Wesen des Menschen als solchen gesprochen. Der Erzähler ist einfach die Verkörperung der naturhaften Vernunft und der Zögling die sich selbst entdeckende Natur. In „Bekenntnissen" ist das „Ich" aber ein Eigenname, der keine Pluralform hat und den man von der einzigartigen und unikalen Persönlichkeit nicht entfremden kann … Das heiβt: Rousseau ging den Weg vom Personalpronomen „ich" bis zum Eigennamen „Ich" durch. Das sind aber die wichtigsten Polen des menschlichen Denkens. … Denn die Struktur des Personalpronomens „ich" ist viel einfacher als die des Eigennamens „Ich". Der Eigenname lässt sich durch kein definites Sprachzeichen ausdrücken." (Lotman 2001, S. 167-168).

Die Popularität von Jean-Jacques Rousseau sei nach Ch.Taylor (2000, S. 35) aber darauf zurückzuführen, dass er die in der Kultur schon stattfindenden Tendenzen ausdrückte. Taylor vertritt den Standpunkt, dass Rousseau mit der Entwicklung der für die moderne Kultur charakteristischen Idee der authentischen Ethik verbunden war. Den Mittelpunkt der Idee der authentischen Ethik bildet die Treue zu sich selbst (die Selbsttreue), die die Treue des Menschen zu seiner eigenen Originalität ausdrückt und die niemand auβer dem Menschen selbst auszudrücken und zu entdecken vermag. Nur der Mensch selber kann, seine ursprünglich eigene Potentialität realisieren (vgl. auch Heidegger 1993). In der deutschen Kultur, so Taylor, setze Herder diese Denkweise mit der Idee fort, dass jeder von uns seine ursprüngliche Art habe, Mensch zu sein. Nach Herder habe jeder sein eigenes Maβ. Auch Taylor deutet hierbei auf das Ich in der lotmanschen Bedeutung des Eigennamens hin. Das heiβt, dass der Eigenname *Ich* im gegebenen Fall die individuelle Originalität des Menschen bezeichnet. „Dass … eine Art des Menschseins existiert, die *meine* Art ist: Ich muss mein Leben auf die für mich eigene Weise leben, ohne dabei die Anderen nachzuahmen. Das alles verleiht der Selbsttreue eine neue Bedeutung. Denn, wenn ich selber mir selbst nicht treu bin, erfasse ich den Sinn meines eigenen Lebens nicht, und ich verpasse das, was für *mich* Menschsein bedeutet.“ (Taylor 2000, S. 36–37). Im Anschluss an Taylor und Lotman konnte man sagen, dass die Selbsttreuen ständig auf der Suche nach ihrer eigenen Ich-Identität, auf dem Weg zu sich selbst sind. Im Kontext der qualitativ denkenden Sozialwissenschaften könnte man hierbei aber fragen, wie der Mensch zu seiner narrativen Koheränz – zur narrativen Ich-Identität – gelangt?

Narrative Ich-Identität

Mit seiner Behandlung der narrativen Identität will Ricœur (1996) eine neue Ich- und Handlungsauffassung begründen, wobei er die in der Metaphysik und auch der Psychologie gewurzelte Tradition der dichotomischen Behandlung der Persönlichkeit und der Handlung überwinden will. Auf die Identität geht Ricœur vom Aspekt der Selbstheit (*ipse*) und der Selbigkeit (*idem*) ein. Die Identität als Selbstheit und Selbigkeit charakterisieren nicht nur die Einzelperson, sondern das sind die Charakteristiken, die auch eine Kulturgemeinschaft charakterisieren können.

Nach Ricæur (1996) ist die Identität eine praktische oder pragmatische Kategorie, denn um die Identität eines Individuums oder einer Gemeinschaft zu bestimmen, muss gefragt werden, wer die Handlung durchführte, wer der Handelnde und wer der Autor war. Die Antwort auf diese Frage deutet gewöhnlich auf den Eigennamen von jemand. Ricæur hält hierbei aber nicht den Eigennamen *Ich* im Sinne von Lotman im Auge, sondern den Vor- und Nachnamen eines Menschen. Der Eigenname bedeutet nach Ricœur etwas Ständiges. Auch der Soziologe Bourdieu (2003) schreibt, dass der Eigenname die soziale Identität eines biologischen Individuums in verschiedenen Sinnbereichen gewährleistet. Deswegen ist es gar nicht zufällig, dass man die Unterschrift für Kennzeichnung der Identität des Menschen gehalten wird.

Bei Bourdieu (2003) weist der Eigenname als Bezeichnung der Identität zugleich aber auch auf die Entfremdung des Menschen von der Gesellschaft. Nach seiner Auffassung ist der Eigenname eine Institution, die von der Zeit und dem Raum und von den Variationen des Raums und der Zeit losgerissen ist. Nur durch die biologische und soziale Veränderung behält der Eigenname seine nominale Konstanz – seine Gleichheit mit sich selbst als der Identität – im Sinne von *constantia sibi*. Die soziale Ordnung verlangt nach der nominalen Konstanz.

Die Pflicht des Menschen zu seinem Eigennamen wird deswegen in vielen sozialen Bereichen zur heiligsten Pflicht, um sich selber als Selbst zu bestimmen. Nach Bourdieu ist der Eigenname ein steifer und anonymer Signifikant, weil er in sich keine Informationen enthält, die die Subjektivität des Menschen zum Ausdruck bringt. Zum Namen als der sozialen Identität gelangt man somit um den Preis des höchsten Abstrahierens. Außerdem ist der Name mit verschiedenen Kontrollmechanismen, wie mit dem Pass, dem Personalausweis, mit *Curriculum vitae*, dem Nekrolog, dem Strafregister usw. verbunden. Auch der deutsche Philosoph Karl Löwith (1981, S. 36) meint, dass der Eigenname ein Fremdname des Menschen sei, der ihm von anderen und für andere gegeben worden ist. Der eigentliche Eigenname des Menschen sei nach ihm das erste Personalpronomen, denn durch die Äußerung „ich bin" drückt der Mensch sich auf die für ihn am meisten authentische Weise aus, indem er auf sich selbst als auf sein biographisches Leben hinweist.

Nach Ricæur verbindet das Narrativ dagegen den Eigennamen des Menschen mit seiner subjektiven Identität. Die Beantwortung der Frage „wer?" weist, so Ricæur, auf die Geschichte des Lebens hin, die an den Eigennamen festgebunden ist. Die erzählte Geschichte bestimmt ihn als eine *Wer*-Handlung. Die *Wer*-Identität existiert nur als eine narrative Identität. Ricæur versteht, dass man bei der Bestimmung der Identität auf solche Weise, auch in Schwierigkeiten geraten kann, denn das mit sich selbst identische Subjekt ist eine substantielle Illusion. Das Dilemma verschwindet jedoch, wenn man die als Selbstheit oder Ichheit verstandene Identität durch eine narrative Selbigkeit ersetzt wird. Im Kontext der narrativen Identität verschwindet der Gegensatz zwischen dem Selbst und dem Anderen, denn das Selbst ist zugleich der Andere. Und da das Narrativ seinem Wesen nach eine zeitliche Struktur ist, kann man hierbei von einem dynamischen Identitätsmodell sprechen. Sich selbst narrativ konfigurierend, refiguriert das Selbst sich ständig. Die Selbigkeit ist auf solche Weise keine abstrakte Substanz mehr, weil die das Selbst konstituierende narrative Identität als das erlebte Leben zu verstehen ist. Die Subjektivität konstituiert sich selbst als Leserin und Schreiberin ihres eigenen Lebens. Ricæur beruht in seinen Auffassungen auf der Analyse der literarischen Autobiographien und behauptet, dass das Leben des Individuums ständig durch wahre und fiktive Geschichten, die das Subjekt von sich selbst erzählt, refiguriert wird. Durch die Refiguration wird das Leben zu einem Muster von erzählten Geschichten, das als der Weg der Selbsterkenntnis interpretierbar ist, auf dem durch Katharsis das Selbst geschaffen wird. Nach Ricœur ist der Begriff der narrativen Identität sehr fruchtbar, indem er es ermöglicht die Mentalität der individuellen Subjektivität und die der Kulturen zu untersuchen. Die narrative Identität ist somit eine Zeit, aus der die Traditionen entwachsen.

Der Leser eines Narrativs kann sich in eine beliebig welche Zeit setzen, indem er sich auf solche Weise in einer Quasi-Gegenwart befindet, die gleichzeitig auch eine Quasi-Vergangenheit oder sogar eine Quasi-Zukunft sein kann. Gerade vom Standpunkt der Erforschung der Zeit als solchen sind, so Ricœur, fiktionale Erzählungen dankbare Forschungsquellen, da die Zeit in ihnen in äuβerst konzentrierter Form vorkommt, indem sie in sich die Vielheit verschiedener Grenzerlebnisse beinhalten, denn in jedem fiktionalen Werk existiert die eigene ganzheitliche Welt, und auf die Weise werden die Sujets von der Zeit zu Zeitsujets oder zu Narrativen – sie widerlegen also die Zeit als solche (Ricæur 1991, S. 395-437). In der fiktionalen Kulturtradition (in der Belletristik), aber auch in den Lebenserzählungen von gewöhnlichen Menschen, fungiert das Narrativ als ein Zeitsujet oder eine Identität – also als das die Traditionen begründende Selbst. Das Narrativ ist somit auch eine ethische Lebenseinstellung, denn jede Erzählung hat immer den Bezug zu einem Anderen. Ihrem Wesen nach ist das keine gegenüberstellende Beziehung, denn die Anderen refigurierend verändert der Mensch sich selbst selber. Indem er das Leben der Anderen erlebt, refiguriert er auch sein eigenes Leben, und so verändern sich seine ästhetischen und ethischen Lebenskriterien. Die narrative Identität von Ricœur kann man, unserer Ansicht nach, ähnlich Roland Barthes, als einen Text interpretieren, der eine symbolische Energie oder Sprache ist, von dem der Signifikant sich ins Unendliche entfernt, denn der Text oder das Narrativ entstehen ja nur durch die Kommunikation mit dem Leser oder dem Zuhörer als solchen. Die Bezeichner, d.h. der Mensch oder die Kulturgemeinschaft, schaffen somit, vom Horizont ihres Verstehens, eine narrative Variation des gelesenen oder gehörten Textes. So betrachtet, kann man die Sprache als ein Narrativ – einen Prozess bezeichnen, der immer innovativ offen ist, denn seine Struktur und sein System hängen vom Leser bzw. von dem Zuhörer ab. Auf diese Weise kann man die Sprache als das Narrativ als eine Struktur oder ein System bestimmen, die nie ihre vollkommene Endlichkeit erreicht. Als ein Erlebnis oder eine Erkenntnis ist das Narrativ ein sich ewig in Richtung der Unendlichkeit bewegendes fiktionales Ereignis (vgl. Barthes 2002, S. 12-139; Ricœur 1991).

Von der Sprechakttheorie ausgehend interpretiert Ricœur die narrative Identität als eine interlokutive, bzw. als eine Ich-du-Beziehung (Ricœur 1996, S. 55–139; so auch Searle 1971). Folglich ist die narrative Identität autonarrativ und als solche ist sie ein aktives und handelndes Selbst. Auf die Frage, wie die Handlung und der Handelnde zueinander stehen, antwortet Ricœur (S. 113) kurz, dass es hierbei nur eine Antwort geben kann: Die Urquelle der Handlung ist der Handelnde selbst. Nach „Ethik von Nikomachos" von Aristoteles würde das bedeuten, dass sich die Handlungen bei dem Handelnden selbst befinden. Also sind das persönliche und das kulturelle Selbst keine Gegenpole.

Zusammenfassend kann man also behaupten, dass das personale Selbst von Ricœur das in einem Text lokalisierte Ich ist, denn der Mensch erreicht seine Selbstheit oder seine ethische Lebenseinstellung, indem er sich vom Aspekt einer Textfigur oder des Protagonisten über sich selbst reflektiert. Das Selbst erwirbt seine narrative Identität oder seinen ethischen Charakter, indem er sich in den semantischen Feldern seines Lebensnarrativs als des Kulturanarrativs bewegt (Ricœur 1996). Auf der Suche nach der Struktur der Selbstheit balanciert der Mensch von Ricœur (S. 425) zwischen der Grenze der Beweglichkeit und der Unbeweglichkeit oder zwischen der Grenze der Quasi-Zeit und der Nicht-Zeitlichkeit, weil er seine Quasi-Lebenssujets immer beenden muss, um mit seiner Selbstverwirklichung immer wieder neu anzufangen (vgl. Lotman 1990). Der neue Anfang ist zugleich aber auf irgendwelche Weise auch eine Wiederholung, denn als Quasi-Zeit ist die personale Identität gleichzeitig auch eine Quasi-Unzeitlichkeit oder eine Lokalisierung in einer Kulturtradition, die man nach Ricœur (1991; 1996), unserer Ansicht nach, auch als eine zur Vollendung gekommene Gegenwart in der griechischen Bedeutung verstehen kann.

Die griechisch aufgefasste Gegenwart kommt nach Tõnu Luik (2002, S. 111) durch das Wort *Geschick* zum Ausdruck. Grammatisch kommt das griechische Wort *Geschick* in der Perfektform vor. Das Schicksal in der Perfektform ist keine vergangene Zeit, sondern etwas, was stattgefunden hat und als Stattgefundenes das Seiende aufbewahren wird. Das Schicksal ist also in Bezug auf das Seiende das Kommende, das vorher schon Stattgefundene, aber zugleich das Noch-nicht-zum-Vorschein-Gekommene. Die vollendete Gegenwart als das Schicksal können nach Luik außerdem noch durch folgende Verben, wie messen, teilen und zuteil werden, charakterisiert werden. Die narrative Geschichte als kulturelle oder soziale Identität von Ricœur kann noch als das Maß des Vor-Seienden, das sich vor der Erkenntnis des Menschen verbirgt, aufgefasst werden. Dem Menschen ist aber die Aufgabe gegeben worden, dieses Sich-Verbergende zu bemerken, indem er auf diese Weise seine Selbigkeit oder seine narrative Ethik erwirbt, die in Bezug auf ihn als das Vor-Seiende da ist. Also, die Zeit des Narrativs von Ricœur ist Zeitlosigkeit. In Bezug zu einem einzelnen Leben existiert das Narrativ als eine bedeutsame Vorgegebenheit. Auch Epp Annus spricht in diesem Zusammenhang vom Narrativ als einem Kulturphänomen oder von einer Art der Welterfahrung, die sie als implizite Narrative oder „große Narrative" bezeichnet, die in der Gesellschaft die Rolle der kulturell etablierten Überzeugungen und Werthaltungen spielen. „Große Narrative" sind mit keinem bestimmten Text verbunden. Doch haben sie textliche Ausdrucksformen. Die Palette dieser

Texte ist breit und verändert sich ständig, indem diese Texte das Verstehen des „großen Narrativs“ beeinflussen. Das „große Narrativ“ entsteht aus der Mitwirkung von vielen kleinen Erzählungen, jedoch kann man es nicht in einem Text zusammenfassen. Ein Kerntext (z.B. die Bibel) kann es wohl verbinden, aber es löst sich von seinen Grenzen los und existiert meistens in einer impliziten Form. Nach Wilhelm Dilthey (1973, S.132–133) könnte man hier auch den Ausdruck die allgemeine Lebenserfahrung gebrauchen, die die individuelle Sichtweise und die persönliche Erfahrung korrigiert und erweitert. Die allgemeine Lebenserfahrung äußert sich in den Lebensbedeutungen, die sich in einer Menschengemeinschaft entwickelt haben und die die Menschen miteinander geistig verbinden, indem sie im Menschen das Gefühl der Behaglichkeit schaffen.

Die Vertreter der nativistischen narrativen Psychologie, wie zum Beispiel Jerome Bruner, sprechen in diesem Kontext vom Narrativ als von einer kulturell-habituellen Veranlagung (Polkinghorne 1998, S. 39). Die Befürworter der sozial-konstruktivistischen Richtung würden hierbei aber sagen, dass das Narrativ das Ergebnis eines soziokulturellen Mitkonstruierens ist (Gergen 1985; 1996). Bruner (1990, S. 64) vertritt zum Beispiel die Ansicht, dass die Menschen für die narrative Wahrnehmung der Welt prädisponiert sind. Sie haben eine vorsprachliche Prädisposition für das Erzählen, das dank den expressiven Äußerungen von anderen Menschen hervorgerufen wird. Donald E. Polinghorne (1998, S. 21) meint dagegen, dass die Fähigkeit, Geschichten zu erzählen und sie zu verstehen, und mit Hilfe von ihnen die Kontingenz des Lebens für sinnvoll zu ordnen, eine anthropologische Charakteristik ist. In diesem Zusammenhang ist Donald E. Polinghorne daran interessiert, ob das Konstruieren der narrativen Erfahrung und die Erfahrung selbst aufeinander zeitlich abgestimmt sind oder ist das ein reflexiver Prozess, d.h. dass das Konstruieren erst nach der Handlung stattfindet. Da die Erfahrung des Menschen aber nie fragmentär oder inkohärent sein kann, könnte man hieran auch von einer pränarrativen Qualität sprechen, die vom Bedürfnis zu erzählen bedingt wird. Dabei ist aber die pränarrative Qualität selbst unvollkommen, weil sie eine unreflektierte Erfahrung ist. Das narrative Wissen kann als eine deutende Interpretation der unreflektierten pränarrativen Qualität verstanden werden. Prä-reflexive unartikulierte Erfahrungen fungieren als Leitfäden einer entstehenden Geschichte, weil nicht aus jeder Erzählung das Prä-Narrative authentisch zum Ausdruck kommt. Die Geschichte wird erst im Laufe der kognitiven Verarbeitung produziert, aus der dann die für das Ich bedeutsame integrierte Identität geschaffen wird. Die narrative Identität ist eine Konstruktion in Form der Geschichten und die kann, muss aber nicht, als eine Ich-Erzählung explizit werden, weil man der Produkte der Selbst-Erzählung nicht bewusst ist.

Vor dem Hintergrund des sozialwissenschaftlichen Kontexts wird, unserer Ansicht nach, aber nicht ersichtlich, ob man, über die narrative Identität sprechend, einen sog. realen Menschen im Auge behält, d.h., dass die Identität des Menschen einem Narrativ ähnlich sein könnte oder ist das Narrativ nur ein heuristisches Mittel zur Interpretation der Subjektivität. J.S.Bruner (1998) scheint zum Beispiel vom Standpunkt auszugehen, dass die Denkweise des Menschen an und für sich narrativ ist. Auch der deutsche Psychologe Jürgen Straub (1998) vertritt diesbezüglich eine konzeptuell nahe Position. Er versteht unter dem narrativen, historischen oder dem Geschichtsbewusstsein die Fähigkeit, die auf spezifischen kognitiven Fertigkeiten beruht. Mit der Entwicklung dieser Fähigkeit beschäftigt sich die Psychologie des zeit-geschichtlichen narrativen Denkens.

Das Bewusstsein oder das Denken von der Zeit-Geschichte ist nach Straub (1998) eine Kompetenz Narrative zu schaffen. Die Kompetenz des Zeit-Wissens ist die Fähigkeit, der Zeit eine Geschichte oder einen Sinn zu geben. Unter dem Zeit-Bewusstsein versteht Straub kein historisches, d.h. auf die Vergangenheit gerichtetes Wissen, sondern das zeit-geschichtliche Bewusstsein. Das Zeit-Bewusstsein ist das Wissen von der Kontinuität, der Differenz und der Kontingenz, von der Unterschiedlichkeit und der Veränderlichkeit. Kurz ausgedrückt – das ist ein durch die Kontinuität und Veränderung konstituiertes Bewusstsein. Als Differenz hat das Zeit-Wissen oder das Zeitbewusstsein eine einheitliche Struktur oder Gestalt in Form des Narrativs, das als Denkweise nicht von einem vereinfachten Zeitmodell, bzw. von der dreidimensionalen Zeit (Vergangenheit-Gegenwart-Zukunft) ausgeht, sondern sie ist dazu fähig, qualitative, d.h. bedeutsame Zeitdifferenzen zu bestimmen und Zeitverbindungen zu konstruieren. Die erzählten Geschichten oder die Narrative sind, so Jürgen Straub (1998, S. 81-116), sprachliche oder textliche Ganzheiten mit einer festen formalen Struktur. Die formale Struktur des Narrativs als der Ganzheit hängt von dem kognitiven Entwicklungsstand des Subjekts ab und ist mit seiner mentalen oder kognitiven Struktur verbunden. Vom theoretischen Aspekt aus betrachtet, ist diese kognitive Struktur ein sehr abstraktes Konstrukt des Bewusstseins und funktioniert unbewusst. Diese Fähigkeit erreicht man erst im Jugendalter. Die Bildung der Kompetenz des Narrativs fängt aber in einem sehr frühen Stadium der sprachlichen Entwicklung an.

Deswegen bestimmt man in den Sozialwissenschaften das Narrativ als eine ontologische Struktur des Menschen. Die narrative Denkweise und der narrative Text sind, so E. Annus (2002, S. 37), miteinander verbunden: „Dem narrativen Text geht das narrative Denken, d.h. die Deutung der Tätigkeit und ihrer Zeitlichkeit, das Konstruieren des Sinns aufgrund der zeitlichen Kausalität, voraus. Der narrative Text beruht auf dem narrativen Denken."

Das oben Gesagte verallgemeinernd, können wir also metaphorisch behaupten, dass wir heutzutage nach der Meinung der Sozialwissenschaftler in einem narrativ strukturierten Zeitraum leben. Das bedeutet nach dem Wissen zu leben, dass unser Leben einen Anfang, einen Mittelpunkt und ein Ende hat. Wir leben folglich wie in Zeitsegmenten, von denen der Lebenssinn erst konstruiert wird. Zum Sinn des Lebens kommen wir aber aus den Genreimperativen und Gattungskanons, die sowohl für die vorigen Epochen als auch für die Gegenwart charakteristisch sind. Indem wir die Lebensnarrative von anderen lesen, deuten wir gleichzeitig auch unser eigenes Leben. Karl Jaspers (1960, S. 363) würde hierbei wahrscheinlich aber widersprechen und sagen, dass wir erzählend überhaupt nicht an das Leben als solches nähern können, denn vom Standpunkt des Lebens führen alle Erzählungen in eine Sackgasse. Indem wir über das Leben erzählen, erfahren wir, dass das Leben an und für sich unerfassbar ist, und dass wir nur nach seinen versteinerten Fertigprodukten greifen. Erzählend werden wir wohl des Wesens vom Leben bewusst, aber wir können es in dessen Ganzheitlichkeit weder erkennen noch bestimmen (Jaspers 1997, S. 204-206; Dilthey 1973).

Nach der narrativen Psychologie und Soziologie ist der Kontakt zur Realität also fraglich geworden, denn es entsteht der Eindruck, als gäbe es keine wirkliche Realität mehr. Es gibt nur eine narrative Realität, die eigentlich wie eine Hyperrealität ist, die nach einem negativistischen Modell von der Gegenwart konstruiert worden ist und die als das positive Vorbild selber die Wirklichkeit modelliert (Baudrillard 1999, S. 8–9). In den Sozialwissenschaften ist auf solche Weise das Narrativ zum Regulierer des verschwindenden oder verschwundenen Zeitraums geworden. Indem das Narrativ sich der Zeit gegenüberstellt, fungiert es selbst wie die Zeit, mit der Absicht, die Zeit als solche loszuwerden. Da die ganze Lebenswelt zeitlich geworden ist, ist aus der Zeit der Feind des Menschen geworden. In diesem Kontext kann der Mensch sich nicht mehr die Frage stellen: „Wer bin ich und zu wem ich gehöre?" Vor dem Hintergrund der zeitlichen Dimension sollte man sich vielleicht fragen: „Wer bin ich gewesen und wer werde ich sein?" Die Zeit des Menschen ist biographisch, denn als subjektive Identität muss der Mensch sich sowohl in der Zeit als auch gegen die Zeit durchsetzen. In der Zeit deswegen, dass die Identität des Subjekts als die Differenz der zeitlichen Fortdauer gesichert werden muss. Gegen die Zeit deswegen, dass von den Differenzen, die die Zeit entstehen lässt, das Bedürfnis nach der Identität oder nach der Erkenntnis der Identität hervorgerufen wird (Nassehi 1994, S. 46). Und so wird in den Sozialwissenschaften von einem fragmentären Ich gesprochen. Oder man spricht vom Subjekt, das unter dem MPS-Syndrom leidet (Kraus 1996; Giddens 1991). Das Subjekt der heutigen Theorien ist ein Mensch, der sich ständig neu kennen lernen, umlernen und mit seinem Leben immer wieder von Anfang an beginnen muss (Hitzler & Honer 1994; Keupp 1994; Gergen 1996), denn man lebt in einer Gesellschaft, die voll von Risiken ist (Beck 1986), weil die Beziehungen zwischen verschiedenen Generationen fern geworden sind. Groβmütter und Groβväter sterben ja in Altersheimen und so muss Einer die Kunst erlernen, die Angst vor dem Tod zu überwinden. Alles, was früher ein normaler Bestandteil des Lebens war – man lebte zusammen, man wurde zusammen alt und man erzog die Kinder auf – kann jetzt mit den Begriffen Landstreichermoral oder Drahtseilbiographie charakterisiert werden (Beck 1986). Wir leben in einer gefährlichen Zeit, in der Markt und Unternehmungskultur für den Menschen zu einem Terror geworden sind (Peter McLaren 1993, S. 201). In solch einer Situation hat die Identität des Menschen keinen Mittelpunkt. Der Mensch ist von sich selbst entfremdet. Seine Identität hat sich fragmentarisiert, und das wird auch vom intellektuellen oder dem wissenschaftlichen Markt begünstigt, weil Marginalisierung und Entfremdung zu einem nutzbringenden Geschäft geworden sind.

Über das Unvermögen der Sozialwissenschaften mit den Begriffen Mensch, Ich, Individualität und Identität umzugehen

In den Sozialwissenschaften hat man eigentlich nie es vermocht, mit den Begriffen Subjekt, Ich, Individualität, Identität etwas anzufangen. Warum? Unserer Ansicht nach könnte man diese Frage mit folgenden Begründungen beantworten:

1.) Die Sozialwissenschaften sind heutzutage zu den in sich selbst geschlossenen theoretischen Systemen geworden. Man verteidigt eifersüchtig ihre eigenen Grenzen, da man Angst vor dem Verlust ihrer wissenschaftlichen Selbstidentität hat. So spricht man zum Beispiel in der sozialwissenschaftlichen Biografie- und narrativen Forschung wohl von der Inter- und Transdisziplinarität. Im Auge werden dabei aber nur die Grenzen des sozialwissenschaftlichen Denkens behalten. Das kommt zum Beispiel sogar in solchen Kleinigkeiten wie in der Frage nach der Benennung der Forschungsrichtung zum Ausdruck: Ob man zum Beispiel den Begriff autobiographisch oder biographisch gebrauchen sollte. Gegen den Namen autobiographische Forschungsrichtung spricht nach Sozialwissenschaftlern vor allem die Tatsache, dass Autobiografik ein literaturwissenschaftlicher Terminus ist. Die Sozialwissenschaften haben sich immer von der Literatur, der Geschichtswissenschaft und von den geisteswissenschaftlichen Disziplinen abgrenzen wollen, um sich so von ihnen äusserst fern zu halten, damit man wie Naturwissenschaften ist. Dabei werden die Naturwissenschaften aber nur nachgeahmt, denn bisher hat man es nicht vermocht, eine Naturwissenschaft von einem sozialen Phänomen als solchen zu kreieren. Wenn die Sozialwissenschaften auf ihre szientifisch-naturwissenschafltiche Orientierung verzichten, geraten sie aber in das Gefahrgebiet der Literatur (siehe Garz 1995), obwohl sie die für sie gefährliche Grenze, unserer Ansicht nach, schon überschritten haben.

Mit Arthur C. Danto (2000) behaupten wir, dass die ganze Psychologie als ein wissenschaftlicher Bereich sich nicht in der neurophilosophischen Richtung entwickeln kann, denn das menschliche Verhalten lässt sich nicht nur in den Termini der Neurochemie erklären. Ein Teil von der Psychologie bleibt verbunden mit der Volkspsychologie. Narrative und konstruktivistische Psychologie sind ja Betrachtungsweisen, die es versuchen eine Verbindung zwischen der naturwissenschaftlichen Psychologie und Volkspsychologie aufzubauen. Das setzt aber den Mut voraus, einzugestehen, dass man – gleich den Geisteswissenschaften – eine Wissenschaft von Texten ist. Hierbei entsteht jedoch unvermeidbar die Frage: von welchen Texten? Auch von literarischen? Und welche könnten die Behandlungsweisen und die Menschauffassungen sein, von denen ausgehend die Texte interpretiert werden. All das setzt unvermeidlich die Bildung eines wirklich interdisziplinären Geistes voraus.

2.) Am Anfang des vorliegenden Unterkapitels bestimmten wir sowohl die biografische als auch die narrative Forschungsrichtung als eine Behandlung, deren Forschungsgegenstand das in Form von Texten beschriebene individuelle Leben ist. Eigentlich gilt diese Definition aber für die Sozialwissenschaft als solche im Allgemeinen, denn mit dem wirklichen Leben oder dem Menschen hat man sich in diesen Wissenschaften nie unmittelbar beschäftigt: Das Individuum und das Soziale hat man immer durch die Texte behandelt. Denn alle verschiedenen Tests, Fragebögen, Interviews usw. sind ja auch eine Art von Texten, aus denen der Mensch als solcher mit Hilfe vom verifizierten Kodieren oder anderer statistischer Verarbeitungsprogramme ständig reproduziert wird. Der auf solche Weise produzierte Mensch oder das Subjekt könnte man als Produktionsprodukt bestimmt werden, das immer wieder in einen neuen Produktionsprozess gerät.

3.) Das Produktionsprodukt selbst identifiziert sich aber nicht, sondern es ist nur das Zu-Identifizierende. Es ist ein Objekt, dessen Subjektivität nur von der wissenschaftlichen Bestimmung abhängt. In den Sozialwissenschaften (in der Psychologie, Soziologie usw.) hat man das Subjekt (den Menschen) bisher, so Zima (2000), als ein einsames Unterworfene aufgefasst, das sich selbst erst dann findet, wenn es sich mit dem Allgemeinen identifiziert. Hierbei entsteht jedoch die Frage, gibt es da überhaupt einen gravierenden

Unterschied zwischen der Bedeutung der neuzeitlichen Semantik und der Semantik des antiken und mittelalterlichen Subjekts. Das lateinische *subjectum* bezeichnet, so Tõnu Luik (2002, S.38), das Unterworfensein, was durch das griechische Wort *hypokeimenon* übersetzt wird: „*Hypokeimenon, subjectum* ist Subjekt, mit dem etwas passiert. Das, was mit ihm passiert, wird vom Prädikat bestimmt."

In dem heutigen sozialwissenschaftlichen Wissen steht das soziale System in der Funktion des Prädikats. Demnach wird der Mensch als eine sozialideologische Funktion aufgefasst (Zima 2000; Giddens 1991). Die nach den wissenschaftlichen Paradigmen modellierten Auffassungen, die in den jeweiligen sozialen Systemen dominierend sind, bestimmen also, wie der Mensch als das Subjekt zu verstehen ist. Zima (2000) weist hierbei darauf hin, dass die westlichen Theorien der Persönlichkeit sozialideologische Konstrukte sind und dass sie monologisch autoritär sind, weil sie sich zu ihrem Subjekt so abseitig objektiv verhalten, als hätten sie vergessen, dass auch sie selbst ein Teil der beschriebenen Realität sind (vgl. Danto 2000, S. 353). Dieses Verhalten impliziert aber unabdingbar eine autoritäre Haltung. Denn man geht von der Metapher der unvereinbaren Gegensätze aus: Helden und Antihelden, Freunde und Feinde, zivilisierte und nichtzivilisierte, kollektivistische und individualistische, primitiv denkende und zum abstrakten Denken fähige Kulturen usw. Peter Zima (2000) nennt solch eine Mentalität für die Krise der Gesellschaft. In verschiedenen biografischen und narrativen, aber auch in geisteswissenschaftlichen Disziplinen wird das vor allem als eine Krise der Persönlichkeit oder der Ich-Identität interpretiert (Klika 2000; Kraus 1996 usw.). Nach uns sind aber eigentlich die sozialwissenschaftlichen Konzeptionen von der Persönlichkeit und von dem Ich in der Krise.

4.) Denn die Auffassung der Sozialwissenschaften vom Menschen (vom Subjekt, vom Ich, vom Selbst, von der Identität) wird, unserer Ansicht nach, nicht mit dem sog. lebendigen Leben untermauert, denn das lebendige Leben ist unvermeidbar kontingent. Und so kann man mit keiner kognitiver Strategie (mit dem Narrativ, der Ich-Erzählung, mit der Autobiographie) das Leben als lebendiges Leben weder erlernen noch vorplanen.
5.) Die Sozialwissenschaft ist folglich ein Signifikant, der wie hinter dem Rücken des Menschen handelt – der nach der wesentlichsten und endlichen Definition sucht. Diese Definition gibt es aber nicht, denn es gibt keine Metasprache, die den Menschen erfassen könnte (Eco 1994, S. 410-411.) Denn das Leben und der Mensch sind in ihrer absoluten Mächtigkeit unfassbar, weil sie als begriffliche Objekte nicht wahrnehmbar sind (Lyotard 2000, S. 1426).

6.) Die Mensch- und Weltauffassung der Sozialwissenschaften bleibt immer für historistisch und vergangenheitsbezogen, denn das Seiende kann nur vom Standpunkt des Vergangenen beurteilt werden. Und deswegen ist es unabdingbar unvermeidlich, dass man – wenn sich alles im Leben sehr schnell verändert – diese Veränderung ihrem Wesen nach nicht zu interpretieren vermag, sondern man deutet sie als etwas Drohendes. Es werden, um diese Veränderung zu verstehen, sehr allgemeine Kategorien begründet. Als ob sie die Geschwindigkeit der Veränderung vermindern könnten, indem sie ferne Epochen, verschiedene Zeiträume, Begriffe und Werte verallgemeinern. Zu einem solchen Begriff ist gegenwärtig Postmoderne geworden die an und für sich eine Kategorie ist, mit der man der Eindruck geschaffen wird, dass alles unveränderlich konstant ist. Das Wort Postmoderne wäre also wie eine Idealsprache für die Umfassung der Realität. Die Einstellung, als ob es eine Idealsprache als eine optimale Ausdrucksform der Realität gäbe, ist einfach eine Illusion. Denn die minimal funktionierende Struktur setzt zwei Sprachen und das voraus, dass keine von den beiden, einzeln genommen, die Welt als solche erfassen können (Lotman 2001, S. 10).

Auch das Narrativ kann als eine oben erwähnte Idealsprache betrachtet werden. In den pädagogischen Auffassungen wird es außerdem noch als eine Lebensdidaktik aufgefasst, denn das Narrativ hat pragmatische Funktionen, es enthält Verhaltensregeln darüber, wie man leben sollte. Folglich verbirgt es in sich das sowohl individuell als auch soziokulturell Typische. Das Narrativ – das ist eine kulturelle Vereinbarung (Bruner 1998 usw.). So verstanden kann man das Narrativ als einen ethischen Imperativ bestimmen und die sozialwissenschaftlichen narrativen Auffassungen als das ABC der Ethik. Außerdem ist das Narrativ ein zeitloser kategorischer Imperativ, denn es selbst ist ja Zeit, und eine seiner Funktionen ist deswegen dem Menschen beizubringen, wie die Gegenwart als ein negatives Narrativ zu stoppen ist (McLaren 1993).

Da man in den Sozialwissenschaften die Fiktion vom „figurativen“ Narrativ-Menschen geschaffen hat, hat man kein Bedürfnis mehr nach den ambivalenten Begriffen wie Identität, Ich, Subjekt, Objekt, weil sie doch alle durch das Narrativ ersetzbar sind. Hierbei entsteht natürlich die Frage, wodurch sich das litaratur-oder sozialwissenschaftliche Narrativ oder die sozial- und literaturwissenschaftliche Textinterpretation voneinander unterscheiden. Denn sie alle sind doch Fiktionen.

Der Forschungsgegenstand der Literaturwissenschaft ist Fiktion als imaginäre Wahrheit, die aus der Erhabenheit der Vorstellung entsteht und die deswegen keine rational zweckmäßige Tätigkeit sein kann (Loytard 2000, S. 1426). Denn durch sie entsteht das Erlebnis als die Seele, nicht aber der Geist, dessen Hauptcharakteristiken Ziel und Zweck sind. Zur Schöpfung der sich erlebenden Seele, muss der Schöpfer aber selbst beseelt werden. Er muss von der intellektuellen Gegenständlichkeit befreit werden, denn nur beseelte Schöpfung vermag es, das Erlebnis als die auf sich selbst konzentrierte Seele in ihrer individuellen innerlichen bildhaften Einheit hervorzurufen. Das konzentrierte Erlebnis als die Seele ist aber ein gegenstandsloser Rhythmus. Der Autor selbst wird passiv, da er in den Anderen (in seinem Werk und in dessen Protagonisten) lebt. Der Autor ist der Rhythmus des Erlebnisses der Anderen. Er wird zum Rhythmus der Stimmen des Werkes, indem er zum Rhythmuschor des Werkes gehört (Bachtin 1986, S. 105-115). Das Bewusstsein des Autors ist somit das Bewusstsein der Anderen und als solche ist es eine potenziell unendliche Realität (Loytard 2000, S. 1428).

Vor dem Hintergrund des oben Gesagten können weder der Mensch noch die Kulturtexte als semiotisch homogen betrachtet werden. Sie sind nur auf der Skala der semiotischen Homogenität oder der Heterogenität zu verstehen, denn sowohl die belletristischen Texte als auch die Kulturtexte entwickeln sich als Semiose im Feld der historisch-kulturellen Kollisionen. Einerseits charakterisiert sie wohl die Ganzheitlichkeit und die immanente Geschlossenheit, andererseits aber auch die semiotische Heterogenität, Widersprüchlichkeit und strukturelle Gegensätzlichkeit (Lotman 1990, S. 275; Zima 1980, S. 12). Mit Lotman kann man also behaupten, dass die Kultur als eine komplexe Ganzheit aus sehr unterschiedlichen Schichten besteht, die sich mit sehr unterschiedlicher Geschwindigkeit entwickeln. Durch die Explosion in einer Schicht entstehen synchron Entwicklungen in den anderen Schichten (Lotman 2001, S. 23). Wir leben also in der Welt verschiedener kontingenter Wechselwirkungen. Und es ist nicht möglich, aus unserer Gegenwart, der Sprache, der Kultur, aus den Institutionen und aus den sozialen Praktiken auszusteigen, um sie alle von einer neutralen Position vergleichen zu können (Rorty 1999, S. 97). Wir sind untrennbar und unmittelbar mit der Welt verbunden.

Die Fiktion in den Sozialwissenschaften verbannt dagegen aus den Texten das Imaginäre als die Potenzialität der Explosion, denn sonst wäre es einfach nicht möglich, zu validen Untersuchungsergebnissen zu kommen. Und so sucht man auch in den qualitativ orientierten Sozialwissenschaften immer wieder nach irgendeiner einheitlich geltenden Idealsprache, auf Grund der man das Ich als den Eigennamen und die von dem Eigennamen geschaffenen Texte typologisieren könnte (Bohnsack 2002). Man kann somit sagen, dass die qualitative „Wende" in den heutigen Sozialwissenschaften nur darin besteht, dass man neue Textsorten zur Untersuchung des Menschen als solchen finden will. Wenn bisher bei der Untersuchung des Menschen verschiedene formalisierte Tests und Interviews dominierten, dann gebraucht man jetzt offene, halboffene, ganz offene usw. Interviews (narrative oder lebensgeschichtliche Gespräche, in freier Form geschriebene Geschichten, Erzählungen, belletristische Texte). Man sucht nach Techniken der Textanalyse, die den Kriterien der Wahrheit der Sozialwissenschaften entsprechen könnten. Wichtig ist somit das Problem der Veränderung der Untersuchungstechnologie, nicht aber die Suche nach neuen Menschauffassungen. Irgendwelche alten Beschreibungen, so Rorty (1999), werden mit Hilfe anderer alter Beschreibungen analysiert, ohne dass man nach neuen Umschreibungen, nach einem neuen Grenzwortschatz sucht (Rorty 1999, S. 131–163).

Die Sozialwissenschaften sind sich aber leider der Nichtkontingnez ihres Grenzwortschatzes sicher. Und so hält man den Forschungsgegenstand als solchen – d.h. die Menschen, die Gesellschaft, die Kultur und die Kulturtexte – für brüchig und kontingent. Deswegen beseitigt die sozialwissenschaftliche Fiktion aus ihren Texten den Menschen als eine Lebensganzheit und auch seinen Seelenrhythmus. Sehr treffend charakterisiert in dieser Verbindung das Wesen der biographischen und narrativen Ausführungen Paul Geyer (1997), indem er behauptet, dass Begriffe Autobiographie, Biografie und Narrativ begriffliche Metaphern sind, die nach alten theoretischen Kriterien konzipiert worden sind und mit denen man die sowohl in der westlichen Metaphysik als auch in den Sozialwissenschaften eingewurzelte Auffassung von der zeitlichen Gespaltenheit des Wesens des Menschen zu verschleiern versucht. Es geht hier also einfach um einen künstlichen Wiederbelebungsversuch des Subjekt-Objektproblems, nicht aber um die Suche nach neuen Metaphern (Rorty 1999, S. 97– 37).

Zum Abschluss des oben Gesagten behaupten wir also, dass der Versuch, den Menschen vom Zeitprinzip ausgehend zu bestimmen in eine Sackgasse führt. Warum? Denn in den bisherigen Behandlungen vom Ich hat der Mensch selbst als Maβ der Zeit fungiert, d.h., dass die Zeit als ein menschliches Bezugsphänomen fungierte. Der Mensch als solcher ist somit zugleich sowohl die Zeit als auch das Maβ der Zeit gewesen. Die Zeit des Menschraums ist aber absolut unveränderlich, denn von welchem Aspekt wird die Veränderung gemessen, wenn die Variable ständig dieselbe ist. Ungelöst bleibt in diesem Kontext eigentlich das Problem, ob die Zeit selbst als solche überhaupt mit dem Parameter der Veränderung verbunden ist. Wir wissen nicht, was oder wer die Zeit als solche ist, die sich auβerhalb der Grenzen der menschlichen Erkenntnis befindet. Das bedeutet aber, dass das Maβ der Zeit an und für sich für den Menschen unbekannt bleibt (Geyer 1997). Wir behaupten, dass das Maβ des Menschen das Dasein ist, denn niemand kann aus dem Menschraum aussteigen. Zum Ort (zur Raumzeit) des Menschen ist und bleibt für immer der Mensch selbst. Das heiβt, dass die Dimension des Menschen immer nur das Menschsein ist, weil es einfach nicht möglich ist, sich von dieser Lokalisiertheit zu entfernen. Die Zeit des Menschen ist folglich zeitlos – also eine Zeit, wo es keine Zeit gibt (Husserl 1998), da der Mensch selber die lokalisierte Zeit ist. Die Subjektivität, das Bewusstsein, das Selbstbewusstsein, das Ich – das alle sind Begriffe, die das Leben als solches zu erfassen versuchen (Jaspers 1960, S. 413), aber das Wesen des Lebens offenbart sich nicht, und wahrscheinlich ist eben deswegen der zeitlose Mensch dazu bestimmt, nach der Zeit als solchen zu suchen.

Literatur

Aichinger, I. (1998). Probleme der Autobiographie als Sprachkunstwerk.- In: G. Niggl (Hrsg.). Autobiographie: zu Form und Geschichte einer literarischen Gattung. Darmstadt: Frotscher Druck, 170-199.

Annus, E. (2002). Kuidas kirjutada aega. Tallinn: Underi ja Tuglase Kirjanduskeskus.

Bachtin, M. (1986). Ėstetika slovesnogo tvorčestva. Moskva: Iskusstvo.

Barthes, R. (2002). Autori surm: Valik kirjandusteoreetilisi esseid. Tartu: Varrak.

Baudrillard, J. (1999). Simulaakrumid ja simulatsioon. Tallinn: Kunst.

Bauman, Z. (1994). Tod, Unsterblichkeit und andere Lebensstrategien. Frankfurt am Main: Fischer Taschenbuchverlag.

Beck, U. (1986). Risikogesellschaft: Auf dem Weg in eine andere Moderne. Frankfurt am Main: Suhrkamp.

Bohnsack, R. (2002). Dokumentarische Methode und die Analyse kollektiver Biographien.- In: G. Jüttemann & H. Thomae (Hrsg.). Biographische Methoden in den Humanwissenschaften. Weinheim & Basel: Beltz, 213–231.

Bourdieu, P. (2003). Praktilised põhjused : Teoteooriast. Tartu: Tänapäev.

Bruner, J. (1996). The Culture of Education. Cambridge: Harvard University Press.

Bruner, J. (1998). Vergangenheit und Gegenwart als narrative Konstruktionen.- In: J. Straub (Hrsg.). Erzählung, Identität und historisches Bewusstsein: Die psychologische Konstruktion von Zeit und Geschichte (Erinnerung, Geschichte, Identität I). Frankfurt am Main: Suhrkamp, 46-81.

Böhme, G. (1990). Lebensgestalt und Zeitgeschichte.- In: BIOS, H.2. Opladen: Leske + Budrich, 135-151.

Danto, A.C. (2000). Ühendused maailmaga: Filosoofia põhimõisted. Tallinn: Hortus Litterarum.

Dilthey, W. (1973). Der Aufbau der geschichtlichen Welt in den Geisteswissenschaften. Stuttgart: Teubner Verlag.

Eco, U. (1994). Einführung in die Semiotik. München: Wilhelm Fink Verlag.

Elias, N. (1990). Über den Prozess der Zivilisation: Soziogenetische und psychogenetische Untersuchungen: Erster Band: Wandlungen des Verhaltens in den weltlichen Oberschichten des Abendlandes. Frankfurt am Main: Suhrkamp.

Elias, N. (1992) Über den Prozess der Zivilisation: Soziogenetische und psychogenetische Untersuchungen: Zweiter Band: Wandlungen der Gesellschaft. Entwurf zu einer Theorie der Zivilisation. Frankfurt am Main: Suhrkamp.

Frank, M. (1991). Selbstbewusstsein und Selbsterkenntnis: Essays zur analytischen Philosophie der Subjektivität. Stuttgart: Philipp Reclam jun.

Garz, D. (1995). Qualitative Sozial- oder Humanforschung – ein anderes Menschen- und Weltbild.- In: E. König & P. Zedler (Hrsg.). Bilanz qualitativer Forschung: Band I: Grundlagen qualitativer Forschung. Weinheim: Deutscher Studien Verlag, 11-33.

Gergen, K. J. (1985). The Social Constructionist Movement in Modern Psychology.- In: American Psychologist, Vol. 40, No.3, 266-275.

Gergen, K. J. (1996). Das übersättigte Selbst: Identitätsprobleme im heutigen Leben. Heidelberg: Carl-Auer Verlag.

Geyer, P. (1997). Die Entdeckung des modernen Subjektes: Anthropologie von Descartes bis Rousseau. Tübingen: Niemeyer Verlag.

Giddens, A. (1991). Modernity and Self-Identity. Cambridge: Polity Press.

Heidegger, M. (1993). Sein und Zeit. Tübingen: Max Niemeyer Verlag.

Henrich, D. (1979). Identität – Begriffe, Probleme, Grenzen.- In: O. Marquard (Hrsg.). Identität. München: Wilhelm Fink Verlag, 169-185.

Hitzler, R. & Honer, A. (1994). Bastelexistenz: Über subjektive Konsequenzen der Individualisierung.- In: U. Beck & E. Beck-Gernsheim (Hrsg.). Riskante Freiheiten. Frankfurt am Main: Suhrkamp, 307-316.

Holdenried, M. (2000). Autobiographie. Stuttgart: Philipp Reclam jun.

Husserl, E. (1998). Phänomenologie der Lebenswelt: Ausgewählte Texte II. Stuttgart: Philipp Reclam Jun.

Jaspers, K. (1960). Psychologie der Weltanschauungen. Berlin & Göttingen & Heidelberg: Springer Verlag.

Jaspers, K. (1997). Aja vaimne situatsioon. Tartu: Ilmamaa.

Keupp, H. (1994). Ambivalenzen postmoderner Identität.- In: U. Beck & E. Beck-Gernsheim (Hrsg.) Riskante Freiheiten. Frankfurt am Main: Suhrkamp, 336-353.

Klika, D. (2000). Identität – ein überholtes Konzept?: Kritische Anmerkungen zu aktuellen Diskursen auβerhalb und innerhalb der Erziehungswissenschaft.- In: Zeitschrift für Erziehungswissenschaft, 285-304.

Kohli, M. (1988). Normalbiographie und Individualität: zur institutionellen Dynamik des gegenwärtigen Lebenslaufregimes.- In: H.-G. Brose & B. Hildenbrand (Hrsg.). Vom Ende des Individuums zur Individualität ohne Ende. Opladen: Leske + Budrich, 33-53.

Kraul, M. & Marotzki, W. (2002). Bildung und Biographische Arbeit – eine Einleitung.- In: M. Kraul & W. Marotzki (Hrsg.). Perspektiven erziehungswissenschaftlicher Biographieforschung. Opladen: Leske + Budrich, 7-22.

Kraus, W. (1996). Das erzählte Selbst: Die narrative Konstruktion von Identität in der Spätmoderne. Pfaffenweiler: Centaurus Verlagsgesellschaft.

Lejeune, P. (1998). Der autobiographische Pakt.- In: G. Niggl (Hrsg.). Autobiographie: zu Form und Geschichte einer literarischen Gattung. Darmstadt: Frotscher Druck, 214-257.

Lotman, J. (1990). Kultuurisemiootika. Tallinn: Olion.

Lotman, J. (1999). Semiosfäärist. Tartu: Vagabund.

Lotman, J. (2001). Kultuur ja plahvatus. Tartu: Varrak.

Luik, T. (2002). Filosoofiast kõnelda. Tartu: Ilmamaa.

Löwith, K. (1981). Mensch und Menschenwelt: Beiträge zur Anthropologie. Stuttgart: J.B. Metzlersche Verlagsbuchhandlung.

Lyotard, J.-F. (2000). Mis on postmodernsus?- In: Akadeemia, 7, 1419-1431.

McLaren, P. (1993) Border Disputes: Multicultural Narrative, Identity Formation, and Critical Pedagogy in Postmodern America. McLaughlin, D. & Tierney, W. G. (Eds.). Naming Silenced Lives. New York & London: Routledge, 201-237.

Nassehi, A. & Weber, G. (1990). Zu einer Theorie biographischer Identität. Episte- mologische und systemtheoretische Argumente. BIOS, H. 2, Opladen: Verlag Leske + Budrich, 154-187.

Nassehi, A. (1994) Die Form der Biographie. Theoretische Überlegungen zur Biographieforschung in methodologischer Absicht. BIOS, H.1, 7. Jg., Verlag Leske + Budrich, 46-63.

Polkinghorne, D.E. (1998). Narrative Psychologie und Geschichtsbewusstsein. Bezie- hungen und Perspektiven.- In: Straub, J. (Hrsg.) Erzählung, Identität und historisches Bewusstsein. Die psychologische Konstruktion von Zeit und Geschichte. Erinnerung, Geschichte, Identität I. Frankfurt am Main: Suhrkamp, 12 - 46.

Ricœur, P. (1991) Zeit und Erzählung Bd. III Die erzählte Zeit. München: Wilhelm Fink Verlag.

Ricœur, P. (1996) Das Selbst als ein Anderer. München: Wilhelm Fink Verlag.

Rorty, R. (1999) Sattumuslikkus, iroonia ja solidaarsus. Tartu: Vagabund.

Röttgers, K. (1988) Die Erzählbarkeit des Lebens. BIOS, H. 1, Verlag Leske + Budrich, 5-17.

Sarbin, Th. R. (Hrsg.) (1986). The Narrative as a Root Metaphor for Psychology. Narrative Psychology. The Storied Nature of Human Conduct. New York: Praeger, 3-22.

Schramke, J. (1974). Zur Theorie des modernen Romans. München: Verlag C.H. Beck.

Schulze, Th. (1993). Biographisch orientierte Pädagogik.- In: Baacke, D. & Schulze, Th. (Hrsg.). Aus Geschichten lernen. Weinheim und München: Juventa, 13-41.

Schulze, Th. (1993a). Autobiographie und Lebensgeschichte. In: Baacke, D. & Schulze, Th. (Hrsg.). Aus Geschichten lernen. Zur Einübung pädagogischen Verstehens. Weinheim, München: Juventa, 126-174.

Schulze, Th. (1993b). Lebenslauf und Lebensgeschichte.- In: Baacke, D. & Schulze, Th. (Hrsg.) Aus Geschichten lernen. Zur Einübung pädagogischen Verstehens. Weinheim, München: Juventa, 174-229.

Schütze, F. (1996) Verlaufskurven des Erleidens als Forschungsgegenstand der interpretativen Soziologie.- In: Krüger, H-H. & Marotzki, W. (Hrsg.). Erziehungswissenschaftliche Biographieforschung. Opladen: Leske+Budrich, 116-158.

Searle J. R. (1971). Sprechakte. Ein sprachphilosophischer Essay. Frankfurt am Main: Suhrkamp Verlag.

Straub, J. (1998). Geschichten erzählen, Geschichte bilden. Grundzüge einer narrativen Psychologie historischer Sinnbildung.- In: Straub, J. (Hrsg.). Erzählung, Identität und historisches Bewusstsein. Die psychologische Konstruktion von Zeit und Geschichte. Erinnerung, Geschichte, Identität I. Frankfurt am Main: Suhrkamp, 81 – 170.

Taylor, Ch. (2000) Autentsuse eetika. Hortus Litterarum.

Zima, P. (2000) Theorie des Subjekts: Subjektivität und Identität zwischen Moderne und Postmoderne. Tübingen, Basel: Francke Verlag.

Zima, P.V. (1980) Textsoziologie. Stuttgart: MCMLXXX J.B. Mezlersche Verlagsbuchhandlung.

Self-Education as Premise and Context for Teacher's Identity Development: The Case of Master Studies in Educational Science

Ligita Dudinskienė, Vilma Žydžiūnaitė

Abstract

The article discloses premises and context of self-education as teacher's identity development. The results of the performed analysis of scientific sources as well as the survey-in-written (in which n = 327 teachers, master students of Lithuanian higher education institutions in the field of educational science took part) have shown that teacher's self-education involves conscious process of learning related to teachers' self-esteem, innovativeness and self-respect. During the research it has become evident that teacher's self-education is inherent from comprehensive personal development, which allows experiencing learning satisfaction. The teachers, who purposefully study in master studies in educational science at the university, deliberately learn from own experience. The identified components (important for teachers' identity development) of learning motivation (personal features, collaboration and learning in action, self-motivation as well as the wish to implement changes), of purposefulness of learning (learner's responsibility, experience, practical value of knowledge abilities and skills, self-analysis, management of learning process) as well as of the analysis of learning activity (reflection on learning process, learning structure, identification of consistency, organization of learning, development) discloses the interaction between self-education and teacher's identity development.

Keywords: identity, teacher, self-education

Introduction

Research relevance

Professional identity is analyzed in educational science (Olesen, 2001; Crisafulli, 2009), psychology (Ikiugu, Rosso, 2003), management (Dobrow, Higgins, 2005), nursing (Eriksen, 2004). Professional identity is not disembodied from person's identity; it is the constituent (integral) of person's identity, thus it is often formed together with social identity. In a sense it is one of the structural parts of social identity, which manifests and is formed in educational situations as well as in social environments of particular professions (Crisafulli, 2009). In this article the attention is paid to self-education as premise to develop teacher's identity, and 'teacher's identity' is not divided into personal and professional by treating this concept as integrating both components – personal and professional.

Foreign scientists' works are rich with research topics related to teachers' identity. Holden (2004) analyzed the involvement of the persons who finished their higher education studies into professional development; he also pointed out that this was the inherent part of the process at university studies. Dobrow and Higgins (2005) investigated how the professional identity of business students' bachelors and master students from the elite school was related to their participation in networks of mentors. The topic of teachers' professional identity was studied by Beijaard et al. (2000); the researchers interviewed teachers of comprehensive schools (N=80). Cattley (2007) strived to reveal the influence of the professional identity of the teachers studying at Flinders University upon the learning environments, which could develop possibilities for personal and professional development. In social work McMichael (2000) identified how the development of professional identity was related to the professional status and image in continuing studies. Chong, Ling and Chuan (2011) highlighted that teachers formed their professional identity by referring to their education / self-education experiences, and Beijjard et al. (2004) analyzed factors and characteristics of teachers' professional identity formation. O'Connor (2006) interpreted the relationship of professional identity and personal spiritual experiences among teachers of secondary schools. Olesen (2001) performed several researches in the field of educational science as he pursued to evaluate the relationship of teachers' personal identity and the process of learning. Roberts (2000) analyzed factors of the professional identity of mathematics bachelor-students and teachers-beginners. Weinberg (2011), Schwartz et al. (2008), Pearce & Morrison (2011), Olsen (2008), Peter and Robert (2002) also analyzed the concept of person's identity in different aspects.

A lot of researches analyzed self-education; however, Lithuanian scientists name *self-education* by different terms. However, still there are few researches on self-education, which is related to teacher's identity development. The article will not separately present and analyze the concept of *self-directed learning*, which is not separated or even identified as self-education in scientific sources (Neale et al., 2008; Starr, 2009). However, it will be mentioned as self-education component in the presentation of the empirical research results by disclosing the complexity of self-education concept and links to teacher's identity development.

Novelty of the research

The article presents the research, which systemizes the conceptions presented by various authors, and the performed empirical research reveals the coherences of self-education and teacher's identity development.

The research problem

Scientific articles (Beijaard et al., 2009) on self-education and teacher's identity disclose separate aspects; thus, when analyzing these concepts together, coherences among teacher's self-education, self-directed learning and teacher's identity development are identified. *Self-education* is related to life-long process of person's development, in which three main characteristics are relevant (Vaivada, 2012): education target, educational tools and forms as well as personality development. The latter disclose main aspects of identity development when a person takes place in different learning activities, which most often are planned by himself / herself. The main factor motivating a person to take part in the process of learning is prospective application of knowledge, abilities and skills in one's life and practice. Thus it is important to understand how a teacher, when studying in the master studies of educational science, which are directly related to practical activity, finds himself / herself in the profession. So far the researches, which would attempt to relate self-education to teacher's identity development at university studies (especially referring to the experiences of the students at master studies level), have not been performed; this article does this. The performed analysis of scientific sources and empirical data allows stating that teachers' identity is not directly developed as 'identity' does not possess precise and clear structure of knowledge and skills. However, it is possible to state that self-education is the premise for developing teacher's identity in higher education. *The search for coherences between self-directed learning and teacher's identity development is the direction of other new continuous research by the authors of this article.*

Thus the formulated question of the research is inseparable from the context of the scientific problem being presented is as follows: What components are relevant for teachers' identity development when studying at the master studies in educational science in the university?

The research object – is coherences of teacher's self-education and identity development.

The research aim – is to substantiate self-education as premise and context for teacher's identity development in the context of master studies in educational science in the university.

The research samples

Considering the research problem the target population – the master students of the field of educational science, who perform pedagogical work – has been chosen. In the research 327 respondents from Vytautas Magnus University (VMU),

Kaunas University of Technology (KTU), Klaipėda (KU), Šiauliai (ŠU), Lithuanian University of Educational Sciences (LEU), Vilnius (VU) and LCC International (LCC) universities have participated. In 2012, when the research was carried out, in total 319 master studies programmes were being implemented in Lithuanian higher education institutions. In 2012 the number of the students studying in master studies in educational science fluctuated from 3 to 15 master students. The population of students at the master studies level consisted of N = 1884 master students of social sciences. The sample of the respondents was appropriately calculated by the spreadsheet of the sample size on the Internet website www.apklausa.lt. According to the achieved result in the spreadsheet, the sample had to consist of not less than n = 319 respondents in order that the survey results would reflect the opinion of the population with the chosen probability (95%) and error (0.5%). In the survey most participants were women (n = 265, this made 81% of the respondents' sample). The part of the men was less n = 62 (19% of the respondents' sample). The age phase of most respondents (82.6%) fluctuated from 23 to 30 years, the least age phase (3.7%) was from 41 to 50 years. The largest part of the survey sample participants studied at VU (n = 20.2 %), VMU (n =19.6%), ŠU (n =17.1%) and KTU (n = 14.7%). *Education Management, Education Quality Management* and *Educational Science* are the most popular master studies programmes in the field of educational science, the branch of social sciences.

The research methods

The matrix method (Klopper et al., 2007) was applied when performing the analysis of science sources. The survey-in-written was implemented by using the structured close-type questionnaire. The research data were processed by using the software SPSS 16.0 for Windows by applying methods of correlation analysis (according to Spearman) and of factorial analysis as well as sizes of descriptive statistics (*Mean*). The check of statistical data suitability was carried out by referring to KMO coefficient (0.853-0.932), in order to evaluate the factor's inner consistency the Cronbach α test was performed. The method or principal components and the Varimax rotation were applied.

The research instrument

The questionnaire consisted of 84 statements, which were divided into three parts: (1) instruction; (2) demography (6 questions); (3) diagnostic (78 statements). The questionnaire values of the Cronbach's α were very high (0.837 – 0.933).

The research results

The research results revealed 57 sufficient strength correlation relationships of self-education as the premise for teacher's identity development, the coefficient of which (according to Spearman) fluctuated from0.401 to 0.639. Three factors were distinguished as the result of the data analysis.

In the first factor (*Self-education, which stimulates motivation of learning, as teacher's identity development context*) the components developing teacher's self-esteem and that are inseparable from identity development became evident (see Fig. 1).

Fig. 1: Self-education, which stimulates motivation of learning, as premise for teacher's identity development

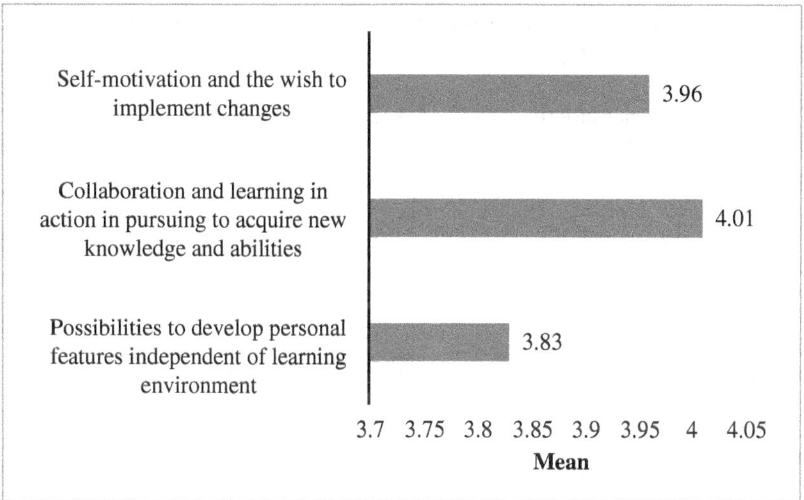

The respondents are more motivated to learn if learning activities are meaningful for them, allow identifying with the teacher's role, and the learning environment stimulates the wish to learn. It is possible to state that the development of teachers' personal factors and behaviour as well as the memorizing of new knowledge by acquiring new experience, discloses in performing different learning activities. The statement *Possibilities to develop personal features without reference to learning environment* (Mean = 4.01) explains that, when the respondents form their personal provisions, there is the possibility to enlarge their motivation of learning. Thus it is possible to presume that the starting point of teachers' self-education motivation for the learning, where the teacher finds his / her second 'I'.

When comparing the results of our research to the research of Beijaard et al. (2009) performed in the Netherlands, similar motives related to development of teachers' identity: self-improvement, communication and learning environment. The authors of the above-mentioned research in the Netherlands disclose that the interaction of identity development and environment, the result of which manifests in teacher's satisfaction with the work, self-education and motivation to develop, is important teachers of comprehensive schools. In summing up it is possible to state that teachers' motivation has direct links with their own positive self-esteem. Thus teachers' self-esteem is one of important components, which are relevant for the identity development of the students' studying in master studies in educational science at the university.

The second factor (*Self-education, which stimulates learning purposefulness, as the context of teacher's identity development*) disclosed the components that stimulate the innovation of teacher's activity and disclose main elements of teachers' identity development (see Fig. 2).

Fig. 2: Self-education, which stimulates learning purposefulness, as the context of teacher's identity development

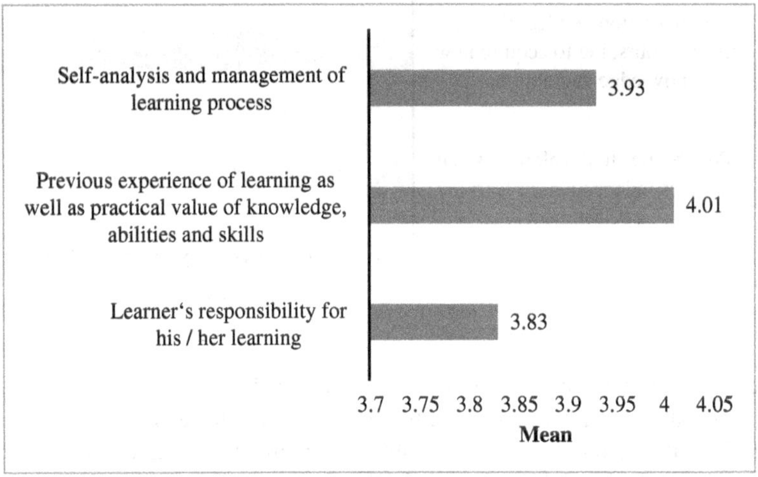

The results show that the teachers studying in master studies in educational science at the university are able to set the direction of their development, and in the purposeful learning a teacher opens as personality. The respondents take their responsibility for learning (Mean = 3.83), which is related to their previous experience of learning and are able to manage their learning. The statement *Previous experience of learning as well as practical value of knowledge, abilities*

and skills (Mean = 4.01) explains that the experience of learning as the self discovery is important for the respondents by purposefully managing the process of own learning and knowing how to purposefully apply knowledge, abilities and skills. The third component of the second factor *I reflect the process of my learning in order to better understand learning ways* shows that by managing the process of their learning (Mean = 3.96) the teachers reflect and analyze the means inseparable from their self-education. The research results also disclosed that the self-education stimulating learning purposefulness becomes evident in the development of teacher's person identity. Adequate research results are presented by Chong et al. (2011), who distinguished five factors: learner's responsibility in the field of learning; self-analysis; professional growth as development of identity (personal and professional), senses; personal needs. As these researchers state, the development of teacher's identity is the continuous process when a teacher can change as well as to self-evaluate in everyday situations. Such components are revealed in the research by Pearce (2011), the results of which showed that teacher's identity development became evident when they purposefully organized their learning as well as the vision and recognition through the scope of the self discovery possibilities and meanings.

In the third factor (*Self-education, which stimulates the analysis of learning, as the context of teacher's identity development*) the components inducing teacher's awareness and self-esteem that unclosed through particular elements of teachers' identity development became evident (see Fig. 3).

Fig 3: Self-education, which stimulates the analysis of learning, as the context of teacher's identity development

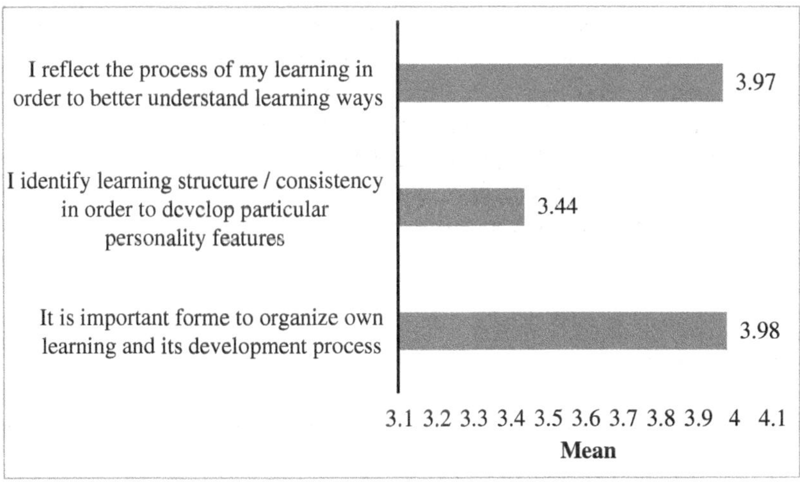

The research results obtained in our research allow stating that components of the learning content, which empower the master students to analyze the learning content, are important for the teachers studying in the master studies in educational science at the university. According to the respondents, the first and the third components are essential for them. Thus, as the teachers state, self-education is inseparable from their independent learning, during which students learn how to solve problems, to make decisions and organize their learning. Such results coincide with the results of the research performed by Cattley (2007), which show that self-education is the premise creating the clear identity of a teacher, which becomes evident when a teacher analyzes the learning content, i.e. self-education is assessed as conscious process. Thus it is possible to presume that the starting point of teachers' identity development is purposeful continuous learning and development, which help a teacher find himself / herself through self-education.

The statement *I identify the learning structure / consistency when pursuing to develop particular features of my personality* (Mean = 3.44) shows that the context of the respondents' self-education development depends on their ability to identify the learning structure and consistency, and here personality features of a teacher play a minor role. Such aspect is adequate to the results of the research performed by Touluse (2007), which showed that person's identity developed in the context of cognition and perception when learning. However, the research by this author the personality gets probably the essential role, which orients to self-esteem, in which the scientists distinguishes several factors, namely: spiritual, physical, emotional-psychological and intellectual. In our research the latter aspect was not included; however, in our further researches we will pay attention to it. However, the results of our research do not contradict the results of the research by O'Connor (2008), which show that the analysis and reflection on the experiences taking place during learning make premises to identify the learning structure as well to continuously learn when teacher's identity develops. Thus in summary it would be possible to state that in self-education the analysis of the learning made purposefully by the person is not inseparable from the self-education consciously implemented; this contributes to teacher's identity development.

Conclusions

Self-education as premise for teacher's identity development is purposeful comprehensive process of learning, during which a person consciously learns from all his / her life activities. Self-education is relevant for teacher's identity development when purposefully organizing own learning, evaluating own needs for learning, i.e. by empowering oneself for self-education.

The identified components (important for teachers' identity development) of learning motivation (personal features, collaboration and learning in action, self-motivation as well as the wish to implement changes), of purposefulness of learning (learner's responsibility, experience, practical value of knowledge abilities and skills, self-analysis, management of learning process) as well as of the analysis of learning activity (reflection on learning process, learning structure, identification of consistency, organization of learning, development) discloses the interaction between self-education and teacher's identity development.

Self-education is the expression of the identity development of the teachers' studying in master studies in educational science at the university. Thus self-education expresses the idea of identity development as a master student in educational science purposefully plans own learning based on personal experience and taking responsibility for own learning. Self-education is relevant discursive process of teacher's identity, which is based on continuous person's life-long learning and comprehensive development.

References

1. Beijaard D., Meijer P. C., Verloop N. (2004). Reconsidering research on teachers' professional identity. *Teaching and Teacher Education*, 20(2), 107-128.

2. Beijaard D., Verloop N., D. Vermunt J. D. (2009). Teachers' perceptions of professional identity: an exploratory study from a personal knowledge perspective. *Teaching and Teacher Education*, 16, 749-764.

3. Cattley C. (2007). Emergence of professional identity for the pre-service teacher. *International Education Journal*, 8(2), 337-347.

4. Chong S., Ling L. E., Chuan G. K. (2011). Developing Student Teachers' Professional Identities – An Exploratory Study. *International Education Studies*, 4(1), 30-38.

5. Crisafulli, S. (2009). *Universiteto studentų profesinė identifikacija socialinių mokslų studijose*. Daktaro disertacija, Socialiniai mokslai, edukologija, 07 S. Kaunas: Kauno technologijos universitetas.

6. Dobrow S. Higgins M. (2005). Developmental networks and Professional identity: a longitudinal study. *Career Development International,* 10(6/7), 567–583.

7. Eriksen T. R. (2004). Gendered professional identity and professional knowledge in female health education-put into perspective by a follow-up study (1987–2002). *Gendered professional identity and professional knowledge, NORA,* 1(12), 20-30.

8. Holden, H. (2004). New graduate engagement with "professional development". *Journal of European Industrial Training,* 28 (2/3/4), 272–282.

9. Ikiugu M. N., Rosso H. M. (2003). Facilitating professional identity in occupational therapy students. *Occupational Therapy International,* 10(3), 206-225.

10. Klopper R. Lubbe S., Rugbeer H. (2007). The Matrix Method of Literature Review. *Alternation,* 14(1), 262-276.

11. Kolbergytė A., Indrašienė V. (2013). Saviugdos formavimosi kontekstas. *Socialinis darbas,* 12(1), 137-150.

12. McMichael, A. (2009). Professional identity and continuing education: a study of social worker in hospital settings. *Social Work Education,* 19 (2), 175–182.

13. O'Connor K. E. (2008). "You choose to care": Teachers, emotions and professional identity. *Teaching and Teacher Education,* 24, 117-126.

14. Olesen H. S. (2001). Professional Identity as Learning Process in Life Histories. *Journal of Workplace Learning,* 13 (7/8), 290–298.

15. Olsen B. (2008). Introducing teacher identity and this volume. *Teacher Education Quarterly,* 35(3), 3-6.

16. Olsen, B. (2008). *Teaching what they learn, learning what they live: How Teachers' Personal Histories Shape Their Professional Development.* Boulder, CO: Paradigm Publishers.

17. Pearce J., Morrison C. (2011). Teacher Identity and Early Career Resilience: Exploring the Links. *Australian Journal of Teacher Education,* 36(1), 48-59.

18. Peter W., Robert J. (2002). The reconstruction of primary teachers‘ identities. *British Journal of Sociology of Education,* 23(1), 89-106.

19. Roberts L. (2000). Shifting Identities: an investigation into student and novice teachers' evolving professional identity. *Journal of Education for Teaching,* 26 (2), 1–17.

20. Schwartz J. S., Zamboanga L. B., Weisskirch S. R. (2008). Broadening the Study of the Self: Integrating the Personal Identity and Cultural Identity. *Social and Personality Psychology Compass,* 2 (2), 635-651.

21. Toulouse P. R. (2011). *Supporting Aboriginal student success: Self-esteem and identity, a living teachings approach.* Laurentian University, p. 1-10 URL: http://www.edu.gov.on.ca/eng/research/toulouse.pdf

22. Vaivada S., Blinstrubas A., Mockevičienė D. (2012). Skirtingų lygių asmeninių žinių fenomenai asmenybės dimensijose: asmenybės saviugdos kontekstas. *Jaunųjų mokslininkų darbai,* 3(1), 255–266.

23. Vaivada, S. (2012). Saviugdos koncepto pagrindimas edukologijos požiūriu. *Jaunųjų mokslininkų darai,* 4(3), 93–100.

24. Weinberg S. (2011). Locke on Personal Identity. *Philosophy Compass,* 6 (6), 398-407.

Education for Sustainable Development in Teacher Training in Baden-Württemberg: From Theory to Implementation

Markus Jahn, Alexander Siegmund

Abstract

The United Nations Commission on Environment and Development highlighted education as a critical force for promoting the guiding principle of sustainable development, and enabling people to address relevant issues. The overall goal of the education for sustainable development (ESD) is to "motivate and empower learners to change their behaviour and take action for sustainable development" (UNESCO 2013).

In the German education system, learning standards and curricula define subject-based skills and abilities to be acquired by pupils. Beyond that, the subject-independent concept of the "Gestaltungskompetenz" (shaping competence) defines learners' interdisciplinary knowledge, skills and competences being necessary to act in a sustainable way. Though these guidelines are important for the implementation of ESD, the teachers and their own competences related to teaching in terms of ESD may be a very influential factor in this regard too.

In their survey, Rieß et al. (2008) asked 2,951 teachers across the federal state of Baden-Württemberg in southwestern Germany about ESD as being a part of their lessons as well as about general conditions promoting or restricting ESD in school. The results of this and other surveys led to a more extensive engagement of the state government and ministries in terms of ESD, especially in the field of teacher training. The goal is to implement ESD at universities more profoundly. As a part of the current measures, the authors conducted a survey to evaluate a.) the status quo of ESD implementation in teacher training, b.) drivers and barriers of implementation as well as c.) the willingness of academics/lecturers to teach about ESD in their courses.

Introduction

For several thousand years, humanity altered the earth's surface due to land reclamation, forest logging, resource mining and urbanization. In 1987, the World Commission on Environment and Development (WCED) came in its momentous report to the conclusion that the development of human activities on our planet overbalanced its natural system:

> *"Many present efforts to guard and maintain human progress, to meet human needs, and to realize human ambitions are simply unsustainable [...]. They draw too heavily, too quickly, on already overdrawn environmental resource accounts to be affordable far into the future without bankrupting those accounts. They may show profit on the balance sheets of our generation, but our children will inherit the losses." (WCED 1987).*

Beyond pointing to the impacts of an unsustainable development on our society's well-being, the WCED referred also to the consequences for future generations, indicating the need for an overall change in the behavior of humankind. Sustainability became the overall goal to be achieved (cf. WCED 1987).

Sustainability and Sustainable Development

The term "sustainability" describes a goal or status to be reached in the future, e.g. a more sustainable world via a long-term maintenance of well-being. Beyond social/cultural issues, sustainability implies also an ecological, economic as well as political dimension, and is often associated with terms such as "productivity", "health", "diversity" and "longevity". In principle, every human or societal activity could be investigated in terms of its sustainability as the underlying concept regards almost every aspect of human livelihood from a local to global perspective. Concerning long-term impacts, the contribution of a human or societal activity to sustainable development could be assessed. The term "sustainable development" (SD) refers to diverse processes and pathways leading to sustainability. Sustainable development is an overarching paradigm of the United Nations (UN) to meet "*the needs of the present without compromising the ability of future generations to meet their own needs*" (WCED 1987).

Education for Sustainable Development

In 1992, the UN Conference on Environment and Development (UNCED) adopted the "Agenda 21", a voluntary plan for global action focusing on SD. In chapter 36 of the agenda, education has been highlighted as an indispensable prerequisite for achieving SD:

> *"Education [...] should be recognized as a process by which human beings and societies can reach their fullest potential. Education is critical for promoting sustainable development and improving the capacity of the people to address environment and development issues." (UNCED 1992).*

According to the chapter, education enables individuals to become aware of the need to act sustainable (UNCED 1992), being "the most effective means that society possesses for confronting the challenges of the future" (UNESCO 1997). The practice of teaching for sustainability is called "education for sustainable development" (ESD), which includes the acquisition of "the knowledge, skills, attitudes and values necessary to shape a sustainable future" (UNESCO 2013). In the context of ESD, teaching and learning processes are oriented towards key SD issues following a participatory and collaborative approach between teachers and learners to "motivate and empower learners to change their behaviour and take action for sustainable development" (UNESCO 2013). Since the concept of ESD goes beyond traditional methods of teaching and learning, its implementation into education systems is closely connected with far-reaching changes "in the way education is often practised today" (UNESCO 2013).

Through launching the international program "Decade of Education for Sustainable Development" (DESD), run by the UNESCO from 2005 to 2014, the UN provided a framework for aligning education systems and practices worldwide – from pre-school training to higher education – to ESD means. In this regard, numerous, if not all, aspects of education are affected, including the training of teachers-to-be (cf. UNESCO 2011).

According to the 2008 survey of the DESD Monitoring and Evaluation Expert Group on the global progress in the implementation of the DESD based on responses from 97 countries, the following 14 learning outcomes for all levels of education had been found (cf. UNESCO 2009):

- *Critical reflective thinking*
- *Understanding complexity/systemic thinking*
- *Futures thinking*
- *Planning and managing change*
- *Understanding inter-relationships across disciplines*
- *Applying learning in a variety of life-wide contexts*
- *Decision-making, including in uncertain situations*
- *Dealing with crises and risks*
- *Acting with responsibility locally and globally*
- *Ability to identify and clarify values*
- *Acting with respect for others*
- *Identifying stakeholders and their interests*
- *Participating in democratic decision-making*
- *Negotiating and consensus building*

The History of ESD in Germany

In 1998, the first guidelines on ESD at all levels of the national education system had been released in Germany, addressing stakeholders in education policy as well as societal actors being engaged in the field of education and training.

Challenges concerning the implementation, educational/didactic principles and key skills as well as concrete activities to be executed by the federal government and the federal states had been formulated (cf. BLK 1998). One year later, the nationwide model program "21", which focused on the implementation of ESD in schools, and the acquisition of "Gestaltungskompetenz" (see explanation below) through pupils, had been launched. Until the end of the program in 2004, about 200 schools participated, and new educational contents related to SD as well as innovative and interdisciplinary ways of learning had been developed (cf. BLK 2005). Those concepts, materials and structures had been transferred and expanded to up to 10 percent of the schools in 14 of the 16 German states in the course of the following program "Transfer-21" (2004 to 2008).

In 2007, the guidelines for the learning part "Global Development" in the context of the ESD had been released. This document emphasized the importance of teaching about societal, economic, political and environmental issues in terms of local-to-global-aspects, development policy and global development. Necessary skills and competences to be acquired by learners as well as performance requirements and class topics had been nominated (cf. BMZ & KMK 2007). In the same year, guidelines for ESD in secondary schools (level 1) focusing on the acquisition of Gestaltungskompetenz by pupils had been published (Programm Transfer-21 2007).

"Gestaltungskompetenz" (Shaping Competence)

The term "Gestaltungskompetenz" (shaping competence) characterizes a comprehensive approach by Gerhard de Haan & Dorothee Harenberg to concentrate all the key competences associated with ESD in a single educational concept. De Haan (2010) summarized the concept's significance as followed:

> "Gestaltungskompetenz means the specific capacity to act and solve problems. Those who possess this competence can help, through active participation, to modify and shape the future of society, and to guide its social, economic, technological and ecological changes along the lines of sustainable development. Gestaltungskompetenz [...] means having the skills, competencies and knowledge to change economic, ecological and social behavior [...]."

Gestaltungskompetenz characterizes a superordinate concept comprising twelve sub-competences, which can be brought in line with the key competences defined by the OECD (2005); each sub-competence consists of a group of several minor ones; some of them recognize attitudes and values related to SD (cf. de Haan 2010; see Table 1). Besides Gestaltungskompetenz, other educational concepts specifying leaners' skills and abilities necessary for shaping the future in a sustainable way have been developed in Germany, Austria and Switzerland, addressing different levels of the education system. Of all those concepts, Gestaltungskompetenz draws the most attention (cf. BMBF 2012).

Tab. 1: Overview of the 12 sub-competences of Gestaltungskompetenz (de Haan (2010).

Subject-based / Methodical Competences
Learners should be able to: • gather knowledge in a spirit of openness to the world, integrating new perspectives • think and act in a forward-looking manner • acquire knowledge and acting in an interdisciplinary manner • deal with incomplete and overly complex information
Social Competences
Learners should be able to: • co-operate in decision-making processes • cope with individual dilemmatic situation of decision-making • participate in collective decision-making processes • motivate oneself as well as others to become active
Personal Competences
Learners should be able to: • reflect upon one's own principles and those of others • refer to the idea of equity in decision-making and planning actions • plan and act autonomously • show empathy for and solidarity with the disadvantaged

Conditions for teaching in terms of ESD in German schools

ESD and the Normative Fundamentals for Teaching in School

Education is manifold as it goes beyond the sheer teaching and learning of subject knowledge and subject-based skills and abilities. However, when it comes to assessing the "quality" of education, especially those aspects take center stage. Owing to modest results of German pupils in international assessments such as PISA, TIMSS and PIRLS, education has been reformed in Germany subsequently. In contrast to the input-based learning of the past, which focused on subject content as well as subject-based methods and techniques, the acquisition of skills and abilities (competences) through pupils is the aim nowadays (output orientation).

Based on the definition of competences by Weinert (2001, p. 27f.), nationwide learning standards have been released for every major subject in school, specifying what pupils should know or be capable of doing/performing at the end of primary and secondary education. As far as December 2013, learning standards are available for Mathematics and German in primary education (1st to 4th grade), for Mathematics, German, English, French, Biology, Chemistry, Physics and Geography in lower secondary education (5th to 10th grade) as well as for Mathematics, German, English and French in upper secondary education (11th to 12th/13th grade). Interestingly, only the learning standards in Biology and Geography refer to the goals of the ESD, and define competences that are comparable or similar to those of the Gestaltungskompetenz.

Due to the federal system in Germany, every state is autonomous regarding its internal organization, including its education system. As a result, every state has its own curricula for the different school subjects. Ultimately, the education guidelines written in those curricula are mandatory for the teaching in school. The nationwide learning standards are, thus, just an approach to harmonize education throughout Germany. However, all states committed themselves to adopt the learning standards. At the time when the first learning standards came out in 2003/2004, the curricula of the state of Baden-Württemberg for example had been adapted towards output orientation too.

Concerning the implementation of ESD into the states' curricula, the conference of German ministers of education (KMK) stated in its report on the situation and perspectives of ESD in schools:

> *"Regarding the implementation of ESD as well as of the connected orientation towards competences into curricula one can say that ESD has found its way – explicitly or implicitly – into the curricula of all federal states meanwhile (e.g. implementation into school laws, as central theme or leitmotif, as educational mandate, in preambles or guidelines, in curricular guidelines, in subjects and subject combinations, and as subject matter)." (Translated from KMK 2012)*

In Baden-Württemberg, it is finally up to the schools and teachers to decide if and to what extent ESD is integrated into lessons and other activities in or outside school. This is likely to change as the responsible state ministry is currently (December 2013) revising its curricula to place more emphasis on ESD as a guiding principle in the state's education system (cf. KMK 2012).

ESD and Teacher Training

Although learning standards and curricula are important normative fundamentals for an effective implementation of ESD in school education, one of the most influential factors may (still) be the teachers and their teaching. Beside personal attitudes, values and interests, the training of prospective teachers in terms of ESD is of crucial significance. In Germany, teacher training is divided into two phases. During academic studies (first phase), the teachers-to-be get to know general pedagogic fundamentals on the one hand as well as subject-didactic teaching methods on the other hand, which can be exercised during hands-on training in schools. After finishing their studies, the students become junior teachers in school for two years, getting advice from senior colleagues (second phase). Afterwards, they could become full-time, in-service teachers.

In 2004, the Federal Ministry of Education and Research (BMBF) stated a lack of evidence concerning the implementation of the guiding principle of SD in teacher training. At that time, in only four (including Baden-Württemberg) of the overall 16 states the examination regulations for teacher training provided by the states specified interdisciplinary studies as a relevant part of the final examination; in general, interdisciplinary studies are considered to be a substantial indicator of the scientific examination of SD. In the ministry's report, only a few, insular good-practice examples of ESD implementation in teacher training at German universities could be given (cf. BMBF 2004).

In 2009, almost all universities and other higher education institutions declared the essential importance of a higher education committed to SD (HRK & DUK 2010). In this context, the integration of ESD into teacher training is of particular importance as teachers do not only pass on their knowledge and competences related to SD to pupils, but also serve as a model in lessons and beyond. Therefore, lectures and courses on ESD are necessary to help future teachers increase their awareness of its importance in school education (DUK 2011) and develop professional teaching competences (KMK & DUK 2006). In this regard, students should experience interdisciplinarity, participation, action orientation and problem orientation on their own during their studies/training, before getting to know teaching methods, forms of learning and organizational structures that are appropriate to teach about SD later in school, and cooperate with stakeholders outside school (cf. DGFE 2004).

A comprehensive concept focused on the competences of lecturers related to teaching in terms of ESD in formal as well as informal education settings has been developed by Franz Rauch, Anna Streissler & Regina Steiner. Based on literature review and interviews with supervisors and participants of training courses on ESD in different European countries, their concept of competences called "KOM-BiNE" differentiates the following aspects related to teaching in terms of ESD (cf. BMUKK 2008; see Table 2):

Tab. 2: Lecturers' competences related to teaching in terms of ESD (BMUKK 2008).

Knowledge
• Ability to acquire basic knowledge on SD and ESD contents e.g. subject-based knowledge), and deal with it • Ability to cross-link knowledge being acquired from different subjects • Ability to catch up up-to-date information on various issues from different sources of information • Awareness of knowledge as being influenced by culture and values • Ability to question the knowledge being acquired critically, and enhance it
Skills
• Methodical knowledge about various teaching methods and pedagogic/didactic strategies to enable learners to shape the future in a sustainable way • Ability to apply and enhance those methods and strategies
Values
• Endeavor to be aware of own presuppositions and values influencing one's own knowledge and action • Endeavor to show solidarity, fairness and respectfulness toward learners and colleagues • Endeavor to honor the heterogeneity and diversity of people and sentiments, avoiding a thinking in extremes such as "right" or "wrong" and "good" or "bad" respectively • Endeavor to honor conclusions made by learners, which deviate from one's own conclusion, and to support learners in making well-founded decisions • Endeavor to appreciate theoretical as well as practical knowledge
Feelings
• Endeavor to be empathetic in terms of listening to learners as well as putting oneself in learners' point of view • Endeavor to be enthusiastic in terms of ESD as well as to enthuse others, e.g. learners • Endeavor to give learners confidence as well as to encourage their engagement to face the challenges associated with SD (empowerment) • Ability (and willingness) to talk about issues that are only rarely discussed or not discussed at all between lecturers and learners
Communication
This ability is closely linked with all the other aspects in terms of: • exchange of knowledge and experiences • dialogue and discussion, e.g. on sensitive issues

• expressing values and feelings • common reflection upon learning contents, values etc. • developing visions as well as planning and organizing • networking
Reflection
• Ability (and willingness) to question oneself as well as one's own knowledge, skills, values and feelings critically • Ability (and willingness) to question one's own networking, development of visions as well as planning and organization
Developing of Visions as well as Planning & Organizing
• Developing of visions, e.g. new strategies, innovative projects or teaching methods • Setting the targets as well as appropriate steps for implementation, e.g. organizing of appropriate conditions for teaching and learning, frame conditions as well as institutional conditions that are favorable to ESD, cooperation with the social environment • Willingness to face challenges while developing visions, planning and organizing
Networking
• Ability (and willingness) to network with colleagues as well as with people and institutions from outside, e.g. contact experts, exchange knowledge, work together in groups or teams, cooperate with people and institutions • Ability (and willingness) to share knowledge, learn together, win allies for one's own cause, get praise and criticism from people with deviating perspectives ("critical friends")

According to Rauch, Streissler & Steiner, lecturers apply these aspects in the following three spheres of activities: teaching setting (e.g. lessons in school), own institution (e.g. the school) and society as the social environment in the institution's surroundings (cf. BMUKK 2008).

Lately, based on the federal states' feedback regarding ESD in their education system, the KMK came to the conclusion that diverse approaches of ESD integration in the first phase of teacher training do already exist in some of the states, e.g. ESD as course topic, basic introduction or course offer, ESD in single areas of the studies, single disciplines, electoral modules or additional qualifications. Concerning the second phase, a similar variety has been reported by the states, e.g. the integration of ESD into the states' regulations for teacher training and admission to teaching in school, as an offer for an additional qualification, as a cross-cutting issue for all teacher training areas, as a component of mandatory or electoral modules, as a part of teacher training courses (KMK 2012).

ESD in Baden-Württemberg: Results from a School Survey

Since ESD had already been integrated into the curricula in Baden-Württemberg, several state ministries concluded a contract to survey the state-wide status quo of ESD in secondary education. Subsequently, Rieß et al. (2008) evaluated the data of a questionnaire being sent to 2,951 teachers across the state during the school year 2006/2007. The central aim of the survey was to explore in which way ESD related contents had been taught, and which conditions promoted or restricted ESD in teaching. Thus, the survey was conducted to find out how much of the teaching in schools at that time actually related to the goals of ESD. The questionnaire focused on different issues of school education, e.g. subject-based aspects, curricula, teaching methods, organizational structures, cooperation structures, advanced training as well as the teachers themselves. Of the 2,951 teachers who had been addressed, 1,835 filled out the questionnaire.

Altogether, the evaluators concluded a general lack of information on SD and ESD by many teachers. In general, ESD would have been quite of significance for many of them, but the knowledge on goals and contents would have been insufficient. They would have known hardly that ESD is already a component in many curricula, and not an additional topic for teaching; ESD topics would be underrepresented in textbooks too. Many teachers would have been rarely familiar with the topics and goals related to ESD, and as a result, many would have taught ESD topics without being aware of it. If possible, curricula should be slimmed and concentrated more on interdisciplinary ESD topics. Teachers should be more introduced to teaching/learning methods and practices as they are recommended by ESD literature. A lack of concrete proposals for projects and lessons would limit the teaching in terms of ESD furthermore. Scientific studies on the efficacy of ESD teaching on pupils' consciousness and behavior, e.g. in comparison to other influential factors, might help to increase the teachers' overall engagement in this context. Examples of best practice concerning (interdisciplinary) cooperation within the school should be given, and possibilities for an active participation of pupils in the teaching/learning process should be pinpointed (cf. Rieß et al. 2008).

Consequences following from the Survey Results

In 2008, the Baden-Württemberg Council on Sustainable Development, an independent advisory board established by the state government, published an expertise on the status of ESD in the state's education system. Though many efforts had already been made by the state government and ministries to implement ESD, the council came to the conclusion that the results of several school surveys, including the one conducted by Rieß et al. (2008), could be summarized all in all as being sobering. The council advocated even more extensive efforts through the state government and ministries while also adverting to the overall importance of education levels other than school education, such as kindergarten, higher education, and advanced and professional training. Due to its basic importance the council emphasized particularly the high priority of teacher training to become more directed to the contents and goals of ESD. The council requested the state government and ministries to adapt the state's curricula, and to establish ESD in teacher training accordingly (NBBW 2008).

In response to the council's expertise, the Ministry of Education, Youth and Sports Baden-Württemberg initiated the state-wide project "Learning beyond the day – Education for a sustainable future" in 2010 (projected ending by 2014). The project's ultimate goal is to implement ESD in school education more successfully through involving and connecting stakeholders from different levels of the education system such as teachers and pupils as well as academics and students. Beside school education, the project focuses also on the implementation of ESD in teacher training (MKJSBW 2011). Since it is ultimately up to the universities, respectively their academics/lecturers, to decide how to implement ESD exactly, a voluntary network of university members on ESD in teacher training (www.bne-hochschulnetzwerk.de) has been established. Though the network is increasing, only a small number of the academics/lecturers across the state involved in teacher training participate so far.

In order to get an overview of current activities related to ESD in teacher training at universities located in Baden-Württemberg, the authors conducted a state-wide survey. Between May and June 2013, about 750 academics from nine universities and six universities of education had been addressed for the completion of an online questionnaire. The addressees had been balanced by university and discipline/subject to allow for a comparison in this respect. The goal of the survey is to give an answer to the following central questions:

- In which way ESD has already been integrated into teacher training?
- Which factors favor or limit the integration of ESD in general?
- How about the willingness of academics to teach about ESD in their courses?

Altogether, three different groups of academics have been asked via e-mail to provide information on the status quo of ESD at their institution. Due to different tasks within the university, each group got its own questionnaire. The collected data will be analyzed statistically, evaluated and reported to the Ministry of Education, Youth and Sports Baden-Württemberg.

ESD in Teacher Training at the Heidelberg University of Education

At the Heidelberg University of Education, the biggest university specialized on teacher training in Baden-Württemberg, numerous courses in various disciplines (subjects) focus on interdisciplinary contents or are given by academics from different disciplines. Especially in Geography, many courses are dealing directly with human-environment-interactions due to the scientific direction of the discipline. Beyond these courses, different educational projects related to ESD have been established in some disciplines, involving the engagement of students.

A few years ago, the teacher training has been reformed completely. Beside the regular courses of the various disciplines, now all students have to attend additional courses belonging to a superordinate field of study. These additional courses are classified into three modules. In module 3, which is for students being in the 6th to 8th semester of their studies, ESD is a major component of the teaching. First courses will be given in summer term 2014.

The Interdisciplinary Institute for Natural Sciences, Engineering and Society (German: NTG) is a think tank for interdisciplinary courses and projects at the university. Its focus is on interdisciplinary topics of relevance to the society, including ESD. Recent colloquia hosted by the institute concentrated on "Sustainability and ESD" and "Sustainability from an Interdisciplinary Point of View". The institute aims at building a center on ESD in cooperation with the Heidelberg Center for the Environment of the Heidelberg University. This center is thought to serve as a starting point for an academy for sustainability.

References

BLK (Bund-Länder-Kommission für Bildungsplanung und Forschungsförderung): Bildung für eine nachhaltige Entwicklung – Orientierungsrahmen. Materialien zur Bildungsplanung und zur Forschungsförderung. Heft 69, Bonn, 1998.

BLK (Bund-Länder-Kommission für Bildungsplanung und Forschungsförderung): Bildung für eine nachhaltige Entwicklung („21"). Abschlussbericht des

Programmträgers zum BLK-Programm. Materialien zur Bildungsplanung und zur Forschungsförderung. Heft 123, Bonn, 2005.

BMBF (Bundesministerium für Bildung und Forschung): UNI 21. Hochschulbildung für eine nachhaltige Entwicklung, Berlin, 2004.

BMBF (Bundesministerium für Bildung und Forschung) (Ed.): Bildung für nachhaltige Entwicklung – Beiträge der Bildungsforschung. Bildungsforschung Band 39, Berlin, 2012.

BMUKK (Bundesministerium für Unterricht, Kunst und Kultur) (Hrsg.): Kompetenzen für Bildung für Nachhaltige Entwicklung (KOM-BiNE). Konzepte und Anregungen für die Praxis, Wien, 2008.

BMZ (Bundesministerium für wirtschaftliche Zusammenarbeit und Entwicklung) & KMK (Ständige Konferenz der Kultusminister der Länder in der Bundesrepublik Deutschland) (Eds.): Orientierungsrahmen für den Lernbereich Globale Entwicklung im Rahmen einer Bildung für nachhaltige Entwicklung, Berlin, 2007.

DGFE (Deutsche Gesellschaft für Erziehungswissenschaft): Memorandum zur Lehrerbildung für eine nachhaltige Entwicklung (BfnE). Kommission „Bildung für eine nachhaltige Entwicklung", Lüneburg/Hannover, 2004.

DUK (Deutsche UNESCO-Kommission): Hochschulen für eine nachhaltige Entwicklung. Nachhaltigkeit in Forschung, Lehre und Betrieb einer Hochschule, Bonn, 2011.

de Haan, Gerhard: The development of ESD-related competencies in supportive institutional frameworks. International Review of Education, Volume 56 (2010), Issue 2-3, 315 – 328.

HRK (Hochschulrektorenkonferenz) & DUK (Deutsche UNESCO-Kommission): Hochschulen für nachhaltige Entwicklung. Erklärung der Hochschulrektorenkonferenz und der Deutschen UNESCO-Kommission zur Hochschulbildung für nachhaltige Entwicklung. Ein Beitrag zur UN-Dekade „Bildung für nachhaltige Entwicklung", Bonn, 2010.

KMK (Ständigen Konferenz der Kultusminister der Länder in der Bundesrepublik Deutschland): Zur Situation und zu Perspektiven der Bildung für nachhaltige Entwicklung. Bericht der Kultusministerkonferenz vom 13.12.2012, Berlin, 2012.

MKJSBW (Ministerium für Kultus, Jugend und Sport Baden-Württemberg) (Ed.): Lernen über den Tag hinaus – Bildung für eine zukunftsfähige Welt. Projektabschlussbericht zur Konzepterstellung – ohne Finanzierungsplan, Stuttgart, 2011.

NBBW (Nachhaltigkeitsbeirat der Landesregierung Baden-Württemberg): "Zukunft gestalten – Nachhaltigkeit lernen". Bildung für Nachhaltige Entwicklung als Aufgabe für das Land Baden-Württemberg, Stuttgart, 2008.

OECD (Organisation for Economic Co-operation and Development): The Definition and Selection of Key Competencies. Executive Summary, Paris, 2005.

Programm Transfer-21 (Ed.): Orientierungshilfe Bildung für nachhaltige Entwicklung in der Sekundarstufe I. Begründungen, Kompetenzen, Lernangebote. Erstellt von der „AG Qualität & Kompetenzen" des Programms Transfer-21. Berlin, 2007.

Rieß, W., Mischo, C., Reinbolz, A., Richter, K., Dobler, C., Seybold, H.: Evaluationsbericht „Bildung für nachhaltige Entwicklung an weiterführenden Schulen in Baden-Württemberg". Teil 1: Befragung der Lehrerschaft. Maßnahme Lfd. 15 im Aktionsplan Baden-Württemberg, Freiburg i.B., 2008.

UNCED (United Nations Conference on Environment and Development): Agenda 21. United Nations Conference on Environment and Development, Rio de Janeiro, Brazil, 3 – 14 June 1992, New York, 1992.

UNESCO (United Nations Educational, Scientific and Cultural Organization): Educating for a Sustainable Future. A Transdisciplinary Vision for Concerted Action, Paris, 1997.

UNESCO (United Nations Educational, Scientific and Cultural Organization): Review of Contexts and Structures for Education for Sustainable Development 2009. Learning for a Sustainable World, Paris, 2009.

UNESCO (United Nations Educational, Scientific and Cultural Organization): Education for Sustainable Development. An Expert Review of Processes and Learning. DESD Monitoring and Evaluation, Paris, 2011.

UNESCO (United Nations Educational, Scientific and Cultural Organization): Education for Sustationable Development (ESD), 2013. [http://www.unesco.

org/new/en/education /themes/leading-the-international-agenda/education-for-sustainable-development/]

WCED (World Commission on Environment and Development): Our Common Future. Report of the World Commission on Environment and Development. Oxford University Press: New York, 1987.

Weinert, F. E.: Concept of competence: A conceptual clarification. In: D. S. Rychen, & L. H. Salganik (Eds.): Defining and selecting key competencies, Seattle, 2001, 45-65.

Die subjektive Bedeutung von Schule und formaler Bildung aus der Perspektive von Schülerinnen und Schülern

Eine empirische Untersuchung an Brennpunktschulen[1]

Carsten Rohlfs

Abstract
In modern society education is of great importance both as an individual resource and category for social orientation. As an effect of the PISA study and the subsequent controversial discourse in Germany it has clearly gained in relevance once more. Yet does the importance attached by students themselves to school and formal education correspond to this objective rise in relevance? This question was at the centre of an empirical survey of N = 1689 pupils in grades 7 and 9 at schools in disadvantaged areas in Bremen, Germany. This article reports on the central findings of an explorative factor and cluster analysis focussing students' attitudes towards school – especially referring to gender, social origin and migration background.

Zusammenfassung
Bildung ist in der modernen Gesellschaft als Orientierungskategorie und Ressource von großer Bedeutung und hat im Zuge des kontrovers geführten Diskurses im Anschluss an PISA noch einmal deutlich an Relevanz gewonnen. Entspricht aber der Stellenwert, den Schülerinnen und Schüler selbst der Schule und der formalisierten Bildung beimessen, dieser objektiven Bedeutungszunahme? Diese Frage stand im Mittelpunkt einer empirischen Untersuchung zu den subjektiven Bedeutungen von Schule aus der Perspektive von N = 1689 Schülerinnen und Schülern der Sekundarstufe I an Schulen in benachteiligter Lage im Land Bremen. Der Beitrag referiert die Befunde einer explorativen Faktoren- und Clusteranalyse zu den schulbezogenen Einstellungen der befragten Jugendlichen und richtet seinen Fokus insbesondere auf die Zusammenhänge mit Geschlecht, Migrationshintergrund und sozialer Herkunft.

1. Einführung

Die Bedeutung von Bildung für die moderne Gesellschaft und das (fach-)öffentliche Interesse an Bildungsprozessen, Bildungschancen und Bildungsreformen haben in den vergangenen Jahren spürbar zugenommen. International wie national vergleichende Schulleistungsstudien haben den Blick auf zum Teil dramatische – und bisweilen dramatisierte – Defizite von Schülerinnen und

1 Der Beitrag ist in ähnlicher Form bereits erschienen in der: Zeitschrift für Bildungsforschung, Heft 3, 2013, Jg. 3.

Schülern an deutschen Schulen in ihren fachlichen Kompetenzen gelenkt und damit einen kontrovers geführten Diskurs auf unterschiedlichsten Ebenen und aus unterschiedlichster Perspektive forciert und „erneut – nach den 60er Jahren des letzten Jahrhunderts – die Rede von einer deutschen Bildungskatastrophe provoziert" (Otto und Rauschenbach 2004, 9). Nur selten – und wenn, dann primär als ein Nebenprodukt leistungsbezogener und kompetenzbereichsspezifischer Studien (vgl. Harazd und Schürer 2006, 208) – wird in dieser Diskussion und der Schulforschung die Frage nach der subjektiven Bedeutung von Schule und formaler Bildung aus der Perspektive der Heranwachsenden in einer Gesellschaft formuliert, die der formalen Bildung objektiv einen sehr hohen Stellenwert beimisst, während sie sich zugleich bemüht, die Herausforderungen der neuen deutschen Bildungskrise zu bewältigen. Der folgende Beitrag rückt eben diese Einstellungen von Schülerinnen und Schülern anhand der Befunde einer empirischen Untersuchung an Brennpunktschulen in den Fokus. Die zentrale Fragestellung lautet demnach: Welche schulbezogenen Einstellungen zeigen sich aus der Perspektive von Schülerinnen und Schülern an Schulen in benachteiligter Lage unter besonderer Berücksichtigung der Merkmale Geschlecht, Migrationshintergrund und sozialer Herkunft und wie lassen sich diese durch eine Typenbildung differenzieren? Der Beitrag konturiert zunächst in einer grundlegenden theoretischen Rahmung das sozialpsychologische Konstrukt der Einstellung (*Abschnitt 2*), liefert in *Abschnitt 3* einen Überblick über den aktuellen Stand der Forschung, worauf in *Abschnitt 4* das methodische Design der empirischen Untersuchung expliziert wird. *Abschnitt 5* stellt die zentralen Befunde zunächst in Form einer Typologie (*5.1*) und folgend im Kontext der Merkmale Geschlecht, Migrationshintergrund und soziale Herkunft (*5.2*) vor, bevor *Abschnitt 6* den Beitrag mit einem Resümee und Ausblick beschließt.

2. Theoretische Rahmung: das sozialpsychologische Konstrukt der Einstellung

Als eine Subjekt-Objekt-Relation bilden sich Einstellungen bereits früh im Kindes- und Jugendalter, sind durch neue Erfahrungen sowie primäre kulturelle und milieubedingte Einflüsse bzw. gezielte (erzieherische) Veränderungsabsichten aber durchaus wandelbar – so auch die Einstellung zur Schule. Eagly und Chaiken (1998) beschreiben die Einstellung als verinnerlichte, also fest zu Eigen gemachte, sehr persönliche und doch stark von sozialen Phänomenen abhängige Haltung – als eine psychische Tendenz: „In formal terms, an attitude is a psychological tendency that is expressed by evaluating a particular entity with some degree of favour or disfavour. The idea that an attitude is a psychological tendency treats attitude as a state that is internal to the person and that lasts for shorter or longer duration. As an internal state, an attitude is not directly observable“ (ebd., 269). Nach Städter (2003) ist die Einstellung eine habitualisierte Reaktionsweise (vgl. ebd., 222), derer sich das Individuum nicht immer bewusst ist (vgl. Gerrig und Zimbardo 2008, 644) und auch deren spezifische Gründe oftmals keinen bewussten Annahmen unterliegen (vgl. Wilson et al. 1989). „Attitudes express passions and hates, attractions and repulsion, likes and dislike. People have attitudes when they love or hate things or people and when they approve or disapprove of them" (Eagly und Chaiken 1998, 269). In der häufig als klassische Definition der Einstellung bezeichneten Begriffsklärung von Allport aus dem Jahre 1935 steht deutlich deren kognitive Dimension im Vordergrund (vgl. Witte 1977, 103; Allport 1935, 810). Thurstone (1931) dagegen reduziert die Einstellung auf den darin liegenden Affekt für oder gegen ein psychologisches Objekt (vgl. auch Witte 1977, 103), und im Kontrast hierzu heben Smith et al. (1956) die verhaltensbezogene Dimension der Einstellung als Wahrnehmungskategorie hervor. Witte (1977) bündelt diese drei unterschiedlichen Perspektiven: „Eine Einstellung eines psychologischen Subjektes gegenüber einem psychologischen Objekt ist seine sozial vermittelte Stellungnahme diesem Objekt gegenüber, durch die in systematischer Weise Prozesse im Bereich des Erkennens, Erlebens und Handelns ausgelöst werden“ (103). Das hier zugrunde liegende Modell der Einstellung als mehrdimensionales System wurde von Rosenberg und Hovland im Jahr 1960 eingeführt und konzipiert die Einstellung somit als ein Gefüge bestehend aus kognitiven, affektiven und handlungsintentionalen Komponenten (vgl. Rosenberg und Hovland 1960). Diesem Begriffsverständnis der Einstellung wird jedoch wiederum bewusst ein eindimensionales Konzept entgegengestellt, um das evaluative Element der Einstellung und deren bilanzierende Eigenschaft (vgl. Thurstone 1931) in den Fokus zu rücken: „Einstellungen basieren auf Bewertungen, da sie aus einer positiven oder negativen Reaktion auf etwas Bestimmtes bestehen. Der Mensch ist kein neutraler Beobachter der Welt, sondern bewertet unablässig das, was er sieht“ (Aronson et al. 2004, 230). Diese eindimensionale Konzeption hat sich zunehmend durchgesetzt (vgl. Mayerl 2008) und erscheint vor allem für die empirische Untersuchung von Einstellungsausprägungen tragfähiger als der dreidimensionale Ansatz, da die Erhebungen hier vorwiegend über Befragungen zur individuellen „evaluation“ eines Einstellungsgegenstandes erfolgen (vgl. Stroebe et al. 1999; Christen 2004, 10) – so auch in vorliegender Studie, die dabei sowohl nach der Einstellung der Heranwachsenden zur Schule insgesamt als auch nach den Einstellungen zu schulischen Einzelaspekten wie dem Lernen, dem

Unterricht, Hausaufgaben, bestimmten Fächern und Kompetenzbereichen, den Lehrerinnen und Lehrern oder den Mitschülerinnen und Mitschülern fragt. Diese spezifischen Einstellungen stehen in einem engen Wechselverhältnis zueinander und beeinflussen zudem die Einstellung zur Schule insgesamt sowie das damit eng verknüpfte Wohlbefinden der Schülerinnen und Schüler in der Schule (vgl. Fend 1998; Hascher und Baillod 2000; Christen 2004, 9).

3. Zum aktuellen Stand der Forschung

Im Fokus unterschiedlicher Kompetenzbereiche attestiert zunächst IGLU den Viertklässlerinnen und Viertklässlern an deutschen Schulen im internationalen Vergleich insgesamt sehr günstige Einstellungen zum Lesen mit einer zunehmend positiven Entwicklung zwischen 2001 und 2006 (vgl. Bos et al. 2003; 2007). Ebenso stellt TIMSS 2007 in der 4. Klasse für die Mathematik eine recht positive (vgl. Walther et al. 2008, 80) und für die Naturwissenschaften eine „hohe positive Einstellung" (Wittwer et al. 2008, 119) fest. Auch PISA 2006 ermittelt günstige „Schülermerkmale" (wie Kompetenzüberzeugungen, motivationale Orientierungen, Interesse und Freude) im Bereich der naturwissenschaftlichen Kompetenz, relativiert diesen Befund jedoch, da hier offenbar interkulturell unterschiedliche Selbstdarstellungsregeln wirken: Deutschland liegt hier deutlich über dem OECD-Durchschnitt, während solche im Leistungstest sehr erfolgreichen Länder wie Korea und Japan auf den letzten Plätzen zu finden sind – aufgrund einer ihrer Kultur eigenen Bescheidenheit (vgl. Schütte et al. 2007, 132ff.). So konstatiert TIMSS: „Gute Testleistungen und eine positive Einstellung zur Mathematik gehen [...] häufig, aber keineswegs immer Hand in Hand" (Walther et al. 2008, 80). Ebenso stellt LAU 5 bei Fünftklässlern weitgehend günstige schul- und unterrichtsbezogene Einstellungen fest, die sich vorwiegend unabhängig vom tatsächlichen Schulerfolg gestalten (vgl. Lehmann und Peek 1997), während im Rahmen von BIJU ein deutlicher wechselseitiger Zusammenhang zwischen Einstellungen zu Schule und Unterricht und den gezeigten Leistungen ermittelt wird (MPIB und IPN 1995). Auch Majoribanks (1992) stellt diesbezüglich signifikante Korrelationen fest, allerdings ist bei der Interpretation Vorsicht geboten: „although there appears to be a relationship between attitude toward school and achievement, this relationship does not suggest or determine any flow of causality between the two variables" (McCoach 2002).

LAU 5 zeigt zudem ein großes Wohlbefinden in der Schule bzw. eine hohe Schulzufriedenheit und große Schulfreude (vgl. Lehmann und Peek 1997), was von zahlreichen Studien insbesondere für die Grundschulzeit – mit bis zu 87 Prozent der Mädchen und 79 Prozent der Jungen, die hier gern in die Schule gehen (vgl. Stecher 2005) – bestätigt wird (vgl. etwa Eder 2007; Leven und Schneekloth 2007). Für die in LAU 7 und 9 befragten Siebt- und Neuntklässler bestätigen sich die insgesamt positiven schul- und unterrichtsbezogenen Einstellungen aus LAU 5. Allerdings fallen die Werte zur Schulzufriedenheit zu Beginn der Klassenstufe 9 niedriger aus als noch in 7 (vgl. Lehmann et al. 1999; 2002). Eine derartige (und noch deutlichere) Abnahme der Schulzufriedenheit und der positiven Bewertung der Schule zeigen zahlreiche empirische Untersuchungen vor allem für die Sekundarstufe auf (vgl. etwa Czerwenka et al. 1988; Fend 1997; Kanders 2000; Büchner und Koch 2001; Eder 2007). Dabei nimmt die Schulunlust in international vergleichender Perspektive in Deutschland im Laufe einer Schulkarriere weitaus stärker zu als etwa in der deutschsprachigen Schweiz, den USA oder Frankreich (vgl. Czerwenka et al. 1990; Fend 1997). Czerwenka et al. (1990) vermuten, die negativere Einstellung von Schülerinnen und Schülern in Deutschland spiegele insbesondere den vergleichsweise stark ausgeprägten kritischen Geist der Schülerschaft wider (vgl. ebd., 428f.).

Geschlechtsspezifische Einstellungsunterschiede werden dabei stets zugunsten der Mädchen konstatiert (vgl. etwa Harazd und Schürer 2006; Valtin et al. 2006; Leven und Schneekloth 2007), bildungsgangbezogen zeigen sich für das Gymnasium die günstigsten Einstellungswerte, vor der Gesamtschule, der Realschule und schließlich der Hauptschule (vgl. Lehmann et al. 1999). Die Befunde zu Differenzen nach sozialer Herkunft erscheinen hingegen ambivalent: Während mehrere Studien in Deutschland für das schulische Wohlbefinden bzw. die Schulzufriedenheit weder sozialschicht- noch herkunftsspezifische Unterschiede ermitteln (vgl. etwa Okun et al. 1990; Klöckner et al. 2007, 33ff.), konnte etwa die 1. World Vision Kinderstudie 2007 für die 8- bis 11-Jährigen konstatieren, dass sich Kinder höherer Schichten tendenziell wohler in der Schule fühlen als ihre Mitschüler niedrigerer Sozialschichten (vgl. Leven und Schneekloth 2007). Darüber hinaus zeigen Schülerinnen und Schüler mit Migrationshintergrund in sämtlichen PISA-Studien eine stärkere Motivation und größeres Interesse an den jeweiligen Inhalten als Jugendliche ohne Migrationshintergrund (vgl. z.B. Walter und Taskinen 2007). Ein angesichts der im Durchschnitt niedrigen Werte von Migrantinnen und Migranten in den Leistungstests bemerkenswertes Ergebnis (Attitude-Achievement-Paradox), das sich in LAU 9 wiederfindet (vgl. Lehmann et al. 2002), aber keineswegs typisch allein für Deutschland wäre. Für die Niederlande bspw. stellen Verkuyten und Thijs (2002) fest, dass sich auch hier Kinder ethnischer Minoritäten in der Schule wohler fühlen als niederländische Kinder und Asakawa und Csikszentmihalyi (1998) zeigen auf, dass Schülerinnen und Schüler asiatischer Herkunft in amerikanischen Schulen größere Schulfreude aufweisen als ihre europäisch-stämmigen Mitschülerinnen und

Mitschüler. Entsprechend konstatiert ein Bericht des BMBF über den Schulerfolg von Jugendlichen mit Migrationshintergrund im internationalen Vergleich auf der Grundlage der Daten aus PISA 2003: „In allen OECD-Erhebungsländern äußern die Schülerinnen und Schüler der ersten und zweiten Generation eine signifikant positivere Einstellung zur Schule als die Schülerinnen und Schüler aus einheimischen Familien" (BMBF 2006).

Als zentral für eine günstige Einstellung zur Schule zeigt sich vielfach eine hohe subjektive Bedeutsamkeit der Lerninhalte (vgl. etwa Giesbrecht 1997; CERI 2006). Allerdings vermögen Schulen in Deutschland die Sinnhaftigkeit von zu vermittelnden Inhalten im internationalen Vergleich nur wenig erfolgreich zu transportieren: „Die Lernprozesse und Lerninhalte erscheinen bei unterschiedlichsten Schülergruppen nur wenig sinnstiftend" (Helsper und Böhme 2002, 582). Aus der Schülerperspektive ist Schule primär eine Veranstaltung, „an der man nicht aus Gründen der lebenspraktischen Bedeutsamkeit teilnimmt, sondern die man vielmehr zum Zweck des Erwerbs formaler, für den Berufseinstieg unabdingbarer Qualifikationen durchlaufen muß" (Wild 2002, 238). Diese vorwiegend extrinsische Form der Motivation erklärt, warum die Relevanz guter schulischer Leistungen und Noten unter Schülerinnen und Schülern an deutschen Schulen ungeachtet der geringen Bedeutsamkeit schulischer Inhalte überaus hoch ist und zum Ende der Schullaufbahn noch einmal deutlich zunimmt (vgl. Oldenburg 2001; Klöckner et al. 2007).

4. Forschungsmethodische Rahmung

Die eigene empirische Untersuchung basierte auf einer standardisierten Fragebogenerhebung[2] aus dem Jahr 2008 an neun Bremer Schulzentren der Sekundarstufe I, die vom Senator für Bildung und Wissenschaft des Landes Bremen als „Brennpunktschulen" ausgewiesen und aus diesem Grund ausgewählt wurden. Es handelte sich somit um Schulen, „die sowohl bezüglich der äußeren Bedingungen in einer kritischen Lage sind (weit überproportionaler Anteil von Migranten, bildungsferne Elternhäuser, schwierige sozio-ökonomische Lage) als auch im Innern Krisensymptome zeigen (schwache Lernergebnisse, Konflikte im Kollegium, eine hohe Zahl von Wiederholern, Rückläufern und Schülerinnen und Schülern, die die Schule ohne Abschluss verlassen, Gewaltvorfälle)"

---

2 Als Teilprojekt des vom Institut für Qualitätsentwicklung im Bildungswesen Berlin (IQB), dem Max-Planck-Institut für Bildungsforschung Berlin (MPIB) und der Friedrich-Alexander-Universität Nürnberg-Erlangen wissenschaftlich begleiteten Projektes „Schule macht sich stark" (SMS).

(Senator für Bildung und Wissenschaft Bremen 2005, 5). Untersucht wurden an diesen Schulen insgesamt N = 1689 Schülerinnen und Schüler sowie deren Eltern, was einem Ausschöpfungsgrad von 89 Prozent auf Schüler- und 50 Prozent auf Elternebene entspricht. 47.7 Prozent der Schülerinnen und Schüler besuchten zum Zeitpunkt der Erhebungen die siebte, 52.3 Prozent die neunte Jahrgangsstufe.[3] 51.8 Prozent waren weiblichen und 48.2 Prozent männlichen Geschlechts und beinahe die Hälfte der Mädchen und Jungen wies einen Migrationshintergrund[4] auf (45.1 Prozent). Nur 7.9 Prozent waren der Gruppe von Familien mit einem hohen sozioökonomischen Status (HISEI[5]) zuzurechnen und etwa 25.0 Prozent der Schülerinnen und Schüler besuchten ein Gymnasium, 37.5 Prozent eine Sekundarschule (integrierte Haupt- und Realschule) und 38.3 Prozent die Gesamtschule, in denen die Schülerinnen und Schüler nicht in getrennte Bildungsgänge aufgeteilt, sondern überwiegend gemeinsam unterrichtet werden.

Für eine differenzierte Untersuchung der schulbezogenen Einstellungen der Schülerinnen und Schüler an diesen Schulen in benachteiligter Lage wurde auf multivariater Ebene eine explorative Faktoren- und Clusteranalyse durchgeführt. Um komplexe Konstrukte, wie die Einstellung zur Schule, empirisch untersuchen zu können, müssen sie „in eine Vielzahl einzelner beobachtbarerer Variablen zerlegt werden, da keine einzelne Variable der Komplexität des Begriffs gerecht werden kann" (vgl. Brosius 2008, 771). Die Faktorenanalyse folgte eben diesen Überlegungen, ging aber genau den umgekehrten Weg, denn den Ausgangspunkt bildete hier eine große Anzahl von Variablen, von denen a priori nicht bekannt war, ob und in welcher Weise sie miteinander zusammenhingen, für die aber vermutet wurde, dass sie bestimmte komplexe Sachverhalte repräsentieren. 36 Variablen zur Erhebung der schulbezogenen Einstellungen gingen in die Faktorenanalyse ein, wurden paarweise miteinander verglichen und für jedes Paar wurde der Korrelationskoeffizient berechnet. Korrelieren Variablen stark miteinander, liegt ihnen, so die Annahme, dieselbe Hintergrundvariable zugrunde. Korrelieren sie nur schwach, wird umgekehrt angenommen, sie

3 Das SMS-Projekt untersucht im Längsschnitt die Jahrgangsstufen 5, 7 und 9. Um eine Fokussierung auf die Lebensphase Jugend vornehmen zu können, beschränkt sich vorliegende Teilstudie auf die Jahrgänge 7 und 9.

4 Definiert analog zu PISA (vgl. Walter und Taskinen 2007, 345).

5 Der HISEI kann Werte von 16 bis 90 annehmen und folgt einer Dreiteilung: Die Punktwerte 16-36 stehen für einen niedrigen, 37-64 für einen mittleren und 65-90 für einen hohen sozio-ökonomischen Status (vgl. BMFSJ 2009, 78). Der Mittelwert in vorliegender Untersuchung liegt bei 39.9 Punkten bei einer Standardabweichung von 13.4.

würden nicht durch einen gemeinsamen Faktor verbunden (vgl. ebd., 774). Die identifizierten Faktoren müssen sich dabei möglichst deutlich von einander unterscheiden. Jeder Faktor sollte also nur von einem Teil der Variablen bestimmt werden (vgl. ebd.). In diesem Zusammenhang ist die Faktorladung ein bedeutsamer Indikator. Denn je höher die Ladung, desto größer die Erklärungskraft und somit die Bedeutung des Faktors für die Variable (vgl. ebd., 785). Entsprechend ist „eine erfolgreiche Faktorenanalyse dadurch gekennzeichnet, dass die Vielzahl der relevanten Variablen auf nur wenige möglichst gut voneinander zu unterscheidende Faktoren zurückgeführt werden kann" (ebd., 774). In vorliegender Analyse wurden die Variablen durch insgesamt zehn komplexe Hintergrundvariablen bzw. Komponenten verknüpft (vgl. Tab. 1).

Tab. 1: Items zur schulbezogenen Einstellung und Ergebnisse der Faktorenanalyse (Rotierte Hauptkomponentenstruktur)

Variablen	**Faktorenladung**	**Faktoren (Komponenten)**
Schule ist mir wichtig.	.682	Subjektive Bedeutungszuschreibungen
Mir ist es wichtig, in der Schule gute Leistungen zu zeigen.	.783	
Mir ist es wichtig, im Unterricht gut mitzuarbeiten.	.779	
Mir ist es wichtig, die Schule mit einem guten Abschluss zu verlassen.	.695	
Ich finde es uncool, wenn jemand gut in der Schule ist.	.540	
Ich lerne, weil ich den Unterrichtsstoff richtig verstehen will.	.671	Intrinsische Motivation
Die Dinge, die wir im Unterricht durchnehmen, interessieren mich.	.700	
Ich lerne, weil ich Lust dazu habe.	.647	
Ich lerne, weil das für mich selbst wichtig ist.	.618	
Ich lerne, weil es mir Freude macht.	.524	
Ich lerne, weil man es von mir verlangt.	.760	Extrinsische Motivation
Ich lerne, weil ich keine Probleme mit meinen Lehrern bekommen will.	.683	
Ich weiß eigentlich nicht, warum ich in der Schule lernen soll.	.561	
Ich lerne, weil ich mich sonst schlecht fühle.	.461	
Ich lerne, weil ich will, dass die Lehrer eine gute Meinung von mir haben.	.516	

Wie oft kommt es im Allgemeinen vor, dass du dich in der Schule langweilst?	.660	Langeweile
Lernen in der Schule ist für mich Langeweile haben.	.590	
Meine Schule ist ein Ort, an dem ich mich oft langweile.	.546	
Lernen in der Schule ist für mich sich geborgen fühlen.	.641	Positive Lernemotionen
Lernen in der Schule ist für mich spannend.	.628	
Ich freue mich, wenn ich in der Schule etwas gut kann.	.778	
Lernen in der Schule ist für mich Spaß haben.	.612	
Lernen in der Schule ist für mich etwas leisten.	.569	
Lernen in der Schule ist für mich Erfolg haben	.470	
Lernen in der Schule ist für mich Enttäuschungen erleben.	.778	Negative Lernemotionen
Lernen in der Schule ist für mich sich blamieren.	.773	
Der Unterricht in der Schule macht mir Spaß.	.482	Schulfreude
Nach den Ferien freue ich mich darauf, wieder zur Schule zu gehen.	.510	
Ich finde es schade, wenn Unterricht ausfällt.	.797	
Ich mache gerne meine Hausaufgaben.	.552	
Meine Schule ist ein Ort, zu dem ich nicht hingehen möchte.	.766	Schulunlust / Unwohlbefinden
Meine Schule ist ein Ort, an dem ich mich oft unwohl und fehl am Platz fühle.	.752	
Wie oft kommt es im Allgemeinen vor, dass du einfach keine Lust hast, zur Schule zu gehen?	.707	
Wie oft kommt es im Allgemeinen vor, dass du dich mit deinen Freunden nachmittags über die Schule unterhältst?	.840	Bedeutung von Schule außerhalb der Schule
Wie oft kommt es im Allgemeinen vor, dass du in deiner Freizeit noch einmal an die Schule denkst?	.810	
Wie oft kommt es vor, dass du im Unterricht heimlich spielst?	.841	Abgelenktsein im Unterricht
Wie oft kommt es vor, dass du im Unterricht Aufgaben für andere Fächer machst?	.786	

Anmerkungen: Extraktionsmethode: Hauptkomponentenanalyse; orthogonale Rotation gemäß Varimax-Kriterium mit Kaiser-Normalisierung, die Rotation ist in neun Iterationen konvergiert. Dargestellt sind aus Gründen der Übersichtlichkeit in der Matrix ausschließlich Faktorenladungen ≥ .300

Dabei wurde ein KMO-Wert (Kaiser-Mayser-Olkin-Maß) von 0.886, erreicht, wodurch eine „rechte gute“ Auswahl indiziert wurde („meritorious“: 0.8 bis unter 0.9) – nahe an einem „fabelhaften“ Wert („marvelous“: 0.9 bis 1.0) (vgl. ebd., 779-780). Die auf diese Weise extrahierten Faktoren fanden im Anschluss als Klassifizierungsmerkmale in der Clusteranalyse ihre Anwendung. Die Clusteranalyse ist ein heuristisches, exploratives Verfahren zur systematischen Klassifizierung der Objekte einer gegebenen Objektmenge, in der vorliegenden Untersuchung der Stichprobenelemente, nach vorher festgelegten Kriterien und bei gleichzeitiger Berücksichtigung mehrerer ausgewählter Klassifizierungsmerkmale. Ziel ist es, die untersuchten Objekte so zu gruppieren, dass die Unterschiede – ermittelt über entsprechende Distanz- bzw. Ähnlichkeitsmaße – hinsichtlich der Kriterien zwischen den Objekten innerhalb einer Gruppe möglichst gering und die Differenzen zwischen den Objekten aus unterschiedlichen Clustern möglichst groß sind (vgl. Bacher 1996, 2; Brosius 2008, 695ff.). Ein grundsätzlicher Vorteil der Clusteranalyse gegenüber anderen multivariaten Verfahren liegt dabei darin begründet, dass sie keine linearen Zusammenhänge zwischen den Ausgangsvariablen der Analyse voraussetzt (vgl. Palentien 2004, 175), also eine nur wenig eingeschränkte wie einschränkende Auswertungsmethode darstellt. Als konkretes Clusterverfahren wurde für die Datenanalyse im Rahmen der vorliegenden Studie die Ward-Methode angewandt, mit deren Hilfe einzelne Objekte zu Gruppen und dann weiter zu größeren Clustern zusammengefasst werden. Dabei werden solche Gruppen fusioniert, deren Zusammenfügen die Varianz innerhalb der Cluster in möglichst geringem Maße ansteigen lässt. Die Ward-Methode stellt damit in Abgrenzung zu divisiven Verfahren, deren Ausgangspunkt die kleinstmögliche Anzahl von Gruppen ist, welche dann aufgeteilt werden, ein agglomeratives Verfahren dar, das von der feinsten Partionierung ausgeht und einen Fusionierungsprozess beschreibt (vgl. Brosius 2008, 720). Bei der Anwendung dieser Methode lag eine entscheidende Aufgabe in der Entwicklung einer für die Stichprobe optimalen Clusterlösung bzw. einer idealen Anzahl von Clustern. Die Entscheidungsfindung orientierte sich an inhaltlichen wie formal-statistischen Kriterien: inverser Scree-Test unter Anwendung des Elbow-Kriteriums, Korrelation von aus z-Werten gebildeten Mittelwerten mit den Clustern und Überprüfung der unterschiedlichen Clusterlösungen im Hinblick auf ihre inhaltliche Plausibilität, Effizienz und Prägnanz (vgl. Backhaus et al. 2011, 430f., Harring 2011). So konnten schließlich fünf stabile Cluster anhand der schulbezogenen Einstellungen der untersuchten Schülerinnen und Schüler gebildet werden, die in ihren Einstellungsprofilen hinreichend differenziert ausfielen.

5. Zentrale Befunde

5.1 Typologie von Einstellungen zur Schule

Die generierten Cluster sind durch eine bemerkenswerte Ungleichverteilung gekennzeichnet: Mit 60 Prozent hat das Cluster 1 der „pragmatisch Leistungsorientierten" den größten Anteil an der Gesamtstichprobe, mit deutlichem Abstand vor den „unzufrieden Gelangweilten" (Cluster 2 mit 13.1 Prozent), den „intrinsisch motiviert Bildungsbegeisterten" (Cluster 5 mit 11.6 Prozent), den „extrinsisch motiviert Angepassten" (Cluster 4 mit 9.1 Prozent) und schließlich den „desinteressiert Frustrierten" (Cluster 3 mit 6.2 Prozent) (vgl. Abb. 1):

Abb. 1: Verteilung der Stichprobe auf die fünf Cluster nach Einstellungen zur Schule

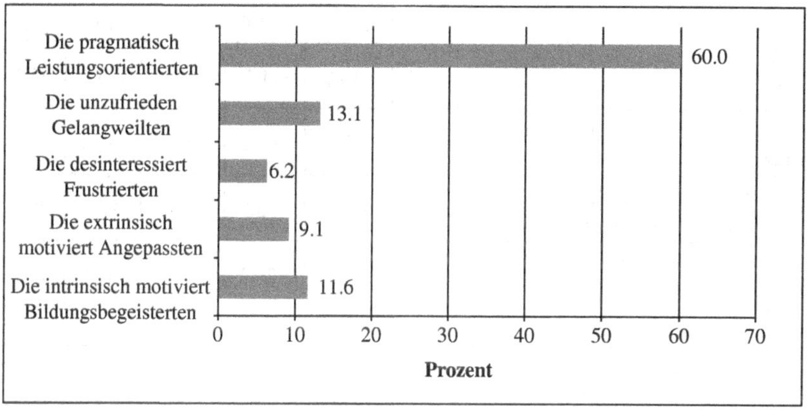

Quelle: Eigene Darstellung

Die Cluster 1 und 5 bilden eine vorwiegend günstige Einstellung zur Schule ab, Cluster 2 und 3 eine weitgehend ungünstige und Cluster 4 ein widersprüchliches inneres Verhältnis der Schule gegenüber. Tabelle 2 zeigt eine Übersicht über die einzelnen Items[6] (geordnet nach den extrahierten Faktoren) und ihre Abweichung von der Gesamtstichprobe für die fünf Cluster sowie soziodemografische und weitere für die Beschreibung der Subgruppen aufschlussreiche Merkmale.

6 Aus den Variablen wurden keine Skalen gebildet, u.a. um die Cluster über die Einzelitems differenzierter charakterisieren zu können, wie in Tab. 2 zu erkennen.

Tab. 2: Übersicht über die fünf Cluster nach Einstellungen zur Schule – Abweichung von Gesamt (Angaben in Prozent)

	Cluster					Ges.
	1	2	3	4	5	
	Die pragmatisch Leistungsorientierten	**Die unzufrieden Gelangweilten**	**Die desinteressiert Frustrierten**	**Die extrinsisch motiviert Angepassten**	**Die intrinsisch motiviert Bildungsbegeisterten**	**Gesamt**
Subjektive Bedeutungszuschreibungen						
Schule ist mir wichtig.	+5.7	+13.2	−41.6	−3.6	+16.0	84.0
Mir ist es wichtig, in der Schule gute Leistungen zu zeigen.	+6.1	−10.6	−50.7	+4.6	+11.2	85.7
Mir ist es wichtig, im Unterricht gut mitzuarbeiten.	+5.9	−9.2	−47.2	+0.8	+12.1	85.1
Mir ist es wichtig, die Schule mit einem guten Abschluss zu verlassen.	+6.6	−2.9	−53.5	−2.4	+9.6	90.4
Ich finde es uncool, wenn jemand gut in der Schule ist.	−10.1	+3.2	+26.2	+43.8	−13.9	16.5
Intrinsische Motivation						
Ich lerne, weil ich den Unterrichtsstoff richtig verstehen will.	+4.7	−15.3	−23.3	−12.1	+18.8	77.6
Die Dinge, die wir im Unterricht durchnehmen, interessieren mich.	−0.9	−16.4	−33.2	+13.4	+26.2	65.9
Ich lerne, weil ich Lust dazu habe.	−4.2	−22.4	+2.0	+2.1	+33.2	50.7
Ich lerne, weil das für mich selbst wichtig ist.	+4.8	−12.7	−27.0	−4.9	+12.4	85.5
Ich lerne, weil es mir Freude macht.	−8.3	−19.3	+6.5	+17.4	+30.9	40.8
Extrinsische Motivation						
Ich lerne, weil man es von mir verlangt.	+0.9	+9.9	+4.4	+21.7	−30.3	46.7
Ich lerne, weil ich keine Probleme mit meinen Lehrern bekommen will.	+2.1	+1.7	−5.5	+17.3	−18.9	44.0

	Cluster					Ges.
	1	2	3	4	5	
Ich weiß eigentlich nicht, warum ich in der Schule lernen soll.	−7.2	+11.2	+25.8	+26.2	−17.8	18.9
Ich lerne, weil ich mich sonst schlecht fühle.	+0.6	−17.7	+6.1	+10.8	+4.2	52.1
Ich lerne, weil ich will, dass die Lehrer eine gute Meinung von mir haben.	+1.0	−11.5	−12.1	+11.4	+4.8	68.0
Langeweile						
Lernen in der Schule ist für mich Langeweile haben.	+4.2	+18.8	+6.3	−2.4	−36.4	41.2
Meine Schule ist ein Ort, an dem ich mich oft langweile.	−4.7	+39.1	+7.9	−10.0	−21.7	35.9
Positive Lernemotionen						
Lernen in der Schule ist für mich sich geborgen fühlen.	+4.5	−18.9	+0.0	−3.7	+5.8	39.0
Lernen in der Schule ist für mich spannend.	−3.8	−22.8	+15.3	+7.2	+29.2	45.7
Ich freue mich, wenn ich in der Schule etwas gut kann.	+6.3	−4.3	−46.1	+0.3	+9.6	88.8
Lernen in der Schule ist für mich Spaß haben.	+0.1	−25.7	+3.3	+8.1	+18.0	74.3
Lernen in der Schule ist für mich etwas leisten.	+5.7	−25.4	−6.2	−4.5	+11.4	84.5
Lernen in der Schule ist für mich Erfolg haben.	+4.6	−24.0	−12.9	−6.0	+10.4	83.6
Negative Lernemotionen						
Lernen in der Schule ist für mich Enttäuschungen erleben.	+0.9	+7.2	+8.0	+4.0	−15.4	23.1
Lernen in der Schule ist für mich sich blamieren.	−1.8	+13.7	+10.3	−0.1	−11.5	12.7
Schulfreude						
Der Unterricht in der Schule macht mir Spaß.	−3.1	−20.3	−26.0	+14.2	+34.9	59.3
Nach den Ferien freue ich mich darauf, wieder zur Schule zu gehen.	−8.6	−11.1	−6.1	+28.8	+23.3	45.3

	Cluster					Ges.
	1	2	3	4	5	
Ich finde es schade, wenn Unterricht ausfällt.	−12.8	−2.5	+24.8	+36.6	+8.3	18.6
Ich mache gerne meine Hausaufgaben.	−10.9	−13.1	+0.3	+29.9	+30.6	44.4
Schulunlust / Unwohlbefinden						
Meine Schule ist ein Ort, zu dem ich nicht hingehen möchte.	−7.0	+37.6	+6.8	−8.3	−11.1	20.3
Meine Schule ist ein Ort, an dem ich mich oft unwohl/fehl am Platz fühle.	−7.7	+33.0	+8.1	−0.8	−9.0	15.1
Wie oft kommt es im Allgemeinen vor, dass du einfach keine Lust hast, zur Schule zu gehen? (oft/manchmal)	+9.8	+18.0	−3.5	−18.0	−37.6	55.5
Bedeutung von Schule außerhalb der Schule						
Wie oft kommt es im Allgemeinen vor, dass du dich mit deinen Freunden nachmittags über die Schule unterhältst? (oft/manchmal)	+1.8	−3.0	+12.1	−17.3	+3.7	48.1
Wie oft kommt es im Allgemeinen vor, dass du in deiner Freizeit noch einmal an die Schule denkst? (oft/manchmal)	+1.1	−12.7	+11.8	−14.0	+13.2	44.8
Abgelenktsein im Unterricht						
Wie oft kommt es, dass du im Unterricht heimlich spielst? (oft/manchmal)	+2.6	−1.1	+7.9	−0.2	−13.2	22.5
Wie oft kommt es vor, dass du im Unterricht Aufgaben für andere Fächer machst? (oft/manchmal)	+5.4	−3.4	+6.4	−5.9	−16.9	34.0
Schulerfolg in Schulnoten abgebildet						
Letzte Deutschnote: gut bis sehr gut	+3.1	−3.4	−15.9	−11.9	+4.8	20.0
Letzte Mathematiknote: gut bis sehr gut	+0.2	−0.8	−6.2	−8.6	+10.1	20.9
Kulturelles Kapital						
Institutionalisiertes Kulturkapital: Höchster Schulabschluss in der Familie: Hochschulreife/Abitur	+2.4	+3.8	−14.1	−12.3	−0.2	18.8

	Cluster					Ges.
	1	2	3	4	5	
Objektiviertes Kulturkapital: Anzahl der Bücher in der Familie über 100	+2.1	+4.7	−4.9	−8.8	−5.4	33.1
Inkorporiertes Kulturkapital: Ich lese mehr als eine Stunde in der Woche	+0.4	+0.0	−10.9	−9.3	+13.4	22.3
Inkorporiertes Kulturkapital: Ich spiele ein Musikinstrument	−1.2	+1.0	+2.2	−0.9	+3.7	8.4
Elterliche Bildungsaspiration						
Erwarteter Schulabschluss: Hochschulreife/Abitur	+3.7	−3.9	−14.3	−16.7	+5.8	50.5
Die hohen Erwartungen meiner Eltern setzen mich unter Druck.	−2.0	+8.1	+3.2	+6.9	−8.4	15.4
Psychologische Grundbedürfnisse						
Erleben von Autonomie: In unserer Schule dürfen wir über viele Dinge mitentscheiden.	+2.3	−9.5	−29.9	+10.6	+9.8	69.7
Erleben fehlender Kompetenz: Kein Mensch kann alles. Für Deutsch habe ich einfach keine Begabung.	−4.3	+2.6	+18.4	+18.4	−10.5	28.5
Erleben fehlender Kompetenz: Kein Mensch kann alles. Für Mathematik habe ich einfach keine Begabung.	−0.4	+0.1	+8.7	+7.1	−9.1	36.8
Erleben von Zugehörigkeit: Meine Schule ist ein Ort, an dem ich mich dazugehörig fühle.	+4.3	−14.0	−14.0	+0.6	+4.8	87.1
Sprache						
In der Familie vorwiegend verwandte Spreche: Deutsch	+1.9	+5.9	−8.6	−0.8	−11.8	79.6
Soziodemografische Merkmale						
Geschlecht	w/m	m	m	w/m	w	w/m
Jahrgangsstufe	7/9	9	7/9	7	7	7/9
Bildungsgang	Gy/ Ges	S	Sek	Sek	Gy/ Ges	Ges/ Sek
Sozioökonomischer Status (HISEI)	MS	OS	US	US	OS/ US	MS/ US

	Cluster					Ges.
	1	2	3	4	5	
Migrationsstatus	ohne Mig.	ohne Mig.	ohne/ mit Mig.	mit Mig.	mit Mig.	ohne Mig.
Gesamt						
N	1014	222	104	153	196	1689
Prozent	60.0	13.1	6.2	9.1	11.6	100

Anmerkungen: für die soziodemografischen Merkmale: vorwiegender Anteil; Gy: Gymnasium, Ges: Gesamtschule, Sek: Sekundarschule; Mig: Migrationshintergrund; w: weiblich, m: männlich; OS: obere, MS: mittlere, US: untere Statusgruppe nach HISEI.

Das *Cluster 1 der „pragmatisch Leistungsorientierten"* ist als eine Subgruppe zu charakterisieren, deren insgesamt als positiv zu bewertende Einstellung insbesondere durch den sehr hohen Stellenwert von Leistung und guten Abschlüssen geprägt wird, weniger durch Lernfreude oder bemerkenswerte Begeisterung – etwa für den schulischen Unterricht. Dabei kennzeichnet vor allem ein realistisch-distanzierter Pragmatismus dieses Cluster, das in etwa gleichem Maße aus Siebt- und Neuntklässlern, Mädchen und Jungen, Gesamtschülern und Gymnasiasten, vorwiegend aber aus Jugendlichen ohne Migrationshintergrund besteht. Für das *Cluster 2 der „unzufrieden Gelangweilten"* zeigen sich insbesondere eine geringe Schulfreude und intrinsische Motivation, ein starke Schulunlust und starkes Unwohlbefinden, ein geringes Gefühl von Geborgenheit, vergleichsweise hoch ausgeprägte negative Lernemotionen sowie Langeweile. Eine in all diesen Dimensionen zu spürende Unzufriedenheit mit der eigenen Situation in den Kontexten formaler Bildung prägt deutlich diese Subgruppe, die sich vorwiegend aus Jungen, Neuntklässlern der Sekundarschulen und Jugendlichen ohne Migrationshintergrund aus der oberen sozioökonomischen Statusgruppe zusammensetzt, die sich in ihren psychologischen Grundbedürfnissen als nur wenig befriedigt, durch elterliche Bildungsaspirationen aber häufiger als in den übrigen Subgruppen unter Druck gesetzt erleben. Möglicherweise handelt es sich hierbei um so genannte „Underachiever", die sich in der Schule unterfordert fühlen. Im Zentrum des *Clusters 3* hingegen steht eine auffällige Distanzierung von den Zielen und Motiven der Schule. Resigniert zeigen die *„desinteressiert Frustrierten"* nur geringe Schulfreude und messen Schule, Leistung und einem guten Abschluss nur wenig Bedeutung bei. Sie sind im Unterricht oft abgelenkt, gelangweilt und wissen vielfach nicht, warum sie eigentlich lernen sollen. Das Desinteresse und die Frustration dieser Jugendlichen, die primär männlich sind, eher die Sekundarschule besuchen, vorwiegend der unteren sozioökonomischen

Statusgruppe zuzuordnen sind und über ein gering ausgeprägtes kulturelles Kapital verfügen (vgl. Bourdieu 1982), sind allerdings differenziert zu betrachten, da sie primär einer Distanzierung vom Gesamtsystem Schule Ausdruck verleihen. Denn es ist die Schule als Abschlüsse vergebende und zu diesem Zweck Leistung einfordernde Institution, welche die schulbezogene Einstellung dieser Schülerinnen und Schüler negativ färbt, jedoch nicht alle übrigen Facetten von Schule gänzlich überlagert, die z.T. positiv bewertet werden und Freude bereiten. So würde es bspw. ein bemerkenswerter Anteil bedauern, wenn Unterricht ausfiele. Dennoch zeigen sie nur geringen – in Schulnoten abgebildeten – Schulerfolg und verbinden das Lernen in der Schule stark mit Blamagen und Enttäuschungen. Die elterlichen Bildungserwartungen sind vergleichsweise niedrig, jedoch fühlen sich die Schülerinnen und Schüler dadurch etwas häufiger unter Druck gesetzt als in den übrigen Subgruppen. *Cluster 4 der „extrinsisch motiviert Angepassten"* ist demgegenüber als eine Gruppe zu beschreiben, deren Einstellungsmerkmale in weiten Teilen zunächst günstig erscheinen – gepaart allerdings mit stark externalisierten Motiven für das Lernen und nur geringem Schulerfolg. Es ist interpretativ zu vermuten, dass ihr Antwortverhalten in der Befragung weit stärker als in den sonstigen Clustern, auch wenn dies nie gänzlich auszuschließen ist, geprägt ist durch eine soziale Erwünschtheit, sodass die Einstellungen hier insgesamt als widersprüchlich zu charakterisieren sind und eine deutliche Außenorientierung und Anpassung an antizipierte Erwartungen – der Eltern, Lehrerinnen, Lehrer, aber auch Mitschülerinnen und Mitschüler (gute Leistungen bspw. werden oftmals als „uncool" bewertet) zeigen. Die Jugendlichen in dieser Subgruppe besuchen zu einem höheren Anteil die siebte Jahrgangstufe der Sekundarschulen, weisen häufiger einen Migrationshintergrund auf, stammen vorwiegend aus der unteren sozialen Statusgruppe und verfügen über ein gering ausgeprägtes kulturelles Kapital. Ihr Einstellungsbild erscheint ambivalent. Auch das *Cluster 5 der „intrinsisch motiviert Bildungsbegeisterten"* besteht mehrheitlich aus Migrantinnen und Migranten, welche vorwiegend weiblich, zu einem höheren Anteil in der Jahrgangstufe 7 und nur selten an den Sekundarschulen zu finden sind. Die positive Einstellung dieser Jugendlichen wird insgesamt durch einen überaus hohen Stellenwert formaler Bildung, durch Lernfreude, Spaß und Interesse – eben durch Bildungsbegeisterung, gepaart mit einer starken intrinsischen Motivation – geprägt. Druck durch elterliche Bildungserwartungen wird vergleichsweise selten wahrgenommen, obwohl deren Aspirationen etwas höher als in den übrigen Clustern ausfallen. Dabei ist das kulturelle Kapital in den Herkunftsfamilien in Teilen stark ausgeprägt, auch wenn diese Schülerinnen und Schüler interessanterweise gleichermaßen der oberen und unteren sozioökonomischen Statusgruppe zuzuordnen sind. Sie sind vergleichsweise erfolgreich in der Schule, auch wenn ein großer Anteil trotz überaus positiver Einstellungen der Schule gegenüber keine guten bis sehr guten Leistungen erbringen kann und gar in den unteren Leistungsbereichen zu verorten ist.

Insgesamt ergeben sich somit durch Korrelation der fünf Cluster mit soziodemografischen und weiteren in diesem Kontext relevanten Merkmalen statistisch relevante Zusammenhänge mit der besuchten Jahrgangsstufe (.303**)[7], dem Geschlecht (.400***), dem besuchten Bildungsgang (.228**), den elterlichen Bildungsaspirationen (.389***), dem Migrationshintergrund (.429***) und der Befriedigung der psychologischen Grundbedürfnisse (Deci und Ryan 1991, 1993) nach Autonomie (.531***), Kompetenz (im Fach Deutsch mit .409***, in Mathematik mit .198*) und sozialer Eingebundenheit (.323***). Auch mit dem in Schulnoten abgebildeten Schulerfolg besteht ein hoch bzw. sehr signifikanter Zusammengang (im Fach Deutsch mit .554***, in Mathematik mit .222**), was jedoch nicht bedeutet, dass eine günstige Einstellung wider Erwarten nicht auch mit geringem Erfolg im Kontext Schule einhergehen kann. Weit weniger stark als erwartet zeigt sich eine Abhängigkeit vom kulturellen Kapital der Herkunftsfamilie (in keiner der untersuchten Dimensionen signifikant) und der sozialen Herkunft (.074^n.s.).

5.2 Die Einstellung zur Schule im Kontext von Geschlecht, Migrationshintergrund und sozialer Herkunft

Die Verteilung der Schülerinnen und Schüler auf die fünf Cluster ergibt durch Korrelation mit soziodemografischen Merkmalen zunächst aus geschlechtsspezifischer Perspektive ein deutliches Bild: In den als günstig zu bewertenden Subgruppen befinden sich vorwiegend Mädchen, während die Jungen primär in den Clustern mit negativen Einstellungen zu verorten sind. Insgesamt bewerten Mädchen hochsignifikant häufiger die Schule (.334***) und einen guten Schulabschluss (.401***)8 als persönlich bedeutsam und lassen eine stärker ausgeprägte Schulfreude erkennen (.366***). Hingegen zeigen deutlich mehr Jungen kein Bewusstsein dafür, warum sie überhaupt in der Schule lernen sollen (.222***),

7 Aufgeführt werden hier der Cramer's V-Wert, der Auskunft über die Stärke, nicht jedoch über die Richtung eines Zusammenhangs gibt, und die Signifikanz (*** = p < .001; ** = p < .01; * = p < .10; n.s. nicht signifikant).

8 Auch hier unter Verwendung des Zusammenhangsmaßes Cramer's V.

sie neigen weit häufiger zur Abwertung guter Schulleistungen (.170***) und berichten eher von Langeweile im Unterricht (.347***). Aufschlussreich ist in diesem Kontext jedoch ein fachspezifischer Zugang: Während im Fach Deutsch keine statistisch relevanten geschlechtsspezifischen Differenzen bezüglich der subjektiven Bedeutung und des Interesses zu erkennen sind, so sind diese in Mathematik beachtenswert: Für die Jungen erscheint dieses Fach weitaus wichtiger (.405***) und interessanter (.114**). Dieses Ergebnis findet sich in noch deutlicherer Ausprägung in zahlreichen Studien bestätigt: Mädchen weisen insgesamt sehr viel weniger Interesse und Freude an Mathematik auf als Jungen. Stattdessen zeigen sie ein sehr viel höheres Niveau an Stress und Hilflosigkeit im Mathematikunterricht sowie Angst vor dem Fach und ungünstige Attributionen. „Es scheint dem deutschen Schulsystem in geringerem Maße als anderen Ländern zu gelingen, Geschlechterunterschiede in diesen Haltungen zur Mathematik auszugleichen" (Pekrun und Zirngibl 2004) bzw. ihr Entstehen zu verhindern. Dennoch erlangen die Mädchen in vorliegender Untersuchung in beiden Fächern bessere Zensuren – in Deutsch ist die Differenz hochsignifikant (.314***). Die besseren Noten von Mädchen im Fach Deutsch finden sich bundesweit vielfach bestätigt, während sich die Befunde zum Fach Mathematik von zahlreichen einschlägigen Studien unterscheiden, die hier leicht bessere Schulnoten für die Jungen identifizieren (vgl. etwa Schurt und Waburg 2007).

Während sich die Jungen mit Migrationshintergrund vorwiegend auf die Cluster 1-4 verteilen, finden sich im positivsten Cluster 5 vor allem Mädchen mit Migrationshintergrund, sodass die Befunde zur nationalen Herkunft hier auch von geschlechtsspezifischer Relevanz sind, denn auch innerhalb der ohnehin sehr heterogenen Gruppe der Migranten zeigen sich erhebliche Geschlechtsunterschiede. Dies gilt es zu berücksichtigen, wenn für die Jugendlichen mit Migrationshintergrund insgesamt deutlich günstigere Einstellungsmerkmale konstatiert werden können: So zeigen sie eine hochsignifikant höhere Schulzufriendenheit (.343***), Lernfreude (.287***) und geringere Langeweile (.299***) sowie ein sehr signifikant stärkeres Interesse an den Unterrichtsinhalten (.188**). Dies bestätigt sich in sämtlichen PISA-Studien. Hier zeigen Migrantinnen und Migranten trotz ihrer geringeren Kompetenzen größeres Interesse verbunden mit einer stärker ausgeprägten Motivation (vgl. Walter und Taskinen 2007). Vorliegende Studie bestätigt dieses Attitude-Achievement-Paradox bei Jugendlichen mit Migrationshintergrund – vor allem für die Mädchen, denn der Schulerfolg steht insbesondere im Fach Deutsch in hochsignifikantem Zusammengang mit der nationalen Herkunft (.404***). Dabei kommt der Beherrschung der deutschen Sprache große Bedeutung zu, denn die in der Familie vorwiegend verwandte Sprache korreliert sehr signifikant mit der letzten Deutschzensur

(.247***) und signifikant mit der Schulnote in Mathematik (.124*). Mädchen und Jungen mit Migrationshintergrund, die zu Hause hauptsächlich Deutsch sprechen, erbringen dabei insbesondere im Fach Deutsch insgesamt bessere Leistungen als Migrantenkinder, bei denen zu Hause vorwiegend eine andere Sprache gesprochen wird.

Im Gegensatz zum Migrationshintergrund stehen die soziale Herkunft und das kulturelle Kapital in keinem statistisch relevanten Zusammenhang mit der Verteilung der Schülerinnen und Schüler auf die Cluster. Allerdings sind die Zusammenhänge zwischen höchstem sozioökonomischen Status in der Familie HISEI und den Schulnoten hochsignifikant für die Fächer Deutsch (.402***) und Mathematik (.344***): Mädchen und Jungen in der höchsten HISEI-Statusgruppe haben in Deutsch bspw. eine fast dreimal so große Chance, die Note „sehr gut" oder „gut" zu erlangen, als Jugendliche der unteren HISEI-Gruppe. Entsprechend ist auch der Zusammenhang zwischen der Bücherzahl in der Familie und den Schulnoten als jeweils hochsignifikant zu bewerten (Deutsch: .458***, Mathematik: .305***). Soziale Herkunft und objektiviertes Kulturkapital erscheinen hier als bedeutsame Determinanten des in Schulnoten abgebildeten Bildungserfolgs. Da sich zudem eine hochsignifikant bzw. sehr signifikant positive Korrelation zwischen der Verteilung auf die Cluster und den Schulnoten im Fach Deutsch (.554***) und Mathematik (.222**) zeigt – wobei von einem wechselseitigem Beziehungsgefüge auszugehen ist –, die Einstellung selbst aber vom sozioökonomischen Status und dem Kulturkapital wie dargestellt in vergleichsweise geringem Maße abhängig ist, eröffnet sich hier eine bedeutsame Perspektive im Kontext der für Deutschland bereits in PISA 2000 ermittelten und noch immer wirkkräftigen überaus straffen Kopplung von sozialer Herkunft und Bildungserfolg (vgl. Baumert und Schümer 2001).

6. Resümee und Ausblick

Die schulbezogenen Einstellungen der untersuchten Schülerinnen und Schüler an Brennpunktschulen lassen sich durch eine große Heterogenität charakterisieren, die sich in den unterschiedlichen Clustern deutlich abbildet. Auch an Schulen in benachteiligter Lage zeigen sich also – entgegen gängiger Stereotypen – nicht etwa vorwiegend negative Einstellungen, sondern sehr vielfältige; und es finden sich Subgruppen (wie etwa die Mädchen mit Migrationshintergrund) mit überaus positiven Einstellungsausprägungen, andere mit ungünstigen – geprägt durch Desinteresse, Distanz oder Frustration. Dieser zentrale Befund der referierten empirischen Untersuchung ist im aktuellen Diskurs um

Bildungsbenachteiligung, Bildungschancen und Schulerfolg von großer Relevanz, denn er differenziert das Bild der „Bildungsverlierer" und der Wirkmechanismen in bildungsfernen Milieus, über die Bildungsmisserfolg und negative Schulkarrieren präformiert werden: Zwar ist die schulbezogene Einstellung mit Schulerfolg eng verkoppelt, nicht aber mit einem spezifischen bildungsnahen oder bildungsfernen Milieu bzw. mit der sozialen Herkunft. So ist sich die deutliche Mehrheit der befragten Mädchen und Jungen offenbar der Schlüsselfunktion guter Schulleistungen und entsprechender Zertifikate für die Eröffnung von Berufs- und Lebensperspektiven bewusst und stellt diese ins Zentrum der eigenen Einstellung (Abschlüsse und Leistung werden vielfach als persönlich überaus wichtig bewertet) – und dies an Brennpunktschulen, die sich in vielfacher Weise in benachteiligter Lage befinden. Die in der theoretischen Rahmung dieses Beitrags konstatierte Milieubedingtheit von Einstellung lässt sich mit vorliegender Untersuchung nicht bestätigen.

Auch die häufig diskutierte enge Verknüpfung von Migrationshintergrund und Bildungsmisserfolg kann auf der Basis der berichteten Ergebnisse differenziert betrachtet werden, denn insbesondere Mädchen mit Migrationshintergrund zeigen oftmals günstige Einstellungsmerkmale, die sie aber nur in geringerem Maße in Schulerfolg umzusetzen vermögen (Attitude-Achievement-Paradox). Dass Jungen im Vergleich zu Mädchen insgesamt deutlich ungünstigere Einstellungen aufweisen und entsprechend schlechtere Leistungen erbringen, bestätigt die geschlechtsspezifischen Befunde einer Vielzahl von Studien (vgl. etwa Harazd und Schürer 2006; Valtin et al. 2006; Leven und Schneekloth 2007). Dabei zeigt sich bei den untersuchten Mädchen und Jungen ein beachtenswerter „Schulpragmatismus" verbunden mit einem ausgeprägten Leistungsethos (das größte Cluster 1 bildet diese Einstellungsausprägung deutlich ab), allerdings auch mit einer fehlenden persönlichen Sinnhaftigkeit der Unterrichtsinhalte, die deutlich wird, wenn nach Gründen für das Lernen oder Interesse an Inhalten gefragt wird. Schule ist, so lässt sich interpretieren, offenbar primär eine Pflichtveranstaltung, die zwar insgesamt von großer Bedeutung, aber von geringem sinnstiftenden Gehalt über die Zertifikatszuteilung hinaus erscheint. Weiterhin zeigt sich, dass spezifische Einstellungen zur Schule insgesamt und zu schulischen Einzelaspekten wie dem Lernen, dem Unterricht, Hausaufgaben etc. in den Clustern eng mit einander verknüpft (vgl. auch Fend 1998; Hascher und Baillod 2000; Christen 2004, 9) und, wenn auch nicht durch das Milieu determiniert, doch durchaus sozial vermittelt sein können (vgl. Witte 1977), wenn man bspw. den engen Zusammenhang zwischen Einstellung und elterlicher Bildungsaspiration betrachtet.

Limitationen der referierten empirischen Untersuchung liegen bspw. in der Erklärung des Attitude-Achievement-Paradoxons, denn auf die Ursachen liefern die Daten – mit Ausnahme des Bezugs zur Sprachbarriere – keine gesicherten Hinweise. So besteht darin ein wichtiger Anknüpfungspunkt für die weitere empirische Bildungsforschung im Kontext von Mixed Methods-Designs (vgl. Gläser-Zikuda et al. 2012), um zu vertiefenden und die subjektiven Bedeutungszuschreibungen tatsächlich erklärenden Befunden zu gelangen. Dies erscheint ebenso sinnvoll, um spezifische Wirkrichtungen einer genaueren Prüfung zu unterziehen. Die vorliegenden Ergebnisse können lediglich ein Wechselverhältnis etwa zwischen der Einstellung zur Schule und dem Schulerfolg, zwischen Einstellung und besuchtem Bildungsgang oder zwischen der Einstellung und der Befriedigung der psychologischen Grundbedürfnisse annehmen. Wie genau aber Wirkprozesse in diesen Kontexten ablaufen, ist auf der Basis der vorliegenden Daten nicht zu erkennen. Eine grundlegende Einschränkung der Befunde liegt weiterhin in der Begrenzung der Stichprobe allein auf Brennpunktschulen und somit auf einen insgesamt niedrigen sozioökonomischen Status und ein nur gering ausgeprägtes kulturelles Kapital in den Herkunftsfamilien begründet. Es stellt sich daher bspw. die Frage, ob das kulturelle Kapital und die soziale Herkunft in einer repräsentativen Stichprobe stärker auf die Einstellungsausprägung wirken. Interessant wäre weiterhin ein Vergleich von Brennpunktschulen mit einer Kontrollgruppe mit Schulen, die sich nicht in benachteiligter Lage befinden, um das Spezifische der Einstellungsmerkmale in beiden Gruppen herauszuarbeiten.

Dessen ungeachtet haben vermutlich bereits das Konstatieren der großen Lernfreude bei Jugendlichen mit Migrationshintergrund durch die breite Rezeption etwa von PISA und die aktuelle fachöffentliche Diskussion um das Attitude-Achievement-Paradox durchaus Bedeutung für eine spezifische Förderung von Migrantenkindern und findet zunehmend Beachtung in entsprechenden Konzeptionen. Womöglich wäre es von ähnlicher Relevanz, wenn die tendenziell günstigen schulbezogenen Einstellungen von Schülerinnen und Schülern an Brennpunktschulen stärker in den Diskurs um Bildungsbenachteiligung Einzug hielten.

Literatur

Allport, G.W. (1935). Attitudes. In Murchison, C.A. (Hrsg.), *A Handbook of Social Psychology* (S. 798-844). Worcester: Clark University Press.

Aronson, E., Wilson, T.D. & Akert, R.M. (2004). *Sozialpsychologie*. 4. Aufl. München: Pearson Studium.

Asakawa, K. & Csikszentmihalyi, M. (1998). The Quality of Experience of Asian American Adolescents in Activities Related to Future Goals. *Journal of Youth and Adolescence 27*, 141-163.

Backhaus, K. et al. (2011). *Multivariate Analysemethoden: Eine anwendungsorientierte Einführung.* Heidelberg et al.: Springer.

Bacher, J. (1996). *Clusteranalyse. Anwendungsorientierte Einführung.* München; Wien: Oldenbourg.

Baumert, J. & Schümer, G. (2001). Familiäre Lebensverhältnisse, Bildungsbeteiligung und Kompetenzerwerb. In Baumert, J. et al. (Hrsg.), *PISA 2000. Basiskompetenzen von Schülerinnen und Schülern im internationalen Vergleich* (S. 323-407). Opladen: Leske+Budrich.

Bos, W. et al. (Hrsg.) (2003). *Erste Ergebnisse aus IGLU. Schülerleistungen am Ende der vierten Jahrgangsstufe im internationalen Vergleich.* Münster et al.: Waxmann.

Bos, W. et al. (Hrsg.) (2007). *IGLU 2006. Lesekompetenzen von Grundschulkindern in Deutschland im internationalen Vergleich.* Münster et al.: Waxmann.

Bourdieu, P. (1983). Ökonomisches Kapital, kulturelles Kapital, soziales Kapital. In Kreckel, R. (Hrsg.), *Soziale Ungleichheit* (Soziale Welt/Sonderband 2) (S. 183-198). Göttingen: Schwartz.

Brosius, F. (2008). *SPSS 16. Das MTIP-Standardwerk.* Bonn: MTIP.

Büchner, P. & Koch, K. (2001). *Von der Grundschule in die Sekundarstufe. Bd. 1: Der Übergang aus Kinder- und Elternsicht.* Opladen: Leske+Budrich.

Bundesministerium für Bildung und Forschung (BMBF) (Hrsg.) (2006). *Schulerfolg von Jugendlichen mit Migrationshintergrund im internationalen Vergleich.* Bildungsforschung Bd. 19. Berlin.

Bundesministerium für Familie, Senioren, Frauen und Jugend (BMFSFJ) (Hrsg.) (2009). *Der Mikrozensus im Schnittpunkt von Geschlecht und Migration. Möglichkeiten und Grenzen einer sekundär-analytischen Auswertung des Mikrozensus 2005.* Baden-Baden: Nomos.

Centre for Educational Research and Innovation (CERI) (Ed.) (2006). *Schooling for Tomorrow.* Paris: OECD Publications.

Christen, F. (2004). Einstellungsausprägungen bei Grundschülern zu Schule und Sachunterricht und der Zusammenhang mit ihrer Interessiertheit. http://www.upress.uni-kassel.de/abstracts_fr/3-89958-062-1.html. Gesehen 05.08.2008.

Czerwenka, K. et al. (1988). Was Schüler von der Schule halten. *Die Deutsche Schule 80*, 132-146.

Czerwenka, K. et al. (1990). *Schülerurteile* über *die Schule. Bericht* über *eine internationale Untersuchung.* Frankfurt a.M.: Lang.

Deci, E. L. & Ryan, R. M. (1991). A Motivational Approach to Self: Integration in Personality. In Dienstbier, R. A. (Hrsg.), *Perspectives on Motivation. Nebraska Symposium on Motivation, 1990* (p. 237-288). Lincoln: University of Nebraska Press.

Deci, E. L. & Ryan, R. M. (1993). Die Selbstbestimmungstheorie der Motivation und ihre Bedeutung für die Pädagogik. *Zeitschrift für Pädagogik 2*, 223-238.

Eagly, A.H. & Chaiken, S. (1998). Attitude Structure and Function. In Gilbert, D., Fiske, S.T. & Lindzey, G. (Eds.), *The Handbook of Social Psychology* (S. 269-322). Vol. 1. 4th Printing. New York: Oxford University Press.

Eder, F. (2007). *Das Befinden von Kindern und Jugendlichen in der österreichischen Schule. Befragung 2005.* Innsbruck et al.: Studien-Verlag.

Fend, H. (1997). *Der Umgang mit Schule in der Adoleszenz. Aufbau und Verlust von Lernmotivation, Selbstachtung und Empathie.* Bern et al.: Huber.

Fend, H. (1998). *Eltern und Freunde. Soziale Entwicklung im Jugendalter.* Bern: Hans Huber.

Gerrig, R.J. & Zimbardo, P.G. (2008): *Psychologie* (18. Aufl.). München: Pearson Studium.

Giesbrecht, K.H. (1997). *Escuela Indigena – Escuela Moderna. Interkulturelle Kommunikation am Beispiel indianischer Schulen in Paraguay.* Frankfurt a.M.: IKO.

Gläser-Zikuda, M. et al. (Hrsg.) (2012). *Mixed Methods in der empirischen Bildungsforschung.* Münster et al.: Waxmann.

Harazd, B. & Schürer, S. (2006). Veränderung der Schulfreude von der Grundschule zur weiterführenden Schule. In Schründer-Lenzen, A. (Hrsg.), *Risikofaktoren kindlicher Entwicklung: Migration – Schulleistung – Emotionalität* (S. 208-222). Wiesbaden: VS Verlag für Sozialwissenschaften.

Harring, M. (2011). *Das Potenzial der Freizeit: Soziales, kulturelles und ökonomisches Kapital im Kontext heterogener Freizeitwelten Jugendlicher.* Wiesbaden: VS Verlag für Sozialwissenschaften.

Hascher, T. & Baillod, J. (2000). Auf der Suche nach dem Wohlbefinden in der Schule. *Schweizer Schule* 3, S. 3-12.

Helsper, W. & Böhme, J. (2002). Jugend und Schule. In Krüger, H.-H. & Grunert, C. (Hrsg.), *Handbuch Kindheits- und Jugendforschung* (S. 567-596). Opladen: Leske+Budrich.

Heymann, H. W. (2008). Lernen inszenieren – Interesse wecken. *Pädagogik* 6, 6-9.

Kanders, M. (2000). *Das Bild der Schule aus der Sicht von Schülern und Lehrern II.* Dortmund: IFS-Verlag.

Klöckner, C., Beisekamp, A. & Hallmann, S. (2007). *LBS-Kinderbarometer Deutschland 2007. Stimmungen, Meinungen, Trends – von Kindern und Jugendlichen in Deutschland.* Berlin: Bundesgeschäftsstelle der Landesbausparkassen

Lehmann, R.H. et al. (2000). *QuaSUM. Qualitätsuntersuchung an Schulen zum Unterricht in Mathematik. Ergebnisse einer repräsentativen Untersuchung im Land Brandenburg.* Reihe Schulforschung in Brandenburg, H. 1. Potsdam.

Lehmann, R.H. et al. (2002). LAU 9. Aspekte der Lernausgangslage und der Lernentwicklung – Klassenstufe 9. Ergebnisse einer längsschnittlichen Untersuchung in Hamburg. http://www.hamburger-bildungsserver.de/welcome.phtml?unten=/schulentwicklung /lau/. Gesehen 23.08.08.

Lehmann, R.H., Gänsfuß, R. & Peek, R. (1999). LAU 7. Aspekte der Lernausgangslage und der Lernentwicklung von Schülerinnen und Schülern, die im Schuljahr 1996/97 eine fünfte Klasse an Hamburger Schulen besuchten. Bericht über die Erhebung im September 1998 (LAU 5). http://www.hamburger-bildungsserver. de/welcome.phtml?unten=/ schulentwicklung/lau/. Gesehen 23.08.08.

Lehmann, R.H. & Peek, R. (1997). LAU 5. Aspekte der Lernausgangslage und der Lernentwicklung. Bericht über die Erhebung im September 1996 (LAU 5). http://www.hamburger-bildungsserver.de/welcome.phtml?unten=/schulentwicklung/lau/lau5/. Gesehen 23.08.08.

Leven, I. & Schneekloth, U. (2007). Die Schule – frühe Vergabe von Lebenschancen. In Hurrelmann, K. & Andresen, S. (Hrsg.), *Kinder 2007. 1. World Vision Kinderstudie* (S. 111-142). Frankfurt a.M.: Fischer.

Majoribanks, K. (1992). The Predictive Validity of an Attitudes Towards School Scale in Relation to Children's Academic Achievement. *Educational and Psychological Measurement 52*, 945-949.

Mayerl, J. (2008). Kognitive Grundlagen sozialen Verhaltens. Theoretische und statistische Analysen zur Modellierung von Einstellungs-Verhaltens-Beziehungen.

http://elib.uni-stuttgart.de/opus/volltexte/2008/3513/pdf/mayerl_2008.pdf. Gesehen 07.08. 2008.

Max-Planck-Institut für Bildungsforschung (MPIB)/Institut für die Pädagogik der Naturwissenschaften (IPN) (Hrsg.) (1995): Bildungsverläufe und psychosoziale Entwicklung im Jugendalter (BIJU). Kurzbericht. Berlin/Kiel: Max-Planck-Institut für Bildungsforschung/Institut für die Pädagogik der Naturwissenschaften.

McCoach, D.B. (2002): A Validity Study of the School Attitude Assessment Survey (SAAS). *Measurement and Evaluation in Counseling and Development 35*, 66-77.

Okun, M.A., Braver, M.W. & Weir, R.M. (1990). Grade Level Differences in School Satisfaction. *Social Indicator Research 22*, 419-427.

Oldenburg, I. (2001). *Schulangst*. Oldenburger Vordrucke. Nr. 436. Oldenburg: Carl-von-Ossietzky-Universität.

Otto, H.-U. & Rauschenbach, T. (Hrsg.) (2004). *Die andere Seite der Bildung. Zum Verhältnis von formellen und informellen Bildungsprozessen*. Wiesbaden: VS Verlag für Sozialwissenschaften.

Palentien, C. (2004). *Kinder- und Jugendarmut in Deutschland*. Wiesbaden: VS Verlag für Sozialwissenschaften.

Pekrun, R.& Zirngibl, A. (2004). Schülermerkmale im Fach Mathematik. In Prenzel, M. et al. (Hrsg.), *PISA 2003. Der Bildungsstand der Jugendlichen in Deutschland – Ergebnisse des zweiten internationalen Vergleichs* (S. 191-210). Münster et al.: Waxmann.

Rosenberg, M. J. & Hovland, C. I. (1960). Cognitive, Affective, and Behavioral Components of Attitudes. In Hoveland, C. I. & Rosenberg, M. J. (Eds.), *Attitude Organization and Change*. New Haven: Yale University Press.

Schurt, V. & Waburg, W. (2007). Geschlechtsspezifik und/oder Fachtypik? Selbstwirksamkeit, Interesse, Stimmung und körperliches (Wohl)Befinden von Schülerinnen in ausgewählten Schulfächern im Vergleich. In Herwartz-Emden, L. (Hrsg.), *Neues aus alten Schulen – empirische Studien in Mädchenschulen* (S. 115-160). Opladen et al.: Budrich.

Schütte, K. et al. (2007). Schülermerkmale, naturwissenschaftliche Kompetenz und Berufserwartung. In Prenzel, M. et al. (Hrsg.), *PISA 2006. Die Ergebnisse der dritten internationalen Vergleichsstudie* (S. 125-146). Münster et al.: Waxmann.

Senator für Bildung und Wissenschaft Bremen (2005). Qualitätsprojekte an Schulen des Landes Bremen. Beschlussfassung. Vorlage Nr. L 165 für die Sitzung der Deputation für Bildung am 24. November 2005. http://www2.bildung.bremen.de /sfb/behoerde/deputation/depu/l165_16.pdf. Gesehen 05.08.2009.

Smith, M.B., Bruner, J.S. & White, R.W. (1956). *Opinions and Personality*. New York: John Wiley & Sons.

Städter, T. (2003). *Lexikon der Psychologie*. Stuttgart: Kröner.

Stecher, L. (2005). Schule als Familienproblem? Wie Eltern und Kinder die Grundschule sehen. In Alt, C. (Hrsg.), *Kinderleben. Bd. 2: Aufwachsen zwischen Freunden und Institutionen* (S. 183-197). Wiesbaden: VS Verlag für Sozialwissenschaften.

Stroebe, W. et al. (1999). *Sozialpsychologie. Eine Einführung*. Berlin: Springer.

Thurstone, L.L. (1931). The Measurement of Social Attidutes. *Journal of Abnormal and Social Psychology 26*, 249-269.

Valtin, R., Wagner, C. & Schwippert, K. (2006). Jungen – benachteiligt? Einige Ergebnisse aus IGLU. *Die Grundschulzeitschrift 194*, 18-19.

Verkuyten, M. & Thijs, J. (2002). School Satisfaction of Elementary School Children: The Role of Performance, Peer Relations, Ethnicity and Gender. *Social Indicator Research 59*, 203-228.

Walter, O. & Taskinen, P. (2007). Kompetenzen und bildungsrelevante Einstellungen von Jugendlichen mit Migrationshintergrund in Deutschland: Ein Vergleich mit ausgewählten OECD-Staaten. In Prenzel, M. et al. (Hrsg.), *PISA 2006. Die Ergebnisse der dritten internationalen Vergleichsstudie* (S. 337-366). Münster et al.: Waxmann.

Walther, G. et al. (2008). Mathematische Kompetenz im internationalen Vergleich: Testkonzeption und Ergebnisse. In Bos, W. et al. (Hrsg.), *TIMSS 2007: Mathematische und naturwissenschaftliche Kompetenzen von Grundschulkindern in Deutschland im internationalen Vergleich* (S. 49-85). Münster et al.: Waxmann.

Wild, E. (2002). Lebensraum Schule. Analysen zum Wohlbefinden von Schülern und ihren Einstellungen zu Schule und Lernen. In LBS-Initiative Junge Familie (Hrsg.), *Kindheit 2001 - Das LBS-Kinderbarometer* (S. 237-256). Opladen: Leske+ Budrich.

Wilson et al. (1989). Introspection, Attitude Change, and Attitude-behaviour Consistency: The Disruptive Effects of Explaining Why We Feel the Way We Do. *Advances of Experimental Social Psychology 22*, 287-343.

Witte, E. (1977). Einstellung. In Herrmann, T. et al. (Hrsg.), *Handbuch psychologischer* Grundbegriffe (S. 103-115). München: Kösel.

Wittwer, J., Saß, S. & Prenzel, M. (2008). Naturwissenschaftliche Kompetenz im internationalen Vergleich: Testkonzeption und Ergebnisse. In Bos, W. et al. (Hrsg.), TIMSS 2007: *Mathematische und naturwissenschaftliche Kompetenzen von Grundschulkindern in Deutschland im internationalen Vergleich* (S. 87-124). Münster et al.: Waxmann.

Improvisational Theatre in the Secondary EFL Classroom: Teacher-Researcher Collaboration as In-Service Teacher-Training

Raphaëlle Beecroft

Abstract

In Germany, English language teaching is becoming increasingly competence and output-oriented, with the promotion of Intercultural Communicative Competence (ICC) firmly established as a main goal of the ELT curriculum. In order to mirror and do justice to these developments in ELT, new, dialogical forms of oral assessment have been introduced at all secondary levels. The contents of these high-stakes examinations as well as the school syllabi leading up to these aim at reflecting the competences and levels detailed in the Common European Framework for Modern Languages (CEFR).

These developments in ELT present a challenge for established teachers, who find themselves having to interpret and adapt European demands, new syllabi and new assessment forms to their own varying educational contexts.

In light of the above, the paper presents a mixed-methods action-research study which investigates the use of Improvisational Theatre in ELT as a means of both promoting ICC in young learners and preparing them for various dialogical assessment forms. After an overview of the theoretical foundations of the project, the paper focuses on the research design and implementation of the study, demonstrating how it acts as a didactic tool, involving the teachers on different levels in a form of in-service-teacher-training by enabling them to investigate one potential method of meeting the demands discussed above.

Introduction

In Germany, English Language Teaching (ELT) is becoming increasingly competence and output-oriented, with the promotion of Intercultural Communicative Competence (ICC) firmly established as a main goal of the ELT curriculum. The model of ICC developed by Byram forms the basis for the secondary EFL curriculum, demanding of students that they acquire both the skills and pragmalinguistic knowledge of an "intercultural speaker" (Byram 1997: 31ff), enabling them ultimately to contribute adequately and appropriately to an intercultural communication situation. This requires a combination of both affective and cognitive-linguistic dimensions of speech production (Kessler and Küppers 2008: 6).

In order to mirror and do justice to these developments in ELT, new, dialogical forms of oral assessment have been introduced at all secondary levels. The contents of these high-stakes examinations as well as the school syllabi leading up to these aim at reflecting the competences and levels detailed in the Common European Framework for Modern Languages (CEFR) (Council of Europe 2001) which itself has, as its foundation, a dynamic, action-oriented understanding of language and which draws heavily, in its model of communicative competence, on Byram's above named concept of ICC. (cf. Vogt 2011: 11, 29)

As an example, the Year 10 curriculum guidelines state that students must be able to hold their own during a conversation in English by reacting spontaneously and demonstrating a range of emotions as required by the current situation. This is then assessed in a high-stakes final exam (students may leave school after Year 10 to embark on vocational training) in the form of presentations, mediation, as well as a dialogue with the examiner or an improvised role-play.

These developments in ELT present a challenge for established teachers, who find themselves having to interpret and adapt European demands, new syllabi and new assessment forms to their own varying educational contexts.

In light of the above, this chapter will present a research project, currently being carried out, which investigates the use of Improvisational Theatre in ELT as a means of promoting ICC in young learners. After an overview of the theoretical foundations of the project, the chapter will focus on the research design and implementation of the study, demonstrating how it acts as a didactic tool, involving the teachers in a form of in-service teacher-training by enabling them to investigate one potential method of meeting the demands discussed above.

Theoretical Assumptions

From a theoretical point of view, Improvisational Theatre, as a form of unscripted theatre based on a framework of games, where the rules are given but the contents are improvised on stage (cf. Spolin 1999 and Johnstone 1999, for an overview of the history of improvisational theatre cf. Sawyer 2011), is considered to hold particular potential for fostering ICC and preparing for assessment in the classroom as it offers parallels to everyday (intercultural) conversational interaction (cf. Sawyer 2011) and can be, due to its flexible framework, adapted didactically to cover different contexts and contents.

The aim of the study is thus to empower teachers to, through improvisational theatre, provide their students with meaningful sociocultural activities for which, in order to master, they have to acquire the appropriate language and affective skills (cf. Zuengler and Miller 2006: 39). These activities should reflect the ICC goals set out in the school syllabi and assessment guidelines. The research study is furthermore based on the work of Kurtz (2001), who first explored the effect of improvisation-based activities on the improvement of students' spontaneous oral expression.

At the core of the study, improvisational theatre is considered to have the potential to fulfil the above-named aims on two levels: At the methodological level, the structure of improvisational theatre and the rules of the games within it offer the opportunity to train the affective-procedural dimension of ICC: The skills required in order to improvise successfully (and engage in successful conversation – see above), such as spontaneity and flexibility, can be deemed to be parallel to the skills laid out by Byram pertaining to an intercultural speaker – for example the ability to "suspend disbelief" (Byram 1997: 34) in interaction. At the content-related level, improvisational theatre is considered to be able to train cognitive-linguistic skills through the open contents of improvisation games, which can be specifically designed to meet textbook, syllabus and assessment demands, and/or to include pragmalinguistic information such as prosodic-cues, intonation-contours, turn-taking, emotions, directness/politeness, interpretation of illocutionary force (LoCastro 2012:81) promoting, as has been called for in the literature, pragmatic competence through drama methods. Furthermore, the contents of the improvisation games can be adapted from intervention to intervention, matching the pace and the needs of the individual students.

Research Questions

The research questions framing the study are based on the investigation of the actual feasibility and potential benefits, for teachers and students, of integrating improvisational theatre in day-to-day secondary English language teaching, bearing in mind the manifold institutional constraints as well as syllabus and assessment demands teachers are faced with:

- Can the direct demands of the syllabus as well as contents promoting the goal of ICC be dealt with in a 45/90 minute improvisational task-cycle? Can this be integrated sustainably into English lessons?
- Do the students, during the improvisation sessions, engage in a different type of oral discourse than they do in their other English lessons?
- Do the students, during the improvisation sessions, speak freely whilst reacting spontaneously and flexibly? Does this develop over the course of the study?
- Do the students, during the improvisation sessions, use pragmalinguistic structures? Does this use develop over the course of the study?
- How do the improvisation sessions impact on the students' oral marks?
- How do the improvisation sessions serve to prepare the students for summative oral assessment? How do the improvisation sessions serve as a formative assessment task?

Research Design

The project is designed as an empirical, mixed-methods intervention study of action-research character. An overview of the -mainly qualitative- data-gathering methods are provided below:

Picture 1: Overview of research methodology

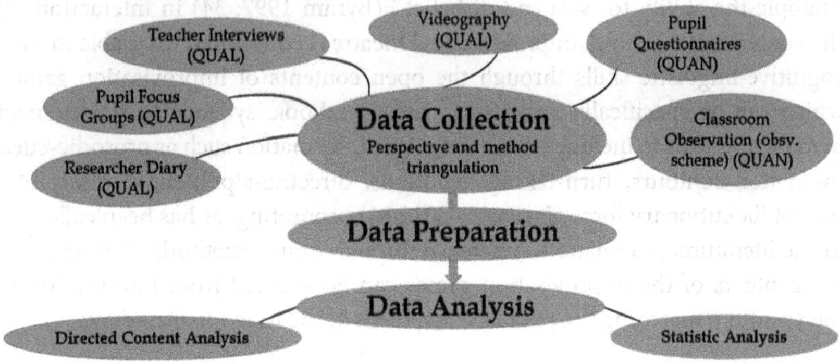

The mixed-methods design depicted above offers the possibility of triangulation on two levels: methodological and perspective triangulation. Methodological-triangulation enables, according to Dörnyei (2008: 43f) the reduction of "the inherent weaknesses of individual methods by offsetting them by the strength of another, thereby maximising both the internal and external validity of research" as well as making it possible to "… [V]alidate one's conclusion by presenting converging results obtained through different methods" (Ebd.: 164)

The triangulation of perspectives (teacher, student, researcher) is considered to be of particular value in this study, as it makes it possible to, whilst investigating the students' development with regard to the research questions, document and cross-relate the teachers' experiences too, in order to enable a didactic process in the vein of in-service teacher-training.

This is reflected in the over-arching framework of the research design, which is of action-research character (according to John Elliot, cf. Altrichter and Posch 2007: 13). Action-research traditionally sees teachers as researchers, carrying out their own studies themselves and not solely as users of knowledge generated by scholarly research (cf. Ebd.: 12 and Kurtz 2001: 62), with the focus of the research being current challenges identified by the teachers themselves within their particular institutional context. However, in this study, a slightly altered form of action-research was chosen in which both the author (a university researcher) and the teachers are involved equally, from different perspectives. The aim of this double-perspective is for a dialogue between theory and practice (which, in a traditional action-research situation may be missing) to ensue, with the researcher providing theoretically founded impulses which can then be developed and implemented by the teachers (cf. Kurtz: 61) and researcher in team-teaching units, reflected upon by the teachers in light of actual needs of the class and redeveloped and implemented in, as in traditional action research, a long-term cycle of action and reflection (Altrichter and Posch 2007: 17) -the distinction being that in this case, teachers and researcher are involved, from their different perspectives, to the same extent. This 'confrontation of different perspectives', according to Altrichter and Posch (2007: 18) serves to compare the different perceptions of teacher and researcher, with 'discrepancies' serving as food for mutual reflection and further action. The aim of this altered action-research concept is for it to thus function as in-service teacher-training, with the goal of, once the researcher-teacher collaboration ends, empowering the teachers to implement improvisational theatre as a means of promoting ICC independently in their different classes, adapting the concept to their various current needs, perhaps even leading on to teacher-led action-research.

Implementation of the Research Design:

The research study is currently being carried out in two *Gesamtschulen* in the federal state of Baden-Wuerttemberg in southern Germany. *Gesamtschulen* house the three school-types (*Werkrealschule, Realschule* and *Gymnasium*) pertaining to the German state-system under one roof. According to their academic performance, students are, depending on the school, streamed into one of the school-types either after the second or third year of secondary school (*Orientierungstufe*), with certain grades having to be achieved in a selection of subjects (one of them being English) in order to be streamed into either the *Realschule* (leaving age: 16) or the *Gymnasium* (leaving age: 19) system.

Through existing contacts with the two schools mentioned above, three classes volunteered to take part in the study during the 2012-2013 school-year. The study is now continuing into its second school-year with two classes thanks to the successful collaboration between teachers and researcher, with one teacher taking on a new class. Despite coming from similar types of school, the classes have different requirements and levels, making it possible to compare the potentials of improvisational theatre between the classes. The following table shows the individual features of each class.

Overview of the individual features of each class involved in the study

Class No.	**I**	**II**	**III**	**IV**
School	"M"		"H"	
English Level	'Top-set' class	'Lower-set' class	Mixed ability 'theatre class'	
English lessons	1 of 5 45-min. lessons per week		2 of 5 45-min. lessons per week	
School year and type	Yr. 6 and 7 *Orientierungsstufe*	Yr. 9 and 10 *Realschule*	Yr. 6 *Orientierungsstufe*	Yr. 5 *Orientierungsstufe*
School stream	Will be streamed into one of the school-types at the end of Yr. 7	Will take their high-stakes final English oral exam in Nov. 2013 and leave school in July 2014.	Were streamed into one of the three school-types at the end of the 2012/13 school-year	Will be streamed into one of the three school-types at the end of the 2014/15 school-year
Students	n=28	n=27	n=27	n=27
Age	11-13 in Yr. 6	14-16 in Yr. 9	11-13	10-12
Teachers	1 teacher 1 student teacher	1 teacher 2 student teacher interns	2 class teachers	2 class teachers 3 theatre students
Focus of the Improv-sessions	Textbook contents, grammar, and pragmalinguistic information	Spontaneous oral production in English, specific exam preparation	Project-work and theatre-skills as well as grammar revision and pragmalinguistic information	

With the study having begun in October 2012 with preliminary intensive workshops, the following structure for the 45/90 minute weekly improvisation sessions with variable contents according to the class requirements was established by the teachers and researcher after a 3-month first research-cycle.

Pre-task warm-up:

- greeting forms, tongue-twister, poem or sentences in varying emotions
- statues: Small groups of students build a "statue" to activate vocabulary on a current (textbook) topic
- small drill-improv: Focus on spontaneous use of a grammatical form, e.g. present progressive: "What Are You Doing?"

Main improvisational task: Improvised scene with a minimum of 2 students, for example:

- an improvised 4-line dialogue in varying emotions
- an improvised dialogue with an opening sentence as a prompt
- an improvised role-play with topical roles (e.g. antisocial teens)
- an improvised scene with role/setting/ object suggestions from the audience or including current vocabulary
- a spontaneous story /dialogue with two pupils using randomly picked current vocabulary.

Post-task cool down:

- whole group "statue" or "association chain" on a current topic
- whole group reflection on lesson

Teacher-Researcher Collaboration as In-Service-Teacher Training

In the following, the forms of teacher-researcher collaboration as an altered form of action-research during the implementation of the research design will be documented and their didactic potential for in-service teacher-training highlighted.

Begin of teacher-researcher collaboration

Through existing contacts between Heidelberg University of Education and the two *Gesamtschulen* mentioned above, it was possible to approach a broad spectrum of teachers. In the action-research vein, the three class teachers who were interested in collaborating on the project had identified particular challenges they were facing and were of the opinion that the concept of improvisational theatre to promote ICC could be helpful in overcoming these.

These challenges included:

- enabling students to speak freely and fluently whilst remaining grammatically and pragmatically accurate
- preparation of a lower-level, unmotivated class for the final high-stakes oral dialogical exam
- the successful integration of theatre methods in everyday English teaching whilst keeping up with syllabus/textbook demands.

These challenges were discussed between teachers and researcher and first "practical theories" (Ebd.: 16) were formulated on the potentials of improvisational theatre regarding these. Furthermore, the institutional parameters within which the collaboration could take place were established. Initially, each of the teachers was willing to use one of their five 45-minute lessons for the project.

Weekly Reflection, Alteration and Planning

In the initial planning phase, it was decided to have an 'intensive workshop phase', during which the researcher would introduce a series of basic improvisation games to the class and the teachers. This phase was considered as a first form of active input from the researcher in the role of a coach, for the teachers to experience the concept of improvisational theatre from a practical perspective. In school M, the intensive phase was made up of one 90-minute lesson, whereas in school H, with the theatre class, it was made up of a week of 5 daily 90-minute lessons. At the end of the intensive workshop phase, the two class teachers at school H decided to dedicate one 90-minute lesson a week instead of 45 minutes, as they were of the opinion that the longer time period would serve their purposes better.

After the intensive phase began, in November 2012, the long-term cyclical process of action and reflection (Ebd.: 17) on the part of both teachers and researcher which constitutes the foundations of the project. Each weekly lesson is planned both by the teachers and researchers in dialogue with each other, leading on from the preceding lesson, which is analysed in a feedback discussion. The various current needs of the respective class are discussed and ways in which improvisational theatre exercises can cater to these needs – either through the introduction of new exercises or through varied content input according to, for example, the textbook contents - are developed. In this way, the teachers are themselves, developing new ideas week to week (Ebd.: 16) which fit into their own institutional and curricular context in dialogue with the researcher, whose theoretical concepts are adapted in practice. The teachers' ideas, which can also be seen as practical theories forming the planning of the lesson, are then 'tested' in the next lesson -carried out in different forms of team-teaching between teachers and researchers- and reflected in continuation of the cycle.

Expert Interviews

The above-mentioned planning and implementation collaboration cycle between researcher and teachers is enriched through "semi-structured expert interviews" (cf. Pfadenhauer 2007), where the teacher is the expert, and which form part of the research methods pertaining to the study. These interviews aim to gain an insight into the teacher's expert opinion on the functioning and development of the study with regards to the specific challenges of the respective class and to the broader goal of promoting ICC in their respective institutional contexts. Carried out at different points in time in the study in addition to the weekly reflection sessions, the questions forming the semi-structured script of the interview focus on the general progression of the study, making it possible to, at the end of the study, trace its development from the teachers' perspectives.

Assessment

In all of the classes involved in the study, the improvisational sessions are/were used by the teachers to assess their students' oral skills both for summative and formative purposes, with two class teachers choosing to grant their students their complete oral mark on the basis of the weekly improvisational sessions as a whole. The third teacher gave the students their oral mark summatively, based on improvised role-plays acted out by groups of students at two points in time during the study which aimed at reflecting the contents of the final high-stakes oral exam. Although an aim of the study, it was not explicitly asked of the teachers to assess their students through the improvisational sessions – this happening much rather as a result of the dialogue and collaboration between the researcher and the teachers.

Internal Workshops

As an example of one further aim of action research, embedding individual research in a professional community context (cf. Altrichter and Posch 2007: 18), the teachers involved in the study organised internal workshops for their colleagues which are recognised as official in-service teacher-training units by the school. These workshops provided a further opportunity for a dialogue between theory and practice on two levels: On the one hand, teachers and researchers once again collaborated in planning and carrying out the workshop. On the other hand, the teachers involved in the project themselves initiated a theoretical discussion with their colleagues, who, during the workshop, in turn formulated practical theories based on their own teaching contexts and requirements which they would then apply in practice.

Outcomes for Teachers

The aim of the study described in this chapter is to develop a means of empowering teachers to investigate ways of dealing with the challenges presented by an increasingly competence-oriented, globalised educational landscape regarding English Language Teaching and, more specifically, in the case of this study, oral communicative skills.

The study has, as its object of research, a method, improvisational theatre, which, it is proposed, can promote ICC, a competence which has been declared as one of the pillars of foreign language teaching and which forms the basis of the CEFR. By implementing this method over a long-term, cyclical period of research in collaboration with the researcher, as described above, the teachers are involved in an altered form of action research, whose double goal, reflection and development (cf. ibid.: 21) is enriched by the teacher-researcher collaboration. This collaboration, and the ensuing dialogue between theory and practice, can be seen as a form of in-service teacher-training which takes the current needs of the teachers as a focal point. By playing an equal part in the implementation and development of the improvisational sessions, in a long-term cycle of action and reflection, the teachers are themselves empowered to carry on the implementation beyond the scope of the current research study independently. This will at school M, take place through one class teacher implementing improvisational sessions in another, younger class – in this way, applying the theoretical concept to another group of learners with differing requirements – and at school H, with the teachers taking on the complete planning and implementation of the improvisational sessions, which now feature as a 'subject' in the students' timetables, demonstrating that, to date, the project has had a successful outcome, with teacher-researcher collaboration yielding a stable concept for teachers to continue independently.

References

Altrichter, Herbert, Posch, Peter: Lehrerinnen und Lehrer erforschen ihren Unterricht 4. Edition, Bad Heilbrunn, Klinkhardt, 2007.

Byram, Michael: Teaching and Assessing Intercultural Communicative Competence, Clevedon, Multilingual Matters, 1997.

Council of Europe: Common European Framework of Reference for Languages: Learning, Teaching, Assessment, Cambridge, Cambridge University Press 2001.

Dörnyei, Zoltan: Research Methods in Applied Linguistics, Oxford, Oxford University Press, 2008.

Johnstone, Keith: Impro for Storytellers, London, Faber and Faber, 1999.

Kessler, Benedikt, Küppers, Almut: 'Dramapädagogik, interkulturelle Kompetenz und holistisches Fremdsprachenlernen', Scenario, Journal for Drama and Theatre in Foreign and Second Language Education 1/ 2008.

Kurtz, Jürgen: Improvisierendes Sprechen im Fremdsprachenunterricht, Eine Untersuchung zur Entwicklung spontansprachlicher Handlungskompetenz in der Zielsprache, Giessener Beiträge zur Fremdsprachendidaktik, Tübingen, Gunter Narr Verlag, 2001.

LoCastro, Virginia: Pragmatics for Language Educators, New York, Routledge, 2012.

Pfadenhauer, Michaela: Das Experteninterview, Ein Gespräch auf Augenhöhe. In: Buber, Renate, Holzmüller, Hartmut. (Eds.): Qualitative Marktforschung, Konzepte - Methoden - Analysen, Wiesbaden, Gabler, 2007, 449-461.

Sawyer, R. Keith: Creating Conversations, Improvisation in Everyday Discourse, Kreskill, Hampton Press Inc, 2001.

Spolin, Viola: Improvisation for the Theatre (Third edition), Evanston, Illinois, Northwestern University Press, 1999.

Vogt, Karin: Fremdsprachliche Kompetenzprofile, Giessener Beiträge zur Fremdsprachendidaktik, Tübingen, Gunter Narr Verlag, 2011.

Zuengler, Jane, Miller, Elizabeth R: 'Cognitive and Sociocultural Perspectives: Two Parallel SLA Worlds?' Tesol Quarterly, 1:14, 2006, 35-58.

War and Peace in Historical Education of Modern Lithuania

Aivas Ragauskas, Raimonda Ragauskienė

Abstract
The article is devoted to experience of teaching peaceful values in history in contemporary Lithuania. The author present tendencies of history teaching in Lithuania from 1990 to present. Lithuanian educational policy documents and historical textbooks are analysed (from 1990 to 2013) in the context of peace education.

Context

There are a lot and various explanations which describe the concept of *peace education* because it is fundamentally dynamic, interdisciplinary and multicultural. Peace education has its roots in the history of peace movements of the 19th and 20th century. The peak of peace education in many countries was in the 1980s, after the European Security Conference which was organized in Helsinki 1975. Later on, the actual peace education was replaced by other trends like human rights education. In the beginning of the 21st century, peace educators which were concerned about civil and domestic forms of violence developed a new form of peace education known as conflict resolutions education. All these and other different forms of peace education about the roots of violence and strategies for peace have teaching and learning in common. "One of the founding principles of peace education initiatives is that learner can develop a sense of possibility that enables them to become agents of social change"[1]. It aims to cultivate the knowledge, skills and attitudes needed to achieve and sustain a global culture of peace, and to create in the human consciousness a commitment to the ways of peace[2].

One of the key roles in constructing that understanding is played by historical education. History science and education has a direct effect on the human ego and social identity and can potentially undertake a more peaceful mission

---

1 M. Bajaj, Introduction, *Encyclopedia of Peace Education*, M. Bajaj (Ed.), Charlote, North Carolina: Information Age Publishing, 2008, p. 1-9.

2 Ian M. Harris, M. Lee Morrison, *Peace Education*, McFarland, 2003; *Critical Peace Education. Difficult Dialogues*, P. P. Trifonas, B. Wright (Eds.), Springer Science, 2013.

in the formation of the cultural codes for communities[3]. And in the last years peace education research has been examining history teaching more often. But the context of peace education programs in different countries is different[4].

We want to present only some results of those research. We have, for example, a study devoted to history teaching in a time of crisis in Spain. Based on the difficult situation of the 20th century, when Spain took part in two world wars and Spanish Civil War, professor Gemma Tribó Traveria suggests not to avoid talking about war. We have to remember that those wars resulted in more than 100 million deaths. But it is necessary to speak about a meaningless of the Spanish Civil War that still shapes their present days. The author suggests replacing some patriotic elements of school history, which evoke negative views of others, by more social contents[5]. Another study written by Mark Langager is about elements of war and peace in America's and Japan's history education. It reveals that teachers in both countries feel expected to teach mostly historical facts and find it difficult to teach about events when their countries had to engage in immoral acts. The difference between those two countries is that Japanese students are taught that the war was an immoral act which their country is historically guilty of, whereas American students are not necessarily taught like this6. There are also more articles suggesting to use 'historical empathy' that includes a peaceful understanding to individuals[7] and so on.

In this article we will try to present our experience of teaching peaceful values in history in contemporary Lithuania. The major aim was to analyse Lithuanian

3 M. Alpargu, E. Sahin, S. Yazici, Teaching History and Its Contribution to Peace, *International Journal of Social Inquiry*, 2009, vol. 2, nr. 2, p. 201.

4 Paul R. Carr, Brad J. Porfilio, Introduction: Framing Peace and War Within the Educational project – Willful Dis(Engagement) and the Meaning (and Cost) of Conflict, *Educating for Peace in a Time of a "Permanent War". Are Schools part of the Solution or the Problem?* P. R. Carr, B. J. Porfilio (Eds.), Routledge, New York, 2012, p. 16; Ian M. Harris, L. J. Fisk, C. Rank, A portrait of university peace studies in North America and Western Europe at the end of the millennium, *The International Journal for Peace Studies*, 1998, January, Vol. 3, Nr. 1.

5 G. Tribó Traveria, History teaching and peace education in a time of crisis. How can we build a shared muture based on dialogue and peace? *Policy Paper*, 2013, No. 9, September, p. 1-6.

6 M. Langager, Elements of war and peace in history education in the US and Japan: a case study comparison, *Journal of Peace Education*, 2009, vol. 6, issue 1, p. 119-136.

7 M. Alpargu, E. Sahin, S. Yazici, Teaching History and Its Contribution to Peace, *International Journal of Social Inquiry*, 2009, vol. 2, nr. 2, p. 199-214.

educational policy documents and historical textbooks (from 1990 to 2013) in the context of peace education. A short view of historical teaching tendencies in Soviet Lithuania will be presented as well. Content analyses methodology was used for gathering data.

War and Peace in Historical Education of Modern Lithuania: Analysis

There is little ***Lithuanian research literature*** on history teaching and peace education. We have several major research: historical and sociological articles. Evaldas Bakonis was one of the first in Lithuania who examined historical textbooks in 2000. He raised a question: "Are Lithuanian historical textbooks peaceful?" Nevertheless, he did not apply a concept of peace education in his article but he proposed to use quantitative as well qualitative methods for deeper analysis. But even that data what he gathered allowed him to make conclusions about militarization of our historical textbooks which were published before 20008. In the other article the same author analyses aspects of multiculturalism in our historical, civic education and geographic textbooks. He suggested for textbooks authors to pay more attention to histories of women, ethnic minorities or religious confessions9. Irene Andersson and Irena Zaleskienė conceptualised peace education from social, historical and educational perspectives and presented case studies from Lithuania and Sweden10. Their article was based on the findings from the comparative research 'Peace Education and teacher training: today in the context of yesterday. A comparative perspective from Finland, Lithuania and Sweden'.

8 E. Bakonis, Ar „taikūs“ lietuviški istorijos vadovėliai? Taikos klausimai lietuviškame istorijos vadovėlyje, *Istorija*, 2000, t. 45, p. 60-65. (Are our Lithuanian history textbooks 'peaceful'? Questions about peace in Lithuanian textbook).

9 E. Bakonis, Socialinių mokslų vadovėliai. *Multikultūrinių vertybių skaida mokykliniuose vadovėliuose*, sudarė T. Tamošiūnas, 2000, accessed 5 February 2014, available from http://skf.osf.lt/old/FK-BL-05.htm. (Textbooks of social sciences. Of multiculturial values in school textbooks).

10 I. Andersson, I. Zaleskienė, Education for peaceful life: case studies from Lithuania and Sweden, *Socialinis ugdymas*, ed. G. Kvieskienė, Vilnius, 2011, nr. 16 (27): *Vaikų ir suaugusiųjų socializacija: tyrimai ir patirtis (Socialization of children and adults: research and experience)*, p. 5-18.

We have much more research devoted to citizenship, civil society education and place of history in those processes[11]. But all of them are only partly related with history teaching or peace education. So in this article we will not talk about them.

History teaching and peace education in Soviet Lithuania (until 1990)

In brief, history teaching and peace education in Soviet Lithuania (until 1990) as a concept did not have strong traditions in education system. Nevertheless, they were mentioned as a general idea in the history programs or textbooks for secondary schools and teacher training institutions. But really Soviet Lithuania education concept was based on an old Roman phrase 'Peace through strength' (si vis parem, para bellum/ if you desire peace, prepare for war)[12]. So history textbooks were rather teaching about war than about peace. Students had to know a lot about heroic aspects of the won wars, to praise military heroes (especially soviet military heroes), places related to military events, other related facts but not to analyse reasons of those wars. The contribution of peace makers was ignored as well. Besides, the information about war events was very often incorrect[13]. Even without a deep analysis it seems clear that education of this time could be named as war education rather than peace.

History teaching and peace education in Lithuania 1990-2002

Practically this period covers a little bit more than a decade of Lithuanian independence. We have chosen this period for a reason. In 1990 Lithuanian statehood was restored and some important political steps towards strengthening peaceful environment for the members of society were taken by Lithuanian politicians.

---

11 *Švietimo problemos analizė*, 2013, may, Vol. 3 (87), Pilietinio ugdymo įgyvendinimo politika: tikslai, priemonės, rezultatai (The realization of the civil society education policy: purposes, measures, results), accessed 20 January 2014, available from http://www.nmva.smm.lt/wp-content/uploads/2012/12/Pilietinio-ugdymo-igyvendinimo-politika.pdf.

12 Ian M. Harris, M. Lee Morrison, op. cit., p. 18.

13 I. Andersson, I. Zaleskienė, op. cit.

1. Educational policy documents. Ministry of Education and Science of Lithuania took some initiatives together with UNESCO to prepare some textbooks and teacher manuals for peace education at secondary schools. The published books were simply called 'Peace Education' (1996)[14]. From 1992 to 2002 there was a new law on education. This law stated that a student had a right: 'to study in a psychologically, emotionally and physically safe environment based on mutual respect' (Article 46). The same ideas are met in our curriculum of history subject (in 1999). It was proclaimed in the introduction that all history teaching must be based on 'ideals of democracy, humanism, tolerance, solidarity and cooperation'. Besides there was a special program 'War and peace'. It focused on preventing aggression and genocide and revealing the role of Lithuania and Lithuanians in peace movements[15]. Content analysis of the curriculum helped me to identify some elements of the curriculum, related to peace education. If you look carefully through all this curriculum, you will notice that the word 'peace' was used more often than the word 'war'.

Table 1: The proportion of the words war/peace in historical curriculum (1999)[16]

Form	War	Peace
V	3	0
VI	3	0
VII	2	0
VIII	2	0
IX	7	0
X	15	3
Total	32	3

2. Historical textbooks. Almost similar situation can be found in our historical textbooks from that period. From the restoration of Lithuanian statehood until 2002 old and newly published history textbooks were rewritten. A lot even 110 new historical textbooks were published during the 13 years (on average

14 *Ibid.*
15 *Bendrojo lavinimo mokyklos programos. Istorija V-XII klasė*, Vilnius, 1999, p. 3, 69. (Secondary school curiccula. History for V-XII forms).
16 E. Bakonis, op. cit., p. 61.

8 historical textbooks per year)[17]. And 4 textbooks written in 1998-1999 were analysed more deeply in terms of historical teaching and peace education.

Historical textbooks analysed in terms of historical teaching and peace education

1. A. Kasperavičius, R. Jokimaitis, *Naujausiųjų laikų istorija*, Vadovėlis 10 klasei, Vilnius: Kronta, 1998; (*The history of modern times*, Textbook for 10 form)
2. J. Brazauskas, S. Jurkevičius, K. Petrauskis, *Naujųjų amžių istorija. Nuo Didžiosios prancūzų revoliucijos iki pirmojo pasaulinio karo pabaigos.* Vadovėlis 9 klasei, Vilnius: Kronta, 1999; (T*he history of modern times. From the great French revolution until the end of the First World War.* Textbook for 9 form).
3. R. Jokimaitis, A. Kasperavičius, E. Manelis, B. Stukienė, P*asaulio ir Lietuvos istorija VI-XVIII amžiai.* Vadovėlis 8 klasei, Vilnius: Kronta, 1999; (*The history of the world and Lithuania in the VI-XVIII centuries.* Textbook for 8 form).
4. B. Šetkus, L. Pobedinska, *Senovės istorija*, Vilnius: Kronta, 1998. (*The ancient history*).

In the textbooks a lot attention was paid to discuss peace agreements; however, peace movements were completely ignored. We have much more topics of peace from the XXth century history. Often in the discussions of these topics words 'but' or 'however' brushed away all attempts to emphasize the importance of peace. For example, in the sentence: 'in 1947 peace treatments were signed, but they changed the political map of Europe very little'. Moreover, war was even praised. We can see the in the following sentence: 'Lithuanian welcomed the War [the Second World War] with joy because it saved them from the soviet terror'. Similar situation can be found in 1783 when Crimea was annexed to Russia. Crimea was named as 'robbers' lair from which Tatars devastated the Grand Duchy of Lithuania for some centuries'.

The proportion of the words war/peace in paragraphs of these textbooks shows that the pro-war education prevailed. We can conclude that historical teaching in that time was still more military.

---

17 A. Svėrienė, Istorijos vadovėlio kultūra (1999–2002 m.), *Knygotyra*, 2004, t. 43, p. 9-24. (Culture of the Lithuanian history textbook, 1990–2002).

Table 2: The proportion of the words war/peace in paragraphs of the textbooks (1998-1999)[18]

Form	Paragraphs	War	Peace
VII	58	7	1
VIII	54	4	0
IX	48	7	3
X	54	19	1
Total	214	37	5

History teaching and peace education in Lithuania 2003-present

1. Educational policy documents. A New national Education Strategy of 2003-2012 and Lithuanian laws on education (2003, 2011) were adopted by Lithuanian Parliament[19]. Revised curricula and the program of history exam (2011) were presented[20]. There are no significant differences between previous and contemporary curriculum in terms of using words 'war' and 'peace'. In general educational policy documents showed that peace education in contemporary Lithuanian society is socially based, i.e.:
   - conflict is seen as the clash of differences (ideologies, ideas, actions, etc.);
   - people are so different that it is not possible to avoid conflicts;
   - people must live together, because of one shared world;
   - people have to learn how to solve conflicts[21].

2. Historical textbooks Since 2003 until today 110 historical textbooks (for all forms) have been published in Lithuania. It is significant that in 2003 the requirements for textbook were approved[22]. So contemporary historical textbooks in context of peace education are much more different from previous textbooks. This is clear from the analysis of some of them. We selected

18 According to E. Bakonis' analysis, E. Bakonis, op. cit., p. 63.
19 Accessed 15 February 2014, available from *http://www3.lrs.lt/pls/inter3/dokpaieska.showdoc_l?p_id=395105&p_query=&p_tr2*
20 Accessed 15 February 2014, available from *http://www.nec.lt/failai/2804_1.11_Istorijos_brandos_ezamino_programos_priedas.pdf*
21 I. Andersson, I. Zaleskienė, op. cit.
22 A. Svėrienė, Teoriniai mokyklinio vadovėlio aspektai, *Knygotyra*, 2006, t. 47, p. 28-46. (Teoretical aspects of school textbook).

5 historical textbooks for 10th and 11th forms, published in the period from 2007 to 2010, and examined their contents, the titles of paragraphs as well as the quality of historical material.

The list of selected Lithuanian historic textbooks published in the period of 2007-2010

1. I. Kapleris, A. Meištas, K. Mickevičius, A. Laužikienė, Ž. Tamkutonutė-Mikailienė, *Laikas*, Istorijos vadovėlis 10 klasei, I dalis, Vilnius: Briedis, 2007. (*Time*, part I, History textbook for 10 form).
2. I. Kapleris, A. Meištas, K. Mickevičius, A. Laužikienė, Ž. Tamkutonutė-Mikailienė, *Laikas*, Istorijos vadovėlis 10 klasei, II dalis, Vilnius: Briedis, 2007. (*Time*, part II, History textbook for 10 form).
3. M. Tamošaitis, *Istorijos vadovėlis 11 klasei*, II dalis, Vilnius: Baltos lankos, 2009. (*History textbook* for 11 form, part II).
4. I. Kapleris, R. Laužikas, A. Meištas, K. Mickevičius, M. Žolynas, *Laikas*, Istorijos vadovėlis 11 klasei, I dalis, Vilnius: Briedis, 2010. (*Time*, part I, History textbook for 11 form).
5. I. Kapleris, R. Laužikas, A. Meištas, K. Mickevičius, M. Žolynas, *Laikas*, Istorijos vadovėlis 11 klasei, II dalis, Vilnius: Briedis, 2010. (*Time*, part II, History textbook for 11 form).

Table 3: The historical textbooks (2004-2013)[23]

Form	2004/05	2005/06	2006/07	2007/08	2008/09	2009/10	2010/11	2011/12	2012/13	Total
V	7	0	0	1	2	1	0	0	0	11
VI	10	0	0	0	1	2	1	0	0	14
VII	9	3	0	0	1	1	2	1	1	18
VIII	9	1	2	0	0	1	3	1	0	17
IX	3	0	1	0	1	2	0	0	1	8
X	8	0	0	0	0	1	2	0	0	11
XI	14	0	0	1	1	1	2	1	1	21
										100

We would like to talk a bit about titles and contents in those textbooks. On the one hand, the words 'peace' and 'war' almost disappeared from them. These

23 This data was gathered by Rūta Šetkuvienė (Lithuanian University of Educational Sciences, Faculty of History, Centre of History didactics).

words were changed by other words with similar meanings, like *conflict, tragedy* or *culture, daily life* etc. On the other hand, in textbooks historical events are still dated according to wars, for example paragraph 'Lithuania in interwar period' instead of 'Lithuania in the period of 1919-1941'; or 'Lithuania in the storm of war' instead of 'Lithuania in the period of 1941-1945'.

Table 4: The proportion of the words war/peace in paragraphs of the textbooks (2007-2010)

Textbooks	1.	2.	3.	4.	5.
The number of the paragraphs	23	23	28	24	24
War	7	1	0	0	0
Peace	0	1	0	0	0

A lot has been done to make historical teaching more peaceful. In the textbooks a lot attention is paid to analyse reasons of wars; however, the importance of peace is emphasized both implicitly and explicitly. For example, the following sentence can illustrate this: 'Millions of Europeans met the end of the First World War with great joy'. Peaceful movements such like Lithuanian resistance movement 'Sajudis', peaceful resistance way to Baltic Independence 'Baltic way' or the movement now known as the Singing Revolution and other similar examples are used.

It is very good that we have much more peaceful and interesting topics in each of these textbooks such as the topic of the favorite game in Lithuania – basketball – and how Lithuania became a country of it or the topic of historical humor. It is important that historical anecdotes or caricatures rather show tragedy of soviet time than tell the situation using mere facts. For example, "It is Sunday morning in Lithuanian city in 1940. Some Red army soldiers are walking near the church and see beggers begging at the church's door. With a big surprise one solder says to another one: "Look Vasia, there are collective farmers in Lithuania as well". (This anecdote shows that collective farmers in the Soviet Union lived like Lithuanian beggers and that was true).

Although it is normal to leave some historical topics on military like dictatorships, crimes against humanity, crusades, revolutions or to speak about two World Wars. It is inevitable and this has been done.

Conclusions

1. In Soviet Lithuania peace education could be title as a war education rather than peace education. A decade after 1990s pro-war education still prevailed.

And only in the last decade we can observe a significant progress in terms of historical teaching and peace education.

2. There are two main reasons why contemporary Lithuanian history textbooks became more peaceful. First of all, stability of Lithuanian political situation and new strategies of national education appeared. There were some changes in Lithuanian history science. We have a lot of a new research devoted not only to political/war history but to social history as well as to national minorities, daily life and so on.
3. Development has no limits. Our historical teaching is no exeption. We still need to introduce women into history, to speak more about daily life in the past, to pay more attention to multicultural process along with other similar topics.

Gewaltlos glücklich

Helmut Wehr

Abstract
Violence appears as the expression of missing personal strength and resilience. This refers to deficits in education and shows the frustration of basic needs: relation/bonding, (self-) control/orientation, positive self-esteem, positive balance of lust and listlessness and elements of healthiness (sense of coherence: manageability, knowability and meaningfulness). Therefore, empowerment appears, like it is offered in a school subject "luck/happiness" in Heidelberg, as a positive possibility to make children and youngsters "strong" that they can renounce on account of own personal strength a violent interest penetration. How Pippi Langstrumpf and Tom Saywer own their differentiated alternative potentials to solve problems and conflicts and they are competent in realising these in everyday life.

Zusammenfassung
Gewalt erscheint als Ausdruck von fehlender personaler Wirkkraft und Resilienz. Dies verweist auf sozialisatorische Defizite und Frustration grundlegender Bedürfnisse: Bindung-Beziehung, (Selbst-)Kontrolle/Orientierung, positives Selbstwertgefühl, positive Lust-Unlust-Balance) und Gesundheitsfaktoren (Kohärenz: Handhabbarkeit, Verstehbarkeit und Sinnhaftigkeit). Demnach erscheint Empowerment, wie es im Schulfach „Glück" in Heidelberg angeboten wird, als positive Möglichkeit, Kinder und Jugendliche „stark zu machen", dass sie aufgrund eigener persönlicher Stärke auf eine gewalttätige Interessendurchsetzung verzichten können. Wie Pippi Langstrumpf und Tom Saywer eigenen sie sich differenzierte alternative Konfliktlösemechanismen an und realisieren diese im Alltag.

I

Zu allererst erscheint der Titel widersprüchlich, unpassend, inhaltlich völlig entgegengesetzt. Die Zusammenhänge möchte ich im Folgenden aufweisen, denn wir kommen nicht umhin, alltägliche Gewalt, vor allem auch an Schulen zur Kenntnis nehmen zu müssen. „Gewalt ist die absichtliche Herbeiführung [...] von physischem, psychischem, materiellem, sozialem, ökologischem Schaden[1]

- Physische Gewalt (Schlagen, Treten,
- Psychische Gewalt (u.a. Schlafentzug, Erniedrigungen, Demütigungen, Drohungen, Mobbing, Cyber-Mobbing, Gerüchte verbreiten, "teilen" [Facebook],)

1 Krüger, Udo Michael: Gewalt in Informationssendungen und Reality-TV. Quantitative und qualitative Unterschiede im öffentlich-rechtlichen und privaten Fernsehen, in: Media Perspektiven 94 (2), 1994, S. 72-85, hier S. *73).*

- Soziale Gewalt (u.a. Einsperrungen, Kontaktverbote, soziale Isolierung, Ignorieren,..)
- Sexualisierte Gewalt (u.a. Vergewaltigung, sexuelle Belästigung, Zwang zu sexuellen Handlungen, sexueller Missbrauch,)
- Ökonomische Gewalt (u.a. kein Zugang zum Familieneinkommen, Verbot der Erwerbstätigkeit, Ausschluss vom Sozialleistungen, Entzug der Leistungen des Bildungspaketes,..
- Strukturelle Gewalt (Verringerung von Chancen, z.B. durch schlechte Noten, diskriminierende Vorurteile, Sitzen lassen, Schulverweis, Lernumgebung: zu laut, reizüberflutend…).

Folgen wir dieser Vorstellung von Gewalt, so wäre der nächste Schritt, den Ursachen gewalttätigen Verhaltens auf die Spur zu kommen. Diese können wir mit Krüger (1994, S. 72-85.) in folgenden acht Schwerpunktmotiven verorten: schlechte Noten und Probleme in der Schule, normative Anomie und soziale Vereisung, intensive Nutzung gewalthaltiger Medien, elterliches Vorbild verstärkt kindliches Aggressionshandeln, permissives und inkonsequentes elterliches Erzieherverhalten oder ein dominant machtbetonter Erziehungsstil führen zur positiven Bewertung von Aggressiven (medialen, sozialen) Vorbildern, das von devianten (kriminellen) Peers stabilisiert wird. Verstärkt wird dieser Effekt durch gesellschaftliche Erfahrungen von Enttraditionalisierung, Risiko der Übergänge, Desintegration und Individualisierung.[2] Diese Sozialisationserfahrungen haben nolens volens psychisch mögliche Defizite in der kindlichen Psyche zur Folge.

So machen viele Autoren darauf aufmerksam, dass gewalttätige Jugendliche durch spezifische Charakteristika auf sich aufmerksam machen. Ins Auge fällt, dass sie häufig physisch überdurchschnittlich stark sind und durchaus einen ausgeprägten Machtwillen zeigen, dem gegenüber steht aber oft ein geringes Selbstbewusstsein und eine wenig ausgeprägte Selbst- und Impulskontrolle. Da sie zugleich auch über eine geringe Anzahl von Konfliktlösungsmuster verfügen, kann aufgrund einer unrealistischen Fremdwahrnehmung eine fast beliebige Äußerung zur "Provokation" werden, auf die mit einem Wutausbruck reagiert und damit inadäquat geantwortet wird. Die Empathie-Schwäche gekoppelt mit einer niedrigen Hemmschwelle sorgt für eine schnelle "Entsicherung" von gewalttätiger Aggressionsimpulsen. Den Kindern/Jugendlichen entgleitet ihre emotionale Aggression, ihre Wut, ein Kontrollverlust tritt ein. Das rationale Ich

---

2 vgl. Keupp, Heiner: Bildung zwischen Anpassung und Lebenskunst. Ressourcen der Lebensbewältigung. Vortrag auf dem „Tag der Weiterbildung" am 03. März 2006 in St. Johann im Pongau.

wird von emotionalen Impulsen überschwemmt, zumindest im Augenblick beherrscht. Hier geht er "aufrechte Gang" verloren.

Dies ist umso gravierender, als gerade (Selbst-) Kontrolle zu den vier Grundbedürfnissen eines gelungenen Lebens gehört. Die folgenden vier psychischen Grundbedürfnisse des Kindes/Jugendlichen müssen im Sinne eines gelungenen, gesunden Lebens befriedigt sein, damit Heranwachsende eine effektive Konflikt- und Problemlösekompetenz zeigen können: Ein Bedürfnis nach Orientierung und (Selbst-)Kontrolle: Hier artikuliert sich das menschliche Bestreben sich selbst und die Welt zu verstehen und sich kompetent und selbstbestimmt in ihr bewegen zu können. Hinzu kommt das Bedürfnis nach Selbstwerterhöhung/-schutz, was nicht anderes bedeutet, als das Bedürfnis nach selbstaufwertenden Wahrnehmungen und einem positiven Selbstbild. Dies schließt ein, dass das Bedürfnis nach Lustgewinn/Unlustvermeidung gestillt wird. Auf jeden Fall intendiert das Kind/der-die Heranwachsende eine positive Lust-Unlust-Bilanz ziehen zu können. Hierzu kann durchaus die erfolgreiche Bedürfnisbefriedigung nach Bindung, d.h. das grundlegende Angewiesenheit des Kindes/Jugendlichen auf Mitmenschen und Nähe eine wesentliche Verwurzelung aufweisen. Die konsistente und positive Befriedigung der vier Grundbedürfnisse erscheint Grawe[3] als die Grundlage eines adäquaten Antwortens auf Stimuli aus der Umwelt. Demnach reagiert ein Kind dann auf eine fragende oder kritische Kommunikation oder Interaktion eines anderen dann angemessen, wenn er die entsprechenden Grundbedürfnisse befriedigt und die entsprechenden Kompetenzen und Potenzen als Ressourcen aufbauen konnte.

Potenzen, um aggressive „Entsicherungen"[4] und Wutausbrüche zu verhindern sind die in der (primären) Sozialisation aufgebauten Ressourcen": d.h. „alle Möglichkeiten [eines] Menschen zur Befriedigung seiner Grundbedürfnisse"

---

3 Grawe, Klaus: Neuropsychotherapie. Göttingen 2004; Rupp, Robert: Bedürfnisorientierung als gesundheitserzieherisches Grundprinzip, in: Knörzer, Wolfgang; Rupp, Robert (Hrsg.): Gesundheit ist nicht alles – was ist sie dann? Baltmannsweiler 2011, S. 20-33, hier S. 7.

4 vgl. Heitmeyer, Wilhelm: Der doppelte Kontrollverlust. Jugendliche Massenmörder: Wer die Ursachen von Amokläufen verstehen will, muss die Bedingungen beachten, unter denen Jugendliche aufwachsen - und die Konsequenzen mangelnder Anerkennung, in: bildung & wissenschaft, Juni 2009, S. 18-20; Heitmeyer, Wilhelm: Entsicherung, Individualisierung und Gewalt. In: Beck, U./Beck-Gernsheim, E. (Hrsg.).: Riskante Freiheiten. Frankfurt a.M. 1994; Heitmeyer, Wilhelm: Freigesetzte Gewalt. Gewalt als Bearbeitungsform einer neuen Unübersichtlichkeit. In: Pädagogik, H. 6, 1994, S. 35-40.

(S. 6) und hierzu gehört via Bedürfnis nach (Selbst-) Kontrolle auch der Impuls, die eigene Aggression und Wut unter Kontrolle des Realitäts-Ich zu halten, um die Lust-Unlust-Balance nicht ins Negative kippen zu lassen. Dies hat insofern für die Psyche und deren Lust-Unlust-Bilanzierung Bedeutung, als die psychischen Prozesse, also auch die aggressiven Impulse sowohl im Inneren einer Person dem Gesundheitsziel zustreben (Konsistenzstreben), als auch das physiologisch-biologische System, das kindlich-jugendliche Individuum, in seiner Interaktion mit der biologischen und sozialen Umwelt nach Kongruenz strebt.

Hier sind Brüche, seien es Wutausbrücke, Gewalt-Entgleisungen, komatöses Zudröhnen[5] dem Streben nach Kohärenz kontraproduktiv entgegengesetzt. Dies wird deutlich, wenn wir uns des Gesundheitsbegriffes von Antonovsky vergewissern. Dieser versteht "Gesundheit" in einem dialektisch-polaren Sinne nicht als einander ausschließende Polarität, sondern als synthetisch im Alltag vorkommende "Mischungen". Demnach ist niemand ganz "gesund" oder ganz "krank". Beides ist immer im anderen enthalten. "Entsicherung" von Gewalthandlungen hieße demnach im Sinne Heitmeyers ein doppelter Kontrollverlust, der Verlust sozialer Anerkennung (Bedürfnis nach Bindung, Selbsterhöhung) und der Zerfall von Selbstkontrolle eigener Lebensgestaltung (Bedürfnis nach Orientierung und Kontrolle), es droht Orientierungsverlust, in gravierenden Fällen durch deviante Peers stabilisierte Desintegration.[6] Heitmeyer radikalisiert dies insofern er davon ausgeht, dass jede Person immer darum bemüht sei, eine positive Identität aufzubauen und zu sichern. Gewaltausübung dagegen stelle eine Ohnmachts- und Unterlegenheitssituation dar, die es im Sinne eigenen gelungenen Lebens zu überwinden gilt. Damit dies ermöglicht werden kann, sind Kompetenzen zur Bewältigung solcher Konflikte aufzubauen – durch Empowerment[7], also den partizipativen Aufbau von Selbstgestaltungskräften im Heranwachsenden, womit den Bedürfnis nach (Selbst-)Kontrolle (Grawe) und dem kohärenten Handeln (Antonowsky) Genüge getan sei. Beide Aspekte sind gleichermaßen zu betrachten.[8]

---

5 Verres, Rolf: Drogen oder die Sehnsucht nach dem Glück, in: Knörzer, Wolfgang; Rupp, Robert (Hrsg.) Gesundheit ist nicht alles – was ist sie dann? Baltmannsweiler 2011, S. 127-136.

6 siehe Anm. 3.

7 Hintermair, Manfred: Empowerment als konzeptioneller Orientierungsrahmen für eine gesundheitsförderliche Lebensführung, in: Knörzer, Wolfgang; Rupp, Robert (Hrsg.): Gesundheit ist nicht alles - was ist sie dann? Baltmannsweiler 2011, S. 46-56.

8 Rupp, Robert: Bedürfnisorientierung als gesundheitserzieherisches Grundprinzip, in: Knörzer, Wolfgang; Rupp, Robert (Hrsg.): Gesundheit ist nicht alles - was ist sie dann? Baltmannsweiler 2011, S. 20-33, bes. S.10

Dieses von Antonovsky beschriebene Kohärenzgefühl wird geprägt durch: Machbarkeit: das Ausmaß, in dem man wahrnimmt, dass man geeignete Ressourcen zur Verfügung hat, um den alltäglichen sozioemotionalen, kognitiven und schulischen Anforderungen effektiv zu begegnen. Sinnhaftigkeit: das Ausmaß, in dem man das eigene Leben als emotional sinnvoll empfindet, was durchaus mit den grundbefriedigten Impulsen nach Orientierung/Selbst-Kontrolle, Bindung und einem positiven Selbstbild im Sinne Grawes zusammenhängt. Als kognitive Dimension schließt sich die Notwendigkeit an, Konsistenz durch Verstehbarkeit zu realisieren: Situationen und Reize als geordnete, strukturierte Informationen verarbeiten zu können. Sind diese drei Dimensionen kohärent-schlüssig im kindlich-jugendlichen Lebensalltag repräsentiert, so ergibt sich das Phänomen der Gesundheit. Diese ist allerdings auf "Kraftquellen" der Resilienz[9], auf prosoziale Teilhabe und damit Entwicklung von psychosozialen Stärken, Kompetenzen angewiesen.

Als externe Ressource erscheint hiermit nur kompatibel eine Kultur der Ermutigung, der Anerkennung als notwendige Antwort auf die Frage nach der Prävention von Gewalt.

II

Kinder/Jugendliche müssen für sich selbst folgende Fragen positiv und ermutigend, förderlich beantworten können[10]: Wer braucht mich? Wer hört mir zu? Wozu, wohin gehöre ich? Bin ich gleich wert? Werde ich gerecht und fair behandelt? Werden meine Gefühle akzeptiert? Wer schätzt und respektiert mich?

Die Anerkennung von Personen und der Respekt von Normen stabilisieren sich gegenseitig. Dieser Prozess ist allerdings äußerst störanfällig, wenn Lehrer oder Eltern das Gerechtigkeitsgefühl verletzen. Die Folgen: Soziale Bindungen, emotionaler Rückhalt können verloren gehen. Hier muss sich das von

---

9 Hintermair, Manfred: Empowerment als konzeptioneller Orientierungsrahmen für eine gesundheitsförderliche Lebensführung, in: Knörzer, Wolfgang; Rupp, Robert (Hrsg.): Gesundheit ist nicht alles – was ist sie dann? Baltmannsweiler 2011, S. 46-56, hier S. 52, und Göppel, Rolf: Resilienzförderung als Aufgabe der Gesundheitspädagogik?, in: Knörzer, Wolfgang; Rupp, Robert (Hrsg.): Gesundheit ist nicht alles - was ist sie dann? Baltmannsweiler 2011, S. 57-68.

10 Heitmeyer, Wilhelm: Der doppelte Kontrollverlust. Jugendliche Massenmörder: Wer die Ursachen von Amokläufen verstehen will, muss die Bedingungen beachten, unter denen Jugendliche aufwachsen - und die Konsequenzen mangelnder Anerkennung, in: bildung & wissenschaft, Juni 2009, S. 18-20.

Hurrelmann als „magisches Dreieck“[11] bezeichnete feinfühlige Erzieherverhalten auch im Schulalltag als anerkennend, anregend und anleitend konkretisieren, denn dieses muss die Kinder/Jugendlichen „stark machen“, sie resilient werden lassen, damit es ihnen gelingt, Gewaltausübung zur Interessen-, Meinungs- und Bedürfnisartikulation eine Absage erteilen zu können, sich von Gewalttätern abgrenzen zu können, selbst zwischen aggressiver und gewalthaltiger Interaktion differenzieren zu können, Boundary-Management-Kompetenzen im Sinne Keupps auszuprägen. Um externe und interne Resilienzfaktoren zu harmonisieren, sind Kindern folgende Punkte wertvoll:

Liste guten Lebens für Kinder, die sich als „glücklich“ einschätzen*[12]*:

---

11 Hurrelmann, Klaus: Zu viel für manche kleine Seele. Trend zu immer mehr psychosomatischen und mentalen Störungen bei Kindern und Jugendlichen. Gefragt sind nicht nur neue Medikamente, sondern vor allem kompetente Eltern, in: http://nbn-resolving.de/urn:nbn:de:0168-ssoar-202197 [10.12.13]. Erschreckend, welche Zahlen die Gewaltstudie 2013: Gewalt- und Missachtungserfahrungen von Kindern und Jugendlichen in Deutschland, empirisch belegt. Hierbei spielen auch sozioökonomische Bedingungen eine wesentliche Rolle, so zeigt sich, dass rund 40% der "sozial benachteiligten Kinder" Gewalterfahrungen in der Familie machen, 45% sich in der Schule missachtet fühlen. Auch die Mobbingerfahrungen sind relevant höher als bei "privilegierten Kindern". Hier setzt die Stiftung Kinderjahre ein, sie verfolgt folgende Ziele: Chancengleichheit für benachteiligte Kinder ….dass sie auch „Glück“ erfahren und erlernen können, z.B. durch das Projekt „Schulfach Glück“. http://www.stiftung-kinderjahre.de/ziele.html [12.12.13].

12 vgl. GEOlino-Kinderwerte-Monitor 2010: Repräsentative Untersuchung des Kindermagazins GEOlino in Zusammenarbeit mit UNICEF, in: http://www.geo.de/_components/GEO/info/presse/files/2010/kinderwerte/Praesentation_PK_final.pdf[10.12.13]

Ähnlichkeiten zur Liste des guten Lebens von Martha Nussbaum sind deutlich sichtbar*):

1. Die Fähigkeit, ein volles Menschenleben bis zum Ende zu führen;
2. Gesundheit insbesondere als Ernährung, Wohnen, Sexualität und Mobilität;
3. Fähigkeit, unnötigen Schmerz zu vermeiden und freudvolle Erlebnisse zu haben;
4. Fähigkeit, fünf Sinne zu benutzen, sich etwas vorstellen und denken zu können;
5. Bindungen zu Dingen und Personen einzugehen, zu lieben, zu trauern, Sehnsucht und Dankbarkeit zu empfinden;
6. Sich Vorstellungen vom Guten zu machen und kritisch über die eigene Lebensplanung nachzudenken;
7. Für andere und bezogen auf andere zu leben, verschiedene Formen familiärer und sozialer Beziehungen einzugehen;
8. Verbundenheit mit Tieren und Pflanzen und der ganzen Natur zu (er-)leben;
9. Fähigkeit zu lachen, zu spielen und Freude an Erholung zu haben;

- Freundschaft 76%
- Familie 75%
- Geborgenheit 58%
- Ehrlichkeit 52%
- Bildung 50%
- Vertrauen 57%
- Zuverlässigkeit 51%
- Gerechtigkeit 46%
- Hilfsbereitschaft (Freunde, Tiere) 78%
- ohne Gewalt aufwachsen 81%

„Freundschaft" und „Familie" sind gegenwärtig die wichtigsten Werte, „Respekt" und „Vertrauen" legen an Bedeutung zu, die Bereitschaft der Kinder zu sozialem Engagement wächst, die Kinder sehen den Beruf der Eltern positiv, wünschen sich aber mehr gemeinsam verbrachte Zeit. Dies zeigt die hohe Relevanz des Konzeptes der „einbettenden Kulturen"[13] für familiäre und außerfamiliäre, also schulische Netzwerke für den Prozess einer gelingenden Identitätsarbeit bei Heranwachsenden. Dies kann im Sinne von Modellen selbstwirksamer Lebensprojekte erfolgen.

Hinzu kommt, dass Keupp mit Bourdieu Bildung als „symbolisches Kapital" anerkennt, als eine Ressource, die Heranwachsende in die Lage versetzt, selbstbewusst und -gesteuert den eigenen Lebensentwurf zu entwickeln und immer wieder zu reflektieren und umzugestalten. In diesem Sinne hat Bildung viel mit (selbst-)kontrollierter Lebensbewältigung, Lebenskompetenzen[14] und

---

10. Das eigene Leben und nicht das eines anderen zu leben;
11. Fähigkeit, sein eigenes Leben in seiner eigenen Umgebung und seinem eigenen Kontext zu leben

(vgl. Nussbaum, Martha: Gerechtigkeit oder Das gute Leben, Frankfurt am Main 1999.) Auch hier ist eine sichere Bindung, die die individuelle Entfaltung fördert erkennbar. vgl. Moser, Michaele: Nachdenken über gutes Leben, in: http://www.armutskonferenz.at/index2.php?option=com_docman&task=doc_view&gid=179&Itemid=69[10.12.13]; auch Gewaltstudie 2013: Gewalt- und Missachtungserfahrungen von Kindern und Jugendlichen in Deutschland, in: http://kinderförderung.bepanthen.de/de/kinderarmut/index.php[10.12.13]. Die Erfahrung guten Lebens ist demnach eng mit der Entwicklung kindlicher Ressourcen und Resilienz-Kompetenzen verwoben.

13 Keupp, Heiner: Bildung zwischen Anpassung und Lebenskunst. Ressourcen der Lebensbewältigung. Vortrag auf dem „Tag der Weiterbildung" am 03. März 2006 in St. Johann im Pongau, S. 1-17, hier S. 8 und S. 10.

14 Keupp, Heiner: Bildung zwischen Anpassung und Lebenskunst. Ressourcen der Lebensbewältigung. Vortrag auf dem „Tag der Weiterbildung" am 03. März 2006 in

Identitätsarbeit zu tun. Entwickelt sich hier doch auch die Fähigkeit, Grenzziehungen zu leisten, Prioritäten des Handelns zu bewerten und reflektiert zu realisieren, also Boundary-Management-Kompetenzen zu entfalten. Denn kognitive Widerstandsquellen sind "symbolisches Kapital", also Intelligenz, Wissen und Bildung. Eine zentrale Widerstandsquelle bezeichnet die Ich-Identität, also eine emotionale Sicherheit in bezug auf die eigene Person. Die Ressourcen einer Person schließen als zentralen Bereich seine zwischenmenschlichen Beziehungen ein, also die Möglichkeit, sich von anderen Menschen soziale Unterstützung zu holen, sich sozial zugehörig und verortet zu fühlen. Beides erscheint notwendig, die emotionale und die kognitive Kompetenz- und Ressourcenentwicklung im Sinne guten Lebens, das der gewalttätigen Interessendurchsetzung entbehren kann. Dieser Doppelstrategie von Bildungs- und Sozialtraining folgt das "Schulfach Glück"[15] an der Willy-Hellpach-Schule in Heidelberg. Wichtig erscheint hier durchaus auch die soziale Vernetzung mit Projektpartnern wie der Stadt Heidelberg, vor allem mit dem Kinder- und Jugendamt, der Universität Heidelberg, dem Institut für Kriminologie und der Polizeidirektion Heidelberg und dem Verein Sicheres Heidelberg e.V. Dabei wird das übergeordnete Lernziel formuliert, die Schülerinnen und Schüler lebensfroh zu sehen und sie für die Realität sowohl des Berufslebens als auch des privaten Umfelds vorzubereiten. Die Schülerinnen und Schüler sollen sich einer gelassenen Heiterkeit hingeben und selbstverantwortlich sowie selbstwirksam handeln (vgl. Fritz-Schubert 2011, S. 157.). So sind die von Fritz-Schubert allgemein formulierten Ziele des Faches Glück beschrieben. Bei der Durchführung im

St. Johann im Pongau, S. 1-17. vgl. auch den Bildungsaspekt bei: Bieri, Peter: Wie wäre es gebildet zu sein? Vortrag Bern 2005, gleichfalls in: Bieri, P.: Wie wäre es, gebildet zu sein? In: Göppel, R./Lenhart, V./Rihm, Th./ Schön, B./Strittmatter- Haubold, V. (Hrsg.): Bildung ist mehr.... Potentiale über PISA hinaus. Heidelberg 2008, S. 13-21 und Brumlik, Micha: Bildung und Glück. Versuch einer Theorie der Tugenden, Berlin; Wien 2002.

15 Fritz -Schubert, Ernst: Schulfach Glück", Herder Verlag, Freiburg, Basel, Wien 2008; Fritz -Schubert, Ernst: Glück kann man lernen: Was Kinder stark fürs Leben macht, Berlin 2011; Märzinger, Karin; Plangl, Vicky; Smolka, Heide-Marie; Teufel, Ingrid: Mein Glückstraining. Ein Portfolio für Kinder zum Lernen des Glücklichseins ab Klasse 5, Augsburg 2012; Bertrams, Alex: Unterricht zum Glücklichsein 2012, in: http://www.uni-mannheim.de/1/presse_uni_medien/pressemitteilungen/2011/august/unterricht _zum_gluecklichsein/index.html[1.10.2012]- kritisch: Burow, Olaf-Axel; Hoyer: Timo: Positive Pädagogik oder: Wie das Glück aus der Schule verschwand – und wie es in die Ganztagsschule zurückkehren könnte, in: Lehren & Lernen Heft 6, 2012, S. 28-33.

Unterricht werden diese Ziele nach einem in vier Einheiten geteilten Schema verfolgt: „*Glück des Augenblicks*“, „*Abenteuer Alltag*“, „*Kultur und Kulturtechniken als Grundlage für soziales Leben*“, „*Das Ich und die soziale Verantwortung*“ (vgl. Fritz-Schubert 2011, S. 157f.). Die erstgenannte Einheit ‚Glück des Augenblicks' enthält vielerlei Übungen, die Konzentration, Bewusstheit, Kommunikation und Wahrnehmung fördern sollen. Die betreuten Konzentrations- und Achtsamkeitsübungen sollen die Schüler/-innen lehren, „aufmerksam und wach zu sein und zu sehen, dass alles eine Bedeutung hat und sorgfältig und behutsam damit umzugehen ist“ (Fritz-Schubert 2011, S. 158). Meditations- und Konzentrationsübungen motivieren die Schüler/-innen, auf Körper und Seele zu achten, Stille im Kopf und außerhalb des Körpers aufzunehmen. Die Übungsreihe vervollständigen Praktiken zur Schulung der Wahrnehmung, also der Sinne. Die Sinne sind, so Fritz-Schubert, das Tor zum Spüren von Glücksgefühlen. In der 4. Einheit „Das Ich und die soziale Verantwortung“ setzen sich die Schüler/-innen mit ihrem Handeln und dessen Auswirkung auf die Beziehung zu ihren Mitmenschen auseinander. Die Jugendlichen konfrontieren sich mit den Anforderungen und problematischen Situationen ihres Umfelds, um die Wichtigkeit und die Konsequenzen ihres eigenen Handelns kennen und ein-schätzen zu lernen. Hierein ist das Projekt: “Gewaltlos glücklich” eingebettet. Das Gewalt-Präventions-Projekt “Gewaltlos glücklich”[16] hat zum Inhalt unterschiedliche Methoden, die der sozio-emotionalen Erziehung und dem Bildungsprozess förderlich sind:

- Coolness-Deeskalationstraining, Rollenspiele,
- Sport
- Erlebnispädagogik

So werden die sozialen Kompetenzen zur gewaltfreien Konfliktregelung und Konfliktschlichtung trainiert und als sinn- und wertvoll erfahren. So verstärkt das gelungene Konfliktmanagement auf der Grundlage einer bindungsorientierten Konfrontationspädagogik die Konfliktlösemethoden und lösungsorientierten -alternativen mit dem Ziel des Erkennens eigener, authentischer Ziele und dem Einschlagen sozialverträglicher Wege zu den je eigenen Zielen. Die Voraussetzung hierzu ist, dass die Heranwachsenden eine Sensibilität für Normakzeptanz schärfen und die Fähigkeiten zur Kommunikation und

16 vgl. http://www.jugendgewalt-vorbeugen.de/projekte/gewaltlos-gluecklich [10.12.13] und http://www.sicherheid.de [10.12.13].

Kooperation effektivieren im Sinne der Durchsetzung eigenen Interessen auf "unverletzende Weise"[17].

Denn untersuchen wir Gewalt-Präventionskonzepte wie Faustlos, Arizona, usw. so stoßen wir immer wieder auf folgende fünf Grund-Fähigkeiten, die eine effektive, gewaltfreie und damit "unverletzende" Konfliktregelung ermöglichen.

- Positive Selbst-Akzeptanz; Lern-Leistungs-und Sozialermutigung im Erziehungsprozess ("Anerkennungskultur", Heitmeyer).
- Realistische Selbst- Wahrnehmung ("Boundary-Management", Keupp; "Machbarkeit", Antonovsky), verhindert Größenphantasien und Aggressionen aufgrund von Bedrohungsgefühlen, die in Minderwertigkeitsgefühlen gründen.
- Rationale Selbst- Kontrolle (Affekt-, Wutkontrolle)(Grawe);
- Empathie durch Perspektive-Wechsel mittels sozialer Phantasie;
- Einüben kreativ-konstruktiver Konfliktlösungsalternativen- dies bedeutet Durchsetzungsfähigkeit fairer und gerechter Konfliktlösungsmodelle und damit Ausbildung sozialer Kompetenzen (Marsal). Diese wiederum benötigen Ich-Stärke und Zivilcourage, diese gründen in Selbst-Akzeptanz.

Hiermit werden die für die (gesunde) Persönlichkeitsentwicklung relevanten Dimensionen effektiv angesprochen. Demnach zeichnet sich eine gelungene Identitätsentwicklung durch folgende fünf große (Basis-)Kompetenzen[18] aus: sich nach außen hin präsentierende Kommunikative Kompetenz, Offenheit für neue persönliche und soziale Erfahrungen, durch emotionale Intelligenz begründete soziale Verträglichkeit, Gewissenhaftigkeit. Dies führt zur ausbalancierten emotionalen Stabilität und damit zu geringem Neurotizismus.

Dies zeigt, wer verbale oder sogar andere konstruktive Alternativen zur physischen oder psychischen Gewalt kennt und anzuwenden vermag, der sieht durchaus deren „Nutzen" auch für sich selbst: soziale Anerkennung, Prestigegewinn, Respekt, Solidarität, Zuneigung, Aufmerksamkeit usw. Die Motive der Schüler/-innen für eine Teilnahme am Projekt gründen in der Ambivalenz cool

---

17 Dölling, Dieter; Beisel, Horst; Hermann, Dieter; Hoge, Ulrike: Bericht über die wissenschaftliche Begleitung des Projekts „Gewaltlos glücklich", Heidelberg, 21. September 2010, http://www.modul100.de/455/files/20101021145242Glueck-Abschlussberichtneu.pdf[22.9.2013] und Marsal, Eva: unverletzende Selbstbehauptung. Das Karlsruher Jugendtraining, Opladen 1997.

18 Pervin, Lawrence A.; Cervone, Daniel; John, Oliver, P: 2005 (5.Auflage), Persönlichkeitstheorien, München, Basel 2005, 5. vollständig überarbeitete und erweiterte Auflage, S. 322.

zu sein, dazuzugehören, sich anzupassen und gleichzeitig sich abzusetzen, etwas Besonderes zu sein. Diese Ambivalenz gilt es im Sinne von konstruktiver Selbstwirksamkeit und Lebenskunst auszubalancieren.

Gelungene Basis-Bedürfnisbefriedigung in der Kindheit und Jugend erscheint demnach die Voraussetzung einer „starken" Identitätsentwicklung, die es wiederum unnötig macht, sich ständig seiner real vorhandenen Ressourcen zu vergewissern und soziale Grenzen ständig zu überschreiten, aufzubrechen und damit gewalttätig zu werden. Eine dialogische Erziehungskultur, die Anerkennung, Anregung und Anleitung verbindet, erscheint als die Basis, auf der eine gewaltfreie, glückliche Kindheit und Jugend aufbauen kann. Diese kann sich durchaus an den Bildern von Pippi Langstrumpf (Astrid Lindgren) und Tom Saywer und Huckleberry Finn[19] (Mark Twain) orientieren.

Literatur

Andresen, Sabine; Hurrelmann, Klaus: Was bedeutet heute "Glück" für Kinder?, in: Aus Politik und Zeitgeschichte 38/2010 vom 20. September 2010, S. 3-8.

Bertrams, Alex: Unterricht zum Glücklichsein, 2012, in: http://www.uni-mannheim.de/1/presse_uni_medien/pressemitteilungen/2011/august/unterricht_zum_gluecklichsein/index.html[1.10.2012]

Bieri, Peter: Wie wäre es gebildet zu sein? Vortrag Bern, gleichfalls in: Bieri, P.: Wie wäre es, gebildet zu sein? In: Göppel, R./Lenhart, V./Rihm, Th./ Schön, B./Strittmatter- Haubold, V. (Hrsg.): Bildung ist mehr... Potentiale über PISA hinaus. Heidelberg 2008, S. 13-21.

Brumlik, Micha: Bildung und Glück. Versuch einer Theorie der Tugenden, Berlin; Wien 2002.

Burow, Olaf-Axel; Hoyer: Timo: Positive Pädagogik oder: Wie das Glück aus der Schule verschwand – und wie es in die Ganztagsschule zurückkehren könnte, in: Lehren & Lernen Heft 6 - 2012, S. 28-33.

19 Schiffer, Eckhard: Warum Huckleberry Finn nicht süchtig wurde. Anstiftung gegen Sucht und Selbstzerstörung bei Kindern und Jugendlichen, Weinheim und Basel 1997. Hier spielt die Salutogenese-Konzeption von Antonovsky eine entscheidende Rolle. Das Kind gewinnt das "Recht auf sein *eigenes* Leben" im Sinne von Korczaks Kinderrechte-Pädagogik.

Csikszentmihalyi, Mihaly: Flow – der Weg zum Glück. Der Entdecker des Flow-Prinzips erklärt seine Lebensphilosophie, Freiburg, Basel, Wien 2010.

Dölling, Dieter; Beisel, Horst; Hermann, Dieter; Hoge, Ulrike: Bericht über die wissenschaftliche Begleitung des Projekts „Gewaltlos glücklich", Heidelberg, 21. September 2010, http://www.modul100.de/455/files/20101021145242Glueck-Abschlussberichtneu.pdf[22.9.2013].

Fritz-Schubert, Ernst: "Schulfach Glück", Freiburg, Basel, Wien 2008.

Fritz-Schubert, Ernst: Glück kann man lernen: Was Kinder stark fürs Leben macht, Berlin 2011.

Fuhr, Thomas: Das Glück des Kindes, in: Zeitschrift für Pädagogik 48 (2002) 4, S. 514-533.

GEOlino-Kinderwerte-Monitor 2010: Repräsentative Untersuchung des Kindermagazins GEOlino in Zusammenarbeit mit UNICEF, in: http://www.geo.de/_components/GEO/info/presse/files/2010/kinderwerte/Praesentation_PK_final.pdf[10.12.13]

Gewaltstudie 2013: Gewalt- und Missachtungserfahrungen von Kindern und Jugendlichen in Deutschland, in: http://kinderförderung.bepanthen.de/de/kinderarmut/index.php [10.12.13].

Gewaltstudie 2013: Gewalt- und Missachtungserfahrungen von Kindern und Jugendlichen in Deutschland, in: http://kinderförderung.bepanthen.de/de/kinderarmut/index.php [10.12.13]

Göppel, Rolf: Resilienzförderung als Aufgabe der Gesundheitspädagogik?, in: Knörzer, Wolfgang; Rupp, Robert (Hrsg.): Gesundheit ist nicht alles – was ist sie dann? Baltmannsweiler 2011, S. 57-68.

Göppel, Rolf: Sind Kinder heute (noch) glücklich? Über den erstaunlichen Kontrast zwischen kulturkritischen Klagen und kindlichen Selbsteinschätzungen, in: Göppel, Rolf: Aufwachsen heute. Veränderungen der Kindheit – Probleme des Jugendalters, Stuttgart 2007, S. 153-179.

Göppel, Rolf: Von der „Leidabwehr" zur „Lebenskunst"? Die Psychoanalytische Pädagogik und die Frage nach dem Glück, in: Göppel, Rolf: Pädagogik und Zeitgeist. Erziehungsmentalitäten und Erziehungsdiskurse im Wandel, Stuttgart 2010, S. 284-300.

Heitmeyer, Wilhelm: Der doppelte Kontrollverlust. Jugendliche Massenmörder: Wer die Ursachen von Amokläufen verstehen will, muss die Bedingungen beachten, unter denen Jugendliche aufwachsen - und die Konsequenzen mangelnder Anerkennung, in: bildung & wissenschaft, Juni 2009, S. 18-20.

Heitmeyer, Wilhelm: Entsicherung, Individualisierung und Gewalt. In: Beck, U./ Beck-Gernsheim, E. (Hrsg.): Riskante Freiheiten. Frankfurt a.M. 1994.

Heitmeyer, Wilhelm: Freigesetzte Gewalt. Gewalt als Bearbeitungsform einer neuen Unübersichtlichkeit. In: Pädagogik, H. 6, 1994, S. 35-40.

Hintermair, Manfred: Empowerment als konzeptioneller Orientierungsrahmen für eine gesundheitsförderliche Lebensführung, in: Knörzer, Wolfgang; Rupp, Robert (Hrsg.): Gesundheit ist nicht alles – was ist sie dann? Baltmannsweiler 2011, S. 46-56.

Hurrelmann, Klaus: Zu viel für manche kleine Seele. Trend zu immer mehr psychosomatischen und mentalen Störungen bei Kindern und Jugendlichen. Gefragt sind nicht nur neue Medikamente, sondern vor allem kompetente Eltern, in: http://nbn-resolving.de/urn:nbn:de:0168-ssoar-202197 [10.12.13].

Knörzer, Wolfgang; Rupp, Robert (Hrsg.): Gesundheit ist nicht alles – was ist sie dann? Baltmannsweiler 2011.

Marsal, Eva: unverletzende Selbstbehauptung. Das Karlsruher Jugendtraining, Opladen 1997.

Märzinger, Karin; Plangl, Vicky; Smolka, Heide-Marie; Teufel, Ingrid: Mein Glückstraining. Ein Portfolio für Kinder zum Lernen des Glücklichseins ab Klasse 5, Augsburg 2012.

Moser, Michaele: Nachdenken über gutes Leben, in: http://www.armutskonferenz.at/index2.php?option=com_docman&task=doc_view&gid=179&Itemid=69 [10.12.13].

Münch, Joachim; Wyrobnik, Irit, Pädagogik des Glücks. Wann, wo und wie wir das Glück lernen, Hohengehren 2011[2].

Nussbaum, Martha: Gerechtigkeit oder Das gute Leben, Frankfurt am Main 1999.

Pervin, Lawrence A.; Cervone, Daniel; John, Oliver, P: (5. Auflage), Persönlichkeitstheorien, München, Basel 2005, 5. vollständig überarbeitete und erweiterte Auflage.

Rupp, Robert: Bedürfnisorientierung als gesundheitserzieherisches Grundprinzip, in: Knörzer, Wolfgang; Rupp, Robert (Hrsg.): Gesundheit ist nicht alles – was ist sie dann? Baltmannsweiler 2011, S. 20-33.

Schiffer, Eckhard: Warum Huckleberry Finn nicht süchtig wurde. Anstiftung gegen Sucht und Selbstzerstörung bei Kindern und Jugendlichen, Weinheim und Basel 1997.

Schmid, Wilhelm: Glück. Alles, was Sie darüber wissen müssen, und warum es nicht das Wichtigste im Leben ist, Frankfurt am Main, Leipzig 2007.

Wehr, Helmut: Wohl-Sein – der bedürfnis- und ressourcenorientierte Gesundheitsbegriff bei Erich Fromm, in: Knörzer, Wolfgang; Rupp, Robert (Hrsg.): Gesundheit ist nicht alles – was ist sie dann? Baltmannsweiler 2011, S. 69-77.

Mathematische Bildung auf der Fährte der Reproduzierbarkeit

Karl Hayo Siemsen, Joachim Schwarz, Hayo Siemsen

Abstract (overview)
The trail of mathematics at the university, which we follow, consists of four sections. The first section shows, how strongly the vivid conceptualization and gestalt are statistically superior to a logicist approach (the ordering of concepts). What follows is a documentation in which two experienced professors in university mathematics are transmitting the topic closely following the logic of the topic. Their students are nevertheless in the end only 35% successful, a meager result. In the third section - after a year of coaching in vivid conceptualization, gestalt and the beginning at the point, which the students surely understand and the "jump of the colleagues into the cold water" -, already up to 65% of the students pass the exam at the first try with the same level of difficulty and the same preparation time. In the repeated test a total of 71% have passed, and one can see, how it is possible to get nearly 100% of success rate. Mathematical instruction does not have to become the horror of a natural-science, technological study. The students immediately reach a higher achievement of learning (and keep it also later), the teachers know after a year of coaching, how they have to adapt their teaching. The understanding of the concepts forming a Machian teaching, in order to consistently replicate it, will still take another one or two years for the teachers. The statistics is leading into this direction. If the measurement goals are not achieved then the vivid conceptualization, the gestalt or the connecting with the students at the point, where they are in learning is inaccurate. There, one can improve, though not necessarily for the current cohort in case that one does not recognize the mistake in time.

Abstract 1: Following Gestalt and Concept in University Teaching of Mathematics
Examples show that mathematics has been using visualization (Anschauung) and gestalts for a long time. Visualizations can be reduced to gestalts, i.e. to something elemental and to link these. From these gestalts (webs of significance (/meaning), "Bedeutungsgewebe"), step by step mathematical concepts develop.

The linked gestalts are identified in an automated way, analyzed with statistical methods regarding their effectivity and each chain is evaluated. The problem of exercizing of concepts is also analyzed. By this, hard schematic work, which can lead to rigidity, is reduced. It is replaced by exercizing at exactly the location, where the error appears.

The foundation of this approach can be traced back to Pestalozzi, regarding the visualization, regarding the gestalts to Ernst Mach, his erkenntnis-theory and his economy of thought. One can see by this the gestaltpsychological problem of university level mathematics education and an at least partial possible solution.

***Keywords*:** visualization, gestalt, statistics, mathematical terms, *Erkenntnis*theory

Anschauung ist das Fundament aller Erkenntnis (Pestalozzi)

Gestalten sind das empirische Fundament aller Erkenntnis (Karl Hayo Siemsen)

(Gestalten sind Anschauungen, die mit wissenschaftlichen Methoden weiter herausgeschält, d.h. auf Wesentliches reduziert sind.)

Lebenslang hat Pestalozzi gegen das 'Maulbrauchen' gekämpft. Er verstand darunter jegliches Sprechen, das nicht durch deutliche Begriffe getragen und nicht von Wahrheit und Erfahrung erfüllt ist. Er hielt die Schulen seiner Zeit insofern als *ver*bildend, als sie das Kind vor dem Lesen und Schreiben nicht sprechen lehrten. Er legte daher, als ihm praktische Schularbeit ermöglicht wurde, auch sofort alle Bücher weg und führte die Kinder hin zu den Dingen. Die Schüler sollten zuerst tragfähige, von klaren Vorstellungen geprägte Begriffe erwerben. Das in den Schulen verbreitete und nicht auf Anschauung gründende Schwatzen über alles und jedes nannte er 'Lirilariwesen'. Er verstand darunter *„alles, was den Kindern so eine Art gibt, mit dem Maul ein Weit und Breites über die Sachen zu machen, hinter denen für sie nichts steckt und die sie nicht verstehen und nicht im Herzen tragen, mit denen man ihnen aber doch die Einbildungskraft und das Gedächtnis so anfüllt, dass das rechte Alltagshirn und der Brauchverstand im menschlichen Leben dadurch zugrunde geht"* (6, 116 f.). Pestalozzi hielt das 'Lirilariwesen' somit nicht nur für unnütz, sondern für schädlich. *„Wo Wörter und Urteile dem Überlegen und Anschauen voreilen, da ist denn Überlegen und unbefangen Anschauen doppelt schwer"* (Br. 3, 142), lesen wir schon 1782 in einem seiner Briefe an den Hauslehrer Petersen. (zitiert nach Arthur Brühlmeier)

Das folgende kann als Nachweis für Pestalozzis Aussage genommen werden, obwohl ihm diese Art der Beweisführung sicherlich fern lag. Aus heutiger Sicht erscheint sie dringend notwendig.

Pestalozzi und Chambers haben ihre Erkenntnisse festgehalten, bevor Ch. Darwin 1859 die Ergebnisse seiner Expedition mit der Beagle veröffentlichte. Dennoch hatten sie die grundsätzlich korrekte Idee, sie verfolgten eine wegweisende Spur, deren Beweis auf sich warten lässt.

Die konkrete Situation im Mathematikunterricht des ersten Semesters bestand darin, die Durchfallquote von über 60 % zu senken, wenn möglich auf unter 10 %. Die ursprüngliche Versuchsplanung war, die Kontrollgruppe aus dem vorausgegangenen ersten Semester zu nehmen und als *Experimentalgruppe das ganze derzeitige Semester*. Im Unterschied zur Vergangenheit sollte die Experimentalgruppe über Gestalten zu Beginn des jeweiligen Abschnittes mit dem gleichen Inhalt wie bisher vertraut gemacht werden. Die Lernleistungen wurden jeweils nach dem Abschluss des jeweiligen Inhaltes am Ende des Semesters von den Lehrenden gemessen. Die Inhalte sind voneinander weitgehend unabhängig. Der Schwierigkeitsgrad beim Messen der jeweiligen Lernleistungen wurde in jedem der Fälle als vergleichbar und schwierig angesehen, d.h. im Rahmen der Versuche gegenüber den Vorjahren aufrecht erhalten.

Dies wurde in den ersten neun Tagen so arrangiert. Doch dann kam es anders. Einige Studenten des ersten Semesters beschwerten sich anschließend gezielt bei einigen am Ablauf beteiligten Personen, der mathematische Inhalt sei mit den Gestalten *zu leicht*. Der für den Einbau der Gestalten verantwortliche Kollege war auf einer Tagung im Ausland und somit nicht in der Lage, in die Entscheidung einzugreifen. So fand der weitere Verlauf des Semesters (14 Wochen) in der klassischen Form statt, allerdings mit Teilen der Inhalte elektronisch nach den Antworten der Studierenden über Fernbedienung (Clicker) ein Vierteljahr später ausgewertet, so auch für den Abschnitt „Exp/Log", der anfangs neun Tage mit Gestalt durchgeführt wurde. Die weiteren Abschnitte „Diff" und „Grenz" wurden ohne Gestalt vorgetragen und auf die gleiche Weise wie in den Vorjahren "geübt".

Eine Statistik über die gesamten Studenten ohne besondere Berücksichtigung von Gestalt wurde erstellt, und das Ergebnis war wie im vorausgehenden 1. Semester. So wundert es nicht, dass die Durchfallquote wie gewohnt oberhalb 60 % lag und auch sonst die Statistik keine Veränderungen gegenüber den Vorjahren zeigte. Dabei hätte es bleiben können.

Der für den Einbau der Gestalten verantwortliche Kollege war mit diesem Ergebnis nicht einverstanden. Er wollte eine eigenständige Auswertung der neun Tage "mit Gestalt" („exp/log") (und damit auch der beiden weiteren Abschnitte). Im Gespräch mit einigen Kollegen meinten die Kollegen, es sei unmöglich, nach über einem Vierteljahr von diesen neun Tagen noch irgendetwas statistisch nachzuweisen. Doch dem war nicht so.

Der Nachweis wird in mehreren Schritten erbracht (Anhang 1, Anhang 2). Zunächst wird nur nachgewiesen, dass die Studierenden „Exp/Log" wesentlich erfolgreicher absolviert haben als „Diff", und zwar mit einer Irrtumswahrscheinlichkeit von 0.0001, d.h. 100 ppm oder 1/10tel Promille (Anhang 1).

Type III Tests of Fixed Effects				
Effect	**Num DF**	**Den DF**	**F Value**	**Pr > F**
Unterrichtseinheit	1	1923	41.23	<.0001

Die Erfolgsquote der Studenten der Unterrichtseinheit „Diff „war nicht einmal halb so groß wie die Erfolgschance der Studenten in der Unterrichtseinheit „Exp/Log". Da letztlich insgesamt über 60 % durchgefallen sind, wären es für den Abschnitt „exp/log" weniger als 30 % gewesen.

Dasselbe Ergebnis $p < .0001$ erhält man, wenn man anstelle von „Diff" stattdessen „Grenz" zum Vergleich heranzieht.

Allerdings bleiben zunächst einige Zweifel, ob es an der Gestalt gelegen hat. Es ist bis zu diesem Punkt nicht auszuschließen, dass zwei andere Variable beitragen: ein möglicher IQ-Unterschied der jeweiligen Studierenden oder unterschiedlicher Leistung des Vortragenden.

Glücklicherweise gibt es gesonderte Datensätze „Grenz" und „Diff", beide ohne Gestalt (Anhang 2), die darauf hinweisen, dass es genau die Gestalt (und weder der studentische IQ noch die unterschiedliche Leistung des Dozenten) gewesen ist, die den Einfluss genommen hat.

Fasst man beide „GrenzwReflekt"-Datensätze zu einem zusammen, ist der Unterschied in den Testergebnissen zum „DiffReflekt"-Datensatz nicht signifikant (p-Wert von 0,67). Insgesamt kann man sicher sagen, dass die Unterschiede zwischen „DiffReflekt" und „GrenzwReflekt" deutlich geringer sind als deren Unterschiede zu „Exp/LogReflekt".

Man kann die „GrenzwReflekt"-Daten gegen die „DiffReflekt"-Daten laufen lassen und stellt fest, dass in diesem Fall annähernd die gleichen Studenten dabei waren, die auch an den Übungen „Exp/Log" teilgenommen hatten. Die Zahl der Studenten liegt in allen drei Fällen zwischen 73 und 83, eine Schwankung im üblichen Rahmen, wenn es den Studenten freigestellt ist, zu erscheinen. Es gibt im Verleich „GrenzwReflekt"-Daten und„DiffReflekt"-keine Signifikanz, sondern Rauschen. Auch in diesem Fall spielte Gestalt in beiden Gruppen keine Rolle (und damit können es ja nur die anderen Effekte gewesen sein, die Signifikanz hätten auslösen können, und die erweisen sich als viel geringer, wenn sie überhaupt nachweisbar sind.

Dies kann im konkreten Fall als ein schöner empirischer Nachweis für Pestalozzis Aussage über Anschauung gesehen werden.

Das Gestaltradio

Es folgt eine weitere Gestalt: Das Radio, sein Drehknopf zur Sendersuche und das magische Auge zur Anzeige des gefundenen Senders. Wir nehmen anstelle von Frequenzen, auf denen wir senden, Gestaltketten, im Beispiel Exp/Log, Diff und Grenz. Ob eine Gestalt trägt, zeigt sich aus der Statistik (ein $p < 0.0001$ ist gegenüber dem Rauschen dramatisch!).

Mathematik 1 besteht grundsätzlich aus vielen solchen Gestaltketten in einzelnen Dateien mit Rohdaten (wie in „Exp/LogReflekt"). Es gibt immer eine Datei "mit (intensiver) Gestalt" und eine "ohne (erkennbare) Gestalt" am Anfang als Grundlage für das Experiment. Anstelle der Rundfunk-Sendersuche im Radio wählen wir die Gestaltketten-Sendersuche. Das Statistikprogramm läuft weitgehend automatisiert (keine manuelle Eingabe nach dem Start der Dateien).

Es gab früher im Radio ein „magisches Auge", das zusammenging, wenn ein Sender gefunden wurde. Man konnte die Sender darauf einstellen. So suchen wir für die einzelnen Gestaltketten nach dem p-Wert. Ist der zu hoch, also z.B. nicht signifikant, muss an der betreffenden Gestaltkette bezüglich der Gestalt nachgebessert werden. (An einem $p < 0.0001$ wird es wenig Bedarf zur Verbesserung zu geben). Das Ergebnis ist für alle Gestaltketten ein Ausdruck des p-Wertes als Hinweis, ob die Gestalt trifft, oder sie als Gestalt nicht wahrgenommen wird, also wie bei der alten Rundfunk-Sendersuche.

Gesetzt, man programmiert in einer übergeordneten Steuersprache ein bestehendes Statistikpaket. Damit kann man zwei unterschiedliche Gestalten in ihrer pädagogischen Wirkung sicher unterscheiden. Damit wird der Umgang mit Gestalten beziehungsweise das Verbessern von Gestalten nach mehreren Durchläufen in mehreren Jahren deutlich vereinfacht, und Lernen wird erheblich präziser umsetzbar.

Pädagogik erwartet, Studierende dort abzuholen, wo sie sich befinden

Man kann anhand der statistischen Daten den Versuchsablauf auch wie folgt interpretieren: in den Übungen 14 Wochen lang wurden die Studierenden nicht dort abgeholt, wo sie sich befanden, sondern „einige Stockwerke darüber". Der Einfluss dieser 14 Wochen für den Lernprozess ist im Verhältnis zum Aufwand sehr gering. Die neun Tage gestaltorientiertes Lernen dagegen tragen im Vergleich erheblich bei.

> *„Allgemein gesehen, zeigt sich immer wieder ein unglaublicher Absturz des "Wissens", (was man darunter auch verstehen mag) […] Ich kann mir schlechterdings nicht vorstellen, daß die Fachlehrer von dieser Hinfälligkeit des Schulwissens genug wissen. Man wird ja auch von dieser Frage abgelenkt, wenn man die Erfolgsmessung für erledigt hält nach Prüfungen[…]."*
>
> (Martin Wagenschein)

Gehen wir davon aus, dass 95 % des Unterrichts vergessen werden. Wir haben Wagenschein dann ein Messer in die Hand gegeben, mit dem er von vornherein die 95 % wegschneidet, die ohnehin nach dem Unterricht vergessen sind. Unterricht wird dann ganz ruhig statt hektisch. Die Details der Inhalte treten zurück und lassen sich aus den Gestalten rekonstruieren. Das Problem war bisher: wie identifiziert man die 95 %, bevor man sie über Generationen mitschleppt. Das Gestaltradio ist ein Ansatz, schon nach einem Durchlauf diese Stellen automatisch zu identifizieren.

Doch wenn man Wagenschein pädagogisch folgt, ergeben sich erhebliche Probleme. Im Gegensatz zu Mach sind seine Gedanken zwar häufig ebenfalls tragfähig, doch an entscheidenden Stellen widersprüchlich.

> „In Erinnerung geblieben ist mir der Stoß, der mich traf, als ich an einer offenen Hörsaaltür vorbeigehend, auf der Wandtafel mathematische Zeichen erblickte. Es waren Gleichungen oder Formeln, die offenbar pädagogische (oder auch nur didaktische) Begriffe, als messbare Größen behandelt, miteinander verbanden. Genaues weiß ich nicht mehr. Nur das eine: Da ich ja mit dem Mathematischen wie mit dem Pädagogischen auf vertrautem Fuß lebte, sah ich diese Verbindung mit Schaudern: Welch eine Mésalliance!“ *(1983, S. 91).*
>
> *(Martin Wagenschein)*

Wagenscheins Abwehr richtet sich gegen das, was er Physikalismus nennt. Machs Empirismus in der Pädagogik und Psychologie in der Nachfolge von Jacobi, Beneke, Hering und Fechner wird von ihm fälschlich unter Physikalismus subsumiert, aber Mach nicht zugeschrieben. Der Kreis der von Wagenschein Akzeptierten umfasst Mach, Planck (welch Widerspruch! Planck hat den Physikalismus explizit eingeführt als Gegenpol zu Machs Empirismus), Wolfgang Metzger (1983, S. 88), Wertheimer, Wittenberg (1983, S. 76), Freudenthal. Wertheimers Dissertation ist eine Arbeit über Statistik (S. 41) und bleibt unerwähnt. Charlotte Bühler und ihre Nachfolger werden ebenfalls nicht genannt.

Ein weitere Gedankengang, der unter Bezug auf Mach dringend einer Erörterung bedarf, ist Wagenscheins Aussage "genetisch - sokratisch - exemplarisch", gelegentlich auch „historisch“ (nur scheinbar Rückgriff auf Mach). Wagenschein versteht unter *Genese* die des Aristoteles und nicht die von Darwin und in der Nachfolge von Mach. Das lässt sich aus dem zweiten Begriff erschließen: sokratische Logik. Machs Genese ist nicht teleologisch, kennt kein Ziel (im Gegensatz zur sokratischen Logik). Es macht einen großen Unterschied, ob in der Pädagogik in den Anfängen der Begriffsbildung Logik enthalten ist. Letzteres führt zu Interferenzen und erschwert den Zugang für den Lernenden erheblich. Logik zu diesem Zeitpunkt ist nicht notwendig. Logikfreie Gestalten sind bei Wagenschein faktisch ausgeschlossen.

Diese beiden Punkte im Denken Wagenscheins müssten geklärt werden, damit sich die Pädagogik von ihren selbst gewählten Restriktionen befreien kann und die Gestalten ihre pädagogische Wirkung entfalten können. Ausschließlich Introspektion ist - wie die Geschichte der Pädagogik zeigt - nicht geeignet, Widerspruchsfreiheit zu erwirken.

Mathematik, ein angeblich schwieriges Fach

Nochmals zur Beschwerde einiger Studierender, eine solche Mathematik 1 sei *zu leicht*. Um es deutlich zu sagen: die Aufgaben, die es am Ende zu lösen gilt, sind die gleichen, ob mit oder ohne Gestalt. Mit Gestalt werden sie als leicht empfunden, ohne Gestalt beliebig schwer. Mit Gestalt wären weniger als 30 % durchgefallen (weitere oder intensiver wirkende Gestalten könnten zu noch geringeren Durchfallquoten führen), ohne Gestalt sind es mehr als 60 %. Ein Teil der Studierenden mit der Aussage *zu leicht* "will offenbar durchfallen", damit Mathematik den von Studierenden selbst gesetzten Standard erfüllt.[1]

Hans Hahn, Student von Ernst Mach, Mitbegründer des Wiener Kreises und Lehrer von Kurt Gödel, Rudolf Carnap, Karl Popper und Karl Menger, bemerkt spöttisch und ignoriert dabei absichtlich die Eleganz der Mathematik:

> „Haben sich die Mathematiker seit den Tagen Euklids damit abgefunden, dass es zu ihrer Wissenschaft keinen Königsweg gibt, so darf doch andrerseits der Zugang zu ihr nicht über die langwierigsten und halsbrecherichsten Hochgebirgspfade führen, sodass die Mehrzahl aller Beschreiter unterwegs scheitern müsste, die wenigen aber, denen es gelänge, deren Schwierigkeiten zu überwinden, zu Tode erschöpft ankämen, und zwar nicht am Ziele, sondern dort, wo nun die eigentliche Mathematik erst beginnen soll."

Der Eindruck der Studierenden während des gestaltorientierten Lernens

Die folgenden Bilder skizzieren die Meinung von Studierenden während eines gestaltorientierten Unterrichts. In Bild 1 kommt zum Ausdruck, dass

1 Im April 2013 demonstrieren Abiturienten im Lande Nordrhein-Westfalen gegen die dortige Abiturklausur in Mathematik (vergleichbarer Inhalt wie Mathematik 1). Sie sei viel *zu schwer*.

Mathematik grundsätzlich als schwer eingestuft wird. Der Eindruck von gestaltorientierten Übungen, die leicht erscheinen, irritiert einige Studierende und sie haben Sorge, dass etwas Wesentliches weggelassen wurde, dass sie mit einem derartigen Vorgehen mehr oder weniger „verschaukelt" worden sind. Da hilft auch nicht, wenn ihnen nachgewiesen wird, dass ihre Vorgänger – auf die gleiche Weise unterwiesen – erfolgreich gewesen sind. Assistenten, die sich dieser Diskussion stellen, können den fehlerhaften Eindruck problemlos ausgleichen. Fehlt aber dieser Ausgleich, kommt es zu Situationen wie der Forderung, dem Wunsch, das Vertraute zu bevorzugen – mit der verdrängten Folge der hohen Durchfallquote. Es kommt sogar der Wunsch nach „Kampfrechnen" auf [vgl. Schwenk].

Bild 2 zeigt, dass 80 % der Studierenden das Vorgehen, mit Gestalten zu lernen, akzeptieren, und Bild 3 weist nach, dass das gestaltorientierte Lernen unter hoher Konzentration abläuft, also nicht zu beanstanden ist. Das Fundament nach Pestalozzi wird mit aller Sorgfalt gelegt.

Da sind dem konventionellen Lernen mit den vielerlei Ablenkungen eher Grenzen gesetzt [vgl. Schwenk]. Als Beispiel beim Vereinfachen von Termen sei von dort nur zitiert: √ (9) bleibt ebenso unaufgelöst wie √(1). Wie selbstverständlich wird ein Term (x-a) gegen $(x\text{-}a)^2$ in Zähler und Nenner gekürzt, und eine ^{2} bleibt „in der Luft hängen".

Bild 1: Die Erwartung von Lernenden an gestaltorientiertem Unterricht hinsichtlich der Schwierigkeit, die sie beim Lernen empfinden, einmal bezogen auf den Inhalt, dann bezogen auf die konkrete Übung mit Gestalt (•) und wie sie es sich wünschen (▪).

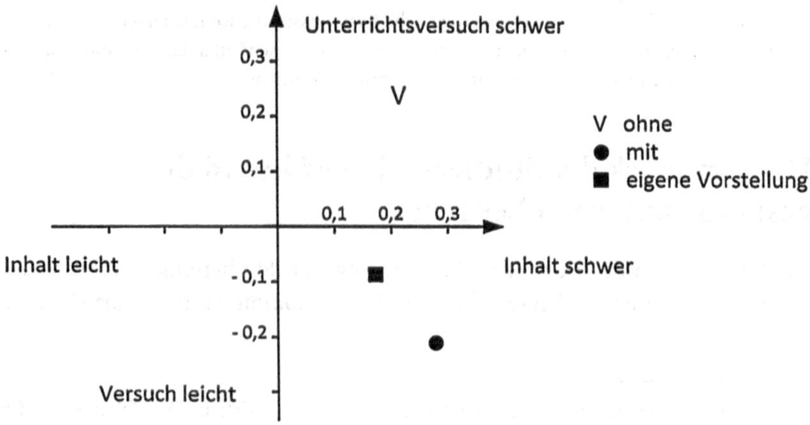

Bild 2: Lernen ohne Leistungsdruck in einem gestaltorientierten Unterricht

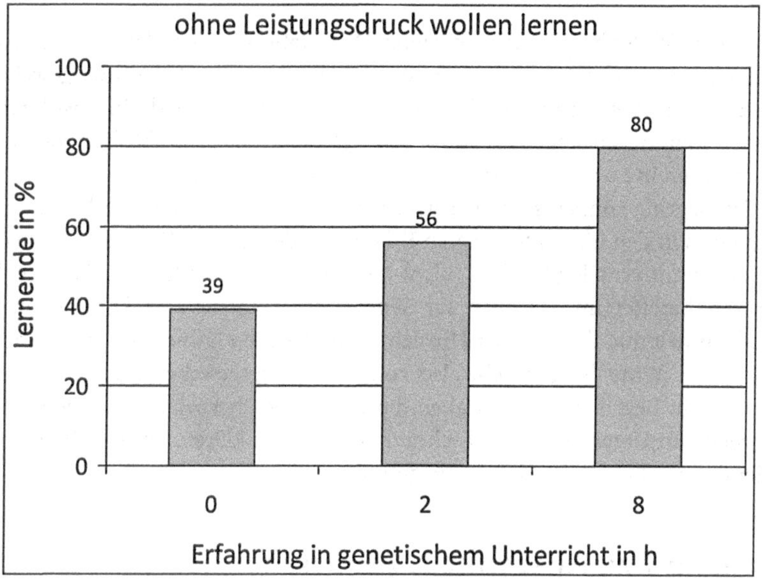

Bild 3: Die Selbsteinschätzung der Konzentration der Lernenden im gestaltorientierten Unterricht nach unterschiedlicher Unterrichtserfahrung von 8 Stunden

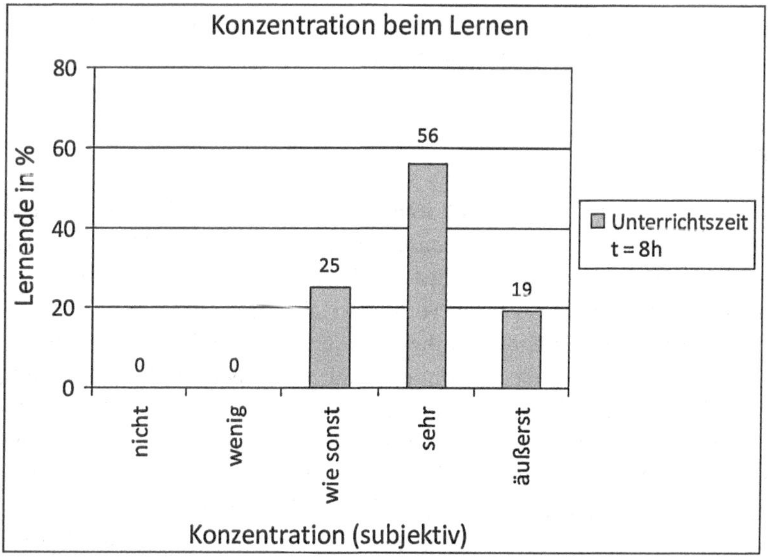

Schlussbemerkung zum Abschnitt 1

Alexander Wittenberg spricht in seiner Dissertation in der Mathematik „Vom Denken in Begriffen" von einer Vorstufe des Denkens, den Bedeutungsgeweben. Gestalten sind vorbegriffliches Denken. Der Unterschied liegt darin, dass Begriffe nach Kant im Entstehen an Logik geknüpft sind (einige Nachfolger von Hans Hahn waren in Logik führend, Hahn selbst war da skeptischer).

Vor-Begriffe können so konstruiert sein, dass sie keine Logik enthalten, und sie sind es in den hier untersuchten Fällen. Mit der Genese Darwins wurde der Weg zu intuitivem, logikfreiem (nicht-teleologischem) Denken als Vorstufe des Denkens bereitet (im Gegensatz zur Genese des Aristoteles 2400 Jahre zuvor, die Logik enthält und die in dieser Hinsicht unser Denken immer noch weitgehend bestimmt). Wittenberg spricht also von Bedeutungsgeweben als Vorstufe des Denkens in Begriffen. Ob Gestalten diese Aufgabe übernehmen können, ist in der weiteren Untersuchung (Abschnitte 3 und 4) zu klären. Empirisch gesehen jedenfalls sind Gestalten dafür eine gute Wahl.

Anhang 1 zu Abschnitt 1: Analysebericht
(Joachim Schwarz)

Untersuchungshypothese

Studenten, die einen Unterricht mit Gestalt absolviert hatten, erzielen bessere Abschlussergebnisse als Studenten ohne einen Unterricht mit Gestalt.

Datenerhebung:

Im 1. Semester wurde der Gestalt orientierte Unterricht im Rahmen einer Mathematikveranstaltung an der Hochschule Emden-Leer erprobt. Dabei wurde zunächst in einer Unterrichtseinheit das Rechnen mit Exponentialfunktionen und Logarithmen gelehrt. Hierbei wurde Gestalt angewandt.

In einer zweiten Unterrichtseinheit wurden die Grundlagen der Differentialrechnung präsentiert, diesmal ohne Gestalt. Nach beiden Unterrichtseinheiten fand jeweils eine Multiple Choice Lernerfolgskontrolle unter Verwendung elektronischer Hilfsmittel (Clicker) statt. Die Studenten bekamen die Fragen online präsentiert und mussten dann mit dem Clicker aus den vier vorgegebenen Antwortmöglichkeiten die jeweils korrekte Antwort auswählen. Diese Antworten wurden bezogen auf den einzelnen Studenten gespeichert. Für beide

Unterrichtseinheiten lag zu jedem Studenten für jede Frage vor, ob sie richtig beantwortet worden ist oder nicht.

Datensatzbeschreibung

An der Unterrichtseinheit zum Logarithmus („Exp/Log“) haben 73 Studenten teilgenommen, an der Unterrichtseinheit zur Differentialrechnung („Diff“) haben 83 Studenten teilgenommen. Die Lernerfolgskontrolle Exp/Log umfasste 17 Fragen, die Lernerfolgskontrolle Diff umfasste 11 Fragen. Für den Fall, dass ein Student eine Frage nicht beantwortet hatte, wurde die entsprechende Datenzeile aus dem Datensatz für diese Frage entfernt. Hatte der gleiche Student andere Fragen beantwortet, so sind diese Antworten im Datensatz verblieben.

Damit ergibt sich für beide Datensätze für die Gesamtheit der Fragen folgendes Bild:

	Exp/Log	Diff	Summe
Falsch	530	451	981
Richtig	664	391	1055
Summe	1194	842	2036

Von den insgesamt 1194 vorliegenden gültigen Antworten des Datensatzes Exp/Log war die Mehrheit der Antworten (664) richtig, während die Mehrheit der 842 Antworten des Datensatzes Diff falsch war.

Das Analyseverfahren

Da die Zielgröße nur die Ausprägungen „richtig“ und „falsch“ annehmen kann, ist diese nicht normal, sondern binomial verteilt. Weitere deskriptive Analysen ergaben:

- Die Verteilung der falschen und richtigen Antworten über alle Fragen war nicht gleich. Es gab offensichtlich leichtere Fragen, die häufiger richtig beantwortet wurden als andere, schwerere Fragen.
- Die Verteilung der falschen und richtigen Antworten über alle teilnehmenden Studenten war ebenfalls nicht gleich. Einige Studenten waren offensichtlich leistungsstärker und beantworteten mehr Fragen richtig als andere.

Beide Effekte müssen in der weiteren Analyse als Störgrößen berücksichtigt werden.

Ein geeignetes Modell, welches sowohl binomial verteilte Zielgrößen als auch Störgrößen berücksichtigt, ist das verallgemeinerte lineare Modell für eine binomial verteilte Zielgröße und der Logit-Funktion als Linkfunktion. Der für die

Untersuchung relevante Effekt ist die Unterrichtseinheit mit den Ausprägungen „Diff" und „Exp/Log", während die anderen beiden Effekte als Störgrößen mitmodelliert wurden. Die Modellanpassung wurde mit der Prozedur Glimmix im Programmpaket SAS durchgeführt.2

Ergebnis und Diskussion

Zunächst wurde die Hypothese geprüft, ob es einen signifikanten Unterschied zwischen den beiden Unterrichtseinheiten Diff und Exp/Log gibt. Das Ergebnis zeigt die folgende Tabelle:

Type III Tests of Fixed Effects				
Effect	**Num DF**	**Den DF**	**F Value**	**Pr > F**
Unterrichtseinheit	1	1923	41.23	<.0001

Unter Kontrolle der beiden genannten Störeffekte gibt es einen signifikanten Unterschied bei den Testergebnissen von der Unterrichtseinheit Exp/Log und der Unterrichtseinheit Diff. Die Regressionskoeffizienten der verallgemeinerten linearen Modelle können allerdings nicht analog der linearen Regression als Steigungsparameter interpretiert werden, weshalb sie hier nicht dargestellt werden. Um nun die Wirkungsrichtung und die Stärke des Unterschieds in beiden Unterrichtseinheiten zu bestimmen, kann hier der Odds Ratio[3] herangezogen werden:

		Schätzwert
Diff	Exp/Log	0,474

Die Chance, in der Unterrichtseinheit Diff die Fragen richtig zu beantworten, ist lediglich 0,474 Mal so groß wie die Chance, in der Unterrichtseinheit Exp/Log die Fragen richtig zu beantworten – also weniger als halb so groß.

Ergebnisinterpretation und Empfehlungen für die weitere Forschung

Die Analyse zeigt, dass Studenten, die an der Unterrichtseinheit Diff teilgenommen haben, die Fragen signifikant schlechter beantwortet haben als Studenten, die an der Unterrichtseinheit Exp/Log teilgenommen haben. Die

2 Dabei wurden die Unterrichtseinheit und die Fragen als feste Effekte modelliert, während die Leistungsstärke der Studenten als zufälliger Effekt modelliert worden ist.

3 Für eine Definition und Interpretation des Odds Ratios vgl. etwa Hosmer/Lemeshow (2000), S. 49 ff. sowie die dort angegebene Literatur. David W. Hosmer / Stanley Lemeshow (2000), „Applied Logistic Regression", 2nd. Edition, Wiley, New York. http://de.wikipedia.org/wiki/Odds_ratio, Abruf am 27.02.2013.

Erfolgschance der Studenten der Unterrichtseinheit Diff war nicht einmal halb so groß wie die Erfolgschance der Studenten in der Unterrichtseinheit Exp/Log.

Allerdings kann aus diesem Ergebnis noch nicht geschlossen werden, dass Gestalt einen Einfluss auf die Testergebnisse hat, da noch nicht alle potenziellen Störgrößen kontrolliert worden sind. Auf der anderen Seite kann ein Einfluss von Gestalt nicht ausgeschlossen werden.

Eine Störgröße, die bisher nicht kontrolliert werden konnte ist, dass die Studenten in den beiden Unterrichtseinheiten aus dem gleichen Personenkreis stammten. Bei der Datenerhebung wurde allerdings nicht darauf geachtet, übergreifend eindeutige Studenten-IDs zu erfassen. Unterschiede im Antwortverhalten einzelner Studenten zwischen den beiden Unterrichtseinheiten konnten so nicht im Modell berücksichtigt werden. Weiter wurden die Unterrichtseinheiten zwar von der gleichen Person, aber an verschiedenen Tagen oder Stunden abgehalten, sodass es auch hier zu Unterschieden beim Dozenten etwa in Form von Stimmungsschwankungen über die verschiedenen Zeiten kommen kann. Diese konnten ebenfalls nicht kontrolliert werden.

Aufgrund der Ergebnisse und hier insbesondere der deutliche Unterschied bei den Erfolgsquoten beider Unterrichtseinheiten ist eine Wiederholung des Versuches, diesmal mit sorgfältiger Planung unter Kontrolle aller möglichen und bekannten Störvariablen, ratsam.

Zur Sicherheit wird die Auswertung ein weiteres Mal durchgeführt, diesmal mit GrenzwReflekt. Einmal werden die beiden Datensätze GrenzwReflekt zu einem zusammengefasst, einmal werden sie getrennt modelliert. In beiden Fällen ist der Unterschied in den Testergebnissen zwischen den einzelnen Veranstaltungen zu Exp/LogReflekt hoch signifikant (jeweils $p < .0001$), und auch die Odds Ratios betragen alle um die 0,5, d.h. die Studenten im Datensatz Exp/LogReflekt haben deutlich häufiger richtig geantwortet.

Anhang 2 zu Abschnitt 1: Diff und Grenz (beide ohne Gestalten)

Die beiden Datensätze DiffReflekt und GrenzwReflekt werden gegeneinander ausgetestet, also ohne Exp/LogReflekt. Fasst man beide GrenzwReflekt-Datensätze zu einem zusammen, ist der Unterschied in den Testergebnissen zum DiffReflekt nicht signifikant (p-Wert von 0,67). Trennt man hingegen die beiden Teildatensätze von GrenzwReflekt auf und testet dann gegen die Ergebnisse von DiffReflekt, so gibt es signifikante Unterschiede (P-Wert von 0,0287). Grund ist, dass die Ergebnisse im ersten Teildatensatz von GrenzwReflekt signifikant besser sind als im zweiten Teildatensatz (P-Wert 0,0141) - (Unterschied zwischen den beiden Teildatensätzen von GrenzwReflekt).

Insgesamt kann man sicher sagen, dass die Unterschiede zwischen DiffReflekt und GrenzwReflekt deutlich geringer sind als deren Unterschiede zu Exp/LogReflekt.

Unser Dank gilt Herrn Rabe und dem Team aus Best4HEL, das uns die Daten zur Verfügung gestellt hat.

2. Der zweite Anlauf, die klassische (lineare) Unterweisung

Ein weiterer Durchlauf (Klausur) in Mathe1 (nach Abschnitt 1) mit seiner Statistik und die zugehörige Interpretation zur neuen Klausur Mathe1 wird dokumentiert. Der Gestalt-Teil ist durch einen klassischen Teil ersetzt, so dass es sich ausschließlich um klassischen Hochschulunterricht handelt, wie er den Professoren vertraut ist. Wir beziehen uns hinsichtlich der Ergebnisse auf die Excel - Tabelle statistik-Mathe1_SS13.xlsx, die von den Professoren aufgestellt worden ist.

Zu Beginn der Mathe-Vorlesung (noch keine Klausur) waren etwa 85 Teilnehmer dabei.

Die jetzige Klausur war vergleichbar schwierig zu den vorausgehenden Klausuren. Es wurde zusätzlich darauf geachtet, dass die Texte keine Stolperstellen (z.B. Fremdwörter, Schachtelsätze) aufwiesen.

Es haben ***32 Wiederholer*** teilgenommen. In den Vorlesungen Mathe1 waren regelmäßig etwa ***23 Wiederholer anwesend***.

9 Wiederholer haben sich also die Vorlesungen und Übungen "erspart", kenntlich in Spalte "T" der Excel-Tabelle: Ich nehme sie aus der eigentlichen Statistik heraus, denn sie halten nach ihrer Auffassung das Wiederholen der Vorlesungen und Übungen nicht für erforderlich. Damit sind sie für welche personell geführte und wie auch immer verbesserte Lehre nicht erreichbar. Es sind dies die Wiederholer 11, 13, 16, 22, 23, 24, 31, 35, 36. Von diesen 9 Wiederholern hat nur einer bestanden, Kandidat Nr. 11 mit 51 Punkten (50 Punkte waren zum Bestehen das Minimum). Die anderen Kandidaten liegen mindestens 10 % unter 50 Punkten.

Die Bestehensrate liegt bei 11 %.

4 dieser Kandidaten waren Wiederholer im 2. Versuch, 5 im 3. Versuch. Der Kandidat, der bestanden hat, hatte seinen 3. Versuch.-

Nun zu den Wiederholern, auf die die Kollegen R. und W. haben Einfluss nehmen können. Es sind die Kandidaten 4, 5, 6, 7, 8, 9, 10, 12, 14, 15, 17, 18, 25, 26, 27, 28, 29, 30, 32, 33, 34, 37, 38. Von diesen Kandidaten haben 7 bestanden: 4,

7, 9, 12, 28 34, 37. *Die Bestehensrate liegt bei 30 %, also dem 3-fachen im Vergleich zu den Kandidaten, die die beiden Kollegen nicht haben erreichen können,* weil die Studies zu den Vorlesungen und Übungen nicht erschienen sind.

Es hat sich also sowohl für die Wiederholer als auch für die Kollegen, die sich der Mühe unterzogen haben, gelohnt. Allerdings sehen die Kollegen das Ergebnis unter Berücksichtigung des Aufwandes in der Lehre - wie wir - nicht als wirklich erfolgreich an.

15 dieser Kandidaten, die an der Wiederholung der Vorlesung teilgenommen haben, waren in der Klausur im 2. Versuch, 8 im 3. Versuch. Von den 15 Kandidaten haben 2 bestanden, von den 8 deren 5 bestanden. von den 3 Kandidaten nach der 3. Wiederholungsprüfung war dann eine mündliche Prüfung mit weiterer Vorbereitung (üben, üben, üben…) fällig. Damit hat es dann bei allen diesen Dreien geklappt, wie mir mündlich mitgeteilt wurde.

Der Anteil, den die Wiederholer im 2. Versuch für das Üben aufgewendet haben, war offenbar für die meisten dieser Kandidaten nicht ausreichend. Wer vor dem 3. Versuch war, hat „33 %" mehr geübt (als in 2 Versuchen). Dann hat es gerade so eben gereicht, zu bestehen: 63 % der Drittwiederholer haben bestanden. Das "gute" Gefühl, das die Zweitwiederholer hinsichtlich des Übens hatten, war offenbar nicht zielführend. Die Fehleinschätzungen sind auffällig.

So merkwürdig das meinen Kollegen auch klingen mag: *63 % Erfolgsquote für Drittwiederholer mit Wiederholen der Vorlesungen und Übungen ist für klassischen Unterricht erfolgreich.* Auch die anderen 3 Drittwiederholer haben 41,5 bis 45,5 Punkte erreicht, sind also nicht vollständig abgeschlagen und haben nach weiterem (offenbar nochmals intensiven) Üben (wie mir berichtet wurde) in einer mündlichen Prüfung noch bestanden. Also *100 % Erfolgsquote nach dem 4. Wiederholen und Üben für diejenigen, die diesen Stress durchgehalten haben.* Trotzdem haben über alle gesehen, die anfangs angetreten sind, nur 35 % Mathe1 geschafft.

Welche Verbesserungen halten wir für möglich?

Offenbar klappt das mit dem Üben im klassischen Unterricht immer noch nicht ausrei-chend, sei es auf der Seite der Studierenden (z.B. als Bauchgefühl), sei es im Angebot der Lehre. Es erscheint insgesamt wie ein endloses Hindernisrennen.

Über die Art des Übens gibt die Statistik keine unmittelbare Auskunft. Die Studierenden unterschätzen den Aufwand, den sie für das Üben in der klassischen Form der Lehre inve-stieren müssen. Beim 2. Versuch fehlen noch mindestens 35 % an herkömmlicher Übungs-zeit, realistisch eher 50 %, um das Erreichen der Ziellinie sicher oberhalb von 50 Punkten zu erreichen. Von Seiten der Lehre muss das Üben effektiver gestaltet werden (und das erscheint mit Hilfe von Anschauung und Gestalten möglich (siehe Abschnitte 3 und 4).

Die Grundsätze der Erkenntnistheorie sind bei den Nachzüglern nicht ausreichend verankert ("Morgen- und Abendandacht" fehlt, wie im Beispiel des Projektstudiums nach 10 Semestern gezeigt). Es geht um Selbstreflexion oder Anleitung zur Reflexion.

Die Nachzügler sind, wie es aussieht, zufrieden, sobald sie den Inhalt zum ersten Mal verstanden haben. Das reicht offenbar nicht.14 Tage später muss es auch noch verstanden und die Vorgehensweise abgesichert und automatisiert ablaufend sein! Das ist ein Teil der Selbstdisziplin und der Erkenntnis, wie das eigene Gedächtnis funktioniert und was es nicht leisten kann, wie man dem gegensteuert. Mit unausgegorenen Vorschlägen, wie man "besser", „zielstrebiger" studiert, ist es nicht getan. Sie führen in die Irre.

Ob die Nachzügler im 2. Versuch das Üben intuitiv wirklich berücksichtigt haben? Und was ist mit dem trügerischen Gefühl der Sicherheit, während das Fundament beim Lernen noch wackelt? Das ist genau so trügerisch, wie das Gefühl, mit Gestalten zu Lernen sei zu einfach und daher sofort abzubrechen.

Den Fehler, den Lon-Capa in sich trägt, haben wir in einer gesonderten Arbeit (Grenz) ausführlich erläutert. Die Zeit kann man viel besser für Anderes investieren. Ein tumbes Bauchgefühl von Studies nach 2 (bis 9) Tagen darf nicht so weitreichend sein, dass eine sicher funktionierende Methode „mit Gestalt" (die übrigens über alles gesehen die weitaus geringste Zeit in Anspruch nimmt, siehe Mathe1-Statistik des ersten Durchlaufs) allen Studies auf Dauer verwehrt wird. Sonst sind die 65% Abbrecherquote in Mathe1 unabänderlich.

Abschließend möchte ich auf das Intensivieren von Gestalten oder zumindest von Anschauung hinweisen. Das hilft dem Gedächtnis, sich zu erinnern. Darauf muss die Lehre mehr Zeit investieren, um das Ausufern des Übens in den Griff zu bekommen.

3. Auf den Spuren von Anschauung, Gestalt und Begriff im Hochschulunterricht Mathematik nach dem Coaching

Die Situation

Nach einem Jahr pädagogischer Betreuung (Coaching) hinsichtlich Anschauung und Gestalt für die Kollegen in Mathe1 findet die Vorlesung Mathe1 mit anschaulichen Elementen und die jetzige Klausur Mathe1 nun ohne eine solche Betreuung statt.

Bild 4: Statistik der Klausur Mathe1 für Erstsemester und Wiederholer Januar 2014

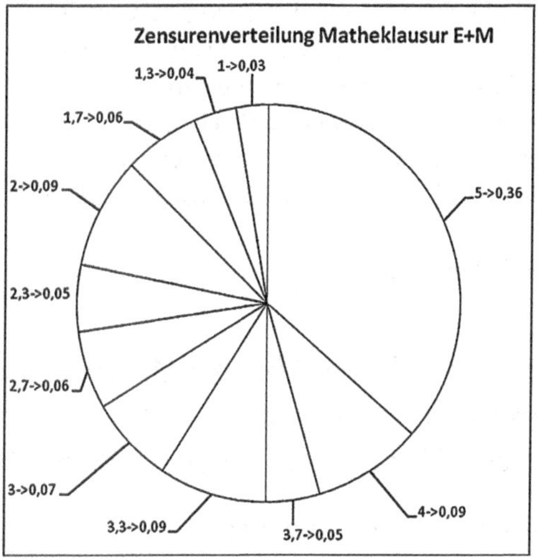

Das Bild ist vom Kollegen Rabe übernommen. Die Aufgabenschwierigkeit der Klausur für externe Beobachter ist gegenüber den Vorjahren vergleichbar (für die Studierenden erschien sie selbstverständlich leichter als den Studierenden in den Vorjahren).

Es haben 110 Teilnehmer an der Klausur Mathe1 Januar 2014 der Studiengänge Elektrotechnik und Medientechnik teilgenommen.

In den vorausgegangenen Semestern nahmen unter den gleichen Kollegen jeweils ca. 85 Teilnehmer mit einer Durchfallrate von ca. 65 % teil.

Fazit

Wir haben diesmal nach dem ersten Versuch 70 Studies mit der Erlaubnis, unbeschwert weiter zu studieren, gegenüber ca. 34 in den Jahren zuvor. ***Die Zahl der Studierenden, die bestanden haben, hat sich also verdoppelt.*** Anstatt 35% der Studierenden haben nunmehr 64% bestanden.

Das sollte bei uns Freude und das Verlangen nach Fortsetzung auslösen!

Anmerkungen

Das bedeutet aber auch, dass die Labore ein wenig überquellen werden (siehe das Projektstudium im 1. Semester 2001 und die damaligen Konsequenzen für 2002).

Dort wurde schon angemerkt (und hier vorausgesagt), dass mit einer intensivierten wissenschaftlich orientierten Pädagogik ("Gestalten") mindestens 90% der Studierenden bestehen können(, und das nicht nur in der Elektrotechnik und Medientechnik, sondern im gesamten technischen Bereich, der in der Vergangenheit die hohen Durchfallquoten aufweist).

Ein Kollege hat das Ergebnis wie folgt kommentiert:

Die Schulen hatten sich gemeinsam entschlossen, uns diesmal "gute" Studenten zu liefern!

Herzlichen Dank. Weiter so!

4. Nachtrag zum dritten Abschnitt: Anschauliche Unterweisung mit Spuren von Gestalt
(mit verfasst von Dirk Rabe, Joachim Wiebe)

Die Situation für die Nachklausur:

Nach einem Jahr pädagogischer Betreuung hinsichtlich Anschauung und Gestalt für die Kollegen in Mathe1 wurden die Vorlesung Mathe1 und die Erstklausur Mathe1 (Abschnitt 3) auf Wunsch der Kollegen also ohne Coaching gehalten. Die Wiederholer (vgl. Bild 5) erschienen zur Nachklausur Mathe1 im zweiten Versuch.

Bild 5: Statistik der Nachklausur Mathe1 für Erstsemester und Wiederholer März 2014

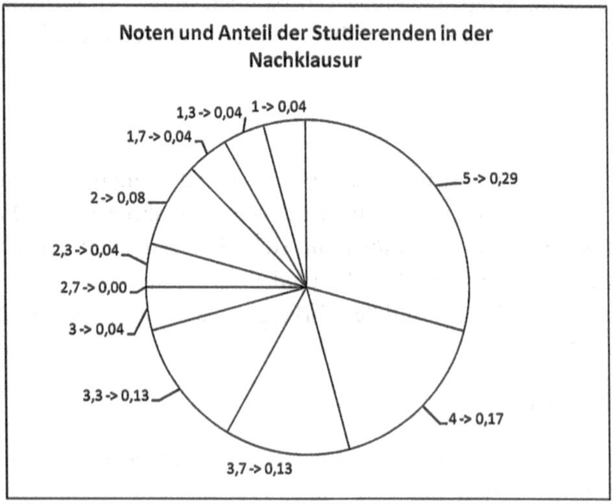

Die Aufgabenschwierigkeit der Nachklausur für den externen Beobachter ist gegenüber den Vorjahren und der vorausgehenden Klausur vergleichbar.

Es haben 110 Teilnehmer an der Klausur Mathe1 Januar 2014 teilgenommen (Abschnitt 3), an der Nachklausur im März (Abschnitt 4) sind es 24 Teilnehmer. Bild 5 zeigt, dass 29% der Teilnehmer die Nachklausur nicht bestanden haben. Das sind bei 24 Wiederholern 7 Teilnehmer. 17 haben bestanden. Im ersten Durchlauf hatten bereits 70 Teilnehmer bestanden (Abschnitt 3).

Damit haben nach dem 2. Durchlauf in der Summe 87 Teilnehmer bestanden, das sind insgesamt 79%. Demgegenüber haben in den Vorjahren nach vier Klausuren 34 Teilnehmer von regelmäßig etwa 100 Teilnehmern bestanden, rund 35 % (z.B. Abschnitt 2). Die weiteren 65 % dieser Teilnehmer in den Vorjahren sind damit endgültig durchgefallen, müssen das Studium abbrechen.

Mit der Klausur und der Nachklausur ergibt sich, dass anstelle der seit Jahren üblichen 34 Studierenden nunmehr zusätzlich 53 Studierende weiter studieren können. Die Hochschule kann also 87 Studierende in zwei Studiengängen des Fachbereichs Technik zum Bachelor oder zu weiteren Studien führen. Ohne die Hilfe dieser statistischen Daten unterschätzen die Lehrenden, die Kollegen in der Umgebung und die Studierenden die Erfolgsquote in getrennten stichprobenartigen Befragungen vor Bekanntgabe der Resultate um typisch mehr als einen Faktor zwei.

Bild 6: Prozentsatz der Studierenden über der erreichten Punktzahl nach zwei Klausuren

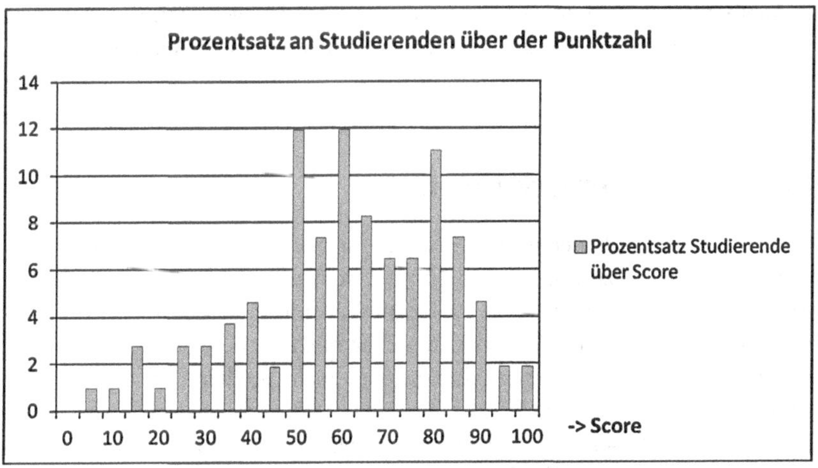

50 % der Punktzahl reichen zum Bestehen. In Bild 6 sind die Studierenden, die in der Nachklausur angetreten sind, mit ihren ersten Klausurergebnissen gelöscht

und stattdessen für sie die Klausurergebnisse der Nachklausur eingetragen. Man erkennt - wie an den PISA-Daten von Finnland bis 2006 - die empirische Harke (Siemsen 2013), mit der die Studierenden aus der linken Hälfte in die rechte Hälfte herübergezogen worden sind (vgl. auch Bild 7 und Bild 8).

Ca. 23 Studierende (21%) von 110 Studierenden haben Mathe1 nach 2 Versuchen noch nicht bestanden. Für sie bietet es sich an, die vorhandenen methodischen Mängel (die im Dechiffrieren von Gestalten und in der Erkenntnistheorie liegen) zu beseitigen und sie in wenigen Tagen für die nächste Prüfung und insgesamt für das Studium methodisch fit zu machen, vorausgesetzt, der Fachbereich ist bereit, die minimale Investition für diesen Zweck bereitzustellen.

Man erkennt, dass – ausgebildete Lehrende vorausgesetzt – der Erfolg sich unmittelbar im selben Semester, in dem die veränderte Lehre zum Tragen kommt, einstellt (siehe auch Siemsen etal. 2013). Aus vorherigen empirischen Untersuchungen ist bekannt, dass einerseits die Lernleistung bereits unmittelbar nach dem Lernprozess abgerufen werden kann, andererseits noch nach einem halben Jahr sicher vorhanden ist (Siemsen 1981). Das ergibt sich aus der Verarbeitung durch Gestalten anstelle durch das Lernen mit Hilfe von Logik.

Die Lehrenden dagegen brauchen zuvor ein Jahr, bis sie auf diese Weise ausgebildet sind. Wenn sie verstanden haben sollen, warum das Vorgehen erfolgreich ist (also nicht nur, dass es erfolgreich ist) brauchen sie mindestens ein weiteres Jahr. Die Begriffe, die sie aus der Vergangenheit mitbringen, behindern ihr Denken, ein Grund für die wesentlich längere Dauer der Ausbildung der Lehrenden.

Allein auf den Fachbereich Technik insgesamt bezogen, könnten auf diese Weise jährlich 250 bis 300 Studierende zusätzlich zu einem Abschluss geführt werden.

Ergänzende Bemerkungen

Was ist mit den 21% der in der Nachklausur Nichterschienenen? Sie werden hier als *Schwund* bezeichnet. In Abschnitt 2 ist ihre Zahl erheblich größer als in Abschnitt 4, sodass man die Vermutung äußern kann: je hürdenreicher und steiniger (längerdauernd) der Weg zum Bestehen noch vor dem Studierenden liegt, je logizistischer die Lehrveranstaltung, umso mehr Studierende, die betroffen sind, werfen das Handtuch.

Einer der Verfasser hat jahrelangen Umgang mit einigen Hundert der in dieser Hinsicht gefährdeten Studierenden und ist der Ansicht, man könne fast alle in kurzer Zeit (zwei Tage bis maximal sechs Wochen) auffangen. Kenntnisse der Psychophysik und der Erkenntnistheorie in der Ausbildung der Lehrenden sind die Schlüssel, die Probleme zu lösen und diese Studierenden zu stabilisieren. Die ehemaligen Nachzügler reihen sich nicht unbedingt am unteren Ende der Statistik ein, sondern überdecken das gesamte neu entstandene Spektrum (siehe Bild 7 und Bild 8; Siemsen 2013). Wohlgemerkt: Aufgabenschwierigkeit und Dauer an Lehre sind bis dahin konstant im Vergleich zu Abschnitt 2. Ob ein Fachbereich oder eine Hochschule diese vergleichsweise zusätzliche geringe Investition gewillt ist, zu tätigen, ist eine andere Frage. Wäre die Hochschule dazu gewillt, kämen zu den 79% (87 Studierende) noch einmal mindestens 15% an Studierenden hinzu, die die Hochschule zu einem Examen führen kann. Beide Ergebnisse wurden prinzipiell vorausgesagt.

Nun zur Frage der Stabilität des Vorgehens, der Nachhaltigkeit. Das finnische Bildungssystem ist dafür ein gutes Beispiel. Der finnische Kollege zusammen mit seiner Frau hat zwei Jahrzehnte Lehrer in einem internetgestützten Kurs ausgebildet, zwischen 200 und 300 Lehrende (Kurki-Suonio 2010).[4] Er hatte sich die Machsche Methodik in der Kailaschen Variante selbst intuitiv angeeignet. Nach zwei Jahrzehnten ging er in den Ruhestand. Die finnischen Bildungsergebnisse blieben stabil, die von ihm ausgebildeten Lehrer arbeiteten weiter nach seinem Verfahren. Allerdings wurden von da an keine neu hinzukommenden Lehrer auf diese Weise ausgebildet, da die Kailasche Theorie unter Philosophen und Physikern heftig umstritten war. Nach 10 weiteren Jahren begann das finnische Bildungssystem zu erodieren: die von Kurki-Suonio ausgebildeten Lehrer gingen nun ebenfalls der Reihe nach in den Ruhestand. Der finnische Erfolg endete „abrupt". Die Experten standen bei der Suche nach der Ursache vor einem Rätsel.

Das sagt uns, dass es ohne eine schriftliche fixierte und widerspruchsfreie Theorie in der Bildung auf Dauer nicht geht. Einige der bisherigen Grundlagen für Bildung sind zudem mit erheblichen Fehlern behaftet, ohne dass die Forscher, die das erfunden hatten, das zum damaligen Zeitpunkt erkennen konnten. Teilweise haben sie sogar die Einschränkungen genannt, unter denen sie ihren Ansatz vertreten haben. Spätere Entdeckungen (z.B. Evolution (Genese), Sensualität oder Empirie) haben die Fehler in den Ansätzen aufgedeckt, aber die Einschränkungen sind durch den gewohnheitsmäßigen, häufigen Gebrauch des veralteten Ansatzes übersehen worden. Sich mehrheitlich von diesen Fehlern zu lösen, ist ein schwieriges und langwieriges Unterfangen über Jahrzehnte, wenn nicht gar Jahrhunderte (Siemsen, H. 2014).

4 Die anderen Lehrenden wurden klassisch ausgebildet (etwa wie in Abschnitt 2 beschrieben).

Die Verfasser haben sich für eine doppelt abgesicherte Vorgehensweise entschieden: sie prüfen jeden Schritt durch Statistik („empirisches Vorgehen", allerdings mit dem damit verbundenen Nachteil, dass der Aufwand für Statistik groß, die Lesbarkeit der Ergebnisse wegen des statistischen Werkzeugs gering ist und man auf diese Weise nur scheinbar langsam voran kommt, dafür aber besser abgesichert gegen die Metaphysik früher entstandener Begriffe) und ziehen als Theorie diejenige von Mach (Vorläufer Jacobi und den Neokantianer Beneke, Nachfolger die Gestaltpsychologen Wertheimer, Ch. Bühler und Kaila) heran (Sinnes- und Gestaltwahrnehmung, denk-ökonomische, daher monistische Weltsicht, Erkenntnistheorie, Psychophysik; dies sind die Säulen der Theorie, wie wir sie sehen; das Entfernen auch nur einer dieser Säulen führt offenbar sofort zu schweren inneren Widersprüchen und zu dramatischem Versagen der erwarteten Ergebnisse durch Statistik, vergleichbar mit Bauten, wenn dort auch nur eine Säule entfernt wird und die Statik nicht mehr stimmt). Dieses theoretische Konzept in seiner Gesamtheit wird von der Statistik bislang nachhaltig unterstützt. Mit anderen Worten: Kon-zept und Statistik gemeinsam sind stabil, haben bislang nicht zu Widersprüchen geführt. Es sind mit diesem Verfahren sogar präzise, nicht mehrheitenfähige Voraussagen über die Wirkung von Bildung zu machen, wie das Beispiel Finnland (Siemsen etal. 2014) und Abschnitt 1 eindrucksvoll zeigen. Sobald ein statistisches Vorgehen selbstverständlich geworden ist, kann es automatisiert werden. Ergebnisse sind dann – vergleichbar mit Werkzeugen wie Excel – ohne Aufwand sofort verfügbar.

Wer bereit ist, trotz der erschwerten Lesbarkeit statistische Daten geduldig zu lesen, wird dem Konzept, wie wir es in den vier Abschnitten angewendet haben, unter Anleitung („Coaching") folgen können. Der Lehrende bringt – wie in Finnland – den Studierenden bei, zu lernen, sie sind in der Lage, Lernen richtig anzupacken. Allerdings braucht man, um das, was man tut, zu verstehen oder gar zu verbessern, die zugehörige Theorie, die sich nach bisherigem Kenntnisstand nicht in wenige Sätze verpacken lässt, leider sich zudem auf Begriffe stützt, die umgangssprachlich ungewohnt sind, weil die heutigen umgangssprachlichen Begriffe nach unseren Erkenntnissen Lernen nicht angemessen beschreiben. Zudem braucht es ein gerüttelt Maß an Statistik.

Der Traum einiger Professoren ist, den „Supereinstein" durch Selektion unmittelbar zu finden. Die Realität zeigt, dass ein einziges Leben nicht ausreicht, um mit dieser Vorgehensweise in dieser Frage auch nur einen Treffer zu landen. Der etwas realistischere Versuch ist, mit Hilfe der Logik Inhalte zu ordnen (Abschnitt 2). Das führt für die Studierenden zu einem schier endlosen Hürdenlauf vorbei an Stolpersteinen, immer nahe am Aufgeben (Zitat Hans Hahn in Abschnitt 1). Die 32 Wiederholer der 2. Klausur (Abschnitt 2) liegen bei einem sehr hohen Prozentsatz und haben, dem logizistischen Ansatz folgend, so gut wie nichts verstanden und bestehen ohne inhaltliche Wiederholung zu nur 11 %, mit erneuter inhaltlicher Wiederholung und nochmaligem Übungsangebot zu 30 % (vgl. Abschnitt 2). Im Falle eines Lehrangebotes mit Bezug zur Anschauung und zu Gestalten *zu Beginn des jeweiligen Lernens* (Abschnitt 3) sieht das im Einklang mit Comenius und Pestalozzi völlig anders aus. Vor der Nachklausur haben die Dozenten keine Wiederholung angeboten. 24 Studierende unterhalb der Bestehensgrenze der ersten Klausur haben sich der Nachklausur gestellt und bis auf sieben bestanden (vgl. Bild 5). In der Nachklausur liegen die Leistungen der 17 Studierenden über das ganze Spektrum verteilt, also nicht nur „knapp über 50 Punkten", wie in Abschnitt 2. 25% dieser Studierenden übertreffen eine Punktzahl von 75. Das entspricht einer Note 2. Ein vergleichbares Ergebnis haben wir in Finnland beobachtet. Die Abszisse ist deshalb der der PISA-Ergebnisse nachempfunden, ohne dass daraus eine vollständige Übereinstimmung abgeleitet werden kann.

Bleibt noch die methodische Ausbildung der sieben in der Nachtragsklausur gescheiterten. Es ist – im Gegensatz zu den Nachzüglern in Abschnitt 2 - nur eine geringe Anzahl, da lohnt sich ein Nachschulen in Projekten auf jeden Fall, und sie werden mit dieser empirischen Harke (vgl. Siemsen 2013) über das ganze Spektrum verstreut. So wird ein intellektueller Raubbau an Begabten sicher vermieden.

Bild 7: Leistungsbild der Nachzügler ***vor*** *der Nachklausur. Bestehensgrenze 50%*

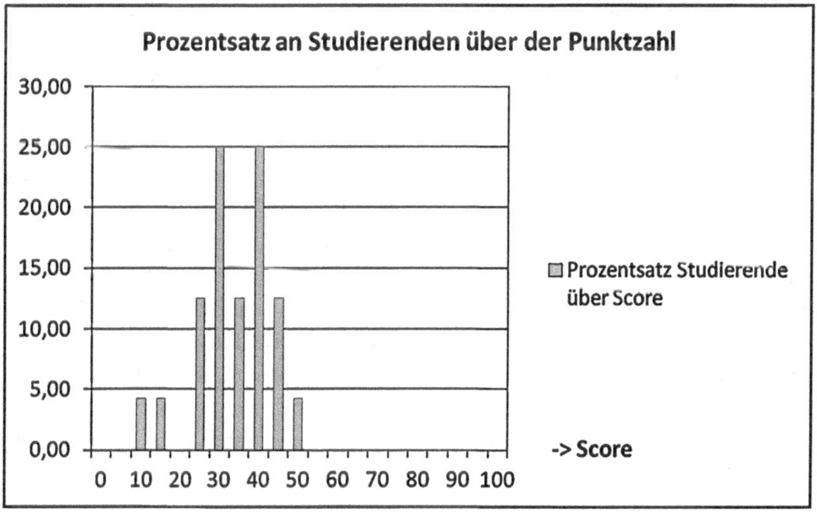

*Bild 8: Leistungsbild derselben Nachzügler **nach** der Nachklausur. Bestehensgrenze 50%*

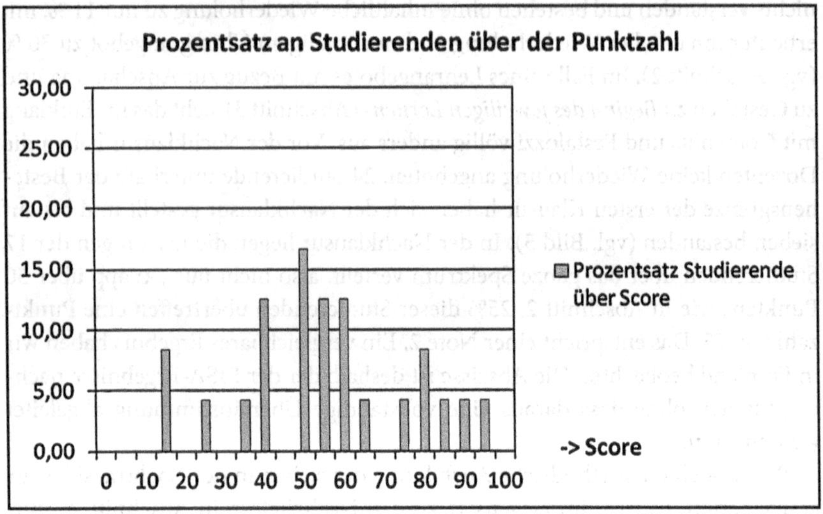

Weiterhin lässt sich das Verfahren empirisch zum *Herausfiltern von besonders Begabten* einsetzen, ein Ansatz, der in Deutschland offenbar hohe Priorität hat. Man verwende das Verfahren der Harke auf die gesamte Kohorte und überarbeitet die Gestalten nochmals, insbesondere hin zu Hapten, am in der Vorgehensweise sinnvollsten über Projekte (Siemsen et al. 2013). Dort wurde der Erfolg in Form von Veröffentlichungen und Dissertationen der Lernenden - also die höhere intellektuelle Leistung - bereits nachgewiesen.

Dramatisch erfolgversprechender ist also ein Vorgehen, das Anschauung, besser noch Gestalten in Form von Hapten und Ikonen und das aktuelle Verstehen der Studierenden in den Vordergrund stellt. Von den in Abschnitt 2 geschilderten „Neben-" Wirkungen (die gleichzeitig ein fester Bestandteil des klassisch logizistischen Vorgehens sind, vgl. Siemsen, H. 2014) ist nichts zu bemerken. Diese Anschauung ist die *Spur* (Chambers 1844), die zu den Gestalten führt und mit denen dann Lernen für Studierende „leicht" zu bewältigen ist (Abschnitt 1, die Statistik über Gestalt). Je stärker die Gestalten den Studierenden in diese Richtung weisen, umso größer ist die Zahl der Studierenden, die schon im ersten Anlauf aus eigener Kraft erfassen, lernen und verstehen (Abschnitt 3, Siemsen 2013). Umso geringer ist der Schwund, d.h. die Anzahl der Studierenden, die vor dem Lernen verzweifeln, weglaufen und das Studium abbrechen. Es lohnt sich, diese erfolgversprechende und historisch beleg-bare Vorgehensweise systematisch anzuwenden.

Dank: Unser Dank gilt Herrn Menno Ternieden für das Ausarbeiten der Zeichnungen.

Literatur

Brühlmeier, A. über Pestalozzi. www.bruehlmeier.info/Anschauung.htm (Zugriff am 10.4.2013)

Chambers, R. (zunächst anonym). Vestiges of the natural history of Creation. 1844. www.archive.org

Hans Hahn. Empirismus, Logik, Mathematik. Besprechung von Alfred Pringsheim. Vorlesungen über Zahlen- und Funktionslehre. S. 68. Suhrkamp, Frankfurt, 1988.

Kurki-Suonio (2010) Principles Supporting the Perceptional Teaching of Physics: A "Practical

Teaching Philosophy". Springer, Science and Education, 22. 8.2010. published online.

Lorenz, K. Vom Weltbild des Verhaltensforschers, Gestaltwahrnehmung als Quelle wissenschaftlicher Erkenntnis. dtv, München 1968.

Luchins, A. S. (1993). On Being Wertheimer's Student. A contribution to the eighth international *Scientific Convention of the Society for Gestalt Theory and its Applications* (GTA) at the University of Cologne, Friday, March 26, 1993.

Schwenk-Schellschmidt, A. Mathematische Fähigkeiten zu Studienbeginn. DNH 1, 2013.

Siemsen (1981) Genetisch-adaptativer rechnergestützter Kleingruppenunterricht. Lang, Frankfurt.

Siemsen, K.H. (2012). Learning by Gestalt. Talk held at the Symposium "100 Years of Gestalt Psychology" 28.-29.09.2012, Helsinki, Finland, herunterladbar unter http://www.lfs.fi/esitelmat/ (accessed 03.01.2013).

Siemsen (2013) Learning by Gestalt. Journal der GTA, Gestalt Theory Vol. 35 No. 2. Krammer, Wien.

Siemsen et al. (2013): Das Projektstudium nach 10 Jahren. Baltische Studien Bd. 26, Lang, Frankfurt.

Siemsen et al. (2014): Finn-ished. Baltische Studien Bd. 27, Lang, Frankfurt.

Siemsen, H (2014): The Genesis of Central Ideas from Ernst Mach. Unpublished, in print.

Wagenschein, M. Was bleibt? In Rumpf, H. (Hrsg.) …zäh am Staunen. Kallmeyer. Seelze-Velber 2002. ISBN 378004196-0

Wagenschein, M. Erinnerungen für morgen. Beltz 752.Weinheim 1983. ISBn 3 407 22752 3

Wertheimer, M. Experimentelle Untersuchungen zur Tatbestandsdiagnostik. Diss. Würzburg 1905, Engelmann, Leipzig.

Wittenberg, A. Vom Denken in Begriffen. Dissertation ETH Zürich 33688-02.

The Pre-service Social Pedagogues'-First Year Students'-Vision of Quality Studies

Gintautė Žibėnienė, Marijona Barkauskaitė

Abstract
Referring to the Law on Science and Studies of the Republic of Lithuania (2009) and the documents of the Bologna Process, it is important to analyze the students' opinion on different aspects related to the quality of studies systematically, analyze the European and national documents and observe that students have become significant participants of the processes assuring the quality of studies. This paper aims to review the pre-service social pedagogues' vision of quality studies. The research instrument is a questionnaire consisting of both open-ended and partially open ended questions. The research was conducted in 2012 and the participants were the first year students of Social Pedagogy from Lithuanian University of Educational Sciences, Mykolas Romeris University and Vilnius College. The survey answered the common methodological requirements, the participants took part in the research on their free will, and the survey was conducted at a convenient time for the students and in line with ethical standards. The analysis of the first year students' opinions revealed that the pre-service social pedagogues relate the vision of *quality studies mostly with the study process (65), study results* (40). Other connections, that were indicated less frequently, include: the psychological environment (16), instructors (14), prestigious and popular institutions and studies (11), different personal considerations (9). The smallest number of the first year students related the quality of studies with material and methodological resources even though they are considered to be a very significant sphere during the external study quality assessment in Lithuania. The analysis of the first year students' opinions, divided into subcategories, demonstrated that most pre-service social pedagogues relate quality studies with practically oriented training which is integrated into the studies, with clear and interesting teaching, with the acquired relevant knowledge, the psychologically comfortable environment and the study results.

Keywords: pre-service social pedagogues, first year students, quality studies, vision of quality studies

Introduction

The European documents induce to involve students into the internal quality assessment of a higher education school. Students are considered to be significant participants of the processes assuring quality studies [10; 11; 12; 13]. The European Union (EU) and Bologna process documents (BOLOGNA beyond 2010) systemically review issues of quality assurance of studies, prompt higher education schools not only to analyze the status of study quality consistently, but also to investigate the opinions of the participants of the study process – students [9].

This European provision is reflected in Lithuania since the national legal acts indicate the need for a school of higher education to systematically announce "… the accurate quantitative and qualitative information about <…> the students', graduates' and other concerned parties' opinion of the quality of studies" [25, article 41 paragraph 2]. Alifanovienė and Jankauskaitė (2009) reviewed the existing research and note that the interest in the quality of studies worldwide prompts discussions and studies across countries, within countries and in schools of higher education [1, p. 71]. Almadani, Reid, and Rodrigues (2011) indicate that the question how a higher quality of studies can be obtained is one of the most topical ones, and suggest focusing on the implementation of goals and the quality of activity with learners in every single stage of the study process [3].

Halsan (2009), Balasooriya et al. (2009) stress the significance of investigating the students' opinion about their studies [4; 14]. Both Lithuanian and foreign researchers are analyzing the opinion of students in different study programmes concerning the issues of quality of their studies, yet, the number of studies (conducted and published) of pre-service pedagogues' opinions is very small. Savelјeva, Petružienė, and Braslauskienė (2011) researched the quality improvement of social pedagogues' training by developing the students' professional vocation and their career vision. The survey of pre-service social pedagogues of different years of study aimed to determine their motivation for vocation decision making, their satisfaction with the chosen profession and the future career plans [27, p. 128]. Alifanovienė and Jankauskaitė (2009) investigated the attitudes of students of Social Pedagogy and Social Work to the chosen study programmes, the motives of choice, the developed skills, the scope of studies and blocks of study subjects, modules and professional practice [1, p. 71].

Shirbagi (2011) reviewed the experience of Iranian universities and notes that in this country the investigation of the quality of studies includes a popular survey "students assess their instructors" [28]. In questionnaires students assess the instructor's teaching, and the development of the general system of quality management. The students' attitude to the quality of teaching, as a significant aspect of quality studies, was analyzed by Allan and Pileičikienė (2010) [2]. Şahin, Zoraloglu, and Sahin Firat, (2010) surveyed the students' opinion about what helps them implement the study goals at the university, and Masic (2013) explored the quality assessment in medical studies at the university of Sarajevo [19; 26]. The opinion of trainee teachers' concerning the issues of quality of studies was reviewed by Baranauskienė, Bukauskienė, and Valatkienė (2011) who focused on the students' attitude to the mission of the university, the goals and the motives of choosing the studies. Bobrova, Grajauskas, and Alūzas (2012) investigated the students' attitude to the quality of teaching; Bobrova, Grajauskas, and Norkus

(2010) analyzed the students' attitudes to the quality of studies evaluating their conception of what learning is and what it means to learn [5; 6; 7]. In 2008 in Lithuania there was a survey, prepared on the basis of a questionnaire designed in the United Kingdom, which explored the students' attitude to teaching, the assessment of students' achievements, the possibilities of personal improvement, academic support and organization of studies, the infrastructure of studies and the external relations of their school of higher education [29].

It is evident that the quality of studies is being analyzed from various perspectives. In the reviewed surveys the students were asked to assess the process of studies, the quality of teaching, the attainment of study goals and results, the material resources and the system of quality management. The surveys also analyzed the students' conception of learning, however, there has been no information found about a research of how students define their vision of quality studies when they start their studies. It is important to determine what the initial expectations of the first year students are, what they relate to quality studies, i.e. what their vision of quality studies is. The European documents state that the pedagogue, together with the school community, is one of the most important figures in the system of assuring the quality of education.

This paper aims to present the first year social pedagogues' conception of quality studies. The analysis of the websites of assessment agencies belonging to INQAAHE[1] and ENQA[2] revealed that in the investigation of the quality of studies the following aspects are important: 1) input (the entrants' level, the criteria of selection and admission, the instructors' qualification); 2) process (the efficiency of realization of study goals, the organization of the study process, the realization of the study curriculum); 3) effect (analysis of the statistical numbers of the graduates, drop outs and successfully employed as well as analysis of the context). Hence, the paper raises a ***problem*** to it is essential to analyze the vision of quality studies of the first year students, especially the pre-service pedagogical workers, as the important future participants of the quality assurance system of studies, from the very beginning of their studies, i.e. the aspect of input.

The research ***object*** is the vision of quality studies of the first year students of social pedagogy. The **goal** is to reveal the vision of quality studies of the first year students, the pre-service social pedagogues.

---

1 INQAAHE, The International Network for Quality Assurance Agencies in Higher Education, http://www.inqaahe.org.

2 ENQA, The European Association for Quality Assurance in Higher Education, http://www.enqa.eu

1. The Organization and Ethical Considerations of the Research

The survey was organized in autumn of 2012; it involved 60 first year students of pedagogical studies: 35 from Mykolas Romeris University, 11 from Lithuanian University of Educational Sciences, 14 from the Faculty of Pedagogy at Vilnius College. The sampling of the research was convenience since the participants of the survey were the students who attended classes, who were not difficult to find on the basis of their target characteristics, i.e. the first year students who had been studying in the study programme "Social Pedagogy" for the first three months.

The research ***methods*** applied: literature and documentation review, a survey of the first year students which included open-ended and partially open ended questions. The data analysis: qualitative content analysis. The students' definition of quality studies was analyzed by means of a traditional (qualitative) text analysis (Krippendorff, 1980, Mayring, 2000), when the analysis is conducted consistently by analyzing and organizing. A multiplex reading of the students' texts was applied, the texts were organized, there were categories distinguished and they were supported by encoding units extracted from the text, and they were interpreted.

The ethical considerations: the survey was conducted following the principle of students' free will, in their study environment, at the time which was convenient for the students. The organizers of the research also obtained the respondents' consent and permission, there was a possibility arranged not to participate in the survey.

2. The Vision of Quality Studies of the First Year Social Pedagogy Students

The Lithuanian legal acts state that the quality of studies is determined by assessing the following spheres: the goals o the study programme and study results, the structure of the programme, the staff, the material resources, the course of studies, the programme management; and the quality of the higher education school is determined by: strategic managements, life-long studies and learning, research and (or) artistic activity, the influence on the development of the region or the whole country [20].

The survey revealed that the majority of first year Social Pedagogy students also relate quality studies with the study process and study results; but only several students consider the material and methodological resources significant. The management of the study programme and other aspects included in the legal acts of Lithuania were not related with the quality of studies by the first year students; however, they indicated new aspects, i.e. the psychological environment, the popularity of the institution and the study programme, prestige and some personal considerations. The analysis of the pre-service social pedagogues' definitions of quality studies distinguished eight categories which were then divided into subcategories (Table 1). The categories were distinguished on the basis of the students' observations and opinions, and the spheres of external quality assessment of studies (referring to the Lithuanian experience). The analysis conducted in relation to the determined categories revealed that the pre-service social pedagogues mostly relate the conception of quality studies with the study process (65), study results (40), the psychological environment (16), the instructors (14), the prestigious and popular institutions and study programmes (11) and different personal observations (9), while the material and methodological resources (3) seem to important for the smallest number of the first year students (Table 1) even though during the external quality assessment of study programmes in Lithuania this aspect is evaluated on the basis of not less than 4 criteria.

The subcategories show that the pre-service social pedagogues mostly relate quality studies with the organization of the study process (42), the quality of teaching (21). The other aspects were mentioned by a smaller number of the first year students, e.g. the acquired knowledge and relevant knowledge (9+9), the psychologically comfortable environment (15), competent instructors (10) (Table 1).

Table 1: Pre-service Social Pedagogues Definition of Quality Studies

Category	Subcategory
STUDY RESULTS (40)	Acquired knowledge (9)
	Relevant, new information (9)
	Acquired knowledge and skills (11)
	Acquired skills (3)
	A good specialist (3)
	Preparation for life (2)
	Versatile preparation of a person (2)
	Successful integration into the labour market (1)
STUDY STRUCTURE (5)	Subjects which are necessary for work (4)
	Useful things (1)

STUDY PROCESS (65)	Teaching (21)
	Organization of the study process (42)
	Study methods (2)
PSYCHOLOGICAL ENVIRONMENT (16)	Psychologically comfortable environment (15)
	Encouragement of cooperation (1)
MATERIAL AND METHODOLOGICAL RESOURCES (3)	Material resources (2)
	Methodological material (1)
TEACHER (14)	Competent (10)
	Developing professionally (2)
	Love their job
	Responsible (1)
PRESTIGIOUS, POPULAR INSTITUTIONS AND STUDIES (11)	Prestige of the institution (5)
	Popularity of studies (3)
	Prestigious studies (3)
PERSONAL OBSERVATIONS (9)	Personal interests (6)
	Personal comfort (3)

2.1. The Conception of Quality of Studies in Relation to Study Results

On the basis of the national documents, the quality assessment of the study process at school of higher education, the preparation of a new study programme or the assessment of the existing one, requires a lot of attention for the clarity of the goals of a study programme and the study results, for the correspondingly granted qualification, for the cycle of studies, as well as for the consistency between the requirements of the labour market and acquired qualification [20]. Nevertheless, the first year students did not consider most of these aspects to be important as they did not even mention them. There is one positive sign, however, that the pre-service social pedagogues (11) mentioned knowledge, skills, and competences. Yet, it was noticed that some first year students still mention only knowledge (18) or skills (3). It can be assumed that at the beginning of the study programme the students did not yet know or did not understand the notion of study result as it is defined in the European documents and scientific literature – it is what learners know, understand and are able to do, what the person will acquire during the study process [15]. There were several opinions that the quality of studies is related with the preparation of a good specialist (3), however, it is not clear yet or it is not indicated what a "good specialist" is (Table 2). Moreover, some first year students believe that a school of higher education has to "prepare a person for life" (2), quality studies prepare "a versatile specialist" (2), "...give as much information as possible...", "when you learn as much as possible what you will need in your work...", or they relate the quality of studies to a person's successful integration into the labour market (1).

Table 2: Interrelation between the Vision of Quality Studies and Study Results

Category	Subcategory	Supporting statements	No.
STUDY RESULTS	**Acquired knowledge**	"....knowledge, information that is necessary at work...", "...knowledge applicable at work...",	**9**
	Relevant, new information	"...newest information...", "...newest knowledge...", "...knowledge of global level...", "...knowledge that is relevant in the labour market..."	**9**
	Acquired knowledge and skills	"...relevant knowledge and skills that are applicable practically...", "...modern knowledge and skills...", "...acquired relevant competences..."	**11**
	Acquired skills	"...acquired skills...", "...practical skills..."	**3**
	A good specialist	"...a good specialist is prepared...", "...a competent specialist..."	**3**
	Preparation for life	"...it is possible to learn different things which will be useful in life...", "...when life skills are acquired..."	**2**
	Preparation of a versatile specialist	"...as much information as possible...", "when you learn as much as possible of what will be necessary in your job..."	**2**
	Successful integration into the labour market	"...after graduation it is possible to successfully integrate into the labour market and the multicultural environment..."	**1**

2.2. The Conception of Quality of Studies in Relation to the Study Structure and Study Process

The analysis of the vision of quality studies of pre-service social pedagogues revealed that most of the students (65) relate it to the study process. The analysis of the students' answers clearly distinguished two subcategories – the organization of the study process (42) and teaching (21) (Table 3).

Table 3: The Interrelation between the Vision of Quality Studies, the Study Process and Structure

Category	Subcategory	Supporting statements	No.
STUDY PROCESS	**Teaching**	"…when it is clearly explained…", "…clearly and slowly…", "…clearly and coherently…, "… teaches in an understandable way…"	**8**
		"…teaches interestingly", "…when it is presented interestingly, it is easily remembered…"	**8**
		"…teaches creatively…", "…quality lectures…"	**3**
		"…demonstrates vividly…", "…uses visualization while explaining…", "visual means are used to illustrate "	**2**
	Organization of the study process	"…practically applicable tasks, not theory-based teaching …", "…a lot of practice…", "…when practical training dominates …"	**10**
		"…create the conditions to improve…", "…the studies prompt improvement…", "…prompt self-education, development…", "…possibilities for self-realization…"	**8**
		"…a possibility to study abroad for some…", "it is possible to have practice abroad"	**5**
		"…the seminars are not boring…", "…organizes seminars in an interesting way…",	**5**
		"…theory and practice are related…", "…after theory it is possible to try things practically…"	**4**
		"…organized creatively and creativity is encouraged…", "…is organized so that creativity is developed…"	**3**
		"…learnt to work in the professional sphere…"	**2**
		"…when there is cooperation with other institutions…", "…real practicians, who explain how things work in practice…"	**2**
		"…while studying a person can determine his strengths, abilities…"	**2**
		"…the tasks are connected with the future work…"	**1**
	Study methods	"…effective methods…"	**1**
		"…group learning…"	**1**

STUDY STRUCTURE	**Subjects which are useful for work**	"...only things which are necessary for work...", "...important for work..."	**4**
	Useful things	"...useful..."	**1**

When the first year students spoke about teaching, they said if it was clear and interesting (8+8) – "...when it is clearly explained...", "...clearly and slowly...", "clearly and coherently...", "teaches in an understandable way...", "explains in an interesting way...", "when it is presented in an interesting way, it is easily memorized..." The students also indicated the visual aspects of teaching (2), three first year students relate the quality of studies with teaching, but they do not explain the idea, merely indicate that "...teaches qualitatively...", "...quality lectures..." These observations relate with the results of a study published in 2009, which focused on the opinions of social pedagogues and social workers [1]. Alifanovienė and Jankauskaitė (2009) determined that "...students most appreciate such forms of study which are based on rendering the knowledge (lectures), and they do not enjoy the forms of study which are related to practical activity, the application of knowledge and skills in research and practical activity" [1, p. 83].

When the vision of quality studies is related to the organization of the study process, the students' opinions diverge. Most of them (10) distinguish the importance of practical training ("...practically applicable tasks, not just theory...", "...a lot of practice...", "...when practical training dominates..."). Slightly fewer first year students (8) relate quality studies with the possibility to develop and improve ("...create the conditions to improve...", "...the studies encourage development...", "...they encourage self-education, development...", "...possibilities for self-realization..."). The first year students find it important that the study process is interesting (5) and there are possibilities to study abroad for some time (5) (Table 3). It is also significant for them that studies are organized "...innovatively...", "in a modern approach...", "...organized in a modern style...", "... studies with multimedia, PowerPoint..."). Some students (9) believe that quality studies are not boring ("...the seminars are interestingly organized...", "...it is interesting to study...", "...interesting studies...").

A lot of researchers (Petty, 2006; 2008, Ramsden, 2000) also emphasize the objective to organize the study process so that it is interesting; they speak about the importance of clear goals and clear explanation, theory and examples, a purposeful selection of tasks and their compatibility [21, 22, 24]. Some first year students relate the vision of quality studies with the correlation between theory and practice (4), and this again answers the key provisions of the pedagogical process [23]. The pre-service social pedagogues relate quality study process with the development of creativity (3), the personal desire to improve (2). These observations are connected with the topicalities of teacher training indicated by Hargreaves (2006): "...teacher training has to be directed at the development of leadership (the ability to transmit knowledge not only in a formal way, but also seeing the real life and the students' needs, the ability to work in a team, problem solving, the ability to involve every pupil and believe in his possibilities)" [16]. Nevertheless, it is interesting to note that even though the selection of evaluation and study methods is an important criterion in the quality assessment of the planned or implemented study programme [20], only 2 first year students related the study methods and the form of organization of studies with quality studies. The assessment of personal achievements was not related to the quality of studies by the interviewed first year students. The results can be explained by the fact that first year students have a limited experience of studying and are not yet able to distinguish the significance of the didactic process.

The Lithuanian national acts regulating the quality of studies state that the structure of studies is a separate sphere which is assessed by experts. In Lithuania it is assessed according to six criteria, accordance with the legal acts, consistency of subjects, accordance with the cycle and type of studies, the volume, content, and connection with the recent achievements of the corresponding sphere, etc. [20]. However, the first year students did not consider the study structure to be related with the quality of studies, this fact was mentioned only by 5 people. Most of them (4) emphasized the following: "...only things which are important for work...", "...important for work..." (Table 3). The first year students paid a lot of attention to the study process, but significantly less to the structure of studies.

2.3. The Conception of Quality of Studies in Relation to Teachers, the Psychological Environment and Material, Methodological Resources

In reference to the Lithuanian national acts regulating the quality of studies, teachers' competence is a distinct sphere of assessment. The analysis of foreign practice concerning the aspects of quality assessment of studies revealed that in all 41[3] reviewed countries human resources, teachers were mentioned.

---

3 Ireland, Albania, Argentina, the Bahamas, Belgium, Czech Republic, Chile, Denmark, Estonia, Philippines, Hong Kong, India, Indonesia, Island, Spain, UAE, Jamaica, Japan, USA, UK, Canada, China, Cyprus, Costa Rika, Latvia, Poland, Lithuania,

The first year students (14 out of 60) relate the quality of studies with teachers. Most first year students (10) mentioned teachers' competence – "...professionals...", "...competent teachers...", "...qualified teachers...", "...experts at their subject..." (Table 4). The first year students also believe that quality studies are possible when teachers develop professionally (2), love their job (1) and are responsible (1). The first year students' relation of quality studies with competent teachers is logical and corresponds to the reseachers' observations. Bogue and Hall (2003), who distinguished the features of qualitatively working educational institutions, think that only the educational institutions which have "impressive" resources (intellectual, human and material) can be of high quality [8, p. 5].

Table 4: Interrelation between the Vision of Quality Studies and Teachers, Psychological Environment and Material, Methodological Resources

Category	Subcategory	Supporting statements	No.
TEACHERS	**Competent**	"...professionals...", "...competent teachers...", "...qualified teachers...", "...experts at their subject...", ".., it can be seen that they really know..."	10
	Developing professionally	"...developing professionally...", "...teachers go abroad to improve their qualification..."	2
	Love their work	"...teachers who love their work..."	1
	Responsible	"...responsible teachers..."	1
PSYCHOLOGICAL ENVIRONMENT	**Psychologically comfortable environment**	"... good relationship between teachers and students...", "...friendly relations...", "... teachers communicate willingly...'	4
		"....the opinion is taken into account...", "...interests are taken into account..."	3
		" ...attention for the student...", "...all kinds of questions are answered..."	3
		"...pleasant environment...", "...friendly environment..."	2
		"...responsible students..."	1
		"...environment enabling a person to adapt well, to answer all the needs, self-expression..."	1

Mexico, New Zealand, Nigeria, Norway, the Netherlands, Oman, South Africa, Russia, Slovakia, Finland, Sweden, Switzerland, Germany

Category	Subcategory	Supporting statements	No.
	Promoting cooperation	"...the environment promotes cooperation..."	1
MATERIAL AND METHOD-OLOGICAL RESOURCES	**Material resources**	"...varied and modern inventory ...", "...modern teaching basis..."	2
	Methodological material	"...all the necessary material for studies, methodological material, the syllabus are easily attained..."	1

The pre-service social pedagogues believe that the psychological environment is an important condition for the quality of studies (16), the great majority relates it with a psychologically comfortable environment (15), and some of the respondents mentioned the environment which promotes cooperation (1). The first year students relate comfort with relationship with teachers (4) "...good teacher-student relationships...", "...friendly relations...", "...teachers socialize willingly...", with the mutual understanding (3) and attention (3). According to Tijūnėlienė (2006), a favourable educational environment can be also facilitated by the teacher's authority [30, p. 58]. The results of this survey of the first year students correspond to the results of the research conducted by Alifanovienė and Jankauskaitė (2009). Students of Social Pedagogy and Social Work wanted a better communication with teachers, individual consultations [1, p. 83].

Least frequently social pedagogues relate quality studies with material and methodological resources (3). While assessing the quality of studies, external assessors pay a lot of attention to the material basis, however, the first year students do not find this aspect important (Table 4). Only two people mentioned material resources ("...varied and modern inventory...", "...modern teaching basis...") and only one person mentioned methodological resources. The comparison of this data with Alifanovienė and Jankauskaitė's (2009) findings concerning the opinions of pre-service social pedagogues and social workers of different years of study reveals ambiguous results: "...most students are currently dissatisfied about the lack of study material: they lack teaching aids, textbooks, educational studies, handouts" [1, p. 83]. This contradiction could have appeared because of several reasons – five years passed after the first survey, during this period a lot of e-books appeared, students now have access to various data bases from any convenient place. On the other hand, the respondents of the study of 2012 were first year students, while Alifanovienė and Jankauskaitė (2009) surveyed students from different years of study, thus, they could have been more demanding.

The Conception of Quality of Studies in Relation to Personal Observations and Prestigious, Popular Institutions and Studies

In their considerations about the vision of quality studies, the first year students mentioned different personal observations which can be grouped into personal interests (6) and personal convenience (3) – "you get what you expected…", "…when it answers your expectations", "…flexible timetable and studies…", "… there are suitable conditions for studies…" (Table 5). These aspects are not included in the external assessment of quality of studies. However, one of the theoretical conceptions of quality of studies is related to the provision "*quality as fulfillment and/or exceeding client's needs*". Competition among schools of higher education prompts the application of models used by institutions working in the sphere of customer service, and thus students are treated as clients. On the basis of this conception, it is significant to identify the users of education services, determine their needs and answer the purposeful ones.

Table 5: Interrelation between the Vision of Quality Studies and Personal Observations, Prestigious and Popular Studies and Institutions

Category	Subcategory	Supporting statements	No.
PERSONAL OBSERVATIONS	**Personal interests**	"you get what you expected…", "… when your expectations are answered"	**2**
		"…it meets my personal needs…", "…I like learning…"	**2**
		"…interesting extra-curricular activities…"	**1**
		"…new acquaintances…"	**1**
	Personal convenience	"…flexible timetable and studies…", "…when there are suitable conditions for studies…"	**2**
		"…the study process takes into account the students' interests…"	**1**
PRESTIGIOUS, POPULAR INSTITUTIONS AND STUDIES	**Prestige of the institution**	"…good name…", "well-known university"	**5**
	Popularity of studies	"…popular studies…", "…popular programme…"	**3**
	Prestigious studies	"… good education…", "…good programme…", "…famous programme…"	**3**

The pre-service social pedagogues determine the vision of quality studies by relating them to prestigious, popular institutions and studies: "…good name…", "well-known university", "...popular studies...", "the programme is in great request...", "...good education...", "...good programme...", "...famous programme...". This opinion of the first year students was not unexpected even though the national documents on the implementation of study programmes do not mention prestige. In the research literature this aspect has been mentioned. Bogue and Hall (2003) related the quality of education with six aspects, one of which says that only the educational institutions which are acknowledged by the society and have a good name can implement quality studies [8, p. 5].

Conclusions

The analysis of the first year students' opinions revealed that the pre-service social pedagogues mostly relate the vision of quality studies with the study process (65), study results (40). The other aspects were mentioned more rarely: the psychological environment (16), teachers (14), prestigious and popular institutions and studies (11), and different personal observations (9).

The smallest number of the first year students relate the quality of studies with material and methodological resources, even though the external quality assessment of studies in Lithuania consider this sphere to be important.

The analysis of the first year students' opinions and the distinguished subcategories revealed that most pre-service social pedagogues relate quality studies with practical training, integrated into the studies, clear and interesting teaching, the acquired relevant knowledge, the psychologically comfortable environment and study results.

References

1. Alifanovienė, D., Jankauskaitė, D. The peculiarities of designing socio-educational study programmes at university type higher educational institutions: students' attitude. *Teacher Education*, 2009, vol. 12 Issue 1, p. 71-84.

2. Allan, A., Pileičikienė N. Teaching quality assessment in university studies: possibilities for application of student survey. *The quality of Higher education*, 2010, no. 7, p. 60-87.

3. Almadani, K., Reid, N., Rodrigues, S. Quality assurance: a pressing problem for education in the 21st century. *Problems of education in the 21st century*, 2011, vol. 32, p. 9-22.

4. Balasooriya, C. D. etc. Impact of a new integrated medicine program on students' approaches to Learning. *Higher Education Research & Development, 2009, vol.* 28, no. 3, p. 289–302.

5. Baranauskienė, I., Bukauskienė, V., Valaikienė, A. Assessment of quality assurance in studies of higher education from the students' point of view. *Pedagogic*, 2011, no. 102, p. 16-24.

6. Bobrova, L., Grajauskas, L., Alūzas, R. Attitudes of teaching and learning quality: students' assessment context in the university physical education studies. *Studies in Modern Society*, 2012, no. 3, p. 30-37.

7. Bobrova, L., Grajauskas, L., Norkus, S. Assessment of quality of university studies of physical education: student's opinion. *Teacher education*, 2010, no. 15 (2), p. 162-176.

8. Bogue, E.G., Hall, K.B. Quality and Accountability in higher education: Improving policy, enhancing Performance. Westport, Connecticut London, 2003.

9. BOLOGNA beyond 2010. Report on the development of the European Higher Education Area, 2010. [accessed 2013-03-01]. <http://www.ehea.info/Uploads/Irina/Bologna%20beyond%202010.pdf >.

10. Budapest_Vienna Declaration on the European Higher Education Area March 12, 2010. [accessed 2012-08-22]. <http://www.ond.vlaanderen.be/hogeronderwijs/bologna/documents/declarations_communiques.htm>.

11. Communiqué of the Conference of Ministres responsible for Higher Education in Berlin on 19 September 2003. [accessed 2011-04-02]. <http://www.ond.vlaanderen.be/hogeronderwijs/bologna/documents/declarations_communiques.htm>.

12. Communiqué of the Conference of European Ministers Responsible for Higher Education, Bergen, 19-20 May 2005. [accessed 2011-04-02]. <http://www.ond.vlaanderen.be/hogeronderwijs/bologna/documents/declarations_communiques.htm>.

13. Communiqué of the Conference of European Ministers Responsible for Higher Education, Leuven and Louvain-la-Neuve, 28-29 April 2009.

[accessed 2011-04-02]. <http://www.ond.vlaanderen.be/hogeronderwijs/bologna/documents/declarations_communiques.htm>.

14. El Hassan, K. Investigating substantive and consequential validity of student ratings of instruction. *Higher Education Research & Development, 2009, vol.* 28, no. 3, p. 319–333.

15. European Commission. Proposal for a recommendation of the European Parliament and of the Council on the establishment of the European qualifications framework for lifelong learning. Brussels: European Commission, 2006. (COM (2006) 459 final).

16. Hargreaves, A. III Tarptautinė Europos ir centrinės Azijos šalių konferencija "Švietimo kokybė ir kontekstualumas", 2006 [accessed 2008-06-09]. <http://www.lyderiulaikas.smm.lt/ll/index.php/nuorodos/autoritetai.html>

17. Krippendorff, K. Content Analysis: An Introduction to its Methodology. Beverly Hills, California: Sage Publications,1980.

18. Mayring, P. Qualitative Content Analysis. Forum: Qualitative Social Research, 2000, No. 1(2). [accessed 2008-04-16]. <http://qualitative-research.net/fqs>.

19. Masisc, I. Quality assessment of Medical education at faculty of Medicine of Sarajevo University - comparison of assessment between students in Bologna Process and old system of Studying. *Acta inform med, 2013*, no. 21(2), p. 76-8.

20. Order No. V-23 of study programme methodic requirements, Quality Assessment Director of the Centre in 22 of April, 2013 (Official Gazette, 2013, No. 49-2469).

21. Petty, G. Šiuolaikinis mokymas. Vilnius, 2006.

22. Petty, G. Įrodymais pagrįstas mokymas. Vilnius: 2008.

23. Rajeckas, V. (1999). Mokymo organizavimas. Vilnius.

24. Ramsden, P. Kaip mokyti aukštojoje mokykloje. Vilnius, 2000.

25. Republic of Lithuania law on Higher education and research, 30 April 2009, No. XI-242, Vilnius (As last amended on 24 April 2012 - No XI-1987).

26. Şahin İ.; Zoraloglu, Y., R.; Sahin Firat, N. 2010. *University Students' Opinions About Factors Affecting Their Achievement Of Educational Goals.* Inonu University journal of the faculty of Education, Vol. 11, Issue. 2, pp. 133-154.

27. Saveljeva, R., Petružienė, S., R. Braslauskienė. Socialinių pedagogų rengimo kokybės gerinimas plėtojant studijuojančiųjų profesinį pašaukimą ir karjeros viziją. *Tiltai,* 2011, vol. 56, issue 3, p. 125-136.

28. Shirbagi N. Iranian university teachers' and students' views on effectiveness of students' evaluation of teaching. *The quality of higher education,* 2011, no. 8, p. 118-131.

29. Studentų mokymosi motyvacija, požiūris į studijų kokybę ir mokestį už studijas, 2008), Mokslo ir studijų stebėsenos ir analizės centras. [accessed 2012-08-10]. <http://www.mosta.lt/images/documents/analize/tyrimai_ir_ataskaitos/2008/studentu_%20poziuris_i_aukstojo_mokslo_sistemos_tobulinima.pdf.

30. Tijūnėlienė O. The problem of the teacher's authoritys a positive factor in education: theory and empirics. *Teacher education,* 2006, no 6, 45-59.

The Controversy of National Education: Towards Depth or Shallowness?

Elvyda Martišauskienė

Abstract
The paper presents the essence and development of national identity in Lithuania. It emphasizes the fact that national education permeates all the spheres of education and fundamentally relates them with humanity; highlights the problems of national education in the context of globalization when the relationships between man and nation are changing, the opposition between the global market and the global society is deepening, the social stratification is growing and the national education is transforming. The paper reviews the national experiences of adolescents which demonstrate the movement of national education towards shallowness.

Keywords: national individuality, national education, problems of national education in the globalization context, adolescents' national experiences

Introduction

In the development of humanity *nation* occupies a special position since it developed, according to Maceina (1990), on the basis of origin (race), residential environment, historic heritage to answer people's essential needs which, just like the inborn emotions that are re-shaped in the process of education, are responsible for the energization and direction of sensations and behavior (Herber & Vásárhelyi, 2003). As it is widely known, Maslow's pyramid of needs (physiological, safety, belonging and love, self-esteem and self-actualization) composes a hierarchic structure. It is important that at least a partial satisfaction of the lower needs enables the realization of higher levels needs. The residential environment and historical heritage have also had its influence on these processes, including the development of a personality. It has to be noted that even the lower level needs have interpersonal relations built in; hence, it is understandable why there developed people's groups of varying unity until there was the level of the *nation* reached and it enabled versatile, sustainable relations. This paper does not aim to analyze those groups of people, but it wants to focus on factors determining national vitality in the modern world and the possibilities and ways of education to affect them.

When the State of Lithuania was restored (1918-1940), the issues of national education became the priority for the most prominent figures, e.g. Šalkauskis, Vydūnas, Maceina and others. The essence of the nation was focused on the national individuality which was approached through the spread of a man, individual, as the undivided whole, since "*national individuality only resides in separate human individuals*"(Maceina, 1990, p.71). Therefore, national individuality was considered to be a mode of being, and affecting at the same time, it was viewed as something which enables the sensation of commonality originating from race, the residential environment and historical heritage: "human individuals are equal *in their essence*. But specifically, in their manifestations, none is *exactly the same*, and none is *totally different* (ibid., p. 72). Just like personal individuality, the national one preserves its indivisibility which comes from the inside, the depths of the nation, and unites the whole community. National identity is characteristic of similarity and diversity. Moreover, it provides the creative impulse, distinctively indicates the direction of activity which sublimates the communal potential, and objectivizes it by means of new versions of science, art, religion, politics etc. The continuum of the national individuality is also changing and acquiring the forms of the Homeland, State and National unions.

Hence, national individuality, as the expression of national spirit, comprises both common human and national springs which can be generalized by Diesterweg's apt observation: "my name is man, and my surname is German" (1988, p. 362). It is important to conceive this in the context of globalization, which is an opposite process of the national crystallization aiming at the emphasis of specific features and otherness. Nowadays, however, the goal is to find common features, similarities which could help nations understand each other and endeavor at sustainable co-existence. No special research is necessary to be able to see the scarce results of these processes when there constantly appear new trouble spots, sometimes even acquiring forms of national fundamentalism. It can be assumed that the interrelation of all the factors (origin, environment and historical heritage) in the modern context does not receive sufficient consideration, and sometimes it is even distorted.

National individuality, as a mode of national existence, evolves irrespective of the level of its perception, but as a way of activity it has to be realized. Hence, there appears national education which was started to develop purposively only in the period of Romanticism, and it has to answer the unique issues presupposed by the environment and the nation's historical destiny. When those issues are well familiar to most nations, national education can assimilate; however, as early as the 19th century it was conceived that moral education, or, using the modern terminology, the ultimate function of education, cannot be globalized as it is rooted too deeply in the national holistic construct, the national spirit since "every nation has a unique system of national education, thus, it is impossible for one nation to borrow the system of moral education from another" (Ušinskis,

1987, p. 86). Therefore, in the context of globalization the issues of national education are becoming even more acute, their solutions, most frequently related with national identity, are controversial, redirecting at the search for conceptual foundations which are the research *problem* of this paper as "in a wider perspective, by embedding the personal and national identity, national education becomes a tool which can enhance resistance to the negative effects of globalization, and consolidate the cultural variety and the significance of the process of civilization" (Grigas, 2007, p. 22).

The research object is national education

The *goal* of the paper is to reveal the essence of national education and the topical issues of national education in the context of *globalization*.

The objectives: to review the development of the concept of national education in Lithuania; to highlight the present-day problems of national education; to determine the adolescents' national experiences.

Research Methodology

The paper is based on:

Existentialism philosophy and existential psychology, which speak about the universality and unchangeability of man's essence, i.e. the common qualities of all people; about the transcendental essence of man's spiritual dimension;

Humanistic psychology, which explores the aspects of people's authenticity, uniqueness and self-expression, as well as the preconditions for improvement;

Pedagogical anthropology which covers the holistic expression of a person (as a subject of physical, mental and spiritual dispersion) in the process of education.

The research methods: systemic – analytical review of scientific literature on national education and its current problems;

Modified inventories of value-related experienced by Dodonov, which enable to determine the purposefulness and stability of adolescents' experiences;

Statistical data analysis (SPSS software) which helped to determine the statistical frequencies of the data.

The Organization of the Research

The survey was conducted in 2009 in the 11th and 12th forms in different types of schools (general education, art schools, humanitarian and practical gymnasiums, Jesuit high school) in Vilnius and Kaunas and included 904 respondents. The survey aimed to preserve a similar distribution of pupils in the 8th, 9th and 10th forms. The gender-based distribution was random.

The Development of the Concept of National Education in Lithuania

The investigation of national education requires the determination of its key definitions. It has to be noted that very frequently the terms *education* and *moral education* are used in parallel or even synonymously. The definition of education in Lithuanian pedagogy was initially proposed by Šalkauskis (1992) who related the term with guiding upwards, just like in the case of the Latin *education* and the German *Erziehung*. The functions of education are upbringing and fostering, training and moral education. The latter aims to help the pupil become the person he is supposed to be. Šalkauskis's disciple Maceina (1996) related moral education with the organization of the person's center, i.e. with the empowerment for humanity expressed by the ultimate values; Jovaiša (2007) associated it with the help and guidance in the formation and maturation of personal relations; Bitinas – with the purposefully regulated individual assimilation of social ideas (2004) etc. Hence, the oldest function of school – moral education – is unanimously understood as the ultimate function which uniquely summarizes the process of teaching / learning: moral education enables a person to support his relationships and activity with the ultimate values. It must be the reason why education and moral education are collated, and cannot exist without one another.

The present-day scholars of educational sciences agree to define *education* as a purposeful development of a person's abilities on the basis of a creative interaction, and *moral education* as a development of a person's spiritual powers based on common human values. The perception of the importance of education is improving and as a result schools are renamed as general education schools, and the fostering of nationality is termed as national education, because the function of education can be separated only for theoretical purposes. In the educational reality those functions intertwine, yet the educator has to be well aware of those mechanisms and be able to plan and implement them. Otherwise, a person's activity and relations become fragmentary, not related into a system, unsubordinated on the value-related basis.

It is important to conceive the interrelation between *education* and *national education*. At the first sight, they are related as the whole and its part. However, this relation is special as all education happens in the sphere of national education. At the first half of the 20th century it was noted that "education is always national, hence, it originates from the national spirit and corresponds to the national essence" (Maceina, 1990, p. 109). Therefore, it is believed that national education is not a separate aspect of education (like the aesthetic aspect among others), but the principle which permeates all the directions of education, fundamentally relates it with humanity, the foundation of education, derives the national expression, deepens the relation with that expression and enhances the cultural significance and uniqueness. It helps to conceive why after the restoration of independence the national school is being created on the basis of the following principles in the given sequence: "humanity, democracy, change and the fourth one, which intertwines all the others, - nationality. It is not perceived superficially"(Lukšienė, 2000, p.312). The idea of the national school faded when there grew the dominance of the expression of humanity, rather than its roots, which, according to Hargreaves, corresponds to the elements of safe imitation (pretense that you have what you really do not have) of the postmodern existence. Those elements "typically prefer what phenomena *look like*, rather than what they really *are* (1999, p. 112).

Undoubtedly, the basis of national education is moral education, and "every person's feeling of nationality is so strong that it dies last, after all the other sacred and noble feelings died" (Ušinskis, 1987, p. 82). In other words, if a person loses his sensation of nationality, there is no use to look for humanity, its deepest root. Maceina views national education as a peculiar continuation of moral education, or, to be more precise, a special objective of social education since "national education in its essence is a special type of moral education of a person" (Maceina, 1990, p. 162), which matures the pupils' attitudes to other people (family, nation, state, humanity) and reveals the "universal content of human spirit" (Šalkauskis, 1954, p.110). Moreover, as it has been mentioned, national education permeates the other spheres of education which develop the mind, physical abilities, aesthetic sensation etc., and thus prepares all the personal powers to implement the national objectives.

The ideas of the national education conception, formulated by Šalkauskis and developed by Maceina, are still relevant in the present-day society. They help to comprehend the problems of the modern world as they were based on the principles of the nation's formation, rise and development, related to the harmony between origin and nature, which can acquire various forms in different periods of history, but cannot lose their interrelation. The aforementioned authors shrewdly distinguished the main objectives of national education: ethnic, patriotic and national. Ethnic education aims to develop the pupils' national identity and lay the grounds for national education. Patriotic education invokes their love of the nation and the home country, and national education aims to develop national attitudes, cultural creation, which integrates into people's life to implement their specific mission, enriching the existence of humanity and protecting from "the constricted nationalism which, being an exclusive kind of chauvinism, stops the nations from joining into a harmonious union of the whole humanity" (Šalkauskis, 1954, p. 110). Everything is logical, consistent and answers the present-day vision of the global world.

Thus, national education emerges as a necessity to help a man spread the individual and group humanity, relying on the common experience and the need to feel a free and equivalent member of the nation and humanity as a union of nations. Therefore, the development of any nation, whatever level it achieves, preserves the *common human* part which determines a person's status, and a miscellaneous *experience of expressions a cultural phenomenon* marked by the historical destiny. Maslow correctly observes that "culture is the sun, food and water; it is not the seed", which existed in the fetus, but rather creates the conditions (prompts, encourages, helps) for the seed to become real and relevant (2011, p. 245). On the other hand, "nationality, in its full sense, is not inborn as race, but it is also not completely acquired as, for instance, education" (Maceina, 1990, p. 169). In other words, the human nature accumulates certain dispositions which facilitate the formation of a person's positive attitudes to the nation. National education aims to foster a person's ultimate humanity on the basis of the national rudiments, expression of the national rewards and the resolution of the human nature concerning nationality. Thus, national education is the bridge to the universal, global unification. Hence, why is national education becoming a problematic issue under the processes of globalization?

The Present-Day Issues of National Education

There is no doubt that they result from the processes of globalization. Though, according to Steger, "globalization is as old as the human history" (2008, p. 45), its modern development is dislodging the borders of nations and countries, invading their cultural world and touching upon everybody's life since "national identity originates from trust in cultural values and loyalty to them: the native language, ethnic traditions and education" "(Krikštopaitis, 2005, p. 90). Naturally, the changing economic, political and ideological dimensions of globalization determine the change of the cultural dimension which is by some authors related to the essence of the present-day globalization as culture is viewed in the form of "symbolic constructions, expressions and meanings" (ibid., p.112), i.e. in the creative dimension that implies the personal and, at the same time, national values which, according to Jokubaitis (2012) can hide serious moral and political oppositions and can permeate all forms of globalization with humanity / inhumanity.

As it has been already mentioned, the interrelation between *a person* and *nation* is special: the cultural forms, polished by centuries, mature humanity, form personal needs and answer them in the most acceptable ways. Therefore, *a person* and *culture* became the foundation of the Conception of Lithuanian Education (1988), which was prepared just before the restoration of the country's independence. The goal was to strive for depth as the occupation government had been pursuing a denationalizing policy for fifty years, which, according to Maceina, is sure to "make injustice for an individual's nature" (1990, p. 169), when in different spheres of life the voice of conscience (truth), which is the moral foundation of the nation's unity, is being suppressed. This process was enhanced by the prohibition of the historical memory as well as by pseudo culture. When freedom does not originate from the personal spiritual sensation of existence, is administered by the government and based on the partial truth, it does not become mature and does not bring the fruit of nationality and citizenship.

After the restoration of independence, national education was primarily oriented to the perception of our roots, the cultural heritage. That is the external side, an obvious, valued part of the education policy, but the target results can only be expected when teachers and pupils experience moral ties with the heritage, comprehend some personal significance, when they want to identify themselves with the heritage and participate in the development of culture. Accessibility is not enough, the freedom made the screen of the "iron shield" fall down which revealed the moral decline and prompted the social stratification, while a necessary precondition for the unfolding of nationality is "socialization across the social classes" (Held et al., 2002, p.371). In that situation the fragile moral and national ideals were weakening, giving way for the shallow contexts of globalization which "invaded" Lithuania by all possible ways.

The processes of globalization, based on the market principles, which, according to philanthropist Soros, are immoral beside their economic significance as they "let people follow their own interests without applying any moral demands for those interests" (From: Hargreaves, 2008, p. 52). On the other hand, if people experience seclusion, globalization "pushes them towards their own culture, religion and ethnicity as the sources of the meaning of life and identity" (ibid, p.53), yet they are typically assessed in the context of fundamentalism as exclusivities. Obviously, morality does not reach the level that enables one to rise to a humane attitude towards all people: those who see the world through the window of communications, information or commerce, as McWorld, and those who confront the group, nation, community to others, in one or another respect, and so intercept the further development of humanity and also stop it in the achieved stage, sometimes even deviating to jihad. The aforementioned Soros says it openly that "a global society cannot be created without crushing the moral motives" (ibid, 55) as the global market does not intrinsically presuppose a global society, without which sustainable harmony is hardly possible at all.

How are these problems being solved? The most frequent tendency is to focus on their most explicit expressions and mitigate them by creating "secular" religions, emphasizing the civic relations, promoting a cosmopolitan, hybrid, universal, or European identity. However, the global practice shows that the results are local; the trouble spots do not disappear. It can be assumed that one of the reasons related to education, as the extent of globalization is growing, is degraded or even eliminated national identity, though it, as it has already been mentioned, creates the best conditions for humanity to unfold, helps to perceive the key values and personally experience their significance and meaning. Only when humanity sprouts are strong, a person is able to accept otherness, transcend it and then see the common human springs. In other words, it would be purposeful to approach citizenship through nationality, and then look for paths to the global society where people's differences prompt not opposition, but rather interest; they cause no threats for security or individuality.

Another obvious reason is the inability of people, belonging to different social groups, to communicate, understand another person and look for ways to lessen the social seclusion. Even though on surface these seem to be social problems, deep in their roots they also refer to the phenomenon of truth which resides in human nature and, thus, belongs to the sphere of teaching / learning. People's spiritual nature creates preconditions for a person, people's groups or communities, nations to decide upon their lifestyle by themselves. When this right, directly or indirectly, through economic, social or other decisions, is usurped, the moral foundation of existence is violated, which, just like the social environment, integrates into education and by its own ways deepens stratification.

The situation of teaching / learning is also quite complicated. In the context of the knowledge-based economy, even though the educational policy declares the focus on education, the dominant function of school is training, knowledge and skills, while the moral education is left aside. Then education does not perform its major function which enhances the moral basis of a personality. This situation is deepened by the spiritual foundations of moral education, which are insignificant for the pragmatic world, and the procedural nature of mechanisms of education is diminished.

Therefore, the curriculum, as the visible part of the process of education, becomes most important and creates the conditions for the existence of pseudo patriotism: the enforced ideas of national education frequently condition the opposite processes. The issues of national teaching / learning are especially relevant in adolescence when young people start looking for authenticity.

Adolescents' National Experiences

The adolescents' existential relation with the life of the nation reveals itself through the search for identity which is typical for this age and whose significant constituent part is national identity. It covers the surrounding events (micro and macro), influences, formal and informal knowledge and other real or virtual factors which are merged together into a coherent chain of attitudes, experiences, evaluations which implicates the past, the present and the future. Experiences are most important. Even if they are not fully conceived, they still reveal an authentic relation with the reality and thus are very important for the practice of education: they reveal the true state of the relationship and enable to create strategies aimed to deepen the former, since "in the process of internalization of values emotions play a role which corresponds to the role of thinking in the process of learning" (Bitinas, 2000, p. 182).

Depending on the level of conception or verbalization, experiences are revealed through emotions, feelings, evaluations and are differently classified: according to the content of experience, stability, the intensity of expression, etc. Aiming to determine the purposefulness of adolescents' experiences, the study employed an adapted technique by Dodonov, which was expanded with the inventories of religious and national experiences, and there appeared natural experiences, which originate from the relation with nature, and aesthetic experiences, which are evoked by artistic manifestations. The presented rating of the descriptions of 13 values according to their *stability* (most frequently, the individually experienced pleasure) determined which experiences are mostly related by the respondents with the authentic experience. There were 7 level distinguished: very high – the experience was given the 1st or 2nd place; high – 3rd – 4th place; sufficient – 5th – 6th; average – 7th; insufficient – 8th – 9th; low – 10-11th; very low – 12th – 13th. Table 1 presents the assessment of national experiences.

Table 1: Distribution of Adolescents' National Experiences

Levels	No.	Percent
Very high	32	3.6
High	63	7.0
Sufficient	93	10.3
Average	74	8.2
Insufficient	146	16.2
Low	244	27.2
Very low	247	27.5
Total:	*903*	*100.0*

The data show that only one tenth of adolescents very frequently or frequently experience positive experiences evoked by national expressions, and only one fifth of the respondents have a favourable attitude to national experiences. However, more than seventy percent of the respondents expressed a negative attitude.

It has been determined that national experiences range among the other feelings that adolescents' experience. Figure 1 presents the data of the highest and lowest levels which reveal the dominant tendencies.

Fig. 1: Distribution of Adolescents' Value-Related Experiences of the Highest and Lowest Levels (2009)

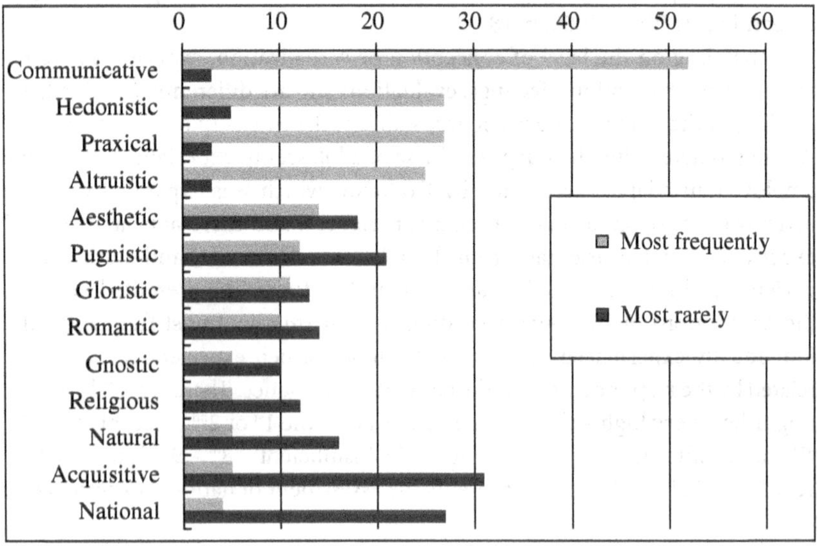

Figure 1 demonstrates that national experiences are put in the final place: only 4 percent of the respondents assigned them to the highest level. It also has to be noted that the *acquisitive, natural, religious*, and *gnostic* experiences are similarly assessed. Adolescents most enjoy socialization, and more than half of the respondents assigned these experiences to the highest level. This reveals a very important need typical of the age group. Yet, it is not necessarily based on moral values. More than one fourth of the adolescents most frequently deal with *hedonistic, praxical* and *altruistic* experiences, the combination of which also reveals their moral instability.

The comparison of the research data with the adolescents' experiences determined by the same methodology ten years ago (1999) reveals similar tendencies, while national experiences decreased: the highest evaluation diminished from 7 to 4 percent, and the lowest slightly increased from 26 to 27 percent.

Conclusions

National identity, citizenship, patriotism, ethnicity, multiculturalism and others are the controversial issues of the present-day world which are frequently discussed on the conceptual level, yet the studies on national education are not prolific. The focus has been on separate aspects: the problems of national education in the context of globalization (Ubartaitė-Vingienė, 2005; Senkus, 2007; Taylor, 2000); the pupils' attitude to the issues of being Lithuanian (Čiužas, Ratkevičienė, 2005); the students' attitudes to the survival of the nation (Kazlauskienė, 2005), to patriotism and cosmopolitism (Antinienė, Merkys, 2007); the development of nationality and citizenship at school (Klimka, 2007; Martišauskienė, 2007, Ranonytė, 2007); the national awareness and identity of pupils of senior classes (Aramavičiūtė, 1999, 2007); the adolescents' attitudes to the nation (Martišauskienė, 2004), etc. The data of this research, revealing the close interrelation between national, religious and moral values, are concerning. The initial idea was that their spread in Lithuania is related to the restored independence. However, when similar tendencies were determined by foreign researchers as well, there appeared a need to look for their deep, natural relation. More and more springs of religious and national fundamentalism are appearing in the world, and it can be assumed that their appearance is related to the loss of connection between religious or national teaching/learning and morality, people's natural powers, which were called by Socrates as an inner oracle. After all, nobody has determined moral fundamentalism since morality cannot be a distinguishing factor between people or nations as in its essence it is the foundation of people's unity. Obviously, the pioneers of national education saw this interrelation and considered national education to be special moral education. The interrelation between national and moral experiences is also revealed in the presented survey of adolescents' experiences: when some are of a low level, the others are also similarly experienced.

For the harmony among people and nations, for the enhancement of their unity, the present-day postmodern world is searching for more external approaches and focusing on national education, hybrid or cosmopolitan identity, etc. However, these approaches cannot give significant results since they do not have a fundamental basis and, despite the noble goals, are directed at shallowness.

In conclusion it can be argued that:

1. In the development of humanity nation occupies a special place as it can answer people's essential needs which, being inborn and by education reformed emotions, are responsible for the energization and direction of sensations and behavior.
2. National education is related to national individuality which is approached through the spread of a separate person, an individual as the indivisible whole, which implicates common human and specific features, when "none is exactly the same, and none is totally different" (Maceina, 1990, p. 72).
3. National teaching / learning has been of interest in Lithuania since the beginning of the 20th century; it was emphasized then that it was not a separate aspect of education (like the aesthetic or moral one), but was a principle which permeates all the spheres of education and fundamentally relates them with humanity, and at the same time it enables the formation of a typically national expression, deepens that relation with the expression and enhances the cultural significance and uniqueness. It is acknowledged that the basis of national education is moral education and the following objectives of national education are distinguished: ethnic, patriotic and national, which function as bridges to the universal, global approximation of people and nations.
4. The processes of globalization are stirring the established relationships between people and the nation, highlighting the effects of occupation, are revealing the opposition of the global market and the global society, deepening the social stratification; they presuppose the transformations of national education which to a certain extent eliminate the moral backgrounds.
5. The adolescents' national experiences occupy the last place among the other feelings and demonstrate a tendency to decrease. The moral feelings are also unstable.
6. National education is heading for shallowness promoting to reconsider this phenomenon anew.

References

1. Antinienė, D., Merkys, G. (2007). Lietuvos akademinis jaunimas patriotizmo ir kosmopolitizmo kryžkelėje. In: *Tautiškumas ir pilietiškumas. Atskirtis ar dermė?* [scientific monograph]. – Vilnius: VPU, p. 200 -221.

2. Aramavičiūtė, V. (1999). Tautinė savimonė: prasmė ir problema vyresniame mokykliniame amžiuje. In: *Acta Pedagogica Vilnensia*. – Vol. 6 p. 23-27.

3. Aramavičiūtė, V. (2007). National Identity and its Change in Senior Pupils' Attitudes. In: *Entwicklung erzieungswissenschaftlicher Paradigmen: Theorie und Praxis*. Band 14. Peter Lang, p. 243 - 255.

4. Bitinas, B. (2000). *Ugdymo filosofija*. Vilnius: Kronta.

5. Bitinas, B. (2004). *Hodegetika. Auklėjimo teorija ir technologija*. Vilnius: Kronta.

6. Čiužas, A., Ratkevičienė, V. (2005). Moksleivių požiūris į lietuvybę. In: *Tautinės tapatybės dramaturgija*. Vilnius:VPU leidykla, p. 234 - 243.

7. Dystervėgas, A. (1988). *Pedagoginiai raštai*. Kaunas: Šviesa.

8. Grigas, R. (2007).Nacionalinis sąmoningumas šiandien: jaunuomenės pasirinkimo drama. In: *Tautiškumas ir pilietiškumas. Atskirtis ar dermė?* [scientific monograph]. – Vilnius: VPU, p. 11-28.

9. Hargreaves, A. (1999). *Keičiasi mokytojai, keičiasi laikai: mokytojų darbas ir kultūra postmoderniajame amžiuje*. Vilnius: Tyto alba.

10. Hargreaves, A. (2008). *Mokymas žinių visuomenėje: švietimas nesaugumo amžiuje*. Vilnius: Homo liber.

11. Held, D., McGrew, A., Doldblatt, D., Perraton, J. (2002). *Globaliniai pokyčiai: politika, ekonomika ir kultūra*. Vilnius: Margi raštai.

12. Herber, H.J. & Vásárhelyi, E. (2003). Moderne Motivationsforschung als Paradigmenverschmelzung? Die Leistungsmotivationsforshung und ihre Wurzeln als Kern einer allgemeinen Motivationstheorie. Available from: www.sbg.ac.at/erz/salzburger.../herber_02_03_sbg.pdf. (25.10.2013).

13. Jokubaitis A. (2012).*Vertybių tironija ir politika*. Vilnius: Vilniaus universiteto leidykla.

14. Jovaiša, L. (2007). *Enciklopedinis edukologijos žodynas.* Vilnius: Gimtasis žodis.

15. Kazlauskienė, V. (2005). Tautos išlikimas studentų nuostatų aspektu. In: *Tautinės tapatybės dramaturgija:* Vilnius:VPU leidykla, p. 244 - 266.

16. Klimka L. (2007). Pilietiškumas ir tautiškumas: laiko dimensijos ir aktualijos. In:*Tautiškumas ir pilietiškumas. Atskirtis ar dermė?* [scientific monograph]. – Vilnius: VPU, p. 127-141.

17. Krikštopaitis, J. A. (2005). Intelektinio potencialo vaidmuo ES narių varžybų arenoje. In: *Tautinės tapatybės dramaturgija:* Vilnius:VPU leidykla, p. 87-98.

18. *Conception of Lithuanian Education.* (1992). Vilnius: Leidybos centras.

19. Lukšienė M. (2000). *Jungtys.*Vilnius: Alma litera.

20. Maceina, A. (1996). Lavinimas ir auklėjimas. In: *Lietuvos pedagogikos istorijos chrestomatija* // ed. M. Karčiauskienė, K. Pečkus. –Vilnius: VPU, p. 4-53.

21. Maceina, A. (1990). *Pedagoginiai raštai.* Kaunas: Šviesa.

22. Martišauskienė, E. (2007). Vertybinis dramatizmas ugdymo procese. In: *Tautiškumas ir pilietiškumas. Atskirtis ar dermė?* [scientific monograph]. – Vilnius: VPU, p. 165-175.

23. Martišauskienė, E. (2004). *Paauglių dvasingumas kaip pedagoginis reiškinys.* Vilnius: VPU leidykla.

24. Maslow, A.H. (2011). *Būties psichologija.*Vilnius: Vaga.

25. Ranonytė, A. (2007). Moksleivių pilietiškumo nuostatos: teorinės prieigos ir realybės atspindžiai. In: *Tautiškumas ir pilietiškumas. Atskirtis ar dermė?* [scientific monograph]. – Vilnius: VPU, p. 176-199.

26. Senkus, V. (2007). Globalizacija ir patriotizmas: tarp realybės ir perspektyvos. In: *Tautiškumas ir pilietiškumas. Atskirtis ar dermė?* [scientific monograph]. – Vilnius: VPU, p. 29-36.

27. Steger, M. B. (2008). *Globalizacija: labai trumpas įvadas.* Vilnius: Eugrimas.

28. Šalkauskis, S. (1992). *Rinktiniai raštai. Pedagoginės studijos.* Vol. I and II. Vilnius: Leidybos centras.

29. Šalkauskis, S. (1954). *Ateitininkų ideologija.* Putnam, Conn.

30. Taylor, M. J. (2000). Values Education: Isuues and Challenges in Policy and School Practice. In: *Education, Culture and Values* / ed. by M. Leicester, C. Modgil and S Modgil. - London and New York, 2000, vol. 2, p. 151-168.

31. Ubartaitė-Vingienė, D. Globalizacija ir tautinio ugdymo problema XXI a. Lietuvos mokykloje. *Mokytojų ugdymas.* 2005. No. 5, p. 13-19.

32. Ušinskis K. (1987). Apie tautiškumą visuomeniniame auklėjime. In: K. Ušinskis. *Pedagoginiai raštai.* Kaunas: Šviesa, p. 71-86.

33. Додонов, Б. (1978). Эмоции человека. Москва: Полииздат.

(Re)conceptionalisation of Curriculum in Contemporary Educational Discourse: International and Lithuanian Perspectives

Tatjana Bulajeva, Šarūnė Nagrockaitė

Abstract
Traditionally curriculum theorists and education philosophers have tried to clarify the concept of curriculum by proceeding from a review of definitions, found in the curriculum research field, to presenting their own definitions. Our paper provides a brief discussion of the issue of defining *curriculum* concept in contemporary educational discourse showing how curriculum is viewed by curricular theorists (traditionalists and reconceptualists) representing the modern and postmodern perspectives.

The discussed problems of Lithuanian curriculum and its terminology based on the analysis of Lithuanian educational discourse show that the main most widely used term „ugdymo turinys" is exclusively defined in terms of content or subject matter. It calls for the necessity of clarification and reconceptualisation of curriculum notion considering its complexity and relatedness to different aspects of educational process.

Introduction

Since the second half of the 20th century the analysis of the academic discourse on education provides evidences that the problems of curriculum and its concept remain in the focus of educational research. In spite of the constantly growing number of publications dealing with this theme on the international and national arenas, there seems to be a confusion still existing within this research field, which to some extent is due to the ambiguity of the curriculum concept in general, and curriculum theory in particular. Marsh (2004) states that a curriculum concept research field at the end of the 20th century was one of the most complicated and less investigated compared to other education aspects (education objectives, methods, management process, assessment etc.) related to curriculum.

A complicated curriculum research situation is also mentioned by other authors (Young, 2013; Pinar, 2004; Pinar et al. 2008; Jackson 1992) stressing the current crisis in curriculum theory, arguing that recent developments in curriculum theory have led to it losing its object – what is taught and learned in school – and its distinctive role in the educational sciences. They speak of the complexity, multiple aspects of the curriculum concept and necessity to understand and define its role in the changing education reality. Metaphorically speaking Pinar (2004) defines this situation as a *complicated conversation.*

It is hardly reasonable to consider curriculum theory as one coherent theory. Rather, it seems more relevant to view it as a number of theories, focusing on different aspects of social phenomena. Due to the usage of many different definitions of the word *curriculum* and competing conceptions appearing in response to social changes, it makes it difficult for both theorists and teachers involved in educational practice of curriculum implementation to come to its common understanding. With reference to curriculum, Berg (1993) states that the different definitions of this word have at least one common coincidental point – they always concern a plan for teaching (Berg, 1993, p. 159).

The developments in the curriculum research field show that curriculum vocabulary is gradually being enriched, that the concept of curriculum and associated terms comprise educational phenomena in nearly all its aspects. Smith and Lovat (1991) illuminate this by presenting twelve definitions of the curriculum-concept, which were further elaborated by a number of Western researchers. Most authors have offered not a conclusive but a stipulative definition. Some theorists argue that a universally acceptable definition is important, that without it communication and conceptual progress is hampered. Ian A. C. Rule (1973), for instance, in his doctoral dissertation „*A Philosophical Inquiry into the Meaning(s) of „Curriculum*“ identifies 119 different definitions of the term *curriculum* (quoted in Portelli, 1987, p. 357).

Henchey (1981) thinks that the main problem is not in the multiplicity of curriculum definitions through which authors try to explain curriculum phenomenon, but in their epistemological position and philosophical base. Since 70-80 ies the critical situation in curricular theory becomes obviuos. The earlier motives used to explain curriculum become irrelevant due to multiple changes taking place in the educational reality. The definitions offered are all problematic in one respect or the other. For when simplified definitions are used to define curriculum, its complex ties with complex educational reality they become distorted. In his „Curriculum as Myth“ Henchey argues that not recognizing the complexity of curriculum leads to reductionism, to creation of another negative myth. The stronger the belief in a definition of curriculum, the stronger the myth (Henchey, 1981, 264-265).

Doherty and Briant (2013) in their overview of the articles on curricular research issues published in the journal of *Theory into Practice* in the last 50 years state that curriculum studies acquired new lines of enquiry. Educators and scholars asked very different questions about official curriculum, the enacted curriculum, the hidden curriculum (Reagan, 1975; L.L.Tyler, 1986), the curriculum-in-use (Barone, 1982), the crowded curriculum (Stevahn, 2004), the measured and unmeasured curriculum, the transactive curriculum (Seidman, 1983), the emerging curriculum (Zaret, 1986), the outside curriculum (Schubert, 1992), the negotiated curriculum (Miller, 1992) and the transformative curriculum (Doherty and Briant, 2013, p. 43).

In 1980s in the USA, the reaction to this critical situation in curricular research was the appearance of *reconceptionalists* - a new curricular school of thought. They are guided by the notion that there is no universally accepted definition of curriculum, because different social, political, cultural and personal contexts influence the formation of curriculum concept (Pinar, 2004). So it is offered to understand the notion of curriculum as the discourse of ideas and words used to speak about the curriculum (Pinar ir kt., 2008).

The existing problem of curriculum conceptions appears not because there are many curriculum definition but because there is no clear and common understanding of the factors influencing the formation of multiple conceptions. According to Kansanen (1999) curriculum definitions depend on different semantic contexts, in which they are being formed, and on educational objectives causing the development of curricula (p. 26). How the curriculum content is understood, how it is changed, how it is implemented in practice – all these research questions are related to the analysis of *conceptualizing curriculum* within the curriculum field of study.

The **aim** of this article is to analyse the problem of Lithuanian curriculum concept presenting it in the broader context of international theoretical discourse.

The focus of this article is to concentrate on theoretical aspects trying to show what changes have taken place within the field of curriculum conceptualisation. In the first part of this article we would discuss the main curriculum paradigms (the modern and the postmodern) and theoretical schools of the 20th century (*traditionalists* and *reconceptualists*), which have influenced the changes in understanding of curriculum in educational theory and practice in the world. In the second part of the article we will discuss the academic curriculum discourse in Lithuania, presenting the problems which are being faced trying to define and understand the changing curriculum concept and the influences made by international descourses.

The Change of Curriculum Paradigms and Contemporary Curricular Schools of Thought

The intensive interchange of paradigmatic ideas, which has begun in the 20th century is often defined by the modern and postmodern thought divide that has influenced the development of curriculum theory. Though different paradigms often coexist, they may have some common characteristics: they act as opponent to each other; are related to definite theoretical schools of thought; the shift of them marks some qualitative changes. The two contrasting paradigms within the field of curriculum, *the modern* and *the postmodern* are most widely analized by researchers (Doll,1993; Behar-Horenstein & Glanz, 2000; Slattery, 2006).

Curriculum in the modernist and traditionalist perspectives

Hartley (1997) in his analysis of curriculum and modernity states that modern period saw the rapid urbanization of the population and increasing mechanization of production process. In his view the development of curriculum at that period was framed by the following: by bureaucratic rationality as a dominant style of thought; by the emegence of structuralist theory (behaviourism in psychology and social systems and marxist theory in sociology); by calls from liberal philosophers for curriculum whose content should be value neutral; by the need of both industrialists and governments for trained and educated workers. (Hartley, 1997:44). In his view

„The modern curriculum is set very much within the Enlightment project. The managerialist and positivistic rhetoric rooted in behaviourism and systems theory suffuses the curricular prespritions of modernity. Aims and objectives, inputs and outputs, standards and specifications – all can be fixed, fast and securely, in the system-serving organization called school. The family is said to be functional for school, the school for the social system, a system marked by a liberal-democratic polity, and by a fordist production process within industrial capitalism" (Hartley, 1997:44).

This quotation presents a clear picture of a modern curriculum understanding which can still be found today. Definitions of modern view presented by Behar-Horenstein & Glanz (2000) are said „to have been influenced by: (1) public perceptions of what school should be like; (2) parental memories that are suggestive of what school experiences should be; (3) societal expectations for cultivating a productive adult citizenry; (4) standardised tests; (5) teacher accountability for student outcomes; (6) metaphors that describe school as a factory; (7) competing views of what knowledge or whose knowledge is most valuable" (p.6). According to them at the heart of postmodern paradigm there are some criticisms against the modernists, for modern view of curriculum has failed to ensure equal opportunities for all students irrespective of class and social status and has promoted the knowledge and values of the dominant class (Behar-Horenstein & Glanz, 2000:6).

As Marsh (2004) states the changed education reality necessitated the development of a new view on curriculum and the construction of new vocabulary, new terminology and metaphors (Marsh, 2004, 199).

Most often the modern and postmodern paradigms are presented using the logic of differences, which is the precondition for the formation of the postmodern paradigm. In Behar-Horenstein and Glanz (2000) view, the postmodernists try to look at conceptions of the modern paradigm from the critical perspective and offer new focuses of view. Doll (1993) uses the term of *the postmodern* referring to a paradigmatic shift in thinking from empiricism and rationalism to one in which modern notions are subject to critical reappraisal and transformation. The deconstruction of meaning becomes the hallmark of the postmodern theory, the main strength of it is the creation of new knowledge and the transformation of learning.

This is also true speaking about the formation of other curricular theoretical schools – *traditionalists* and *reconceptionalists*. Each paradigm, as well as a theoretical school of thought related to it, has the vocabulary of its own and the terminology which can be seen in definitions of curriculum developed by theorists representing these schools.

Though the formation of a traditionalist school of thought is dated back to the first half of the 20th century (it obviuosly has much in common with the modernist view), the critique of traditionalist views on curriculum expressed by *reconceptionalists* (Jackson, 1992; Wraga, 2003; Pinar, 2004; Hlebowich, 2005; Westbury, 2005, Pinar et al., 2008; Pacheco, 2012) activated curricular discussions.

Portelli (1987) commenting on the main definitions of curriculum classifies them under these categories: (1) *curriculum* defined in terms of content, (2) *curriculum* defined in terms of experience (activities), (3) *curriculum* defined in terms of a plan. According to him the traditional vision of *curriculum* as content identifies curriculum with a course of studies which lists content, subjects or subject matter. So curriculum is understood as what students should be taught. This view is propounded in the work of theorists belonging to *perennialism* (the movement in philosophy of education that objects to some of the main tennets of progressivism, main representatives are R.M. Hutchins and M. J. Adler) and *essencialism* (another educational philosophical perspective put forth by educationalists W. C. Bagley, T. Briggs and H. H. Horne who state that there are certain essential subject matters that one should know to become educated). Identifying curriculum with content assumes that no distinction between subject and curriculum is made as well as there is no clear distinction between instruction and curriculum. Traditionally educators think that if one knows a subject, one can teach it. The problem is that *content* and *subject matter* are still used synonymously with *curriculum* (Portelli, 1987, 357-359).

The definition of curriculum as experiences (ranging from student experiences to school and life experiences) arose as critical reaction to curriculum as content. It is ussually found in progressivist discourse this appreach was dominant from early 1930s till 1970s. The criticism most commonly raised with this approach is that the definition of curriculum as experiences is too broad, whereas the definition of curriculum as content is too narrow (Portelli, 1987, 360).

The attempt to define curriculum as a plan first appeared in the 1960s. This approach gained popularity, those who favour it, state that the definition of curriculum as a plan helps to avoid problems posed by the other two (curriculum as content and curriculum as experiences) approaches (Portelli, 1987, 361). There are also some other definitions of curriculum related to a traditional and modern view found in curricular discourse: curriculum in terms of a teaching/ learning programme, curriculum in terms of learning outcomes, curriculum in terms of the educational process. The variuos types of curriculum definitions discussed in this paper and in curricular theory are problematic, they show that curriculum is a complex notion and separate definitions cannot capture its full meaning, that is why, no coherent curriculum theory has yet emerged, able to unite the desparate trends.

The traditional and modern curriculum conception is still widely used in official documents regulating school curricular issues. It is practice oriented and has a clearly cut and planned model. This curriculum notion makes teachers' everyday work easier for traditional definitions of curriculum are simple and easy to understand, setting clear subject-specific objectives, methods and assessment of learning outcomes achieved by students.

No doubt that modern paradigm and traditionalists made an important contribution to a better understanding of curriculum as a planned and implemented programme (Benhar-Horenstein, 2000). This conception of curriculum gained populiarity due to well defined learning outcomes, which were introduced into the curriculum, and their assessment helped in measurement of education quality. Evaluating this conception from the critical perspective most criticisms were addressed to a standardised curriculum, which was used for mechanistic teacher work strictly following the instructional recommendations, and for teacher control.

The main principles of traditional and modern curriculum can be summarised as follows:

- The curriculum is used for reproduction of existing political and and social systems including education;
- Curriculum inquiry is based on qualitative research methods and it is aimed at practical curriculum application in schools;

- The main problemic question related to curriculum formation and implementation – how to develop, improve, assess the curriculum;
- Educational objectives based on behaviouristic approach help to form the definite behaviour according to defined skills;
- Clearly defined subject matter as a knowledge system, presented as content, as a plan, as a programme, is transmitted;
- Curriculum (plans, programmes, education standards) is related to institutionalisation of official curriculum.

Curriculum in the postmodernist and reconceptionalist perspectives

A post modern currucular theory brings to view some other aspects of education and its management which are in close relation with curriculum concept. W. E Doll (1993) offers his own vision and describes a set of curriculum concepts derived from a postmodern perspective. He makes the point that postmodernism (comparing it with the modernism) as a philosophical school of thought and as a base for curriculum analysis is in its infancy and difficult to define (Doll, 1993, 158). He acknowledges that in societies characterised by advanced technologies and complex social dynamics, transforming curriculum from its deep roots in modernism to a post modern paradigm requires new and different ways of thinking. A linear, step-based Tyler's rationale for curriculum planning upon which educators have depended upon for a long time, should be replaced by a postmodern frame of open, relational systems. The post modern curriculum perspective represents a shift in focus from descrete to the complex, from the predictive to the relational, and from linear and determinant to the self-generating and indeterminant. Empiricism and rationalism are giving way to a post modern reappraisal and reconceptualisation (Slattery, 2006).

From a critical perspective of Aronowitz and Giroux (1991) educators having a postmodern view believe that the curriculum can best inspire learning only when school knowledge is built upon tacit knowledge derived from cultural resources that students already possess. For example, electronically mediated popular culture is treated by postmodern education as a legitimate object of knowledge. In contrast to practices that uncritically transmit apparently unrelated disciplines of math, literature, geography, and history, organized in discrete time periods of the school day, the teacher attempts to integrate these knowledges within a series of projects chosen jointly with students (Aronowitz and Giroux, 1991).

Reconceptionalists (an American school of thought), though criticising the modern views on curriculum actually use theoretical insights and vocabulary of postmodernism in their discourse. In their understanding the curriculum is a symbolic idea, concept, which expresses what the older generation chooses to say to the younger generation, so the curriculum is a communicative act, „a complicated conversation", taking place in a definite context in which it is created and in which the conflict of generations is embedded (Pinar, 2004).

Reconceptionalists define the curriculum discource as historical, political, gendered, racial, phenomenological, autobiographical, esthetical, theological and international (Pinar et al. 2008, p. 848). Curriculum is also related to the education processes of individuation and socialisation. It is related to a present life-world and is constructed in lived spaces (Pinar et al., 2008).

This curriculum notion does not fit into curriculum models of traditionalists, which reconceptionalists name as institutional, developed on the basis of international and national agreements and commitments to develop textbooks suitable for teaching according to the programmes of the state, the so called official curriculum. The main characteristics of the postmodern curriculum defined by reconceptionalists are as follows:

- Curriculum definition is not final, it does not have stable models;
- It is often expressed as an idea, as a vision, as orientations, as a discourse;
- It is constantly changing and depends on changing contexts;
- It is not only institutional, it comprises all the processes of personal lifelong development, a hidden curriculum can be part of it;
- It is formed by different (social, cultural, political, personal) contexts and educational realities;
- It is *subjectivised*, in it a personal (subjectivist) experience is seen as having biographical and phenomenological aspects which become the objects of post positivistic, qualitative curriculum enquiry.

Curriculum in Lithuanian Educational Discourse

Paradigmatic shift having taken place in Lithuanian education theory is related to the process of regaining of national independence and reestablishment of Lithuanian statehood, which have greatly changed the political, economic and social context of education in Lithuania.

The first curricular changes were introduced in *The Concept of National School* (1989) (Tautinė mokykla. Lietuvos vidurinės bendrojo lavinimo mokyklos koncepcija, 1989) and other strategic documents, which have made the legal foundation for the reform of school education. Bruzgelevičienė (2013) states, that though the education reform aimed at paradigmatic change, which started more than two decades ago, there is still no clear divide and understanding of different educational and curricular paradigms (p. 49). She presents different paradigm descriptions used by Lithuanian curricular theorists. Usually these paradigms are defined as oppositions (classical and free education, modern and postmodern, traditional and contemporary) representing confronting positions. Classical, traditional and/or modernist paradigms are often related with the educational experience of the past, whereas free / emancipatory, post modern are understood as contemporary and more appropriate for the Lithuanian education of the present. The consolidation of this contemporary thinking has to appear in a newly developed education policy, its strategies, new national school curriculum, new textbooks and new education terminology.

Lithuanian term „ugdymo turinys" most widely used in Lithuanian educational discourse (it is an equvalent of *curriculum* term, though its word-by-word translation means „education content") went through several stages of development.

At the first stage the term „ugdymas"(equivalent of an *education* term) has entered the Lithuanian educational discourse. It was used in the period between World War I and World War II by Lithuanian researchers St. Šalkauskis, A. Maceina, J. Laužikas, but it was not the main pedagogical term (the term „mokymas" - teaching – was the main term then). According to Bitinas (2011) in the post-Soviet period this term and its meaning have undergone further changes. For a long time it was used together with psychological terms when speaking about character education or emotional education. Only in 1980 J. Laužikas suggested that this term should replace the term „auklėjimas" (moral education) and become the basic pedagogical term (Bitinas, 2011, p. 11). This way the term „ ugdymo turinys" began to mean „education curriculum"in its broadest sense.

In the second stage of development the process of identification of this concept in the international discourse began. The *curriculum* term entered Lithuanian educational discourse, though it was translated, explained and interpreted differently by different Lithuanian authors (Pukelis and Sajienė, 2000; Dabriškienė, 2001; Adaškevičienė, 2003; Saugėnienė, 2003; Musneckienė, 2004; Kliminskas, 2009; Rupainienė, 2008; Katiliūtė, 2008; Laužackas, 2008). The problems and difficulties faced in translation of the term *curriculum* into Lithuanian was widely discussed.

Analising the Lithuanian curricular debates we have discovered that two translation strategies were offered to be employed. The first strategy – to find the right Lithuanian equivalent to the term *curriculum*. There were several terms offered: „Ugdymo programa“ [education programme] (Saugėnienė, 2003; Bitinas, 2011) „Ugdymo turinys“ [education content] (Dabriškienė, 2001; Adaškevičienė, 2003; Musneckienė, 2004; Monkevičienė et al., 2012); „Mokymo programa“ [teaching programme] (Laužackas, 2008); „Mokymo turinys„ [teaching content] (Pukelis, Sajienė, 2000). The results of the usage of the first strategy have shown that there are many terms in Lithuanian terminology which can be used to define and explain the *curriculum* concept. It is obvious that all variants of terminology offered are closely related with the modernist paradigm defining curriculum in terms of content, or in terms of a programme. They also show that Lithuanian curricular theorists have a modernist view of the notion of curriculum.

The other group of authors, who used a different strategy, aimed at separation of Lithuanian terms from the internationally used term of *curriculum.* Scholars, which employed this strategy, state that Lithuanian terms „ugdymo/mokymo turinys“ [education / teaching content] and „ugdymo/mokymo programa“ [education /teaching programme] do not fully express the notion of *curriculum* (Kliminskas, 2009; Rupainienė, 2008; Katiliūtė, 2008), so they suggest that it should not be translated but the lituanised version of *curriculum*, the term „kurikulumas“ should be used instead (Musneckienė, 2004). The second strategy evidently shows that we do not have enough Lithuanian terms or the ones that exist can not fully express the postmodern aspects of *curriculum* concept as it is being (re)conceptualised by Doll (1993); Behar-Horenstein & Glanz (2000). For the Lithuanian term „ugdymo turinys“ [education content], which is most often used in Lithuanian Curricular discourse today, is too narrow and has no identification with the postmodern understanding of curriculum and curriculum concept which is supported by reconceptionalists. There is no doubt that his problem needs further discussions of theorists and practicians and has to be solved (Laužackas, 2008, p. 45).

Conclusions

The analysis of international and Lithuanian curricular discourses have shown that the problems related to the conceptualisation of *curriculum* exist in both. The theorists working in the field of curriculum studies for more than five decades, representing different schools of thought have made many attempts to clarify the notion of curriculum and provide an internationally acceptable definition of curriculum concept, which were not successful. There are many competing definitions and curricular rationales today. The analysed discourses show that a modern, traditional view of curriculum prevails and coexists with a postmodern view of curriculum, which is being employed by reconceptionalists.

The discussed problems of Lithuanian curriculum conceptualisation and its terminology based on the analysis of Lithuanian educational discourse show that the main most widely used term „ugdymo turinys" is exclusively defined in modernistic terms of content or subject matter. It calls for the further development of curriculum enquiry and necessity of further clarification, reconceptualisation and academic discussions of curriculum concept and its notion considering all its complexity and relatedness to different aspects of educational process and rapidly changing world. We think that in the future school curriculum will continue to be the battle field for curricular theorists in international and national arenas. And hope that new knowledge about the curriculum not only makes our understanding of this phenomenon better, but also will improve the everyday practice of schooling.

References

Adaškevičienė V. Europinio identiteto kaip švietimo dimensijos raiška Lietuvos bendrojo lavinimo mokyklos ugdymo turinyje. Daktaro disertacija. Socialiniai mokslai, edukologija (07S), Kaunas, 2003.

Aronowitz, Stanley, and Henry A. Giroux. *Postmodern education: politics, culture, and social criticism.* Minneapolis: University of Minnesota Press, 1991.

Benhar-Horenstein S. L. Can the Modern View of Curriculum Be Refined by Postmodern Criticism? // Debates in Curriculum and Supervision: Modern and Postmodern Perspectives, [edited by] Glanz J., Behar-Horenstein S. L. Publisher: Bergen & Garvey, Westport, Connecticut – London, 2000.

Behar-Horenstein S. L. & Glanz J. Paradigm Debates in Curriculum and Supervision:

Modern and Postmodern Perspectives, [edited by] Glanz J., Behar-Horenstein S. L. Publisher: Bergen & Garvey, Westport, Connecticut – London, 2000.

Berg, G. Curriculum and state schools organisations. A Scandinavian view.// Uppsala Reports on Education, 1993, 30. University of Uppsala.

Bitinas B. Edukologijos terminija: kokybė ir problemos: studija / Bronislovas Bitinas. – Klaipėda: Klaipėdos universiteto leidykla, 2011.

Barone, T. Insinauted theory from curricula –in-use. //*Theory in Practice, 21*, p. 38-43, 1982.

Bruzgelevičienė R. Būsimų pedagogų tapatinimosi su ugdymo paradigma tendencijos // Pedagogika, 2013, T. 109, p. 49-56.

Carr W. Reconstructing the Curriculum Debate: an editorial introduction // Curriculum Studies, 1993, Vol. 1, No. 1, p. 5-9.

Doll, W. E., Jr. Prigogine: A new sense of order, a new curriculum. //*Theory in Practice, 25, p. 10-16, 1986.*

Doll W. E. A Post-Modern Perspective on Curriculum, New York: Teachers College Press, 1993.

Dabriškienė V. Pedagogo veikla individualizuojant ugdymo programas Lietuvos bendrojo lavinimo vidurinėje mokykloje. Daktaro disertacija. Socialiniai mokslai, edukologija (07S), Kaunas, 2001.

Doherty, C. and Briant, E. Unnatural selection: Curricular Conversations in *TIP* 1962-2012. //Theory into Practice, 52, p. 42-51, 2013.

Grimmett P.P. & Halvorson M., From Understanding to Creating Curriculum: The Case for the Co-Evolution of Re-Conceptualized Design with Re-Conceptualized Curriculum // Curriculum inquiry, 2010, Vol. 40, issue 2, p. 241- 262.

Hartley D. (1997). Re-schooling society. London: Falmer Press.

Henchey N. (1981). Curriculum as Myth // McGill Journal of Education, Vol. 16, No. 003, 258-266.

Hlebowich, P.S. More on "Generational Ideas" (A Rejoinder to Ian Westbury and Handel Kashope Wright) Author(s) // Curriculum Inquiry, 2005, Vol. 35, No. 1, p. 119-122.

Jackson, P. W. Conceptions of curriculum and curriculum specialists. In Handbook of research on curriculum, ed. P. W. Jackson, New York: Macmillan, 1992, p. 3-40.

Kansanen P. (1999) The *Deutsche Didaktik* and the American Research on Teaching // TNTEE Publications, Vol. 2, No 1, 21-35.

Katiliūtė, E. Švietimo politika ir jos tyrimo metodologija: monografija, Kaunas: Lietuvos kūno kultūros akademija, 2008.

Kliminskas R. Konstruktyvistinis požiūris į mokyklos lygmens *curriculum* realizavimo sąlygas. Daktaro disertacija. Socialiniai mokslai, edukologija (07S), Kaunas, 2009.

Laužackas R. Mokymo turinio projektavimas. Standartai ir programos profesiniame rengime. Kaunas: VDU, 2008.

Musneckienė E. Edukacinių paradigmų kaita ir meninio ugdymo turinio planavimo perspektyvos. //Mokytojų ugdymas, 2004, Nr. 3, p. 97-106. Šiaulių universitetas.

Marsh K. J. Key concepts for understanding curriculum. Third Edition. Publisher: Routledge Falmer. London and New York, 2004.

Miller, J.L. Shifting the baundaries: Teachers challenge contemporary curriculum thought.// *Theory in Practice, 31*, p. 245-251, 1992.

Pacheco, A. J. Curriculum Studies: What is The Field Today? // Journal of the American Association for the Advancement of Curriculum Studies, 2013, Volume 8, 2012, p. 1-18.

Pinar W. F., Reynolds W. M., Slattery P., Taubman P. M. Understanding curriculum.

An Introduction to the study of historical and contemporary curriculum discourses, Publisher: Peter Lang, 2008.

Pinar W. F. What is curriculum theory? Publisher: Peter Lang, 2004.

Portelli J. P. (1987) Perspectives and imperatives of defining curriculum // Journal of Curriculum and Supervision, 1987, Vol 2. No. 4, p. 354-367.

Pukelis K., Sajienė L. *Curriculum* sampratos problema lietuvių pedagoginės kultūros kontekste // Pedagogika, 2000, T. 40, p. 14-27.

Reagan, G. M. Moral education in theory and practice.// *Theory in Practice,* 14, p. 221-223, 1975.

Rupainienė V. Pradinės mokyklos bendruomenės diegiant *curriculum* inovacijas skirtingų edukacinių paradigmų aspektu. Daktaro disertacija. Socialiniai mokslai, edukologija (07S), Kaunas, 2008.

Saidman, W. H. Transactive curriculum: A practical guide to curriculum adaptation.// *Theory in Practice, 22,* p. 216-223, 1983.

Saugėnienė N. Ugdymo programų planavimas ir realizavimas Mokomoji knyga. – Kaunas: Technologija, 2003.

Schubert, W. H. Practicioners influence curriculum theory: Autobiographical reflections.// *Theory in Practice,31*, 236-244, 1992.

Slattery, P. Curriculum development in the postmodern era. 2Ed. London, UK: Routledge, 2006.

Smith, D.L. and Lovat, T.J. Curriculum. Action on Reflection. Australia:Social Sciene Press, 1991.

Stevahn, L. Integrating conflict resolution training into the curriculum.// *Theory in Practice,* 43, p. 50-58, 2004.

Tautinė mokykla. Lietuvos vidurinės bendrojo lavinimo mokyklos koncepcija (1989). Vilnius: Žinija.

Tyler, L.L. Meaning and schooling.// *Theory in Practice*, 25, p. 53-57, 1986.

Westbury, M. Reconsidering Schwab's "Practicals": A Response to Peter Hlebowitsh's "Generational Ideas in Curriculum: A Historical Triangulation" // Curriculum Inquiry, 2005,Vol. 35, No. 1, 89-101.

Wraga W. & Hlebowitsh P. Toward a renaissance in curriculum theory and development in the USA // Journal of Curriculum Studies, 2003, Vol. 35, No. 4, 425-437.

Young M. Overcoming the crisis in curriculum theory: a knowledge based approach, Journal of Curriculum Studies, 2013, Vol.45, No. 2, 101-118.

Zaret, E. The uncertanty principle in curriculum planning.// *Theory in Practice,* 25, p. 46-52, 1986.

Identitätsentwicklung bei Schwächen in der Wahrnehmungsorganisation

Aufwachsen unter erschwerten Bedingungen mit besonderer Berücksichtigung der Affektregulations- und Bindungstheorie

Eva Rass

Abstract
The development of a human individual is embedded in complex psychosocial conditions. In this paper the emphasis will be on the importance of hidden and undetected neurological deficits which cause psychological problems in the interaction with the environment. The mutual influencing of the individual's disposition and his environment and the consequences for the emotional development will be discussed. Children and adolescents and even adults with neurologically caused disabilities in their motor and perceptual skills are a challenge for their environment – especially for parents, educators and therapists. These deficits and their consequences for development have been given little attention to date. The paper describes the impaired child's experience of everyday life. In order to make it comprehensible it will illustrate in a rough schema the most important perceptual areas that have proven useful and successful in the corporation with doctors, teachers, parents and therapists. In this context it is noteworthy to pay attention to the shame experience of the individual and to create an empathic environment.

Zusammenfassung
Sowohl der Prozess des Heranwachsens als auch das Leben des Erwachsenen sind in ein vielschichtiges psychosoziales Geschehen, d. h. in Interaktionen mit der Umwelt eingebettet. Neuropsychologisch bedingte Schwächen in der Wahrnehmung bedeuten somit eine Herausforderung an den betroffenen Menschen und an das ihn umgebende, betreuende und behandelnde Umfeld.

Diesen Defiziten und den daraus resultierenden Folgeerscheinungen wurde bislang zu wenig Beachtung geschenkt, was häufig zu das ganze Leben hindurchziehenden Belastungen führte. Diese Probleme sind nicht ungewöhnlich, da ein nicht unerheblicher Prozentsatz der Menschen (ca. 30 %) unter diesen Störungen leidet. Da die Beeinträchtigungen nicht leicht zu erkennen sind und bislang zu wenige Begriffe, die die Beobachtung hätten strukturieren können, zur Verfügung stehen, erfahren die davon Betroffenen bezüglich ihrer Schwierigkeit konstitutioneller Genese nur unzureichende Einfühlung, was jedoch für die gesunde Entwicklung einer stabilen psychischen Konfiguration unerlässlich wäre. Dargestellt werden die Alltagserfahrungen eines davon betroffenen Menschen und die wichtigsten Wahrnehmungsbereiche, die für die Interaktionen mit dem Lebensumfeld von Bedeutung sind, da sie eine vitale Rolle im Bereich des sozialen Informationsflusses, im Affektaustausch und im Bindungsdialog spielen. Besondere Berücksichtigung gilt dabei dem Schamerleben.

Schlüsselwörter: Wahrnehmungsorganisation; Wahrnehmungsstörungen; Wahrnehmungsbereiche; Rhythmuserkennung; sensorische Integration; Identitätsentwicklung; Schamerleben

Der Prozess des Heranwachsens ist in ein komplexes psychosoziales Geschehen eingebettet. Neben den psychobiologischen, den psychologischen und emotionalen Bedingungen beeinflussen basale neuropsychologische Vorgegebenheiten das Interaktionsgeschehen. Es sind sowohl Anlagen als auch Umgebungsfaktoren, die das jeweilige Endresultat und somit auch das Identitätserleben bestimmen. Von besonderer Bedeutung ist dabei die Wahrnehmungsorganisation (vgl. Rass 2002, 2008).

Anhand eines Beispiels soll verdeutlicht werden, was intra- und interpersonell unter „Wahrnehmungsorganisation" zu verstehen ist. Man führe sich ein mit Vehemenz durchgeführtes Völkerball-Spiel vor Augen: der Ball wird schnell unter den Mitspielern hin- und hergeworfen, um dann letztendlich mit großer Wucht einen Spieler „abzuschlagen". Das individuelle Mitschwingen im Spielverlauf erfordert ein schnelles visuelles und visuomotorisches Differenzieren der Spielzüge, der eigene Körper muss sich in Raum und Zeit auf dieses fortwährende interpersonelle Geschehen zeitlich passend einstimmen und Visuelles muss mit Motorik synchronisiert werden, um zur richten Zeit am richtigen Ort zu sein. Der Bewegungsfluss der anderen muss „gelesen werden", zumal auch Tricks und Täuschungsmanöver zum Spiel gehören. Der mit Macht geworfene Ball sollte auf einen Körper treffen, der über ein gutes Gleichgewichtssystem verfügt, um nicht aus der Balance zu geraten. Die Arme haben sich dem anfliegenden Ball im richtigen Moment geöffnet zu nähern und der Händegriff muss fest genug sein, damit der Ball nicht zu Boden fällt. Bei dem Gesamt der Abläufe braucht nur ein einziger *Aspekt nicht stimmig zu laufen, und schon gelingt das* Gesamt *der ganzen Handlung nicht. (Schon 1982 sprach der herausragende Kinderpsychiater Reinhart Lempp in diesem Zusammenhang von einer „Kette", die so stabil ist wie ihr schwächstes Glied.)*

Ein Mensch, dessen Wahrnehmungsorganisation langsamer und/oder ungenau arbeitet, wird bei einer derartigen Aktivität Unsicherheit, Beschämung und keine Freude erleben und sich – wenn möglich – herausnehmen.

Die Welt trifft über den Gleichgewichtssinn, den Hautsinn, den Stellungs- und Spannungssinn, den Gesichts-, den Gehör-, den Geruch- und den Geschmackssinn auf das Verarbeitungssystem des Menschen; erst danach ist die Schwelle von außen und innen überschritten. Diese Aufnahme über die Sinnesorgane kann sehr unterschiedlich erfolgen und das jeweilige Outcome beeinflusst ganz maßgeblich unser psychologisches Erleben. Dazu Ratey von der Harvard-University: „Wenn ein Therapeut die Ausrüstung erforscht, mit der ein Mensch das Leben wahrnimmt, dann beginnt er gewissermaßen am Ursprung … Störungen in der Wahrnehmung …können den Patienten in trostloses Elend stürzen." (Ratey 2001, S. 14). Das frühere Leben verlief im Vergleich mit dem heutigen bei weitem nicht so schnell und ohne Reizüberflutung. Zudem war der Mensch von früh bis spät auf den Beinen und das Leben war von Motorik geprägt, was die Sinne – auch die schwächeren – schulte und aufpolierte. Auf der Harvard-Spring-Conference 2013 in Boston „The Therapeutic Action of Psychodynamic Psychotherapy: Current Concepts of Cure" – eine alljährliche psychotherapeutisch-psychoanalytische Veranstaltung vom Feinsten – wurde von den beiden Veranstaltungstagen ein halber Tag der Körperlichkeit gewidmet, da ein regelmäßiges aktives Körperprogramm ungeheure Auswirkungen auf den hirnorganischen Stoffwechsel und die synaptischen Vernetzungen hat, was psychische Prozesse beeinflusst und prägt. Regelmäßige körperliche Aktivität beeinflusst die Chemie des Gehirns, dient der Regulation der Emotionen und optimiert das intellektuelle Funktionieren. Das heutige Leben bietet einem Menschen mit körperlichen Unzulänglichkeiten enorme Versuchungen zur Passivität, und so ist es nicht einfach, gerade einen Heranwachsenden mit sensorischen Unzulänglichkeiten zum Erwerb dieser „Körperfremdsprache" zu verführen. Wird sich nicht bewegt, nimmt nicht nur die Entwicklung der sensorischen Integration Schaden, sondern entwickelt sich in der Folge auch eine andere hirnorganische Chemie und ein damit einhergehendes Netzwerk (Ratey 2008). Sinnvoll aktiviert, wird man zweifelsohne kein rhythmisch-flexbiler Sportler, jedoch kommt es nachweislich zu einer hirnorganischen synaptischen Verdichtung, die koordinierteres Funktionieren erleichtert. Die Weisheit des Spruches „Wer rastet, der rostet" findet sich im hirnorganischen Prinzip: „use it or loose it" wieder

Die Wahrnehmungsorganisation ist *das* Schlüsselloch, das die von außen eintreffenden Reize passieren müssen. Und diese Verarbeitung bestimmt wesentlich das erst danach mögliche reaktive Verhalten. Diese Erkenntnis ist nicht neu. Schon 1633 formulierte der große Pädagoge Comenius: „Es gelangt nichts in den Verstand, was nicht zuvor in den Sinnen war." Nur mit der Erforschung dieser Ausrüstung, mit der der Mensch das Leben in sich aufnimmt, holt man ihn gewissermaßen dort ab, wo er steht. Unausweichlich prägt somit die individuelle sensorische Verarbeitung den Prozess der Identitätsfindung.

Der Mensch versucht die Welt mit seinem Körper und den darin eingebetteten beschriebenen Sinnesorganen zu fassen. Je genauer, je differenzierter, je schmeidiger und zeitangemessener diese Verarbeitung ist, umso schneller und sicherer kann man auf die von außen eintreffenden Stimuli antworten – d.h. sich zeitnah darauf einstellen um mitschwingen. Gelingt dies nicht, hat der Mensch schlechte Karten, da er die jeweiligen Abläufe „nicht schnell genug blickt" „mal wieder nicht hinhört" oder in Bezug auf sein Bewegungs- und Gleichgewichtssystem nicht „hieb- und stichfest" ist. Wer Probleme in der Feinmotorik hat, hat scheinbar „zwei linke Hände" und „es geht alles kaputt, was dieser Mensch in die Finger bekommt“. Die Liste derartiger *entwertender* Beschreibungen könnte fortgesetzt werden und es ist vermutlich just jene damit einhergehende Beschämung, was die Annäherung an dieses Thema so problematisch macht.

Wie geht es aber einem Menschen, der vieles nicht schnell genug „blickt“, der häufig immer wieder „auf dem „Schlauch“ oder wie der berühmte „Ochs vorm Berg“ steht? Wie geht es einem Menschen, wenn er die ihn umgebende Geräusch- und Sprachkulisse nicht zügig genug zu differenzieren vermag und vieles einem Brei ähnlich an ihm viel zu schnell vorbeirauscht – ähnlich der Erfahrung, wenn man in einem fremden Land die Sprache nicht versteht, die Menschen aber fortwährend darauf drängen, dass man zeitnah reagiert. Wie geht es einem Kind, das nicht „hieb und stichfest“ ist, das am Klettergerüst einer Rutsche steht, gerne hinaufsteigen würde, aber andere robuste und körperlich vital sicherere Kinder drängeln und schubsen? Wie positiv kann unter diesen Umständen eine Selbstbesetzung aussehen, wie die Bausteine einer sich-selbst-bejahenden Identität?

Wenn man den Bewegungsablauf des angeführten Spiels auf die allgemein menschliche Kommunikationsgeschehen überträgt, wo Bewegungen, Sätze, Worte, Wortfetzen im Verbund mit Blicken hin und her geworfen werden, kann man mit etwaigen Unzulänglichkeiten in diesem Geschehen nicht oder nur mit großer Mühe und Anspannung mithalten.

Die zuvor angeführte Weisheit des großen Comenius zeichnet sich auch in der deutschen Sprache nach, wo mit dem anfänglichen „Begreifen“ – die Welt mit den Händen erfassen – sich allmählich mit der Reifung der mentale „Begriff“ bis hin zum „Wort“ und Schriftbild entwickelt. Wenn im primären Be-Greifen im Kontakt der Hände des krabbelnden Säuglings mit den Gegenständen das Erfasste in seiner Eigentümlichkeit taktil nur wenig, unsicher und eventuell sogar mit Unbehagen erspürt wird, wird nachvollziehbar, dass die lineare Entwicklung ganz anders verläuft wie bei einem Krabbelkind, das sich mit feinstem sensiblen und lustvollen Erspüren eine Vorstellung von der Welt erarbeitet. Ist die basale Sinnesverarbeitung von neuropsychologischen Unzulänglichkeiten geprägt, spricht man von primären Wahrnehmungsschwächen (vgl. Warnke 2000, S. 197 ff.). Diese konstitutionelle Schwäche darf nicht mit einer Wahrnehmungsschwierigkeit, die aus einer durch eine Veränderung des Gemütszustandes beeinflussten Wahrnehmung, aus Empathieverzerrungen oder aus einer psychologischen Dissoziation herrührt, verwechselt werden.

Schon 1985 beschrieb Daniel Stern in seinem zukunftsweisenden Buch „Die Lebenserfahrung des Säuglings“ differenziert die konstitutionellen Unterschiede: “Die Fähigkeiten, die es dem Säugling ermöglichen, seine vielfältigen sozialen Erfahrungen miteinander zu verbinden, sind zu einem sehr großen Teil konstitutionell – und das heißt genetisch – determiniert. Entweder sind sie von Anfang an vorhanden, oder sie entfalten sich unter der Voraussetzung, dass das Zentralnervensystem ebenso wie die Umwelt intakt ist, nach einem angeborenen Zeitplan. Für die nahe Zukunft könnte sich die Untersuchung der individuellen Unterschiede dieser Fähigkeiten im Hinblick auf *psychopathologische* (Hervorhebung E. R.) Entwicklungen sehr kleiner Kinder als das *fruchtbarste* (Hervorhebung E. R.) klinische Forschungsfeld erweisen. (…) Man hofft, dass die frühesten Abweichungen in den sozialen und intellektuellen Funktionen sich auf Anomalitäten dieser Fähigkeiten zurückführen lassen. (…) Sehen wir uns kurz die klinischen Konsequenzen an, die Defizite dieser Fähigkeiten zur Folge haben können. Die Fähigkeit, Informationen aus einer Sinnesmodalität in eine andere zu übertragen, ist für die Integration des Wahrnehmungserlebens von so zentraler Bedeutung, dass die Probleme, die aus einem Defizit dieser Fähigkeit erwachsen können, fast unbegrenzt sind. Lernschwierigkeiten sind eines der ersten Probleme, an die man sogleich denkt, da das Lernen zu einem so großen Teil auf der Informationsübertragung von einem Sinnesmodus in einen anderen, vor allem auf Übertragung zwischen Gesichtssinn und Gehör, angewiesen ist. (…) Auch in sozialer und emotionaler Hinsicht *könnten* (Hervorhebung E. R.) Kinder durch solche Defizite beeinträchtigt werden; ein reibungsloser intermodaler Informationsfluss trägt in hohem Maße dazu bei, das Sozialverhalten anderer Menschen sowie die integrierten Handlungen, Empfindungen, Affekte usw. des eigenen Selbst *richtig zu erfassen* (Hervorhebung E. R.). Diese Ergebnisse sind einstweilen noch zu neu, als dass man sich von den Grenzen ihrer psychopathologischen Relevanz schon eine Vorstellung machen könnte" (dtsch. 1992, S. 265f).

Das Leben des heutigen Menschen und natürlich auch das des Kindes ist von einem hohen Anpassungsdruck geprägt, der vom Zeitgeist diktiert wird. Alles hat flexibel, im Team und kommunikativ zu geschehen – so als ob alle Menschen diesbezüglich gleich ausgestattet seien. Kaum eine Zeitschrift oder Fachzeitschrift beschäftigt sich seit geraumer Zeit nicht mit dem Phänomen des „Burnout" und der „Depression". Nicht wenige Menschen sind mit einer nicht optimalen Wahrnehmungsorganisation ausgestattet: in einer Studie des Gesundheitsministeriums Baden-Württemberg wurden im Juni 2001 600 unausgelesene Kinder bei einer Einschulungsuntersuchung (d.h. keine *ausführliche* testpsychologische Untersuchung) auf Wahrnehmungsstörungen überprüft und die Ergebnisse zeigten, dass 23% der Kinder in diesem sehr guten Kurzverfahren unter optisch-graphomotorischen, aktustisch-phonematischen und kinästhetisch-artikulatorischen Verarbeitungs- und Wahrnehmungsstörungen litten (Deutsches Ärzteblatt 6/2001). Es ist m. E. nicht auszuschließen, dass ein Zusammenhang mit den später auftretenden Erschöpfungszuständen besteht, da das Leben der Erwachsenen eine hohe Anforderung an die individuelle Flexibilität und Mobilität stellt.

Die Basiskompetenz der Rhythmuserkennung (Jäncke 2008, S. 144; Sacks 2008) scheint für die Reizverarbeitung eine Grundlagenfähigkeit darzustellen. Die meisten Abläufe im Leben stehen damit in Zusammenhang. Einen Rhythmus zu erkennen ermöglicht, ihn biomechanisch und neurophysiologisch nachzuempfinden, was notwendig ist, um sich mit den von außen eintreffenden Reizen zu synchronisieren. Jedes menschliche Zusammensein erfordert dieses Mitschwingen, sei es in der Bewegung, in der Sprache oder im Miteinander-Handeln. Sprachliche geschmeidige Kommunikation ist nur möglich, wenn sich die Sprachrhythmen der Partner zeitnah begegnen. Gemeinsames motorisches Miteinander verlangt auf der sensomotorischen Ebene einen ähnlichen Synchronisationsprozess, und so ist es nicht verwunderlich, dass in allen Kulturen wichtige Ereignisse, wo es gilt, die Menschen miteinander abzustimmen, von Musik und rhythmischen Ritualen begleitet sind. Es handelt sich um eine elementare Kommunikationsebene, die Einfluss auf Emotion, Motorik und kognitive Funktionen nimmt. Rhythmischer Gleichschritt ermöglicht ein Wir-Gefühl, das es wiederum ermöglicht, aus der Einsamkeit auszusteigen. Synchrones Handeln ist die Voraussetzung für prosoziales Verhalten, was das Gefühl von Zusammengehörigkeit unterstützt. Jaffe/Beebe et al. erforschten 2001 die „Rhythms of Dialogue in Infancy" und arbeiteten die Notwendigkeit der zeitlich-rhythmischen Abstimmung im frühen körperlichen Mutter-Kind-Dialog heraus. Erwähnenswert sind in diesem Zusammenhang die Untersuchungen von Condon/Sander (1974) und Condon (1975), wo gesunde Neugeborene in den ersten zwei Wochen ihres Lebens beobachtet und ihnen Tonbandkassetten mit amerikanischer und chinesischer Sprache, vereinzelte Vokale und Schlaggeräusche angeboten wurden. Die Babys bewegten sich völlig synchron (Augen, Kopf, Schulter, Augenbrauen, Ellbogen, Füße) mit der vom Erwachsenen vorgegebenen Sprachstruktur. Sie waren somit von Anfang an intensive Teilnehmer an diesem Prozess und bewegten sich im Rhythmus der Sprache. Ein einziges Baby – ein Risikokind – bewegte sich nicht synchron, was damals als Einzelfall klinisch aber noch keine Bewertung erfuhr.

Das Wissen, wie stark Rhythmik die Entwicklung beeinflusst und auch fördert, scheint in allen Kulturen im Umgang mit Säuglingen und Kleinkindern implizit vorhanden zu sein: das Kind wird zur Beruhigung auf den Armen oder in der Wiege geschaukelt; die ihm vorgesungenen Lieder bestehen aus einfachen sich wiederholenden Rhythmen und Reimen und rhythmische Gebärden werden benutzt (Hoppe Hoppe Reiter ...), was je nach Einsatz und Ausdruck belebt, beruhigt und vereint. Wenn die rhythmische Verarbeitung erschwert ist, machen später rhythmische Tätigkeiten wie z.B. Laufen und Schwimmen wenig Freude und müssen wie eine Fremdsprache erlernt werden – was, wie bei der Sprache, in der Kindheit leichter fällt. Dieser Sachverhalt erklärt, warum in dieser Hinsicht „unbegabte" Kinder (später auch die Erwachsenen) diese Tätigkeiten eher vermeiden. Dies wiederum hat zur Folge, dass der Körper nur bedingt oder gar nicht in den Zustand des „Runner's High" gelangen kann, der dann auftritt, wenn der Körper regelmäßig zumindest zwanzig Minuten kontinuierlich rhythmisch bewegt wird. Durch dieses Training wird der hirnorganische Stoffwechsel wie durch ein antidepressiv wirkendes Medikament angekurbelt, und so kann selbstwirksam eine bessere Stimmung herbeigeführt werden. Daneben wird aber auch noch die Neubildung von Nervenzellen angeregt, die bei chronischem Stress und Depression eingeschränkt ist (Rüegg 2009).

Der Rhythmus des Gehens, Tanzens, Joggens, Schwimmens, Reitens u.a. stabilisiert aber nicht nur das Gehirn in Bezug auf eine bessere Stimmung. Die Nervenzellen erzeugen ihre elektrischen Impulse im Gleichtakt, was sich positiv auf die Wahrnehmung, auf die Erinnerungs- und die Denkleistung auswirkt. Im Ergebnis der Studie der Universitätsklinik Ulm formulierte der Direktor der Klinik für Psychiatrie: „Sport macht das Gehirn effektiver". Intensives Lauftraining führt zu deutlichen Verbesserungen im visuell-räumlichen Gedächtnis und in der Konzentration. Rhythmisches Training macht somit nicht nur „schlau", sondern sorgt auch für „gute Laune" (In: Rhein-Neckar-Zeitung 19./20.04.2008, S. 17; Frankfurter Rundschau – WISSEN & BILDUNG 31.10.2006, S. 28).

Diese Erkenntnisfortschritte belegen, wie weitreichend die Auswirkungen körperlich-rhythmischer Aktivitäten und daraus resultierenden Kompetenzen sind. Bei primärem sensomotorischem Mangel – der früher durch den bewegungsgeprägten Alltag ausgeglichen wurde – können diese Kompetenzen mit viel Selbstüberwindung und Ausdauer in gewissem Umfang erarbeitet werden – wozu ein erhebliches Maß an Motivation, Fleiß und Disziplin vorhanden sein muss. Tief beeindruckende Sätze zu „Rhythmus" und dessen archaischer Auswirkung und Bedeutung sind im Film „Rhythm is it" zu hören, wo sich Sir Simon Rattle, der Dirigent der Berliner Philharmoniker, in ergreifender Form dazu äußert (2005).

Wenden wir uns nun der Wahrnehmungswelt des Kindes zu:

Eine intakte sensorische Verarbeitung ist keine Angelegenheit von Entweder-Oder. Man hat nicht eine optimale Wahrnehmungsorganisation oder gar keine. Die Übergänge zum Normalen sind fließend. Nach statistischer Wahrscheinlichkeit muss angenommen werden, dass die sensorischen Reaktionsmöglichkeiten bei einer großen Zahl von Menschen zwischen guter und manifest schlechter sensorischer Integration streuen (vgl. Lempp 1970/1975/1982). Eltern von Kindern und später deren Partner, die von derartigen Unzulänglichkeiten nichts wissen, können sich nur unzureichend auf die Bedürfnisse ihres diesbezüglich beeinträchtigten Mitmenschen einstellen.

Die Folgen derartiger Schwächen werden in der frühen Kindheit in einem überschaubaren, kleineren menschlichen Interaktionsrahmen oft kaum spürbar, da sich in der Zweier- oder Dreier-Situation die Menschen leichter aufeinander abstimmen können und das Gesamt der interaktionellen Stimuli überschaubar und beantwortbar ist. Dies gilt auch später z.B. im Zweierkontakt der Lehrer-Schüler-Interaktion und im therapeutischen Einzel-Setting, wo diese Probleme meist kaum spürbar sind. In einem komplexeren Geschehen kommt es aber zu schnellen, zu rivalisierenden und vielschichtigen Interaktionen, deren Bewältigung für einen sensorisch schwächer ausgestattete Menschen sehr schwierig ist. Dass das daraus resultierende Erleben *in der Folge* zu psychischen Beeinträchtigungen führen kann, ergibt sich von selbst, doch darf dieses Outcome nicht als das Primäre erachtet werden. „Setzt die Therapie bei seinem (des Menschen, Anm. E. R.) Unglücklich sein an, führt das zu nichts. Unsere heutige Welt ist voll von Menschen mit ... Wahrnehmungsstörungen, die sich fragen, warum all die Antidepressiva und Analysen nichts gefruchtet haben" (Ratey 2001, S. 14). Größte Aufmerksamkeit sollte daher dem sensomotorischen interpersonellen Geschehen zukommen, da sich im Lebensalltag durch die entsprechenden Unzulänglichkeiten mehr oder weniger gravierende Misserfolge einstellen, die *in der Folge* zu psychologischen Belastungen werden können.

Grob skizziert sollen im weiteren Verlauf die Unzulänglichkeiten in der taktil-kinästhetischen, auditiven und visuellen Wahrnehmungsverarbeitung beschrieben werden:

Von Lebensbeginn an spielt die taktil-kinästhetische Wahrnehmung, d.h. die Verarbeitung der Informationen aus dem Tastsinn und den Lagerezeptoren, eine herausragende Rolle (Ayres 1979), da alle weiteren Verarbeitungsschemen davon beeinflusst sind. Der Fötus erlebt die Bewegungen der Mutter, er berührt die Nabelschnur, er lutscht am Daumen und mit dem Größerwerden erspürt er den eigenen Körper durch die Enge im Mutterleib immer deutlicher. In der Außenwelt angekommen, versucht die Mutter ihr Baby über ihren Körper, d.h. durch Wiegen, Streicheln, zärtliche Berührungen und Körperwärme in seiner Reizverarbeitung zu unterstützen. Das Berührt-Werden und das beruhigende Sich-Selbst-Berühren (self-touch) ermöglichen ein Nachlassen von Spannung. Vielleicht werden diese mütterlichen Bemühungen, um dem Kind das Gefühl von Sicherheit und Bindung zu geben, von einem diesbezüglich konstitutionell schwächer ausgestatteten Baby abgewehrt, da es nicht in der Lage ist, durch die wiegende rhythmische Bewegung der Mutter und durch ihr Streicheln körperliche Zentrierung, Ruhe und affektive Regulierung seiner Spannung zu erfahren. Vielleicht erlebt es die Bewegung als ängstigend und die zärtliche Berührung eher als unbehaglich und reagiert aversiv, weil es sie als übererregend oder ungenau empfindet. Dieser misslingende Dialog zwischen Mutter und Kind prägt den affektiven Zustand beider und damit auch die künftige Entwicklung der Beziehung (vgl. Rass 2002/2008). Das taktil- kinästhetisch schwächere Kind erlebt die Welt quasi „wie mit Boxhandschuhen" und es hat für manches wenig Gespür oder es reagiert hypersensibel. Die Koordination seiner Bewegungen und sein sicherheitsgebendes Gleichgewichtssystem, mit dem es den Raum um sich herum erfährt, sind ebenfalls wenig sicherheitsgebend ausgeprägt. Wenn „Raum" nicht mit körperlicher Sicherheit erlebt wird, was nur mit einer sicheren taktil-kinästhetisch-vestibulären Körperwahrnehmung gelingt, leiden darunter alle spontanen und freudvollen Bewegungen (vgl. Ayres 1979). Spielerischer Kontakt unter Kindern ist meist körperlich. Wenn ein Kind nicht in der Lage ist, dieses Rivalisieren und den Wettbewerb mit Freude zu erleben, reagiert es verstört. Diese Verstörung beeinträchtigt in der Folge sein Streben nach Unabhängigkeit, denn Sicherheit und Exploration liegen auf einem Kontinuum. Wenn ein Kind unter Kindern mit derartigen Beeinträchtigungen zu kämpfen hat, nützt ihm in diesem Augenblick auch die beste Beziehung zu den Eltern nichts (vgl. ebd.). Es wird vorsichtig und skeptisch und versucht, diese an sich gesunde Rivalität zu vermeiden. Dies wiederum bedeutet eine schwere Verletzung des aufkeimenden kindlichen Selbstwerterlebens. Ein nächster fataler Schritt in dieser unseligen Entwicklung ist das Auftauchen von Scham. Zuweilen wird ein derartiges Kind als ängstlich oder schüchtern beschrieben, was aber nur den psychischen Aspekt erfasst, während die körperlichen Unsicherheiten, die zu diesen Reaktionen führen, meist unbemerkt bleiben. All zu oft nehmen Eltern, Ärzte, Lehrer und Therapeuten das Ausmaß der Unzulänglichkeit nicht wahr (Rass 2000/ 2008). Während das Kind in der ruhigen vertrauten Zweierbeziehung im Erwachsenenkontakt unauffällig erscheint, bereiten ihm Gruppenspiele, feinmotorische Anforderungen unter Zeitdruck und später schulisches Lernen keine Freude, vielmehr ermüdet die große Anstrengung. Oft wird das Nachlassen der Spannkraft mit „Konzentrationsschwäche" beschrieben. In der Schule verlangen Denkprozesse, Rechnen und Aufsätze-Schreiben die Vorstellung eines sicheren mentalen Raumes, der sich nur entwickeln konnte, wenn der Körper des Kindes in der Lage war, den tatsächlichen Raum, in dem es lebt, mit adäquater Sicherheit zu erobern. Selbstorganisation – ein machtvolles Wort unseres Zeitgeistes – ist schwer, wenn es einem aus der Sensomotorik heraus nicht gelingt, etwas alleine „auf die Reihe zu bekommen". Der rote Faden einer sinnvollen Handlung taucht nicht auf, der seriale Ablauf wird nicht erkannt und das Kind „zäumt das Pferd am Schwanze auf". Statt Hilfe erfolgen von außen meist eher Ermahnung und Kritik.

Ein weiterer Wahrnehmungsbereich ist der der auditiven Wahrnehmungsorganisation: Akustische Informationen werden großteils – in weit stärkerem Maße als bei anderen Sinnesorganen – lange bevor wir uns ihrer bewusst werden, verarbeitet. Auf der Zwischenstation entlang der Hörbahn vom Ohr bis an die Stelle, wo sie in Bewusstsein vordringt, werden die Lautsignale differenziert und fein angestimmt (Ratey 2001, S. 112). Im Hirnstamm werden die Geräusche in ihrer Tonqualität unterschieden und entsprechend ihren Klangeigenschaften in Einheiten zerlegt. Ähnlich wie die rhythmische und taktil-kinästhetische Wahrnehmung ist auch die auditive Wahrnehmung ein sehr komplexes Geschehen. Sie besteht u.a. aus Aufnahme, Selektion, Differenzierung, Analyse, Synthese, Ergänzung, Integration und Raum- Lagedifferenzierung und Speicherung (vgl. Rosenkötter 2003). Der Weg vom akustischen Signal und zur phonologischen Bewusstheit ist somit ein sehr komplexer, der nur zeitnah und differenziert benutzt werden kann, wenn die verschiedenen Aspekte integrativ miteinander arbeiten (Ptok et al. 2000). Am besten können ältere Erwachsene, die in jüngeren Jahren sehr gut hören konnten und nun auf ein Hörgerät angewiesen sind, nachvollziehen – da sie den Vergleich kennen – , welch komplexes Phänomen die Sprachverarbeitung ist, und wie schwierig die Programmierung selbst des sensibelsten Hightech-Computers des Hörgerätes ist, der fortwährend auf die unterschiedlichen Umgebungsfaktoren (z.B. Nähe, Ferne, glatte oder raue Umgebungsgegebenheiten, räumliche Akustik) programmiert werden muss. All diese Leistungen – und natürlich noch viel viel besser – übernimmt ein funktionstüchtiges auditives Verarbeitungssystem. Um einen Menschen mit einer intakten auditiven Wahrnehmungsorganisation an das Erleben einer unzulänglichen auditiven Wahrnehmungsverarbeitung heranzuführen, ist anzuraten, dass er sich vorstellen soll, in einem Land zu sein, wo er die ihn umgebende Sprache nicht voll beherrscht und das Umfeld keine Rücksicht auf die ungenaue, langsamere und mühsamere Verarbeitung nimmt und ein schnelles Reagieren verlangt – was noch schwerer ist, wenn trotz intensivsten Bemühens auch noch Störgeräusche auftreten, wie z.B. Papierrascheln, Hüsteln, Straßenlärm, Radiomusik, usw. Dem Sprachfluss eines anderen Menschen zeitangemessen und flexibel zu folgen, wird unter diesen Umständen zu einer Schwerstarbeit. Während der schwerhörig gewordene Erwachsene im Alter eventuell genervt, unwirsch und schnell missgelaunt ist, ist das Kind in seiner Entwicklung eher unsicher, irritiert und – da sein Hörorgan vom Arzt ja als intakt bezeichnet wird – häufig in der Kritik. In der Schule taucht in diesem Zusammenhang meist das Wort „Konzentrationsschwäche" bis hin eventuell zur Diagnostik einer „ADHS" auf.

Natürlich sind solche Verarbeitungsschwierigkeiten auch dazu prädestiniert, den Lese-Rechtschreiberwerb zu erschweren (vgl. Firnhaber 2005). Bei nur *einem* Gegenüber ist es weniger problematisch, einzelne akustische Aspekte heraus zu filtern. In einer ruhigen Situation – z.B. in einer Kindertherapiestunde – tauchen diese Probleme daher häufig auch nicht auf, zumal sich im Zweierbezug die Partner leichter auf den jeweiligen Verarbeitungsrhythmus und die Verarbeitungsgeschwindigkeit des anderen einstellen.

Ähnlich wie bei den beiden vorausgegangenen Wahrnehmungsbereichen ist auch die visuelle Verarbeitung ein komplexes Geschehen. Nach der visuellen Aufnahme müssen die unterschiedlichsten Differenzierungen vorgenommen werden, wobei dem Wahrnehmungs*raum*, den es motorisch zu erfassen gilt, noch zusätzliches Gewicht beigemessen wird. Neben den Feinheiten des „Durchschauens" spielen Aspekte des Vorder- und des Hintergrunds, der Raum-Lage-Qualität, aber auch der Zeitaspekt eine herausragende Rolle. Ein gelungenes „Durchschauen" einer Situation und ein schnelles Bedeutung-Beimessen erlauben ein ganz anderes Sicherheitsgefühl, als wenn der „Film des Lebens" in Windeseile vorbeirauscht und die Einzelheiten nicht differenziert werden können. Unter solchen Umständen erzeugen neue und unvertraute Situationen eine hohe Anspannung und Angst und die Interaktion mit anderen Menschen ist dadurch erschwert. Da die anderen, insbesondere im Gruppengeschehen – das sich fortwährend ändert – nicht darauf warten, bis der davon Betroffene die Situation mit ihren jeweils wichtigen Details zeitnah differenziert und realisiert hat, ist die subjektive Erfahrung meist folgende: „Keiner mag mich, keiner will mich dabei haben" – ohne dass für das Scheitern subjektiv ein Grund gefunden werden kann. Natürlich erschwert dies mit der Zeit auch die Entwicklung eines stabilen Selbstwertgefühls. Auch hier bedeutet Gruppenarbeit in der Schule, jedes Team im Berufsleben, jeder Mannschaftssport, wo es fortwährend gilt, den Ablauf schnell und strukturiert zu durchschauen, um sich zeitnah mit den anderen zu koordinieren und zu synchronisieren, Schwerstarbeit. Schwächen in der visuellen Wahrnehmungsorganisation beeinträchtigen somit viele Lebenssituationen, denn dieses „Durchschauen" ist in jeder menschlichen Gruppierung beim Bewältigen komplexer Situationen notwendig (vgl. Rass 2002/2008).

Das bisher Beschriebene macht nachvollziehbar, wie wichtig der Umgang der Umwelt mit einem davon betroffenen Menschen – insbesondere mit einem Heranwachsenden – ist. Im familiären Rahmen der frühen Kindheit tauchen in der vertrauten Umgebung diese Schwierigkeiten zunächst meist nicht auf, zumal wenn die Eltern intuitiv dem Kind komplementär und abgestimmt helfen. Sie schmiegen sich somit in die defizitäre Verarbeitung ein und ermöglichen dadurch einen kontinuierlichen Begegnungsfluss (Palombo 2006/2011). Mit diesen

Erfahrungen ziehen es Kinder häufig vor, mit Erwachsenen zusammen zu sein, z.B. in der Nähe der Erzieherin, mit der zudem körperliches Rivalisieren keine wichtige Rolle spielt. Problematisch wird es in der frühen Kindheit, wenn die Eltern die Unzulänglichkeiten intuitiv nicht erspüren und sich nicht mit ihrem Kind komplementär abstimmen können, sodass der für das Kind so notwendige synchronisierte affektive und somit beruhigende Dialog entgleist. Ein hektischer und unruhiger Bewegungsfluss, eine eigene hohe innere Anspannung der Eltern, eine nicht auf das Kind abgestimmte Prosodie werden von einem intersensorisch schwächeren Kind nicht nur als Empathiefehler, sondern als ein Scheitern im basalen Kommunikationsgeschehen erlebt. Das Problem der Anpassungsstörungen bei Säuglingen sollte auch unter dieser Perspektive betrachtet werden, da – wie uns die Säugling/Kleinkind-Eltern-Therapie zeigt – wir als Behandler den Eltern helfen sollen, ihr Kind „zu lesen". Immer wieder sind dabei auch Kinder, die s.g. high-need-babies, die viel schwieriger „zu lesen" sind, was in der Fürsorgeperson hohe innere Anspannung aufkommen lässt, was es ihr erst recht erschwert, sich auf dieses Kind abgestimmt einzulassen. Komplementarität ist somit wesentlich für die Aufrechterhaltung eines integrierten Selbstgefühls beim Kind (Palombo 2011).

Wenn dem noch sehr kleinen Kind die reaktive Rhythmik der nonverbalen Kommunikation – das ist der Austausch von Mimik, Gestik, Prosodie – nicht gelingt, es z.B. das veräußerte Verhalten der Eltern falsch „liest" und wenn die kontinuierliche Fürsorgeperson das kleine Kind nicht komplementär dort abholt, wo es in seiner Verarbeitungsmöglichkeit steht, dann ist vorhersagbar, dass der affektive Dialog von Nicht-Abgestimmtheit und Dissonanzen geprägt ist. Problematisch und entwicklungsbeeinträchtigend ist vor allem, dass das allgemeine basale Stressniveau des Kindes dadurch sehr hoch ist und jeder hinzukommende Alltags- und Entwicklungsstress auf ein hohes Anspannungsniveau trifft und jenes Kind/jener Mensch völlig anders reagiert als jemand mit einer inneren Gemütsruhe.

Die Muster und die Geschwindigkeit, d.h. die Rhythmik der verbalen Kommunikation sowie des Blickkontaktes und des Bewegungsflusses, sind wesentliche Elemente dieser subliminalen Kommunikation und es sind diese drei Botschaften, die die Natur der Interaktion bestimmen. Nur eine abgestimmte psychobiologische Kommunikation mit der Fürsorgeperson vermittelt dem Kind Sicherheit. Störungen und Fehlanpassungen der intuitiven Kompetenz auf Seiten der Betreuungsperson entstehen jedoch häufig, wenn das individuelle Kind in seinen Auslöse- und Rückkoppelungssignalen schwer verständlich ist. Die kindliche Responsivität kann sowohl auf motorischer Seite als auch auf Seiten der Wahrnehmung soweit eingeschränkt sein, dass die elterlichen Bemühungen um einen dialogischen Austausch allmählich erlöschen oder in einen direktiven zudringlichen Interaktionsstil einmünden (Papousek et al. 1994). Es fällt Eltern unglaublich schwer, sich – wenn sie über eine geschmeidige und flinke Wahrnehmungsorganisation verfügen – in die langsamere und zähere ihres Kindes einzufühlen. Je abgestimmter die Fürsorgepersonen ihr Aktivitätsniveau an das des Kindes während Perioden des sozialen Miteinanders anpassen können, desto synchronisierter sind ihre Interaktionen. Sensitive Eltern erleichtern somit die Informationsverarbeitung ihres Kindes. Diese „high-need"-Kinder verlangen meist von ihren Eltern mehr als „good enough parents" zu leisten vermögen. Auch im Zusammenhang mit der Entwicklung des *Erscheinungsbildes* von ADHS ist zu bedenken, dass für die Herausbildung synaptischer Angebote zu Stabilisierung komplexer Verschaltungsmuster im frühen Kortex ungünstige Bedingungen wie Irritationen, Stress und psychische Belastung während der frühen Phase der Hirnentwicklung von Bedeutung sind (Hüther et al. 1996). Deswegen treffen wir bei unruhigen Kindern häufig auf solche mit Wahrnehmungsstörungen, da ungünstige Bedingungen in notwendigen Phasen der Hirnentwicklung den Aufbau und die Stabilisierung entsprechender Muster verhindern.

Da Störungen in der sensorischen Integration sich nicht auswachsen, erschweren sie den gesamten Prozess des Heranwachsens, und die Entwicklung kann in ein davon sehr belastetes Erwachsenenselbst münden, womit sich der Kreis zum Beginn meiner Ausführungen – wo es auch um gestresste und erschöpfte Erwachsene ging – schließen kann. Auch der Ablösungs- und Individuationsprozess in der Adoleszenz kann durch diese Problematik stark beeinträchtigt sein, zumal sich der Jugendliche gerade in dieser Zeit ständig mit den Gleichaltrigen vergleicht und bei entsprechend negativem Ergebnis in hohem Grad in seiner Selbstachtung und in seinem Selbstwertgefühl verletzt ist, was depressiven Rückzug, narzisstische Wut und/oder psychosomatische Beschwerden nach sich ziehen kann. Der ohnehin bestehende enorme Sog der medialen Welt wird unter diesen Umständen noch mächtiger.

Ein Kind, dessen verschiedene Sinnesmodalitäten anders arbeiten als die der „Norm", das deswegen „aneckt", macht es seiner betreuenden Umgebung schwer, es richtig einzuschätzen. „Good enough parents" reichen – wie schon erwähnt – bei einem solchen Kind nicht. Seine Bedürfnisse werden häufig nicht richtig und rechtzeitig erkannt und falsch gedeutet, was die Herstellung der inneren Homöostase und eines mutuellen Anpassungssystems mit der Pflegeperson erschwert. Selbst- und Fremdregulation sind hohen Belastungen ausgesetzt, was allzu oft zu malignen Regelkreisen führt. Säuglinge mit Regulationsschwierigkeiten und Anpassungsproblemen an die Rhythmen des Lebens erschöpfen häufig das Responsivitätsvermögen der erwachsenen Beziehungspartner. Den

Eltern muss daher das Erleben ihrer Kinder n°°ahe gebracht werden und das anders geartete Verhalten und Erleben muss ihnen „übersetzt" werden, um auf diese Art und Weise den Eltern zu einem besseren Kontakt mit der inneren Welt ihres Kindes (Ornstein, A. 1977) zu verhelfen. Die Motivationsforschung zeigt, dass sich ein Kind schon ab dem 3. Lebensjahr mit den anderen vergleichen kann (Heckhausen 1971; Rheinberg 1997; Oerter 1995). Das Scheitern, das keinen Namen hat, beeinflusst aufgrund der Unzulänglichkeiten die sich entwickelnde Persönlichkeit und die Leistungserwartung, sodass die Selbsteinschätzung in Bezug auf „erfolgreich" oder „nicht erfolgreich" schon stark das kindliche Selbstwerterleben prägen – also lange bevor die Schule beginnt. Wie zuvor beschrieben, ist der Prozentsatz der Kinder, die nicht zeitnah und rhythmisch mitschwingen können, nicht gering, und die Vorverlegung des Eintritts in die institutionelle Erziehung bedeutet gerade für die Kinder eine noch problematischere Schwellensituation.

Da das Leben eines Kindes unter diesen Voraussetzungen häufig auch mit schweren Schamgefühlen einhergeht – „was ist nur falsch an mir?" – gilt es auch diesem Erleben unter dieser Perspektive Aufmerksamkeit zu schenken. Die Bindungs- und Kleinkindforschung belegt, dass das sich entwickelnde Selbst des Säuglings von Anfang an sozial bezogen ist, und dass die Erfahrung des auftauchenden Selbst nur im Umgang mit anderen möglich ist. Somit spielen Affekte im Austausch mit sich selbst und den anderen eine soziale Rolle. Schwierigkeiten bei der Abstimmung von Affekten und im mangelhaften Zusammenspiel mit den wichtigsten Pflegepersonen rufen beim Kind erhebliche Irritationen hervor, wenn seine Aktionen nicht angemessen begleitet werden. Da sich bis zum 2. Lebensjahr das Selbstkonzept zu entwickeln begonnen hat, gibt es auch schon ein Schamerleben, das durch schmerzliche Erkenntnisse des – vermeintlich – eigenen Versagens ausgelöst wird.

Emotionaler und sozialer Rückzug sind oft die Folge dieser Wahrnehmungsdefizite, was wiederum zu einem Mangel an Lebenserfahrung führt, der insbesondere heute, wo Teamwork und Flexibilität angesagt sind, von höchster Wichtigkeit ist. Schon 1991 belegten die Untersuchungen der Universität Mannheim zu umschriebenen Entwicklungsstörungen, dass die Gruppe der beeinträchtigten Heranwachsenden, die meist in der Schule und in ihren Ausbildungen erhebliche Probleme hatten, in der Gruppe der Delinquenten, Suchtkranken oder seelisch kranken Erwachsenen, die am Leben scheiterten, signifikant stark vertreten waren. Als Kinder wurden sie nicht dort abgeholt, wo sie in ihrer Entwicklung wirklich standen.

Und genau hier liegt ein wichtiger Ansatz der psychoedukativen und/oder psychodynamischen Behandlung. Ähnlich wie bei der Säugling/Kleinkind-Mutter/

Eltern-Therapie liegt es am Behandler, das Kind mit in seinem Verhalten einzuordnen – sei es durch geschulte Beobachtung oder auch mit Hilfe von Testverfahren – um dann aus der Perspektive des Kindes einen stimmigeren Beziehungsdialog mit den Eltern einzuleiten. Differenziert müssen die sensorischen von den sozioemotionalen Defiziten diagnostiziert werden, um nicht zu der alltäglichen Kumulation von Minimaltraumen durch Ängste und Scheitern, weitere hinzuzufügen. Angelehnt an Winnicott (1974), der eine Kinderbehandlung primär aus der Betrachtung der Entwicklung konzeptualisierte, müssen wir die beeinträchtigenden Hindernisse verstehen und sie zur Seite räumen, um progressives Wachstum einzuleiten. Das von den Eltern unter dem Einfluss des Therapeuten Verstandene muss von den Eltern umgesetzt werden, da sie in der unmittelbaren sozialen Umgebung die wesentlichen Verantwortlichen sind. Schon Winnicott maß der „sozialen Situation" große Bedeutung bei, und gerade ein dergestalt belastetes Kind muss von einem "therapeutischen Milieu" umgeben sein, in das auch die professionellen Erzieher mit eingebunden sein sollten. Wenn wir bedenken, dass durch den frühen Krippen- und den zeitlich intensiven Kita-Besuch die Kinder mit diesen Schwächen schon sehr früh in Gruppen integriert werden sollen und die Erzieher von sensorischen Integrationsproblemen meist nichts wissen, ergibt sich hier ein großes Belastungsfeld für den sich entwickelnden Menschen. Ein sicherheitsgebender Umgang mit dem kleinen aber auch dem größeren Kind sollte sowohl in der familiären als auch in der professionellen Erziehung gewährleistet sein, um die sich entwickelnden Repräsentanzen und Identitätsbausteinen durch die beschriebenen Unzulänglichkeiten nicht negativ zu tönen. Jugendliche, deren Selbstbild ohnehin in der Adoleszenz einer Baustelle gleichkommt, können sich einer derartigen Diagnostik nicht oder nur mit großen Schmerzen stellen – und auch hier spielt die unterstützende Einstellung der Eltern eine maßgebliche Rolle. Daher sollte der Früherfassung größte Aufmerksamkeit zukommen, um zum einen die notwendige Förderung einzuleiten und gleichzeitig eine Integration dieses Persönlichkeitsanteiles zu leisten. Wenn elterliches und professionelles Beziehungsgeschehen auf individuelle Besonderheiten abgestimmt und der passende Schlüssel gefunden ist, müssen diese neuropsychologischen Besonderheiten das Entwicklungspotenzial eines Kindes nicht zwangsläufig eingrenzen oder determinieren. Der Radarschirm der therapeutischen Aufmerksamkeit sollte breit ausgefaltet sein, damit diese Heranwachsenden – in einer Gesellschaft, die keine Unzulänglichkeiten duldet – dennoch ihren Platz finden. Einer meiner früheren männlichen Patienten, der sowohl unter auditiven als auch unter kinästhetisch-propriozeptiven Problemen leidet, die den Sprachkontakt erschweren und den Bewegungsfluss etwas langsamer gestalten, der aber im Laufe der Schulzeit (Förderbereich) durch vielfältige

Förderung zu guten Ergebnissen gelangte, was die basale Problematik überdeckte, geriet durch eine falsche berufliche Einschätzung und in der Folge in eine berufliche Überforderung. Seine Angst war sehr stark und er wurde mir als junger Erwachsener mit einer „Panikstörung" (Händezittern u. ä.) erneut zugewiesen. Bald konnten wir im vertrauten Gespräch das Dilemma erkennen. Und so kam er zu folgendem Schluss: „Ich weiß es schon immer: Ich möchte Gärtner werden; die Pflanzen reden nicht und außerdem wachsen sie langsam".

Literatur

Ayres, J. (1979): Bausteine der kindlichen Entwicklung. Springer, Berlin/Heidelberg.

Condon, W. (1975): Speech makes babies move. In: Lewin, R. (Hg.): Child Alive. London 1975, S. 81-90.

Condon, W./Sander, L. (1974): Neonate Movement Is Synchronized With Adult Speech. In: Science, 183, S. 99-101.

Deutsches Ärzteblatt: Vorschuluntersuchungen. Schäden ausgleichen. 98, 2001, 25, S. P. 403.

dpa (2008): Laufen macht schlau und sorgt für gute Laune. In: Rhein-Neckar-Zeitung, 19./20.04.2008.

Esser, G. (1991): Was wird aus Kindern mit Teilleistungsschwächen? Ferdinand Enke, Stutt-gart.

Firnhaber, M. (2005): Legasthenie und andere Wahrnehmungsstörungen. Fischer, Frankfurt a.M.

Heckhausen, H. (1971): Förderung der Lernmotivation und der intellektuellen Tüchtigkeiten. In: Roth, H. (Hg.): Begabung und Lernen. Klett-Cotta, Stuttgart, S. 193-228.

Hüther, G./Doering, St./Rüger, U./Rüther, E./Schüßler, G. (1996): Psychische Belastungen und neuronale Plastizität. In: Zeitschrift für psychosomatische Medizin und Psychoanalyse, 42, 1996, S. 107-127.

Jäncke, L. (2008): Macht Musik schlau? Hans Huber, Bern.

Jaffe, J./Beebe, B. et al. (2001): Rhythms of Dialogue in Infancy. Monographs of the Society for Research in Child Development. S. Nr. 265, Vol. 66, Nr. 2, 2001. Blackwell, Bos-ton/Oxford.

Lempp, R. (1970): Frühkindliche Hirnschädigung und Neurose. Hans Huber, Bern, Stuttgart, Wien.

Lempp, R. (1975): Eine Pathologie der psychischen Entwicklung. Hans Huber, Bern, Stutt-gart, Wien.

Lempp, R. (1982): Wenn das schwache Glied in der Intelligenzkette nicht früh genug erkannt wird. In: Musik und Medizin. 12/1982, S. 26-36.

Mertens, M. (2006): Sport macht schlau. In: Frankfurter Rundschau - Wissen & Bildung, 31.10.2006, S. 28.

Oerter, R. (1995): Motivation und Handlungssteuerung. In: Oerter/Montada (Hg.): Entwick-lungspsychologie. Beltz, Weinheim, S. 758-822.

Ornstein, A. (1977): Kontakt mit der inneren Welt des Kindes. In: Familiendynamik 2, 1977, 4, S. 282-315.

Palombo, J. (2006): Nonverbal Learning Disabilities. Norton, New York/London.

Palombo, J. (2011): Executive Function Disorders and Self-Deficits. In: Gitterman, A./Heller, N: (Hg.): Mental Health and Social Problems: A Social Work Perspective. Routledge, N.Y.

Papousek, ./Hofacker, N./v./Malinowski, M./Jacubeit, T./Cosmovici, B. (1994): Münchener Sprechstunde für Schreibabies. Erste Ergebnisse zur Früherken nung und Prävention ..... von Störungen in der Verhaltensregulation und der Eltern-Kind-Beziehungen. Sozial- .... pädiatrie in der Pädiatrie für Praxis und Klinik, 16, 1994, S. 680-686.

Ptok, M./Berger, R./von Deuster, Chr./Gross, M./ Lamprecht-Dinnesen, A./ Nickisch, A./Radü, H.J./Uttenweiler, V. (2000): Auditive Verarbeitungs- und Wahrnehmungs-störungen. Konsensus-Statement.

Rass, E. (2002): Kindliches Erleben bei Wahrnehmungsproblemen. Peter Lang, Frankfurt/M.

Rass, E. (2008): Kontaktaufnahme mit der Wahrnehmungswelt des Kindes: (Unerkannte) Störungen in der Wahrnehmungsorganisation und deren Auswirkungen auf die psychi-sche Entwicklung. Books on Demand, Norderstedt.

Ratey, J. (2001): Das menschliche Gehirn. Walter, Düsseldorf/Zürich.

Ratey, J. (2008): Spark. The Revolutionary new Science of Exercise and the Brain. Little, Brown and Company, NY/Boston/London.

Rattle, Simon (2005): Rhythm is it. DVD.

Rheinberg, F. (1997): Motivation. Kohlhammer, Stuttgart/Berlin/Köln.

Rosenkötter, H. (2003): Auditive Wahrnehmungsstörungen. Klett-Cotta, Stuttgart.

Rüegg, J.C. (2009): Dem Trübsinn davonlaufen. In: PSYCHOLOGIE HEUTE, 36. Jg. Heft 4, April 2009, S. 56-57.

Sacks, O. (2008): Der einarmige Pianist. Rohwohlt, Reinbek bei Hamburg.

Stern, D. (1993): Die Lebenserfahrung des Säuglings. Klett-Cotta, Stuttgart.

Warnke, A. (2000): Wahrnehmungsprobleme bei Kindern und Jugendlichen. In: Bundschuh, K. (Hg.): Wahrnehmen, Verstehen, Handeln. Klinkhardt, Bad Heilbrunn.

Winnicott, D.W. (1974): The maturational process and the facilitating environment. NY 1962. Dtsch.: Reifungsprozesse und fördernde Umwelt. Kindler, München.

Personal Competence in the Structure of Model of Mentor Competences and its Expression

Birutė Autukevičienė, Ona Monkevičienė

Abstract
The article focuses on the problem of personal competence of mentor. The problematic place of personal competence of a teacher-trainee as well as mentor in strategic education documents is analysed. The authors discuss the most important components of mentor's competence and his/her personal competence. The empiric research aims to identify the place of personal competence of a mentor in the structure of competence model considering its importance and expression of competences.

Keywords: *mentor, personal competences, competence model, pre-school education*

Introduction

Mentor is an experienced educator, who facilitates professional development of a student and a novice teacher. The Lithuanian Regulations of Teacher Training (2007) defines a mentor as a supervisor of a student and a teacher to be: 'Mentor is a pedagogue practitioner, who has accumulated experience and has completed trainings organised following the procedure approved by the Ministry of Education and Science; mentor provides methodological assistance to students during their teaching practice'. Only an experienced teacher working in the same level of education may be appointed a mentor of a pre-school and pre-primary teacher.

To provide a novice teacher with an efficient support, a mentor needs certain competences. The researchers refer to ***various competences that are of importance to a mentor***: personal, social, information and communication, evaluation and expert assessment, managerial, adult teaching and consulting, professional practice organisation and reflection, didactic (Clutterbuck D., 2003; Podsen I. J., Denmark V. M., 2000; Norman Dr., Cohen H., 1999; Elliott B., 1995, Mentoriaus kompetencijos, 2007).

Teacher's personal competence (including that of an experienced teacher, i.e., mentor) has been extensively discussed in Lithuania. 'The Descriptor of Professional Competences of Teachers' (2007) provide for common-cultural competence, professional and generic competences; however, personal competence is not included into the group of latter competences. Discussing with the developers of Lithuanian strategic education documents, E. Martišauskienė emphasises the importance of the personal competence and points out that this competence is provided for only in 'the Standard of Teacher's Profession' (Martišauskienė E., 2008).

Preparing the descriptor of mentor competences, mentor's personal competence was included into the model of mentor competences (Monkevičienė O., Glebuvienė V.S., 2007).

Mentor's personal competence is defined as an active position of an experienced teacher, his/her attitude and ability to establish an emphatic contact, to maintain friendship and to supervise and provide emotional support to a student. The main components of this competence embrace personal characteristics of a mentor, his/her value based attitudes, abilities, knowledge and experience.

Mentor's personal qualities enable a mentor to successfully perform the assumed role of a mentor. A mentor has to be friendly, good-willing, sensitive, understanding, tolerant, supportive, possessing a sense of humour but professional, experienced, organised, courageous and responsible (Jonson K. F., 2002; Podsen I.J., Denmark V.M., 2000). Such personal qualities of a mentor as reticence, criticism, autocracy, arrogance, indecision, selflessness, instability and passiveness (Clutterbuck D., 2003) impede communication with a mentee. The aforesaid features of a mentor stimulate defensive behaviour of a mentee, result in distrust and tension as well as distance in mutual relations.

Mentor's value-based attitudes are of equal importance: a positive attitude towards nurturance of a novice educator as a teacher, a personality and a professional sharing experience; a principle to facilitate revelation of a mentee and development of his/her professional skills; a principle to establish relations based on equal partnership, etc. Establishment and maintenance of relations based on trust and respect create conditions for a mentor and a mentee to attain goals of professional development together. It is necessary for a mentor to want to assume this responsibility and to be devoted to mentoring (Hobson A. J., Ashby P., Malderez A., Tomlinson P. D., 2009).

One of the most essential components of mentor's personal competence includes abilities: an ability to analyse and understand the motives and goals of own behaviour; an ability to establish and maintain positive communication with a mentee; an ability to help a mentee to establish personal relations with learners, other teachers and administration; an ability to solve communication problems on time and in a constructive way (mentor and mentee, mentee and learners, mentee and community) and others. (Jonson K. F., 2002; Podsen I.J., Denmark V.M., 2000).

The following knowledge related to personal competence is of importance to a mentor: knowledge of stages of mentoring relations and dynamics of relation with a mentee; awareness of ways for establishment and maintenance of warm personal relations; awareness of personal qualities of a novice teacher, possible emotional and behavioural problems, knowledge of defensive behaviour models and methods of coping with difficulties, etc. (Clutterbuck, 2003; Reinforcement Techniques, 2006).

The scientific research reveal that students see a mentor as ***an example*** (model), who is followed to self-improve or improve their practical pedagogical work; therefore, mentors have to be experienced practitioners (Jonson K. F., 2002). A mentor becomes a model, when s/he guides own activity employing positive values, is in search for new trends in education, demonstrates professional work with learners and is open to changes and improvement. A student trainee has to observe the educational activity that is modelled by a mentor and to learn from mentor's experience thus applying the acquired experience in independent activities.

The personal competence of a mentor is based and explained by a number of theories.

Empathy theory

J. MacCallum, S. Beltman (1995) and E. F. Clifford (1999) pointed out that emotional empathy facilitates establishment of successful relations between the mentor and the mentee. The researches of E. F. Clifford (1999), J. P. Young and M. Sato (1988) (in Reynolds R., Grushka K., 2007) show that empathy strengthens the mentor's willingness to help others, to establish close contacts avoiding formal communication. Empathy comprises understanding of others, encouragement of their development, orientation to service provision, constructive employment of differences and appropriate reaction to group moods.

Emotional intellect theory

According to D. Goleman (2003), personal competence consists of self-awareness, self-regulation and motivation, which ensure expedience of personal behaviour and provides strength to achieve the set goals, to acquire good self-feeling and favourable emotional climate. To achieve directiveness of the personality an individual needs self-regulation, reliability, honesty, adaptability and openness to innovations. At first sight all these qualities can be considered to be internal; however, they have a particularly strong effect on external behaviour.

Individuals with high emotional intellect are oriented to high professional achievements, are ready to estimate risk factors, apply undertakings, they are proactive and optimists. A mentor possessing the aforesaid qualities is a good example and authority to a mentee.

Social interaction (cooperative learning) theory

Following the theories developed by R. Gallimore (1992) and L. Vygotsky (1962, 1978) (in Reynolds R., Grushka K., 2007), it is argued that success of the novice pedagogue's learning is predetermined by social links and collaboration with a more experienced colleague and work in creative groups. Thus, a mentor is a person, who helps the mentee to create a field of personal and professional relations, which favourable for young specialist's development.

It may be concluded that the personal competence of a mentor is theoretically substantiated. However, it is necessary to clarify:

*What is the place of the personal competence in the structure of model of mentor competences?

*What are the most important components of mentor's personal competence?

*What peculiarities of expression of mentor personal qualities are revealed supervising students during their teaching practice?

Striving for the answers to the raised problematic questions, the empiric research on the structure of personal competence of pre-school and pre-primary teachers and its expression was carried out.

The goal of the research

To identify the place of personal competence in the structure of mentor competence model and its expression during teaching practice.

The objectives of the research:

1. to identify the place of mentor's personal competence in the structure of the model of competences necessary for a mentor in terms of their importance.
2. to identify the place of the mentor's personal competence in the structure of the model of competences necessary for a mentor from the point of view competence expression.
3. to reveal the importance of separate components of mentor's personal competence and their expression during teaching practice of students.

The methods and the sample of the research

Carrying out the research, the quantitative and qualitative methods were combined:

1. The questionnaire survey of experts (qualified mentors) employing the descriptor of mentor competences. The sample included 71 mentors of pre-school and pre-primary education, i.e. 87 % of all the specialists, holders of mentor's qualification certificate in this field, in Lithuania. The experts evaluated importance of all the mentor competences and their components to mentor activities.
2. Students' survey employing the descriptor of mentor competences. The sample of the research included 184 students of the study programmes of pre-school and pre-primary education. The students, who had their practice under supervision of qualified mentors. They submitted their evaluations about the expression of mentor's personal and other competences during practice.
3. Students' discussion groups. Five discussion groups of students from the study programmes of pre-school and pre-primary education, who were supervised by experienced mentors, were organised and 46 students took part in discussions.

Methods of data analysis

Statistical analysis of research data was conducted applying the statistical data processing programme (SPSS for Windows 17).

The factorial analysis was conducted on the basis of correlation matrix and Pearson's correlation coefficient was applied and *VARIMAX* rotation was used carrying out the factorial analysis.

The method of correlation *pleiades* was employed. *Index of popularity – PI* was used.

The results of the research

1. The place of mentor's personal competence in the structure of the model of competences necessary for a mentor.

Striving for identification of the place of mentor's personal competence in the structure of the model of competences necessary for a mentor, the factorial analysis was conducted.

During the first phase of the research, the factorial analysis of the data on the attitude of qualified mentors towards the importance of competences necessary for a mentor was conducted.

The results of scale factor analysis are presented in Table 1. Having overviewed the results, it can be concluded that the correlation coefficients indicate *a strong link* of variables (r ranges from 0.62 to 0.80) and validity of connecting variables into scales *(B. Bitinas, 2006, p. 311)*.

The *VARIMAX* rotation showed that according to factor weight mentor competences were grouped into three factors, which explain 79.2 % of all the data dispersion: Factor 1 - 32.7 %, Factor 2 - 23.2 %, Factor 3 - 23.3 %. As it can be seen, all the empiric variables have a sufficient factor weight (> or = 0.4), therefore, it can be stated that the identified factors express the essential manifestations of mentor competences. Moreover, a significant grouping of components inside of factors was also revealed.

Table 1: Results of factorial analysis of competence scales (after VARIMAX rotation), mentor values

No.	**Competence**	**features**	**1**	**2**	**3**	**4**	**5**	**6**	**7**	**8**	**Factor 1**	**Factor 2**	**Factor 3**
1	Personal	1-9	1.0	0.73	0.67	0.67	0.62	0.62	0.63	0.71	0.35	**0.80**	0.26
2	Social	10-19	0.73	1.0	0.77	0.73	0.72	0.71	0.72	0.72	**0.59**	0.54	0.35
3	Information-communication	20-28	0.67	0.77	1.0	0.79	0.74	0.73	0.73	0.74	**0.70**	0.43	0.34
4	Evaluation and expert assessment	29-37	0.67	0.73	0.79	1.0	0.76	0.77	0.74	0.76	**0.69**	0.39	0.39
5	Managerial	38-48	0.62	0.72	0.74	0.76	1.0	0.75	0.80	0.73	**0.60**	0.32	0.54
6	Adult teaching and consulting	49-57	0.62	0.71	0.73	0.77	0.75	1.0	0.75	0.74	**0.60**	0.34	0.51
7	Practice organisation and reflection	58-70	0.62	0.72	0.73	0.74	0.80	0.80	1.0	0.78	0.45	0.33	**0.77**
8	Didactic	71-97	0.71	0.72	0.74	0.76	0.73	0.74	0.78	1.0	**0.51**	**0.49**	**0.49**
	Dispersion explained										32.7	23.2	23.3

The first factor embraces mentor competences, which are based on communicative, managerial and adult teaching skills. This factor may be referred to as *communicative-managerial* factor. It connects social; information-communication; managerial; adult teaching and consulting; evaluation and expert assessment competences. The correlation of competences with the factor remains significant (from 0.59 to 0.70).

The third factor unites variables of competences of professional practice organisation and practice reflection (r = 0.77). This factor may be regarded as *administrational.* Didactic competence of a mentor is linked with all the three factors.

Mentor's personal competence is related with the second factor. It may be called a ***personality*** factor. Personal competence strongly correlates with the personality factor (r = 0.80). Thus, the personal competence in the structure of mentor's competences emerges as a separate essential competence. This competence is linked to the attitudes and abilities of a mentor to educate and develop a young colleague, to establish an emphatic contact and maintain friendship as well as to provide an emotional support.

During the second phase, the factorial analysis of students' attitude towards expression of mentor competences during teaching practice was carried out. It allowed establishing the structure of mentor competences according to the data on its expression as well as to evaluate the place of the personal competence in it.

The results of scale factor analysis are presented in Table 2. Correlation coefficients reveal *a significant correlation* among variables (r varies from 0.59 to 0.82) and validity of connecting variables into the scales. The factors, which explain 77.9 % of the total dispersion of the data are identified: Factor 1 - 34.6 %, Factor 2 - 25.3 % and Factor 3 - 18.0 %. All the empiric variables have a sufficient factor weight (> or = 0.4).

Table 2: Results of factorial analysis of competence scales (after VARIMAX rotation), student values

No.	Competence	features	1	2	3	4	5	6	7	8	Factor 1	Factor 2	Factor 3
1	Personal	1-9	1.0	0.73	0.66	0.73	0.60	0.65	0.59	0.68	**0.79**	0.27	0.23
2	Social	10-19	0.73	1.0	0.70	0.71	0.67	0.65	0.69	0.73	**0.69**	0.37	0.32
3	Information-communication	20-28	0.66	0.70	1.0	0.71	0.61	0.67	0.67	0.76	**0.63**	0.31	0.43
4	Evaluation and expert assessment	29-37	0.73	0.71	0.71	1.0	0.76	0.72	0.72	0.72	**0.67**	0.55	0.22
5	Managerial	38-48	0.60	0.67	0.61	0.76	1.0	0.74	0.78	0.75	0.41	**0.75**	0.30
6	Adult teaching and consulting	49-57	0.66	0.65	0.67	0.72	0.74	1.0	0.77	0.82	0.48	**0.52**	0.50
7	Practice organisation and reflection	58-70	0.59	0.69	0.67	0.72	0.78	0.77	1.0	0.80	0.41	**0.62**	0.48
8	Didactic	71-97	0.68	0.73	0.76	0.72	0.75	0.82	0.80	1.0	0.50	0.43	**0.70**
	Dispersion explained										34.6	25.3	18.0

The grouping of mentor competences according to the indicator of expression is slightly different compared to the grouping according to their importance, which was conducted by qualified mentors.

Mentor's personal competence was included into the group of competences, which are based on human relations with other human beings, social environment and communication skills, i.e., they make up *interpersonal–communication factor*. This factor unites personal; social; information and communication; evaluation and expert assessment competences. However, the personal competence has strongest correlation with its factor (r = 0.79) and, therefore, is an essential one.

The group of competences, which was formed on the basis of expression of competences, relates to managerial and organisational factor. Thus, the group of competences united by the *managerial-administrative* factor was identified consisting of managerial; adult teaching; professional practice organisation and reflexion competences. It is clear that this factor embraces competences that require managerial excellence of a mentor.

In terms of expression of mentor competences, the didactic competence is related to a separate factor, which may be relatively referred to as *didactic factor.* The correlation of competence with the didactic factor is strong (r = 0.70), whereas according to the importance of mentor competences evaluated by qualified mentors, the didactic competence was associated with all the factors including the personal one as well.

Evaluating the place of mentor's personal competence in the structure of model of mentor competences, it can be concluded that mentors evaluated the importance of this competence subjectively, i.e., from the position 'I', whereas students evaluated it in the context of interpersonal relations. This would explain its isolation and correlation with a personality factor from one side and links with other competences as well as correlation with interpersonal-communication factor.

I. The most important components of mentor's personal competences from the point of view of qualified mentors

Applying the popularity index (PI), an attempt was made to identify the most important and less relevant components of mentor's personal competence and employing the method of correlation pleiades aimed to more explicitly reveal links among components of mentor's personal competence and to test one-dimensionality of a scale. Table 3 presents evaluation of qualified mentors and the index of popularity.

Table 3: Evaluation of structural components of mentor's personal competence (opinion of qualified mentors)

No.	Personal competence (PI=67)	important component according to mentors		Index of popularity
		N	%	PI
1.	**Such personal qualities as empathy, sincerity, a sense of humour, reliability, honesty are characteristic of a mentor**	**49**	**69.0**	**66**
2.	**A mentor guides his/her personal life and professional activity by common human values**	**49**	**69.0**	**66**
3.	**A mentor is prepared to nurture a young teacher as a personality and a professional**	**60**	**84.5**	**83**
4.	A mentor is aware of stages of novice teacher's professional development, and methods of personality maturation	40	56.3	52
5.	**A mentor is able to cognise and control own emotions, to reflect own behaviour and to change it, to assume a personal responsibility**	**58**	**81.7**	**81**
6.	**A mentor is able to be an example to a pre-school education student-trainee**	**56**	**78.9**	**78**
7.	A mentor is able to evaluate the situation without preconceived prejudices from own position and that of a student	45	63.4	60
8.	A mentor is able to recognise signs that indicate that a student needs support	38	53.5	53
9.	**A mentor is able to apply various measures of personal effect and support (encouragement, persuasion, motivation, etc.)**	47	66.2	**65**

According to the mentors, who supervise the teaching practice, the most important structural components of mentor's personal competence are mentor's *determination to mature a young teacher as a personality and a professional* (PI=83) and *an ability to identify and control own emotions, to reflect own behaviour and to change it; to assume a personal responsibility* (PI=81). Mentor's *conforming to common human values both in professional life and in personal life* (PI=66), *possession of personal features that are important for a teacher* (PI=66) are also considered as important. Thus, the maturity of mentor's personality and altruistic mood to help less experienced teachers to be are emphasised. A mature teacher *is able to be an example to a student, who is to become a pre-school teacher,* s/he contributes to nurturance of student's personality, therefore, the aforesaid component is also of importance (PI=78). Development of a teacher-personality is the main challenge both to the mentor and to the university, and, according to the respondents, such a teacher may be educated only by another teacher-personality.

The correlation pleiade (Pic. 1) shows internal links among components of mentor's personal competence. It was established that the statement *'a mentor is able to be an example to a pre-school education student-trainee'* (6) unifies the majority of statements of the scale. Relatively strong links are observed regarding the statement that a mentor *'conforms to common human values both in professional life and in personal life'* (2). The correlation pleiade method revealed that the statements with high but not the highest PI are concentrated in the centre of mentor's personal competence. This includes mentor's ***conforming to values and an ability to be an example to a student during his/her teaching practice and to contribute to maturation of his/her personality.*** These components of mentor's personal competence make up the core of it.

The component *'a mentor is aware of stages of novice teacher's professional development and methods of personality maturation'* (4) is isolated. Weak links are observed between components *'a mentor is able to evaluate the situation without preconceived prejudices from own position and that of a student'* (7) and *'a mentor is able to recognise signs that indicate that a student needs support'* (8). According to the respondents, these structural components of personal competence are not most relevant.

Pic. 1: Correlation pleiade of structural components of mentor's personal competence

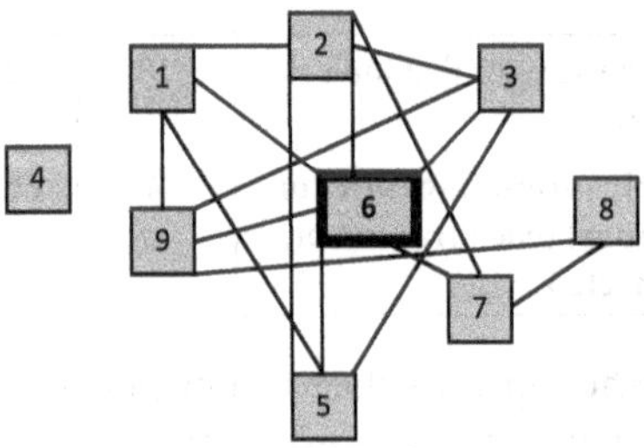

II. Mentor's personal competences and expression of its components

Applying the PI of students' evaluations, an attempt was made to identify expression of mentor's personal competence and its elements supervising students' pedagogical practice.

The analysis of the research data showed that, according to students opinion, mentor's *personal competence* (PI=53) revealed itself most frequently during teaching practice, while other competences were less observable: *social* (PI=51), *didactic* (PI=50), *managerial* (PI-46), *adult teaching and consulting* (PI-42). During teaching practice, mentors demonstrated their abilities to establish a good-natured contact, to provide emotional support and to maintain friendship. They were willing to mature a young teacher as a personality and a professional. According to students, mentors' competences of *information and communication* (PI=35), *evaluation and expert assessment* (PI=38) and *professional practice organisation and reflection* (PI=39) were least frequently observed.

These data allows concluding that mentor's personal competence is particularly important to students during their teaching practice and it serves as a prerequisite for a good contact and professional success of teachers to be.

Pursuing a more thorough analysis, the comparative analysis of importance and expression of components of mentor's personal competences was conducted (the data accumulated during qualitative research on opinions of experts (holders of mentor's qualification certificates) and students were compared).

Table 4 presents research data that the following components of mentor's *personal competence* were best observed by students during their teaching practice: *empathy, sincerity, sense of humour, reliability, honesty (PI=76), ability to recognise and control own emotions, to reflect own behaviour and to change it (PI=67), an ability to be an example to students-trainees (PI=67), conformance to common human values in life and professional activity (PI=62).*

It can be stated that students envisaged and pointed to mentor's conformance to values, positive personal features and an ability to be an example to a student – trainee as to the most essential components of personal competence, i.e. singled out the core components in the correlation pleiade.

Table 4: Evaluation of expression of components in the mentor's personal competence, students' point of view

No.	**Personal competence** (PI=53)	**Students: frequent manifestation of the component**		**Index of popularity**
		N	**%**	**PI**
1.	**Such personal qualities as empathy, sincerity, a sense of humour, reliability, honesty are characteristic of a mentor**	140	76.1	76

2.	**A mentor guides his/her personal life and professional activity by common human values**	**120**	**65.2**	**62**
3.	**A mentor is able to cognise and control own emotions, to reflect own behaviour and to change it, to assume a personal responsibility**	**128**	**69.6**	**67**
4.	**A mentor is prepared to nurture a young teacher as a personality and a professional**	**103**	**56.0**	51
5.	**A mentor is able to be an example to a pre-school education student-trainee**	**128**	**69.6**	**67**
6.	A mentor is aware of stages of novice teacher's professional development, and methods of personality maturation;	85	46.2	38
7.	A mentor is able to evaluate the situation without preconceived prejudices from own position and that of a student;	84	45.7	39
8.	A mentor is able to recognise signs that indicate that a student needs support	81	44.0	33
9.	A mentor is able to apply various measures of personal effect and support (encouragement, persuasion, motivation, etc.)	95	51.6	44

Comparing the results of qualified mentors' and students' surveys, it can be noticed that there is no considerable difference between what components of mentor's personal competence are regarded relevant by mentors and how these components reveal themselves during teaching practice. The most substantial difference between evaluations of importance and expression of mentor's competence is observed in mentor's *determination to nurture a young teacher as a personality and a professional* (mentors - PI= 83; students - PI=51). Mentors see this sand as the most important in the mentoring activities but students observe practical influence of this component on their personality less frequently.

Making an attempt to analyse peculiarities of expression of mentor's personal competence, the focus group research of students was conducted.

The established sub-categories are presented in Table 5. The analysis of students' statements allowed identifying a range of variables of expression of most important components of mentor's personal competence (marked by sub-categories). Though the majority of these variables are positive (i.e., of positive valency), negative variables (i.e., of negative valency) were also identified.

Referring to mentor's conformance to values, students also expressed their negative experience. The following sub-categories were identified: ***dishonesty***

(MS, Group 2-4, '*... she had different beliefs, different personal features, which were not always appropriate to a teacher, there occurred cases of dishonest behaviour'*), ***irresponsibility*** (A, Group 1, *'The mentor said: come whenever you want, do whatever you want and I won't control you at all...'*).

Discussing the category

A mentor showed a disposition to mature a young teacher as a personality and a professional, the following sub-categories were established: unfavourable attitude to student's ideas, lack of communication, lack of support. Students remembered the cases, when their mentors did not support their ideas *('...too much regulation from teacher's point of view – do this, do this in that way, do not do this – perhaps it is not needed'*- In S, Group 3), they did not communicate *('She used to say: this is your practice, so you have to work with children and I go for a cup of coffee'*- JK, Group 5). The informant RS (Group 3) did not feel equal and open relations *('When an educator builds a wall, you are not able to push it').* It can be assumed that such mentors did not have any positive attitudes towards nurturance of a young teacher.

The following sub-category of negative valency was identified during discussion about a mentor as an example to a student: *Mentor was a negative example to a student,* i.e. demonstrated neglect of duty, amateurishness. The informant JK (Group 5) pointed out: *'...I got used to the situation that when she was at work, I had to work with children not only these five academic hours as obligatory but longer because this teacher would not educate children and play with them not to speak about support to me'.*

Discussing the methods of mentor's support to a student, the respondents presented their negative experiences as well. The following sub-category was singled out: ***No support.*** *('Nobody showed me how to plan and what to plan. I would plan everything myself'*- BK, Group 5).

Table 5: Sub-categories that reveal expression of mentor's personal competence

Category	**Sub-categories (positive valency)**	**Sub-categories (negative valency)**
Personal characteristics of mentor	Empathy Joviality, youthfulness Sincerity Communicability Open communication	

Category	Sub-categories (positive valency)	Sub-categories (negative valency)
Guidance by common human values in personal life and professional activity	Respect Love for children, students Responsibility Benevolence Honesty	Dishonesty Irresponsibility
Disposition to nurture a young teacher as a personality and a professional	Emotional support, warm relations Good-heated exchange of experience Encouragement of trainee's self-dependence Attitude to a student as to a partner	Unfavourable attitude towards students' ideas Lack of communication Lack of support
Example to a pre-school student-trainee	Inspirational example	Mentor was a negative example to a student
Application of various methods and techniques of personal effect and support	Emotional support Advice, offering of own help Encouragement, motivation Acknowledgement Demonstration through own example Immediate practical help	Withdrawal of support

The qualitative analysis of student's experience revealed a relatively rich expression of essential components of the personal competence, when a mentor supervises students' practice. However, the research highlighted negative features of essential components of personal competence such as irresponsibility, dishonesty, lack of support, help and insufficient backing of students' ideas.

Conclusions

1. In terms of importance of mentor's competences (qualified mentors' point of view), the personal competence is a separate competence in the structure of mentor's competences related with the personality factor. Considering the aspect of expression of mentor competences (according to the students), the personal competence belongs to the group of competences, which are related to interpersonal –communicative factor. In both cases, this competence is essential in the structure of mentor's competences.

2. According to the qualified mentors, the most important components of mentor's personal competence are: mentor's values, personal qualities, disposition to mature a young teacher as a personality and a professional, to be an

example to students. These components are concentrated in the centre of the correlation pleiade and are linked with other components.

3. Mentor's personal competence was expressed strongest supervising students' teaching practice. The analysis of students' experience proved expression of essential components of mentor's personal competence. Insufficient personal competence of a number of mentors supervising students' teaching practice is proved by the problems that students faced: mentor's irresponsibility, dishonesty, lack of support and help to a student and insufficient encouragement to students' ideas.

References

1. Clifford E. F. (1999), A Descriptive Study of Mentor-protégé Relationships, Mentors' 1. Emotional Empathetic Tendency, and Protégés' Teacher Self-efficacy Belief, *Early Child Development and Care*, Vol. 156, p. 143–154.

2. Clutterbuck. D. (2003). *The Competencies of Mentors and Mentees, Mentoring and Coaching Research Group*. European Mentoring and Coaching Council: Clutterbuck Associates.

3. Available at: <http://72.14.207.104/search?q=cache:8uhQyhfv04MJ:www.temple.edu/martec/onlinet>

4. Elliott B. (1995), Developing Relationships: Significant Episodes in Professional Development, 3. *Teachers and Teaching: Theory and Practice*, vol. 1 (2).

5. Goleman D. (2003). *Emocinis intelektas*. - Vilnius: Smaltija.

6. Hobson A. J., Ashby P., Malderez A., Tomlinson P. D. (2009). Mentoring beginning teachers: What we know and what we don't. / *Teaching and Teacher Education* No. 25, p. 207–216 [accessed 6 October 2011]. Available at EBSCO Publishing.

7. Johnson S. K. (1999). The Mentoring Model Theory: Dimensions in Mentoring Protocols. In: *Career Development International*. [accessed 8 May 2009]. Available at: <http://proquest.umi.com/pqdweb?did=86921101&sid=9&Fmt=3&clientId=57766&RQT=309&VName=PQD.

8. Jonson K. F. (2002). Being an Effective Mentor. CA, Corwin Press.

9. MacCallum J., Beltman, S. (1999), 7. *International Year of Older Persons. Mentoring Research Project. Canberra: Department of Education, Training and Youth Affairs.*

10. Martišauskienė E. (2008). Pedagogo kompetencijos samprata ir modeliai. / Pedagogo kompetencijos. Kaunas, Technologija, 61-77 p.

11. Mentoriaus kompetencijos. (2007). Kaunas, Technologija.

12. *Mokytojų profesijos kompetencijos aprašas.* (2007). The Order of the Minister of Education and Science of the Republic of Lithuania No. ISAK-54 (Žin., 2007, No. 12-511) of 15 January 2007. Available at <http://www.smm.lt/prtm/docs/mkt/2007-01-15-ISAK-54.pdf>.

13. Monkevičienė O., Glebuvienė V.S. (2007). Asmeninė ir socialinė kompetencija. / Mentoriaus kompetencijos. Kaunas, Technologija.

14. Norman Dr., Cohen H. (1999). *The Mentee's Guide to Mentoring.* Canada, HRD Press.

15. *Pedagogų rengimo reglamentas.* (2010). The Order of the Minister of Education and Science of the Republic of Lithuania No.V-54 of 8 January 2010. Available at <http://www.smm.lt/prtm/docs/mkt/Pedagogu_rengimo_reglamentas_20100108.pdf>.

16. Podsen I.J., Denmark V.M. (2000). *Coaching and Mentoring First-Year and Student Teachers.* US, Eye of Education.

17. Reinforcement Techniques (2006). Available at: <8. http://www.massteacher.org/career/new_members/strategies/reinforcement.cfm>.

18. Reynolds R., Grushka K. (2007), 9. *Mentoring in a University / senior High School collegiate.* A Study of the Learning Environment and the Mentor / Mentee Relationship. School of Education: The University of Newcastle. Available at: <www.springerlink.com/index/j163g17563521461.pdf>.

Professional Identity Characteristics of a Teacher Implementing Inclusive Education: Pupil and Teacher Experience

Alvyra Galkienė

Abstract

The article looks at the characteristics of teacher professional identity through the comparison of the reflections of teacher and pupil experiences in a school implementing inclusive education. Recognizing the diversity of the pupils, willingness to lead every pupil towards his/her personal success, expertise in the didactics of heterogeneous groups, inclination to build educational environment that reduces the need for assistance to a pupil, and, in case of necessity, the ability to provide such assistance are all fields of competence, through which the professional identity of a teacher implementing inclusive education reveals itself.

Keywords: professional identity, inclusive education, values, didactics of heterogeneous groups

Introduction

Continuous socio-cultural processes taking place in society and reflecting globalization phenomena, scientific progress, economic and political developments result in a professional activity environment filled with constantly changing requirements for a teacher. Therefore, it is important to review the meanings of a teacher's activity in the process of educating the young generation at different aspects: self-perception in the profession (O. Tijūnėlienė, V. Kavaliauskienė, 2009), expression of competencies (E. Urnėžienė, 2012, A. Norvilienė, V. Zuzavičiūtė, 2011, E. Martišauskienė, 2009, J. Lasauskienė, 2008, E. Rodzevičiūtė, 2006), and reflection of teachers' experience (O. Tijūnėlienė, 2011, R. Paulauskas, A. Gumuliauskienė, H. Volodka, A. Širiakovienė, 2010). The research aims at reviewing teacher professional identity markers from a relatively new prospective of inclusive education.

In social and cultural anthropology, *identity* is perceived not only as identifying oneself with something but also as a difference, as otherness. Thus, according to V. Čiubrinskas (2008, p. 13), identity limits are drawn through certain identification markers. Identity can be analysed as a social construct, in which the limits of one's identity are set through interpersonal relationships and by the person's self-consciousness and signs sent by others that allow recognizing certain commonness and differences.

In the processes of self-perception, a person recognizes the meanings of being in society, thus discovering his/her identity. E. Martišauskienė (2012) links these processes with the distinctiveness of a human as a spiritual being who gives meaning to his/her life through certain values. The search for one's own identity, in order to be oneself in various contexts (those of a nation, religion, profession), continues throughout a life time. Professional identity, according to B. Jatkauskienė (2010, p. 62), encompasses the processes of self-perception, preservation of one's individuality as well as identifying oneself with a professional group, in the context of interaction where the group affects the individual and the individual affects the group. V. Žydžiūnaitė (2012), discussing the professional identity formation process, emphasizes the course of time, during which experience is acquired, and meaningful feedback, which helps develop attitudes, priorities and values.

The identity of a pedagogue is an inseparable part of the identity of the educational science. Looking at the development of the identity of educational science in history, one would see that educational science, according to J. Bengtsson (2006),developed its identity in philosophical, psychological, theological, and sociological contexts. The fundamental marker of the identity of educational science is the practice where the (self-)education-oriented interaction between the educator and the educated takes place. As educational goals change, educational practice changes accordingly; however, new markers of educational science identity appear, presenting new challenges to the teacher's professional identity.

A significant change in the development of educational science took place in 1970s with the arrival of a shift from segregated education system towards inclusive education open to every child. The significance of such education to all learners, including those with and without special needs, is proven by the results of research by Ch. Kliewer (1998), who claims that this is an attitude to the world where fundamental education is provided to all people ensuring full participation, full membership, and fully-pledged civic rights. R. Opertti, J. Brady, L. Duncombe (2010) maintain that accepting pupils with different possibilities and cultural backgrounds to the same classrooms serves as a clear sign of integration and cohesion in society, attesting a much greater value to society that heterogeneity in education might bringthan learning results only. The expression of competencies of a teacher implementing inclusive education was analysed by a group of researchers and practitioners from 28 EU member states, composed under the initiative of the European Agency for Development in Special Needs Education (2012). Fields of competencies were identified, which are a starting point for the development of teacher competence:

1. Respect for pupil diversity.
2. Assistance to all the pupils.
3. Work together with the others.
4. Personal professional development.

Significant changes in pedagogical practice doubtlessly bring changes to the professional identity markers of a teacher as well. In order to achieve high-quality education and professional satisfaction of the pedagogues, it is worthwhile to specify professional identity characteristics of a teacher in the context of inclusive education.

Research problem is presented in the form of a question: "What socio-educational characteristics do inclusive education participants, both pupils and teachers, use to define professional identity of a teacher?"

Research object: Aspects of pupil and teacher experience that reveal professional identity of a teacher in the context of inclusive education.

Aim of the research: to identify socio-educational characteristics that pupils and teachers use to define the professional identity of a teacher in inclusive education.

Research methods

Considering that a teacher implementing inclusive education builds his/her activity on respect to pupil diversity, assistance to all the pupils, and strive for co-operation and personal development, the research structure is based on the idea by AurelijusAugustinas (1999) that it is not the words a teacher utters that teach, but rather the truth that the teacher reveals through his/her outlook, knowledge and behaviour.

By applying the qualitative content analysis method based on the methodological paradigm of phenomenology, aspects of experience reflected on by pupils and teachers are analysed, since a teacher's profession is realized in close interaction between him/her and the pupil. Through the comparison of the reflected experience of pupils and teachers, the search for points in common and differences is carried out, which would allow drawing conclusions on the essential professional identity characteristics of a teacher implementing inclusive education.

Data collection methods. The pupils reflected on their experiences by composing essays "A Teacher Whom I would Like to Follow, and Whose Work Inspires Me." The teachers' experience was revealed through the method of in-depth individual interview. Transcribed interview material was analysed.

Data interpretation

When interpreting the empirical research data, hermeneutical cycle was applied: from subjective perception of the experience of education participants, towards objective and generalized experience (S. M. Laverty, 2003). The texts were read in stages. During the first reading, the texts were analysed in their entirety and insights were made for more in-depth analysis; during the second reading, separate parts of texts were looked into, dimensions of the expression of the professional identity of a teacher were identified and topics formulated; during the third reading, the revealed meanings were reviewed, and topics formulated were specified; during the fourth reading, generalized insights were made and conclusions drawn.

During the research, the perception of pupils educated under the conditions of inclusive education about a good teacher that meets their expectations was compared to the deliberations of teachers working in the same school on aspects of their professional competence.

Research environment and respondent group

The research was carried out in a school where inclusive education has been realized for 20 years. The educational system in the school is based on clearly stated values, a flexible curriculum planning system is applied, and aassistance team for a child works for the sake of every pupil's success. In every general education class in the school, which contains 25 - 28 pupils each, 4 – 5 pupils with disabilities learn together with those without any special needs.

The research included 9 teachers with work experience in the school ranging from 9 to 20 years (the average is 15.2 years). Essays were composed by 182 pupils aged 11 to 17, learning in the 5th – 11th forms.

Research results discussion

The construct of a teacher's professional identity is revealed by comparing the pupils' expectations towards a teacher and prevailing attitudes and values regarding the teachers' professional activity. When analysing the pupils' essays, twotrends clearly emerge, reflecting the pupils' expectations towards a good teacher. On the basis of these trends, two professional identity dimensions of a teacher were defined: 1) A teacher's personal traits and attitudes; 2) A teacher's educational skill.

Dimension 1: Teacher's personal traits and attitudes

The teachers' personal traits enjoy particular emphasis in the pupils' essays. When discussing the image of a teacher they would like to follow, the pupils emphasize his/her personal traits in various situations of educational reality and in various aspects. When reading the texts of the essays, the statements of the pupils were grouped into four categories of characteristics. The attitudes revealed in the deliberations by the teachers also reflect features of a teacher's personality. Comparison of the pupils' expectations and the teachers' deliberations with regard to successful pedagogical activity is used in order to identify professional identity characteristics.

Table 1.1: Category group "Reflection of Teacher's Ability to Build Relationships with Pupils in Professional Identity Construct"

<table>
<tr><th colspan="3">Pupils' expectations</th><th colspan="3">Teachers' attitudes</th></tr>
<tr><th>Category</th><th>Subcategory 2</th><th>Subcategory 1</th><th>Category</th><th>Subcategory 2</th><th>Subcategory 1</th></tr>
<tr><td rowspan="4">Emphatic teacher, responding to pupil's needs</td><td rowspan="2">Taking pupils' feelings into consideration</td><td>Emphatic</td><td rowspan="4">Recognizing the pupil's uniqueness in pedagogical interaction as a precondition for professional satisfaction</td><td rowspan="2">Professional satisfaction in community with children</td><td>Loving children in professional love</td></tr>
<tr><td>Attentive</td><td>Enjoying being with children</td></tr>
<tr><td rowspan="2">Acknowledging pupils' needs</td><td>Respectful</td><td rowspan="2">Recognizing pupil's uniqueness</td><td>Acknowledging a child's personality</td></tr>
<tr><td>Youthful</td><td>Understanding a child's possibilities and difficulties</td></tr>
</table>

The first category of characteristics reflecting personality traits of a teacher implementing inclusive education emphasizes his/her competence to recognize the diversity of pupils (Table 1.1). When discussing teacher's characteristics, the pupils underline his/her empathy and ability to respond to the pupil's needs. The pupils attach significant importance not only to the fact that teachers are able to understand their needs: "*Understand pupils whatever they are*," but also that they go through the successes and failures together with the pupils: "*Is happy to write a 10 (the best mark, - transl.)*", "*Is interested in the pupils' lives.* "The pupils find it particularly important to feel the teacher's recognition and respect: "*... treat them as equal to themselves.*" "*Talk to them as to their colleagues.*" When emphasizing high level of ability to feel another person in a teacher, the pupils also underline the ability to get involved in the pupil's matters: "*Able to discuss various topics.*"

The teachers' deliberations clearly show their personal relationship with the pupil. The pupils' wish to receive empathy and recognition on behalf of the teacher is fully expressed in the teachers' attitudes. When reflecting on their experiences, the teachers emphasize not only the ability to acknowledge the uniqueness of every pupil but they also express the attitude towards a child through the value of love to the child: "*You must love children, all of them; this is the most important*" and the professional satisfaction experienced: "*A teacher must love their profession, what the children are makes no difference to him/her.* "The teachers' deliberations clearly show open relationship with a child, which brings enjoyable experiences: "*Lessons and time spent with children must be a celebration to the teacher.*" The teachers make a clear division between a child's special needs and his/her person, expressing equal and respectful attitude towards the personality of these children as to all the other ones: "*Respect towards a child with special needs as a personality.*" Although overseeing and ignoring difficulties a child faces due to his/her needs is impossible, the teachers' compassion is rational, it does not restrict the child's independence nor reduce his/her efficiency: "*Compassion. But compassion must be fair. You cannot only feel sorry, there'll be no work if you do.*" The relationships of teachers implementing inclusive education with the pupils manifest sensitive openness to every one of them, which strengthens the personality of the teacher him/herself. As K.A. Trimakas puts it (1997), the more a person gives himself to others, the more he becomes himself both as a human and as a person.

The pupils, when defining professional identity of a teacher, see in his/her person a human being able to experience success and failure together with the pupils, feel their needs, and build a relationship of partnership with them.

The teachers, when indicating the markers of their professional identity, identify the profession of a pedagogue with love to a child, they emphasize knowledge and recognition of interpersonal differences as an indicator of pedagogical success, and compassion as a component in pedagogical solutions but not as a basis for interpersonal interaction.

The second category of characteristics describes the character of a teacher, which plays a significant role in building educational environment (Table 1.2).

Table 1.2: Category group "Reflection of the Character of a Teacher in Professional Identity Construct"

<table>
<tr><th colspan="3">Pupils' expectations</th><th colspan="3">Teachers' attitudes</th></tr>
<tr><th>Category</th><th>Subcategory 2</th><th>Subcategory 1</th><th>Category</th><th>Subcategory 2</th><th>Subcategory 1</th></tr>
<tr><td rowspan="4">Teacher who builds joyful microclimate professionally</td><td rowspan="2">Joyful sociability</td><td>Having innate joyfulness</td><td rowspan="4">Cheerful and well-rounded teacher</td><td rowspan="4">Well-roundedness and good mood of a teacher</td><td rowspan="4">Funny, creative, having artistic abilities</td></tr>
<tr><td>Acquired joyfulness</td></tr>
<tr><td rowspan="2">Joyful professionalism</td><td>Unaffectable joyfulness</td></tr>
<tr><td>Business-like joyfulness</td></tr>
</table>

While analysing the pupils' essays and summarizing the teachers' deliberations, the second category of characteristics was distinguished, which reveals the importance of character traits of a teacher implementing inclusive education in his/her professional identity construct. The pupils' essays present a clear expectation of emotionally elevated and joyful educational environment that can be created by a teacher characterized by innate joyfulness: "*With a good sense of humour,*" able of creating elevated mood: "*Keen on joking and quipping,*" able to understand the motives of the pupils' acts and control the situation while preserving good mood: "*Understanding our jokes when we're tired.*" The pupils understand that educational reality also brings tough and complicated situations. A teacher appeals to them who is able to manage such situationswithout surrendering to negative emotions nor raising tension: "*Not surrendering to bad mood,*" on the other hand, the pupils find it important that the educational environment filled with elevated mood was sufficiently work-encouraging: "*Cheerful but taking his/her subject seriously.*"

Whereas the teachers' deliberations present a significantly smaller number of direct references to building joyful educational environment; however, they emphasize the teacher's creativity and resourcefulness: "*…all: a painter, a* *performer, a "handler."* The inclusive education environment itself calls for it. The factor that ensures good mood and self-confidence to a teacher educating pupils with different possibilities is the ability to understand the pupil, the motives for his/herbehaviour, and the reasons for his/her success and failure: "*Always smile and understand the pupil.*" The teachers confirm that pedagogical vocation helps create environment favourable to learning: "*Loving his/her job.*"

In *the pupils'* experience, the image of a good teacher that meets the pupils' expectations is revealed through the relationship with a pedagogue who is constantly building elevated and work-encouraging atmosphere. The pupils indicate the competence of creating favourable emotional environment as the second most important one in the definition of professional identity of a teacher.

The teachers, when discussing professional identification, do not attach relevance to the ability to make jokes; however, they reveal the competence of creating favourable emotional environment expected by the pupils through the teacher's creative, artistic abilities, his/her well-roundedness and inclination to always preserve good mood.

The third category of characteristics reflects a teacher's inclination to help the pupils, and provide them with necessary and rational assistance (Table 1.3).

Table 1.3: Category group "Reflection of Teacher's Ability to Provide Assistance to Pupils in Professional Identity Construct"

Pupils' expectations			**Teachers' attitudes**		
Category	**Subcategory 2**	**Subcategory 1**	**Category**	**Subcategory 2**	**Subcategory 1**
Teacher as assistance provider to a pupil	Helping the pupil to experience personal success	Helping	Teacher, aiming and able to understand the needs of every pupil and to respond to them	Focusing attempts to adapt teaching methods to the pupil's needs	Willing to know the pupil's individuality
		Meaning well			Focussing attempts to find teaching methods suitable to the pupil
	Compassionate	Inclined to understand and sympathize			

The pupils' essays show a clear expectation of a helping teacher. The pupils expect help from the teacher when they face difficulties in learning: "*You can address him/her when you don't understand something and you can be sure he/she will help you.*" They expect assistance from the teacher in other situations of school life too, perhaps even outside school: "*Helping to solve problems.*" However, it is not only assistance from the teacher that the pupils find important. They also look forward to understanding from the teacher: "*Attentive to the pupil,*" sincere attitude towards the pupil's success: "*Advising,*" "*Meaning well.*" In case of difficulties, the pupils look forward to and expect understanding and compassion on behalf of the teacher: "*Able to sympathize, comforting;*" however, it is very important to the pupils that compassion should not violate their dignity: "*Not denouncing those who don't succeed.*" The pupils trust such a teacher and are open to him/her: "*A good friend to the child.*"

In the reflections of the teachers' experience, the inclination to try to understand the needs of every pupil and to respond to them was revealed as one of the greatest values. Firstly, the teacher must recognize the diversity of pupils' possibilities: "*...knowing that not everyone can be the same*" and to try to accept the pupil as he/she is: "...willingness to get to know another person." This means that the teacher, when organizing the pupils' education, will take into consideration the individuality of every pupil and will endeavour to create learning conditions for him/her that will allow the pupil to experience success: "*...the children are different; if the method has been suitable to one of them, it may not be suitable to another.*" When the teacher builds learning environment favourable to every pupil, the need for direct assistance to the child decreases significantly. Assistance to the pupil becomes indirect: "*...willingness to look for ever different ways of education.*" In order to help the child, the teacher must accumulate a rather vast collection of educational methods, teaching methods, and means that in some cases allow individualizing the work to the maximum and in other cases, on the contrary, bringing pupils with different possibilities together for joint activity and ensuring learning success to every pupil.

The pupils define professional identity of a teacher through the competence of assistance to the pupil. What they find important is the teacher's openness to the pupil's failure, understanding and readiness not only to help solve arising problems but also to lead the pupil towards success.

The teachers, when discussing their professional identification, put greater focus on the knowledge of the pupil and the individualization of educational methods; thus creating preconditions for the pupil's success and prevention of the need for assistance.

The fourth category of characteristics reflects, in the deliberations by the pupils and the teachers, the teacher's inclination to work towards the maximum learning result of every pupil (Table 1.4).

Table 1.4: Category group "Reflection of Teachers' Insistence in Professional Identity Construct"

Pupils' expectations			Teachers' attitudes		
Category	**Subcategory 2**	**Subcategory 1**	**Category**	**Subcategory 2**	**Subcategory 1**
Teacher, able to educate the pupils objectively and demandingly but with love	Educator that demands strictly but loves pupils Making fair and objective decisions with regard to the pupils	Educating in a demanding and responsible manner	Teacher, able to demand of everyone according to their individual possibilities	Demanding but taking individual possibilities of the pupils into account	Differentiating requirements
		Demanding strictly but understanding pupils			Not demonstrating pity
		Assessing achievements objectively			
		Taking equal decisions with regard to everyone			

It might seem strange that the pupils, when discussing the image of a good teacher that meets their expectations, emphasize his/her insistence next to assistance, compassion and understanding. The pupils acknowledge the mistakes they make and misunderstandings that take place and they assign the role of an assessor and manager of such situations to the teacher: "*...applying discipline on us, teenagers, when we behave wrongly.*" However, the demanding stance of a teacher should be based on fairness: "*...strict but just,*" sincerity: "*...strict but warm character.*" The teacher should not always be strictly inclined. According to the pupils, he/she should be "*...strict only when it is really needed.*" The image of a strict teacher is expressed in the pupils' deliberations quite intensively; however, equal emphasis is put on the teacher's justness towards the pupils: "*...evaluates fairly,*" "*...does not ruin young people through marks,*" "*...does not have favourites,*" and, on his/her behalf, "*is capable of acknowledging his/her mistakes and correcting them.*" These characteristics of a teacher are certainly important in educating homogeneous groups as well; however, they gain particular relevance to the pupils because the difficulties frequently faced by some pupils might encourage making concessions or create an impression of making concessions to them. The observations by the pupils, stemming from the processes taking place in heterogeneous environment, show that the pupils notice it.

In teachers' deliberations, the demanding attitude towards the pupils is expressed to a significantly smaller extent than love towards a child, attempt to recognize every child's needs and respond to them; however, the teachers are clearly against sycophancy towards children with disabilities, and unjustified lowering of requirements: "*...you must require of them what they can and must do.*" Inclusive education environment must serve for the preparation of a child with disability for an independent life; therefore, it is obligatory to accustom the child to using his/her possibilities fully and to working towards his/her maximum result: "*...you cannot catch the fish for him, you must teach him how to fish.*" The teachers, while educating children with disabilities and particularly having an opportunity to compare their capabilities with those of others, observing the difficulties and limits to their capabilities that emerge, naturally avoid the arising sense of pity. However, they understand that pity cannot prevail in interpersonal relationships. The teachers believe that the way out is "... *not demonstrating pity.*"

The pupils, emphasizing sensitive demandingness and fairness of the teacher, which is one of the features of pedagogical activity that they find important, highlight the competence of upbringing the young as an important component of the teacher's identity.

The teachers indicate insistence as part of the definition of professional identity in a more reserved manner but they do it accurately. Their insights show the inclination to require of everyone according to their personal abilities, which does not only confirm the competence of education differentiation that they have mastered but also reveals their ability to develop fairness in pupils through their personal example.

Dimension 2: Teacher's educational skill

The pupils, when discussing the characteristics of a good teacher, also emphasize the professionalism and educational skill of the teacher in their essays (Table 1.5).

Table 1.5: Category group "Reflection of Teachers' Ability to Build Favourable Educational Environment in Professional Identity Construct"

Pupils' expectations			Teachers' attitudes		
Category	**Subcategory 2**	**Subcategory 1**	**Category**	**Subcategory 2**	**Subcategory 1**
Teacher, creating attractive learning activity by means of harmony between the pupil's learning style and methods	Creating attractive learning activity	Raising interest in learning activity	Teacher, directing the educational process towards the pupil's success in a targeted and flexible manner	Putting focused attempts to achieve the pupil's success through knowing the pupil and meeting his/her needs	In constant search for the most suitable teaching methods
		Creating conditions for the pupil's self-expression			Observing the pupils and making decisions
	Responsively adapting ways and methods of teaching to the pupil's individual needs	Taking responsibility for the pupil's success			Planning the pupil's success
		Using many various teaching methods that respond to individual needs		Planning the educational process consistently but applying the plan flexibly	Planning the educational process consistently
					Organizing educational process flexibly

The teacher's ability to organize educational process is less relevant to the pupils as his/her personal traits. The features showing the teacher's skill are summarized in two criteria groups. However, discussing their educational experiences, the pupils are very accurate in reflecting the fundamental didactic principles of heterogeneous group education.

The analysis of the pupil's essays revealed that they find it important that the educational process organized by the teacher should encourage aiming at high results. Although heterogeneous class includes pupils with different abilities, the intrigue when learning, created by the teacher, is still relevant: "*…raises questions that force us to think*," "*…work is hard but interesting*." It is obvious that the teacher, when organizing the educational process, differentiates the tasks in a way that makes every pupil aim at his/her best personal result: "*…keeping operational tension*." This encourages the pupils' motivation to learn and the learning activity becomes interesting. The reflections on the pupils' experiences clearly show the importance of curriculum differentiation. The pupils learn successfully with a teacher who "*...presents material in such a way that makes it understandable to everyone*," thus creating favourable conditions for the pupils' self-expression, without unifying neither educational activity, nor the limits of potential achievement. When learning in such educational environment, the pupils feel that the teacher "*...gives space for the pupil's abilities*." The reflections on the pupils' experiences clearly reveal that the decisions of the teachers working in heterogeneous classes are based on an inclination to aim at the success of every pupil, by creating flexible, child-oriented educational environment. In case of pupil's failures, the teacher "*...focuses on every pupil's knowledge gaps and problems*" not only by adapting curriculum for him/her. The pupils notice that the teacher "*...knows how to adapt teaching methods*." In case of a pupil's failure, it is aimed at finding ways and responding to the pupil's needs: "*...if the pupil doesn't learn, he/she changes their teaching style*." In the pupils' deliberations, the teacher's ability to create favourable educational environment is reflected through the teacher's focus on every pupil's success, flexible adaptation of the curriculum and ways of activity to the pupils' individual needs.

The teachers, when discussing their work experience under the conditions of inclusive education, highlight the same aspects of educational reality as the pupils. Their deliberations show clear inclination towards constant search, creativity and flexibility in creating the educational environment favourable to the pupils: "*...you search, make mistakes and search again*." The reflections of the teachers, as well as those of the pupils, show clear orientation towards experiences of success in learning for every pupil: "*...the pupil must feel that he/she can and how much he/she can, the pupil must be proud of him/herself*". Such attitude inevitably means the necessity to differentiate educational methods: "*...the children are different; if the method has been suitable to one of them, it may not be suitable to another*." The teacher implementing inclusive education makes decision by observing the children. The main resource for the teacher is the pupil him/herself: "*...everything must take a natural flow. You see a child and you know how you must act*." The diversity of children in the classroom, the aim at responding to the needs of every pupil, and opening space for successful learning of every pupil requires the teacher's particular flexibility, and ability to make decisions quickly bearing in mind the current situations: "*...the teacher must be an improviser because you never know what might happen in a lesson, and you should not be afraid of it*." It is obvious that a teacher implementing inclusive education successfully must know his/her pupils well, constantly observe them and respond to occurring circumstances.

Another characteristics criteria group identified during the research reflects the teacher's didactic skill (Table 1.6).

Table 1.6: Category group "Reflection of Teacher Skill in Professional Identity Construct"

<table>
<tr><th colspan="3">Pupil's expectations</th><th colspan="3">Teachers' attitudes</th></tr>
<tr><th>Category</th><th>Subcategory 2</th><th>Subcategory 1</th><th>Category</th><th>Subcategory 2</th><th>Subcategory 1</th></tr>
<tr><td rowspan="4">Teacher as a master that involves pupils in learning activity</td><td rowspan="2">Skilfully teaching pupils to learn</td><td>Presenting various learning strategies to pupils</td><td rowspan="4">Teacher, constantly looking for innovative educational methods that involve every pupil</td><td rowspan="2">Constantly aiming at pedagogical professionalismand that in his/her field</td><td>Constantly developing, looking for efficient and innovative methods</td></tr>
<tr><td>Teaching his/her subject skilfully</td><td>Professional in his/her field, building partnership networks</td></tr>
<tr><td rowspan="2">Filling the educational process with innovation and expression</td><td>Putting effort to raise pupils' interest</td><td rowspan="2">Putting effort to create interesting learning environment to every pupil</td><td>Able to involve every pupil in the activity</td></tr>
<tr><td>Using innovative teaching methods</td><td>Planning the learning activity interesting to every pupil</td></tr>
</table>

When discussing their learning experiences, the pupils give relevance not to knowledge that the teacher conveys to the pupil but to the teacher's didactic solutions that help the pupil learn to learn: "*...teaches ways to learn more easily and simply*," "...teaching to systemize knowledge." The pupils understand that the learning process is full of challenges and they are inclined to accept them; therefore, they expect of the teacher a space for their independent decisions and a right to take responsibility for them: "*...allows the pupil to make mistakes*." A professional teacher, according to the pupils, does not only provide pupils with certain knowledge and experience but also helps them discover themselves and choose their field of interests: "*...revealing the pupil's abilities*," provides opportunities to express themselves and defend their opinion: "*...allows us to freely express our thoughts*." The pupils also notice the teacher's pedagogical expertise that helps the teacher create favourable learning conditions: "*...knowing all the subtleties of education*." Although the pupils appreciate a possibility to feel independent in the learning process, and to experience self-expression, they also expect methodically provided professional assistance from the teacher: "*...explains everything in a way that no-one else can, it then sets me on fire*." The pupils do not speak much of the teacher's deep expertise in the field. They find the teacher's ability to convey the knowledge on the subject in a manner acceptable to the pupils particularly important. The pupils' expectations reflect high level of pedagogical skill in the teacher, his/her creativity in conveying the content taught to the pupils: "*...finds ways to present boring topics in an interesting way.*" There is a clear need for the teacher to be able to structure his/her ideas in a manner acceptable to the pupils, taking their age, abilities, and experience into consideration, since the pupils feel good when they are able to understand the educational material presented to them: "*...presents the topic in such a way that leaves no questions to the pupils.*" The pupils notice the teacher's efforts and responsibility in his/her work, his/her inclination to create learning activity attractive to the pupils: "*...is prepared for every lesson to make it interesting.*" The pupils attach great importance to attractive emotional atmosphere in the lesson: "*...the lessons are funny and full of challenges.*" The pupilsemphasize the teacher's didactic preparation and their ability to apply suitable teaching methods, which are particularly efficient in educating heterogeneous groups: "*...allowing to work in groups, and discuss; then the pupils learn to communicate.*" New, modern teaching techniques appeal to the pupils and they expect them to be applied in their educational environment, they voice the requirement for the teachers to follow the latest developments and to improve in their fields: "*...work by modern methods.*" A teacher's ability to also invent and combine him/herself, as well as to apply ways of education in their own unique manner is important: "*...unique teaching;*" thus a possibility emerges to respond to real needs of the pupils in the classroom.

The teachers, when discussing their work experience, respond to the pupils' expectations. The category of characteristics summarizing their experience highlights the teacher's innovativeness that allows involving every pupil in the educational process. The teachers implementing inclusive education give particular emphasis to personal development, constant learning and following the latest scientific achievements and technologies: "*...to read and follow what's new in the world.*" According to the teachers, self-reflection is important in the path of professional development. One can learn not only from books; practical experience, in combination with acquired theoretic knowledge, is an excellent school of professionalism: "*...learning to work in such cases, to enrich your knowledge, to look for what to do in the situations that emerge.*" The aim to learn should be a teacher's inner inclination since inclusive education environment presents numerous challenges, individual requirements, and it hardly fits into general standards: "*...a person in search because there are no specific recipes.*" Such conditions call on the teacher to constantly analyse, evaluate situations and make independent decisions: "*...you must always step forward, you must always think.*" The teachers implementing inclusive education, when discussing lesson organization principles, focus on the ability to divide their attention individually in the classroom: "... *the teacher must see the class as a whole and every single one in the class separately.*" Presenting educational material is not sufficient, it is very important to make it "... *interesting to everyone.*" Thus, what is important is "*...the search for new methods of how to do it better.*" The teachers believe that it is not sufficient to have a good command of teaching techniques, know the children and continuously monitor their progress; the teacher will not manage to achieve good results if he/she is not an expert in his/her field: "*to know your subject well.*" Although the pupils put greater emphasis on the teacher's pedagogical skill, the teachers believe that expertise in the field is of equal importance. In inclusive education environment, high results can be achieved through co-operation with other teachers and assistance providers to the pupil; therefore, the teacher must be able to work in a team, "*...to work together with other teachers.*" The teachers, as the pupils, highlight the fact that the teacher cannot come to a heterogeneous classroom without sufficient preparation, "*one must have time to prepare for every lesson.*" The aim to respond to every pupil's needs at the same time presents a requirement to review curriculum, the course of the lesson, and educational goals for every pupil in advance.

D. Brackenreed (2008), who researched factors that cause the greatest tension to teachers working under the conditions of inclusive education, proves that pupils with special needs do not cause any stress to teachers. Stress results from the fact that the teachers educating these pupils in general classrooms do not have sufficient prior preparation for such work. In most cases, experience should come naturally: how to understand and manage the behaviour of these children, how to meet the parents' expectations, which in many cases exceed the actual possibilities of the child. Often the teachers lack special assistance needed in their classrooms.

The pupils emphasize the teacher's pedagogical skill that fills the educational process with innovation and expression.

The teachers identify themselves with a pedagogue who continuously learns and looks for innovation, who is able not only to respond to the pupil's needs but also to inspire a willingness to learn and to work towards a result according to their abilities in everyone.

Conclusions

In order to respond to a problem question "What socio-educational characteristics do participants of inclusive education, both pupils and teachers, use to define professional identity of a teacher?", the comparative analysis of the pupils' expectations and reflections on the teachers' experiences revealed that inclusive education reality manifests clear features of modern didactics based on ideas of constructivism, and a shift from the paradigm of effect to that of interaction is taking place. A clear need to personalize the educational processpresents the necessity to build learning environment favourable to the participation of all pupils, it brings relevance to the realization of back-up feedback and the application of innovative educational techniques. Activity in inclusive education reality encourages the teacher to constantly reflect on his/her personal experience, to share it with others, and aim at professional development. The results of the research allow identifying the socio-educational characteristics of the professional identity of a teacher.

In the professional identity construct of a teacher implementing inclusive education, the pupils put particular emphasis on his/her personal traits that reveal themselves through the teacher's high level of empathy, his/her ability to thoroughly understand the pupil's situation and aim for successful learning together with them. An extrovert teacher capable of creating vivid, emotionally elevated educational environment that at the same time invites to work appeals to the pupils. The teacher's inclination and ability to provide help to the pupil not only when learning but also in other life situations is important. The pupils perceive the teacher's help as a sincere accompaniment towards success in learning, while respecting the pupil's dignity. What appeals to the pupils is a demanding teacher, capable of clearly stating the requirements to every pupil and keeping to them, who at the same time is warm, sincere and fair to the pupils and him/herself.

The pupils identify the teacher's professional skill with the ability to organize education, maintain a motivating intrigue that encourages to discover, to learn, and, at the same time, is adapted to personal possibilities and learning style of every pupil. The pupils attach less significance to the teacher's knowledge on the subject but they highly appreciate his/her skill in proposing learning methods that help every pupil to experience success in learning.

The teachers, when reflecting on their experience of activity under the conditions of inclusive education, also put strong emphasis on establishing personal relationship with the pupils. They find knowing the uniqueness of every pupil and acknowledging their interpersonal differences particularly important. The teachers strongly highlight the value of love for the pupil that leads to professional satisfaction. The teachers recognize that inclusive education environment, containing many interpersonal differences of the pupils encourages the teacher's creativity, well-roundedness and ability to make decisions having understood the reasons for the pupils success or failure as well as their behavioural motives.

The teachers, as well as the pupils, emphasize the skills of assisting the pupil in professional identity construct. However, the teachers' perception of assistance is based on the inclination towards prevention of failure. By applying heterogeneous group education principles, such as personalization of education, and building activity networks, the teachers aim at creating equal conditions for all pupils to experience success in learning. The inclination of the teachers working under the conditions of inclusive education not to make concessions to the pupils with disabilities and to encourage them to aim at the maximum personal result in preparation for independent lifereveals rational insistence of the teachers.

The teachers identify pedagogical professionalism with the ability to organize educational process in a flexible and creative manner by making decisions inspired by individual needs of the pupils and allowing every pupil to trust in their abilities and encouraging their self-confidence. The indicator of the teacher's professional success is the inner inclination to continuously learn and reflect on his/her experience. In order to achieve professional development, the teacher should gain expertise in getting to know the pupils, curriculum and educational process differentiation, co-operation methods and techniques, and constantly improve his/her subject knowledge.

- Summing up the reflection on the pupils' and teachers' experiences in inclusive education reality, it can be stated that the main socio-educational characteristics in the construct of professional identity of a teacher are revealed in the following fields of teacher competence;
- internalization of the value of recognizing interpersonal differences and respecting them in their own attitudes and those of the pupils;
- creation of educational environment favourable to all the pupils, emphatic and emotionally elevated;
- assistance prevention activity, and in case of necessity, providing it to every pupil;
- pupil motivation by creating environment for successful learning;
- teaching process management, skilfully involving every pupil in the educational activity;
- continuous learning through the reflection of one's experience, co-operation, and keeping interest in the latest scientific and practical achievements.

The results of the research correspond to the conclusions of the results of a research carried out in the framework of a project "Teacher Education for Inclusion" organized by the European Agency for Development in Special Needs Education (2012), which served as a basis for giving relevance to the following fields of teacher competence: respect for pupil diversity; assistance to all the pupils; work together with the others; personal professional development.

References

1. Aurelijus Augustinas. (O. Daukšienė, vert.) Dialogai apie mokytoją. Vilnius: Katalikų pasaulio l-kla, 1999.

2. Brackenreed, D. *Inclusive Education: Identifying Teachers' Perceived Stressors.* Exceptionality Education International, Vol. 18, No 3, 2008, p. 131 - 147.

3. Čiubrinskas,V., *Tautinio identiteto antropologinio tyrinėjimo klausimai.* Čiubrinskas, V. Kuznecovienė, J. (sud.) Lietuviškojo identiteto trajektorijos. Kaunas: Vytauto Didžiojo universitetas, 2008, p. 13 – 26.

4. European Agency for Development in Special Needs Education *Teacher Education for Inclusion: Profile of Inclusive Teachers.* Profile of Inclusive Teachers, Odense, Denmark: European Agency for Development in Special Needs education, 2012. [accessible by http://www.european-agency.org/sites/default/files/Profile-of-Inclusive-Teachers.pdf]

5. J. Bengtsson. *The Many Identities of Pedagogics as a Challenge: Towards an Ontology of Pedagogical Research as Pedagogical Practice.* Educational Philosophy and Theory, Vol. 38, Issue 2, 2006, p. 115 – 128.

6. Jatkauskienė, B. *Andragogo profesinio identiteto kultūrų raiška profesionalizacijos kontekste.* Mokytojų gdymas. 2010, Nr. 15 (2), p. 59 – 77.

7. Kliewer, Ch. The Meaning of inclusion. *Mental Retardation,* Vol. 36 Issue 4, 1998, p. 317 - 322.

8. Lasauskienė, J. *Muzikos mokytojo dalykinės kompetencijos kokybinė diagnostika.* Vocational Education: Reserarch & Reality. 2008, Issue 15, p. 60 – 73.

9. Laverty, S. M. *Hermeneutic Phenomenology and Phenomenology: A Comparison of Historical and Methodological Considerations.* International Journal of Qualitative Methods,Vol. 2, Issue 3, p. 1 – 29.

10. Martišauskienė E. *Tautinio tapatumo žadinimas – pedagogo misijos šerdis (metodologinis aspektas).* Pedagogika. Nr. 108, 2012, p. 64 – 70.

11. Martišauskienė, E. *Pedagogų požiūris į mokytojo kompetencijų raišką*. Acta Peadagogica Vilnensija, Vol. 22, 2009, p. 88 - 101.

12. Norvilienė, A., Zuzavičiūtė, V. *Mokytojų požiūris į tarpkultūrinės kompetencijos raišką*. Bridges. 2011, Vol. 56, Issue 3, p. 111 - 124.

13. Opertti, R., Brady, J., Duncombe, L. *Moving Forward: Inclusive Education as the Core of Education for All*. Prospects: Quarterly Review of Comparative Education, Vol. 39, No 3, 2010, p. 205 - 2014.

14. Paulauskas, R., Gumuliauskienė, A., Volodka, H., Širiakovienė, A. *Mokytojų mokytojas: refleksijos apie profesorių Jovaišą*. Teacher Education. Vol. 14, Issue 1, 2010, p. 208 - 216.

15. Ridzevičiūtė, E. *Mokytojų organizavimo kompetencijos raiška edukacinėje praktikoje*. Pedagogika. Issue 81, 2006, p. 42 - 46.

16. Šalkauskis, S. *Rinktiniai Raštai: Pedagoginės studijos*, I knyga. Vilnius: Leidybos centras, 1992.

17. Tijūnėlienė, O. *Gabrielė Petkevičaitė-Bitė - lietuvių tautos mokytoja ir auklėtoja*. Acta Peadagogica Vilnensija, Vol. 27, 2011, p. 149 - 151.

18. Tijūnėlienė, O. Kavaliauskienė, V. *Juozo Vaitkevičiaus kvietimas mokytojui pažinti save*. Acta Peadagogica Vilnensija, Vol. 22, 2009, p. 174 - 183.

19. Trimakas, K. A. *Asmenybės raida gyvenime*. Kaunas: Tarpdiacezinės katechetikos komisijos leidykla, 1997.

20. Urnėžienė, E. *Mokytojų nuolatinio tobulinimosi kompetencija, švietimo dalyvių požiūriu*. Pedagogika. Issue 107, 2012, p. 85 - 91.

21. Židžiūnaitė, V. *Profesinio tapatumo vystymasis ir curriculum struktūros komponentų įgyvendinimas universitete: student patirtis*. Acta Peadagogica Vilnensija, Vol. 29, 2012, p. 123 - 132.

The Creative Originality and Technological Inclinations of female Trainee Teachers of Technologies

Palmira Pečiuliauskienė, Ilona Valantinaitė

Abstract
The paper investigates the creative originality of female trainee teachers of Technologies which was determined by means of Torrance's technique of personal creativity diagnostics (a verbal expression questionnaire). The paper adheres to a sociological attitude to the target issue, and explores the creative originality of Generation Y female trainee teachers of Technologies, and the interrelation between creative originality and inclinations for "female" (nutrition, textile) and "male" spheres of technologies (constructive materials, electronics).

Keywords: creative originality, nutrition, textile, constructive materials, electronics

1. Relevance

The typical features of a creative society are the rapid increase of technological, the development of economy, expansion of social relations, the growth of people's mobility, and the appearance of new social structures, all of which make life unstable, crises become more frequent, and these problems can be dealt with only by enhancing people's creativity (Vaicekauskienė, 2009; Petrulytė, 1995; 2001; Rowlands, 2011). Creativity has become not only part of art, but also an increasingly important contribution to all sectors where cultural creative material functions as a basis for a competitive advantage in the global market economy (Levickaitė, 2011; Rowlands, 2011).

The investigation of the phenomenon of creativity requires a solution to a key problem – what is creativity? For more than a century there have been numerous discussions concerning the roles of knowledge, skills and abilities in creation (Torrance, 1971; Hennessey, Amabile, 1987; Jonynienė, 1987; Rowlands, 2011) as well as the interrelation between creativity and intellect (Guilford, 1950; Belcher, Davis, 1971; Torrance, 1971; 1974; 1977; 1987; Gardner, 1993; Gage, Berliner, 1994; Almonaitienė, 1997; Petrulytė, 2001; Стернберг, 2002; Sternberg, 2003; Sternberg ir kt., 2005; Vaicekauskienė, 2009).

Some theoreticians adhere to a theory which states that creativity is part of intellect, others object to this theory, and still others agree with the theory which argues that intellect and creativity partially coincide. Psychologist Torrance

(1977) formulated a "threshold" theory which determines the "threshold" of intellect. If the intellect is lower than the determined "threshold", a person's creative abilities are limited (Torrance, 1977; Petrulytė, 2001).

The essence of creativity is revealed by the main features of creativity: *originality, fluency, flexibility* and *elaboration* (Guilford, 1950; Felker, Treffinger, 1971; Torrance, 1974; Jonynienė, 1987; Petrulytė, 1995; 2001; 2007; Butkienė, 1996; Butkienė, Kepalaitė, 1996; Ragulienė, 1999; Stankevičienė, 2001; Treffinger ir kt., 2002; Grakauskaitė-Karkockienė, 2006; Girdzijauskienė, 2004; Grudžinskytė ir kt., 2009; Rowlands, 2011 et al.). *Originality* is an important feature of creative thinking which first of all related to novelty. Originality is also defined as an ability to find unique, unusual, untypical, unstereotypic, rare decisions, present rare, untypical ideas, based on distant associations.

Some researchers identify creative originality with creativity itself. The key to identifying instances of creativity is identifying instances or originality (King, 2011). King considers whether instances of originality occur in human works of art and/or human inventions; he concludes that instances or originality only occur in human inventions. Having reached this conclusion King then considers whether there are instances of originality in the non-human realm (John King, 2013). It is assumed that the 21st century lacks originality. "For most of us, truly original ideas come around about as often as winning lottery tickets. Our lives are designed around routines <…> Original ideas are not usually thunderbolts from out of the blue. Originality must be nurtured and stimulated. You cannot force a new thought, but you can create the right conditions to generate the sparks and seeds of a new idea"(Key, 2012; Beghetto & Kaufman).

The phenomenon of creativity can be investigated from the perspective of time and changing environment. What does it mean to be creative in the 21st century? How did creativity manifest itself in the middle of the 20th century? The analysis of the phenomenon of creativity in relation to the time period and the young generation inevitably requires stepping over the boundaries of education and referring to the sociologists' insights about the peculiarities of the present-day generation. Western sociologists (McCrindle, Wolfinger, 2010; Armour, 2008; Huntley, 2006) grouped people into generations: the generation of the Great Depression (born in 1912-1921); the generation of WWII (born in 1922-1927); the postwar generation (born in 1928-1945); the Baby Boom generation (born in 1946-1954); the generation of the 2nd Baby Boom (born in 1955-1965); Generation X (born in 1966-1976); Generation Y (born in 1977-1994); and Generation Z (born in 1995-2012). At present the schools of general education of the country are attended by learners of Generation Z, and universities train teachers who in the majority of cases belong to Generation Y. Technology teachers of Generation Y have to be trained so that they could not only help learners of Generation Z use technologies, but also teach them to create new technologies. The creativity and technological breakthrough of learners of Generation Z depends on the creativity of teachers of Generation Y.

Creativity can also investigated from a gender perspective. In Lithuanian schools of general education most teachers are female (Lithuanian Education in Numbers, 2012). Hence, the insights into female creativity are really topical. Reflections on what has been learned about creativity of women during the last 50 years reveal a gap in one major area. Little research has been completed and little is known about diverse, creative women, the choices they make, and the decisions they face about creative productivity in their lives (Reis, 2002). Consistently with Reis's model of talent development (that includes multiple avenues to creative achievement) the women in this research followed diverse paths to creative achievement. Psychologists still have a long way to go before they come anywhere close to understanding creativity in women and minorities (Simonton, 2000). Reis found factors related to women's talent, to include intelligence, personality, environment, and a women's perception of the social importance of using her abilities (Keller-Mathers, 2004).

The lack of research on female creativity is also confirmed by the recent studies. Despite four decades of scholarly interest in gender and creativity, no clear picture has emerged regarding the relationship between these two important and complex phenomena (Stoltzfus, Nibbelink, Vredenburg, & Thyrum, 2011).

The research problem is the creative originality of female trainee teachers of Technologies and its relation to the inclination for separate spheres of Technologies: nutrition, textile, constructive materials and electronics.

The object is the creative originality of trainee Technology teachers.

The goal is to evaluate the creative originality of female trainee teachers of Technologies and reveal its relation to the girls' inclination for different technological spheres.

The research methods: scientific literature and document review, a qualitative analysis based on Torrance's methodology.

The research questions:

1. What is the creative originality of female students of Technology Education?
2. What are the female students' inclinations for different spheres of Technological Education: electronics and constructive materials, nutrition and textile?
3. Is the creative originality of female trainee teachers of Technologies related to their inclinations for electronics, constructive materials, nutrition and textile?

2. Research Instrument

2.1. Research methodology

The assessment of creativity involves a number of criteria such as novelty, relevance, effectiveness, usefulness, and surprise (Boden, 2004; Cropley, 1999; Plucker, Beghetto, & Dow, 2004; Sternberg & Lubart, 1996,). Research on the assessment of creativity has been criticized for not having adequate criterion measures as well as for relying on subjective judgments and for using creativity tests that have theoretically too general or unimportant items to measure such a complex construct Sak & Ayas.

The instrument for the research on the creative originality of the trainee Technology teachers was composed on the basis of Torrance's test technique (TTCT). The study also used Torrance's (1995) and Kim's (2006) works.

One of the features of creativity, i.e. originality, is determined according to the number of unexpected, non-standard, unusual answers. An answer is considered to be original when it is unique, exclusive in the group of the informants. Every original answer is assessed with 5 points. Different groups provide different original variants of the answer.

The questionnaire for the analysis of creativity features consists of two parts: verbal (*TTCT – Verbal*) and nonverbal (*TTCT – Figural*). This paper presents the investigation of creative originality on the basis of only one – *verbal* – part of the questionnaire (Fig. 1). When the informants were completing the tasks in the *verbal part*, they had to *give answers in textual form*. The verbal part of the questionnaire consists of four different tasks in which the research participants have to reveal the variety of objects / phenomena (Table 1); describe objects / phenomena (Table 2); foresee the possibilities of using the objects (Table 3); foresee the possible consequences of using the objects (Table 4). Every task has to be completed in 3 minutes. The verbal tasks are composed on the basis of the curriculum of Technological Education which consists of four spheres: nutrition, textile, constructive materials, and electronics (General Programmes, 2008). The following example presents how the curriculum of nutrition is reflected in the verbal tasks. In the first task the informants have to define nutrition, in the second – to indicate the possible ways of using eggs, in the third – to list as many dishes which include dairy products as possible and say if they meet the principles of healthy nutrition; in the fourth task the informants have to explain the consequences of eating only thermally unprocessed food.

Fig.1: Kinds of Verbal Tasks in the Study on the Expression of Creativity Factors

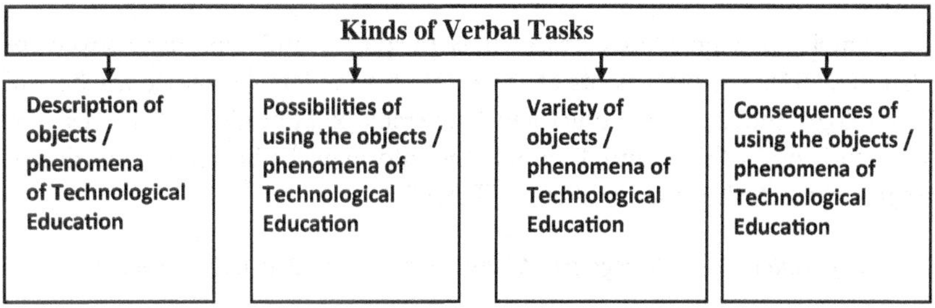

Table 1: Description of Objects / Phenomena of Technological Education

The sphere of textile	The sphere of nutrition	The sphere of constructive materials	The sphere of electronics
Pollution	Nutrition	Plastic	Economy of electricity

Table 2: Variety of the Possible Use of Specific Objects of Technological Education

The sphere of textile	The sphere of nutrition	The sphere of constructive materials	The sphere of electronics
Wool	Egg	Nail	Wire

Table 3: Possibilities of Using the Objects / Phenomena of Technological Education

The sphere of textile	The sphere of nutrition	The sphere of constructive materials	The sphere of electronics
List as many textile articles used 80 years ago as possible	List as many products including dairy products which meet the principles of healthy nutrition as possible	List as many drillable objects as possible	List as many electronic devices which use relatively little electricity as possible

Table 4: Consequences of Using Situations/Objects/Phenomena of Technological Education

The sphere of textile	The sphere of nutrition	The sphere of constructive materials	The sphere of electronics
Imagine what would happen if the last woman able to hand knit, crochet, stitch, sew and weave died in Lithuania	Imagine what would happen if we ate only thermally unprocessed food	Imagine what would happen if it was prohibited to use metal nails	Imagine what would happen if the price of electricity rose at ten times

2.2. Characteristics of the Informants

The sample of a qualitative research is purposive and typically convenient (Patton, 2002). The informants of the research were female trainee teachers of Technology Education at Lithuanian University of Educational Sciences who were studying in the 3rd and 4th years. The informants' educational and demographic characteristics were different (Table 5).

Table 5: Educational and Demographic Characteristics of the Qualitative Research Participants

No	Nationality	Average grade of the last (4th) term	Age	No	Nationality	Average grade of the last (6th) term	Age
Third year students					***Fourth year students***		
1	Lithuanian	9.5	21	1	Russian	9.07	22
2	Lithuanian	10.0	21	2	Polish	8	21
3	Lithuanian	9.03	21	3	Lithuanian	7	22
4	Lithuanian	10.0	21	4	Lithuanian	8	22
5	Lithuanian	9.4	20	5	Polish	9.2	22
6	Lithuanian	9.0	21				
7	Lithuanian	9.7	22				

The research participants were all the students in the programme of Technology Education. It has to be noted that in the third and fourth years this programme is studied only by female students. The curriculum of the study programme covers all the spheres of technologies: nutrition, textile, electronics and constructive materials. The average age of the informants is 20-22, and the average grades of the third and fourth year students vary (Table 5).

In Lithuanian schools girls have been taught handicraft since 1918 and a bit later this subject was also introduced to boys. Since the introduction of this subject, the curriculum of the general education programme has been divided into "female" technologies (nutrition and textile) and "male" technologies (woodwork and metal work). Handicraft was taught on the basis of the learners' gender, and in most cases the teachers for the subject were chosen on the same basis. Since 2003 the educational practice has used the concept *technologies* whose introduction coincided with the rejection of the practice of dividing technological education on the basis of gender. In general education schools girls and boys in the 5th – 8th forms learn in mixed groups and they study all four technological spheres:

textile, nutrition, constructive materials and electronics. From the 9th form learners individually choose the sphere of technologies they are most interested in, which can be later changed on demand. The changing needs of schools are answered by Technology teacher training at the university. Nowadays young people who want to become teachers of Technologies choose the study programme of Technology Education and study all four spheres of technologies and design.

3. Research results

3.1. The Inclination of Female Trainee Teachers of Technologies for "Female" and "Male" Technological Spheres

The tasks in the creativity establishment questionnaire were prepared on the basis on of the Programmes of General Education for Technologies (2008). Every task in the verbal part of the questionnaire was a multiple choice with four options related to different spheres of the curriculum of Technologies: nutrition, textile, constructive materials and electronics.

Half (24 out of 48) of the options in the tasks for the 3rd and 4th year trainee Technology teachers were related to nutrition and textile, and the other half – to constructive materials and electronics (Table 6). Thus, half of the trainee Technology teachers who participated in the research have an inclination for "female" spheres of technologies (nutrition, textile) and the other half – for "male" spheres of technologies (constructive materials and electronics) (Table 7). The data of the qualitative analysis show that the girls' inclination for "female" spheres of technologies – nutrition and textile – is not dominant. It can be assumed that this choice can be determined by some external factors, the technologically rich environment of Generation Y. On the other hand, the similar inclination for "male" technological spheres (constructive materials and electronics) can be conditioned by some internal personal factors.

Table 6: The Informants' Choice of Technology Tasks According to Different Spheres

Year	Informant	Description of objects/ phenomena	Possibilities of using the objects	Variety of objects/ phenomena	Possible consequences
III	1	Nutrition	Nutrition	Textile	Textile
	2	Nutrition	Nutrition	Constructive materials	Textile
	3	Textile	Constructive materials	Nutrition	Textile

	4	Constructive materials	Electronics	Textile	Constructive materials
	5	Constructive materials	Textile	Nutrition	Electronics
	6	Constructive materials	Textile	Constructive materials	Electronics
	7	Electronics	Electronics	Constructive materials	Electronics
IV	8	Nutrition	Electronics	Electronics	Electronics
	9	Nutrition	Nutrition	Constructive materials	Electronics
	10	Constructive materials	Nutrition	Electronics	Electronics
	11	Constructive materials	Nutrition	Nutrition	Electronics
	12	Nutrition	Nutrition	Nutrition	Nutrition

Table 7: The Choice of Technology Tasks According to "Male" and "Female"

Spheres of Technological Education	**Description of objects/ phenomena**	**Possibilities of using the objects**	**Variety of objects/ phenomena**	**Possible consequences**	**Total score**	**Total score according to "male" and "female" spheres**
Nutrition	5	6	4	1	16	24
Textile	1	2	2	**3**	8	
Constructive materials	5	1	4	1	11	24
Electronics	1	3	2	7	13	

The analysis of the interrelation between the informants' study year and their choice of tasks according to technological spheres revealed the following results. Most 3rd year students (6 out of 7) at least once chose a task whose content was related to textile or constructive materials (Table 6). While all 4th year students at least in one activity chose a nutrition task and 4 out of 5 – a task based on electronics. 8 out of 28 choices of the 3rd year informants were in the sphere of textile technologies, while none of the 4th year informants chose a textile-based task. The 4th students expressed a priority to nutrition-based tasks (10 out of 20) (Table 6).

It has been determined that in the group of tasks about the *Possible consequences* the informants most frequently chose tasks from the sphere of electronics (Table 7). Such tasks require a deep and versatile analysis of the given situation, and a creative anticipation of the *possible consequences* of using some object or of some phenomenon. Let us now discuss a task from the sphere of electronics, which is related to consumption culture and economy, given in the group of the Possible consequences which was worded in the following way: *Imagine what would happen if the price of electricity rose at ten times.* The 3rd and 4th year students of Technology Education indicated the possible original consequences of the situation / object. Some answers predict that "*some people would try to pay the price*", and other answers reveal initiative – "*people would protest about it near the Parliament or somewhere else*". Some consequences were that "*people would have no hot water*", "*nothing, absolutely nothing would function*", "*everything would stop in the world*", "everybody would be unhappy", as a result of which "*the birthrate would significantly decrease*", or in some other variants "*people would use electricity more economically, they would become more ecologically-minded*", "*children would try to do their homework in the daytime, at the sunlight, rather than work in the evening using artificial light*", "*all the stock of candles would be bought out*" and "*the newly built houses would have to be wired up*". Even though "*some people would get behind with politics and news since they are used to watching TV*", "*there would appear more educated scientists*". "*People would start looking for an alternative to replace electricity with*", "*I would employ hamsters who spin a wheel, would attach a generator to the wheel and it would produce electricity. Then I would only have to pay for hamster food;)*".

Summarizing the informants' answers about the consequences of more expensive electricity, only one physically correct solution was found: "*I would employ hamsters who spin a wheel, would attach a generator to the wheel and it would produce electricity. Then I would only have to pay for hamster food;)*". Electricity is produced when steam turns a turbine. All the other answers were not related to the solution of the given problem, rather they spoke about adjustment to the predicted situation: "*nothing, absolutely nothing would function*", "*everything would stop in the world*", "*everybody would be unhappy*".

Another frequently chosen task in the group of the *Possible consequences* was from the sphere of textile technologies. This task was chosen by 3 out of 7 3rd year students. That is the most frequent choice of textile-based task of the 3rd year students. It has to be noted that textile technologies were not chosen by the 4th year students even once. The fourth task of the question about textile asked to indicate the possible consequences *if the last woman able to hand knit, crochet, stitch, sew and weave died in Lithuania.* The 3rd year students of Technology Education related the extinction of female handicraft to the danger for the national identity: "*there would be no national traditions, customs left*"; "*part of nationality would disappear*"; "*the greatest Christmas present – woolen socks would disappear*". The original answers also touched upon the gender aspect: "*men would learn these crafts*"; "*there would be nothing to teach girls about*"; "*grandmothers (in the future) would have nothing to do*". The informants' original answers also revealed a positive attitude to the changes of technologies ("*people would be more interested in the executive techniques*"; "*people would create more new technologies which would do the work instead of women*"; *they would start producing, weaving etc. themselves*"), including the use of alternative materials ("*we would replace yarn with different materials: bird feathers, tree leaves, tree bark*"). There were also some possible social-political consequences mentioned: "*families, companies and even the state would go bankrupt*"; "*people would wear their clothing out, they would be ragged and eventually they would walk about naked*"; "*we would buy all the articles at a high price from some foreign countries*". Some of the original answers were rather indifferent ("*most people would not even notice that since there are different factories, workshops whose profit would increase*"), pessimistic ("*eventually the society would start living just like they did in the distant past*") and optimistic ("*people would run about covering themselves with frying pans*").

The generalization of the informants' answers to the task about textile technologies on the possible consequences if hand knitting and sewing technologies disappeared reveals a very wide spectrum of the indicated consequences: cultural, social-political and even emotional consequences. It is important to note that the informants chose tasks based on the "male" sphere of electronics from the most complicated group of tasks that requires original generalizations, i.e. the *Possible consequences*. From the perspective of creative originality there were simpler tasks: describe a phenomenon (Description of objects / phenomena), reveal the variety of phenomena (Variety of objects / phenomena), and indicate the possible uses of objects / phenomena (Possibilities of using objects). However, the informants chose the task based on the "male" sphere of constructive materials from the easiest group of tasks, i.e. *Description of objects / phenomena.*

3.2. The Creative Originality of Female Trainee Teachers of Technologies and their Inclination for Different Spheres of Technologies

The highest scores for creative originality were determined in the 3rd year students' answers to the first task about the description of objects / phenomena of Technological Education: 85 in the task related to the sphere of electronics, and 75 in the task on nutrition technologies (Table 8). Economy of electricity in the variant of the first task related to the sphere electronics was originally described as: *"inconceivable"*, *"smart"*, *"ingenious"*, *"green"*, *"fashion""*, *"ecological"*, *"useless"*, *"contagious"*, *"antique"* etc. And here are some original variants about the description of nutrition: *"balanced"*, *"dangerous"*, *"consumerist"*, *"shoddy"*, *"regular"*, *"punctual"*, *"fast"*, *"dated"*, *"liquid"*, *"scentless"*, *"long-lasting"*, *"colourful"*, *"home-made"* etc.

Table 8: The Female Trainee Technology Teachers' Score for Creative Originality Expression

Year	Informant	Description of objects/ phenomena	Possibilities of using the objects	Variety of objects/ phenomena	Possible consequences
III	1	60 (Nutrition)	5 (Nutrition)	40 (Textile)	30 (Textile)
	2	75 (Nutrition)	15 (Nutrition)	60 (Constructive materials)	25 (Textile)
	3	55 (Textile)	40 (Constructive materials)	15 (Nutrition)	35 (Textile)
	4	45 (Constructive materials)	45 (Electronics)	40 (Textile)	30 (Constructive materials)
	5	55 (Constructive materials)	45 (Textile)	20 (Nutrition)	15 (Electronics)
	6	35 (Constructive materials)	55 (Textile)	25 (Constructive materials)	20 (Electronics)
	7	85 (Electronics)	65 (Electronics)	50 (Constructive materials)	30 (Electronics)
IV	8	45 (Nutrition)	20 (Electronics)	20 (Electronics)	5 (Electronics)
	9	45 (Nutrition)	0 (Nutrition)	20	15 (Electronics)
	10	70 (Constructive materials)	25 (Nutrition)	15 (Electronics)	5 (Electronics)
	11	45 (Constructive materials)	20 (Nutrition)	0 (Nutrition)	20 (Electronics)
	12	60 (Nutrition)	30 (Nutrition)	40 (Nutrition)	50 (Nutrition)
Total score		675	365	345	280

The minimal score for creative originality (which equals zero) was determined in the 4th year students' assessment of the second task (Possibility of using objects) and the third task (Variety of objects / phenomena) (Table 7). The highest score for the sphere of nutrition was given in the assessment of the first task (Description of objects / phenomena). The total score for creative originality in all the nutrition tasks, in comparison to the other spheres of technologies (textile, constructive materials, electronics), is the highest – 505, and the statistical score for creative originality expression in the task about nutrition is 31.53 (Table 9).

It was determined that the statistical scores for creative originality expression are the lowest in the "male" productive task related to electronics – 27.7 and the "female" task related to nutrition – 31.53. The aforementioned facts make it possible to argue that the inclination for a certain technology does not determine a high score for creative originality expression in tasks related to it. The highest score for creative originality expression was determined in the rarely chosen spheres of constructive materials – **43.18** and textile – 40.63 (Table 9). The comparison of the informants' creative originality assessment according to separate spheres of technologies (textile, nutrition, constructive materials and electronics) showed that the highest statistical score for creative originality (43.18) was achieved in the "male" sphere of constructive materials.

Table 9: Assessment of Tasks from Different Technological Spheres: Total and Statistical Scores for Creative Originality

Sphere of technologies	Total score for creative originality	Total score for creative originality according to "male" and "female" spheres	Statistical[1] score for creative originality	The mean of statistical[2] score for creative originality
Textile	325	830	40.63	35.98
Nutrition	**505**		31.53	
Constructive materials	475	**835**	**43.18**	**35.44**
Electronics	360		27.7	

4. Discussion

Research on creativity phenomenon from the perspective of gender does not give a homogeneous answer about female and male creativity. In many studies, no discernible gender differences have been found (Kaufman, Baer, & Gentile, 2004). Some researchers believe that women are more creative than men (Reuter et al., 2005; Wolfradt & Pretz, 2001), other researchers state that men outperform women (Cox, 2002; Dollinger, Dollinger, & Centeno, 2005). Similarly, in studies of creative men and women's personalities, some researchers have found similarities (Chavez-Eakle, Lara, & Cruz-Fuentes, 2006; Szobiova, 2006), while others have found personality differences (Labouvie-Vief, 1994).

Baer & Kaufman (2008), generalizing the scientific discussion, claim that "there are studies that report that girls and women score higher than boys and men, and there are studies that report the opposite. The former (that is, studies in which girls and women score higher) are more numerous so it would be hard to make a case for overall male advantage" (Baer & Kaufman, 2008, p. 79). Generalizing this brief review of research on creativity it has to be noted that most frequently they aim to compare female and male creativity.

We investigated only female creativity. Answering the research objectives, we aimed to reveal the creative originality of female students of Technology Education and its interrelation with female teachers' of Generation Y inclination for electronics, constructive materials. Electronics, constructive materials have been considered to be a sphere of male activity for a long time, while textile and nutrition have been viewed as female activity. Our investigation proves that the division between male and female spheres of activity is disappearing, and the female social-productive role is changing. Social-productive changes can be determined by new gender roles. Stoltzfus, Nibbelink, Vredenburg, & Thyrum (2011) conducted a research which shows that there exists the relationship between gender role and creativity, and in this respect creativity was clearly influenced by gender. Stoltzfus, Nibbelink, Vredenburg, & Thyrum (2011), explored the influence of gender roles on creativity and distinguished the following kinds of gender roles: androgynous (high on both masculinity and femininity), masculine (high on masculinity and low on femininity), feminine (low on masculinity and high on femininity), and undifferentiated (low on both masculinity and femininity). In their research Stoltzfus, Nibbelink, Vredenburg, & Thyrum (2011) consider the influence of gender roles on creativity and their conclusions confirm the results of the previous investigations. Getzels and Csikszentmihalyi (1976) concluded that the psychology of creative men is a feminine psychology and the psychology of creative women is a masculine psychology. The same conclusion about the interrelation between gender roles and creativity was determined by Torrance (1963). He remarked that creative boys are perceived as more feminine than other boys and creative girls possess more traditionally masculine characteristics than other girls do. The studies of Stoltzfus, Nibbelink, Vredenburg, and Thyrum (2011) show that the highest overall picture construction scores were obtained by men who identified strongly with feminine gender role characteristics. The highest female performance, and the second highest picture construction creativity scores overall, were obtained by androgynous women.

Generalizing the studies on the influence of gender roles on creativity, it has to be noted that creativity is a typical characteristic of androgynous and masculine women. Our research revealed that the highest score for creative originality was given for a "male" task related to constructive materials (43.18 points). Nevertheless, the comparison of the informants' statistical score for creative originality according to the male (35.44) and female (35.98) spheres of productive activity demonstrated that they are equivalent. Hence, the informants of Generation Y, conditioned by the technological environment and the rapid technological progress, have acquired an inclination for new social-productive roles (in our research – an inclination for electronics, constructive materials) and possess creative originality in these spheres. Therefore, it is possible to assume that female gender roles are changing and the informants possess an androgynous and masculine gender role. This insight requires new investigations.

Conclusions

1. A rapid change of technologies, an increasing repletion of the household environment with modern technologies are changing the female trainee Technology teachers' inclinations for "male" spheres of activity (electronics, constructive materials). The female trainee teachers of Technologies did not distinguish one specific group of tasks when choosing activities related to different technological spheres (textile, nutrition, electronics, and constructive materials). The frequency of choosing the tasks based on electronics and constructive materials coincides with the frequency of choosing the tasks related to nutrition and textile.
2. The creative originality of the female trainee teachers of Technologies is equivalent in their completion of tasks "male" character (constructive materials and electronics) and "female" character (nutrition, textile). The trainee Technology teachers demonstrated the highest creative originality in the tasks related to constructive materials (creativity score – 43.18) and to textile (40.63). Hence, the highest score for creative originality were gained both in "male" tasks (constructive materials) and "female" tasks (textile).
3. The creative originality of the female trainee Technology teachers is inversely proportional to their choices of tasks related to various technological spheres (nutrition, textile, constructive materials, and electronics). The highest scores for creative originality were achieved in the most rarely chosen spheres of technologies. The highest score for creative originality (creative score – 43.18) was attained in the task on constructive materials; however, this task was the rarest choice (8 choices).
4. Creative originality is determined by gender role androgynous (high on both masculinity and femininity), and masculine (high on masculinity and

low on femininity). The influence of gender roles on women's inclinations for "male" spheres of activity and expression of creativity in them requires new investigations.

References

1. Almonaitienė, J. (1997). *Motyvacija kaip psichologinis kūrybiškumo veiksnys.* Psichologija. Mokslo darbai, 16, 65–84.

2. Armour, S. (2008). *Generation Y: They've arrived at work with a new attitude. USA Today.* Interaktyvus. Prieiga per Inernetą adresu: <http://www.usatoday.com/money/workplace/2005-11-06-gen-y_x.htm>.

3. Baer, J.; Kaufman, J. C. (2008). *Gender differences in creativity*, Journal of Creative Behavior, 42(2), 75-105.

4. Beghetto, R. A.; Kaufman, J. C. (Eds.) (2010). *Nurturing creativity in the classroom.* Cambridge, England: Cambridge University Press.

5. Belcher, T. L.; Davis G. A. (1971). *Interrelationships Among Three Standardized Creativity Tests and IQ.* New York: University of Wisconsin.

6. *Bendrosios ugdymo programos: pagrindinis ugdyma/ Technologijos (2008).* Interaktyvus. Prieiga per Internetą adresu: http://portalas.emokykla.lt/bup/Documents/Pradinis%20ir%20pagrindinis%20ugdymas/Technologijos.pdf.

7. Boden, M. A. (2004). *The Creative Mind: Myth and mechanisms.* London, New York: Routlendge: Taylor & Francus Group.

8. Butkienė, O. G. (1996). *Mokymosi psichologija.* Vilnius: VPU leidykla.

9. Butkienė, O. G.; Kepalaitė A. (1996). *Mokymasis ir asmenybės brendimas: pedagoginės psichologijos įvadas studentams, mokytojams, tėvams.* Vilnius: Margi raštai.

10. Cathy Key, C. (2012). *What is Originality? How to bring fresh thinking to your events.* Interaktyvus. Prieiga per Internetą adresu: <http://www.onlineregistrationreview.com/what-is-originality-how-to-bring-fresh-thinking-to-your-events/>.

11. Chavez-Eakle, R. A.; Lara, M. C.; Cruz-Fuentes, C. (2006). *Personality: A possible bridge between creativity and psychopathology? Creativity Research Journal, 18(1), 27-38.*

12. Cox, B. F. (2002). *The relationship between creativity and self-directed learning among adult community college students. Unpublished doctoral dissertation, University of Tennessee, Knoxville.*

13. Cropley, A. J. (1999). *Creativity in Education and Learning: A Guide for Teachers and Educators.* UK: Kogan Page.

14. Dollinger, S. J., Dollinger, S. M. C., & Centeno, L. (2005). *Identity and creativity. Identity: An International Journal of Theory and Research, 5(4), 315-339.*

15. Felker, D. W.; Treffinger D. J. (1971). *Self Concept, Divergent Thinking Abilities, and Attitudes About Creativity and Problem Solving.* (ERIC Document Reproduction Service No. ED 047 347). New York: Purdue University.

16. Gage, N. L.; Berliner D. C. (1994). *Pedagoginė psichologija.* Vilnius: Alma litera.

17. Gardner, H. (1993). *Multiple Intelligences: The Theory in Practice.* NY: Basic Books.

18. Getzels, J. W.; & Csikszentmihalyi, M. (1976). *The creative vision: A longitudinal study of problem finding in art.* New York: Wiley.

19. Girdzijauskienė, R. (2004). *Jaunesniojo mokyklinio amžiaus vaikų kūrybiškumo ugdymas muzikine veikla: monografija.* Klaipėda: Klaipėdos universiteto leidykla.

20. Grakauskaitė-Karkockienė, D. (2006). *Kur dingsta kodėlčiukai?: kūrybiškumo ugdymo pagrindai.* Vilnius: Logotipas.

21. Grudžinskytė, A.; Norvilienė, A.; Grudžinskienė, V. (2009). *5–6 metų vaikų kūrybiškumas: pedagogų ir tėvų požiūris.* Pedagogika, 96, 95–102.

22. Guilford, J. P. (1950). *The Nature of Human Intelligence.* New York: McGraw-Hill.

23. Hennessey, B. A.; Amabile T. M. (1987). *Creativity and Learning: What Research Says to the Teacher.* Washington D. C.: Nationat Education Association.

24. Huntley, R. (2006). *The World According to Y: Inside the New Adult Generation.* Interaktyvus. Prieiga per Inernetą adresu: http://books.google.lt/books?id=b7RV5njJ3zcC&lpg=PP1&dq=generation+y&pg=PA10&hl=lt#v=onepage&q&f=false.

25. Jonynienė, V. (1987). *Jaunesniųjų moksleivių kūrybinio mąstymo ugdymas: metodinės rekomendacijos.* Vilnius: PMTI.

26. Kaufman, J. C.; Baer, J.; Gentile, C. A. (2004). *Differences in gender and ethnicity as measured by ratings of three writing tasks.* Journal of Creative Behavior, 38(1), 56-69.

27. Keller-Mathers, S. (2004). *A Qualitative Study of Women of Extraordinary Creative Achievemen.* A Dissertation. Submitted to the Faculty of Argosy University/Sarasota, Florida.

28. Kim, K. H. (2006). *Is Creativity Unidimensional or Multidimensional? Analyses of the Torrance Tests of Creative Thinking.* Creativity Research Journal, 18 (3), 251–259.

29. King, J. (2011). *What is Creativity?: Originality, Art & Invention.* Light Press.

30. Labouvie-Vief, G. (1994). *Women's creativity and images of gender.* In B. F. Turner & L. E. Troll (Eds.), *Women growing older:* Psychological perspectives, pp. 140-168. Thousand Oaks, CA: Sage.

31. Levickaitė, R. (2011). *Creative Industries: Classification Systems and Models in the Context of a Creative Economy of the 21st Century. Kūrybinės industrijos: klasifikavimo sistemos ir modeliai XXI a. kūrybos ekonomikos kontekste.* Science – Future of Lithuania, 3, 21-27.

32. *Lietuvos švietimas skaičiais* (2012). Interaktyvus. Prieiga per Internetą adresu: http://www.sac.smm.lt/images/file/e_biblioteka/Lietuvos%20svietimas%20skaiciais%202012_Bendrasis%20ugdymas_pdf.

33. McCrindle, M.; Wolfinger, E. (2010). *The ABC of XYZ: Understanding the Global Generation.* Australia: University of New South Wales Press Ltd.

34. Patton, M. Q. (2002). *Qualitative research & evaluation methods. Thousand Oaks (Calif.) [etc.]: Sage Publications.*

35. Petrulytė, A. (1995). *Kūrybiškumo ugdymo aktualijos*. Vilnius: Leidybos Centras.

36. Petrulytė, A. (2001). *Kūrybiškumo ugdymas mokant*. Vilnius: Presvika.

37. Petrulytė, A. (2007). *Vidurinės mokyklos sustiprinto mokymosi profilio mokinių kūrybiškumo ypatumai*. Pedagogika, 86, 90-98.

38. Plucker, J. A.; Beghetto, R. A.; Dow, G. T. (2004). *Why Isn't Creativity More Important to Educational Psychologists? Potentials, Pitfalls, and Future Directions in Creativity Research. Educational Psychologist, 39(2), 83–96.*

39. Ragulienė, L. (1999). Jaunųjų fizikų mokykla "Fotonas" - moksleivių kūrybingumo ugdymo priemonė: rankraštis: daktaro disertacija: socialiniai mokslai, edukologija (07S). Šiauliai.

40. Reis, S. M. (2002). *Toward a Theory of Creativity in Diverse Creative Women*. Creativity Research Journal. 14, Nos. 3 & 4, 305-316.

41. Reuter, M.; Panskepp, J.; Schnabel, N.; Kellerhoff, N.; Kempel, P.; Hennig, J. (2005). *Personality and biological markers of creativity. European Journal of Personality, 19(2), 83-95.*

42. Rowlands, S. (2011). Discussion Article: Disciplinary Boundaries for Creativity. *Creative Education. 2.* 47-55.

43. Sak, U.; Ayas, M. B. (2013). *Creative Scientific Ability Test (C-SAT): A new measure of scientific creativity. Psychological Test and Assessment Modeling, Volume 55 (3), 316-329.*

44. Simonton, D. K. (2000). *Creativity: Cognitive, Personal, Developmentas and Social Asopects*. American psichologist, 55, 151-158.

45. Stankevičienė, K. (2001). Ikimokyklinio amžiaus vaikų kūrybingumo ugdymas tautodaile: rankraštis: daktaro disertacija: socialiniai mokslai, ugdymas (edukologija) (07S). Vilniaus pedagoginis universitetas, Vilnius.

46. Sternberg, R. J. (2003). *The psychology of problem solving*. Cambridge; New York: Cambridge University Press.

47. Sternberg, R. J.; Lubart, T. I. (1996). *Investing in Creativity*. American Psychologist, *57*, 7, 677-688.

48. Sternberg, R. J.; Lubart, T. I.; Kaufman, J. C.; Pretz, J. E. (2005). *Creativity.* The Cambridge handbook of thinking and reasoning, ed. K. J. Holyoak, G. Morrison. New York: Cambridge University Press, 351-369.

49. Stoltzfus, G.; Nibbelink, B. L.; Vredenburg, D.; Thyrum, E. (2011).*Gender, Gender Role, and Creativity. Social Behavior and Personality: an international journal, 39 (3), 425-432(8).*

50. Szobiova, E. (2006). *The five personality dimensions in relation to creative thinking of adolescents. Studia Psychologica, 48(3), 241-249.*

51. Torrance, E. P. (1971). *Creativity in Communication with Young Children.* Georgia University.

52. Torrance, E. P. (1974). *Norms-technical manual, Torrance tests of creative thinking.* Bensenville, IL: Scholastic Testing service.

53. Torrance, E. P. (1977). *Creativity in the Classroom: What Research Says to the Teacher.* Washington D. C.: Nationat Education Association.

54. Torrance, E. P. (1987). *The nature of creativity as mainifest in its testing. The blazing deive: The creative personality.* Buffalo, N. Y, 43-76.

55. Torrance, E. P. (1995). *Insights About Creativity: Questioned, Rejected, Ridiculed, Ignored.* Educational Psychology Review, 7 (3), 313-322.

56. Torrance, E. P. (1998). *Torrance tests of creative thinking: Norms-technical manual: Figural (streamlined) forms A & B.* Bensenville, IL: Scholastic Testing Service.

57. Treffinger, D. J.; Young, G. C.; Selby, E. C.; Shepardson, C. (2002). *Assesing Creativity*: [A Guide for Educators]. Research Monograph Series. National Research Center ont the Gifted and Talanted, Stors, CT: University of Connecticut, University of Virginia, Yale University.

58. Vaicekauskienė, V. (2009). *Kūrybingumo (ne)ugdymas mokykloje.* Vilnius: Petro ofsetas.

59. Wolfradt, U.; Pretz, J. E. (2001). *Individual differences in creativity: Personality, story writing, and hobbies.* European Journal of Personality, 15(4), 297-310.

60. Стернберг, Р. J. (2002). *Практический интеллект.* Санкт-Петербург: Питер.

Technological Education as a Platform for Developing Global and National Identity

Živilė Sederevičiūtė-Pačiauskienė

Abstract
The paper reviews the curriculum of technologies at Lithuanian schools of general education and its correspondence to the Lithuanian and European strategic and legal documents. The study of technologies today is relevant not only because it aims to develop the learners' technological literacy and creative abilities, but also because it serves as a platform for developing nationality identity and tolerance for other cultures. The curriculum content of technologies is appropriate for the implementation of goals of globalization and has to be developed envisaging the acquisition of learners' multicultural competence. The objectives of technological education meet the provisions presented in Lithuania's Progress Strategy (2012) – to reinforce the citizens' historic awareness and self-respect by promoting a variety of cultural and artistic expression, updating the national cultural and educational programmes as well as the repertoire and symbolism of public state events, and by guaranteeing the preservation of objects of cultural and national heritage and their skillful use for public needs. The curriculum of technologies corresponds with the Programme of the Principles of Lithuanian Ethnic Culture for schools of general education (2000) and this subject remains the main practical teaching subject which upholds ethnoculture, crafts and develops national identity.

Keywords: national identity, globalization, technological education

Introduction

Lithuania's Progress Strategy "Lithuania 2030" (2012) names the following problems of a modern society: identity crisis, the power of stereotypes, emigration, the reticence of society, the lack of tolerance and trust, the lack of concern about one another and the surrounding environment, and a feeble belief in the country's success. The document also states that "the changes of Lithuanian citizens' culture, thinking and behavior as well as the values of the society will determine the successful development of the country and will help to prepare for a resolute acceptance of the challenges of global competition". In the modern society people face a challenge to be citizens of the global society who are tolerant, able to manage the unpredictable processes of change and competition, and who are at the same time able to preserve their Self and national identity. Education in Lithuania also faces the same challenges: "In the age of globalization it is important to reconsider one's national identity and perceive what unites us (including the emigrants from Lithuania and immigrants) and bonds with the State of Lithuania" (The National Strategy of Education of 2013-2022).

Identity, defined as "being oneself", "staying who one is", today is more topical both on the personal and national planes than it has ever been. The modern lifestyle, affected by globalization, enables young people to cognize other cultures which often seem more attractive and so leave no place for a person's cultural identity. A person's cultural identity is related to the cultural environment where one lives and matures as a personality, where one develops and learns the achievements of the culture he bears, consumes and creates (Kuzmickas, 2011: 66). As Kavolis said in 1993: "People always defined and were defined primarily with the most evident entities: their dependence to a country, traditions, religion" (Kavolis: 89). However, in the global society people can live in various countries, hence, there appears a natural threat for their identity: what do I belong to, what country am I a citizen of? Aiming to resemble a global citizen, a person gradually loses his/her national identity. Paradoxically, the global world is interested only in people who distinguish themselves by their national identity. "Globalization activates a person's opening up to the world and expands the horizons of his/her relations and actions, however, at the same time it alienates a person from the higher values and ideals which function as the main sources of personal maturity. <...> In the context of heterogeneous challenges of globalization there first of all appears national culture which experiences numerous significant transformations due to the rapidly spreading multiculturalism (Aramavičiūtė, Martišauskienė, 2010: 19). Therefore, it is "important to reconsider one's national identity and realize what unities people in the modern world. We should find strength in our history and be a modern society that is open to the present (Lithuania's Progress Strategy "Lithuania 2030", 2012).

The goal of the paper is to reveal the correlation between the expressions of global and national identity in technological education.

The object of the study is global and national identity.

The paper uses the ***method*** of content analysis of education documents and academic literature.

Typically globalization is understood as an economic, political and cultural process which changes the role of most nations and states in relation to the global markets, conventions and traditions. Under the processes of globalization the formation of a multicultural society becomes an inevitable phenomenon in modern countries. In the multicultural society there grows the significance of the global education which has a number of benefits (Salo-Lee, 2012: 75). Nevertheless, on the plane of research discussion there appears insight into the negative aspects of globalized education. The recent attempts to analyze and understand the multiple and complex influence of globalization on education show that there is no single, unanimous treatment of the influence of the process of globalization on teaching and learning at school and other institutions of education (Carnoy, 1999; Burbules & Tores, 2000; Hargreaves, 2003).

The challenges of the changing society are primarily faced by the school and its community. The traditional education systems of different nations are equalized by European standards and the shared systems of learner achievement assessment. Aiming to compare students' achievements, there are different surveys conducted (e.g. Trends in International Mathematics and Science Study, TIMSS), the data are compared (Program for International Student Assessment, PISA), and that inevitably leads to the unification of achievement evaluation systems (Sahlberg, 2009). According to Hargreaves (2003), the need to compare results is one of the strongest pretexts to reform schools in the majority of countries, including most countries of transitional economy. Hence, globalization is becoming the force affecting the social reforms of nations and states since the sector of education adjusts to the new flexible global environment.

The Objectives of Lithuanian Technological Education

Aiming to guarantee the quality of education and to answer the society's needs and the main strategic documents of Lithuania, general education is facing great demands. The teaching subjects are being continuously updated as they have to implement both the goals of global education and the national priorities. One of the general education curriculum subjects is technologies and the improvement of its content is especially dependent on the society's needs and development tendencies. The purpose of technological education is to develop the general and technological competences – the whole of knowledge, awareness, skills and attitudes which help students acquire the backgrounds of technological literacy, essential for every person in the continually changing socio-cultural environment (General Programmes of Secondary Education, 2011). Technological education has an objective to provide students with the conditions to cognize the development of technologies in the historical context and the interrelations of technologies with other subjects, to assess the influence of technologies on culture and the change of technologies in the social environment. Technological education is conceived as the whole of a creation and production process (Primary and Basic Education General Programmes of Technologies, 2008). Thus, the objectives of technological education answer the provisions stated in Lithuania's Progress Strategy (2012), namely, to enhance the citizens' historic awareness and self-respect by promoting a variety of cultural and artistic expression, by updating the national cultural and educational programmes as well as the repertoire and symbolism of public state events, and by guaranteeing the preservation of objects of cultural and national heritage and their skillful use for public needs.

The conception of technological education, the general programme of technologies and the education standards were endorsed in Lithuania in 2003 and replaced the former subject of household chores and culture. The range of the subject of technologies was defined; compulsory programmes were created for primary and basic education. Technological education in Lithuania originates from vocational education which is analyzed by researchers as a precondition for upholding traditions, vocational culture and aesthetics (Galkauskas, 2001). Even though in the last two decades in the academic discourse there has developed a new approach reflecting the orientation of technological education to the reality of post-industrial trend which has a clear direction towards technological creation (Statauskienė, 2006), the school subject of technologies has remained the plane for the development of national traditions which upholds the national heritage and ecological awareness. These competences are also included in the technology teacher training programme at Lithuanian University of Educational Sciences – *technology education*. One of the goals of the programme is "to be able to create a safe physical and psycho-emotional environment of technological education that would support the students' emotional, social, intellectual and spiritual development and promote upholding the national cultural traditions, moral responsibility, respect and tolerance for other cultures". Students become familiar with the national traditions, the cultural heritage, and the acquired competences become the background for teaching technologies.

Nevertheless, it has to be noted that the subject of technologies, which has a clear productive – creative orientation, is losing the function of developing learners' national identity and upholding the cultural heritage, and, in comparison with the curriculum that was in force up to 2003, now is more oriented to the development of consumer – producer competences rather than cultural traditional values (General Programme of the Integrated Course of Technologies and Methodological Guidelines). Due to the rapid development of the society as well as scientific and technological achievements, the second trend of technological education is distinguished as the new teaching paradigm of the 21st century which aims to determine the target skills and the needs of the society, to raise and optimize ideas and implement them. Hence, it can be said that design and technologies have a common goal – to create and produce. This tendency can also mean the total elimination of the aspect of national identity within the curriculum which is observed in the results of foreign research on the change of the curriculum of analogous subjects: the content of the subject of technologies is eventually "purified" and the craft dimension disappears since it does not answer the spirit of the time. The content of technological education loses the family and interpersonal relations in the multicultural society, folk traditions and customs (Sederevičiūtė, 2005). As it is stated in the Lithuanian National Education Strategy of 2013-2022, "the need for cultural self-expression is now creating significant pressure on formal education which is frequently unable to answer all the needs and expectations without overloading the formal curriculum". This a significant point of reference for the revision of the content of education and the development of the concept of technological education taking into account the state policy and the need of the modern society to preserve something most nations are desperately trying to retrieve – namely, national identity.

The Development of National Identity in Technology Classes

According to Kuzmicas (2011), a person's identity is defined not only by the way he conceives and sees himself, but also by a number of objective factors which depend on the socio-cultural environment the person lives in. In Lithuania, which has been affected by the processes of globalization, there no longer is a clear Lithuanian identity, "there appear a lot of individuals of dual or multiple cultural identity and their increase is conditioned by migration of various directions" (Kuzmickas, 2011). According to Lukšienė (2000), "some people fail to creatively master and implant the surrounding information on the basis of the model of their own culture, and thus they submit to the influence of the worldview brought by the new information. The core of their national culture shrinks and they become either indifferent or alien to their culture. If the number of such members of the society grows, the national culture starts declining, changing, it stops following its inherent laws, and becomes an echo of the spirit of other cultures" (Lukšienė, 2000, 19-20). Therefore, with the increasing establishment of the processes of globalization and universality, the nation, nationality and identity are becoming the issues of new contemplations, interpretations, discussions and even clashing hypotheses (Aramavičiūtė, 2005). The visible side of national identity which manifests itself in the forms of tradition, crafts, culinary heritage, has traditionally found its place in the general education programmes of technologies in Lithuania.

The Lithuanian Republic Law on the Principles of State Protection of Ethnic Culture (1999, No. 82-2414) defines the significance of ethnic culture for the preservation of national identity: "ethnic culture is the essence of the nation's existence, survival and stability, the background of the national culture". The objectives of the state protection for ethnic culture presented in this law emphasize the educational role of ethnic culture: "to develop a personality with a mature national self-awareness by integrating ethnic culture into the system of education". In the Programme of the Principles of Ethnic Culture for Schools of General Education (2000), approved by the Ministry of Education and Science of the Republic of Lithuania, the following goals of ethnic culture are indicated: to develop national and civic consciousness, to promote the creative succession of traditions, to form the moral norms and environmental skills, and to develop aesthetic culture. Thus, ethnic culture should be given a lot of attention in general education; nevertheless, the success of implementing these educational objectives on the personal level of students will depend on the specific tools and forms used to impart those target values. The curriculum of technologies covers the Programme of the Principles of Ethnic Culture for Schools of General Education (2000) and remains the main practical subject which upholds ethno-culture, crafts and develops national identity.

Technologies are a subject which aims to develop the practical skills of students. In the first (5th - 6th forms) and second (7th – 8th forms) cycles of the curriculum learners look for information about the historical, cultural traditions of the national crafts and trade, while in the third cycle (9th – 10th forms) students are provided with a possibility to choose subjects from all the spheres of general education programmes, they are allowed to test their possibilities in different spheres of learning and activity, and to familiarize themselves with real workplaces (The Guidelines for Enhancing the Possibilities of Choosing the Learning Programmes for 14-19-Year-Old Students, 2008). This is especially important for students who have difficulty achieving a satisfactory level of achievements in other subjects (mathematics, Lithuanian, history, foreign languages, natural sciences, etc.). They are provided with a possibility to choose modules of these subjects with orientation to practical activity (General Programme of the Integrated Course of Technologies and Methodological Guidelines, 2007). Hence, the subject of technologies is not only an educational plane for the technological process of creation, but also a significant point for imparting values and education where the producer's and consumer's values and interests meet, and the business ethics is revealed. This is the context where a person develops by choosing to consume in an educated way and create wisely. Technologies still are and should remain the environment for the intercultural development of national identity. There appears one of the most complicated objectives – to determine the principle values of every learner, and respect them while trying to adjust the curriculum according to the needs of the majority on the personal, institutional and national levels. Answering the national strategic goals, the subject of technologies has to introduce learners to cultures existing in the society, initiate a deeper cognition of cultures as well as the promotion of tolerance, understanding and respect. The school guarantees the transfer of national values through educational projects and events which reveal Lithuanian traditions and their place among the other cultures. Thus, this is how a cultural environment, a civilizational plane is created, which, according to Kuzmickas (2011) helps the individual to mature as a personality, to develop and educate himself in culture, "the culture which he is a bearer, consumer and creator of". When the society and its expectations change, the intercultural aspect becomes an integral part of activity at the school which guarantees equal possibilities, where teachers and students are encouraged to be interested in each other, to create a sustainable and democratic community, i.e. to create intercultural relations and develop an intercultural dialogue. It is the dialogue whose pivot is not rational arguments, but rather respect for others and otherness that is the key precondition for interculturalism (Bielskienė, Duobienė, Tamulionytė, 2012).

The development of learners' national identity in the context of technological education requires teachers or their trainers to uphold the national uniqueness by appreciating the traditional crafts, techniques, emphasizing the importance of the national heritage, developing pride in their national uniqueness, by learning tolerance together, by emphasizing the historical aspect of technologies and crafts, by responsibly developing their own and their learners' culture of consumption. These are the goals listed in Lithuania's Progress Strategy: "to develop the society's ecological consciousness, promote sparing consumption and a responsible approach to the development of economy <...> we have to develop technologies which would cause minimal damage to the environment and would guarantee a sustainable, sparing growth". The technological education still has a topical goal to develop the correct notion of quality of life.

The Challenges of Globalization for Technological Education

The general programmes of technologies also answer the challenges of globalization – getting ready for the labour market. In the 9th - 10th forms, learners, who aim at motivated decisions concerning what technologies to study and where to continue their studies, can test their creative and practical abilities, familiarize themselves with technological processes, professions, branches of industry (in order to increase the possibilities of choosing learning programmes). The general programmes of technologies for secondary education (2011) state that in order to answer the challenges of the modern society, learners have to understand the influence of people's decisions, they have to learn to "function" in the *system nature-man-material environment* <...>, perceive the influence of technologies on nature, people and the material environment as well as to understand how one should assess the benefit of technologies and their possible negative effect for a person's and the society's health, security and welfare.

Nevertheless, the curriculum of technologies covers not only technological processes and their influence on the environment, but also the social consequences of globalization. Technologies, which are conquering the world markets and young people's minds, can become a point of reference and a practical foundation for one's knowledge of cultural diversity. They can encourage learning from each other, teaching sensitivity to the problems of the period of globalization. While teaching technologies, we have to discuss about the issues of global ethics, sustainable development, as well as to instill respect and trust in one another. The development of a multicultural society has to be viewed not as a background for problems, but as the appearance of possibilities for intercultural education. Such possibilities are opened up by the growth of new technologies: while sitting at a school desk, learners can cognize other cultures, their lifestyles, can speak with students from various countries about different issues, including globalization and national identity.

If we speak about the teachers' technological literacy and teachers' competences to develop it, we must also speak about the teachers' and learners' cultural literacy as a background for the development of intercultural competence. This is the key objective of global education accompanying multiculturalism which is a new phenomenon for the Lithuanian society and which will seriously challenge the Lithuanian education. The multicultural environment can be conceived as a possibility to learn from one another, as a plane for creativity development. Nevertheless, this environment can be a cause of problems or a conflict. Intercultural education aims to guarantee a respectful and socially correct relation of cultures, it prioritizes common human ethics rather than cultural differences (Van Dreil, Gundara, 2012).

Teachers have to be able to convey and interpret the influence of culture on creation, to emphasize its significance, and direct the learners to other cultures, teach them to learn from others and otherness. The modern teacher cannot do without the intercultural competence which was defined by Wood, Landrey and Bloomfield (2006: 20) in the following way: "The ability to read, understand and find the significance of diverse cultures and, as a consequence, to be able to evaluate, compare and decode the varied cultures that are interwoven in a place. It allows one to attribute meaning and significance to anything seen and produces. It is a form of cultural capital that enables us to act sensitively and effectively in a world of differences"(Wood, Landrey, Bloomfield (2006; 20)

The most general points of reference for intercultural education are UNESCO Guidelines for Intercultural Education (Guidelines for Intercultural Education, UNESCO, 2006) which emphasize the cultural dialogue encompassing such segments of education as "knowing", "acting", "interacting", "being". The guidelines speak not only about the upholding of cultural heritage, but also about the cultural diffusion. Hence, the teacher faces an educational paradox: for one thing, teachers have to develop a citizen of the global society, a member of a homogenous society, for another, they have to pay more attention to the recognition of cultural diversity and differences (Bielskienė et al., 2012). Even though some authors define the current educational and pedagogic situation in the context of national identity and multiculturalism as transitional, "fluctuating" between two extremities (Duoblienė, 2009), in general education the subject of technologies is just the plane for the intersection of these two extremities and the expression of their content.

The subject of technologies enables learners to fulfill themselves in their creation, discover and express themselves as personalities and as members of the country / culture they identify themselves with. Creation, crafts always reveal a nation's exclusivity, irrespective of the modern tones a creation might possess. In this way, by "identifying oneself with certain spheres of culture, one integrates into the culture of his/her country or nation and there finds his/her place and role either as a creator or as an amateur, a consumer, a conformist or a rebel" (Kuzmickas, 2011). The mission of technologies is to arouse a wish to look for connections or discrepancies among these roles as well as to help learners to find their own selves and establish themselves with that clearly defined identity in the global culture.

Conclusions

The subject of technologies is still relevant not only because of its goals to develop the learners' technological literacy, but also because it is a platform to develop one's national identity and tolerance for other cultures. The curriculum of technologies is suitable for the implementation of goals raised in the context of globalization, and it has to be developed in order to cultivate the learners' intercultural competence. The objectives of technological education answer the provisions presented in Lithuania's Progress Strategy (2012) – to enhance the citizens' historic awareness and self-respect by promoting a variety of cultural and artistic expression, updating the national cultural and educational programmes as well as the repertoire and symbolism of public state events, and by guaranteeing the preservation of objects of cultural and national heritage and their skillful use for public needs. The curriculum of technologies corresponds with the Programme of the Principles of Lithuanian Ethnic Culture for schools of general education (2000) and this subject remains the main practical subject which upholds ethnoculture, crafts and develops national identity.

References

1. Aramavičiūtė V. Tautinis tapatumas ir jo kaita vyresniame mokykliniame amžiuje. Pedagogika 2005, T. 79, p. 40-45.

2. Aramavičiūtė V., Martišauskienė E. Leono Jovaišos auklėjimo mokslas: fundamentalumo ir inovatyvumo sinergija. Acta Paedagogica Vilnensia, 2010, T. 25, p. 19- 29.

3. Bielskienė J., Duoblienė L., Tamulionytė E. Tarpkultūrinis ugdymas Lietuvos mokykloje: į pagalbą mokytojui. Methodological aid. UAB Lodvila, 2012. 54 p.

4. Burbules N., Torres C. A. Globalization and education: Critical perspective. New York: Routledge. 2000, p. 1-27

5. Carnoy, M. 1999. Globalization and educational reform: What planners need to know? Paris: Unesco and IIEP.

6. Duoblienė L. Ar pavyks išvengti hibridinio tapatumo formavimo(si) informacinėje visuomenėje? Acta Paedagogica Vilnensia, 2009. T. 23, p. 79-91.

7. Galkauskas J.K. Technologijų dalyko patrauklumas. Pedagogika. 2001. T. 55, p. 45-48.

8. Guidelines for Intercultural Education, UNESCO, 2006. Available from: http://unesdoc.unesco.org/images/0014/001478/147878e.pdf.

9. Hargreaves A. Teaching in the knowledge society: Education in the age of insecurity. New York: Teachers College, 2003. 230 p.

10. General Programme of the Integrated Course of Technologies and Methodological Guidelines. Švietimo aprūpinimo centras, 2007, Vilnius. Available from: http://www.upc.smm.lt/ekspertavimas/biblioteka/failai/Integruoto_gramos_mr.pdf

11. Kavolis V. Lietuviai, komunizmas ir tautinis charakteris, 1993. In: Kuzmickas Individo kultūrinis tapatumas. Problemos, 2011, p. 65-74.

12. Kuzmickas B. Individo kultūrinis tapatumas. Problemos, 2011. T. 80, p. 65-74.

13. Lithuania's Progress Strategy "Lithuania 2030". Available from: http://www.lrv.lt/bylos/veikla/2030.pdf

14. The Lithuanian Republic Law on the Principles of State Protection of Ethnic Culture. Endorsed on September 21, 1999, by decree No. VIII-1328, by the Seimas of the Republic of Lithuania. Available from: http://www3.lrs.lt/pls/inter3/dokpaieska.showdoc_l?p_id=370909

15. Lukšienė M. Jungtys. Vilnius: Alma Litera, 2000. p. 559

16. The Guidelines for Enhancing the Possibilities of Choosing the Learning Programmes for 14-19-Year-Old Students. Endorsed by the Minister of Education and Science of the Republic of Lithuania on March 15, 2008. Decree No. ISAK-715. Available from: http://www.upc.smm.lt/ekspertavimas/biblioteka/failai/2008-03-15-ISAK-715.pdf

17. Primary and Basic Education General Programmes of Technologies / Technologies, 2008 Available from: http://www.upc.smm.lt/ekspertavimas/biblioteka/failai/programos.pdf

18. Salbergas P. Mokymasis ir globalizacija. (online document).

19. Salo-Lee L. Towards Cultural Literacy. Education for Global Responsibility. Finish perspectives. Publication of the Ministry of Education 2007: 31. Ed. By T. Kaivolan, M. Melen-Passo, p. 74-82.

20. Sederevičiūtė Ž. Technologijos ar namų ūkis? Dalyko raidos Europoje lyginamoji analizė. Pedagogika. 2005. T. 76, p. 67-72.

21. Statauskienė L. Technologinis ugdymas. Realių pokyčių galimybės. Pedagogika. 2006. T. 82, p. 85-95.

22. The National Strategy of Education of 2013-2022 (project). Available from: http://www3.lrs.lt/pls/inter3/dokpaieska.showdoc_l?p_id=438859
23. Van Dreil B. & Gundara J. International association for intercultural education. In. J. Banks. Encyclopedia of diversity in education. Thousand Oaks, CA: SAGE Publications. 2012, p. 1237-1240.
24. General Programmes of Secondary Education, 2011. Available from: http://www.upc.smm.lt/suzinokime/bp/2011/Technologijos_8_priedas.pdf.
25. Wood P. C., Landry & J. Bloomfi ed. Cultural diversity in Britain. A toolkit for cross-cultural cooperation. Josept Rowntree Foundation, York. 2006; 20. p. 75.

Being Oneself: Changes in the Lithuanian Prospective Teachers' National Identity Building

Ona Tijūnėlienė, Janina Bukantienė

Abstract

The article aims to reveal the significance of formation of the contemporary national identity of the prospective teachers in Lithuania in specific interactions of personal life experiences and social and cultural contexts. The national identity is defined as a constantly changing and regularly revised multi-layer construct. One of important factors in the change of the national identity is considered to be a socio-cultural context of higher school which creates prerequisites for the reflection of the experience accumulated in the family, comprehensive school, and university, and for its transformation and synthesis.

A qualitative empiric survey was conducted. The applied research method was a non-structured biographical interview. The interpretation of the obtained outcomes disclosed the trends of the change in the national identity of Lithuanian prospective teachers preconditioned by the holistic self-training of an individual.

Keywords: The context of national identity building, social and cultural environment, self-building, trends of change

Introduction

The necessity of analysis of the issues of national identity in teacher training is caused by the concern about Lithuanian culture being alive, so that the future generations, on the basis of that culture and by its further fostering, could create a meaningful and dignified life for themselves. The survival of viable Lithuanianness, Lithuanian culture, and being oneself in any historical culture, including the present, could never have been and can not be taken for granted: it calls for conscious and purposeful efforts of all the society, and especially teachers. Under the conditions of contemporary globalization, two trends in the behaviour of Lithuanians increasingly stand out: with the opening of the door to the diversity of the world cultures, one part of the Lithuanians is becoing more indifferent to their own identity and tends not recognize it as a value, while the other part tends to withdraw just into their own culture. Both post-modernist trends pose a threat for the preservation of national identity. National education is understood by the authors as an individual's introduction into the world of the national language, culture, and history, and when discussing the viability of the national identity, they assume that the formation of modern identity can be only ensured by the common efforts of the family and the school, in order to implement their vocation: to actively foster the culture (of the state, the language, traditions and customs, respect, and harmony) and to set a personal example to students when pursuing the goal. Focused and spirited activity of public education and training institutions would contribute to the fostering of the kind of national identity that would enable an individual of the 21st c. to grow free, responsible, and creative, and would encourage the appreciation of one‘s own culture and care for its continuity [8].

Public attention to the need of the building of national identity is drawn by the education documents of the European Union [6] and the Republic of Lithuania. The updated version of *General Curricula of Secondary Education* (2012) emphasizes the totality of culture and creates the image of Lithuania‘s multiculturalism. The aim of implementation of the said document is to strengthen the Lithuaniannism, to cultivate patriotism, and to raise the nation‘s self-esteem. Therefore, teacher, when carrying out the social commission, should actively help students to get to know the objective values of the education content so that they would become their integral part and that young people would be willing to live in accordance with them and would accept the responsibility of being successors of culture, as a link between the past and the present [37].

The latest national document *The National Education Strategy 2013-2022* [35] places emphasis on the tasks of preservation of the nation‘s integrity, Lithuania‘s modernization, and its viability whose ways of execution might be the building of the identity of a modern nation, of self-respect, and self-confidence. The documents accentuate that Lithuania needs patriotically-minded people who have respect for national values as creators of its future.

The isssue of national identity building in post-modernist Lithuania from the theoretical and practical viepoints have been analyzed by a number of researchers (philosophers and educologists): R. Grigas,1995 10, 2011, 2013 [12, 13, 14]; B. Kuzmickas, 2001 [22]; E. Krukauskienė, 2003 [21]; V. Pruskus, 2005 [30]; J. Srėbalius, 2006 [33]; L. Duoblienė, 2011 [8], and others. The national attitudes of academic youth as potential creator of the future society have been discussed by M. Barkauskaitė, 2008, 2009 [2,3]; O. Tijūnėlienė, J. Bukantienė, 2010 [34]. Teacher‘s mission to rekindle national identity has been examined by E. Martišauskienė, 2007, 2012 [24, 25].

The research in the language as the basis of identity have been conducted by foreign researchers J. W. Edwards, 1983 [9], R. G. Rumbaut,1994 [32]; J. S. Phinneyet al., 2001 [28]; and J. Argelaguet, 2006 [1] et al. N. Piattoeva, 2009 [29], given the transformation of the political context in Finland and Russia, carried out comparative research in the situation of national and civic education in those countries. D. Faas, 2011 [11] compared the content of geography, history, and civic education in Greece, Germany, and England, while A. Osler, 2011 [27] made public the concepts of the relationship of national and civic education in European countries and the data of empiric research in the implementation of the civic education programme in England.

The issues of the building of national identity of teachers trained in Lithuania have so far received little attention. Therefore, **the problem of the present research** is formulated as follows: what are the factors of change in the national identity of prospective teachers trained in Lithuanian universities and what are the trends of its building.

The object of the research

The experience of prospective teachers' national identity building.

The aim of the research

To reveal the factors of building the national identity of prospective teachers trained in Lithuanian universities.

The objectives of the research:

1. To discuss the nature of national identity.
2. To bring out the role of external factors in the change of prospective teachers' national identity.
3. To identify the trends of change in the building of prospective Lithuanian teachers' national identity.

Theoretical and methodological attitudes

1. "If you do not know, and do not want to know, the history of your own nation, if your heart does not feel its painful echo, you are losing part of your human value. You are no longer interesting either to an Irish, or an Englishman, or a Muslim from Dubai. Your perception and the range of your feelings become poor. You fail to understand the true meaning of your human existence"[R. Grigas].
2. "Tradition will never become obsolete, as it is the expression of the nation's creativity which arises spontaneously, and is recreated spontaneously, and therefore every time it is relevant and viable. <…> Tradition is nation's experience embodied in the shapes of culture" [A. Maceina].
3. Natural sciences "explain" things and processes as recurrent "cases" in the causally linked nature, while spiritual sciences try to "understand" historically unique things and events related by multiple spiritual meaningful links" [A. Halder].

4. When reflecting on one's biographical experience in the fields of existence and culture, one considers and, by means of critical analysis, becomes aware of, and makes meaningful, the content, forms, and presumptions of one's own further activity and the structure and characteristics of one's own spiritual world. Reflection both allows one to know oneself and to improve one's own activity and relations with the environment" [L. Jovaiša].

'The conception of national identity

The most characteristic feature of an individual is his identity and a simultaneous need is to have identity. The nature of identity from the ontological viewpoint was disclosed by Lithuanian philosopher, classic of pedagogy Antanas Maceina. He noted that a human being was always identical to himself, that everything around us was changing, and everything was changing inside us, but we ourselves alawys remained the same. "No change is able to turn me into somebody else. A personality sticks through the changes of nature and the human nature as a rock witnessing one's own identity. <…> Identity is an individual's state <…>, and a normal state of an individual is being, but not becoming. And the one who perceives the becoming must be non-becoming. I only know that everything changes both next to me and in me, as I do not change. I am the bearer of the change, or its subject, but in no way a changing being or the object of the change" [23].

In science, several approaches to the nature of nationality are known. The most common one is the interpretation of the relationship between the language and nationality, and it is related to the names of the representatives of German Romanticism Johann Herder, Vilhelm von Humboldt, and Johann Fichte. The conceptual arguments of the interpretation are the following: the relationship betwen the language and the nationality is essential; the language is the spirit of the nation which provides it with life; the language ensures the development of national culture; the identity of the nation is revealed via the language; and the language by its looks is a sign of the nation which distinguishes it from other nations. When an individual masters the language of another nation, it acquires the identity of that nation; and the nation will survive as long as it keeps its language alive.

The German Romanticism explained the language and the nationality as essential attributes of the human community that distinguished it from other communities. The language and the nation were an unchanging foundation of the community. The community could fully develop only on its basis. The approach resulted in the understanding of an individual's relationship with his own nation:

once an individual became a member of the nation, he could not abandon it, as he possesed the features of the nation. To deny one's nationality would mean to offend oneself. In accordance with that approach, nationality acquired a moral value [26].

Another interpretation of the nature of nationality was affected by the evolution theory and an approach that emphasized the evolutionary change: nationality was not a stable thing, but an ongoing process which depended on the self-consciousness of people making up a national unit.The self- consciousness could be based on different factors: the language, the religion, or people's wish to stay the members of their community. The natural change of the national self-consciousness was stimulated not by the fostering of culture or appreciation of the language, but rather by simple pragmatic calculations: the pursuit of economic welfare and social harmony. Thus, in the process of change of national self-consciousness, the possessed identity was not lost, but merely changed. Therefore, the process should not be artificially affected or pushed in the "right" direction, or restrict people's desire to build such a national identity that would best meet their aspirations and desires. The said position of J.R.Edwards was presented by A. Norvilas [26].

Pope John Paul II related identity to the preservation of culture. He raised a question how to preserve the common property of the humanity, the common property of all cultures so that it would keep growing. John Paul II was interested in finding an answer as to how to retain the correct ratio between economics and culture in order to save the good which was greater and more humane from being destroyed by the money society and the growing needs of the economic power. In his speech given to UNESCO, John Paul II outlined his thoughts on the role of culture in human life: man lives an authentic life due to the culture; the culture is a unique way of human existence; culture is what makes man stronger and more important; a nation is a large community of people who are linked by different relationships, and especially by culture; the nation exists for the culture, and the culture exists for the nation, as the culture is a great teacher that makes them to be more important in the community; the nation saves itself from the foreign aggression and self-destruction just by its culture [19]. "What I am saying about the right of the nation to the foundation of culture for the sake of their own future is not an echo of any "nationalism", it just reflects a sound and promising element of human experience and human development. We can talk about independence when a human being becomes increasingly independent. <...> In the UNESCO meeting, my idas about the the building of the national identity via culture reflected the most vital needs of young nations looking for ways to consolidate their independence" [19].

In the contemporary theories of identity, P. Ricoeur, 2001 [31], by developing the ideas of the classics of phenomenology M. Heidegger, 1992 [17], and E. Husserl, 2005 [18], emphasizes the perception of identity in relation to otherness. Therefore, the identity is proposed to be realized not as a static essence of being, not as certain contents, but rather as relationships with other cultures and the forms dicatated by them. Therefore, an individual's identity turns into a reflexive process. It forms by reflecting, by perceiving oneself, by acting: it inspires certain social and cultural activity of an individual that possesses meaning. Conscious construction of identity, to quote R.M.Vabalaite, 2006 [36], presupposes the mastering of spiritual values and promotes the deep thinking and the worldview development.

Research methodology

The research applied the phenomenological strategy of the object cognition whose aim was to investigate the phenomena and to disclose the ways of human experiences. The researcher studied the daily routine of an individual, was interested in the ways of their daily experiences, and tried to understand the experiences. She disclosed her own understanding on the basis of the data collected from the respondents and tried to understand what was common and what was unique. Different narratives of the respondents encouraged to reveal the basic structure of the phenomenon and its specific characteristics. The individual who decided on the said object cognition theory was guided by an assumption that the reality was a social construct, and the object of cognition was the respondent's internal experience which was given meaning and construed in the process of the respondent's interaction with the researcher. The researcher became a co-author of the narrator, therefore, the information collected during the research was the product of the moral experience of both the respondent and the researcher [36].

Data collection method: a non-standardized biographical essay.

The research was conducted in June to September 2013 with the students of Lithuanian universities. 32 final year students, prospective pedagogues, shared their experience.

In the process of research conducting, target homogeneous student group selection was applied. The groups were selected which shared some common characteristics: they had male and female students; all of them were similar in terms of age (20 to 30); all of them took study programmes in the field of educology; all of them had the experience of the personal identity self-training and could disclose whether the family, comprehensive school, or university made an impact on the change in their national identity and could share the experiences of the trends of changes in the investigated phenomenon. The respondents' essays were catalogued, and the numbers were indicated in the text.

In the explanation and summarizing of the research data, attempts were made to consider and present the most typical respondent statements that characterized the experiences of the self-training of national identity in the family, secondary school, and university. Precision and completeness in the case of the investigated phenomenon was tried to achieve by discerning in student narratives the basic meanings by means of isolating oneself from the known theories. Moreover, the attitude was observed that "the qualitative research "encoded sensitiveness", and the latter automatically created validity, as the researcher was in the field ("territory") which was controlled by the respondents (in a figurative sense) more strongly than by the researcher, i.e. the researcher was in the context of the views presented by the respondents" [5].

Research outcomes

The empiric research revealed how the prospective teachers understood the process of the self-building of national identity. In non-standardized biographical essays, one can discern the links found by the respondents between the personal identity (one's exclusiveness, being oneself) and the national identity. In the definition of the national identity, the respondents named a positive relationship of an individual primarily with the national language, then with the culture, history, customs, and responsibility for the fate of the nation, and emphasized the resolution to create and to protect national values. Their responses did not reflect a personal relationship with the land or territory. We may assume that the academic youth recognized a contemporary modern state without boundaries or limits. That is the result of globalization. From that viewpoint, there is a distinct watershed between the present generation of intellectuals (the prospective teacher respondents) and the developers of the Lithuanian national state (V. Kudirka, J. Basanavičius, and other) [25].

The prospective teachers identified several external factors that inculcated the feeling of national identity, i.e. family, educational institutions, and the closest environment, most frequently, the community; the impact of the distant environment (travel and life abroad) was also distinct).

Almost all the respondents (97%) considered family to be a strong factor that made a special impact on national education in childhood. After an in-depth analysis of the family impact, one can identify the division of the responsibility for the building of child's national identity: from one person (mother) to all the kin. *Mum used to be the closest...(7); I was brought up not only by parents and grandparents, but also by godmother...(14); Close relatives – aunts and uncles – all together we attended the most memorable and honourable places of Motherland, such as Kernave, Trakai, the Hill of Crosses ...(9).*

National education in the family manifested itself in different forms and ways. The respondents mainly named socializing in the family and kin, celebrations that got people together, observation of the national customs and Lithuanian traditions, more seldom, more intimate communication in the family and land studies: *I was taught and encouraged to observe traditions and customs during different festivals, we attended ethnic events...(21); I was told about the national history, taken to museums and different events (e.g., St. John's Festival on Rambynas Hill). Grandparents told about their experiences and grievances, their deportation to Siberia, and explained why and how they held their nation dear... (4) I was educated... by stories and folklore (5). Grandparents inculcated values and explained that one needs, and why one needs, to love one's own nation, land, culture, and customs (6). Parents and grandparents made an impact on my national spirit by their stories, memories, behaviour, and customs (7). The adults did not mind speaking in a dialect, I was taught folk songs and tales, as well as love and respect for the Lithuanian language. They encouraged me to express my Motherland-related feelings. My sister and I were taken around Lithuania (even now I am invited) and shown its beauty and exclusiveness (1). Granny and other members of the family knew a lot of ancient Lithuanian folk songs: we would sing them, especially encouraged by my Granny, and she intended to pass that on to her children (3).* The respondents' essays reflect deep respect for the national history of the elder generation (especially of grandparents, to a lesser degree, of parents), the need to educate the historical memory of the children, and to pass on the language (dialect) and the tradition of land studies and custom observation. On the basis of the materials of biographical essays, one can state that the national identity feeling of the elder generation is strong, and the respondents correctly understand the values adopted from the adults and reflect on them.

Among the educational institutions that played a meaningful role in the consolidation of national identity, the respondents named ***kindergarten and school.*** The educational environment, the forms, methods, ways, and educational materials, all the factors were acceptable and important for the respondents. The respondents assess them from the present-day positions, as they already have their own professional pedagogical theoretical and practical experience: *At primary school, the teacher used to relate different assignments to Lithuania, even if they were not in the textbooks, not only during Lithuanian lessons, but also during the lessons of art, technologies, maths, and even physical culture... (25). We used to draw flags and cut from the paper, build our native town from different materials, sing many songs,* *attend the most beautiful places of Lithuania in excursions, and attend events devoted to the important national dates. Each assignment was commented, discussed, and all the conditions were created for us to feel, to experience, and to share (11). The school held different events devoted to the themes of Lithuania, the country, and the nation. The invited guests would speak about old times. We learnt a lot of poems and stories, and had to speak about our own town or village and family customs. We learnt from the materials in the lessons of nature, geography, and history (3). All the school participated in the project "Our Culture and Customs."It revealed the history of our land, our customs, and our, as pupils';values (4). In a number of national and calendar festivals, we used to sing "The National Hymn", we respected the flag. Our teachers used to tell us different stories about the past of the country, book smugglers, and dukes(5). During state festivals, events were held, national songs and hymns, as well as dances, performed. I have a positive view of pupils'mandatory participation: later they start to like it, and motivation appears (6).* The quoted works witness the respondents'evaluative aspect and prove the competences of prospective teachers to think, to analyze, and to assess the former educational process, however, the impact of the former events, lessons, and experiences on the strengthening/ weakening of the feeeling of national identity is missing.

The school is treated by the respondents as an essential fosterer of the national identity. However, in a number of biographical essays, one can feel a formal, external or outer impact of the school: *The school held events devoted to the national festivals (7); Different events, commemorations, and talks during lessons, of course, may have made an impact (9); We had events devoted to the language preservation and commemorations (1); The lessons emphasized what was dear and unique for Lithuanians (17).*

From a number of essays, it becomes obvious that the respondents miss the sincerity of their former teachers in the inculcation of national identity: creativity, consistency, systematic inculcation of national identity, and reaching pupils' hearts seem to have been missing.

The self-building of national identity in ***higher school*** could be differentiated: the respondents primarily emphasized the ***content of the courses.*** They named the content of the lectures of educology, Lithuanian speech and expression, hodegetics, psychology, and folklore. The respondents commented on the impact of the courses on them: *In the taken courses, different relevant themes are touched upon, which are intertwined with the values, history, and responsibility for the nation's culture. When writing a research essay, I enrich my experience and convey my knowledge of the national history (2). In the lectures of some of the teachers, I learn more and more new things about my native land, its people who sacrificed their works, and sometimes lives, for the sake of Lithuania. I am happy to have an opportunity* *to speak and to hear correct Lithuanian (13). We had lectures on folklore and learnt folk tales, songs, chaants, lullabies, and riddles. We read about dialects and ethnic groups (3).* Second, the respondents indicated that the teachers presented the theories of national identity not sufficiently deeply and thoroughly and did not offer thorough analysis of the data: *Only in several lectures we heard about the national identity (dialects, writing, reading...(2): In the scientific literature recommended in some of the lectures (of psychology, professional speech and expression) I read about man's responsibility for the nation and personal national values (5).*

In higher school, the prospective teachers miss activities that would directly involve them and trigger national identity: *We had a fun folklore lecture in which we sang Lithuanian hymns; I was self-building by hosting different events (4). We were encouraged to read books, played dramas and held events, and sang Lithuanian songs, but only for a short time (6).*

The teachers make an impact on students by their personal example and professionalism, as revealed in the biographical essays: *Teachers often tell us about the national traditions and how they themselves respect them, and I adopt that spontaneously and think what I do and what my family does. Then I understand how much I care for my nation and its fate (3). Part of the teachers convey the content of their courses in a very interesting way and reveal the beauty of the native language, as well as its complexity and exclusivity, and studies become even more interesting (1). The greatest impact was made by teachers of the courses of educology and psychology who inculcated the feeling of responsibility for oneself and for the nation in our minds (5).* The analysis of the data presented by the respondents proved that there was a shortage of teachers' personal examples in academic activity, however, the teachers' personal example was very important for students.

Individual essays of students state that teachers purposefully direct students to build the feeling of national identity: *The teachers recommend literature for independent reading (6). Events are held, and teachers remind us how one or another date is important (the year of dialects) (7).* One can assume that most of the respondents believe in-depth studies of national history and language to be the essence of national education.

The feeling of national identity can only be built when an individual participates in the activities, thinks about national values, and reflects on them. By participating in the building of national culture and its fostering, an individual acquires general cultural, professional, and personal competences of national identity. Some respondents "absorb" the studied materials, others prefer to passively "float". As we noticed, the number of active students is not great. Thus, we can identify two respondent groups: first, the open ones whose national activity forms are diverse, they are ready to take the initiative and make use of every opportunity of self-expression, the voicing of their feelings, and infecting others with their ideas: *We held different events related to Lithuania, its population, and presentation of traditions, customs, and values for students from abroad (1). We would participate in different events and made presentations (2). I hosted events for pupils to commemorate the days of March 11, February 16 (3).With Erasmus students we danced Lithuanian folk dances and played games (7). I participate in the events of the village community, dance folk dances, and sometimes sing with the choir in Song Festivals (4). I participated in some nationality-related events: on the press recovery, on the language and dialects (5).* The other group, the passive ones, state: *I am not active in the building and fostering of national culture and seldom attend events related to that. The only case is discussions of Lithuanian sports with other students and support for national sportsmen (1). I do not participate in the Faculty events (the Lent or other), as I am busy or work (2). I did not participate in any cultural event related to the culture of my nation (6).* When analysing the works of the latter respondents, we saw isolation from national values, indifference to them, and the choice of other priorities (not of national or general cultural self-education). We may assume that the process of studies is not full-fledged for them, it is rather formal and one-sided.

Among the ideas in the respondents'reflections, the researchers looked for ones that would witness the change in the feeling of national identity. Around 92% of the students who wrote biographical essays stated that their national identity was getting stronger. We identified the factors of the strengthening: ***communication activity*** (*The feeling is always getting stronger whenever you meet people who talk about that, who are telling you stories, and who are open. It is important to love not only ourselves, but also our nation, nature, and the native land);* ***emigration*** *(The feeling is strengthening. When I look at the emigrating people, I am acquiring strength to to foster our language and culture and to prevent it from extinction, so that my children and grandchildren would speak the language*); the ***dissemination* of *Lithuania's name in the world*** *(The process is promoted by sportsmen's achievements: such a small state, and it is known in the world. We are small, but wonderful and gifted (3); I think my national identity is only strengthening as, especially after having worked abroad, I understood how much important for me our country and our culture is. Abroad, I was proud to be able to tell everybody I was from Lithuania, I was an ambassador of the country, and would tell people about Lithuania (5).*

Some of the respondents stated that their feeling of national identity was weakening. The researchers discerned the following reasons of weakening: passivity, indifference, and globalization: *The feeling is weakening, as I participate in national events increasingly seldom (8). That feeling is weakening, as the young people in my environment pay increasingly less atention to national identity, and indifference emerges; I keep analyzing the history and customs of the nation and find nothing worth world's attention(6).*

To establish the ***trends in the change of national identity***, the researchers looked for key words that would disclose what the knowledge that one belongs to the nation gives to one.

In the biographical essays of the respondents, one may discern several trends of the change in national identity. The most distinct are the trends of *strengthening* of the national identity. The respondents proved it by talking about their spiritual experience and by emphasizing their pride in their identity and in their culture with which they have a close relationship (see Table 1).

Table 1: Arguments for the trends of the change in national identity

Defines the situation	Promotes trust	Develops pride	Unites
I know what I am, I know I have my own language, my native land, and the songs and dances of my nation. For me, that is important.	*For me, it is pleasant to feel belonging to the Lithuanian nation. It strengthens my self-confidence and the awareness of my uniqueness.*	*For my personality, the knowledge of national identity gives pride, as I believe I am a patriot and would not exchange my land for anything.*	*National identity means unity. We all are united by the same land, the same traditions, cultural heritage, and national achievements which echo all over the world. It is good to know that such a small poppy seed country as Lithuania is my home.*

Almost all the biographical essays (98%) distinctly disclose the changes in the self-building of national identity. Having done an in-depth analysis, the authors identified several factors of change and impact: personal maturity, teachers, events, and literature: *The feeling is strengthened by teachers, peers, or people in the environment, as well as literature reading (1); The feeling is strengthened by studies, by a new approach. The same thing can be looked at from different viepoints and each time deeper and stronger. It is also strengthened by different events (4). The strengthening is effected by events and different commemorations (it seems I will always remember what happened or what will happen), but when you look at the photo in which you see that when they were sitting they were weaving or singing you think that even now you can do something of the kind in order to preserve your identity as long as possible (17). An excess of western information makes an impact (28).*

When looking into the respondents'experiences for their present relationship with the national identity, the answers were found to the questions *what they are feeling to be and how they are feeling.* The answers witness a positive atitude of the respondents: *I am, was, and will always be myself. I know for sure what I want and what I intend to achieve. I want to do good and nice deeds and want to start promoting Lithuania's self-image from the roots, i.e. from children.I feel very well as I can prove all the emigrants that I can realize my dreams, wishes, and desires here, in Lithuania. And I am succeeding in that (4). I feel a citizen of my country who loves his native land (5). I see a lot of injustice - it makes me sad, and simultaneously sets new goals and ambitions (27).*

On the other hand, some of the ideas make one doubt about a young individual's self-determination in the case of national identity preservation: he does not know what he is and looks for himself outside his native land: *I am neither a nationalist or a patriot, I just live in that nation, use its language, but do not feel anything exceptional. It is good to hear Lithuanian words, to have one's own customs, however, I am not relating that to my own nation, as, should I be of a different nationality, I would feel the same (6).*

However, the prospective teachers relate their future visions to meaningful works and plan to meaningfully use the accumulated experience, to find their own place in life, to implement their dreams, and to work in the fields in compliance with their vocation. The respondents' visions include also the fostering of the national cultural values, consequently, the preservation of their own national identity: *In the future, I would like to organize events that would encourage young people to love their country, its history, and the present, to love it for what we have now, but not to slander or criticize. (2). I would like to organize cultural educational events which would emphasize how important it is to have national identity (3). For the prospective pupils, I am going to be an example of bright mind and pure heart and I'll try to inculcate the love and respect for their nation in the youngest minds (1). I will try to educate nation-loving young people (2). I would like to start a Lithuanian school somewhere abroad, so that Lithuanian children would not forget Lithuania, its language and customs...(3).*

Conclusions

1. The concept of national identity is changing under the impact of post-modernist trends. National education is understood by the authors of the article as an individual's introduction to the world of the nation's language, culture, and history, while the viability of the national identity and modern identity can only be ensured by joint effors of the family, secondary school, and university aiming to implement their vocation: to be active fosterers of culture (of the state, the language, the preservation of customs and traditions, respect, and harmony) and to build modern identity by creative and attractive means.

2. The change in the national identity of prospective teachers is affected by external factors: family, school, university, and socio-cultural environment. An individual's national identity turns into a reflective process. It forms in the process of reflection, self-awareness, and activity and inspires certain social and cultural activity possessing meaning.
3. The interpretation of non-standardized biographical essays discloses rather shallow and superficial building of the national identity feelings of prospective teachers. One can discern strengthening trends of the building of contemporary national identity. The absence of structure, of the system of individual (self-) education links, and of university teachers' activities to be emulated weakens the national identity of prospective teachers.

References

Argelaguet J. Subjective national identities in Catalonia. *Nationalism and Ethnic, 2006, No. 12, p. 431-454.*

Barkauskaitė M, Tijūnėlienė O. Lietuvos universitetų studentų dvasinių poreikių raiška. *Pedagogika*, 2008, t. 91, p. 7-13.

Barkauskaitė M, Tijūnėlienė O. Akademinio jaunimo dvasinio tobulėjimo galimybės studijų metais. *Pedagogika*, 2009, t. 93, p. 19-24.

Bitinas B., Rupšienė L., Žydžiūnaitė V. *Kokybinių tyrimų metodologija, I dalis.* Klaipėda: S. Jokužio leidykla. 2008.

Bitinas B., Rupšienė L., Žydžiūnaitė V. *Kokybinių tyrimų metodologija, II dalis.* Klaipėda: S. Jokužio leidykla, 2008, p. 122, 62.

Constitution of the European Union [Entry 22]. Roma, 2004.

Data of respondent`s nonstandart biographical essay. *Personal archive of authors. Klaipėda, 2013.*

Duoblienė L. *Ideologizuotos švietimo kaitos teorijos.* Vilnius: Vilniaus universiteto leidykla, 2011.

Edwards, J. R. *The Irish language: An annotated bibliography of sociolinguistic publications 1771–1977.* New York: Garland, 1983.

Eigminienė D. Kiekvieno atskirai ir visų kartu. In: *Švietimo naujienos*. 2012. Nr. 2/313, p. 17-18.

Faas, D. The Nation, Europe, and Migration: A comparison of geography, history, and citizenship education curricula in Greece, Germany, and England. *Journal of curriculum studies Vol. 43 No. 4:* 471-492. (2011).

Grigas R. *Tautos likimas. Vilnius:* Rosma, 1995.

Grigas R. *Tautinė savivoka*. Vilnius: Rosma, 2011.

Grigas R. Nunykstančios lietuvių šaknys ir jų želmenys. *Mokslo Lietuva*. 2013 m. gegužės 2 d., Nr. 9(497), p. 5.

Grigas R. Nunykstančios lietuvių šaknys ir jų želmenys. *Mokslo Lietuva*. 2013 m. gegužės 16 d., Nr. 10(498), p. 4.

Halder A. *Filosofijos žodynas*. Vilnius: Alma Littera, 2002, p. 84.

Heidegger M. Būtis ir laikas. *In: Rinktiniai raštai*. Vilnius: Mintis, 1992.

Husserl E. *Karteziškos meditacijos*. Vilnius: Aidai, 2005, p. 83-85.

Jonas Paulius II. *Atmintis ir tapatybė*. Vilnius: UAB Katalikų pasaulio leidiniai. 2005, p. 97.

Jovaiša L. *Enciklopedinis edukologijos žodynas*. Vilnius: Gimtasis žodis, 2007, p. 245.

Krukauskienė E., Trinkūnienė I., Žilinskaitė V. *Jaunimo kultūrinis identitetas: prioritetai, nuostatos, etninė kultūra*. Vilnius: Eugrimas, 2003.

Kuzmickas B. *Laimė, asmenybė, vertybės*. Vilnius: Lietuvos teisės universitetas, 2001.

Maceina A. *Mintys, apmąstymai*. Vilnius: Eugrimas, 1998, p. 417.

Martišauskienė E. Mokytojų vertybinės nuostatos ir jų sklaida pedagoginėje veikloje. *Acta padagogica Vilnensia*, 2007, nr. 19, p. 135-146.

Martišauskienė E. Tautinio tapatumo žadinimas – pedagogo misijos šerdis. *Pedagogika*, 2012. T. 108, p. 64-70.

Norvilas A. *Tauta, kalba ir tapatybė*. Vilnius: Vilniaus universiteto leidykla, 2012, p. 79 – 81.

Osler, A. Teacher interpretations of citizenship education: national identity, cosmopolitan ideals, and political realities. *Journal of curriculum studies*, 2011, Vol. 43 No. 1: 1-24.

Phinney J. S., Romero I., Nava M., Huang D. The role of language, parents, and peersin ethnic identity among adolescents in immigrant families. *Journal of Youth and Adolescence*, 2001. Nr. 30, p. 135–153.

Piattoeva N. Citizenship and nationality in changing Europe: a comparative study of the aims of citizenship education in Russian and Finnish national education policy texts. *Journal of curriculum studies* 2009. Vol. 41 No. 6: 723-744.

Pruskus V. Tautinė tapatybė ir jos raiška: Laiko iššūkiai ir atsako galimybės. In: *Tautinės tapatybės dramaturgija: monografija*. Vilnius: VPU leidykla, 2005.

Ricoeur P. *Egzistencija ir hermeneutika. Interpretacijų konfliktas*. Vilnius: Baltos lankos, 2001.

Rumbaut R. G. The crucible within: Ethnic identity, self-eswteem, and semented assimilation among child ren ofof immigrants. *International Migration Review*, 1994, No. 28, p. 748-794.

Srėbalius J. Globalizacijos iššūkiai tautiškumui. In: *Asmenybė ir kultūra*. Vilnius: VGTU leidykla TECHNIKA, 2006, p. 5-18.

Tijūnėlienė O., Bukantienė J. Universiteto studentų santykis su tautos dvasinio paveldo vertybėmis – tautinio tapatumo išlaikymo pamatas. In: *Kultūrų dialogas ir asmenybė*. Klaipėda: KU l – kla, 2010, p. 225-239.

Valstybinė švietimo strategija 2013 – 2022 m. 2012, spalis Nr. 17(81). Prieiga internetu http://www. nmva.smm.lt [30-09-2013], p. 6-12.

Vabalaitė R. M. Modernioji tapatybės samprata. In: *Lietuva globalėjančiame pasaulyje*. Vilnius: Logos, 2006.

Vidurinio ugdymo bendrosios programos. 2012. Prieiga internetu http://www. upc.smm.lt/veikime/turinys/ [16-08-2013], p. 2-8.

The Challenges of Globalization for the National Identity of Trainee Teachers

Emilija Urnėžienė

Abstract
The global society is characteristic of fast change which conditions the importance of the ability to adapt both for the individual and the society. The issues of preserving the national identity, historic succession and commitment to national culture are becoming increasingly relevant on a global scale. The transformation brought about by rapid change gives an incentive to preserve national identity and open the mind of the nation to the challenges of globalization. The paper aims to reveal the peculiarities of the expression of national identity of teacher trainees in the context of globalization. Ultimately it is the trainee teachers who will educate the new generation and instill the values of national identity.

***Keywords*:** teacher, national identity, nationality, globalization

Introduction

Modern society is characteristic of rapid change. Globalization, the development of information technologies, political, social, economic, organizational changes condition innovations in various spheres, hence cause the alterations of conventional norms and rules. Globalization is defined as a phenomenon happening on a social plane and encompassing various spheres of activity of the society, state and other (social) bodies as well as their environment (Tomslinson, 2002). The features of globalization – the intensification of interrelations, motion, speed, universality, etc. – determine the importance of adaptation both for the individual and the society without losing their national identity and upholding it. The upholding of Lithuanian culture is conditioned by the developing ties with other national cultures. The issue of *preserving the national identity, historic succession and commitment to national culture* are becoming increasingly relevant on a global scale. The transformation brought about by rapid change gives an incentive to preserve national identity and open the mind of the nation to the challenges of globalization. According to Bauman (2002), "we are becoming more and more responsible for the substantially modified identity".

Identity is defined as people's ability to describe themselves in answer to the question "*Who am I?*" and combine the varying answers together (Antinienė,

2002). Identity is connected with self-reflection as one aims to find his/her place and role in the world (Kroger, 2000, Blank, 2003). In Marcia's (1993) view, identity is "the core of an individual that endows his/her world with meaning and significance". The expression of national identity is nationality that defines the individual's relation with the world, other nations, his/her belonging to human community which is distinct by its unique features as well as spiritual rather than pragmatic relations. The conception of Lithuanian national identity is based not only on tradition, but also on a philosophical and educational approach. Nationality is the self-awareness of being part of a nation, the expression of national individuality. It consists of using and upholding the national language, succession and preservation of the national culture, sensation of the national spirit, cognition of the national history and respect for the nation's past, creative disposition to create the future of a national state, and national activity (Grigas, 2001, Kuzmickas, 2007).

The issues of national identity in the global society have been analyzed by philosophers, politicians, psychologists and the public (Taljūnaitė, 2006). The questions of the development of national identity have been a matter of interest to both foreign (Kroger, 2000; Huddy, 2001; Dekker, Malova, Hoogendorn, 2003; Blank, 2003; Transue, 2007 etc.) and Lithuanian researchers (Antinienė, 2002, 2011; Aramavičiūtė, 2005; Ranonytė, 2006; Taljūnaitė, 2006, 2009; Čiužaitė, 2007; Duobliene, 2009; Čiužas, 2013; Mazolevskienė, Montvilaitė, 2013 etc.). Nevertheless, there are no studies focusing on the expression of national identity of trainee teachers. Such research would help to solve the problem of improving the development of national identity of pre-service teachers. The significance of forming one's national identity, guaranteeing the conditions for a person's mature national self-awareness, is also emphasized in the National Strategy of Education of 2013-2022.

The aforementioned factors determined ***the research problem*** – what is the expression of national identity of trainee teachers in the context of globalization? After all, it is the trainee teachers who will educate the young generation and instill the values of national identity.

The goal of the research is to reveal the peculiarities of expressing national identity among trainee teachers in the context of globalization.

The object of the research is the national identity of trainee teachers.

The research methods: 1) literature review, 2) interview, 3) statistical analysis of research data (descriptive statistics, factor analysis).

Research methodology

309 students – the trainee teachers from Lithuanian University of Educational Sciences – took part in the survey. The respondents were selected by means of

the clustering method. The clusters are composed of faculties representing the humanities, social and natural sciences.

In the opinion of the trainee teachers (Fig. 1), people's nationality is mostly determined by their parents' nationality or origin (91.9%), education in the spirit of national traditions (76.7%), their language (67.6%). The less significant factors determining people's nationality in the respondents' view were the emotional connections with the nation (62.8%), orientation to the cultural traditions of the nation (60.5%) and the place of residence (59.5%). Concern about the country (44%) and participation in the political life of the nation (30.7%) were in the respondents' opinion the least significant factors determining a person's nationality.

Fig. 1: Factors Determining People's Nationality in the Trainee Teachers' Opinion (%)

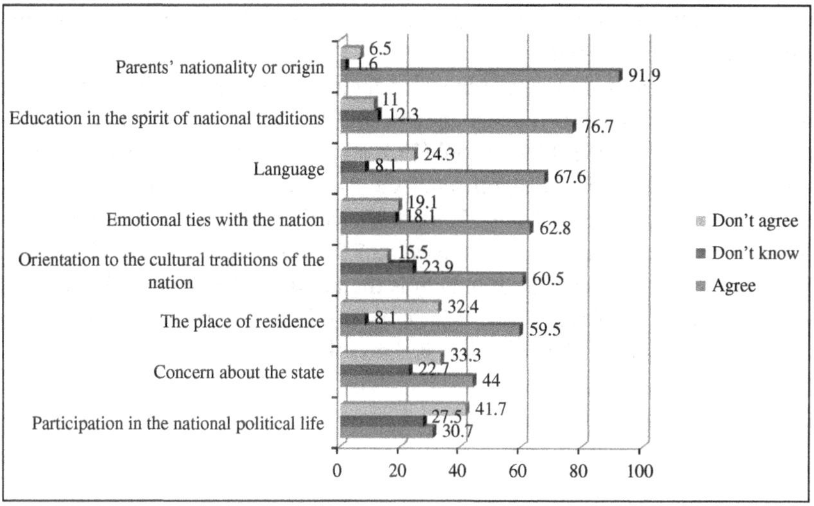

Hence, the trainee teachers are most concerned about their origin, traditions, customs, language, and this subjective evaluation of factors determining a person's nationality, according to most researchers, has most influence on one's national self-determination.

Aiming to determine the trainee teachers' attitude to the features of nationality (Fig. 2), the respondents indicated the mother tongue (88.2 %), holidays, customs and traditions (70.6 %) as the most significant features of nationality. Such results confirm the students' attitude towards the importance of inherent entities.

Fig. 2: The Features of Nationality in the Trainee Teachers' Opinion (%)

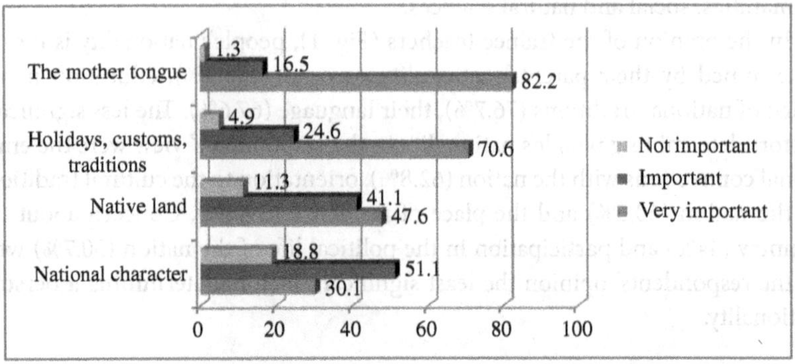

The trainee teachers are mostly proud of their homeland on national holidays (70.6 %), a little less – during sports competitions (69.3 %) and the international shows of the national performers (56.3 %).

The research aimed to determine what factors condition the development of national identity (Fig. 3). The respondents said that the main factors are parents and closest relatives (93.9 %), school (76.7 %) and language (75.7 %). In the respondents' opinion, the less significant factors are peers (60.8 %), the mass media (46.9 %) and the church (41.7 %). The trainee teachers acquired national feelings from their parents and closest relatives (88.7 %) and school (67 %). Hence, the foundation of instilment of nationality is the family that upholds values, traditions and customs. Nevertheless, educators also have a significant influence on the development of one's nationality. It shows that trainee teachers have to reinforce their own nationality identity since it is them who will develop the young generation's values of national identity.

Fig. 3: The Trainee Teachers' Opinion about the Development of Nationality (%)

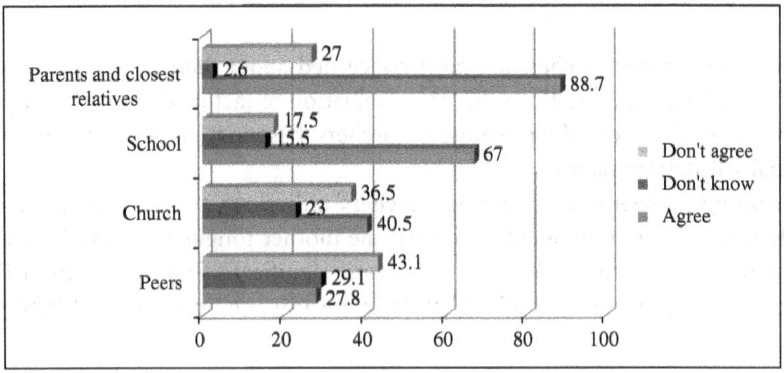

The results of the research show that the trainee teachers distinguished the following factors as a risk for losing one's nationality: emmigration (89.6 %), the loss of customs and traditions (83.2 %) and dangers for the mother tongue (64.4 %).

Fig. 4: The Attitudes to Nationality in the Trainee Teachers' Opinion

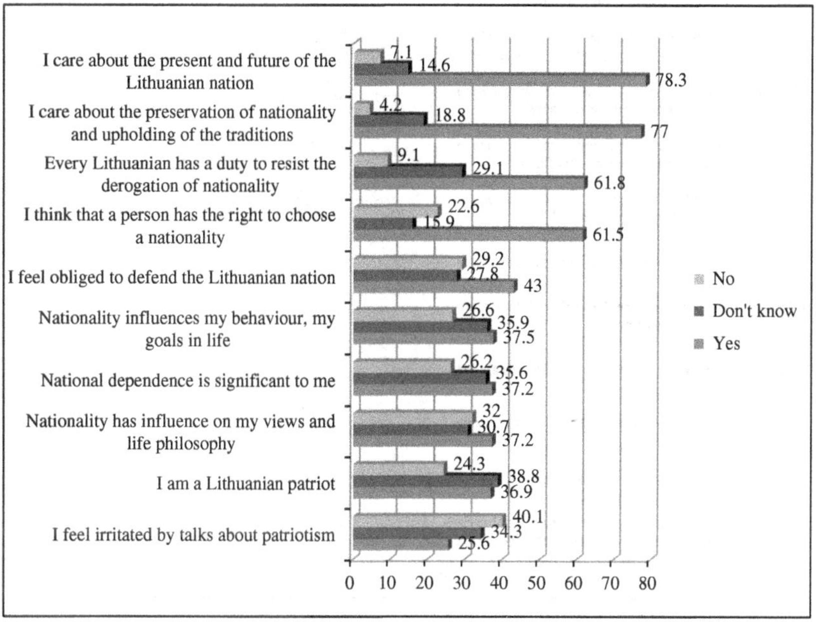

Figure 4 reveals the respondents' attitudes to nationality. Most of the trainee teachers expresses a concern about the present and future of the Lithuanian nation (78.3 %) as well as the preservation of nationality and upholding of traditions (77 %). In the respondents' opinion, it is the duty of every Lithuanian to resist the derogation of nationality (61.8 %). A bit lesss than a half of the respondents feel obliged to defend their homeland (43 %). Hence, a personal national obligation dominates. The research data show that the trainee teachers care about the future of their nation, its customs and traditions, and they feel obliged to develop nationality. However, only a bit more than one third of the respondents consider themselves to be patriots (36.9 %), and one quarter of them (25.6 %) are unwilling to speak about patriotism. The latter are inclined to a negative treatment of nationality. More than a third of the respondents find national dependance significant (37.2 %) and nationality influences their behaviour, goals in life (37.5 %) as well as their views and life philosophy (37.2 %). The results reveal a small influence of nationality on a person.

Fig. 5: The Trainee Teachers' Attitudes to Emigration

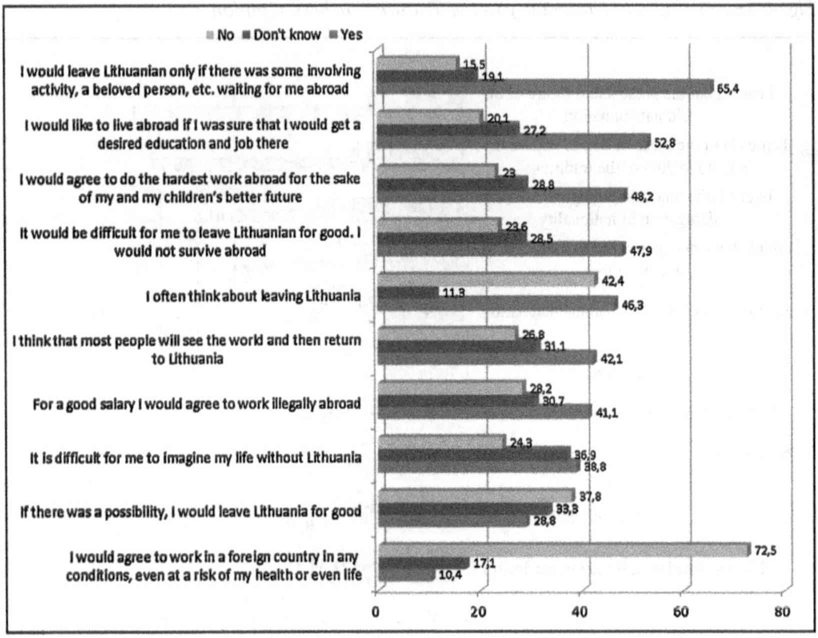

The research also revealed the trainee teachers' attitudes to emigration (Fig. 5). The analysis of results showed that more than a half (65.4 %) of the respondents would leave Lithuanian only if there was some interesting activity, a beloved person, etc. waiting for them abroad. Half of the future teachers (52.8 %) said that they would want to live abroad if they were sure that they would get a desired education or job there. A little bit less than a half would find it difficult to leave Lithuania for good (46.3 %) and they believe that most people will see the world and eventually return to Lithuania (42.1 %). 48.2 percent of the trainee teachers are ready to leave Lithuania for the sake of a better future, they would agree to work really hard abroad, and a significant part of the respondents (41.1 %) would agree to work illegally abroad for a good salary. 10.4 percent of the respondents would agree to go to a foreign country to work in any conditions, even at a risk of their health and even life. About one third of the trainee teachers (28.8 %) are ready to Lithuania for good as soon as they have this opportunity. Yet a bit more than a third (38.8 %) of the respondents are unwilling to leave their homeland, they find it difficult to imagine their life without Lithuania. The research data reveal a rational attitude to emigration and one of the main motives for emigration is the economic one.

It can be seen that both the ethnocentric and the economically motivated attitudes to emigration are typical to a similar percentage of the respondents.

Conclusions

1. National identity can be preserved when the young generation develops an awareness of their nationality, relate it to the historic experience, present and future. The challenges of globalization should be viewed as the consolidation of nationality, not as a threat.
2. Most trainee teachers care about the future of the Lithuanian nation, the development of nationality and its preservation; they feel obliged to resist the derogation of nationality. Nevertheless, only a little bit more than a third of the respondents consider themselves to be patriots of Lithuania.
3. Only a third of the respondents find national dependence significant, and nationality influences their views, life philosophy and their goals and behavior.
4. The respondents possess a rational attitude to emigration (they would agree to go abroad and live there only if there was a beloved person, some interesting activity or they would get a desired education, job and a good salary there).
5. Both the ethnocentric and the economically motivated attitude to emigration is typical to a similar percentage of the respondents. Almost one half of the respondents demonstrated an ethnocentric attitude to emigration – the students would find it difficult to leave Lithuania for good and they believe that most emigrants will return to their homeland. A similar part of the respondents consider leaving Lithuania, they would agree to do the hardest work for the sake of a better future for themselves and their children, yet most of the respondents are not inclined to risk their health and life.
6. It is recommended to develop the students' national identity during pedagogical studies, emphasizing respect for the homeland and upholding the customs and traditions. The trainee teachers should envisage the possibilities to develop their own and their future pupils' national identity.

References

1. Antinienė D. (2002). Asmens tautinio tapatumo tapsmas. Sociopsichologinės šio proceso interpretacijos // Sociologija. Mintis ir veiksmas, Nr. 2, p. 100-107.

2. Antinienė D. (2011). Lietuvos akademinio jaunimo tautinis tapatumas / Mokslo monografija: KTU leidykla „Technologija".

3. Aramavičiūtė V. (2005). Tautinis tapatumas ir jo kaita vyresniame mokykliniame amžiuje // Pedagogika, T. 79, p. 40-45.

4. Bauman Z. (2002). Globalizacija: pasekmės žmogui.

5. Blank T. (2003). Determinants of National Identity in East and West Germany: An Empirical Comparison of Theories on the Significance of Authoritarianism, Anomie, and General Self-Esteem // Political Psychology 24, p. 259-288.

6. Čiužaitė M. (2007). Studentų tautinės savivokos bruožai: VPU atvejis // Tautiškumas ir pilietiškumas: atskirtis ar dermė? / sud. R. Grigas. – Vilnius: Vilniaus pedagoginio universiteto leidykla, p. 222-244.

7. Čiužas R. (2013). Mokytojo kompetencijų svarba mokinių tautinio tapatumo formavimuisi // Pedagogika, T. 109, p. 27-32.

8. Dekker H., Malova D., Hoogendorn S. (2003). Nationalism and Its Explanations // Political Psychology 24 (2).

9. Duobliene L. (2009). Ar pavyks išvengti hibridinio tapatumo formavimo(si) informacinėje visuomenėje? // Acta paedagogica Vilnensia, Nr. 23, p. 79-91.

10. Grigas R. (2001). Tautinė savivoka. – Vilnius: Rosma.

11. Huddy L. (2001). From social to political identity: A Critical Examination of Social Identity Theory // Political Psychology 22, p. 127-156.

12. Kroger J. (2000). Identity Development: Adolescence Through Adulthood. – London: Sage Publications.

13. Kuzmickas B. (2007). Tautos tapatumo savimonė: lietuvių savimonės bruožai / Monografija. Vilnius: Mykolo Romerio universiteto Leidybos centras.

14. Marcia J. (1993). The ego identity status approach to ego identity // Ego Identity: A Handbook for Psychosocial Research / Eds. J. Marcia, A. Waterman, D. Matteson, S. Archer, J. Orlofsky. NY: Sprynger-Verlag, p. 3-21.

15. Mazolevskienė A., Montvilaitė S. (2013). Ikimokyklinio amžiaus vaikų tautinio identiteto formavimo (si) ypatumai emigracijoje // Pedagogika, T. 109, p. 33-38.

16. Ranonytė A. (2006). Lietuvos moksleivių pilietiškumo ir tautinio tapatumo formavimosi prielaidos // Acta paedagogica Vilnensia, T. 16, p. 63-72.

17. Taljūnaitė M. (2009). Tautinio tapatumo politika: teorinės paradigmos paieškos ir modeliai // Filosofija. Sociologija, T. 20, Nr. 4, p. 294-301.

18. Taljūnaitė M. (2006). Lietuvių tautinio tapatumo konstravimas sociologiniuose tyrimuose // Filosofija. Sociologija, Nr. 2, p. 8-17.

19. Tomlinson J. (2002). Globalizacija ir kultūra. - Vilnius: Mintis.

20. Transue J. E. (2007). Identity Salience, Identity Acceptance, and Racial Policy Attitudes: American National Identity as a Uniting Force // American Journal of Political Science 51 (1), p. 78-91.

21. The National Education Strategy of 2013-2022 [online document]. Available from: <http://www3.lrs.lt/pls/inter3/dokpaieska.showdoc_l?p_id=438859>.

Peculiarities of Learning and Teaching History to 8th – Formers on the Basis of Gender Distinctions

Benediktas Šetkus

Abstract
The paper analyzes the idea of the beginning of the 20th century that boys and girls possess distinct learning peculiarities and different interests in the field of history. On the basis of the results of the 8th-formers' test that took place in 2013, the research aims to determine whether history topics have any influence on the level and learning results of boys and girls. The paper explores the validity of the statement that boys are more interested in politics, sovereigns and war, while girls are more concerned with culture and religion – hence, have a better knowledge of these topics. The paper also analyzes some other peculiarities of boys' and girls' learning of history that influence their learning outcomes.

Keywords: learning of history, boys, girls, differences

Introduction

In the 1929–1930 school year, Pranas Penkauskas taught a course of "The Methodology of Teaching History" at the University of Lithuania (Vytautas Magnus University since 1930) and emphasized that, when designing the curriculum programmes of history, it is essential to consider the differences between boys and girls. In his opinion, when teaching history to girls, it is important to take into account the features of female character as well as the position of women in the family and state. Hence he proposed to exclude the topics of the history of state politics and wars at girls' schools and pay more attention to the history of art, poetry, fine arts, and music. When teaching the history of culture, one should focus on the life of family and local inhabitants, citizenship, i.e. inventions, national traditions, the routine of family and public life, its beauty, stability, usefulness, etc. Penkauskas suggested putting emphasis on the cases when women from different social groups distinguished themselves by their diligence or honour and thus could serve as good examples for schoolgirls. He claimed that girls have to be focused on the issues of religion and morality [9, 24].

This case shows that the peculiarities of teaching (learning) history on the basis of gender distinctions were analyzed as far back as the beginning of the 20th century. It can be presupposed that Penkauskas borrowed the ideas from books on the methodology of history published in Germany at the time as the titles are listed in his "Methodology of Teaching History" ("Istorijos dėstymo metodika"). He states having used 21 books, the majority of which were published in Germany, and some – in Austria. Those were methodologies of teaching history, published in the period from 1889 till 1927 in Berlin, Munich, Leipzig, Regensburg, Tubingen, Breslau, Vienna, Innsbruck. It has to be noted that in 1922-1923 Penkauskas studied at Munich University and that must be the reason why he was well acquainted with the works by German authors [9, 7].

In the period of the Republic of Lithuania (1918-1940) and the Soviet Lithuania (1940-1941; 1944-1990) there were numerous publications of articles and dozens of books on methodology of teaching history (mostly translated from Russian), yet they did not focus on the peculiarities of teaching history to boys and girls. It was only in about 2000 when there appeared several articles developing the idea that history course books should include more information on the role of women [1]. This idea triggered a moderate discussion among the teachers of history in Lithuania since the opinions diverged.

It is apparent that the peculiarities of boys' and girls' teaching and learning were initially emphasized by pedagogues and psychologists in Western countries. Their example was followed by pedagogues in Lithuania, for example, Liudmila Rupšienė. In 2001 she conducted a survey of teachers aiming to explore the teachers' opinion about the motivation of boys and the reasons for their unwillingness to learn [10]. The interest in this issue has been growing gradually and now this kind of research is widely spread.

On the basis of Bill McBride's personal experience it is possible to assume that this issue has also been quite recent in the United States as well. He claims that in the 1970s, when he was a student at university, he was a great supporter of feminism. However, having first heard of the peculiarities of teaching children of different genders, he grew skeptical of feminism [7]. It has to be noted here that in Lithuania there were also teachers of similar beliefs who thought that this issue is irrelevant.

In the 1990s in Lithuania there were surveys of schoolchildren conducted aiming to determine their attitude to history in general and the topics that interest them most of all. However, at the time gender distinctions were not taken into consideration. In 2004 there was a survey of 7th-12th-formers conducted which aimed to compare the boys' and girls' interest in history. The results showed that the political events of the past interested 16.4% of the interviewed boys and only 2% of the girls. The history of daily lifestyle was interesting to 12.3% of the boys and 21.2% of the girls. The gender-based interests were also revealed through their answers to the other questions of the survey. When the interviewed pupils were asked to indicate which personalities interest them most, it was revealed that girls are five times more interested in artists than boys, and they are also interested in explorers. Yet boys are interested in scientists twice as much as girls, and in war leaders – three times more than girls [12, 67].

The goal of this paper is to determine the peculiarities of teaching history to boys and girls as well as the peculiarities of their learning.

The research objectives:

1. To determine whether boys possess a better understanding of wars, sovereigns and politics which are assumed to be the history topics that interest boys more than girls;
2. To determine whether girls have a better understanding of culture and religion which are presupposed to be the history topics that interest girls more than boys;
3. To determine other peculiarities of learning history which influence the pupils' learning results.

The research methods: analysis of research works and methodological publications, the interview method, the comparative historic method, the summary of professional experience.

The research context and procedure

The research was conducted in 2013 when the standardized history tests were being prepared. A standardized test is a test that is prepared, administered and assessed on the basis of the established regulations. The test consists of standardized tasks which correspond to the proportions provided in the test specification, and is intended to assess the levels of students' achievements [11, 10].

Since 2009 it has been a common practice in Lithuania to administer the 8th formers' standardized tests on history which are prepared by the National Examination Centre, and the author of this paper takes part in the preparation of the test. The essence of the standardized tests can be briefly summarized in the following way. First of all, several tests on history are prepared and the tests are taken by the 8th-formers in Lithuanian schools. Then the test results are analyzed and the efficiency of each task is assessed. The too complicated and too simple questions are eliminated, likewise the tasks with poor distinctive features (i.e. the completion of such tasks shows that the results of strong and weak students barely differ) are rejected. The verified efficient tasks are included in a standardized test. It is designed according to the established proportions – one part of the tasks is related to the topics of world history, another part – to the topics of

Lithuanian history, there are some questions (10-15%) on the period of ancient history, and much more questions (85-90%) on the Middle Ages. Besides, it is regulated how many tasks are related to the causes and consequences of historic events, personalities, history concepts, etc., how many tasks assess the students' understanding of chronology and historical space, their ability to analyze and interpret historic sources. The tasks are also selected on the basis of cognitive skills: knowledge and understanding comprise 30-40%, application – 40-50%, and the higher mental abilities – 15-20% of the tasks. The tasks are also chosen in accordance with the levels of students' achievements: the tasks of minimal level (easy tasks) constitute 25-35%, the basic level tasks (of average difficulty) – 45-55%, the tasks of higher level (complicated tasks) – 15-25% of all the tasks [11, 102-104]. In our opinion, this kind of test is appropriate and can be used to objectively assess all students.

Such a test is given to the teachers of history at schools. They are also given the scoring instructions and the data on each pre-assessed task all over Lithuania [11, 97-126]. Thus every school can administer the test with 8th-formers and then compare the achievements in history of their school, specific class or separate pupils in the context of the achievements of all learners in Lithuania.

When the completed tasks are graded during the stage of initial testing, there is also analysis of boys' and girls' answers. In the recent years there has been three tests, i.e. 150 tasks, piloted yearly, which makes it possible to draw certain conclusions about the peculiarities of boys' and girls' learning of history.

This paper is limited to the data of three history tests piloted in 2013 [4, 5, 6]. The tests were administered in 166 classes of 154 schools in Lithuania. The first test was taken by 282 pupils, the second – by 276, and the third – by 266 pupils. The test was conducted in cities (52 schools), towns (62 schools) and villages (42 schools). Likewise, in order to maintain proportion, the administrators of the test took into consideration the type of school: the tests were piloted in 60 lower secondary or basic schools, in 31 progymnasiums, 28 secondary schools, and 37 gymnasiums.

There is a slight difference between the number of boys and girls who took the test. The first test was taken by 149 boys and 133 girls. The second was taken by 141 boys and 135 girls. The third was taken by 153 boys and 113 girls. The boys' and girls' results achieved in the test are presented in the following way. On the basis of the pupils' achievements they are divided into four levels – from the lowest to the highest. The results of students of highest achievement (i.e. the group of the best pupils) are termed "Level A" in this paper. The students who are considered to be good are called "Level B". The learners who got satisfactory results are named "Level C", and the weakest students (who received the lowest passing grade) are termed as "Level D". The results of each group are presented in percentage.

With the purpose to reduce the amount of the information acquired during the research, this paper mostly presents the results of Level A boys' and girls' answers. In some cases they are compared with the answers of students of Level C or any other level.

War, Sovereigns and Politics: Are These Topics Favourable for Boys?

Having in mind Penkauskas's idea, expressed as early as the beginning of the 20th century, when he suggested that boys are more interested in political history, as well as the analogous survey conducted in 2004 when schoolboys confirmed this statement, it was decided to verify this viewpoint and compare it with the results of tests of 2013. Therefore, from the three tests we selected ten tasks which should be more relevant to boys, since the questions deal with the topics of war, sovereigns and politics. We will now present the questions or the tasks and the acquired results on the basis of gender distinction [4, 5, 6].

Question 1: *Who is considered to be the founder of the Frankish Empire?* (The possible answers are: Alexander of Macedonia, Otto the Great, Charlemagne, Octavius Augustus.)

Results: this question was correctly answered by 72% of Level A boys and 56% of Level A girls. It has to be noted that the question was correctly answered by as many as 34% of Level C boys, while only 9% of girls of the same level managed to answer correctly. Hence the assumption that boys are more interested in famous war leaders and sovereigns, and have a better knowledge of them is supported.

Question 2: *Indicate the name of the tribes that came by sea in the 8th–10th* centuries and attacked the Baltic tribes. (an open-ended question; the students are provided with a map "The Baltic Tribes in the 12th century" which does not mention the Vikings)

Results: this question was answered by 97% of Level A boys and 97% of Level A girls. It has to be noted that this question was answered by 68% of Level C boys and only 32% of girls of the same level. Thus Vikings are known to twice as many boys of satisfactory learning results as to girls of the same level. Therefore, it is possible to draw the conclusion that boys are more interested in Vikings than girls.

Question 3: *The first Lithuanian location mentioned in the written sources is ...* (The following possibilities are given: Apuolė, Kernavė, Trakai, Voruta.)

Results: this question was answered by 47% of Level A boys and 10% of girls. It has to be noted that the first indicated location is Apuolė which is famous because in 853 A.D. it was attacked by Vikings and the local inhabitants defended themselves fiercely. Hence almost half of Level A boys know the fact, while the girls' answers are random (statistically). The answers of Level C boys and girls do not exhibit any differences based on gender.

Question 4: *Describe the changes that had any influence on the development of the Baltic tribes in the 18th century by inserting the missing words.*

"In the 18th century in the neighbourhood of the Baltic tribes there settled down the [...] Order in the west and the Livonian Order in the north. This initiated the alliance of the Baltic tribes into [...]"

Results: this question was answered by 88% of Level A boys and 55% of girls. In this case the pupils had to insert "The Teutonic Order" and "The Livonian Order" which were fought against by the inhabitants of Lithuania. It has to be emphasized here that the answers of Level C boys and girls barely differ, while Level B boys significantly outnumber the girls (41% of the boys and 28% of the girls).

Question 5: *Which Baltic tribe, under the lead of Herkus Monte, fought against an enemy from the west and failed to survive later?* (an open-ended question)

Results: this question was answered by 81% of Level A boys and 69% of girls. In this case the pupils had to indicate that the tribe was the Prussians. Hence this fact is better known to boys. Nevertheless, as far as Level C students are concerned, more correct answers were given by girls – 30%, while only 18% of the same-level boys managed to answer it.

Question 6: *What joint battle relates Jogaila with Vytautas the Great?* (an open-ended question)

Results: this question was answered by 89% of Level A boys and 97% of girls of the same level. In this case the students were supposed to write "The Battle of Grunewald", and girls were better at this than boys. However, it is interesting to note that Level C boys surpassed the girls of the same level (71% of boys and 55% of girls). It is complicated to account for the contradictory results.

Question 7: *Who was defeated by the joint Lithuanian-Polish army in the Battle of Grunewald?* (an open-ended question)

Results: this question was answered by 94% of Level C boys and 70% of girls of the same level; by 97% of Level B boys and 86% of the girls; by 93% of Level A boys and 94% of the girls. It can be seen that boys of the lower level know better that the army beat Germans (as it is typically said). But in the group of the best students the girls surpass the boys by 1%. The reasons are hard to explain – maybe the boys of Level A considered this question to be too simple? Or it is possible that some of them simply refused to answer the question. There are always boys like that.

Question 8: *Which chronological sequence of Lithuanian sovereigns is correct?* (There were several options given, all including the following rulers: Algirdas, Gediminas, Jogaila, Mindaugas.)

Results: this question was answered by 84% of Level A boys and 66% of girls. It has to be noted that Level C girls gave more correct answers than the boys of the same level. It could be explained by the fact that girls are always more responsible about their tests. Therefore, in this case some boys of Level C simply did not want to solve the "puzzle" which consists of several lines listing the same sovereigns in a different sequence.

Question 9: *Which of the indicated historic personalities could have visited Vilnius University?* (There were several options given including the following rulers: Vytautas the Great, Stephen Bathory, Gediminas).

Results: this question was answered by 80% of Level A boys and 64% of girls. All the above-mentioned historical personalities were famous Lithuanian sovereigns, war leaders, politicians. This could be the reason why boys know more about them – in this case about the historical period they lived in – than girls. It has to be indicated that the correct answers of boys of all the levels outnumbered those of the girls by 8–12%.

Question 10: *What is characteristic of Vytautas' the Great rule?* (There were several options given: the end of battles against the German Order, the Christening of Lithuania, the development of the state, the end of battles against Vikings.)

Results: this question was answered by 93% of Level A boys and 71% of girls. In the group of Level C the boys surpassed the girls by as many as 25%. Thus it is possible to draw the conclusion that boys know much more about the most significant Lithuanian sovereign and war leader than girls do.

Summing up the above listed questions and tasks, which are all related to politics and war, it can be stated that these topics are better understood by boys. Sometimes the girls' answers were better than those of boys, but such cases were rare. Yet, it is possible to draw a conclusion that girls can perform better in the tasks which require more concentration, thinking, and, as a result, more time. In this case, the character qualities of girls have more significance than the possessed knowledge.

It is important to note that the absolute majority of the listed questions and tasks deal with the history of Lithuania. It is difficult to assess the extent to which this circumstance can be significant in relation to boys' and girls' different interests in history. It just happened so that it was not planned to study the pupils' results on the basis of our research objectives during the preparation of the tasks.

Culture and Religion: Are These Topics Favourable for Girls?

As it was stated above, in 1930 Penkauskas claimed that it is important to pay more attention to culture, religion and lifestyle when teaching history to girls. In 2004 there was a survey of 7th–12th-formers conducted which showed that girls are more interested in culture, religion and the history of lifestyle. Aiming to determine whether it influences their learning results, ten questions covering the girls' favourite topics were selected and the results of the research are presented below [4, 5, 6].

Question 1: *What architectural style is represented by this building?* (There were several possible options given: Renaissance, Baroque, Gothic, and Romanesque; there was a photo of a fragment of a baroque building.)

Results: this question was answered by 60% of Level A girls and 44% of boys. The differences between the girls' and boys' answers in lower levels are practically insignificant.

Question 2: *What language was the language of education in European universities of the Middle Ages?* (There were several possible answers given: English, Latin, the mother tongue, French.)

Results: this question was answered by 100% of Level A girls and 86% of boys. The differences between the answers of boys and girls of lower levels are insignificant, but the girls surpass the boys.

Question 3: *How were the copies of books multiplied in the Middle Ages?* (an open-ended question)

Results: this question was answered by 97% of Level A girls and 67% of boys. This question was better answered by Level B girls; however, Level C boys were better at this question than girls of the same level (44% of boys, 28% of girls).

Question 4: *The legend of the foundation of Rome is related to a she-wolf. The legend of which Lithuanian town is also related to a wolf?* (An open-ended question: there is a photo of the Capitoline wolf.)

Results: this question was answered by 97% of Level A girls and 94% of boys. It has to be noted that the legend is about the foundation of Vilnius which is familiar to the majority of students. Level B and C girls surpassed the boys by 1%-2%.

Question 5: Who wrote the first book published in the Lithuanian language? (There were several options given: Martynas Mažvydas, Abraomas Kulvietis, Stanislovas Rapolionis, Mikalojus Radvila.)

Results: this question was answered by 100% of Level A girls and 98% of boys. It has to be noted that this topic is also a subject of Lithuanian literature classes. On the whole, the question is very easy. A slightly bigger difference is observed in Level C as the girls surpassed the boys by 6%.

Question 6: *Which book was the first to be published in Europe?* (There were several options given: the Bible, the Koran, the Torah, and Catechism.)

Results: this question was answered by 84% of Level A girls and 84% of boys. It has to be noted that Level B girls yielded to boys by 9%, while Level C girls – by 8%. The reasons for this are difficult to explain.

Question 7: *Who copied books in the Middle Ages?* (an open-ended question)

Results: this question was answered by 94% of Level A girls and 98% of boys. This question does not give clear information about gender distinctions as in Level C boys gave better answers than girls (the difference is 17%). Nevertheless, in Level B girls were better than boys (by 4%).

Question 8: *Who was the originator of Islam?* (an open-ended question)

Results: this question was answered by 91% of Level A girls and 84% of boys. However, the answers of pupils of Level C diverged more: 65% of girls and only 31% of boys answered correctly.

Question 9: *How do we call a holy picture, painted on a wooden board?* (There were several options given: mosaic, stained-glass, icon, fresco.)

Results: this question was answered by 66% of Level A girls and 50% of boys.

Question 10: *Indicate the names of two religions that are related by the city of Jerusalem.* (an open-ended question)

Results: this question was answered by 52% of Level A girls and 47% of boys. It has to be noted that Level B girls surpassed the boys of the same level by as many as 20%, while the answers of Level C boys are slightly better than those of girls.

Summing up the results of culture and religion questions we could say that virtually the girls showed better results than boys. In some cases the results slightly differ, and the reason for this divergence is not clear. There is a possibility that it can be accounted by the factors which are going to be discussed in the next section of the article.

Some Insights Concerning the Other Differences

The discussion of the peculiarities of 8th-formers' learning and teaching on the basis of gender distinctions can evoke ambiguous opinions concerning the objectivity of this research. First of all, there can appear doubts whether the choice of the questions and the results was not biased.

It has to be noted that there are a lot of questions which are for some unknown reasons answered better by boys or by girls. There can be speculations concerning the role of the subject matter of questions – whether they are more "masculine" or more "feminine". If the test had been designed purposefully to study the gender distinctions, we would have had more questions for the research. For example, there would have been more questions about the history of lifestyle, the position of women in the society, etc. In this case the research was of reverse character: first of all, the tasks were designed and tested, and only then there arose an idea to select them on the basis of boys' and girls' interests and compare them.

The analysis and comparison of the results of history tasks accomplished by boys and girls highlighted certain gender differences. It is obvious that boys, more often than girls, do not try to answer all the questions, especially if they are more complicated and require more attempts. It has to be noted that in this specific test students were not given marks, and this circumstance could have been significant. Nevertheless, the conditions for taking the test were identical for both boys and girls.

As far as the comparison of the numbers of students who did not answer the questions is concerned, the results of the history tests taken in 2010 are presented as an example. The first part of the test consisted of 15 closed-ended questions [3, 6-8]. The table below demonstrates the first ten questions and the number of students who did not answer them (Table 1).

Table 1: The students who did not answer the questions in relation to their gender

Question number, its wording	Number of boys who did not answer (%)	Number of girls who did not answer (%)
1. Where did the Renaissance appear?	1.8	0.9
2. What did ancient Greeks call by the name of Acropolis?	0.9	-
3. Under whose rule did Vilnius become the capital of the Grand Duchy of Lithuania?	1.8	-
4. When was Lithuania mentioned for the first time in written sources?	0.4	0.4
5. What does this sign symbolize?	0.9	0
6. In what historic period was this sculpture created?	0.9	0
7. From what country did Lithuania take christening in 1387?	0.9	0
8. Which sequence of events is correct?	1.8	0.4
9. Which line presents the correct sequence of events?	7.9	3.1
10. What is characteristic of a town possessing Magdeburg rights?	3.5	1.8

The table presents the results of the survey of 452 students. As it can be seen, the majority of the students did not answer question 9 – as many as 25 students failed to answer this. A closed-ended question includes the possible answers; hence it may seem strange why some pupils did not mark any of the given options. This quality is more characteristic of boys.

The second part of the test consisted of historical sources [3, 9-11]. The students were supposed to answer more complicated tasks. Here we list the first ten tasks and the number of the students who did not answer them (Table 2).

Table 2: The students who did not answer the questions in relation to their gender

Question number, its wording	Number of boys who did not answer (%)	Number of girls who did not answer (%)
1. Indicate the historic periods that are reflected in sources A, B, C, D.	39.6	27.7
2. What common topic unites sources A, B, C, D?	16.7	6.3
3. Sources A, B, C speak about two religions. Indicate the titles of those religions and write their originators.	16.7	10.3

4. What is the title of the book which is there in the hand of the man in source C?	11.9	6.3
5. What is the title of the book pictured in source D?	16.3	11.2
6. Indicate the common feature of these religions.	23.8	13.8
7. What battle is pictured in sources A and B?	2.6	1.8
8. Using sources A, B and C, indicate the warring sides.	10.1	7.6
9. Indicate one cause of the battle described in the sources.	25.1	23.1
10. Using sources A and C, indicate in what state and place of the time the battle took place.	16.3	8.9

As it can be seen, the more complicated questions are more frequently unanswered by boys rather than girls. This discrepancy is evident. It has to be noted that questions No.1 to 6 are related to religion, which should be more interesting to girls, and questions No.7 to 10 are related to the Battle of Grunewald, which should be more interesting to boys. Nevertheless, in both the cases the boys did not do the tasks more frequently than the girls. In other words, girls attempt to answer the questions more often, they are more responsible. Thus this factor is also significant for determining the students' achievements in history.

The research established that girls can concentrate better while analyzing historical sources. When they have to work on a historical document, they notice the key words better and understand what significant information they reveal. As a result, girls give better answers than boys. The following example could support this statement [4, 4].

Indicate the causes and effects of geographical discoveries by inserting the options in the table.

***A** The goal to trade with Eastern countries without intermediaries.*
***B** It was proved that the Earth is round.*
***C** The wish to find a shipping lane to India.*
***D** The discovery of the continent of America.*

Causes	*Effects*

As it can be seen from the wording of the above-given task, the question can be answered without any knowledge of geographical discoveries. The task can be accomplished by paying attention to the choice of words. It is obvious that the words "wish" and "goal" indicate the causes. And the other two variants show the effects. However, boys pay less attention to those key words; hence the girls' results were much better than those of the boys: the question was answered by 65% of Level D girls and 28% of boys of the same level; by 76% of Level C girls and 47% of boys; by 80% of Level B girls and 69% of boys, by 89% of Level A girls and 90% of boys.

Pic. 1: The medal dedicated to the Christening of Lithuania

1

In the cases when there are illustrations given, girls are better at noticing the details. An example of this could be the case when there was a picture of a medal given in the test. On the medal there was the coat of arms of Lithuania – "Vytis", i.e. an armoured knight with a sword and shield (Pic. 1). The question was: *What is pictured as number 1?* [3,7]. This task was accomplished by 83% of Level D girls and 50% of boys; by 91% of Level A girls and 93% of boys.

The given examples show that in both the cases girls were better at the tasks that required more attention. However, in the group of the best learners, the boys slightly surpassed the girls. This conclusion can be drawn not only on the basis of the two mentioned cases.

The results of the research conducted in 2013 can be compared with the main conclusions about the differences of girls' and boys' learning drawn in the national research on pupils' achievements carried out in 2003-2008. Since the national research did not investigate the pupils' achievements in history, only certain tendencies can be compared.

The research established that in all the surveyed classes among the learners with better learning outcomes there were more girls, and conversely among the weaker learners there were more boys than girls. The higher the class the better results the girls had than the boys in the same class. The girls mostly surpassed the boys by the results of tests on the Lithuanian language. Boys are frequently assessed more strictly and get lower grades. From 2004 to 2008 the learning aspiration of boys declined much more significantly than that of girls [2, 1].

Similar tendencies and insights are presented by many foreign authors, including Lisa Lawell who claims that the learning results of boys gradually decline, and boys and girls like to learn the same subjects in different ways, have different inclinations for the same activities – reading, writing, etc. [8, 30-33].

It would have been useful to compare the results of the research on history learning with the analogous studies conducted in other countries. However, the author of the paper failed to find publications on the differences of boys' and girls' learning of history; hence the comparison was not done.

Conclusions

1. In his "Methodology of Teaching History" (1930) Penkauskas claimed that boys and girls should be taught different topics of history. This opinion was supported by the survey of 7th–12th-formers conducted in 2004 in Lithuania. The study revealed that boys like learning about politics, sovereigns and wars better. Girls are more interested in culture, religion and the history of mode of life. The validity of these statements was the object of the history test taken by 8th-formers in all types of schools in Lithuania in 2013.
2. The ten questions, selected from three history tests, were related to politics, sovereigns and war history, and showed that the boys' answers were better than those of girls. The majority of the questions were about Lithuanian dukes, their fights with enemies, about Vikings. In some groups of different achievements the differences were really significant, while in others they were barely noticeable. In some cases the girls' answers were better. The summary of the answers to those 10 questions in the group of the best learners revealed that the average number of boys who answered the questions was 65.6%, while the average number of girls was 52%. The difference is 13.6% and it is possible to state that the discrepancy is considerable.
3. The ten selected questions that were related to culture and religion showed that girls answered them better than boys. The majority of the questions were about the books of the Middle Ages, some – about Christianity, Islam,

universities. In some groups of different achievements the differences between boys and girls are considerable, while in others - of little significance. The summary of the answers to those 10 questions in the group of the best learners revealed that the average number of girls who answered the questions was 84.1%, while the average number of boys was 75.2%. The difference is 8.9% and it is possible to state that the discrepancy is noticeable.

4. The conducted research shows that there were more boys than girls in the group of pupils who did not answer the questions. That was typical even in the cases when it was a multiple choice task. Much more boys do not do more complicated tasks - when they have to apply knowledge in practice, to apply higher thinking skills, when the task requires more time. Moreover, girls are more observant; hence they get better results in the analysis of historical sources.

References

1. Bakonis, E. (2000) Apie moteris istorijos vadovėliuose. Kokiais šaltiniais galima parodyti istorinį moters pasaulį? *Informacinis biuletenis*, Vilnius: VTIC,1(4), pp. 11-13.

2. Gudynas, P. et.al. (2009). Ar Lietuvoje užtikrinamos lygios privalomojo švietimo galimybės berniukams ir mergaitėms? *Švietimo problemos analizė*, 3 (34), pp. 12.

3. *Istorija. 4 sąsiuvinis. 8 klasė.* (2010), Vilnius: Ugdymo plėtotės centras, pp. 1-16.

4. *Istorija. 1 sąsiuvinis. 8 klasė.* (2013), Vilnius: Nacionalinis egzaminų centras, pp. 1-9.

5. *Istorija. 2 sąsiuvinis. 8 klasė.* (2013), Vilnius: Nacionalinis egzaminų centras, pp. 1-9.

6. *Istorija. 3 sąsiuvinis. 8 klasė.* (2013), Vilnius: Nacionalinis egzaminų centras, pp. 1-8.

7. McBride, B. *Closing the Achievement Gap: Teaching to Gender Differences.* Retrieved November 5, 2013, from http://www.nassp.org/tabid/3788/default.aspx?topic=Closing_the_Achievement_Gap_Teaching_to_Gender_Differences

8. Lawell L. K. (2010) *Gender Differences in Education: Teaching All Middle School Students in a Gender-Responsive Way*. Retrieved November 3, 2013, from http://csus-dspace.calstate.edu/bitstream/handle/10211.9/609/Lisa%20 Lawell%20-%20PROJECT%20FINAL%20-%20Copy.pdf?sequence=1

9. Penkauskas, P. (1930) *Istorijos dėstymo metodika: paskaitos, skaitytos 1929-30 mokslo metais Lietuvos universitete*. Kaunas.

10. Rupšienė, L. (2001) Berniukų mokymosi motyvacijos ypatumai. *Pedagogika*, 53, pp. 117-121.

11. *Standartizavimo procedūrų aprašas. II dalis. 8 klasės. Istorijos standartizuotos programos ir testų pavyzdžiai*. (2012). Vilnius: Ugdymo plėtotės centras, pp. 97-126.

12. Šetkus, B. (2004) Istorija bendrojo lavinimo mokykloje: moksleivių požiūris. *Istorija*, 61, pp. 65-68.

Human beliefs, values and social – education analysis; of the philosophy, of theology and psychology pisture

Kęstutis Ralys

Abstract
Surrounding environment, social, political situation has affected our lives. Thus, the all-around individual processes have some impact on its value system. Human social behavior and motives, his control at all times, the researchers were interested in the practical and theoretical point of view. Value orientation system, reflecting the lifestyle characteristics of the public consciousness through the prism, is one of the world of the individual components. Generally speaking, it is found that human behavior is determined the person inside, and psychological characteristics, experience, situational and environmental condition in which the action or interaction, the conditions. Of course, that the internal, psychological characteristics of individual behavior is an important value orientation (dispositions) system. Requests to the overall success of dynamic and sometimes hostile environment, they have formed the world view of the basic provisions of its 'I' conservation needs. On the other hand, modern humanism itself is not limited to preservation of humanity, and to disclose personal practices that preserve the humanity of another human being (Buber, 2000), (Rev.dr.K.Ralys 2013).

Key words: society, philosophy, religion, psychology, psychoanalysis, family.

Introduction

Lithuanian scientists, on the basis of European values and global studies, their research examined the cultural changes in Lithuanian society. Published work analyzing changes in the values of religion, family, work, politics and other fields. Scientific work focuses on age of the population impact of value orientations. In various areas of life observed differences between the securities generations.

Values are the desired objectives or fundamental principles of individual life with a different value for a specific individual. They formed human socialization process of internalisation of the social environment, family and societal values. Therefore, internal internalize values (value orientations) in a given group of people depends on the values that are prevalent in society. Man is based on the life experiences to grow and learn. Experience emerges in certain mental structures - attitudes, dispositions, which determine the direction of personal conduct. It is logical to assume that a person's value orientation is chaotic, and a framework. Human value orientation shows that he tends to do with their limited time and energy.

Value - an abstract, the general principle of standard of conduct to justify specific actions and goals. Values of the person taking over the socialization process, identifies himself with them emotionally - internalize. They are general standards of conduct, which is expressed in terms of social norms. Individuals accepting the same values of the uncertain nature of the property may disagree on specific rules, embodied those values. They lead to the choice of treatment facilities and, therefore, the study also include the attitudes, behavior, interaction and social structure of the investigation. Value orientation - is an internal personal characteristics, determining his behavior. The same value is not necessarily equally important for all members of the group (Theodorson, 1969, p. 455-457).

J. V. Uzdila (2003) believes that the value of a trinomial and personal self, hierarchical structure of values of orientation (lower level), values (more significant - the middle tier) and value (internalize the purposes for which the entity is willing to "put down your head" - level).

This article aims and objectives, discuss the theoretical reflections about the security of public faith in the light of developments. Value - an abstract and general principle of conduct. Where to look in philosophy, religion, psychology, education, and an explanation of interaction based on changes in public securities. Compliance with the provisions of that traditional values are interdependent world elements characteristic of the whole culture in the broad sense. Substantive distinction between society and the values of property, money, etc.

Method: *philosophical, religious, psychological, literary analysis, statistical analysis of the philosophy of family and eduction terms*

The object of investigation

The twenty-first century family values: philosophical, religious and psychological approach to family values education, and religious context.

Society can identify those areas of value: money, friendship, family, religion and morality, politics and social organization, love and sex, leisure, maturity, character traits and work (Rathaus St. Louis, 1966).

Social values of nature can be tangible and intangible. They cover every subject in which the individual can respond to treatment form (orientation). Social values include beliefs, activities, opinions, aspirations, moral principles, philosophical doctrines, education, scientific theories and traditions of self, imagination and public opinion, even longer list of more objective and tangible things. This list could continue ad infinitum, because it encompass everything that you can give to the provision or attitude. Values for different individuals are different interest, but there are similarities in the recognition of equal value.

Family values change analysis revealed that younger people are looking at the more liberal family and marriage. Labor relations of the young people supported the provisions that must be the choice to work or not to operate in freedom. The dynamics of religious values in 1990-1999 Analysis showed that younger residents of Lithuania are less religious, their religion is often perceived as a cultural heritage, the most serious scholars of Christian life in the absence of base (Uzdila, 1993). According to JA. According to the re (2005), Lithuania stands out from other European countries in terms most favorable to the Church's moral teaching, family and spiritual needs, and even social challenges. Catholics identify themselves Lithuania more than 80 percent. population. But this may simply show a common cultural attitude that the Church does not teach bad things, because the assessment of its effective role in public life, as many as 27 percent. Catholics themselves bearing in 1999 believed that the Church's role in our society is too large. Formal catholicity and show that more than half of Catholics themselves bearing personal prayer life is practically nonexistent. Half the Catholic self-supporting people of their faith in no way linked to the Church, because it does not feel the need to go, or feel that their religious beliefs to practice Church of the total needed. Church attendance once a week or more in 1999 Lithuania was only 18 percent. These data confirm that much behind in Ireland, Poland, Italy, Spain and Belgium.

Fundamental Christian confession of faith actually Lithuania more or less different from the average, but up to now in Lithuania many people profess to be a sin, or that he is God. In the Catholic countries, we greatest confession of reincarnation (27 percent). Attention is drawn to the fact that a very high level of baptism, marriage and funeral rites of ecclesiastical importance of evaluation, as well as participation in them, so far untapped evangelization and ecclesial sense of community strengthened. 1999 religious communities were increasingly turn on (at least once a week), only 8 percent. people and 14 per cent. at least once or twice a month. It should be noted that even in Estonia and Latvia in particular, people's participation in religious communities is higher, although the confessional addiction rate is lower.

It is clear that our general rule not just Catholicism is based on the ownership of cultural traditions, now and situational social conformism. So far, the ongoing acceptance of Church authority yet does not lead to real and personal belonging to the Church and the Christian community identity is confirmed by our public and social value orientation behavior analysis. The most painful manifestation of the biggest challenges is profound vulnerability of the human person - a wide-ranging personal self-esteem and loss of human dignity. Decay exactly what is and should be the concern of the Church. However, the studies provide very interesting and important information about the human desire to join a believer in the Church's spiritual environment, even globally through the sacraments and tradition. Recent studies on the family situation of the Church, as a couple for marriage preparation for the Sacrament of both scientific as well as philosophical, religious and psychological points of view shows how a man should be God.

***Research:** couple marriage preparation process measurement results clearly indicate their willingness to mingle with the church spiritually.*

Recently noticed that more and more couples decide to marry his testimony to the Church, etc. Y. take the sacrament of marriage. Persons seeking to marry in the Church, faith, this may be very different. The first pastoral duty - to help reveal and explain why the church tends to authorize rites for those whose sense of humor is perfect. Often on social views, but not for religious reasons. But we must not refuse an application-based reasons, if couples are integrated into the baptism of Christ's covenant with the Church. It can not be allowed to perform ceremonies where the spouses expressly deny what the Church by the sacrament of marriage.

Figure 1 shows the trend of adoption of the sacrament of marriage

Figure 1: Grant St. sacrament of marriage. Peter and Paul Church (2009-2013) Vilnius

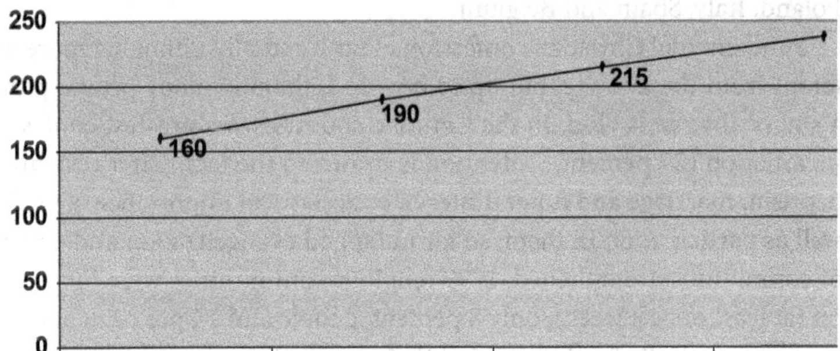

	2009	2010	2011	2012
Sacrament of Marriage	160	190	215	238

The analysis of literary sources, the question arose as how the processes of globalization and the cultural and human values. According to the researchers of globalization (Dion, 1988; Ekerte-Jafe, 1996; Kuzmickas, 2003), we have to take into account all forms of human activities that provide meaningful life. Man perceives himself as a practice entity. Similarly, he is aware of their actions and choices. Culturologists E. Hatfield (1986), A. Andrijauskas (2000), I. Geciene

(2000) believe that human activity is an important communication and creativity. They focus more on human values and lifestyles. Modern western society, according to A. Andrijauskas (2000), associated with the spiritual, moral perfection, which can not be separated from moral values as a whole.

Good and evil are two very basic categories of moral evaluation, since each begins with a description of an ethical theory. Believing the highest human good - God. In view of this, as he understands God, and understand and good. Since ancient times, mythological gods and their behavior at the forefront of mercy and punishment was the main man guidance, as he must do. During long periods of human history gradually conquering the minds of philosophical theory, in which goodness is associated with one of the principle. Humans are very complex creature, his mind-set up and manage a closely related genetic, biological and psychological factors. A man living in the outside world, but the latter often distorts the inner world of his childhood experiences shaped the relationship between the parents and relatives. In addition, the inner world may change and new relationships with people.

Concise presentation of the author mentioned in the article to a theoretical insights based on philosophy, religion, psychology interfaces with the securities change, visible in many areas. The article on the presumption that traditional values are interdependent world elements characteristic of the whole culture in the broad sense.

Philosophical changes in mind

Philosophical thought is directly related to the spiritual culture of humanity, reflects the search for the human mind to know the whole world and its phenomena, to understand the meaning of human existence. Twentieth century intensive mental health, psychology, evolution science nurtured a variety of theories have become the classic tradition and directly related research areas, their strategies and methodologies. In addition to flourishing philosophical, invoking the rhetoric of the scientific conception of psychoanalysis.

Western culture, specifically the concept of an important personality. How many a man introduces himself as a person, and he expresses himself in the world. Therefore, to understand ourselves, it is important to consciously respond to the questions: Who am I? What does your life? what people want to be? what kind of world I want to live? Human emotions, aspirations, thoughts, depends on human effort to overcome the main constraints on the conflict between social and biological impulses. Someone is aggressive, pleasure seeking biological impulses and internalized social constraints, so the main man is to live - to overcome these constraints. Modern European philosophy remains the same problems being most fundamental. In search of existence, based on the philosophy of science, religion and art. Exploring, interpreting existence, but the philosophy and shows its uniqueness, which can be aggregated to the knowledge nor the faith. Being identical with philosophical reflection than rational abstract structures, or mystical intuition (although the moments as they form part of philosophical knowledge). Being fraught of, philosophy reveals the sensitive nature of this thinking, which would be just and open. And this is not a consistent set of logical operations than the poetic imagination. Be thinking - a distinctive way of life. This is a special duty of thought, whose aim - to keep our existence from slipping into meaninglessness. It is rare, and only the most extreme of all human efforts to the maintenance of spiritual forces in a moment, and which gives meaning to our existence in the world, eternity is a kind of interim assessment of everyday life.

The most important factors affecting an individual interested in people since ancient times. Representatives of the Ancient Philosophy person was treated as a slave in destiny, as God's look-alike as a political animal, as a part of nature, and so on. Antique philosophy egzistencelizm speaking philosophers believed that man is suffering because it is distracting emotions, so he needs to strengthen the powers of the mind. Considered unfortunate that the happy life is peace of mind, which they called Ataraxia, Stoics and happy life as depicted in the absence of desire (apatheia). Euclid argued that there are two of the principles by which cells bind and dissociate. It brings together elements of love and breaking - hatred. Aristotle's famous work "On the soul", wrote that the inherent nature of mind is ensured. He said that the soul lies the reasons for the movement of living beings. Aristotle, which can be described as the first psychologist, did not seek to divide the world. He dismissed the heart of abstraction and its separation from the bodily dimension. According to the philosopher, the soul is a perfect expression of living entities, the essence of physical existence. Z. Freud, after more than 2,000 years in a similar conspiracy.

Christian St. Augustine interpreted the subjective emotional experience. He said that a man going through a constant struggle between and control. St. Augustine believes that human nature dualism: body and soul. Therefore, this man is, earthly, heavenly world and the term 'citizen. The essential fact is that a person has two kinds of interests: secular and benefits associated with the body. Other, ulterior, is related to the human soul. It brings a man to life and death secrets. The main mass of the will – love. Love is above the mind and senses. St. Augustine will head the force is love.

Humanistic ethics has alleged that a man is by nature good and that the destructiveness is not an integral part of (integral) part. However, this approach is considered controversial. Socrates was a source of evil is not innate human tendency, and ignorance. Socrates considered the failure error. According to the Old Testament, the human story begins with an act of sin and human desire from an early age is bad.

The late Middle Ages, has faith in human dignity, power, and native goodness. Renaissance thinkers and theologians have argued that the man inside is bad, but the biggest obstacle to the salvation of human pride, which can be overcome only by believing in obedience to God and God's mercy.

I. Kant, a person - a moral entity, able to rise above the empirical world. This means that a person has the ability to act independently of external circumstances. One can argue that he is able to obey the laws of the mind. Preached the idea of human greatness in education philosophy. Bevertiškumo human vanity and the idea of radical preaching, F. Nietzsche putting superman morality, which measured force and victory.

According to the philosopher S. Kanišauskas (2004), life expectancy is the soul of excellence or perfection of its criteria. In addition to the physical, spiritual and even the death, much worse than the physical. Experienced a spiritual death can be described as "living corpses." They are not madmen. They are no longer living, it feels great. They could be called a soulless people. Of course, and they have a spirit and soul. However, the spirit and tightly surrounded by a thick earth, emotions, desires, instincts coating, so that they. „The living dead "may be physically very strong, because they are primarily concerned with the body, matter, but they are alien concepts such as love, even close to man, forgiveness, devotion, not to speak about God.

Psychoanalysis and philosophy of interaction

Psychoanalysis - self-awareness and treatment method, developed by the nineteenth and twentieth century junction. This approach led to some understanding of the human psyche revolution has entered a new era not only in medicine but also in culture, philosophy, and therefore it is natural that a man who wants to understand ourselves and the world, there are a number of questions. Psychoanalysis as the founder and creator of Z. Freud, Austrian psychiatrist.

Jekentaitė (2007) suggests that the preconditions for the emergence of psychoanalysis led to the nineteenth century second-hand turn from the philosophical thought for centuries the dominant systematic rationalism to A. Schopenhauer

(Schopenhauer), F. Nietzsche (Nietzsche) theories. The work of philosophers, human rights, while remaining proud of their minds, a rational handling of its subject's life, the idea changed the confused, suffering individual who is a regular choice of lead and carry the responsibility for the burden of their choice. With respect to F. Nietzsche's philosophy No. Jekentaitė (2007) states: "Nietzshe's insights were the first lessons of psychoanalysis deep-European spirit to clarify the division."

Psychoanalytic Z. Freud's theory of positivism is guided close to the scientific knowledge of human perception and interpretation of the spirit world in search of its natural biological mechanisms. At the same time he approached the inracionalizm as subjective as the world knew instinctively irrational, unconscious the whole event behind the human conscious thoughts, words and actions. The core element of Z. Freud's theory of principle - the libido (sexual drive of human energy, unconscious mental activity of human source) can be redirected. These are all inclinations can be a creative energy. Human moral conscience, its core - of conscience, guilt, sense of responsibility - the appetite and the reality effect of the conflict.

In many cases, Z. Freud was a typical representative of the philosophy of education in the hope of the human mind and the right to protect their natural needs of the social conventions and oppression. Z. Freud's theory is dualistic. He sees a man basically evil, but is controlled by two equally strong opposing forces of life and death, love and strife, day and night, black and white. The choice between good and evil is a basic human way. It is choosing between productivity and destructiveness, and inability to ability, virtues and vices. According to Z. Freud (2009), destructiveness inherent to all human beings.

Destructive forces of human life comes from human sensory, emotional, physical, intellectual abilities of expression blocked. If you are prevented from living increase is blocked energy changes and transforms the life of destructive energy. Destructiveness is a subsidiary of human potency, which occurs only when it fails to achieve its original potency.

Z. Freud focused on the social and cultural purposes. He argues that both the individual and the whole society (or humanity), life is a difficult challenge. Therefore, people think it is looking for faith. Religion, Z. Freud was a culturally sanctioned the illusion of social control. In essence, all forms of religious practice - faith, rituals, mystical experiences, ethical standards - Z. Freud was treated as an expression of neurosis and Infantilism. He describes himself as "nobody tics" an individual, not communicate with the spirit and based on its mind and has not experienced the wonders of (Freud, 1992). Culture refers to the formation of constraints and lean to the irrational anxiety, fear, guilt, respect for authority.

Hostility to religion may be linked to the historical roots of cultural psychoanalysis. Sacred sense of that time in Europe have decreased rapidly. Industrial revolution was destroying traditional forms of communal life and values in science is unquestionably established itself positivist ideology.

Concept of Analytical Psychology

Q G. Jung was, Z. Freud's disciple and collaborator. He explored the dynamic drive impact of human activities and behavior. Q G. Jung believed that the unconscious lies much more than sexuality or aggression drive. He was convinced that the person is unconscious images, which come from human evolution and history. They are innate. A more comprehensive explanation of Q G. Jung used the term collective unconscious.

C.G. Jung identified three partially related systems of personality: ego, personal unconscious and collective unconscious. Ego – a center of consciousness, connecting thoughts, feelings, memories, sensations, which helps us to understand ourselves as a human being as a whole. Ego includes not only internal but also external perception of the world and knowledge about it. Basically - it is consciousness. Personal (proprietorship), the subconscious mind is aware of past experiences, trends, achievements that have been forgotten. Depends on the personal unconscious and the experiences that are unpleasant, present pain, and were driven out. Personal unconscious contents can easily switch back to consciousness. Ego and personal unconscious is a living person are acquired. The situation is different with the collective unconscious. It comes from individual mental limits. Collective represents the subconscious mind of mankind and the evolution of the elements common to all people. This is a deeper layer of the unconscious, which is a kind of mental field that connects all of humanity and the existence of the individual limits. The collective unconscious and the secret lies in the tendency that way, they do understand the reality. Q Jung I do not present the actual psychiatric center, as it seems to us today. " According to R. Grigas (1995), natural, naturalistic human psychiatric centers manifests itself "it is a subconscious way to inherit Anna, archetypal and cultural elements."

Q.G. Combines all the religious events move to the soul behind the scenes' psychology: "Religion is a living relationship with the mental processes, independent of consciousness taking place across the soul, behind the scenes" (Wuchterl, 1994).

C.G. Jung describes in detail some of the archetypes: persona, the shadow of the soul (woman and man image) identity. Q G. Connect the archetypes are the common humanity of the collective unconscious. R. Grigas (1995) states: "A man by his nature, inherited through the genes and cells to inherit a social (as well as national) orientation, KA G. Young's mother tongues - collective - tribal subconscious mind. That nature, upon the normal conditions for the development and man, and people may give greater resilience, creativity, expressions (of course, if the artist was able to get closer to human nature, its existence secret). "Similarly, the collective unconscious explains the importance of other Lithuanian philosophers A. J. Graham (1991), VA Kavolis (1994). Lithuanian philosophers have devoted considerable attention to the relationship between an individual and personal interpretation. Irrelevant to the nation and its language a person is helpless.

However, as already mentioned, the phenomenon of nationalism philosopher understood only as a second stage of existence. Humanity is like a nation-wide individual people, because all people are important, they need to compete in the cultural sphere, free from violence and abuse. Q G. Jung emphasized the importance of balance and harmony. He cautioned that modern people rely too much on science and logic and that the best way to combine science and spirituality.

Conclusions

Science, and religion to interpret the world that surrounds man. Only going in different directions. Science relies on the power of human reason, it is based on the experiment, measurement. More religious knowledge - insight, intuition, moving the material reality (transcendentalism) path. The former is based on the materiality of the world and his knowledge. Second - focus on spiritual phenomena in the human physical senses or mind, even for potential dormant in the world. These different paths led to fundamental understanding of human consciousness forms of incompatibility, which has dealt with since the ancient philosophers.

The researchers note that personal relationship with religion are closely linked with other more traditional values of family, the authority of the community. Religiosity, faith belonging to the family guaranteed acceptance in the community. Church ordered the people to preserve the tradition intact and training your sermon, Bishop. Grušas G. (2013), refers. Religion not only gives people a sense of community in a broad sense, but also fosters a sense of sociability on these fundamental aspects of social responsibility, fear of retribution. Thus, religious dogmas set by the socialization process in connecting people, and what actions the public tolerance for his actions, encouraged, and what not, what can be expected to breach the general rules for living together.

Today, the human consciousness of the moral pollution of instability, particularly important to find peace within myself. This finding is of God, faith and life in him. Philosophers of all ages based on the eternal life of God and through God. Pope Francis, and Pop. Benedict XVI, in his speeches on science and spirituality in education as well as convergence points, and service to man and society as a whole, eternal God, creator of a man and man.

References

1. Advilonienė Ž. Religinis tapatumas ir religingumas posovietinėje Lietuvoje. Sociologinis žvilgsnis. *Filosofija. Sociologija*, 2005, Nr. 4, p. 12-20.

2. Andrijauskas A. Kultūros univeraslumas: morfologinės analizės prolegomenai. In: *Kultūrologija*: straipsnių rinkinys. Sud. Arvydas Matulionis. Vilnius: Kultūros, filosofijos ir meno institutas, 2000, t. 6, p. 12–165.

3. Aristotelis. *Apie sielą*. Vilnius: Valstybinė politinės ir mokslinės literatūros leidykla, 1959, 203 p.

4. Baranova J. Literatūra ir psichoanalizė. *Literatūra ir menas*, 2005, Nr. 5.

5. Buber M. „Aš ir tu", 2000 m.

6. Dion K. Romantic Love: Individual and Cul-tural Perspectives.. In: *Psychology of Love*. Eds. R. Sternberg, M. Barnes. New Haven: Yale UP, 1988.

7. Ekert-Jafe O., Sofer C. Formal Versus Informal Marriage: Explaining Factor. In: *Evolution or Revolution in Europe*. European Population Conference. Milano, 1996.

8. Gečienė I. Gyvenimo kokybė Lietuvoje. In: *Kultūrologija:* straipsnių rinkinys. Vilnius: Kultūros, filosofijos ir meno institutas, 2000, t. 6, p. 278–294.

9. Greimas A. J. *Iš arti ir toli*. Vilnius: Baltos lankos, 1991. 232 p.

10. Grigas R. *Tautos likimas*. Vilnius: Rosma, 1995. 450 p.

11. Grušas G. *Lūžis tarp bažnyčios ir eilinio kataliko: autoritetas, sąžinė ir tikėjimo iššūkiai*. Prieiga per internetą: <http://ateitis.org/saat/js2005grusas>.

12. Ralys K. *Rengimo šeimai pagrindai*. Vilnius: Vilniaus pedagoginio universiteto leidykla. 130 p.

13. Raths L. Compendium the word 1966 m.

14. Hatfield E., Sprecher S. Measuring passionate love in intimate relations. *Journal of Adolescence*, 1986, Vol. 9, p. 383–410.

15. Froidas Z. *Kasdieninio gyvenimo psichopatologija*. Vilnius: Mokslas, 1992. p. 128.

16. Freud S. *Psichoanalizės įvadas*. Paskaitos. Vilnius: Vaga, 2009. p. 488.

17. Jekentaitė L. *De profundis. Psichoanalitinės filosofijos žvilgsniu apie mąstytojus ir menininkus*. Vilnius: Lietuvių literatūros ir tautosakos institutas, 2007. p. 519.

18. Jekentaitė L. *Froidizmas ir humanistinė psichoanalizė*. Vilnius: Mintis, 1992. p. 132.

19. Jekentaitė L. *Etikos ir moralės pagrindai*. Kvalifikacijos kėlimo kursų medžiaga.

20. Johnson P. *Žydų istorija*. Vilnius: Aidai, 1999. p. 872.

21. Jung C. K. *Psichoanalizė ir filosofija*. Vilnius: Pradai, 1999. p. 438.

22. Kanišauskas S. *Likimo anatomija: nestebuklingi stebuklai*. Vilnius: Gardenija, 2004.

23. Kavolis V. *Žmogus istorijoje*. Vilnius: Vaga, 1994. p. 565.

24. Kuzmickas B. Globalizacija – priklausomybės ir tapatybės. In: *Globalizacija*. Vilnius: Lietuvos teisės universitetas, 2003.

25. Kvieskienė G. Socializacija ir vaiko gerovė. Vilnius: Baltijos kopija, 2003. p. 249 .

26. Theodorson M. *Pastor in the word and family*. 1969.

27. Uzdila J.V. *Dorovinis asmenybės ugdymas šeimoje*. Vilnius Academia, 1993.

The Present-Day School and Teacher's Professionalism: the Parents' Evaluation

Vilija Grincevičienė

Abstract

The paper aims to reveal how the pupils' parents / guardians evaluate the present-day school as an educational institution and the teacher as an educator of a developing personality and a *professional*. Once the characteristics of family and school interaction are determined, it is possible to establish feedback concerning the efficiency of the developed model of education change as it faces the reality of education.

The data of long-term research (school year 2002/2003 and school year 2012/2013) make it possible to state that the dynamics of the interaction between the pupils' parents / guardians and the school is marginally positive: during the surveyed period the number of families that frequently communicate with school increased. Moreover, the number of pupils unwilling to attend school decreased. The expectations of pupils' parents in regard to the teacher's professionalism are still high: only personalities can educate personalities. The research revealed that the pupils' families are not indifferent to the processes of the education system reform and wish the organizers of the education process to hear out their ideas and suggestions. Nevertheless, the parents have no intention of taking over the role of initiators in the enhancement of partnership between schools and parents: the community of educators, that still possesses the status of the leader, has to design the model of interaction between the active main institutes of socialization (family and school) in the realm of primary socialization.

Keywords: school, pupils' parents, pupils, the teacher's professionalism, institutes of socialization, identity

Problem

Georg Wilhelm Friedrich Hegel (1770-1831) identified school with the mirror of the epoch. It means that the establishment and assessment of the socio-dynamics of a specific society enables the highlighting of the trajectory which characterizes the development of schools in that country.

When the Lithuanian system of education integrated into the realm of education of the European Union, there began rapid changes of life structure as well as of the dimensions of time and space. The social changes conditioned changes in the whole education system of the country, but the most obvious changes took place in the sphere of general education. Challenges in the life of school became a routine: the model of the change of the process of education designed by the policy-makers transformed into a new quality after encountering the reality of education. The analysis of the changes of general education schools from a pedagogical-sociological point of view helps to determine the scope and trends of the changes. One of them is the sphere of changes in the processes of education; and its trends are the interaction of the participants of education, the teacher's social status in the society, the prestige of the teachers' profession and the teacher's professionalism.

The reform of the national system of education, especially of its primary segment – *general education schools* – aims at the development of an independent, creative personality who is able to reveal his/her abilities to the maximum at school. For this purpose the curriculum (*what should be taught?*) was respectively designed, its design, the problems of organizing teaching (*how things should be taught?*) were approached and democratic relations between the teachers and pupils were formed. The goal was to make the pupils' parents directly participate in the process of education and become a constituent part of the school community.

The State Strategy for Education of 2013-2022 presents an ambitious goal: *to attain that the educators' communities would consist of continuously developing and efficiently working professional teachers and instructors* (8). Professional teachers (educators of personalities) and the continuous of their qualification are a necessity which appears in the context of the rapid social changes. Vaitkevičius argues that *Lithuania will not be reborn until school is reborn, and the school will not be reborn until the teacher is reborn* (6). On the basis of this idea, educational and social policies have to be responsibly designed, and teacher training has to be based on the paradigm where teachers are oriented to the pupils' personality and its main idea is that the major factor in the process of school education is the learners' motivation. The teacher and the pupil are partners aiming at a common goal.

The system of education is the key social and socialization institute. When the system changes (general education schools are reformed), in due time it is relevant to adequately assess the attitude of all the subjects of the education system (teachers, pupils and their parents), their interests and expectations in regard to the process of education since this is feedback concerning the efficiency of the continuously developing model of education change after it encountered the reality of education. The determined negative phenomena can be treated as the trajectories of the further development of the system of education, hence, it can be assumed that the reform of our national system of education can be become a nationwide project which involves the active and creative cooperation of all the participants of education.

The key normative documents regulating the process of changing the national system of education (The Law on Education of the Republic of Lithuania, The Provisions of the State Strategies for Education of 2003-2012, the State

Strategy for Education of 2013-2022, etc.) emphasize the fact that parents are a constituent part of the school community (4,7,8). On a theoretical plane there arises no difficulty concerning the curriculum of general education school which includes the parents' participation; nevertheless, research shows that the reality of education is different: there is an evident discrepancy between the theoretical and real models of school community (2). A close interaction between the school and family, real partnership is still a goal to be attained (5).

The problem has not been yet sufficiently investigated: there are relatively few exhaustive scientific studies which would reveal the opinions of pupils and their parents, their attitudes and expectations concerning school as an institution for developing personalities and concerning the professionalism of teachers (educators of personalities). While only a partially determined attitude of the participants of education (teachers, pupils, their parents or guardians) to the present-day school impedes the generation of insights in regard to feedback which is especially important for the design and construct of theoretical statements underlying the further trends of modernization of general education schools.

The research object is the pupils' parents' (guardians) attitudes and evaluation in regard to school and the teacher's professionalism.

The research goal is to reveal how the pupils' parents (guardians) evaluate the present-day school and the professionalism of teachers who educate their children.

The research objectives:

1) to investigate the strength of the interaction between the pupils' parents (guardians) and the school;
2) to reveal the pupils' parents' (guardians) attitudes to their children's motivation to attend school;
3) to examine how the pupils' parents (guardians) evaluate the change of educational processes in the context of general education.

Methodology

The study is based on: a) the paradigm of structural functionalism: the society is a solid, integral, social system which is composed of many different elements and each of them has its own specific function; all the elements function on the basis of strict interaction and constitute a coherent system which guarantees a normal function of the whole organism; b) humanistic theory of personality development (Rogers) which emphasizes man's freedom, his goal to actualize himself, develop his innate abilities; there is a focus on the principal provisions of Rogers' theory:

– a person is oriented to his goal and is able to reach for it actively; and that is a contradiction with the idea that people's behavior depends on uncontrollable forces;

– when nothing hinders the realization of a person's nature, he acts in a positive way – searches for harmony with himself and others;

– humankind tends to move towards social responsibility, creativity and maturity.

The Research Methods

Theoretical: review of sociological, pedagogical, psychological, philosophical literature, normative acts which regulate the system of education and data of the educational process in regard to a chosen aspect;

A contrastive analysis of philosophies, conceptions and attitudes;

Empirical: interviewing the respondents (the pupils' parents / guardians) aiming to determine their opinions, attitudes and expectations, as well as the analysis of their letters.

The Research Technique

The initial empirical data were collected by means of the quantitative method. The research was conducted in the school years 2002/2003 and 2012/2013. The research geography covered all the regions of the country. The sample volume: the pupils parents (guardians): 30-50-year-old residents of Lithuania. The design of the sample refers to the data basis of the Lithuanian Department of Statistics: the residents in regard to their age and gender. The quantitative scope (the residents from 30 to 50 years old) was relatively equated with the whole of the general population:

http://db.stat.gov.lt/sips/Dialog/varval.asp?ma=gs_dem05lt&ti=Gyventojai+pagal+am%FEi%F8+ir+lyt%E1&path=../Database/cen_lt/s71lt/1.%20demografija/&lang=1.

When the total of the general population is about 100000, the sample volume should be > or = 398 (3. p.322). Since the research results are applied not only in regard to the research group, but also in regard to the whole researched population, aiming at representability, the sample included 700 respondents in 2002/2003 and 600 respondents in 2012/2013.

The Design of the Random Sample

The random sample was designed aiming at representability. The error level is 0.05 since it is the typical measure in social research. The sample volume was based on the "snowball" principle. First of all, from all the regions of the country there were people selected who have school-aged children. Then, these respondents indicated others, and the latter – still other, etc. In such a way the sample size grew:

- up to 700 units (in 2002/2003). 641 filled-in questionnaires returned;
- up to 600 units (in 2012/2013). 502 filled-in questionnaires returned.

The initial data was encoded and statistically processed with the help of SPSS 10.0 / Windows software for statistical data analysis. The data was analyzed by means of descriptive statistics.

The study applied methods of mathematical statistics confidence intervals. Assuming that the reliability is 95 %, the interval then is:

$$[(a-1{,}960\sqrt{\frac{a(1-a)}{m}}\,\%;\ (a+1.960\sqrt{\frac{a(1-a)}{m}}\,\%]$$

here m – the number of the respondents (sample size);

$$a = \frac{m_A}{m} \bullet 100\% \text{ - percentage of the respondents who have quality (A);}$$

m_A – the number of the respondents who said that they have quality (A).

Information about the Respondents

The research involved >87 % of women. The distribution of the respondents according to their age is as follows:

- Younger than 40 – 65 %;
- 41 – 50 years old – 31 %;
- 51 - 60 years old – 4 %.

According to their education the parents represented all the levels:

- Secondary – 2 %;
- Vocational – 16 %;
- Advanced vocational – 44 %;
- Higher – 38 %.

Most of the respondents (78 %) lived in the five biggest cities of Lithuania, 14% - in towns (regional centers) and 8 % - in villages (small towns, settlements).

According to their adherence to social groups the pupils' parents distributed in the following way:

* housewives – 8 %;
* non-manufacturing specialists (specialists of culture, education, healthcare system, lawyers) – 13 %;
* manufacturing specialists (engineers, technologists) – 13 %;
* workers – 31 %;
* farmers – 1 %;
* businessmen (having their own business) – 14 %;
* clerks (providers of different services, hired specialists) – 20 %.

The Research Results

The analysis of the research results revealed that the interaction between parents (guardians) and school during the target period was not close: every second respondent visited the schools of their children only when they attended parental meetings (Table 1).

Table 1: The Interaction between the Pupils' Parents (Guardians) and School

School year 2002/2003 **Statements:**	**Percentage and confidence intervals of the percentage, reliability - 95 %**
I have hardly any contacts with the school and I don't look for them	2.0 [0.9 – 3.1]
I visit the school only if there is a necessity	19.0 [16.0 – 22.0]
I visit the school only during the parental meetings	56.0 [52.2 – 59.8]
I frequently contact the school	22.0 [18.8 – 25.2]
Did not respond	1.0 [0.2 - 1.8]
School year 2012/2013 Statements:	Percentage and confidence intervals of the percentage, reliability - 95 %
I have hardly any contacts with the school and I don't look for them	1.0 [0.1 - 1.9]
I visit the school only if there is a necessity	17.0 [13.7 – 20.3]
I visit the school only during the parental meetings	53.0 [48.6 – 57.4]
I frequently contact the school	28.0 [24.1 – 31.9]
Did not respond	1.0 [0.1 – 1.9]

The analysis of the data demonstrates that only every third or fourth family with school-aged children can enjoy a close cooperation with the school, i.e. the teachers and children's parents follow the principles of good partnership. Every fifth-sixth parent/guardian was quite self-critical and said that they visit school only if there is a necessity. The research results revealed the existing relation among the respondent's gender, age, education and the intensity of interaction with the school: women cooperate with teachers more closely than men; young parents (up to 40 years old) visit their children's schools more frequently than older parents; the respondents possessing higher and advanced vocational education have more frequent contacts with the school than parents (guardians) with secondary or vocational education. The parents of the primary school pupils are a distinctive group of the respondents. They possess the strongest partnership with the school. The parents of primary school children are especially attentive to their children's well-being at school; they care about the children's motivation to get good results and their wish to attend school.

The comparison of the research conducted in 2002/2003 and 2012/2013 revealed that the interaction dynamics of the pupils' parents (guardians) and the school is only marginally positive: in the recent period almost every third family (28 %) frequently contact their school. Nevertheless, the stereotype that parents should only visit school only to attend parental meetings has survived up to date: every second parent (guardian) was self-critical and said that visit school only for parental meetings. The analysis of the respondents' letters revealed that the pupils' parents (guardians) would want to take more active part in the design and modeling of the key aspects of the process of education, i.e. they would want to become full-rate members of the school community. Families are not indifferent to the processes of the reform of the system of education and they would want the organizers of the education process to take their ideas and suggestions into consideration. Obviously, the parents (guardians) have no intention of taking over the role of initiators to enhance the partnership between the family and school: the teachers' community, who possesses the status of leadership, will have to continue looking for ways and methods to improve the interaction between the school and family (society).

The long-term research revealed the dynamics of the pupils' parents' (guardians) attitudes to the strength of children's motivation to attend school: the growing prestige of science and research in the country marginally affected the young generation's (children and adolescents) motivation to pursue education. The question ***Would your children attend school if it was not compulsory?*** Was answered affirmatively by the majority of the respondents (65.0 – 68.1 %). The comparison of the research data reveals a marginal increase of the dynamics of pupils' motivation to attend school: the number of the respondents who believed that their children would probably not attend school decreased (5.2 – 3.8 %)

or would definitely not attend (2.8 – 1.4 %). Nevertheless, the problem of not attending school still exists in the country. Only every third respondent (27.0 – 26.7 %) stated that their children would probably attend school. The social changes had hardly influence on this result (Table 2).

Table 2: The Parents' (Guardians) Attitudes to their Children's Voluntary School Attendance

School year 2002/2003 Statements:	**Percentage and confidence intervals of the percentage, reliability - 95 %**
Would definitely attend	65.0 [61.3 – 68.7]
Would probably attend	27.0 [23.6 – 30.4]
Would probably not attend	5.2 [3.5 – 6.9]
Would definitely not attend	2.8 [1.5 – 4.1]
School year 2012/2013 Statements:	Percentage and confidence intervals of the percentage, reliability - 95 %
Would definitely attend	68.1 [64.0 – 72.2]
Would probably attend	26.7 [22.8 – 30.6]
Would probably not attend	3.8 [2.1 – 5.5]
Would definitely not attend	1.4 [0.4 – 2.4]

The pupils' refusal to attend school is a multifaceted problem. The reasons can be socio-cultural, pedagogical and psychological. According to Targamadzė, the adolescents' problems of learning motivation have to be solved from different perspectives (pedagogical, psychological, social, cultural, medical, etc.) and there should be a programme prepared for the purpose (1, p. 227). Cold relationships with teachers could also be one of the reasons why children do not want to attend school. A professional teacher should be able not only to diagnose this problem, but also to act energetically, professionally and in due time to find its possible solutions.

The research aimed to determine how the pupils' parents (guardians) evaluate the expression of the changing educational processes in the present-day context of general education. The questionnaire also asked "***What positive changes have you noticed in your children's school?***" However, only half of the respondents answered this question, hence, it is complicated to assess the present situation at length and in great detail. The pupils' parents must have been confused when they read this question: every third emphasized the fact that there are no positive changes in their children's schools. The respondents who answered this question enumerated the positive aspects of their children's education:

- There appeared more extra-curricular activities (dance, music, theatre, fine arts);
- The premises have been renovated (classrooms, cloakrooms, corridors, toilets, etc.);
- There appeared more young teachers;
- More attention is paid to the maturity of the child's personality;
- The children's clothing was unified (uniforms);
- There are more excursions organized (more interesting after-class activity);
- Competent teachers work at school;
- There appeared a possibility to contact teachers via e-mail;
- More attention is paid to preventing smoking and other harmful habits;
- The children's motivation to attain good study results is developed.

Only a deeper situation analysis could reveal what positive features of the changing school the children's parents (guardians) tend to emphasize and evaluate positively in the dynamic context of the present-day general education.

The research also aimed to reveal the negative changes observed by the pupils' parents (guardians) at school: the questionnaire also asked ***What negative changes have you noticed in your children's school?*** Unfortunately, this question was answered by every second respondent. Every fourth respondent (25 %) emphasized that there are no negative changes at school. The rest of the respondents enumerated the issues they find concerning at their children's schools:

- Inconsistent programme of the Lithuanian language;
- Problematic relationships between the pupils and teachers;
- Problematic relationships among pupils;
- The parents are indifferent to their children's behavior;
- The workload problem (the children are overloaded with tasks);
- The problem of drugs, smoking;
- The lack of competent teachers;
- High tension before examinations;
- The problem of violence;
- Social and national inequality;
- The programmes of the school subjects are inappropriately prepared.

The fact that this question was answered only by a half of the respondents hinders the exhaustive interpretation of the situation – a deeper analysis is necessary. However, a proper consideration of the parents' (guardians) opinions enables the analysis and adequate corrections of the process of general education school change. This is a part of feedback concerning the efficiency of the model of education change which constantly develops as it encounters the reality of education.

The research data also reveal that the pupils' parents did not really emphasize the problems of national identity and integrity in changing (global) world when they were evaluating the changes in the present-day school. Nevertheless, the respondents' observation that in their children's schools there exist social and national inequality is of great concern and presupposes the initiation of an exhaustive study aiming to disclose the situation. But primarily this is an incentive for teachers to consider the guidelines of developing the model of national, civic and patriotic education while cooperating with the pupils' parents.

Conclusions

In the realm of primary socialization there interact two institutes of socialization: families and schools. Their strong partnership is one of the guarantees that the process of socialization of a maturing personality will be consistent and sustainable since socialization preconditions the expression of the potential abilities and skills of a personality.

The analysis of the long-term research data revealed that in our country the interaction between school and family was not strong in the target period: only every third family of a general education school pupil consistently communicates and cooperates with teachers; most parents (guardians) socialize with educators only episodically, i.e. during parental meetings or in case of necessity; the parents' (guardians') input in modeling and organizing the process of education has recently been episodic; partnership with school is of higher quality in the group of young (up to 40 years old) parents (guardians) who have higher or advanced vocational education.

In the context of the changing national system of education (the reform of general education) the pupils' motivation to attend school marginally grew: the number of parents who believed that their children would probably not attend school or would definitely not attend it decreased. Nonetheless, the problem of children and adolescents who do not attend school has not been solved in our country yet. This is a multifaceted problem whose reasons and solutions should be reached for consistently by the education policy makers, strategists, researchers and all the participants of the process of education.

The results of the long-term research reveal that the pupils' parents (guardians) ambiguously assess the change of educational processes in the context of general education. It is self-evident that in the target period the expectations of the pupils' parents (guardians) in regard to the school as an institution of education and the teacher's personality and professionalism were high. This is also a challenge for teacher training schools of higher education. The parents' (guardians) observation that in their children's schools there exist *social and national inequality* prompts the organizers of the educational process and teachers reconsider the issue of national identity and integrity in the changing (global) world and focus on the development of the model of national, civic and patriotic education.

References

1. Dereškevičius P., Rimkevičienė V., Targamadzė V. (2000). Mokyklos nelankymo priežastys. – Vilnius: Žuvėdra.

2. Grincevičienė V. (2010). Mokyklos kaitos erdvės ir linkmės: pedagoginis-sociologinis aspektas. – VPU leidykla.

3. Kardelis K. (2002). Mokslinių tyrimų metodologija ir metodai. 2-asis pataisytas ir papildytas leidimas. - Kaunas.

4. The Amended Law on Education of the Republic of Lithuania: March 17, 2011. No. XI-1281.

5. Merfeldaitė O. (2009). Tėvų ir mokyklos bendradarbiavimo stiprinimas sprendžiant vaikų socializacijos problemas. Socialinis ugdymas. Vol. 19, p. 93 – 103. ISSN 1392-9569. Available from: *http://www.i-anager.lt/upload/201004/Socialinis.ugdymas.Nr8(19).-Socialine.pedagogine.pagalba.[2009].pdf.*

6. Vaitkevičius J. (1995). Socialinės pedagogikos pagrindai. – Egaldos leidykla.

7. The State Strategy for Education of 2003-2012.

8. The State Strategy for Education of 2013–2022.

The Development of College Teachers' Intercultural Competences and their Application in Educational Activity

Vida Navickienė

Abstract
The paper aims to reveal the significance of the (self-) development of college teachers' intercultural competence and the necessity of its relation to international mobility. The paper presents the results of the analysis of college missions and visions which demonstrate their intention to expand the limits of internationality, as well the attitude of the college academic community to the possibilities of participating in international mobility.

Keywords**:** college teacher, intercultural competence, international mobility

Introduction

Globalization makes a considerable influence of the participants of higher education –students and teachers who organize various educational activities. The teachers' intercultural competence has become one of the major priorities and this situation is determined by the increasingly frequent interaction with students from different cultures and other participants of higher education (social partners). College teachers are facing a necessity of (self-) development of intercultural competence in order to be able to qualitatively organize the study process and other educational activities – scientific conferences and seminars, forums; to create or pursue the joint study programmes with other institutions of European or other countries, international scientific applied research and projects; to work with Erasmus students, etc.

Hence, the ***goal*** of the paper is to analyze the possibilities of developing the college teachers' intercultural competence and their application in educational activity with the emphasis on the aspects of international mobility.

The research methods: Scientific literature review; document analysis; group, in-depth, semi-structured interview.

Bitinas (2006) argues that the goal of a group interview is to reveal not only the experience of the participants of the process of education, their attitudes or any other target features, but also the group characteristics of the respondents – their unity, dominant values, the public opinion, the peculiarities of the social environment of education, satisfaction with the process of education and its drawbacks, etc. Thus, group, in-depth, semi-structured interview was chosen because of the possibility for the researchers to directly contact the respondents. The group, in-depth, semi-structured interview was based on the strategy of selecting the homogenous cases which is frequently applied aiming at a better well-being of the group participants (5-7 respondents) and a greater focus on the research problem. Colleges were chosen by means of criterion-based selection. In the private and state colleges in different regions of Lithuania there were seven groups of members of the academic community formed: the deputy director for academic activity, deans, vice-deans, heads of departments and different divisions and centers, and teachers.

The Conception of Intercultural Competence

Aiming to define the concept of intercultural competence, primarily it is important to emphasize that intercultural competence is most frequently analyzed in the professional context and is becoming an inseparable part of the professional competence (Paurienė, 2011). Hence, in the discussion of the college teachers' intercultural competence it has to be noted that most Lithuanian and foreign researchers define intercultural competence emphasizing the *ability to work in the miscellaneous / intercultural environment*, for example: "A person's cultural sensitivity, cultural knowledge and skills enabling efficient operation in a miscellaneous environment. Intercultural competence includes the interpersonal, intercultural and professional abilities (Grigaliauskienė, 2009); "Cultural competence is a combination of certain components which includes knowledge, skills, qualities and attitudes of different origin and manifestations. They affect the cognitive and behavioral processes in the cultural context. Competence is perceived as enabling individuals (organizations) to learn and acquire experience, helping to combine various skills determining the activity of organization on the international scale, technologies of activity and successful operation in the present-day situation" (Palaimaitė, Radzevičienė, 2009)

> *"The ability to recognize the limits of behavior expanded to include the aspects of other cultures, the differences and similarities of cultures, and the ability to apply all of this for successful interaction in the situations of intercultural intersections. A strong behavioral orientation towards task accomplishment is also characteristic" (Geistmann, 2002).*

Hence, while developing their intercultural competence college teachers should acquire skills to work with students from different cultures in the miscellaneous cultural environment – both in their own institution of higher education and in some foreign schools of higher education.

The discussion of the college teachers' intercultural competences is purposefully based on Norvilienė's (2011) analysis of the teachers' intercultural competences. In the author's opinion, *intercultural competence* is a combination of certain components which includes attitudes and personal qualities, knowledge, skills and cultural awareness. It affects the cognitive, affective and behavioral processes and enables efficient operation in intercultural situations. Intercultural competence is not distinguished as a separate competence. It is highlighted as the ability to apply all competences essential for a specialist in culturally diverse contexts of activity. Thus, it is possible to assume that the college teachers' *intercultural competence includes general and specific / professional skills which they have to be able to apply in the intercultural environment.*

It has to be noted that, in Virgailaitė-Mečkauskaitė's (2011) opinion, the influence of the general competence and professional context are essential for the development / formation of intercultural competence – hence, the umbrella conception of professional intercultural competence. According to Geistmann (2002), professional competence consists of the specific content of activity (in German *Fachliche Kompetenz*), competence and methodological competence (in German *Metodiche Kompetenz*). Such combination of content, methodological and intercultural competences constitute the competence of professional participation in international activity which is characteristic of any modern profession (businessmen, police officers, teachers, social workers).

Intercultural competence can be defined across three dimensions: the *cognitive* (knowledge about cultural differences), *affective* (motivation and desire to operate in intercultural situations) and *behavioral* dimensions (skills, abilities related to intercultural situations) (Deardorff, 2004; Williams, 2009).

The knowledge dimension of the teachers' intercultural competence covers a wide spectrum of knowledge: information about the general cultural peculiarities and systems of education, science institutions, the curriculum of the study subject, the methods, etc. The influence of the professional and subject-specific context is also revealed in the other dimensions of the teachers' intercultural competence: the skills to discover and interact, to interpret and relate. Hence, the appearance of new elements of the intercultural competence: the ability to communicate with the audience from another country and of a different cultural experience, the ability to "use" the audience's potential of cultural diversity, the ability to interpret the European and other documents regulating education in general and higher education (Mažeikienė, Loher, 2008).

On the institutional level in the Lithuanian schools of higher education the internationality / interculturalism of higher education is most evidently observed through the international mobility of the teaching and research staff and the students coordinated by the Erasmus programme (Žydžiūnaitė, Lepaitė, Bubnys, Čepienė, 2010).

The Analysis of Missions and Visions of the Lithuanian Colleges

The intercultural competence of college teachers is most frequently related to the internationality and international mobility of schools of higher education. Hence, aiming to reveal the significance of intercultural competence for college teachers, the missions and visions (the strategic vision of the future) of the colleges, published on their websites, were primarily analyzed. The analysis aimed to reveal the present activities and future objectives of the colleges, and the environment for the (self-) development of intercultural competence that is important for college teachers.

Mission and vision are the underlying components of planning the strategic activity of every organization, irrespective of their function. According to Želvys (2003), the mission is the point of reference in the organization for everyday activity aiming at the vision of the future. As institutions of professional higher education are developing organizations, the most important element of their strategic planning is a common vision since it "mobilizes energy" (Fullan, 1998, p. 48). Hence, the determination and forethought of the future goals directs everyday activity towards the implementation of the objectives.

The conducted contrastive analysis and systemization of the acquired data on the missions of the colleges, presented in their websites, revealed more diverse tendencies than those which were determined in the group, in-depth, semi-structured interviews (the results are presented in the following chapter). The purposeful comparison of the present-day missions of the colleges with the ones they had at the time of their establishment more than a decade ago revealed that some institutions of higher professional education formulate a too narrow mission statement or do not pay sufficient attention for this issue, while others – conversely – include the concepts and activities which are aimed at and pursues by universities. In the binary system, which is defined by a dynamic rather than static character, the mission cannot stay the same. It is also determined by the Bologna Process which requires developing and activating the international mobility of students and teachers (The Major Documents of the Bologna Process, 2005; the Major Documents of the Bologna Process, 2012). Colleges, as full active members of the Bologna Process, have expanded their mission to include the aspect of internationality, and now the purpose of some colleges is related not only to the needs of the region or country, but also to the "needs of the European labour market". Thus, after a decade since the establishment of the first colleges, these schools of higher education do not limit themselves to the regional level, but rather aim to be acknowledged both on the national and the international levels by participating in international projects and integrating into diverse networks of international partnership. In the specialized colleges the aspect of internationality is especially well-pronounced, as well as their goal to meet "the international requirements of organizations", to prepare and pursue activities which meet "the standards of the European Union", "to take over the global experience in the field", even though these institutions are not big in terms of the number of students of the academic staff. Nevertheless, the missions of some of the colleges lack the aspect of internationality either because of their inactive participation in the international exchange, or because of their careless attitude to the formulation of their mission statement and its presentation for the society.

First of all, the mission is not conceived as a concept / phenomenon or as one of the components of strategic planning. Drucker (2009) argues that the contribution of every member of an organization slightly differs; however, the contribution of all of them has to serve one common goal. Their attempts have to be pointed at the same direction, and their contributions have to match and combine a coherent whole. The scientific literature of management (Drucker, 2009; Jucevičius, 2003; Želvys, 2003 et al.) defines the mission as the overall present-day purpose of an organization.

It is important to emphasize the tendency that none of the missions of private colleges claim their studies are oriented towards answering the needs of the society of the region – the principle of the regional character has disappeared. Hence, private colleges do not relate their activity with the regions they are established in (all the private colleges are established in the bigger cities of Lithuania – Vilnius, Kaunas, Klaipėda, Šiauliai), but rather they aim to attract students from different parts of the country and thus guarantee the continuity of their activity.

The key tendencies of college change are limited by the traditional / present-day system of higher education in Lithuania: universities implement studies of three cycles, while colleges – higher education studies of one cycle. Even though in most European countries (for example, Germany, Finland, Norway, France, etc.) colleges organize studies of professional Master degree, the Lithuanian system of higher education is not yet ready for such transformation. Hence, the tendencies of development of the colleges can be presented only through the prism of their vision – the goals and achievements of the academic communities of the colleges.

According to Želvys (2003), in the rapidly changing world, visions are typically oriented to the period of 3-5 years. The analysis of the visions aims to focus only on the activities which are not involved in the present-day missions to highlight their goals better. On a general scale, it can be stated that due to the dynamism characteristic of the binary system and the attempts of the colleges to improve, the major goals of most state and private colleges are the promotion of internationality and the development of research applied activity. Nevertheless, it has to be noted that Mažeikienė et al. (2008) introduce the concept of research competence which is necessary for international cooperation. According to the authors, the new, broad research competence requires joining the practical competence of conducting research and the general competences. Such specific juncture of the practical research and general competences (*where intercultural competences have a primary role*) in the context of internationalization of higher education and the labour market initiates a discussion on the new, broader umbrella conception of research competences required for international cooperation which answers the peculiar challenges of the global labour market. Hence, the goals of the schools of higher professional education presuppose the unconditional development / self-development of the college teachers' intercultural competence.

The Attitude of the College Academic Communities to Mission, Vision and International Mobility

The group, in-depth, semi-structured interview with the academic communities of the colleges returned little information about their conception and attitude to the pursued mission. The most exhaustive definitions of the mission pursued by their institutions were presented by the respondents who had been involved in the process of formulating the mission, while the others were only able to describe the mission in general terms, referring to the purpose of higher professional education in general. Nonetheless, a certain tendency can be observed – while pursuing the goal for international mobility presented in the documents of the Bologna Process, the colleges have adopted a broader attitude to their missions. Thus, most of the colleges included the goals for internationality in their pursued activity: "we have to be modern, mobile and attractive to others", "to develop internationality, expand the teaching methods, forms..." and justified this as the introduction of new ideas, innovations and higher competences of the college teachers. Aiming at the international aspect of the mission, the academic communities of the colleges prepare study programmes in cooperation with their foreign partners and implement them in English as the language of instruction.

It has to be noted that colleges engaged in international mobility as full active participants when in 2007 they were granted the right to award the Professional Bachelor's Degree. It is related not only with the requirements of a system of easily comparable degrees issued in the documents of the Bologna Process, but also with some other reasons. In order to actively pursue the international exchange of students, the colleges had to establish the offices for international relations with active and sufficiently competent employers and to find contacts, and all of that required significant time and human resources.

The analysis of the international mobility of the academic community revealed different problems that impede a more active involvement in the process. Even though it could be assumed that the main factor preventing from participation in the exchange programme is the insufficient knowledge of a foreign language, the research data revealed other reasons.

International mobility is problematic in a big private college which possesses a diverse academic community. Even though most of the respondents claim that the main reason preventing from their active participation in international exchange is their inability to communicate in a foreign language, in the college there are a lot of teachers of foreign languages, who know the languages perfectly well and still do not want to participate in international exchange. Hence, the professional knowledge of a foreign language is not the main obstacle preventing the college teachers from international mobility.

One of the features of a big private college is that there are a lot of hired lecturers from other educational institutions who engage in international mobility in their main workplace; hence, most of them are not interested in representing another institution. Another feature is that there are a lot of teachers-practitioners, hence, the respondents said that: "practitioners have good positions in their enterprises, so they cannot interrupt their professional activity by leaving even for a short period time" (moreover, teachers-practitioners cannot be full-time employers even though the specificity of these situations requires such professionals in their staff). Thus, the lack of full-time college-teachers impedes active international exchange, and most frequently the same members of the academic community participate in the programme. The informants also believed that the international exchange of the college teachers would become more active with the help of financial motivation. Naturally, this would be possible only for the institutions, most likely the private ones, which can freely administer their financial resources.

Hence, one of the main reasons preventing the academic communities of institutions of higher professional education from active participation in international exchange is their lack of internal motivation (or **affective** dimension (motivation and desire to operate in intercultural situations) (Deardorff, 2004; Williams, 2009) which prompts looking for various forms of excuses why they cannot leave, for example, their fear of cultural difference, inability to leave their families for a week or two, the older teachers emphasize their age, etc.

Speaking about the international teacher exchange pursued by the smaller colleges – both state and private – similar tendencies can be observed: their active participation is impeded by the poor knowledge of foreign languages; it is evident that usually the same teachers participate in the programme. The international mobility of teachers from specialized colleges is stable and this is determined by the traditions established in vocational schools as well as by the young specialists working there. The heads of the private colleges are trying to solve the problems and motivate the academic community in different ways, for example, they organize free-of-charge (or very reasonably-priced) language courses, trips to European institutions of higher education, conduct international projects (their participants are paid for participation), organize weeks of international mobility during which the classes are taught by foreign teachers from colleges and universities. Naturally, this enhances international mobility as contacts are made, disciplinary relations are established, foreign language skills are improved, self-confidence is boosted, etc. Aiming at the enhancement of the rates of international mobility, the networks of international partners start including institutions of higher education of the neighboring countries where teachers could deliver their classes in Russian.

The goal of all the institutions of higher professional education to activate the international mobility of the academic community is impeded by the teachers' workload. The respondents claim that this causes additional problems for the college administration and the participants of the international exchange: "when the teachers leave, the timetables have to be changes, and when they return, they have to make up for the missed classes". Before leaving the teachers also need a good, exhaustive subject-specific and methodological preparation to be able to teach foreign students, this requires a lot of time which is not included into their yearly workload and is not paid for, hence, only highly motivated teachers can resolve to take part in the programme.

A different situation is observed in big state colleges where the international mobility of the academic community is increasingly active. The following response could illustrate the statement well: "in 2007-2008 the plan for teachers' mobility was implemented by 120 percent, there are new projects announced." Naturally, when there are more teachers of different study programmes it is easier to conduct international exchange; there are disciplinary relations established, hence, the teachers need and want to return to the same foreign institutions of higher education. In this type of colleges the priority criterion for participants of international exchange programmes is given for the teachers who have never applied before. Hence, there has to be an ongoing motivation, the spread of information and sharing the good experience of the teachers who have already returned. The analysis of the acquired data revealed one of the most efficient tools of motivating – the introduction of the teachers' certification system (which includes international mobility) into the state and some of the private colleges. The provisions of the certification process list the requirements and their accomplishment gives points for the teachers which promote competition among the teachers, activate them, help to defeat certain personal psychological barriers, for example, their lack of self-confidence, and even initiate the growth of their internal motivation.

Thus, in summary it can be stated that the international mobility of the academic community is one of the main objectives of all the institutions of higher professional education. Aiming at their implementation, there are different ways and forms of motivating the academic community introduced. It is important that the international mobility of students and the academic community is increasing (despite all the existing obstacles), and it is becoming one of the major goals of most colleges. It has to be emphasized that one of the conducted projects created an instrument for the college teachers' self-assessment of their intercultural competence (Žydžiūnaitė, Bubnys, 2010) (there are 13 scales of measurement generalized). The scholars measure the college teachers' international competence in terms of three directions: the organization – the teacher as a provider of teaching facilities at the college and as a member of the academic community of the college; the curriculum – the teacher as a participant of the process of the curriculum change; the components of intercultural competence – the teachers as the assessor of his own trends of qualification development.

Conclusions

1. The (self-) development of intercultural competence is one of the components of the constant development of a personality which is guaranteed through the dimension of life-long learning. The development of intercultural competence should be systematic and cyclic: the development of the competence, sharing the experience, reflection (self-assessment).
2. Due to the dynamic character of the binary system and the attempts of the colleges to improve, the major goals of most of the state and private colleges

are the enhancement of internationality and the development of research applied activity. Hence, the goals of the institutions of higher professional education determine the unconditional (self-) development of the college teachers' intercultural competence.

3. One of the major reasons impeding the active participation of the college teachers in international exchange programmes is their lack of internal motivation (the affective dimension) which is supported by searching for different forms of excuses why they cannot leave, for example, the teachers' fear of cultural difference, their inability to leave their families for a week or two, the older teachers emphasize their age, etc. The academic community of the private institutions of higher professional education applies more diverse forms of developing intercultural competence than the state colleges since they can administer their financial resources more easily.
4. The development of the college teachers' intercultural competence should be promoted and supported on the institutional level by looking for various ways and forms: there should be seminars with foreign lecturers (both for new knowledge and the experience of interaction with people of different cultures); one of the points of the college teachers' certification process should be their participation in international activities; international mobility should be included into their self-assessment questionnaires (or the institutions could start using the already prepared instrument); the institutions should get more involved into Erasmus students' activities, for example, the presentations of their countries could be organized which could be based on the interaction not only among the students, but also among all the members of the academic community; there could be training sessions organized and every year different members of the academic community could be involved; mutual training sessions, self-education, the development of the teachers' competences, etc. are also the possible options.

References

1. Bitinas, B (2006). Edukologinis tyrimas: sistema ir procesas. Vilnius: Kronta.

2. Deardorff, K. D. (2004). The Identification and Assessment of Intercultural Competence as a Student Outcome of internationalization at Institutions of Higher Education in the United States. A dissertation for the Degree of Doctor of Education. Raleigh, North Carolina [accessed on 29-08- 2013]. Available from: <http://www.lib.ncsu.edu/theses/available/etd- 06162004-000223/unrestricted/etd.pdf>.

3. Drucker, P. F. (2009). Drukerio mokymo pagrindai. Vilnius: UAB „Rgrupė".

4. Fullan, M. Pokyčio jėgos: skverbimasis į ugdymo reformos gelmes. - Vilnius: Tyto alba, 1998.

5. Geistmann,Chr. Interkulturelle Kompetenz - Eine wichtige förderbare Fähigkeit in der internationalen Zusammenarbeit. Norderstedt: Books on Demand GmbH. 2002.

6. Grigaliauskienė, E. (2009). Įstaigų, institucijų, formuojančių ir įgyvendinančių valstybės politiką, bei darbdavių žinių ir kompetencijų poreikiai darbui su imigrantais bei pačių imigrantų poreikiai sėkmingai integracijai. Vilnius. Available from: http://www.google.lt/url?sa=t&rct=j&q=&esrc=s&source=web&cd=1&ved=0CCwQFjAA&url=http%3A%2F%2Fwww.chambers.lt%2Flt%2Ffailai%2F26_Tyrimo%2520isvados.doc.doc&ei=_orWUsfUD46RhQem5ICgAw&usg=AFQjCNESX7GawulzAHu46Vz7_tSl0lKI1g.

7. Jucevičius, R., Jucevičienė, P., Janiūnaitė, B. ir kt. (2003). Mokyklos strategija: strateginio vystymo vadovas. - Kaunas: Žinių visuomenės institutas.

8. Mažeikienė, N., Loher, D. (2008). Teachers Intercultural Competence in Mobility Programmes. Socialiniai mokslai, 2(60).

9. Mažeikienė, N., Zubilina, O., Virgailaitė-Mečkauskaitė, E., Ruškus, J. (2008). Research Competence for International Collaboration in the Process of Higher Education and Labour Market Internationalization. Socialiniai mokslai, 2 (60).

10. The Major Documents of the Bologna Process. The Bologna-Bergen Period, 1999-2005 (Svarbiausi Bolonijos proceso dokumentai. Bolonijos–Bergeno laikotarpis 1999–2005 m.), 2005. Vilnius: LR švietimo ir mokslo ministerija.

11. The Major Documents of the Bologna Process. EHEA documents confirmed in the Bucharest conference of ministers, 2012. (Svarbiausi Bolonijos proceso dokumentai. EAME dokumentai, priimti Bukarešto ministrų konferencijoje 2012 m.). Available from: http://www.smpf.lt/lt/programos/bolonija/bolonijos_procesas_1/kokybes_uztikrinimas

12. Želvys, R. Švietimo organizacijų vadyba. Vilnius: Vilniaus universiteto leidykla, 2003.

13. Norvilienė, A. (2011). Mokytojų požiūris į tarpkultūrinės kompetencijos raišką. Tiltai: socialiniai mokslai. Klaipėda. ISSN 1392-3137. 2011, No. 3 (56).

14. Palaimaitė, A., Radzevičienė, A. (2009). Kultūrinė kompetencija jaunimo nevyriausybinių organizacijų tarptautinės veiklos kontekste. Mokslas – Lietuvos ateitis = Science – future of Lithuania: Verslas XXI amžiuje. Vilnius: Technika. ISSN 2029-2341. Vol. 1, No. 3.

15. Paurienė, G. (2011). Tarpkultūrinė kompetencija ir jos ugdymas pareigūnų rengime. Visuomenės saugumas ir viešoji tvarka (5): mokslinių straipsnių rinkinys = Public Security and public Order: scientific articles (5) [electronic resource] / Mykolo Romerio universitetas. Viešojo saugumo fakultetas. Kaunas: Mykolo Romerio universitetas. ISSN 2029-1701. 2011, vol. 5.

16. Virgailaitė-Mečkauskaitė, E. (2011). Tarpkultūrinės kompetencijos ugdymas aukštojo mokslo internacionalizacijos kontekste (magistrantūros studijų aspektas). Doctoral dissertation. Šiaulių universitetas.

17. Williams, T. R. (2009). The Reflective Model of Intercultural Competency: A Multidimensional, Qualitative Approach to Study Abroad Assessment. Frontiers: the Interdisciplinary Journal of Study Abroad, 18, 289-306. Available from: http://www.frontiersjournal.com/documents/frontiersxviii-fall09-twilliams.pdf

18. Žydžiūnaitė, V., Bubnys, R. (2010). Savęs vertinimo apie turimą tarpkultūrinę kompetenciją instrumentas. Vilnius: Vilniaus kooperacijos kolegija.

19. Žydžiūnaitė, V., Lepaitė, D., Bubnys, R., Čepienė, A. (2010). Tarpkultūrinės kompetencijos ugdymo/si metodologija. Vilnius: Vilniaus kooperacijos kolegija.

The Ethnic Identity and Future Plans of the Multinational Student Community

Aldona Vaičiulienė

Abstract
After graduating from the university, some students look for their place in Lithuania, while others are attracted by the wide world. What are the grounds of such decisions? The research aims to determine how the maturity of ethnic identity of the undergraduates of Lithuanian University of Educational Sciences is connected with their future plans. The research was conducted by means of the ethnic identity measure MEIM-R and a questionnaire on the undergraduates' future plans. The research revealed that the maturity of ethnic identity is positively related to their plans of staying in Lithuania.

Keywords: ethnic identity, ethnic group, future plans

Introduction

While studying at schools of higher education, hardly all students clearly imagine their future. After graduation they frequently find themselves at the crossroads. What should they do next – study or work, stay in Lithuania or look for happiness abroad? After the restoration of independence, the young people of Lithuania have the whole world ahead. Now they have a possibility to choose and a significant number of young people choose their studies and, frequently, work abroad. Nevertheless, there still are alarming questions why some young people leave Lithuania after acquiring their education here, what conditions their decision? Maybe there are some consistent patterns? Maybe this category of young people has some common features? In order not to let emigration become an elemental phenomenon, in order to control it, it is essential to analyze the factors determining the process.

Ethnic identity is one of the aspects of the Self. The maturity of ethnic identity defines a person's maturity to a certain extent, determines the psychological welfare within a community and has influence on the social and intersectional relations. This research reveals the fact that such a psychological formation as ethnic identity should be significant when solving the question of one's relation with Lithuania.

From the ethnic perspective, Lithuania is a heterogeneous country. Lithuanians have lived together with a variety of ethnic minorities for ages. The largest communities are Polish and Russian. Hence, there naturally appears a question concerning attachment to Lithuania in the dominant population of Lithuanians and in the communities of ethnic minorities.

There are a lot of studies conducted by foreign researchers concerning the experiences of people affected by migration (emigration) [14]. The states, that were founded quite recently, for example Israel, find this issue of great concern [6; 15]. There are a lot of studies prepared by American researchers who investigate the relation between the ethnic and national (American) identity [12; 1; 2]. In Lithuania there have recently been many publications of significant studies on the issues of ethnic and national identity and migration. However, in the field of psychology these topics are not very popular. The psychological aspects of ethnic identity, the experiences of people affected by migration (emigration), and the psychological welfare of ethnic minorities have been analyzes only episodically [17; 10]. There is a great need for systemic studies. Thus, the present-day situation makes the paper topical and novel.

Ethnic Identity: Theoretical and Methodological Paradigm

Ethnic identity is one of the aspects of identity. Psychologists give different definitions of ethnic identity. From the perspective of development, ethnic identity is the awareness of self as a member of the group which evolves in time due to active inquiry, learning, responsibility and commitment to the ethnic group [13, p. 271]. The formation of ethnic identity, as well as identity in general, is especially important in one's adolescence and youth. The sane and mature identity is the essence of a person's psychological well-being; it is related to a variety of indicators of psychological welfare [4, p. 8-10]. Developmental psychology is concerned with the formation of identity. The roots of research on ethnic identity lie in Erikson's ideas [3, p. 50]. According to Erikson, identity is related to the individuals' inner need to be their real selves. Identity is related to the experience of continuation in space and time which gives an individual a sense of stability and serves as a point of reference in the most important situations of life. Erikson analyzed a person in his historic and cultural context, searched for the interconnections between individual and group identity.

Social psychology [16, p. 33] analyses ethnic identity with a focus on the feeling of belonging to a group and the outcomes of the individual identifications with a social group. In this respect ethnic identity is an aspect of social identity. The theory of social identity says that being a member of a group gives an individual a feeling of community which is an important investment in one's positive Self-Concept. This approach explains ethnic identity as a multidimensional construct which covers such dimensions as self-definition and attribution to a certain category, emotional commitment, a feeling of belonging to a group, ethnic behavior (using an ethnic language, eating certain food, socialization with group members), values and beliefs, etc. [13, p. 271].

Jaspal and Cinnirella [5, p.509] analyze the principles of constructing ethnic identity from the theoretical perspective of the identity process, make assumptions and search for support in the works of other researchers' as they want to prove that identification with an ethnic group has its own advantages and enables a person to function more efficiently.

The word "ethnic" reveals a feeling of dependence and community, and this can also be understood as an organizational principle followed by a group of people in order to distinguish themselves from other groups in terms of race, kinship, language, customs, lifestyle, culture, religion, etc. The claim to a certain ethnic identity can have positive results in meeting the needs for dependence. Membership in an ethnic group enables a member experience acceptance and a feeling of intimacy. This is one of the basic needs for most people. Hence, in this case identification with one's ethnic group will have positive consequences as a person's need for community will be met. The feeling of community, related with membership in an ethnic group, can also function as protection from the imaginary rejection on behalf of other social groups, for example, the dominant ethnic majority. The possible consequence of a real or imaginary rejection is the enhancement of the sense of community.

Moreover, when an individual is forming his/her identity, his/her feeling of exclusivity becomes important. It can be reinforced by a conception that one belongs to a group which is different in its cultural differences. From a psychological perspective, exclusivity, uniqueness in the context of other groups is an important key principle which is connected with the development of ethnic identity. Thus, identification with an ethnic group enhances exclusivity.

When analyzing the historic traditions related to ethnicity, people can start romanticizing ethnic history. This tendency was determined not only by psychologists, but also social anthropologists and sociologists who speak about the "invention of traditions" and creation of national myths and stories. It is important that this constructed history would give group members a positive sense of ethnic identity. When members of an ethnic group create a positive ethnic identity, positive social representations, related to the group history, typically become psychologically distinctive, and negative historic aspects of the group become psychologically latent. A created story can give the group members a positive sense of "succession in time". For example, it was noted that people want to psychologically distance themselves from such negative aspects of their past as violence among groups, or cruelties within a group. Such dissociation can enhance a positive sense of succession.

A group member can conceive himself not as an independent person in the world, but as related to others through their common heritage. This gives the members of an ethnic group a psychological connection between the past, present and future, which is essential in order to guarantee the feeling of succession. Ethnicity is "locked" in traditions typical to ethnic groups. Traditions create a "shared thread", which joins the past, present and future of an ethnic group and so creates conditions for the members of an ethnic group to reconnect themselves with the past, express themselves "ethnically" in the present, and conceive a durable ethnic identity in the future. Therefore, a person tries to sustain the compatibility and succession in time and space. Contemplation of "unchangeable" traditions, related to one's ethnic identity, reinforces the feeling of continuation.

Such behavior creates the connection between generations, gives ancestry to the most humble member of the group, allows a person to identify himself with heroic times, great deeds, with the history of origin of human, cultural and spiritual entities. Ethnic identity can give a sense of experiencing something immortal. This positive feeling of succession can be also useful for the feeling of self-respect which meets a person's need to retain and reinforce the positive Self-Concept. Indeed, group members are motivated to present themselves in a positive light and that helps to improve their self-respect. Thus, the feeling of succession, which stems from a person's ethnic identity, has a positive influence on his/her self-respect.

Summarizing the significance of ethnic identity for a person's efficient functioning, it is possible to make an assumption that the maturity of this psychological formation plays an important role in a person's decision making process in the creation of one's future.

Marcia's [8, p. 159; 9, p. 3] work laid foundation for empirical research on identity in psychology. This study also employs Marcia's model of identity according to which the development of identity consists of two processes: 1) the alternative exploration of identity and 2) commitment, responsibility for important spheres of identity, implementation of one's decisions. The individual explores the possibilities of his own and his environment, on their basis he makes decisions and implements them in reality. If he is not involved in any process – if he does not explore and commit, it shows the person's identity diffusion. If the individual explores the key issues of identity, makes decisions, implements them and takes on responsibility, it reveals the person's achievements in the sphere of personal identity and maturity. Marcia did not study ethnicity himself, but his empirical model is successfully applied in the studies on ethnic identity. From the perspective of development, an individual proceeds from the immature ethnic identity towards the more mature one when a person's exploration and experience determine a stable and firm sense of belonging to an ethnic group, when a person takes on certain responsibility for his/her ethnicity. In adulthood most people acquire a relatively stable and strong feeling of being a member of an ethnic group. However, the question of identity can still be further investigated.

The goal of the research is determine how the maturity of ethnic identity of Lithuanian University of Educational Sciences undergraduates of Lithuanian, Russian and Polish nationality is related to their future plans.

Research objectives:

1. To assess the maturity of ethnic identity of students of Lithuanian, Russian and Polish nationality.
2. To investigate the future plans of students of Lithuanian, Russian and Polish nationality.
3. To reveal the peculiarities of the relation between the ethnic identity maturity and future plans of students of Lithuanian, Russian and Polish nationality.

Methodology

The study objects. The research involved the undergraduates of Lithuanian University of Educational Sciences – 133 respondents. There were 99 females and 34 males. The age of the respondents was from 21 to 50 years old. The average age was 23.69.

The methods. Ethnic identity was investigated by means of the methodology prepared by Phinney et al. [13, p. 276] – namely, MEIM-R (Multigroup Ethnic Identity Measure-Revised). It was translated to Lithuanian by Vaičiulienė [19, p. 15]. It is a modified MEIM methodology [11, p. 156]. MEIM-R is the measurement tool for the process of ethnic identity. The questionnaire has got two scales. One assesses the investigation of ethnic identity (e.g. "I frequently talked to other people to find out more about my ethnic group"), and the other assesses commitment to the ethnic group (e.g. "I feel that I am very attached to my ethnic group"). The answer can be marked in the five-point scale from total disagreement (1) to complete agreement (5). Before answering the questions, the respondents have to indicate what ethnic groups they belong to. In order to be able to mark their ethnic origin, the respondents are provided with a list of ethnic groups. In the MEIM-R questionnaire Cronbach's alpha equals 0.87. In order to assess the students' future plans, they received a questionnaire on their future plans. The respondents were asked what they intend to do after completing their bachelor degree studies – will they stay in Lithuania or go abroad to study or work.

The research process. The students were given the MEIM-R questionnaire and a questionnaire on their future plans. The respondents were asked to read the questions and fill in the questionnaires. Anonymity was guaranteed.

Research results

The respondents were the undergraduate students since it was expected that their ethnic identity would be more mature than that of the students of the first, second or third years, and it would be possible to determine the differences between ethnic groups. The previous research [18, p. 134] shows that students of the first, second and third year are more concerned about interpersonal relations and that is the main sphere where their identity intensively develops. The immaturity of the ethnic identity of the first and second year students was indirectly determined by some other studies [7, p. 37]. Moreover, students of this age are not yet concerned about their plans after graduation. Thus, as it was expected, the results presented in Table 1 reveal the differences in ethnic identity rates of undergraduates belonging to ethnic groups. The average rates show that the ethnic identity of Polish students is higher than that of Lithuanian and Russian ones and statistically significantly differs from the latter two groups. The rates of ethnic identity in the groups of Lithuanian and Russian students were practically almost identical.

Table 1: The Differences of Ethnic Identity Rates among Ethnic Groups

Ethnic group	Ethnic identity		Difference (ANOVA)
	Average	SD	
Lithuanians	20.34	6.30	p = 0.049
Russians	20.58	5.33	
Poles	24.05	4.74	

The analysis of the relation of ethnic identity with future plans (Table 2) gave unambiguous results – the maturity of ethnic identity is positively related with the students' plans to stay Lithuania after graduation and continue their studies or work. The general assessment of ethnic identity is positively and statistically significantly connected with intentions to stay in Lithuania. The undergraduate students who are in the process of exploration, who feel a need to explore different possibilities (their abilities and interests as well as the possibilities provided in the society) are more inclined to choose further studies in Lithuania. However, the undergraduate students who feel that they have sufficiently considered their possibilities and have made corresponding decisions are more committed to Lithuania and intend to work here. More frequently they possess a negative attitude concerning the possibility to work abroad.

Table 2: The Interrelation between Ethnic Identity Results and Further Plans (Pearson)

Ethnic identity scales	**Student's further plans**			
	Studies in Lithuania	Work in Lithuania	Studies abroad	Work abroad
Exploration	**0.228****	0.128	0.095	0.011
Commitment	0.159	**0.338*****	–0.138	**–0.224****
Overall rating	**0.207***	**0.267****	–0.037	–0.130

* $p \leq 0.05$; ** $p \leq 0.01$; *** $p \leq 0.001$.

The results of the research raise questions and promote a more detailed investigation of the ethnic identity of Lithuanians and members of other ethnic communities in Lithuania, the interrelations of ethnic groups, their psychological welfare, their ties with Lithuania and their commitment. This research did not aim to determine what conditions the different results of ethnic identity with the Polish ethnic group and other groups. Now only theoretical assumptions concerning the greater commitment of the Polish youth to their ethnic group are possible. However, in the future such studies are essential.

Conclusions

1. The rates of ethnic identity of undergraduate students of Polish nationality are higher than those of Lithuanian and Russian ethnic groups and the differences are statistically significant.
2. The research results reveal the positive relations between the greater maturity of students' ethnic identity and their plans to settle in Lithuania after graduation.

Summary

From the ethnic perspective, Lithuania is not a homogeneous state. Lithuanians have lived together with a variety of ethnic minorities for ages. The largest communities are Polish and Russian. Hence, in the context of mass migration of young people, there naturally appears a question concerning attachment to Lithuania in the dominant population of Lithuanians and in the communities of ethnic minorities.

Ethnic identity is one of the aspects of the Self. The maturity of ethnic identity defines a person's maturity to a certain extent, determines the psychological welfare within a community and has influence on the social and intersectional relations. This research discusses the theoretical and methodological paradigm of ethnic identity, the benefits of ethnic identity which are experienced by an individual committed to his ethnic group. The summarized material makes it possible to assume that ethnic identity is significant for a person's efficient functioning. Hence, an assumption is made that the maturity of this psychological formation plays an important role in the decision making process when the individual creates future plans or considers his commitment to Lithuania.

The aim of this research was to determine how the maturity of ethnic identity of the bachelor degree undergraduates of Lithuanian University of Educational Sciences is connected with their future plans.

Ethnic identity was analyzed using MEIM-R methodology (Multigroup Ethnic Identity Measure-Revised) prepared by Phinney et al. [13, p. 276]. The methodology was prepared on the basis of Marcia's model of identity [8, p. 159; 9, p. 3] according to which the development of identity includes two processes: 1) the alternative exploration of identity and 2) commitment, responsibility for important spheres of identity, implementation of one's decisions. The respondents also got a questionnaire on their future plans.

The research results showed that the rates of ethnic identity of undergraduate students of Polish nationality are higher than those of Lithuanian and Russian ethnic groups and the differences are statistically significant. The research results reveal the positive relations between the greater maturity of students' ethnic identity and their plans to settle in Lithuania after graduation.

The results of the research raise a lot of questions and promote a more detailed investigation of the ethnic identity of Lithuanians and members of other ethnic communities in Lithuania, the interrelations of ethnic groups, their psychological welfare, and their ties with Lithuania.

References

1. Berry J. W., Phinney J., Sam D. L., Vedder P. (Eds.), *Immigrant youth in cultural transition: Acculturation, identity and adaptation across national contexts.* 2006, Mahwah, NJ: Lawrence Erlbaum Associates.

2. Devos T., Gavin K., Quintana F. J. Say "adios" to the American dream? The interplay between ethnic and national identity among Latino and Caucasian Americans // *Cultural Diversity & Ethnic Minority Psychology*. 2010, 16, 1, 37-49.

3. Erikson E. H. *Identity: Youth and Crisis*. New York: W. W. Norton & Company, 1968.

4. French S. E., Seidman E., Allen L., Aber J. L. The development of ethnic identity during adolescence // *Developmental Psychology*, 2006, 42, 1, p. 1-10.

5. Jaspal R., Cinnirella M. The construction of ethnic identity: Insights from identity process theory // *Ethnicities*. October 2012, Vol. 12 Issue 5, p 503-530. DOI: 10.1177/1468796811432689.

6. Kelman H. C. Anti-Semitism and Zionism in the debate on the Palestinian issue. In M. Polner and S. Merken (Eds.), *Peace, Justice, and Jews: Reclaiming our Tradition*. 2007, New York and Charlottetown, Canada: Bunim & Bannigan Ltd.

7. Laurinavičiūtė J., Vaičiulienė A. Etninio tapatumo ir asmenybės tapatumo ryšio ypatumai // *Psichologija sveikatai ir gerovei: Lietuvos psichologų kongresas: kongreso pranešimų santraukų leidinys*: 2013 m., balandžio 25-27 d., Vilnius, Lietuva.

8. Marcia J. E. Identity in adolescence // J. Adelson (Ed.), *Handbook of Adolescent Psychology*. New York: Wiley, 1980, p. 159-187.

9. Marcia J. E. The ego identity status approach to ego identity // J. E. Marcia, A. S. Waterman, D. R. Matteson, S. L. Archer, J. L. Orlofsky (Eds.), *Ego Identity: A Handbook for Psychosocial Research*. New York: Spring Verlag, 1993, p. 3-21.

10. Norvilas A., Valdkytė-Janušaitienė I. Tautybė ir kalba trijų tautinių grupių sankryžoje // *Psichologija*, 2007, 36, 7-21.

11. Phinney J. The Multigroup Ethnic Identity Measure: A new scale for use with diverse groups // *Journal of Adolescent Research*, 1992, 7, p. 156-176.

12. Phinney J., Devich-Navarro M. Variations in bicultural identification among African-American and Mexican-American adolescents // *Journal of Research on Adolescence*. 1997, 7, 3-32.

13. Phinney J. S., Ong A. D. Conceptualization and measurement of ethnic identity: current status and future directions // *Journal of Counseling Psychology*, 2007, 54, 3, p. 271-281.

14. Rutland A., Cinnirella M., Simpson R. Stability and variability in national and European self-identification // *European Psychologist*. 2008, 13, 4, 267-276.

15. Sonnenschein N., Bekerman Z, Horenczyk G. Threat and the majority identity // *Group Dynamics: Theory, Research, and Practice*, 2010, 14, 1, 47-65.

16. Tajfel H., Turner J. C. An integrative theory of intergroup relations // W. G. Austin, S. Worchel (Eds.). *The social psychology of intergroup relations*. Monterey, CA: Brooks/Cole, 1979, p. 33–47.

17. Užgiris J., Rimkutė E. Conception of Nationality in Self-Identity Discourse. *Abstract of Poster Presented at the Meeting of the International Society for the Study of Behavioral Development*. 1994, Amsterdam.

18. Vaičiulienė A. Paauglystės uždaviniai ir problemos: Studentų ir moksleivių asmenybės tapatumo ypatumų palyginimas // *Ugdymo Psichologija*. 2004, 11-12, 134-139.

19. Vaičiulienė A., Laurinavičiūtė J. Etninio tapatumo tyrimas: MEIM-R klausimyno lietuviškos versijos psichometrinės savybės // *Ugdymo Psichologija*. 2012, 23, 15-21.

Student Support Needs in Individual Work Tasks: Student-Centred Approach

Rasa Tamulienė

Abstract
The aim of the research presented in this article was to reveal student support needs when performing individual tasks. 162 students from Lithuanian state colleges took part in the research which was carried out in the form of a written survey. The results revealed that when performing individual tasks students require five types of support: library services, consultations, methodical support, emotional support and learning to learn support.

Introduction

The implementation of the Bologna Process aims at European higher education institutions requires shifting from traditional teacher-centred studies to student-centred and learning outcomes based studies (Bucharest Communiqué, 2012; Budapest-Vienna Declaration, 2010). The implementation of student-centred studies is based on reforming curriculum in accordance with the European Credit Transfer and Accumulation System (hereinafter referred to as ECTS). One of the key features of this system is to focus on devoting more time to student self-study when designing and delivering study programmes at higher education institutions (ECTS Users' Guide, 2009). Following the Bologna Process recommendations, since September 2011 the curriculum of study programmes delivered in Lithuanian higher education institutions has also been reorganised in accordance with the ECTS logic. As a result, a larger part of student's workload now consists of individual work (Lietuvos Respublikos mokslo ir studijų įstatymas, 2009). Researchers (Prebble et al., 2004) note that even thoughmany advantages of self-directed learning have already been discovered, students face many challenges and difficulties when performing individual work tasks. The lack of independent learning skills often becomes an obstacle when performing these tasks (Danford, 2006). Provision of required student support at higher education institutions is thought to be one of the means that could successfully help students to overcome difficulties that they face when performing individual work tasks (Prebbleet al., 2004).The following question was formulated in the research: "What support students need when performing individual work tasks?" The aim of the research was to reveal student support needs when performing individual work tasks. In order to achieve the aim of the research the following tasks were set:1) to discuss the paradigm of student-centred approach; 2) to review student support provided in foreign higher education institutions; 3) to identify Lithuanian college students' support needs when performing individual work tasks. The object of the research was the opinion of students from Lithuanian state colleges on support needs that they face when performing individual work tasks.

The Link between the Paradigm of Student-Centred Studies and Student Support Needs

Rogers is believed to be the initiator of the idea of student-centred studies. Rogers' (1969) idea of student-centred studies was based on the assumption that the process of education should be based on two main categories, i.e. freedom and trust. According to Rogers (1969), a learner who constantly gets directions while learning cannot develop the feeling of internal freedom. As a result, a learner who grows up in the paradigm of traditional education cannot make decisions independently and does not develop the feeling that a person is responsible for his/her own life. Rogers (1969) notes that learner's internal freedom can only be developed if an educator trusts his/her learner, i.e. believes that a person can find a solution himself/herself.

Although the ideas of student-centred studies were started to be developed some time ago, nowadays the development of student-centred studies is mainly identified with the Bologna Process. It is emphasised in the documents that provide guidelines for the implementation of the Bologna Process aims (Budapest-Vienna Declaration, 2010; LeuvenCommunique, 2009; LondonCommunique, 2007; BerlinCommunique, 2003; PragueCommunique, 2001; Bologna Declaration, 1999) that the development of student-centred studies in higher education institutions should be based on reforming curriculum in accordance with the European Credit Transfer and Accumulation System. Following this system, student's workload should be reorganised in a way that a part of classroom activities were changed into independently performed tasks. However, successful implementation of student-centred studies should be based on not only "mechanical" reallocation of hours in a study plan but also on the changes in student and teacher rolesin this process. Brush and Saye (2000) note that student-centred studies should focus on changing the role of "a teacher who supervises"to "a teacher who helps", and the role of a student from a passive "receiver of information" to an active manager of one's own studies (Adam,

2007). According to Adam (2007),on one hand, student-centred studies give students plenty of independence in planning their workload but, on the other hand, encourage them to take more responsibility for their studies. However, a student can take responsibility for his/her studies only when he is ready, i.e. has sufficient skills to perform independent study tasks. Thus when shifting from traditional to student-centred studies higher education institutions have to not only mathematically recalculate national credits into ECTS credits and mechanically reallocate student workload hours but also create alternative systems that could compensate the reduced number of contact hours in a study plan.Prebble et al. (2004) note that a system that could help to ensure fluent progress of a study process when the amount of contact hours is reduced would be the development of student support.

Student support is defined as the system of services provided by a higher education institution, which fulfils students' emotional, academic and social needs and is a precondition for increasing student's individual welfare and academic success (SajienėandTamulienė, 2012). Scientific literature (Morgan, 2012; Prebble et al., 2004) distinguishes two types of student support services: academic and non-academic. Academic student-support involves such services as career designing, academic student information and counselling services, first year students' orientation courses, support for international students, foreign language centre services, library services, and distance study centre services.(Morgan, 2012; Thompson and Mazer, 2009; Prebbleet al., 2004, SavickienėandPukelis, 2004; La Padula, 2003). Non-academic support involves psychological counselling services, accommodation services, financial support and legal counselling, services that promote a sense of community, health services, and support for students with disability (Morgan, 2012; Prebble et al., 2004; Savickienė and Pukelis, 2004; Report on the Quality Culture Project, 2005). However, even though scientific literature emphasises that academic and non-academic student support services are important for the implementation of student-centred studies, the analysis of the reviewed scientific researches did not reveal what student support services are the most relevant when performing individual work tasks. This question was addressed in the empirical research described below.

Research Methodology and Methods

A qualitative methodology was applied in the research and a written survey that was carried out through a web-based survey system was used as a data collection method. Content analysis was used for the analysis of data.

Research Organisation, Process and Participants

Research participants were students at Lithuanian state colleges who had started their studies not earlier that in 2011 (in 2011 Lithuanian higher education institutions started delivering study programmes reorganised in accordance with the ECTS logic). An invitation to participate in the research was distributed by students' unions at Lithuanian state colleges.162 students from 13 Lithuanian state colleges participated in the research.

Student Support Needs when Performing Individual Work Tasks

The results of the empirical research revealed that student support needs when performing individual work tasks can be described through five categories: library services, consultation services, methodical support, emotional support and learning to learn services. The following sections provide a detailed description of different subcategories that form these categories.

Need for Library Services when Performing Individual Work Tasks

The analysis of the need for library services when performing individual work tasks revealed that students require more comprehensive information about library services, more flexibility in provided services and broader access to library resources (Table 1).

Table 1: Subcategories describing the category of the need for library services when performing individual work tasks

Subcategory	Examples that illustrate subcategory
Informing about library services	*"a simple saying that college students have the right to use libraries of other higher education institutions"*; *"comprehensive information about how to use library services appropriately and efficiently"*; *"to be explained how to use databases and where to get the most recent information"*.
Flexibility of services	*"library services online"*; *"I would like to get books from my library for at least a week and not for just a day"*;

	"I think they could let students to stay at libraries longer"; *"to be able to take required books home".*
Access to library resources	*"to have more books at college libraries and to be able to read various books and articles online";* *"to have more copies of books provided in the list of references"* *"access to scientific literature is not good enough. You have to pay for the majority of articles";* *"access to material of conducted research";* *"need literature that could be available for students when performing individual tasks.* *"for libraries to have enough books for all students and not just have several copies for one group".*

When highlighting the need for more information related to library services, research participants mentioned that they lack "*a simple saying that college students have the right to use libraries of other higher education institutions*", or "*comprehensive information about how to use library services appropriately and efficiently*". Research participants also emphasised that it is important for students "*to be explained how to use databases and where to get the most recent information*" (Table 1). Students mentioned that in order to make library services more flexible, it is important to expand library services online, to be able to loan books for a longer period of time, to extend library opening hoursand to be able to take required books home (Table 1). Students stressed that in order to ensure access to library resources it is important "*to have more books in college libraries and to have free access to read various books and articles online*", "*for libraries to have enough books for all students and not just have several copies for one group*". When emphasising the importance of access to library resources, research participants noted that "*access to scientific literature is not good enough. You have to pay for the majority of articles*". They also mentioned the lack of "*access to material of conducted research*".

Need for Consultation Services when Performing Individual Work Tasks

Upon generalisation of the need for consultation services, four subcategories were distinguished:individual aspects in consultations, the importance of distance consultations, giving feedback and being able to get consultations from particular field specialists (Table 2).

Table 2: Subcategories describing the category of the need for consultation services when performing individual work tasks

Subcategory	Examples that illustrate subcategory
Individual aspects in consultations	*"consultations with each student individually"; "individual consultations, clear explanations"; "individual support"; "individual teachers' consultations"; "individual consultations and flexible schedule (of course, with certain limits) <…>taking into consideration student's lifestyle"; "consultations so that a student knew if he's on a right track".*
Distance consultations	*"more distant communication with teachers when questions arise"; "more frequent online consultations, not only face-to-face ones"; "theability to consult with teachers online about issues related to given tasks".*
Being able to get consultations from particular field specialists	*"in my opinion, what we need most is more varied consultations and not only from that teacher who gives a task"; "it is very important to get consultations from a specific field specialists"; "whena college gives certain tasks, they could mention companies that are willing to give students information about themselves".*
Giving feedback	*"to explain a student his mistakes so that he could correct them and improve. You learn from your own mistakes"; "highlight inaccuracies in one's work, emphasise how to improve one's skills, what to read individually, what exhibition or training to attend"; "correction, explanation from a teacher".*

Research participants emphasised that when performing individual tasks it is very important for a student to get individual consultations from his/her teacher, to be allocated a specific time for consultations "*taking into consideration student's lifestyle*". Students would like to have "*more distant communication with teachers when questions arise*", and "*more frequent online consultations, and not only face-to-face ones*" (Table 2). Research participants also said that it would be useful to get consultations from not only those teachers who give an individual task but also from specific field specialists: "*when a college gives certain tasks, they could mention companies that are willing to give students information about themselves*", "*it is very important to get consultations from a specific field specialists*" (Table 2).Research participants also noted that during consultations they should be given feedback about the tasks they perform:"*to explain a student his mistakes so that he could correct them and improve. You learn from your own mistakes*", "*highlight inaccuracies in one's work, emphasise how to improve one's skills, what to read individually, what exhibition or training to attend*" (Table 2).

Need for Methodical Support when Performing Individual Work Tasks

After generalising the results of the research, a conclusion was made that ensuring methodical support while performing individual tasks requires defining an individual work task, providing examples and developing a clear conception of individually performed work (Table 3).

Table 3: Subcategories describing the category of the need for methodological support when performing individual work tasks

Subcategory	Examples that illustrate subcategory
Clarity of an individual work task	*"that individual work tasks were clear and that it was clearly defined what has to be done and what the purpose of these tasks are"; "clear explanation from a teacher about what s/he requires"; "specific requirements", "specific description of a task"; "teachers should understand that not all students understand a task the same way"; "very specific tasks, requirements, amount of information and time given to complete a task"; "purposeful and clearly defined individual work tasks"; "more accurate information from teachers about what they really need"; "accurate information from teachers about what they expect to find in individual work tasks, what has to be provided in them"; "accurate instructions and list of references as it is ridiculous when you have to correct work maybe 3 times as teachers don't fully explain how to do it and where to find literature and sometimes it's even impossible to find it…"; "accurate descriptions about what a teacher wants from a student".*
Providing examples	*"provide samples of similar tasks as after analysing other works a student might find it easier to understand what has to be done and he gets more ideas how to do it"; "give examples that could be followed when doing one's tasks"; "examples how to do a given task"; "individual work samples that could get the highest score".*
Developing the conception of individual work	*"teachers should pay more attention to the development of the individual work conception and then teach a student"; "the most important thing is that before each individual task teachers discussed its content with each student as certain cases show that students sometimes misunderstand the topic"; "I think that teachers should help and slightly narrow tasks if students require that. Tasks are sometimes too broadand students worry that they might misunderstand them"; "In my opinion, teachers should specify what a student should reveal in that task and what to pay attention to".*

Research participants emphasised the importance of clear requirements for individual work tasks provided by teachers: "*that individual work tasks were clear and that it was clearly defined what has to be done and what the purpose of these tasks are*", "*specific requirements*", "*specific task description*", and "*accurate information from teachers about what they expect to find in individual work tasks, what has to be provided in them*" (Table 3). Students also mentioned that it might be easier for them to complete their individual tasks if they were given samples of other highly-scored works: "*individual work samples that could get the highest score*", "*provide samples of similar tasks as after analysing other works a student might find it easier to understand what has to be done and he gets more ideas how to do it*" (Table 3). Research participants noted that it is important to not only provide "mechanical requirements" related to a given task but also discuss the conception of work with a student: "*the most important thing is that before each individual task teachers discussed its content with each student as certain cases show that students sometimes misunderstand the topic*". Students also said that teachers should narrow the topic of their task when they develop the conception of individual work: "*I think that teachers should help and slightly narrow tasks if students require that. Tasks are sometimes too broad and students worry that they might misunderstand them.*" (Table 3)

Need for Emotional Support when Performing Individual Work Tasks

The analysis of the research data also revealed the need for emotional support when students perform individual work tasks. Students' responses to the research questions can be divided into two subcategories: emotional support from teachers and emotional support from peer students (Table 4).

Table 4: Subcategories describing the category of the need for emotional support when performing individual work tasks

Subcategory	Examples that illustrate subcategory
Support from teachers	*"that teachers understood that students need help with one or another question and not only said "do it yourself, it's your work""*; *"teachers' tolerance, understanding and peaceful and open opinion"*; *"the most valuable support is accurate and informative advice, consultations related to relevant questions. And these questions should not necessarily be related to the taught subject. Students face lots of questions and issues that a clever advice from a teacher might help to solve"*; *"teachers should be more tolerant as individual tasks are understood differently by different students"*;

	"teachers' desire and ability to have time to solve issues"; "more attention from teachers, more listening"; "more cooperation with students".
Support from peer students	*"sharing experience with older students"; "advice from other students who are good at that subject"; "I ask my friends' help".*

Research participants highlighted the importance of teachers' understanding, tolerance and their will to help students when performing individual work tasks: "*teachers' tolerance, understanding and peaceful and open opinion*", "*more attention from teachers, more listening*", and "*that teachers understood that students need help with one or another question and not only said "do it yourself, it's your work*". Students also noted that they would like to get teachers' advice on various topics, not only on the subject that they teach: "*the most valuable support is accurate and informative advice, consultations related to relevant questions. And these questions should not necessarily be related to the taught subject. Students face lots of questions and issues that a clever advice from a teacher might help to solve*" (Table 4). A part of respondents mentioned that support from peer students is important for them when performing their individual work tasks: "*sharing experience with older students*", "*advice from other students who are good at that subject*" (Table 4).

Need for Learning to Learn Services when Performing Individual Work Tasks

Data conducted in the research revealed that in order to facilitate performance of individual work tasks, students should also be provided with learning to learn services (Table 5).

Table 5: Subcategories describing the category of the need for learning to learn support when performing individual work tasks

Subcategory	Examples that illustrate subcategory
Developing information search abilities	*"teachers' help in choosing information"; "help in solving literature search related issues"; "help in finding sufficient amount of references".*
Developing team work abilities	*"training related to public speaking, group work etc. that would enable students to communicate easier and work more efficiently. Thus it's important to show that you can move outside your comfort zone".*
Time planning abilities	*"I think that the most important thing is to get advice on how to manage your time and be able to do everything"; "evenly distribute time as often work is being done on the last evening".*

The analysis of the research data showed that learning to learn services should be oriented towards the development of information search, team work and time planning abilities (Table 5). The lack of information search abilities is illustrated by those students' responses where they emphasise the need for "*teachers' help in choosing information*", "*help in solving literature search related issues*", and "*help in finding sufficient amount of references*" (Table 5). Research participants also highlighted the importance of the development of teamwork abilities: "*training related to public speaking, group work etc. that would enable students to communicate easier and work more efficiently*" (Table 5). According to research respondents, one of problems that students face when performing individual work tasks is the lack of time planning abilities: "*I think that the most important thing is to get advice on how to manage your time and be able to do everything*", "*evenly distribute time as often work is being done on the last evening*" (Table 5).

After summarising the results of this empirical research, a conclusion can be made that students require five types of supportwhen they perform individual work tasks: *library services* (this category involves such subcategories as informing about library services, flexibility of services and access to library resources), *consultation services* (this category involves such subcategories as individual aspects in consultations, distance consultations, being able to get consultations from particular field specialists, giving feedback), *methodical support* (this category involves such subcategories asclarity of individual work task, providing examples and creating the conception of individual work), *emotional support* (this subcategory involves such subcategories as support from teachers and support from peer students) and *learning to learn services* (this category involves such subcategories as developing information search, teamwork and time planning abilities).

Conclusion

1. The paradigm of student-centred studies is based on two main categories – freedom and trust. One of the key features of student-centred studies that are developed in the contemporary European higher education area is reforming curriculum in accordance with the European Credit Transfer and Accumulation System. One of its features is to focus on devoting more time to student self-study when designing and delivering study programmes at higher education institutions.
2. The analysis of scientific literature revealed that higher education institutions provide two types of student support: academic and non-academic. Academic student-support involves such services as career designing, academic student

information and counselling services, first year students' orientation courses, support for international students, foreign language centre services, library services, and distance study centre services.Non-academic support involves psychological counselling services, accommodation services, financial support and legal counselling, services that promote a sense of community, health services, and support for students with disability.

3. The results of the empirical research revealed that students require five types of support when they perform individual work tasks: library services, consultation services, methodical support, emotional support and learning to learn services.When performing individual work tasks it is important that students were sufficiently informed about library services, that library services were flexible and that access to library services was ensured. Provision of consultations when performing individual work tasks should meet the following criteria: consultations should be individual and distant, students should beable to get consultations from particular field specialists and they should get feedback about their completed tasks. Provision of methodical support should be oriented towards the development of the conception of individual work, the provision of examples and clarity of individual work tasks. When performing individual work tasks it is important for students to get two types of support, i.e. emotional support from teachers and emotional support from peer students. The provision of learning to learn services should focus on the development of information search, teamwork and time planning abilities.

References

Adam, S. (2007). An introduction to learning outcomes: A consideration of the natute, function and position of learning outcomes in the creation of the European Higher Education Area. Introducing Bologna Objectives and Tools, B2.3-1.Surname, First name: complete main title of cited work, complete subheading of cited work, place of publication, year of publication.

Berlyn Communique (2003).Communiqué of the Conference of Ministers responsible for Higher Education, Berlin, 19 September, 2003. Retrieved from: http://www.ond.vlaanderen.be/hogeronderwijs/bologna/documents/MDC/Berlin_Communique1.pdf.

Brush, T., Saye, J. (2000). Implementation and Evaluation of a Student-Centered Learning Unit: A Case Study. Educational Technology Research and Development. Association for Educational Communications & Technology, 3, 79-100.

Bucharest Communique (2012). Making the Most of Our Potential: Consolidating the European Higher Education Area. Retrieved from: http://www.smpf.lt/uploads/documents/docs/868_a4d3e86ee4fefe419b0d8ddcddc070f6.pdf.

Budapest-Vienna Declaration on the European Higher Education Area. (2010). Retrieved from: http://www.ond.vlaanderen.be/hogeronderwijs/bologna/2010_conference/documents/Budapest-Vienna_Declaration.pdf.

Danford, G. L. (2006). Project-based Learning and International Business Education.Journal of Teaching in International Business. 1, 7-25.

ECTS Users' Guide. (2009). Luxembourg: Office for Official Publications of the European Communities. Retrieved from: http://ec.europa.eu/education/lifelong-learning-policy/doc/ects/guide_en.pdf.

LaPadula, M. (2003). A Comprehensive Look at Online Student Support Services for Distance Learners// American Journal of Distance Education, 17 (2), 119–128.

Leuven Communique (2009).Communiqué of the Conference of European Ministers Responsible for Higher Education. Leuven and Louvain-la-Neuve, 28–29 April 2009. http://www.ond.vlaanderen.be/hogeronderwijs/bologna/conference/documents/leuven_louvain-la-neuve_communiqu%C3%A9_april_2009.pdf.

Lietuvos Respublikos Mokslo ir Studijų įstatymas. (2009). Nr. XI-242, Vilnius. Retrieved: from: http://www3.lrs.lt/pls/inter3/dokpaieska.showdoc_l?p_id=343430&p_query=&p_tr2=

London Communique (2007). Towards the European Higher Education Area: responding to challenges in a globalised world. London, 18 May, 2007. Retrieved from:

Morgan M. (2012). The evolution of student services in the UK // Perspectives: Policy and Practice in Higher Education, 1, 1–8.

Prague Communiqué (2001). Towards the European higher education area Communiqué of the meeting of European Ministers in charge of Higher Education, Prague, 19 May, 2001. Retrieved from: http://www.smpf.lt/uploads/documents/docs/295_56731fc9186ce4bacc64b9c4f4ce14f4.pdf.

Prebble T., Hargraves H., Leach L., Naidoo K., Suddaby G., Zepke N. (2004). Impact of Student Support Services and Academic Development Programmes on Student Outcomes in Undergraduate Tertiary Study: A Synthesis of the Research // New Zealand: Ministry of Education. Retrieved from: http://edcounts.squiz.net.nz/__data/assets/pdf_file/0013/7321/ugradstudentoutcomes.pdf.

Report on the Quality Culture Project: Developing an Internal Quality Culture in European Universities. (2005). European University Association: Brussels. Retrieved from: http://www.eua.be/eua/jsp/en/upload/QC1_full.1111487662479.pdf.

Rogers, C. R. (1969). Freedom to learn// United States of America: Charler E. Merrill Publishing Company.

Sajienė, L., Tamulienė, R. (2012). Quality assessment parameters for student support at higher Education institutions//The Quality of Higher Education, 9, 120-139.

Savickienė I., Pukelis K. (2004). Institutional Quality Assessment of Higher Education: Dimensions, Criteria and Indicators// The Quality of Higher Education, 1, 26–37.

The Bologna Declaration on the European Space for Higher Education: An Explanation. (1999). Retrieved from: http://ec.europa.eu/education/policies/educ/bologna/bologna.pdf.

Thompson B., Mazer J. P. (2009). College Student Ratings of Student Academic Support: Frequency, Importance, and Modes of Communication // Communication Education, 58 (3), 433–458.

Innovation and its Modeling in Educational Environment

Vilija Targamadzė, Egle Celiešienė

Abstract

Socio-economic, cultural, political and other changes, the trends of globalisation presupposed the necessity of continuous changes in education. The rate of formation of the knowledge society, ensuring dissemination of information and its accessibility raises significant questions for politicians, practitioners and scientists in education, relating to migration and demographical changes, understanding of the identity of the individual and the public spirit, reciprocal understanding and tolerance towards diversity and other. In response to the mentioned challenges legal acts and other documents of Lithuania and of the European Union are being improved, where special attention can be noticed to innovations in educational environment. However it is important to underline that innovation have become important drivers of economic growth and social improvement, especially in educational environment.

Definition of keyterms

Daugiakultūrinėje visuomenėje, kurioje globalizacijos tendencijos apima nuolatinę informacijos kaitą, skirtingų kultūrų ir subkultūrų sanglaudoje atsiranda palankios prielaidos rastis edukacinėms inovacijoms, užtikrinančioms pozityvias galimybes atsirasti naujų kompetencijų poreikiui bei jas ugdyti. The changes taking place in rapidly changing society, very often described as alternations, reforms or innovations, presume the necessity of a more specific description of the educational innovation content. In Lithuania educational changes were analysed by Aramavičiūtė, V. Martišauskienė, E. 2009; Barkauskaitė, M. 2009; Janiūnaitė, B. Kliminskas, R. 2009; Motiejūnienė, E. Žadeikaitė, L. 2009; Rutkienė, A. Zuzevičiūtė, V. 2009; Targamadzė, V. 2009; Želvys, R. 2009, 1999; Bruzgelevičienė, R. 2007; Gečienė, E. 2003 and other scientists who investigated different aspects of the education reform: the role of the teacher, changes of education and values, potentials for development of pedagogues' competences, school learners, students, peculiarities of the future school and other relevant issues. At the same time innovations, based on specific examples from the education field, or educational innovations are rather rarely realised in the works of Lithuanian scientists: Gudauskas, R. Kaušylienė A. (2010), Janiūnaitė, B. Jezerskytė,

E. (2009), Janiūnaitė, B. (2008, 2007, 2005, 2002, 2000, 1999), Jucevičienė, P. (2008, 2006) and others. In the works of foreign authors educational innovations are often related to technological novelties (Shear, Moorthy, 2010; Freeman, Bell,

Comerton-Forde, Pickering, Blayney, 2007; Coll, Rochera, Mayordomo, Naranjo, 2007 and etc.), teacher education (Bakkenes, Vermunt, Wubbels, 2010; Henze, Driel, Verloop, 2009), new learning or methods of didactics (Murray, 2008; Lamie, 2005; Kennedy, Doyle, Goh, 1999; Markee, 1997 and etc.).

Furthermore the innovations in educalional invorement plays an important role in EU policies as well. The educational innovations are revealed in the Pact on Stability and Growth of the European Union (EU) (C236, 1997) and in the strategy "Europe 2020" (EC „Communication from the Commission: Europe 2020, A Strategy for Smart, Sustainable and Inclusive Growth", COM (2010) 2020), where it is said that programs of education and training systems should be focused on creativity, innovation and entrepreneurship. The latter objective is based on the anticipated growth of demand for medium and highly qualified employees, presuming that in 2020 in the EU there will be 15.6 million new jobs for high education graduates and 3.7 million jobs for secondary school graduates, meanwhile, the number of low qualification jobs will go down by 12 million.

The educational innovations are targeted in other documents as well: the strategic framework for European cooperation in education and training (2009/C 119/02), where it is emphasised that it is important "to efficiently invest in human capital through education and training systems at same time as promoting personal fulfilment, social cohesion and active citizenship"; EU renewed social agenda (2010/C 212 E/05), which "encourage the knowledge of ICT and new forms of work organisation", the EU Strategy for Youth (2009/2159(INI), where member states are urged "to facilitate access to new technologies in order to boost young people's creativity and capacity for innovation and generate interest in culture, the arts and science".

Summarising the mentioned legal acts and other documents it is important to note that in the strategic documents innovations and their expression in cohesion with improvement of key competences are dominating: "Research and innovation are crucial to addressing some of the major issues facing the European Union and upholding an EU model based on economic growth, social responsibility and sustainable development". <…> Workers need higherlevel skills and the capacity to adapt to change. Investing in human capital is therefore of crucial importance in order to increase research and innovation efforts". European Commission Communication „A new impetus for European cooperation in Vocational Education and Training to support the Europe 2020 strategy" KOM(2010) 296. In the Communication from the European Commission on competitive European regions through research and innovation: a contribution on more growth and more and better jobs (COM/2007/474).

It should be emphasized that educational innovations in the creation of new communication and cooperation relations, execution of new activities or demonstration of competences, introduction of the new technologies and implementation of the new learning paradigms, acquisition of the skills necessary to successfully live and work in the society of high technologies, information and at the junction of founding of new educational environment. Educational innovations are defined and classified in different ways, their definitions reveal the nature of the novelty, so educational innovation is understood as the difference of thoughts or behaviour with a higher quality; fundamental changes that provide a new content for further development and progress; planned efforts to improve the practice in achieving the objectives and including all new educational systems at the level of pedagogical system.

Innovation Modeling: case studies

Summarizing the educational innovations J. M. Higgins (2002, p. 35) provides the classification of educational innovations where they are distinguished by 5 features: place of creation and nature; objective; subject, external attribute, introduction nature. J. M. Higgins identifies and specifies each feature. This classification is supplemented by one more feature of impact volume – such classification is presented in the works of B. Janiūnaitė (2000) (the author called this impact „The impact of educational innovations according to the extent"). This supplement is necessary because the educational innovations may be implemented in the education system by different extent: local, modular or systematic.

The impact of educational innovation by local, modular or systematic level in assessing the motivation to communicate and collaborate may be based on (civil) moral values (E. Kvieskaitė, 2011).

The local impact of educational innovations consolidating the civic moral values may be disclosed by implementing the educational activities of democratic values and using the innovative methods. An example of this could be the project implemented by Lithuanian College of Democracy, Lithuanian University of Educational Sciences and Modern Didactics Centre[1]. This partnership aimed to improve the professional competence of the teachers seeking to develop the civil, democratic and other values of the students – to help the class teachers and all other teachers to create the educational plenitude by examples of good practice. An instructional book was developed for the teacher. It is easily adapted

1 The project funded by European Union „Education of citizenship and democratic values through innovative methods" (ESF/2004/2.4.0-03-302/BPD-125/1).

to different classes (L.Navickienė, S.Karalienė, A.Navickas, A.Piličiauskas, 2007) and thus the field of local educational innovations impact is distinguished.

The broader field of impact is revealed through civic moral values by the extent of modular educational innovations. This case may be illustrated by the methodology „Project CITIZEN" developed by Centre for Civic Education (USA). This methodology aims to address the issues of civil community by integration into all learning subjects of the school (E.Celiešienė, 2013). The impact of educational innovations in forming the civic moral values by systematic level is actualized in the long-term programme of civic and national education of the Republic of Lithuania (2006-09-26, No. 102-3939). The results of this programme show the change of public moral values. During the implementation of this programme in 2007-2011 it was decided to carry out the monitoring and evaluate the effectiveness of the measures. It is worth to note there were no conducted researches at the national level helping to assess the contribution of the programme to the change of public moral values.

The research on Creativity, Innovation and their synonyms in primary and secondary school curricula in EU27 (European Communities, 2010) shall be emphasized in this context. The results of the research reveal (p. 19) that many teachers still continue to work on the educational paradigm and used mostly classical methods (such a as explaining). In other hand the students underline interest in the usage of alternative methods (project, problem based learning, learning by doing etc.). These results correspond to the results of the research conducted in Lithuania (V. Targamadzė et al., 2010; V. Targamadzė 2011). They indicate that pedagogical paradigm of impact (training) basically prevails at the secondary school. The research revealed that teachers commonly use the methods of impact paradigm, the essence of which is information rendering, i.e. orientation to the teacher's activity instead of a student. This is fundamentally contrary to the essence of interaction and learning paradigms.

The impact of educational innovations at the systematic level may be disclosed by establishing the implementation of national or international law during the analysis of knowledge of citizenship and government. For example, it was aimed by approving the long-term civic and national education programme (2006-09-26, No. 102-3939). The programme raised three objectives, two of them aimed to improve the conditions of civic and national education of Lithuanian and foreign Lithuanian children and youth by the means of formal and non-formal education and improve the civic and national education system of Lithuanian and foreign Lithuanian adults. On the other hand, the effect of systematic educational innovations during the analysis of competences to communicate and cooperate in internal and external networks may be based on *public-private-partnership* and universal multifunction networks or socio-educational cluster outlined in education system (G. Kvieskienė, V. Kvieska, 2012). The report of the research on the universal multi-functional centre (UMC) conducted by the order of Ministry of Education describes the centre as an institution providing the educational, cultural, socio-cultural and recreational services to the children and adults oriented in the needs of people living in the region. The report also stated that the objective of UMC is to help to satisfy the social and educational needs of different interest groups of specific area by mobilizing the public, private, national and local resources and its main functions shall include the activeness, welfare, nurture, education, development, improvement and formation of community members. This is important because in Lithuania it is only started to talk about social entrepreneurship in the educational space which is analysed as the innovative model of welfare state (Kvieskienė, Kvieska, 2012) by enabling the business models which can also be analysed in the context of systematic educational innovations impact and solve the social or environmental issues.

Design of the research

In order to clarify the opportunities and prevalence rates of innovations application in the field of education, group discussions in the scientific literature also referred to as „focus" or „focus group interview", „focused group interview" or „group interview" (Babbie, 2009; Krueger, Casey, 2009; Stewart, Shamdasani, Rook, 2007; Hennink, 2007; Litosseliti, 2003; Edmunds, 2000; Greenbaum, 1998; B.Bitinas, L.Rupšienė, V.Žydžiūnaitė, 2008 et al.) were organized. Lithuanian scientists of methodology (B.Bitinas, L.Rupšienė, V.Žydžiūnaitė, 2008, pages 21-22) indicate that although so far the group discussions were rarely applied in education, this approach helps: „to understand the behaviour, know the approach to study problem and course of events". The open coding program of qualitative research „Kokybis" was invoked to analyse the data.

Subject and sampling

The study consisted of one focus groups from 10 international experts (from 7 Europe country): 6 men and 4 women. All of them were experienced in the usage of online social networking platforms (web.20 tools) and were involved in various kind of the best practice sharing in the mentioned area. Initially, the research asked about all types of online social networking platforms (web.20 tools) used by the participants and its adaptation or usage in education environment.

Research findings and discussion

Interpretation of research results revealed the main trends: innovation value, development of cooperation based on social partnership and innovation opportunities for education. So every area actualized by the informants is discussed by highlighting the key concepts and questions of discussion.

Value of innovation

Informants during the analysis of web.20 tools, paid great attention to their functionality and analysis of additional values to education. It should be emphasized that web.20 tools create the opportunities for personal interaction between students and teachers. The learning system that involves the students of different motivation and skills is often criticized for ignoring the individual needs of the learner. Therefore, the adaptation of innovative systems can allow for individualised learning process. The learners should have the opportunities for individualized learning process in accordance to pace, motivation and career projections of each person academic achievement improvement. Together these opportunities allow learning in a comfortable place and at convenient time, so it can become a substitute for traditional learning at home. It is important to notice that these innovations can reduce the cost of education so they may be attractive to both taxpayers and politicians.

Experts also noted that web.20 tools create conditions to form new complex solutions, particularly in analysing marginalized groups. At the same time, it is pointed out that when the learning activities are organized by this way the strategy of lifelong learning is actually implemented and sustainable learning is consolidated. It should be noted that web.20 tools enable to improve the competence in the chosen field at any time; therefore, according to the experts the formalization of this system and links to formal education is inevitable and at the same time opens the new opportunities and values.

Development of cooperation based on social partnership

The informants in assessing their personal experiences and environments observed that web.20 tools were often invoked by the business to achieve for its goals, mainly – product popularization and sale. Although the field of education still underuses web.20 tools. Informants stressed the need to expand teacher's knowledge on web.20 tools possibilities and its integration to educational process. Namely, the competence development of education participants may create an extra spurt in this field by using web.20 tools in education process. It should be emphasized that web.20 tools also create solutions that could be adapted instantly and presented not only regionally, but also nationally. Furthermore, beneficial partnerships and social networking development, to which web.20 tools open up new possibilities to cooperate in the development of social economy, shall be emphasized in this context.

Innovation opportunities to education

Informants analysing the additional possibilities of web.20 tools, stressed the visibility, development of social partnership and creation of innovative and creative environment in the educational system. Experts said that the participants of both formal and non-formal education use web.20 tools, but noticed there are no estimates of their systematization, efficiency and performance. According to experts, one of the missed opportunities is sharing the best practice experience in web.20 usage in educational processes. In this way, specific methods or techniques may be easily taken over and adapted in different countries of the world and that creates an opportunity to develop the education system. On the other hand the informants emphasized it is necessary to collaborate between public, private sectors by presenting innovative project ideas and disseminating best practices. web.20 tools, that may be involved in both the search for partners for specific projects (for example, by using Linkesid) and in the presentation of results (for example, YouTube) may also serve in the development of this partnership. Generally, informants noted the links between innovation and creativity and their interaction points by claiming that it is necessary to establish the creative and open for innovations environment in the education system allowing for innovative experience and idea communication.

Conclusion

In the summary, it should be emphasized that innovation and its modelling in educational environment rise new possibilities for meaningful personal interaction between students and teachers and ensure individualizations possibilities in the sustainable learning processes. These changes lead to inevitable change of educational system that creates the opportunities of lifelong learning according to individual needs and abilities and developing the educational services, reporting formats and volumes. Innovations gradually consolidating in the field of education are usually identified in local and modular volumes, but their spurt guarantees the innovation development at the national and global level.

Sharing the networking best practice experience is crucial for learning environment stakeholders' competencies development. At this stage of innovative transformation where local and modular innovations rapidly spread in the different parts of the world it is essential to develop the traditions of open culture and promote the dissemination of best practice. Successful cooperation and partnership of business and public sector, including the education, may raise the education for its renaissance. An open, innovative and creative environment where all participants are equally important and their cooperation is based on partnership may become an excellent medium for creation and dissemination of innovations.

Innovation and its modelling in educational environment can play an important role involving the mutual interaction of the different groups and individuals in the organizations. When every user of innovation is enabled to further develop and improve the innovation, the empowerment of each individual stimulating the individual motivation to create and participate is stressed. On the other hand, it is important to note that the innovations would create the real conditions of equal opportunities when marginalized groups have an equal opportunities, conditions and field of activities.

The communication across cultural boundaries becomes one of the basic preconditions for the development of an innovative in educational environment. The perception of educational content requires the appropriate linguistic competence the development of which becomes crucial for the development of innovations. In this case, a linguistic competence may become an important barrier to the spread of innovations in different contexts, therefore it is necessary to develop language tools able to translate and adapt the content to understandable language for each individual.

References

1. Aromavičiūtė, V., Martišauskienė, E. Vertybės: švietimo strategijos ir realybė. Iš: Pedagogika. Mokslo darbai. Vilnius: Vilniaus pedagoginio universiteto leidykla, 2009, T.95, p. 28–34. ISS N 1392-0340.

2. Babbie, E. Twelfth Edition. The Practice of Social Research. 2009, 624 p.

3. Barkauskaitė, M. Skirtingų visuomenės grupių lūkesčiai dėl ateities mokyklos. Iš: Pedagogika. Mokslo darbai. Vilnius: Vilniaus pedagoginio universiteto leidykla, 2009, T. 95, p. 94, 101. ISS N 1392-0340.

4. Bakkenes, I., Vermunt, J. D., Wubbels, T. Teacher learning in the context of educational innovation: Learning activities and learning outcomes of experienced teachers. Iš: Learning and instruction, 2010, Nr. 20(6), p. 533–548. ISS N 0959-4752.

5. Bitinas, B., Rupšienė, L., ir Žydžiūnaitė, V. Kokybinių tyrimų metodologija. Klaipėda: S. Jokužio leidykla-spaustuvė, 2008.

6. Bruzgelevičienė, R. Lietuvos švietimo reforma ugdymo paradigmų kaitos aspektu 1988–1997. Daktaro disertacija. Socialiniai mokslai, edukologija (07 S). Vilnius: Vilniaus pedagoginio universiteto leidykla, 2007, 216 p. [žiūrėta 2011 m. sausio 7 d.]. Prieiga per internetą: <http://vddb.library.lt/fedora/get/LT-eLABa-0001:E.02~2007~D_200 91223_ 134 355-43353/DS.005.0.01.ETD>

7. Coll, C., Rochera, M. J., Mayordomo, R. M., Naranjo, M.. Continuous Assessment and Support for Learning: An Experience in Educational Innovation with ICT Support in Higher Education. Iš: Electronic Journal of Research in Educational Psychology, 2007, T. 5, Nr. 3, p. 783–804.. Publisher: University of Almeria, Education & Psychology I+D+i. Faculty of Psychology, Department of Educational and Developmental Psychology, Carretera de Sacramento s/n, 04120 LaCanada de San. Education and Training Monitor 2012. This publication is based on document SWD(2012)373.

8. Edmunds, T. 'Searching for a Nationalism: Foci for Kazakh Identity in the 1990's', Central Asia Monitor, No.2 2000.

9. European Parliament resolution of 6 May 2009 on the Renewed social agenda (2008/2330(INI)).

10. Europos Komisijos komunikate moksliniais tyrimais ir inovacijomis grindžiamas Europos regionųkonkurencingumas. Pasiūlymas: augimo skatinimas ir naujų bei geresnių darbo vietų kūrimas (COM/2007/474), priimtas 2008 metų balandžio 22 d. Prieiga per internetą: http://eur-lex.europa.eu/LexUriServ/LexUriServ.do?uri=CELEX:52008AE0751:LT:NOT.

11. Europos Sąjungos bendradarbiavimo švietimo ir mokymo srityje strateginė programa („ET 2020"), priimta 2009 metų gegužės 12 d. Prieiga per internetą: <http://eur-lex.europa.eu/LexUriServ/LexUriServ.do?uri=OJ:C:2009:119:0002:0010:LT:PDF>.

12. Europos Sąjungos strategija jaunimui. Investicijos ir galimybių suteikimas, 2010 m. gegužės 18 d. Europos Parlamento rezoliucija dėl ES jaunimo strategijos „Investicijų ir galimybių suteikimas" (2009/2159(INI)). Prieiga per internetą: <http://www.europarl.eur opa. eu/sides/getDoc.do?pubRef=-//EP//NONSGML+TA+P7-TA-20100166+0+DOC+PDF+V0/ /LT>

13. Freeman, M., Bell, A., Comerton-Forde, C., Pickering, J., and Blayney, P. Factors affecting educational innovation with in class electronic response systems. Iš: Australasian Journal of Educational Technology, 2007, Nr. 23 (2), p. 149–170 [žiūrėta 2011 m. sausio 11 d.]. Available under: <http://www.ascilite.org.au/ ajet /ajet23/freeman.html>

14. Gečienė, E.. Pokyčių visuomenėje sąsajų su švietimo kaita paieška. Vilnius: Ciklonas, 2003, 28 p. ISBN 9955-497-18-1.

15. Greenbaum. Thomas L. The Handbook of Focus Group Research. Second Edition. Sage, 1998, p. 261

16. Gudauskas, R., Kaušylienė, A. Žinių vadyba ir edukacinės inovacijos: metodinė priemonė. Vilnius: Vilniaus pedagoginio universiteto leidykla, 2010, 47 p. ISBN 978-9955-20-569-2.

17. Higgins, S., Moseley, d. Raising achievement in Literacy through ICT. Iš: Teaching Primary Literacy with ICT, Monteith, M. Buckingham: Open University Press, 2002, p. 30–45. ISSN/ISBN: 0335 21246-8.

18. Henze, I., Driel, J. H., van, Verloop, N. Experienced Science Teachers' Learning in the context of Educational Innovation. Iš: Journal of Teacher Education, 2009, Nr.60 (2), p.184–199. Online ISS N 1552-7816.

19. Hennink, Monique M. International focus group research: a handbook for the health and social sciences. Cambridge: Cambridge University Press, 2007. 248pp. ISBN: 978-0-521-60780-3.

20. Heilmann, G., Werner B. K (empirica GmbH). [10:47:39] Migle M: The Role of Creativity and Innovation in School Curricula in the EU27A content analysis of curricula documents. Luxembourg: Publications Office of the European Union European Communities, 2010.

21. Janiūnaitė, B. Edukacinių novacijų diegimo tyrimo modelio pagrindimas. Iš: Socialiniai mokslai, 1999, Nr. 4 (21), p. 25–33. ISS N 1392-0758.

22. Janiūnaitė, B. Edukacinės novacijos ir jų diegimas: monografija. Kaunas: Technologija, 2004, 100 p.ISBN 9955-25-162-X.

23. Janiūnaitė, B. Edukacinių novacijų diegimo ypatumai Lietuvos švietimo sistemos modernizavimo kontekste: metodologiniai aspektai. Iš: Socialiniai mokslai, 2002, Nr. 3 (24), p. 97–112. ISS N 1392-0758.

24. Janiūnaitė, B. Educational innovations as a means of educational changes: problematic aspects of conceptualizing, locating and assessing educational innovations. Iš: Baltische Studien zur Erziehungs und Sozialwissenschaft. Frankfurt am Main: Peter Lang, 2002, Bd.7, p. 171–186. ISBN 978-3-631-39869-2.

25. Janiūnaitė, B. Methodology of evaluating educational innovation during the process of implementation. Iš: Socialiniai mokslai, 2002, Nr. 4 (36), p. 70–80. ISNN 1392-0758.

26. Janiūnaitė, B., Tumėnienė, V. Pedagogo veiklos pokyčiai pasaulinių ir europinių švietimo dimensijų kontekste. Iš: Socialiniai mokslai, 2000, Nr. 2 (23), p. 24–34. ISSN 1392-0758.

27. Janiūnaitė, B. Edukacinės novacijos ir jų diegimas: monografija. Kaunas: Technologija, 2004, 100 p. ISBN 9955-25-162-X.

28. Janiūnaitė, B., Tumėnienė, V. Pedagogo veiklos pokyčiai pasaulinių ir europinių švietimo dimensijų kontekste. Iš: Socialiniai mokslai, 2000, Nr. 2 (23), p. 24–34. ISSN 1392-0758.

29. Janiūnaitė, B. Jezerskytė, E. Dėstytojo inovacinės veiklos bruožai tobulinant studijų programas: besikeičiančios edukacinės paradigmos kontekstas. Iš: Jaunųjų mokslininkų darbai, 2009, Nr. 1 (22), p. 73–82. ISS N 1648-8776

30. Janiūnaitė, B. Piliečių inovacinė kultūra: mokslo studija. Kaunas: Technologija, 2007, 100 p. ISBN 9955-25-162-X.

31. Janiūnaitė, B. Edukacinių dimensijų charakteristikos atvirųjų inovacijų paradigmos kontekste. Iš: Socialiniai mokslai, 2008, Nr. 1(59), 2008, p. 42–49. ISS N 1392-0758.

32. Jucevičienė, P., Čiužas, R. Lietuvos mokytojų didaktinė kompetencija. 2006. Availableunder: <http://www.smm.lt/svietimo_bukle/docs/pr_analize/Mokyto judidaktiekomp ten ija.pdf >.

33. Jucevičienė, P. Edukacinės ir mokymosi aplinkos – inovacijos socioedukacinio įgalinimo veiksnys. Iš: Socialiniai mokslai, 2008, Nr. 1 (59), p. 58–70. ISS N 1392-0758.

34. Jucevičienė, P. Teaching in higher education: teacher as learning. Iš: Socialiniai mokslai, 2006, Nr. 3 (53), p. 72–79. ISNN 1392-0758.

35. Kennedy, C., Doyle, P., Goh, C. M. (eds.). Innovation and change in English language teaching: Over adecade of shared perspectives. Exploring change in English language teaching. Oxford: Macmillan Heinemann,1999, 118 p. ISBN 0-435-24053-6.

36. Kliminskas, R., Janiūnaitė, B.. Bendrojo lavinimo mokyklos ugdymo programos kaitos sąlygos Lietuvoje: konstruktyvistinio požiūrio aspektas. Iš: Pedagogika. Mokslo darbai, T.95, p. 58–57. ISS N 1392-0340.

37. Kvieskaitė, Eglė, Tarpkultūrinės komunikacijos ir edukacinių inovacijų sąsajų raiška: [rankraštis]: daktaro disertacija: socialiniai mokslai, edukologija (07S) / 2011 Vilnius: Vilniaus pedagoginio universiteto leidykla, 2011. p. 133

38. Kvieskienė G., Kvieska V. Socialinės partnerystės įtaka inovacijoms. Monografija. Vilnius: Edukologija, 2012. ISBN978-9955-20-797-9.

39. Lamie, J. M. Evaluating change in English language teaching. New York: Palgrave Macmillan, 2005, 254 p. ISBN 1-4039-3975-6.

40. Litosseliti, L. Research Methods in Linguistics.London Se1 7NX. New Yourk, NY 10038, 2010 p. 227.

41. Navickienė, L., Karalienė, S., Navickas, A., Piličiauskas, A. Grožio ir Gėrio link / Pilietiškumo ir demokratinių vertybių ugdymas taikant inovatyvius metodus. Klasės auklėtojo knyga.- Vilnius: Ciklonas.- 2007.- 388 p.

42. Markee, N. Managing curricular innovation. Cambridge: Cambridge University Press, 1997, 227 p. ISBN 0-521-55524-8.

43. Motiejūnienė, E., Žadeikaitė, L. Kompetencijų ugdymas: iššūkiai ir galimybės. Iš: Pedagogika. Mokslo darbai, T 95. Vilnius: Vilniaus pedagoginio universiteto leidykla, 2009, p. 86. ISS N 1392-0340.

44. Murray, D. (ed.). Planning change, changing plans. Ann Arbor, MI: The University of Michigan Press, 2008, 264 p. ISBN 978-0-472-03278-5.

45. Richard, A., Krueger. 4th edition. Focus Groups: A Practical Guide for Applied Research. 2009, p. 219.

46. Resolution of the European Council on the Stability and Growth Pact Amsterdam, 17 June 1997. Official Journal C 236, 02/08/1997 P. 0001 - 0002. Available under: http://eur-lex.europa.eu/LexUriServ/LexUriServ.do?uri=CELEX:31997Y0802%2801%29:EN:HTML

47. Rutkienė, A., Zuzevičiūtė, V. Mokytojo vaidmens kaita: bendruomenės įtraukimas siekiant mokymąsi daryti veiksmingesnį. Iš: Pedagogika. Mokslo darbai, 2009, T. 95, p. 53. ISS N 1392-0340.

48. Shear, L., Moorthy, S. Innovation Around the World: ICT-Supported Educational Innovation in Four Countries. Iš: Z. Abas et al. (Eds.). Proceedings of Global Learn Asia Pacific, 2010, p. 618–623. ISBN 1-880094-79-7.

49. Knight B. Reseach in focus: using focus groups in institutional research. Bowling Green State University OAIRP Spring Conference March 25, 2011.

50. Targamadzė, V.; Nauckūnaitė, Z.; Stankuvienė, I. ir kiti (2010). Galimybių studija 12–14 metų mokinių mokymo(si) didaktinės problemos ir jų sprendimo galimybės. Vilnius: LR ŠMM ŠAC. Prieiga per internetą: http://www.alternatyvusugdymas.lt/wp-content/uploads/2009/12/10_09_03_AU_galimybiu_studija.pdf.]

51. Targamadzė, V. Produktyvus mokymasis – pedagoginės sistemos bendrojo lavinimo mokykloje kaitos akstinas. Acta pedagogica Vilnensia 2011 27. ISSN 1392-5016.

52. Želvys, R. Švietimo vadyba ir kaita. Vilnius: Garnelis, 1999, p. 290

53. Želvys, R. Visuomenės požiūris į švietimo reformą: kas pasikeitė per 20 metų. Public Opinion About Educational Reform: What Has Changed During the 20 Years. Iš: Pedagogika. Mokslo darbai, 2009, p. 27. ISS N 1392-0340.

Identity and Inequality: Artistic Activity as the Precondition for Social Integration

Aldona Vilkelienė

Abstract
The paper discusses the aspects of perceiving and experiencing personal identity and inequality in artistic activity and by means of the phenomenological analysis highlights the features of social integration. The phenomenological analysis of the research data revealed artistic activity as a precondition of personal cognition and interpersonal communication, and the significance of negative experience in artistic activity as a personal and social challenge which is typical for social reality.

Keywords: identity, inequality, artistic activity, social integration

Introduction

Young people have always asked themselves questions like "Who am I? What is my purpose? Are my surroundings the best place for me?" Identity issues were initially investigated by Erikson [7] who determined that the formation of identity results from the interrelation between a person and the society. The later researches stated that the formation of identity takes place in the realm of the integration of opposing qualities and is affected by the family and the whole mankind as well as its morality development [13]. Researchers noted that identity and ethnic culture are inseparable from the perception of individual identity [17]. Hargreaves [10] conducted a survey determining how important the art of music is for personal identity and its enhancement. The phenomenon which is by modern philosophy, sociology and other sciences called a "postmodern society" reveals a variety of attitudes, cultural assimilation and ambiguity of values. The postmodern society is losing the elements of the main agreements: belief in progress, rationalism, long-lived values – they are becoming pluralistic and chaotic [4]. In the authors' opinion, the classical rules are disappearing, and the attitudes to social roles are changing. These researchers also revealed the role of cultural and social phenomena in the formation of identity. They also emphasized the fact that education by means of artistic activity is an appealing and efficient way which discloses the level of identity perception and the possibilities of dialogue in the context of different cultures.

Nevertheless, the development of research in the context of identity and artistic education exhibits a discrepancy between the educational goals. Some goals are raised by the postmodern age, or rather postmodern ideology, which opens the way for a personality, but at the same time eliminates the limits protecting a personality. The national identity is weak; the features of commitment to a person disappear, just as the outline of his identity [5, 2]. Modern philosophers [8] call this a personality identity crisis which makes a negative influence on the development of a young man's personality. Why are the problems of spiritual decline so vital, why are values disappearing? – These are the questions rhetorically asked by researchers of educational sciences [1]. They claim that for one thing information technologies expand communication, but for another – they constrict live, non-virtual relationships and distance a personality from the society. The scientists say that the image of a "freed" man, created by the media, leads to a crisis just as the influence of TV shows and films on a young person, absence of critical thinking and inability to choose the right values. The problem of personal identity in this situation is especially relevant: the image of a delocalized and depersonalized world, which is depicted by postmodern ideologists, causes a personality identity crisis [5]. Dollof [6] states that identity is changing, dynamic, fragmented, however, she believes that we construct our identities on the basis of what we experience and get from the environment. Lithuanian researchers have ambiguous positions concerning identity construction. One of them argues [11] that the perception of reality under the influence of postmodernism is becoming fragmented, and people's involvement in information technologies immerses them in virtual reality, breaks them away from natural culture and creates an environment of cultural and social paradox which doubts the possibility of cognizing the process of personal identity. However, Daujotytė–Pakerienė [5] believes that identity covers the sense of people's independence and ownership which includes the practical sphere of life and the spiritual existence. The need for stability in the process of identity growth was justified by Regelskis [16] who depicts this process as a pyramid on whose stable foundation a person's identity is formed. In the most general sense, these are the unique individual characteristics (age, gender, status) which are reflected in time, social space or some specific context. This aspect sometimes contrasts with the inner aspect of a personality "I know who I am". In the subjective, phenomenological sense we can speak about a personal conception of identity: "I am like this", "I am different". According to Klimka [12], cosmopolitization, happening in the process of globalization, has no perspective; hence, the EU strategy for culture promotes the encouragement of cultural dialogue and multicultural diversity. The investigation of different researchers' educational studies in the context of identity and cultural diversity shows that there are not many analytical works on the intercultural issues which would reveal the significance of artistic activity, as social integration, for the development of personal identity. There are yet no works which would reveal the interrelation of artistic education and intercultural context, hence, this paper raises the following problem question: What does an individual experience in a different socio-cultural context (different culture, unusual experience during artistic activity)? What is the significance of intercultural context in the perception of personal identity? Can practical artistic activity open up new possibilities to experience the cultural variety as a means to enrich oneself spiritually, to make the integration of different cultures and the social environment more productive and to enhance people's perception of their personal and cultural identity?

The object of the research is personal identity and socio-cultural identity in the context of artistic education.

The goal of the research is to reveal the features of expression of personal and socio-cultural identity, which justifies an individual's social integration, in artistic activity.

The research employs a qualitative ***research method***. In such a research it is significant to refer to the theory of social constructivism which states that reality exists in relation to individual experience. Social phenomena are constructed by people who explore such phenomena and participate in social reality [3]. In such a way a qualitative research constructs situations and meanings which are analyzed through the views of the target subjects from their own perspective. One of the ways that lets to cognize the object of a qualitative study is phenomenological analysis. The object of a qualitative study based on this strategy is the data about the study subjects' experience, thus, the research is based on the interview data supplied by the informants themselves, and the phenomenological analysis is directed at their experience on the basis of which the informants construct social reality [14, 18]. In the process of education, artistic activity can be employed in many different ways. This research chose several artistic activities: theatre, music and arts. In the group artistic activity was conducted with students who have no special artistic preparation in Lithuania, Germany, Austria, Northern Ireland (the United Kingdom), and Turkey. The musical tasks were applied in Lithuania; art was employed in Germany, and theatre – in Northern Ireland and Turkey. The choice was determined by what arts teachers (teachers of theatre, music or art) agreed to participate in the research (to conduct artistic activity with the study subjects) and by the possibilities to conduct artistic activity in different cultural situations and locations. Theatre activity took place in Turkey in the authentic place of Trojan events in the amphitheatre of Troy. Other theatric activities took place in Northern Ireland in a studio in Derry. This town, just as the rest of Northern Ireland, is known for its strong political and religious conflict, therefore, the topics related to these conflicts and their dramatic solutions are painful and emotionally affect the local residents since they are closely connected with the people's personal losses, injustice and violence. Some other social and cooperative issues were being solved in Austria in a play about theatre situations that was presented in the anthroposophical academy in Vienna. Musical activity took place in Lithuania by using traditional music instruments, the old folk songs – multipart songs and children's games which prompt social relations. Art activities took place in Germany, in Alanus University and they involved students from different countries (immigrants from Lebanon, Brazil, Russia and Turkey).

The sampling of the research is purposeful. The sampling criteria are 19-25-year-old research participants who represent three kinds of art, have no professional backgrounds of artistic preparation, but are motivated to participate in artistic activity and research. After they completed artistic activity, the participants were asked to answer two open-ended questions which were oriented to their experience during artistic activity. The total number of the research participants was 9 informants from different countries.

The Phenomenological Analysis of the Research Data

In every research, artistic activity is presented and interpreted on two levels – as a creative process and as a symbolic meaning for a personality and at the same time for the society. As it is stated by Brater and Wagner [4], the goal of artistic education is not to try to convince learners or make them interested. The aim is much more quickly attained, especially in artistic activity, by means of *insight*, consciousness and inner awareness. Only such a consensus makes the change of behavior or activity meaningful. According to the aforementioned researchers, only when experience intertwines with cognition and emotions, the rigid patterns of habits, which used to function as the limits of an individual's safe (or supposedly safe) existence, start expanding and the world is seen from a new perspective. The authors argue that if art is used as a means of emotional influence for educational purposes, the process is not a manipulation with emotions.

In this study artistic activity is understood as education based on the active methods, which employs different kinds of art, and the process activates the learners. Yet, there is still some space left for him to relate the activity with his own experience, personal significance and value-related attitudes.

The research participants had to say what they experienced in artistic activity and how the context or situation determined their experience. The data acquired in the partially structured interview were analyzed by means of the interpretative phenomenological analysis. The interview was transcribed, read, analyzed, there were distinct topics singled out, they were grouped, and a descriptive analysis was conducted. When the text was divided into topics and subtopics, it was again returned to the initial data context, the vague meanings were selected, they were explained and the initial interpretations were related to the whole [18]. The research results highlight the features which enable to determine whether the interviewees relate themselves to their nation and its traditions, and thus the study foregrounds the individual's self-worth, the cases of its enhancement, or, conversely, its derogation (Table 1).

Table 1: The Transfer of Biographical Situations and Self into Artistic Form

Topic	Subtopic	Quotation
Sense of self-worth	Relation to the past traditions, historic events	When I heard the Lithuanian multipart songs, their even rhythm and recurrent words, my heart was touched by something like the ancestors' call, I should say. In our country folk songs are viewed differently. Some people of the older generation associate them with the period when Nazis came to govern the state. Unfortunately, it is like this…
	Self-image	When you improvise in the group without a leader, you start feeling yourself. Initially you experience some kind of chaos, but gradually begin to perceive separate fragments.
Possibilities to create a safe environment	Self-help	<…listening to music is different from playing it. When you listen to music, you try to understand what the composer had in mind and compare his ideas with your own perception and sensations. Do we have to speak about how you created a painting, a picture? What if another person starts assessing you, "this is good", and "I don't know what you wanted to express here"? Won't it humiliate a person? You are creative, and he is not?
	Therapy	When you act and tell your story, it breaks away from you and becomes a part of "another", bigger picture or the world.
	Connection with the group	It was help for me. Even when I acted a real story which was not very nice. It helped me to forget some unpleasant things. Even the fact that others are hearing my story, and they are not judging me as a bad or unjust person.
Establishment of Self	Staying true to yourself	I understood that for me it is important to stay true to myself in any situation. The joint painting helped me cognize myself. I was drawing goldfish. I chose the orange colour because I wanted to protect the goldfish from the touches of the other painters.

Topic	Subtopic	Quotation
		I went round and all around the big sheet of paper I left a school of orange fish.
	Sense of leadership	Even though I knew that it was just a play, I enjoyed being the lead. I liked that the head of the theatre group assigned me to be the lead. I do have leadership qualities.

In the subjective, phenomenological sense the interviewees speak about the personal perception of their identity: "I am like that", "I am different". Their experience in the artistic context shows that the roots of identity lie in the internal dialogue when the consciousness controls a persons' actions, memories and fantasies: "when I was singing my nation's folk song, I felt exalted…remembered my home, childhood"; "I would have never thought of painting while listening to music"; "this composer (Čiurlionis) is very talented. We also have a composer who related music to colours and movement". These statements, presenting the informants' experience, reveal the fact that it is impossible to deny the existence of the primary, inborn identity and the significance of activity, knowledge, internalized values and adequate communication as it anchors itself in the society; and that all of this corresponds to the creative possibilities of Regelski's [16] stable foundation which enables a personality to grow and reveal itself.

Another part of the informants' ideas about their experience in artistic activity revealed the features of their relationships with the participants of artistic activity. Table 2 presents the informants' experiences of merging with the group or socializing in the context of different social cultures, and using the language of art symbols in communication.

Table 2: Integration in the Artistic Activity Group

Topic	Subtopic	Quotation
Development of social skills	Artistic activity within a group	At first I wanted to play as loudly as possible to be heard. Later I understood that I will not win since everyone was playing their own melodies. I started according with them, searching for a contact… I got involved as soon as I heard the rhythm and started improvising. She turned back at me right away. A very interesting connection. I like joint creation when you can complement and develop your friend's creative idea. I like watching how colours mix together, the contours disappear, new shapes appear. I even like chaos but only because I want to "correct" it, to make it nicer. Initially I wanted that there was a leader, somebody in charge who would tell us what to do, when to start and stop.

	The differences of the context of artistic communication	Music is a very social art; it immediately involves a person into the process. It is more about emotions rather than intellectual thinking. For example, when we were drawing, it was possible to retreat from the work, to talk about it. Music is a very strong impulse. It is impossible to retreat and reflect. You can dance, sing and enjoy the experience, but afterwards it is impossible to find words to describe it.
	Perception of social and cultural differences	This theatre project helped to conceive myself better. I know what I am, why I am and where I am. I saw that the others sometimes felt lonely and misunderstood. I even made friends here. Now we are keeping in touch. We, Irish people, love rhythmic dances. They raise your spirits. This gives you a sense of community. When I acted, I learnt a lot of historic facts. When I was speaking to the dead, I raised my hands up to the sky, but my mentor remarked that in the period of Antiquity the posthumous world was underground.

Group artistic activity was differently assessed in the informants' answers. After drawing a joint painting, they noted: "*wonderful, I felt a part of a small community. I am not alone. The joint painting is impressive. You could never make anything like that by yourself – so many colours, surprises…*"; "*When you are drawing or making a work of art and a colleague approaches you, it shows that your creation is interesting, appreciated and maybe respected.*" Not all the participants had positive experiences; there were other, different opinions: "*When my group-mate saw flouncing about in my drawing, at first I did not agree with that in my heart. I did not think that was a flounced about image. I saw it as tenderness; I wanted to protect the fish from the other painters. I was walking round and all around the big sheet of paper I left a school of orange fish. In half an hour, as a result of joint drawing, my fish remained only in a couple of places in the painting*".

These statements show that artistic activity and its different forms (individual, group activity) reveal not only the differences of people and cultures, but also the individual's experience of those differences. "*I cannot mix colours. I feel bad about it. I've learnt that I like only primary colours. Maybe this is the reason why I found it difficult to draw a joint painting? It's just like life…*"; "*Initially I wanted that there was a leader, somebody in charge who would tell us what to do, when to start and stop*". The informants' ideas reveal the perception of their own culture and traditions which are evaluated by means of comparison: "*…you have such a strong national culture. Well, you had to protect yourselves from a foreign invasion all the time. We did not have such a problem and maybe this is why our ethnic* *culture is disappearing?"; "I learnt that in the ancient times the posthumous world was underground..."*

The researchers' studies confirm the fact that personal values are inseparable from culture, to which one or another person belongs, and thus they form an attitude that the formation of identity depends on personal, social and cultural factors, while culture motivates behavior, cognition of values, creation and development [4, 10, 17]. In the process of education these factors are also mentioned by Lithuanian researchers: they emphasize the significance of social interactions in education and argue that people can cognize themselves only when seeing others [2, 5]. Personal identity, or the Self, covers such features as the characteristics of their physique, abilities, skills, behavior and many other qualities typical for that personality which are constructed on the basis of the individual's ideas, expectations and social relations [13, 15]. Hence, identity is a personal and social phenomenon. On the other hand, culture, in which personalities develop, also has some significance. It emphasizes the importance of educational interaction and educational environment for the excitation of learners' personality powers.

The fact that the informants' experience in artistic activity was not always positive was not unexpected for the researcher as an art education specialist, since in a phenomenological analysis the researcher rejects all the preconceptions. He proved the theoretical grounds only at the end of a phenomenological data analysis in order to be able to compare the acquired data with the studies of other researchers. In the investigation of a selected phenomenon it is important not only determine what has been already foreseen by raising hypothetical questions or referring to theories, but also discover elements of "otherness" [18]. In this research during the deep unstructured interview the participants frequently spoke about the feeling of unexpectedness, defined artistic activity as a personal challenge. The phenomenological analysis distinguished a topic of negative experience in artistic activity which can be subdivided into five subtopics (Table 3).

Table 3: Social and Pedagogical Challenges in Artistic Activity

Topic	Subtopic	Quotation
Negative experience in artistic activity	Sophisticated precaution	I don't like drawing a joint painting. I could not begin for a long time, I was standing apart till I was able to make out the contours of the drawing. I don't like destroying other people's works. As if I was bossing around in their homes. I was trying to find my own place and draw there.

	Social confinement	I was late for the beginning of the class and was very excited. When I approached the classroom door, I heard music and wondered how I could enter. Everybody will turn silent and look at me. But then I made up my mind and opened the door.
	Cultural distinctions	I could not understand what the teacher was singing as the words were not Lithuanian, but the refrain had the same recurrent motif, so I was singing with the others and, just like the others, I imitated the greeting gesture (holding out my hand).
	Psychological problems	At first the task seemed a bit strange… Since I had my roller-skates with me, I chose them, or rather one roller-skate. I was speaking from his perspective and said that I was very fast and the other roller-skate, the one with a smaller heel, stopped me when necessary. I don't know why I chose a roller-skate. Maybe because I have a brother, but don't communicate much…
	Fear of artistic expression and employment of tools	When I had to play a musical instrument, I got scared. I felt vulnerable and helpless. I thought I would know how to handle it. It is good that we were given some time to test the strings (of the zither). When we were playing together, I was trying to heat what I was playing. I was glad that it was a group improvisation and nobody could hear me. I had never drawn. I felt like I was in a kindergarten. I had a single idea – they will all see my drawing. I felt embarrassed. I can't draw.

The research data demonstrates that some informants felt uncomfortable about the artistic activity, it revealed their psychological problems: "*I saw my colleagues in the woods, meditating. I felt sad. Maybe I did not find any connection with nature. Or maybe with myself? I felt really uncomfortable, I was alone*"; uncovered their frustrations and the undefined or preconceived resistance to the use of artistic tools, musical instruments or improvisation: "*for a second, when I had to play the entrance of an elderly person and his sitting on the chair, I did not know how I looked, how old I was, what my pains were, or maybe I had none, and if I was just like all the other elderly people*". This experience of the informants' relates to the researchers' theories which argue that the tasks of artistic activity and artistic education are special since they require to cognize oneself and be able to listen to the answers of other people or the group participating in the process of artistic activity [4, 19]. Moreover, even the social context is perceived by the informants as a negative environment since here artistic activity is discussed and the individual differences are revealed. On the other hand, every artistic activity is new, unfamiliar, open to experience and that experience is equivalent to every individual's daily meeting with different groups, cultures, activity; thus, the process of a person-group-society is not only positive, but also full of challenges, frustrations and affecting powers, yet, it is always changing and open to the new construct of social reality.

Conclusions

The analysis of scientific sources shows that the problem of identity in the present-day context is a relevant issue. The education paradigm of the postmodern age opens freedom for a personality; however, at the same time it eliminates the limits protecting the person: it levels values and aesthetic tastes, cultures assimilate, the belief in progress is lost. The authors of the analyzed sources agree that there are clear features of identity crisis, yet, they acknowledge the act that one of the most stable grounds for personality development is the enhancement of Self and national identity as well as the development of cultural dialogue. This makes it possible to state that the morality development of a person, his/her family and the whole mankind affects the formation of identity; hence, art, which renders the peculiarities of the development of society, is inseparable from the conception of individual identity.

The data of the qualitative analysis show that artistic activity reveals the possibilities of personality cognition and identity formation. Despite the variety of artistic expression, there exists a common perception of artistic experience which discloses the individual's biographic moments and the characteristics of world cognition by means of artistic instruments. The determined topic of "the significance of self-worth" showed that the informants perceive their identity as nationality (relation to the past traditions and historic events), individual strength (the establishment of one's Self, the sense of leadership), and psychological security (the possibilities of creating a safe environment).

The context of group artistic activity gives possibilities to reveal the social skills of an individual, to communicate by means of the body language and, while comparing oneself with other cultures or working with the other members of group artistic activity, to determine the cultural or personal differences and, at the same, integrate into the group.

The research data also revealed the informants' negative experience when the artistic activity itself became a challenge and so approached the unpredictability of daily life. On the other hand, the fact that artistic activity discarded with the social surroundings, when the research participants had to choose between adapting or expressing themselves in the joint drawing or musical work, helped the informants conceive the image of "otherness" in a multicultural society and search for the best ways of integration.

Summary

The authors who studied the issues of identity (Erikson, 1963; Marcia, 1980 et al.) argued that the formation of identity happens with the integration of two opposite spheres. This process is influenced by the mankind and the development of its culture and morality. The perception of individual identity is inseparable from the perception of a person's nationality. Even though the postmodern age is characterized by the disappearance of boundaries between nations and social roles, by the cultural assimilation, by the changing attitudes to progress and the stability of values, commitment to a person, connection with one's community and the ability to integrate into the multi-cultural world have remained the foundation of personality growth. The problem question about the nature of artistic activity, which creates the preconditions for social integration and enhances the perception of people's identity and inequality, initiated a qualitative research. Nine representatives from different countries, who have no professional preparation in artistic activity, took part in musical, theatre and art classes. In this case artistic activity was an educational means which was meant to activate the informant, yet, left some space for their experience, personal meaning and expression of value-related attitudes. The research participants had to answer what they experienced during artistic activity and how their experience was affected by the context or situation. The data of the qualitative research and their phenomenological analysis show that artistic activity can have influence on the personality cognition and their identity development. Despite the variety of artistic expression, there exists a common perception of artistic experience which reveals the individual's biographical moments and the peculiarities of cognizing the world by means of artistic instruments. The research also uncovered the participants' negative experience when artistic activity discarded with the social surroundings and the informant had to choose either to adapt, or express himself in the joint drawing or musical work. This helped the research participants to perceive the image of "otherness" in the multicultural society and search for the best ways to integrate into it.

References

1. Aleknaitė-Bieliauskienė R. Tautinė tapatybė ir meninis ugdymas globalaus pasaulio akivaizdoje. Iš: *Tautiškumas ir pilietiškumas. Atskirtis ar dermė?* Ed. R. Grigas. Vilnius: Vilniaus pedagoginio universiteto leidykla, 2007, p.69-79.

2. Barkauskaitė M. Valstybingumo pagrindų ugdymas pedagoginiame procese. *Tautiškumas ir pilietiškumas. Atskirtis ar dermė?* Ed. R.Grigas. Vilnius: Vilniaus pedagoginio universiteto leidykla, 2007, p.246-275.

3. Bitinas, B., Rupšienė, L., Žydžiūnaitė, V. *Kokybinių tyrimų metodologija: vadovėlis vadybos ir administravimo studentams.* Klaipėda: S. Jokužio leidykla-spaustuvė, 2008.

4. Brater M., Wagner J. *The Enhancement of Adult Education by Artistic Activities.* Munchen: Gesellschaf fur Ausbildungsforschung und Berufsentwicklung, 2011.

5. Daujotytė – Pakerienė V. Tapatybės, ne(ta)patybės, be(ta)patybės trajektorijos. In: *Tautiškumas ir pilietiškumas. Atskirtis ar dermė.* Ed. R.Grigas. Vilnius: Vilniaus pedagoginis universitetas, 2007, p.55-69.

6. Dolloff, L. Imagining Ourselves as Teachers: The development of Teacher Identity in music teacher education. In *Music Education Research*, 1, 191-207, 1999.

7. Erikson E. Identity and the life cycle. In: *Psychological issues*, 1959, Vol.1, p.1-71.

8. Gaižutis A. *Estetika: tarp tobulumo ir mirties: textbook.* – Vilnius: VDA leidykla, 2004.

9. Giddens, A. *Modernity and self identity. Self and society in the late modern age.* Cambridge: Polity Press, 1991.

10. Hargreaves D., Miell, D. McDonalds, R. *Musical identities.* London: Oxford university press, 2002.

11. Juodaitytė A., Šiaučiulienė R. Postmodernizmo iššūkiai mokytojų rengimui: kintantys ugdymo kontekstai ir tapatumų įvairovė. *Mokytojų ugdymas.* 2012. No. 19 (2), 158-171.

12. Klimka L. Pilietiškumas ir tautiškumas: laiko dimensijos ir aktualijos. *Tautiškumas ir pilietiškumas. Atskirtis ar dermė?* Ed. R.Grigas. Vilnius: Vilniaus pedagoginio universiteto leidykla, 2007, p.127-140.

13. Marcia, J. E. Identity in adolescence. In J. Adelson (Ed.), *Handbook of Adolescent Psychology.* New York: Wiley.1980, p. 109-137.

14. Matulaitė A. *Kai "tavo kūnas tiesiog išprotėja": įkūnytas nėštumo patyrimas".* Doctoral dissertation. Manuscript. Vilnius, 2013.

15. Owens, T. Self and Identity. In J. Delamater (Ed.) *Handbook of social psychology*, (pp. 205–232). New York: Kluwer Academic/Plenum Publishers.2006.

16. Regelski, T. Music **teacher** - Meaning and practice, **identity**, and position. In *Action, Criticism, and Theory for Music Education*, 6(2). Retrieved March 12, 2008, from http://act.maydaygroup.org/articles/Regelski6%5f2.pdf

17. Ruud, E.Music and identity. In *Nordic journal of music therapy*, Vol. 5, p.3-13, 1997.

18. Viluckienė, J. *Judėjimo negalią turinčių asmenų fenomenologinė gyvenamojo pasaulio analizė.* Doctoral dissertation. Manuscript. Vilnius, 2011.

19. Vilkelienė A. *Ypatingųjų mokinių meninis ugdymas.* Vilnius: Edukologija, 2013.

The Experiences of Students of Fine Arts Education During the First Period of Teacher Practice

Birutė Vitytė

Abstract
The paper reviews the reflections of bachelor's degree first year students of Fine Arts Education on the first period of teaching practice – observation practice. The paper analyzes the dispersion of the students' experiences in teaching practice in order to reveal the most interesting and concerning issues for students during the practice period in the context of fine arts lessons.

Keywords**:** pre-service fine arts teachers, teaching practice, reflection on the teaching practice

Introduction

Pečiuliauskienė and Barkauskaitė define teaching practice in the following way: "Teaching practice is an integral part of teacher training studies, intended for the development of the trainee teachers' practical skills. This is a study period when students make the first steps in their professional activity, when they acquire a new approach to the teacher's job, rights, duties and responsibility." [1, p. 7].

Teaching practice is a special period since a learner turns into a teaching person, thus, it is a transition to the other side of the "barricades". The experience has to be especially peculiar for the first year students who were still pupils half a year ago and then become teachers themselves. The inquisitiveness about those first experiences and supervision of fine arts teaching practices induced the idea to investigate the first year students' of Fine Arts Education experiences during the teaching practice. The significance of teaching practice and the increasing attention to this stage of studies can be observed in the *Regulations of Teacher Training*. The review and comparison of the Teacher Training Regulations of 2008, 2010 and 2012 revealed the changes of the teaching practice workload by credits. In 2008 teaching practice was allocated 20 credits, while in 2010 and 2012 – 30 credits. Moreover, since 2010 the Regulations present a scheme concerning the changing roles of the students during the teaching practice: *at the beginning of the teaching practice – observer and teacher's assistant, working with the mentor's and supervisor's help, at the end of the teaching practice – working independently under the mentor's and / or supervisor's supervision*. This scheme appeared when the workload of teaching practice increased, when one independent teaching practice was replaced by four distinct periods of teaching practice: observation practice, pedagogical assistant practice, practice under the mentor's supervision, independent teaching practice, which all have distinct objectives corresponding to the students' preparation level.

The Regulation of Teaching Practice at Lithuanian University of Educational Sciences (2013) raises the following objectives for the first observation practice: to acquire practical knowledge about education and the activity of institution of practice by relating theoretical knowledge with practice; to observe, analyze and reflect upon educational activity taking into account the specificity of every study programme; to promote the trainee teacher's professional motivation. The purpose of this observation practice is to develop practical, common cultural, general, professional and special competences by means of observation, analysis and reflection of the pedagogical process [2, 3, 4].

The review of all the documents regulating teaching practice gave a good understanding of the significance of this stage of studies for trainee teachers; however, there was no information or conducted studies on how students managed to implement all the requirements during the teaching practice, how they felt and what they experienced, what fascinated them most and what was most difficult. Hence, aiming to reveal ***the dispersion of students' experience during the first teaching practice*** twelve first year students of Fine Arts Education got a task to write a reflection of their first teaching practice of fine arts.

The research object is the reflected experiences of the trainee Fine Arts teachers in their teaching practice.

The research goal is to reveal the trainee Fine Arts teachers' experiences of the first teaching practice described in their reflections.

The research methods: literature review, the content analysis of reflections of the teaching practice.

The Characteristics of Recording Experiences of Teaching Practice

The initial analysis of the students' reflections revealed that the first year students of Fine Arts had recorded their experiences during the teaching practice in two ways. One way was characteristic of all the students who wrote reflections – i.e. the ability to *recognize and record* different problems arising in the process of education. The second way of recording experience was typical only of some students and it was termed as "*the deep approach*" since this method of recording experiences required several stages (Figure 1). The scheme presented below illustrates the stages of the "deep" approach and their characteristics are presented further in the paper.

Figure 1: Stages of the "Deep" Recording of Experiences

The first stage (which was typical for all the students) was identification of the problems arising in the process of education, i.e. students of Fine Arts described a problem, a situation, the teacher's or pupils' activity in the process of education which affected the other participants of the process. For example: "*The teacher performs a lot of different activities*"; "*Some pupils took out their textbooks and started doing their homework, but it was not fine arts, of course*". These examples show that some students merely recorded certain situations without analyzing them. It can also be noted that all the students would reach the next stage – analysis. Nevertheless, they analyzed only some of the identified problems. Moreover, in all the reflections there was no single or several problems which had been tendentiously avoided to analyze and reach the next stage of the "deep" experience. The characteristic feature of this stage is that the student does not merely recognize the problem, but also describes it and analyzes its possible causes, e.g.: "*The reason for the pupils' misbehavior was the teacher's complaisance and inability to control the situation.*"; "*The main problem was the inability to keep the pupils' attention; we decided that the tasks were too easy for the topic.*" The given examples show that the students did not only record a problem, but also described one or another possible reason which determined the process of education.

The third stage was searching for a solution and its identification, e.g.: "*... the teacher was observing the process of the activity, walking around the classroom and giving praise for nice work. The tasks had to be performed in the time set by the teacher, and this made the pupils disciplined. I understood that if the time of the lesson is well planned and distributed, it is possible to preserve order and peace in class.*"; "*To be honest, after the practice I was thinking about the teacher's activity. I was surfing the internet looking for the ways how classes can be managed.*" The quoted examples show that the students were able not only to recognize problems, but also to identify their possible solutions. As soon as the reason was determined, the students reflected upon the activity which, in their opinion, positively affected the process of education, solved the problem, while others even tried looking for the possible solution individually. It has to be noted that the distinction of these stages is relative since they are all interrelated and sometimes it is difficult to determine which of the stages was attained by the student. Yet, it was noticed that the depth of most students' analyses limited itself with this stage, and only some descriptions of problems included in the students' reflections had the features of the following stages.

The fourth stage was the reflection and consideration of the solved problems, activity and drawing conclusions. Here are some examples from the students' reflections: "*When I was speaking with pupils, I understood that the teacher did not give strictly defined rules, but always told what she was expecting from the pupils. Good relations with pupils and the class give a positive characterization of the teacher, though personally I sometimes found her methods not understandable, yet the mutual trust proves that her specific methods served the purpose.*" In the rest of this reflection the student critically assessed the teacher's activity; however, this statement proves that even though some of the teacher's methods were not understandable for the student, she still managed to acknowledge their efficiency. Thus, the given quotation illustrates the student's considerations and the conclusions drawn. If the students reach this stage, they consider the actions taken to solve the problem and decide if they were efficient.

The fifth stage was designing a model for solving the problem, e.g.: "*The teaching practice shocked and disappointed me. Yet, it raised a lot of questions, so I found it really useful. I was thinking a lot about what I would do in one or another situation, how the children could be involved in activity, how their attention and inquisitiveness could be promoted. How such a minor detail as rearrangement of desks in the classroom can encourage the pupils to learn and cooperate. I was surfing the internet and found a lot of information. I don't think I know enough now, but the information was really useful*"; "*In order to make the children feel more relaxed, I would arrange more group activities and I would like to play instrumental music in my lessons.*" The students of Fine Arts, who have reached this stage of analysis, contemplate and design the solutions for different problems, imagine how things could work in practice, search for additional information. The peculiarities of this stage of analysis were typical only for some students involved in the reflection writing process.

Hence, all the first year students of Fine Arts, who wrote their teaching practice reflections, recorded their experiences in two ways. In the first case, they recognized and described a wide spectrum of problems, while in the second case the described problems were analyzed deeper and in greater detail, such analysis included several stages. The first three stages of analysis were typical for most students of Fine Arts, but the other two – only for some of them.

The Variety of Problems Described in the Reflections of First Year Students of Fine Arts Education

The analysis of the problems distinguished by the students revealed their possible grouping. All the students who wrote the reflections indicated and described more than three / four problems.

The first and most frequently mentioned issue of the students was the fine arts teacher's activity, its features and versatility. The students noticed the wide and varied activity of the teacher which some of the students assessed positively, and some – negatively. The students distinguished the following activities of the teacher: *managing the class, organizing events, conducting extra-curricular activities, participating in project activity.* A part of the students considered such activities as organizing activities, conducting extra-curricular activities and participating in project activity to be a burden enforced upon a teacher. When the students were describing these activities, they used the following words: *obligation; must; demand; a lot of energy is wanted; for the sake of school; was responsible; very busy; fine arts was in the background; you have to; it is important for the school community; lots of projects; various*, etc. The students' descriptions make it possible to assume that they were surprised by such a varied activity of the teacher; none of them mentioned that the teacher's work was limited to conducting lessons. "*...the teacher of fine arts also has to decorate the school, we experienced it ourselves... we made paper kites, decorated the corridors, the school hall and helped to prepare the dance costumes...*" This extract from a student's reflection is a good illustration of most students' experiences related to the variety of fine arts teacher's activity.

In their reflections the first year students paid most attention to the **features of class management** and described it in great detail. Such words as *chaos, noise, taming, standing behind the back, the pupils' excesses, to calm the class down, noisy, aroused chaos*, etc. were used to describe the atmosphere of fine arts lessons and the pupils' behavior in the lessons. The analysis of the students' experience of the problem of class management revealed the fact that the students assessed the features of class management either positively or negatively. The following examples illustrate the positive assessment: "*he did not even try to silence them as they can work and chatter at the same time, therefore neither I nor the teacher saw any problems here...*"; "*but the teacher is already used to that, I was confused by such chaos and noise, but later I also got accustomed and realized that there is no other way with younger pupils...*"; "*...very active and noisy, as soon as they entered the classroom they aroused a huge chaos, but the teacher was used to such behavior and he was not confused at all*"; "*...I learnt that there is no need to pay* *attention to all the learners' excesses or pranks, of course, when it is not necessary..."; "...The teacher knows her pupils very well. She calmed down only the noisy one, but everybody calmed down even though it seemed that they were all making noise..."; "...the pupils were very calm in the lesson. I talked with the teacher and we were both surprised that they can be so calm..."; "both of us decided that children are very sensitive, agile, active, it is difficult for them to stay in one place for a long time..."* It was noticed that the students who gave positive examples of class management tended to identify themselves with the teacher and used such words as: *neither the teacher nor I saw a problem; we were both surprised; both of us decided.*

The following examples illustrate negative management of the class: "*The teacher had to quiet the class all the time, temper the naughty ones, and try to keep their attention..."; "...she disciplined her students with grades and will punish them at the end of the term, at the credit..."; "Unfortunately, a lot of pupils do not listen to the teacher, they were busy with some extrinsic activity (texting on their phones); they were noisy; they do not feel the limits of educated behavior; they don't listen to the teacher's instructions; they laugh loudly..."; "...the pupils whom the teacher did not manage to control were threatened with notes for the parents or bad marks so that they got frightened and started to work."; "the teacher had to do something, yet all she did was threaten them with notes for their parents and the social worker...*" The analysis of the students' reflections revealed that class management sometimes reminded the students of natural disasters which are difficult to fight with and conquer. It was noted that when the students were describing the teacher's unsuccessful attempts to manage the class and maintain order, they did not use such words that were used in the descriptions of the successful examples, i.e. *we, both of us*, etc.

One of the key features revealed in the analysis of the first year students' reflections was that all the distinguished and analyzed problems, all the participants of the process, the curriculum of fine arts, the observed and described methodology were presented in the context of class management and control. No matter if the students were analyzing the teacher's character qualities, the pupils' attitude to fine arts lessons or the given assignments, nevertheless, they returned to one and the same issue of class management and order. It can be assumed that after the first teaching practice most of the students would gladly attend a course on the peculiarities of class management. Such presuppositions were confirmed by the students in their writing: "*...Maybe I am disappointed now that the teacher's job is not as easy and fun as I thought. I frequently think about one thing: how will I maintain discipline in the lessons?"; "...to tell the truth, after the teaching practice all I could think about was the teacher's activity. I was surfing the internet and looking for information how a class can be managed...";*

"...I understood that a pedagogue first of all has to be a good psychologist or at least should be one. Psychology should help to understand the pupils' behavior and solve the problems."; "...most of the time I spent thinking not about the teaching process or the programme, but contemplating on how a teacher should behave if he wants to create a cozy atmosphere and have sustainable relations with the pupils."; "I can say that the fine arts teacher's work did not disappoint me, I was not frightened that sometimes pupils do not listen to the teacher in art lessons or that they cannot concentrate..."

Hence, the pre-service teachers of fine arts were most concerned by the issue of class management, yet they also noticed that frequently the pupils' improper behavior was instigated by some other factors, such as ***the choice of topics or their arrangement in the lesson***. The following extracts from the students' reflections vividly reveal the students' opinion about the interrelation between the content of the lesson and pupils' behavior: *"Greek art was not affected by any factors and this style lasted till the year 100 BC. There is an Ionic, Doric and Corinthian column." This short sentence finished the explanation about Greek art for the seventh formers. The children were noisy, it was not possible to control them... The teacher provoked their behavior with such an explanation."* Another student wrote about a situation when the teacher told the pupils to make festive decorations for the school for the second lesson in a row: "*...this class was going to make kites for the second time, the pupils were dissatisfied as they were already bored with the task. The teacher had to discipline the children a lot of times and closely observe them...*" The first year students of Fine Arts also presented positive examples when a well chosen task, an interesting activity conditioned the pupils' good behavior: "*the pupils were working calmly and said that they really liked that activity...*" The students also gave such examples as: "*...there was chaos in the classroom. Most pupils were chatting, they did not follow the activity, were inattentive... trying to evade the practical activities even though the topic was really interesting...the pupils were not interested even in the attractively presented version...*" Hence, the students' reflections can be generalized and three schemes distinguished: *a good activity / topic – pupils' good behavior in the lesson; unsuitable activity / topic – pupils' bad behavior in the classroom; a suitable activity / topic – pupils' bad behavior in the lesson.* Unfortunately, the students only recorded the different results of the interrelation between the task / topic and the pupils' behavior, but did not provide any analysis.

The first year students of Fine Arts Education also recorded the problem of ***using templates and stencils in fine arts lessons***, e.g.: "*...the teacher showed an example with a triangular box – and then most of the pupils drew a comic-strip with a triangular box...*"; "*...the teacher distributed stencils, delineate the egg, the* *pupils start working as soon as they get the stencil…it is noisy since they have nothing to do, they are bored and impatient…"; "…the teacher was disappointed with the boys' work since they works are obvious copies of the classmates' drawings."* When recording this problem, the students again only mentioned the fact that fine arts lessons employed stencils, there were copies promoted and said then it was noisy in the class, however, the students did not express their personal opinion whether the actual use of templates or unsuitable examples really provoked the bad behavior.

The formulation of tasks in fine arts lessons is another problem described by the first year students. The following is an example of the students' idea of an unsuccessful task in the lesson of fine arts: *"…the teacher was silencing the class and asked them to listen to her. She told the pupils to measure 14.8 cm and draw a line. From that line the pupils have to measure 5 cm on both sides. The pupils get to work, they measure and draw lines. After 20 minutes it turned out that the lines were measured and drawn correctly, but the pupils had brought irregularly sized sheets of paper and some of them did not draw the line in the centre of the page… I think the teacher is to blame since she failed to present the information clearly. She did not check the size of paper sheets; she was focusing on class management and specification of the task and did not emphasize the fact that the line has to be drawn in the centre of the page. I think the sixth formers are already able to mathematically find the centre of the page individually…"* And here is one more vivid example: *"The teacher told the class to take an orange sheet of paper and rub it with wax. She did not say what would follow. The children were asking if they could take a yellow page. The teacher said yes. – Could they take a brown one? – Yes. – Could they take a red one? – Yes. – And could they take a green one? – Yes.…And so on and so forth. Eventually the teacher said that the sheet of paper could be of any colour, most importantly it had to be light…Moreover, the teacher had said that the children could rub the page not only with wax, but also with wax crayons. The task became even more complicated and there appear more chances of mistake. The children had brought oil crayons. The teacher was running about, wasting the time, and looking for the necessary tools. The children felt confused and turned noisy. The teacher spent a lot of time on silencing the pupils and taming them…Because of the lack of time and her inability, the teacher limited her explanations to the minimum and made the task too specific. The children did not understand it and started asking additional questions, which could have been avoided if the teacher had provided clearer explanations. Another case – the teacher's speaking initiated senseless discussions and strange situations…"* It is interesting that the students analyzed the problems related to the formulation of tasks; they expressed their personal opinions about the teacher's mistakes and the incorrect behavior, or compared the teacher's actions with the teachers of fine arts they used to have themselves. The following examples show that the unconditional "favorites in the sphere of formulating tasks" were the students' teachers of fine arts at school. Student C wrote in his reflection: "*...on the contrary, our teacher of fine arts was able to present the requirements clearly and specifically. When she was giving a task, first of all she explained everything, indicated all the steps we would have to do and why, and then repeated everything when we had started to work...*" The students of Fine Arts frequently remembered their school teachers in their reflections and those reminiscences were exceptionally positive.

"*...the teacher observed the process of work...she was walking around the class... gave praise for nice work... she assigned the time limits and that disciplined the students...the well planned and distributed time in the lesson helped her to maintain order and discipline in the classroom...*; "*...the teacher wanted to involve everybody in class so she would look for the necessary tools: pencils, paper, rulers. The strict structure of the lesson was ruined, the children turned noisy and confused. Such a behavior does not prompt the pupils' responsibility and creates the conditions for mess...*"; "*...the problems would most frequently arise because the pupils had not brought the necessary tools and were not ready for the class. Sometimes the teacher did not notice those problems right away, and sometimes she would waste the precious time on looking for the necessary tools in the closet or cupboard...*" These extracts revealed one more problem observed by the first year students, i.e. ***the significance of planning time in the lesson of fine arts.*** Planning the process of the lesson is very important because, in the students' opinion, it conditions the pupils' behavior. Following the plan consistently lets both the teacher and the pupils work peacefully. According to the students, the situation when the plan is not followed, even in the cases when the teacher has to change the lesson structure because the pupils misbehave or are not prepared for the class, can be assessed only negatively.

One more problem that was described in the reflections of the first year students of Fine Arts Education, which was analyzed in the context of class management, just like all the others, was ***the general atmosphere in the lesson.*** The pre-service teachers of fine arts in their reflections gave examples which proved that the pupils' behavior in the classroom was determined by some factors, e.g. "*...in fine arts lessons it is usually very noisy because in these lessons the children are relaxed and do not obey the strict order.*" The words "*in these lessons*" make it possible to assume that children do not behave like this in other lessons. The other examples provided by the students show that the observed teachers successfully deal with this problem and indicate the things which help the pupils relax and create a positive environment in the lesson of fine arts. On the basis of the students' reflections it is possible to distinguish several factors which influence the general classroom atmosphere and the pupils' behavior – i.e. music and the teacher's immediate interaction with the pupils. For example: "*There was a harmonious environment in the lesson, there was soft music playing in the background, the pupils were concentrated and working, the teacher was walking around the class and giving individual consultations…*"; "*The teacher wanted to silence the pupils so she turned on the music – the pupils calmed down, they were listening to their favorite songs and got really involved in the task.*"; "*The teacher was trying to create a cozy atmosphere in the lesson, she turned on the pupils' favourite music, the pupils relaxed, they felt free and got involved in the task more.*"; "*The most distinct thing that the teacher was using in the lessons was music. The pupils picked the music they wanted to listen to and the work became more pleasant.*"; "*…there was music on, nobody bothered anybody, it was easier for the teacher to work with them.*"; "*…and some pupils were listening to music with their headphones and were working peacefully without bothering anybody…*" Thus, according to the students, music is a special element in the lessons of fine arts, it creates a harmonious atmosphere, music calms the pupils down, and lets them feel free. All the students, who spoke about music in the lesson, unanimously focused on its positive influence. Another positive element, in the students' opinion, which determined a good working atmosphere in the lesson, was the teacher's immediate interaction with the students. "*…in my opinion, it would be possible to try something different. Even though the teacher was performing all her duties and was not violating the pupils' rights, she was avoiding immediacy, was hiding her emotions and did not say much about herself. There was a psychological barrier between the teacher and pupils which created a too serious and cold atmosphere.*" Observations like this show that the students believe that the teacher's more personal and intimate interaction with the pupils could create a better classroom atmosphere. It was noticed that these suggestions were given by the students who had matter-of-fact relations with the teachers during the teaching practice, while the students, who had a friendly and immediate relationship with the teacher, did not give such observations and did not argue that warmer relationships between the teacher and pupils can improve the general classroom atmosphere.

"*A latecomer ran into the classroom, a very impolite pupil who was absolutely unmotivated to work and he started making a terrible din…the teacher had to tame him, and eventually he was turned out of the classroom…*"; :*The pupils' motivation for work was not high as most of them had not brought the necessary tools…*". ***The pupils' motivation*** is another problem observed and described by the students. According to the students, the internal motivation is directly related to their behavior in the classroom; unmotivated pupils express their dissatisfaction, don't listen to the teacher, and don't bring the necessary tools. One more problem indicated in the students' reflections was directly related to motivation and the classroom management problem is ***the pupils' attitude to the lessons of fine arts*** as a school subject, e.g.: *"...some students took out their textbooks and started doing homework (not fine arts, of course)..."; "...the pupils are busy with their own things: they are doing mathematics homework; texting on their phones; drinking coffee..."; "The teacher's job in fine arts lessons is more difficult as the children feel more relaxed than in the other lessons and that conditions noise in the classroom, children's walking around and a smaller motivation for work and creation"; "Moreover, they have an opinion that lessons of fine arts are insignificant and underrate them."; "... a class of stereotypical geeks who are seemingly perfect at the exact sciences – they believe that fine arts is a waste of time, however, I cannot judge them as I had a similar opinion when I was younger. Ironic since I am studying Fine Arts at the moment..."* Similar descriptions and observations were typical for the students not only when they spoke about the lessons of fine arts, but also about teachers of fine arts, e.g.: *"I was talking to one schoolgirl about her class, I asked why it was so undisciplined. She said that they don't consider fine arts to be as important as the lessons of mathematics or the Lithuanian language. And they don't consider the teacher merely a teacher, she is more like a friend to them, therefore, they do not show respect."*

Even though the students were inclined to analyze the aspects of the teaching / learning process in the context of classroom management, the examples given in their reflections demonstrate that their experiences are not limited to a mere observation of the problem, its recording and analyzing. As different statements in the students' reflections show, they successfully and sensitively noticed the other aspects of the process of education and the problems arising in the lessons of fine arts. Hence, the spectrum of the problems presented in the reflections of the first teaching practice is very wide. The students consider the activity of the teacher, the features of class managements in the lessons of fine arts, the selection of topics, the characteristics of their arrangement, the use of templates and stencils, the formulation of tasks, the significance of planning time, the creation of the general classroom atmosphere, the pupils' motivation and attitude to the lesson of fine arts.

The Students' Personal Experiences during the First Teaching Practice

A separate part of the students' experiences in the reflections of the first teaching practice was related to their internal reactions which, contrary to the aforementioned ones, were oriented to their personal feelings and experiences rather than the aspects of the process of education.

"I really enjoyed this practice, I didn't even notice how fast those two weeks passed. It was nice to return to my old school."; "It was nice to speak to my teacher again who had noticed that vein of fine arts in me"; "As I had left this school before the university studies, I had no difficulty adapting to the new work environment..." Statements like these which speak about the quickly passed time, absence of communication or adaption problems at school were typical of the students who had their first teaching practice at their former teacher of fine arts. The following statements confirm the students' strong emotional connection with their former school and teachers: *"I really appreciate this teacher because he also taught me at school, helped me a lot, and gave advice when I had questions. I am very glad that I had my first teaching practice with him since it is always nice to return to your former teacher"; "When I was observing classes, I remembered my own school and lessons"; "I had my observation practice in Gymnasium A. I chose this school as I had attended it myself and finished it last year. It is always nice to return to your former school, meet the teachers again, return to the place where you had spent so much time."; "...I saw one of my paintings made for the creative task of the examination on fine arts, and I felt upsurge with good reminiscences."* As it can be seen in these statements, the first year students of Fine Arts Education find their former school and teacher very important, and remembered them frequently during their teaching practice. The observed teacher was compared with the students' former teacher. The statements show how important the example of their former teacher is, and it is especially favorably assessed if the student is having teaching practice with his former teacher of fine arts.

Another topic of personal experiences revealed in the reflections of teaching practice was the self-assessment of their own abilities and skills: *"I learned a lot, all the time I observed myself, assessed my knowledge, skills and the level of socialization..."; "...after the lessons I understood that I have to learn a lot to become a good specialist, and it is really important to be interested in everything related to fine arts, its teaching methodology, history, tasks and techniques.";"...I missed psychological knowledge in my teaching practice, thus, I had to analyze a couple of books on psychology."; "I perceived that the teacher's work requires a good physical preparation and great patience."; "During my teaching practice I saw that I knew very little about the teacher's job."; "I experienced a constant lack of educational knowledge, I felt not ready to assess a teacher."; "I know that I did my tasks unprofessionally or even unacceptably, nevertheless, I tried to evaluate everything on the basis of my conception and attitude."; "In the lessons I felt short of knowledge about fine arts and pedagogy, I did not know how to keep the class interested, how to communicate and silence the class...*" The students gave a critical assessment of their abilities in the teaching practice and use a lot of negative words to describe them: *I have to learn a great deal, I felt a lack of knowledge, I experienced a lack of knowledge, I felt unprepared, I acted unprofessionally* etc. Nevertheless, the following statements show that even though the students assessed themselves critically in the practice, their motivation increased: "*...I still want to become a teacher of fine arts, my motivation grew during the teaching practice.*"; "*...I made friends with the teacher and pupils, I didn't want to leave, I wanted to stay for a couple of days.*"; "*My professional motivation grew, I envied the teacher her cozy classroom, good pupils and good school.*"; "*During the teaching practice my motivation to become a good teacher was growing...*"; "*I realized that I really wanted to be a teacher of fine arts...*"; "*I am looking forward to the teaching practice next year, when I am able to experience a lot of new things and spend a useful time with the pupils...*"; "*At the end of the teaching practice I was a bit disappointed. I would gladly stay at school for a couple of more weeks.*"; "*That was a unique experience – I loved it...*"; "*I would like to say that I do regret having chosen this profession. This experience convinced me that I get on well with younger people. Moreover, I acquired motivation to develop and find new methods of teaching. The final verdict: the teaching practice surpassed all expectations...*" The aforementioned statements show that the first observation practice motivated the first year students of Fine Arts Education. It has to be noted that none of the students said that after the teaching practice he understood that this activity was not for him. After their critical assessment of the lessons of fine arts and the observation of numerous problems in the process of education, the students assessed their motivation really positively.

In their reflections the students distinguished several factors which could be termed as the most motivating ones, while the students called them the most appealing moments of the teaching practice: "*...but I liked that the teacher did not place me in the distant corner of the classroom and did not tell me to observe the lesson passively. She let me help the pupils and give advice, evaluate and praise them...*"; "*...the teacher had the most difficult job – to assess the pupils' drawings... and he asked me to do it.*"; "*I liked the classes I could work with, and not just be a passive observer...*"; "*...the teacher asked us to help the children, and I liked it more than observation...*"; "*...this class was special because I was the teacher...*"; "*I conducted the lesson...that was the most pleasant part of the practice because I was able to apply the knowledge acquired at the university..*"; "*I was very moved when a girl approached me and asked to explain how she should draw lips correctly. I felt like an expert. I really liked this part of my teaching practice...*" These statements show that the students enjoyed being not only the passive observers of the lesson, but also its active participants, they were glad to help the teachers assess the pupils' works, conduct a part of the lesson, formulate the task or even conduct the whole of the lesson. The following statement is especially eloquent: *"this class was special because I was conducting the lesson."* The students of Fine Arts Education had no preparation to conduct lessons, they did not get this task in their teaching practice assignments, but the teachers who let them do that motivated the students by showing their confidence and accepting them as co-workers and colleagues in the lesson. None of the students was confused when the teacher suggested conducting a class and did not consider this task to be very complicated. The other statements from the students' reflections again prove how important it is for the students to feel like a real teacher, not just a student observing a lesson of fine arts: *"The pupils were in a good mood when they came to class and even greeted me: "Hello, teacher", I felt pleased."; "I enjoyed that the pupils started greeting me: "Good day, teacher" or bidding goodbye: "Goodbye, teacher".* These experiences are essential as it provided the students of Fine Arts Education with a possibility to feel like real teachers.

Hence, the students' personal experiences during the first teaching practice involved the following topics: the relation with the former teacher of fine arts or the former school, the self-assessment of personal skills during the practice, the increase of motivation and the most pleasant moments or the moments that motivated the students most.

Conclusions

1. The analysis of the first year Fine Arts Education students' reflections of the first teaching practice revealed that the students recorded their experiences in two ways. In the first case, they described different educational problems. The spectrum of these problems is very wide: the activity of the teacher of fine arts, the characteristics of class management in the lesson, the selection of topics – the peculiarities of their arrangement, the use of templates and stencils, the formulation of tasks, the importance of planning time, the creation of the general classroom atmosphere, the pupils' motivation and attitude to the lesson of fine arts. One of the key aspects that was determined in the review of the problems was that the students were inclined to analyze everything in the context of class management.
2. It was determined that the students analyzed some of these experiences deeper and in greater detail, there were several stages in these analyses, hence, this way of recording experience can be termed the "deep" one.

3. A separate part of the students' experiences in the reflections of the first teaching practice was related to their personal feelings and reactions concerning: the relation with the former teacher of fine arts or the former school, the self-assessment of their personal abilities, the growth of motivation during the teaching practice and the most pleasant moment or the moments that motivated the students most.

References

1. Pečiuliauskienė, M. Barkauskaitė. 2011. Pedagoginės praktikos mokykloje vadovas: vadovėlis, skirtas pedagoginių studijų bakalaurantams, mokytojams mentoriams, pradedantiesiems mokytojams. Vilnius: Edukologija.

2. Regulations of Teaching Practice. Lithuanian University of Educational Sciences, 2008.

3. Regulations of Teaching Practice. Lithuanian University of Educational Sciences, 2010.

4. Regulations of Teaching Practice. Lithuanian University of Educational Sciences, 2012.

5. Regulations of Teacher Training. 2008, 2010, 2012.

Cultural Competence as Keystone of Training Teachers of Foreign Languages

Gerda Mazlaveckienė

Abstract

Globalization processes and their determining factors raise new challenges for training teachers of foreign languages. A teacher is expected to both cherish own culture and transmit the culture of the taught foreign language; therefore, while training a specialist of a foreign language it is essential to develop his/her cultural competence.

The current article deals with the expression of cultural competence of pre-service teachers of foreign languages. It emphasizes the effects of globalization on the training of foreign language teachers as well as highlights the necessity of raising their cultural competence. With this aim, a quantitative study was conducted in spring 2013. The questionnaire consisted of several sections, which aimed at identifying the current level of the cultural competence of pre-service teachers of foreign languages, namely their knowledge about culture in general and country-specific culture facts, emotions evoked by different cultural events, abilities and experience of functioning in different cultural environments, as well as other factors determining their cultural competence.

Keywords: pre-service teacher of a foreign language, culture, cultural competence, foreign language, globalization

Introduction

The rapid and firm establishment of globalization processes, mass migration and diversity of political, economic, cultural and linguistic processes throw down challenges for the governmental, economic, cultural, educational and other institutions in the world. European and world leaders are expected to ensure the acknowledgement and respect for all cultures, cherishing of national identities and successful overcoming of cultural differences. Hence, a number of European and Lithuanian documents set goals to develop abilities of coexistence, mutual understanding, intercultural dialogue, solidarity and harmonious relations [9; 13; 14; 27; 29; 30; 41 and others].

Researchers of Lithuania and the world analyze the consequences of globalization on the development of humanity, focusing on cultural changes in particular [1; 2; 17; 24; 25; 48 and others). They note that the effect of globalization is twofold: constructive and desctructive. One the one hand, information revolution and the compliant tendencies of culture globalization depolarize nations' spiritual experience, principles of world cognition, systems of values and cultural symbols; as well as help to perceive the cultural diversity of the world. On the other hand, globalization can produce a destructive effect on human life: cultural leveling destroys the individuality of nations, communities and individuals, which results in social and spiritual crises, and is 'one of the factors of internal instability and disintegration' [25, p. 30]. The Lithuanian philosopher B. Kuzmickas notes that only the nation itself can decide which of the two seeds of globalization to grow.

The establishment of the term *globalization* conditions changes in the notion *culture*: in the past few decades such terms as *global culture, interculture, cross culture, multiculturalism* and others have been widely used. All of them presuppose cultural equality opt out of geographical, historical or national cultural differences and permitting all nations to perform in intercultural interaction on an equal basis. In the last decades of the 20th century, national cultures were perceived as individual islands, whereas the world was considered as a mosaic of those islands. However, globalization processes have revealed that homogeneous culture groups do not exist any longer; they have been changed by heterogeneous groups: 'Culture in this sense is perceived not as a static, hermetically sealed system, but as a current of ever changing meanings' [11, p. 6]. Hence, culture is a dynamic process, which embraces norms, values and lifestyles, which are globally intertwined.

Such a changed approach to understanding culture sets new objectives to education, including teacher training. One of the key goals of the Council of Europe for teaching/learning foreign languages is to 'achieve a wider and deeper understanding of the way of life and forms of thought of other peoples and of their cultural heritage' [8, p. 6], which is implemented through the development of cultural competence of pre-service and in-service teachers of foreign languages.

The significance of cultural education is also highlighted in Lithuanian documents on education. *Law on Education* of the Republic of Lithuania (2011) provides for 'conveying national and ethnic fundamentals, and traditions and values of European and world humanist culture, ensuring conditions for a person's mature national identity, moral, aesthetic and academic culture, developing own world outlook, as well as guaranteeing continuity of the national culture, preserving its identity and continuous development of values, cherishing the openness and dialogic nature of the state' as well as 'providing possibilities for a person to acquire fundamentals of democracy-oriented civil and political culture, developing abilities and experience essential for a person to develop as a competent citizen of Lithuania and a member of the European and global multi-cultural community' [28, p. 15-16]. Topicality of cultural education is strengthened in the *National Description of Teacher Competences* (2007). It distinguishes the common cultural competence, which is defined as 'knowledge, skills, abilities, value-based provisions and other personal qualities determining human activity in specific culture(s)' [34, p. 1].

A lot of research of (inter)cultural competence has been conducted by European and American scholars, who explored intercultural communicative competence and its components, developed, verified and assessed its theoretical models, identified its manifestation in training specialists of healthcare, business, economy and management, university teachers and students [3; 4; 10; 11; 18; 19; 20; 21; 43; 44; 49 and others]. Lithuanian scholars have also extensively explored different aspects of (inter)cultural competence and its dimensions in the context of business and management [33; 36], internationalization of university studies [50], teacher training [7] and others.

Despite extensive studies of intercultural competence, Lithuania lacks research of the effect of teaching/learning foreign languages on the development of students' cultural competence. The Lithuanian scholar V. Šernas notes that 'language is an important constituent part of a nation, state and its culture, which reflects its history, tradition, geography, culture and other spheres of life' [45, p. 188]. Thus, while learning a foreign language a student assimilates the culture conveyed through the target language. On the other hand, learning a foreign language permits seeing own culture in the wider context, i.e. in the light of the target language, which involves its evaluation and comparison with other cultures, as well as learning from and communication with other cultures. All the aforesaid factors contribute to a student's personal experience, social and political awareness and tolerance.

The Australian educationalist M. Pegrum (2008) stresses the significance of critical intercultural literacy, which involves abilities and skills 'to read' cultural information in different information sources, critically assess it in terms of personal knowledge and experience, and at the same time critically evaluate the old knowledge in the light of new experience. Such skills are essential for a contemporary student, as they contribute to one's integration into the global world.

Due to the changing role of the student, the teacher's position undergoes certain changes as well. A teacher, especially the one of foreign languages, becomes 'a teacher of culture' and is expected to assume new tasks: his/her attention is focused not on mere transfer of knowledge, but rather to the development of new attitudes, skills and critical awareness. In other words, the new task of the teacher of a foreign language is to develop students' competence, which would encourage them to link and compare the cultural values, attitudes and behaviour with the target culture as reflected in the language.

Training of foreign language teachers is explored in terms of the development of linguistic competence, communicative and metacognitive skills, and even moral values [6; 22; 23; 38; 42; 43; 46 and others]; however, the development of cultural competence of pre-service teachers of foreign languages through target language acquisition has not yet been extensively investigated. Hence, ***the problem of the research*** is as follows: what is the expression of the cultural competence of pre-service teachers of foreign languages and how can it be developed at the university level?

The object of the research is the expression of cultural competence of pre-service teachers of foreign languages.

The goal of the research is to disclose the ways of expression of the cultural competence of pre-service teachers of foreign languages and opportunities of its development at the university.

The objectives of the research are as follows:

a) to define the structural components of the cultural competence of pre-service teachers of foreign languages;
b) to determine the ways of the expression of individual components of pre-service teachers' cultural competence;
c) assess the current level of the cultural competence of pre-service teacher of Lithuanian University of Educational Sciences.

Research methodology

The research is based on the methodological assumptions of:

- *Cultural anthropology*, which implements holistic surveys of humanity and its cultural differences. Anthropologists (Tylor, Frazer, Morgan, Malinowski, Boas) maintain that all people are able to classify and encode their experience, which they transfer to others. They state that culture is transmitted through learning, which results in the evolvement of different cultures in different places and conditions.
- *Theory of structuralism*, which explores the ties between language, culture and society. The main principle of structuralism is based on the assumption that all substances consist of particular structures, i.e. cognitive models of the reality, through which different cultures perceive the surrounding world and express themselves using particular linguistic structures (Jakobson, Levi-Strauss, Foucault, Barthes, Lacan).
- *Philosophy of culture*, in which the conception of culture is linked to development, improvement and creation (Maceina, Šalkauskis, Cohn).
- *Critical pedagogy*, which analyzes different aspects of human identity, expression of multiculturalism and advocates the idea of global culture. Representa-

tives of critical pedagogy promote intercultural education based the ideas of democracy and cultural dialogism.
- *Theory of linguistic activity*, which covers combination of learners' knowledge, skills, abilities, philological and cultural experience as a way to ensure harmony of the internal and external characteristics of communication [45].

Research methods include

Theoretical methods: analysis of scientific literature on training teachers of foreign languages and the conception of cultural competence.

Empirical: a survey, including open-ended and closed-ended questions, which help to reveal the attitude of pre-service teachers of foreign languages to cultural competence and opportunities of its development through target language studies at the university.

The research sample

The research sample included undergraduate students of foreign languages of Lithuanian University of Educational Sciences that will acquire a teacher's qualification. The survey was conducted in spring 2013 and involved 250 respondents. The demographic characteristics of respondents are provided in Table 1:

Table 1: Distribution of respondents according to demographic characteristics

Criterion	Distribution of respondents	Absolute number	Percentage
Year of studies	I	68	27.2%
	II	65	26%
	III	60	24%
	IV	57	22.8%
Gender	Females	224	89.6%
	Males	26	10.4%
Language of studies	English	181	72.4%
	German	19	7.6%
	French	14	5.6%
	Russian	21	8.4%
	Polish	15	6%

As it is seen from the table, distribution of respondents according to year of studies is relatively equal. Sufficient discrepancies can be observed in the distribution of respondents in terms of gender and language of studies. This accounts for the fact that philological studies are mostly preferred by females and the popularity of the English language in comparison to other foreign languages commonly studied in Lithuania.

Theoretical background of the research

Scientific literature and education documents define the term *competence* is a number of different ways. Generally speaking, *competence* is defined as a combination of knowledge, skills and attitudes appropriate to the context [40], which are complemented by values one holds because of one's belonging to a number of social groups [4].

The foundation of *cultural competence* is in the attitudes, which include curiosity and openness, readiness to suspend disbelief about other cultures and belief about one's own. This means that a learner should not assume that his/her values, beliefs and behaviour are the only possible and naturally correct. They have to be able to see how they might look from an outsider's perspective, i.e. they have to develop an ability to 'decentre'.

Another crucial factor is knowledge and awareness, primarily not about a specific culture, but rather knowledge on how social groups and identities function and what is involved in (inter)cultural interaction. Cultural awareness comprises information about how people behave both linguistically and extra-linguistically in common everyday situations, e.g. greetings, excusing oneself, leave taking, forms of address, asking for or giving directions, expressing respect, thanking, inquiry, giving or receiving compliments, etc., which are complemented by paralinguistic behaviour, such as gestures, touching, eye contact and the like [32]. Skills are just as important as attitudes and knowledge. Byram (2002) distinguishes between the skills of interpreting and relating, which refer to the ability to interpret a document or event from another culture, to explain it and relate it to documents or events from one's own; and skills of discovery and interaction, which cover the ability to acquire new knowledge of a culture and cultural practices and the ability to operate knowledge, attitudes and skills under the constraints of real-time communication and interaction.

Finally, however open towards, curious about and tolerant of other people's beliefs, values and behaviours learners are, their own beliefs, values and behaviours are deeply embedded and can create reaction and rejection. Therefore, intercultural mediators need some critical awareness or literacy: an ability to evaluate critically and on the basis of explicit criteria, perspectives, practices and products one's own and other cultures [2; 35]. In other words, a learner needs experience of communicating and operating in a target culture.

In this regard, a four-dimensional model of cultural competence of pre-service teachers of foreign languages has been designed, which includes the following components:

- *knowledge and awareness of cultural phenomena*, which is characterized by the amplitude (the number of provided arguments) and profoundness (identification of fundamental properties) of knowledge and awareness;
- the sphere of knowledge and awareness is closely linked to *emotional perception of cultural phenomena (situations)*. It is assumed that this component helps to reveal a student's emotional reaction and evaluation of a cultural object (phenomenon) as positive, negative or indifferent;
- *cultural behaviour or involvement in cultural activity*, which is characterized by a person's readiness and ability to act in different culture-specific situations;
- *cultural experience,* which helps to reveal and assess a pre-service teacher's reflection on own cultural and linguistic expression in different cultural contexts.

On the basis of the aforesaid model, a diagnostic survey was conducted in spring 2013, which included 250 undergraduate students of English, German, French, Russian and Polish Philology at Lithuanian University of Educational Sciences.

Results of the research

Culture is a very broad concept, so to get to know a given culture means to gain extensive knowledge and awareness of it. Hence, pursuing to explore the dimension of *knowledge and awareness of cultural phenomena* (processes), four main forms of culture were distinguished: elite (or high culture), ethnic (or folk), mass (popular) culture and subcultures.

Elite (high) culture is often attributed to aristocracy or upper social class, which is characterized as the ruling culture, which should be observed by everybody, or a privileged culture marked by peculiar spiritual, economic, political and other properties [47]. On the other hand, elite (or high) culture embraces world-famous and acknowledged works of art, as well as achievements in sports and science [1; 16; 26].

Ethnic (folk) culture, on the other hand, comprises a wide variety of aspects, many of which are inter-connected, including attitudes, assumptions, beliefs, perceptions, norms and values, social relationships, customs, celebrations, rituals, politeness conventions, patterns of interaction and discourse organization, the use of time in communication and other aspects characteristic of a particular nation or country. This form of culture embraces folk myths, rituals, customs, oral, musical and visual heritage, which are determined by common history, territory, race and language [1; 5; 16].

Mass (popular) culture is based on the stereotypes of mass consciousness, which seeks to imitate celebrities, satisfy hedonistic needs, conforms to the changing conditions and neglects own individuality and distinction. Consumer art, fashion and entertainment are attributed to this form of culture [1; 15].

The conception of contemporary culture is fragmented: the society breaks down into individual groups, which have peculiar culture, repudiate the prevailing norms and create alternative approaches to different aspects of life, including identity, religion, society and others. Subcultures involve independently functioning groups, which choose different stylistic forms to express their individuality and distinctness. Very often they are defined as either highly educated part of society, which appears due to professional differentiation, or a completely uneducated marginal group [1; 39].

Seeking to assess the first component of cultural competence, namely **knowledge and awareness of cultural phenomena**, pre-service teachers of foreign languages were asked to attribute the provided features to an appropriate form of culture. The replies of the respondents are provided in Picture 1.

Picture 1: Distribution of respondents' knowledge and awareness of culture forms.

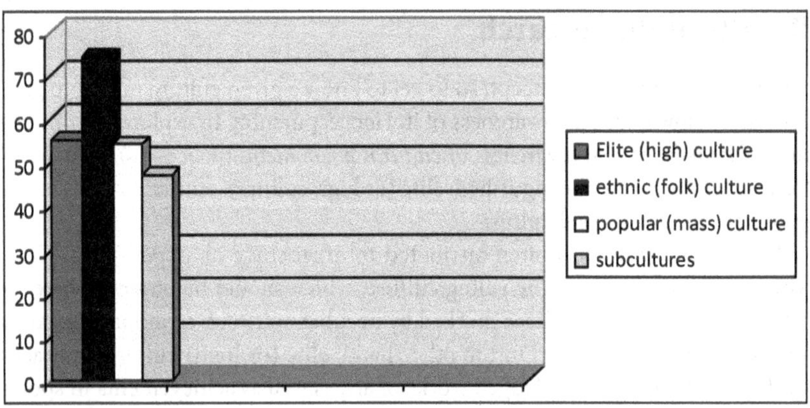

As it is seen from the picture, the knowledge and awareness of pre-service teachers of foreign language was average. The knowledge about elite (high), popular (mass) and subcultures focused around 50% of attributing the outlined features to appropriate culture form, the least recognizable being subcultures (47%). An exception was ethnic (folk) culture: the respondents managed to correctly attribute 74% of the outlined features.

Pre-service teachers of foreign languages were also asked to provide own examples illustrating the four culture forms as represented in own country and the country/countries of the studied foreign language. The replies were assessed with regard to the following criteria: 1. The provided examples fully comply with the selected form of culture and represent the country; 2. The provided examples comply with the form of culture but are not country-specific; 3. The provided examples do not comply with the selected form of culture; and 4. No examples are provided. The obtained results are demonstrated in Picture 2.

Picture 2: Distribution of the provided examples regarding culture forms and their representation in the native and target countries.

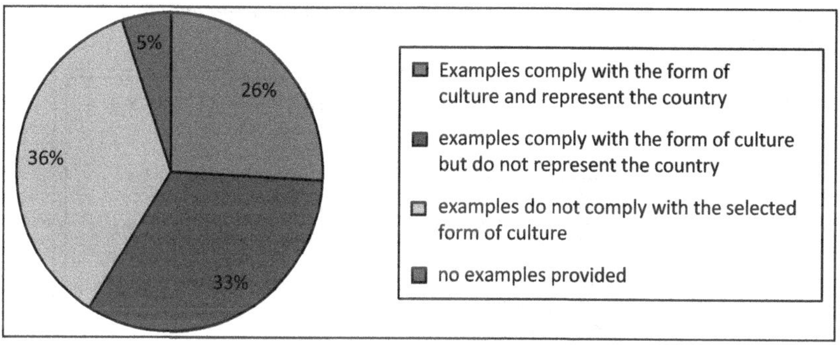

As it is seen from the diagram, only about a quarter of pre-service teachers of foreign languages provided examples that both complied with the form of culture and represented the selected culture. A third of future teachers showed good knowledge of culture forms; however, they were not able to attribute it to a specific context of a country. Approximately the same percentage of students (slightly more than a third) showed no awareness of culture forms. Consequently, it is assumed that in general the knowledge and awareness of pre-service teachers of foreign languages was average: about a half of the respondents showed considerable understanding of culture forms, but only a fourth could illustrate it with country-specific examples.

Knowledge and awareness of cultural phenomena is closely related to the second dimension of cultural competence, namely *emotional perception of cultural phenomena.* It is assumed that emotions and feelings can either promote or suppress the acquisition of received information; thus, this dimension reveals the respondents' emotional evaluation of a cultural phenomenon [12; 53]. Pursuing to assess emotional perception of cultural phenomena, the pre-service teachers of foreign languages were given a list of seven emotions (pride, joy/admiration, concern, responsibility, anger/dissatisfaction, shame and fault). According to Deardorff (2004) and Žemaitis (2005), these emotions are most frequently experienced with regard to own or others' culture. The respondents were asked to define what cultural phenomena (objects, personalities, works of arts, literature, music, etc) raise the aforesaid emotions within the context of native and target culture. Their replies were assessed in accordance with the following criteria: 1. Significant and country-specific cultural phenomena are indicated; 2. Significant aspects are indicated, but they do not relate with culture; 3. Insignificant or trivial phenomena are indicated; and 4. No cultural phenomena are provided. The obtained results are provided in Picture 3.

Picture 3: Distribution of the indicated cultural phenomena in accordance to their significance

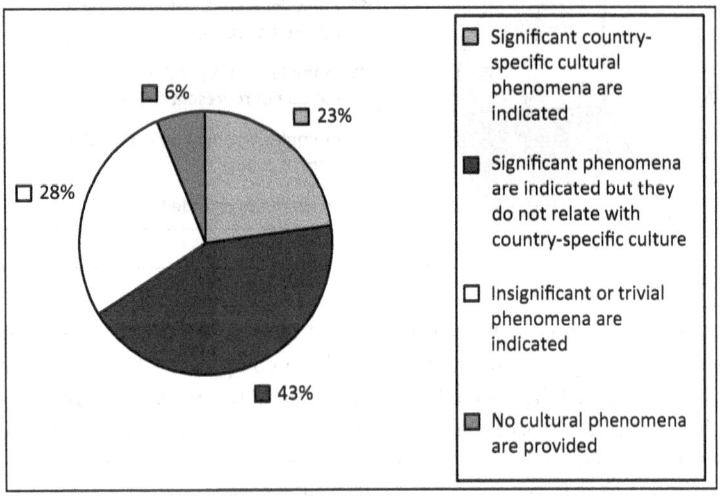

As it is obvious from the picture, less than a quarter of respondents referred to cultural country-specific phenomena when describing their feelings. Among the provided phenomena most frequently mentioned were artists and their works, preservation of objects of cultural inheritance, cultural education of children, etc. Almost a half of pre-service teachers of foreign languages attributed their emotions to different significant phenomena; however, these did not relate with the country-specific culture. They included different personal assumptions, like success of studies, consideration for family members, own health, friendship, protection of nature as well as different political and economic situations. Approximately a third of respondents mentioned trivial or insignificant things that evoked the outlined emotions, whereas about 6% of respondents provided no answers or their answers were too broad to be attributed to one of the afore-said groups. As a result, it is assumed that pre-service teachers of foreign languages did not generally tend to attribute their emotions to cultural phenomena. However, they experienced adequate emotions with regard to other significant phenomena.

The third dimension of cultural competence, *cultural behaviour or involvement in cultural activity*, demonstrates a person's capacity to perform specific actions in a certain culture-related situation. Pursuing to explore this dimension, students were asked to indicate their activity in different groups. They were offered a list of different groups, which included art groups, sports, political youth organization, religious organizations, volunteer organizations, non-traditional cultural groups. The respondents also had an opportunity to expand the provided list by indicating other groups in which they participated. However, none of them took the chance.

The obtained results revealed that a greater part of pre-service teachers of foreign languages (59%) did not participate in the activity of any of the afore-said groups. In terms of the involvement in the activity of art groups, around a quarter of the respondents indicated that they participated in different art groups (music bands, folklore groups, drama circles, etc). 12% of pre-service teachers maintained participating in sports activities. Religious, volunteer and political youth organizations received the least amount of replies, which varied from one to two per cent; no replies were provided for non-traditional cultural groups. Consequently, it is assumed that pre-service teachers of foreign languages were not keen participants in cultural events and activities. Nevertheless, as far as 37% of respondents (more than a quarter) maintained that they participated in cultural activities, such as arts and sports.

Pre-service teachers' cultural behaviour was also explored in terms of their capacity to perform culturally with regard to oneself and other people. A number of researchers [31; 52; 54] maintain that cultural behaviour is manifested via a person's cultural intelligence. An intelligent person (*homo intelligens*) is well-read and highly educated, possesses substantial knowledge and demonstrates abilities of global thinking. Therefore, such a person takes keen interest in different fields of science, new ideas, arts, religion, philosophy, law, history and others. Another important feature is knowledge of the etiquette, which embraces well-mannered behaviour with others (speech, posture, politeness conventions, etc). R. Assagioli (2002) highlights the moral sense and personal responsibility of an intelligent person, which is manifested through such qualities as strong will, clear conscience, generosity, dignity and others. A truly intelligent person shows extensive abilities of self-criticism and self-reflection.

The features of an intelligent person outlined in scientific literature are summarized in a description of an intelligent person, which embraces three aspects: intellect, etiquette and morality, as well as their empirical properties identified. As a result, the respondents were provided with this description and asked to assess their behaviour achievements in each of the afore-said aspects by indicating their capacity (I can't do it yet; I rarely do it; I sometimes do it; I can often do it). Their replies were measured in terms of four ranks: 1. A student is able to mechanically perform certain activities; 2. A student maintains positive relationships with others; 3. A student is able to creatively interpret own cultural activity and defend own position when interacting with others; and 4. A student reflects on own cultural expression, which is considered to show the highest level of a person's cultural intelligence. The obtained results are provided in Picture 4.

Picture 4: Pre-service teachers' assessment of own cultural behaviour

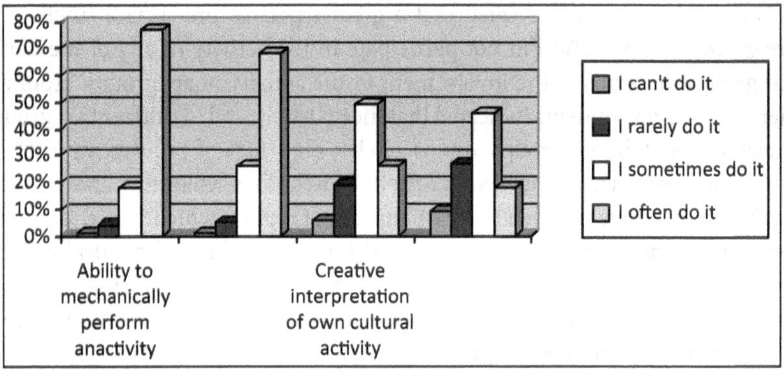

The obtained results demonstrate that as far as 79% of the respondents attributed most frequent occurrences of their cultural behaviour to mechanical performance of activities; whereas 68% maintained being able to frequently establish positive relationships with other people, and 26% stated that they sometimes established positive relationships with others. In terms of the ability to creatively interpret own cultural activity and defend own position interacting with other people, pre-service teachers of foreign languages appeared to be more self-critical: only about a quarter of the respondents (26%) stated that they frequently interpreted their cultural activity, and about a half of the respondents (49%) declared being able to occasionally interpret their cultural behaviour and defend personal stand when communicating with others. Another quarter of pre-service teachers admitted that they rarely or never interpreted their cultural behaviour or defended their position (19% and 6% respectively). As far as pre-service teachers' reflection on own cultural expression is concerned, only 18% claimed doing it frequently, whereas less than a half of the respondents (46%) stated that they reflected on their cultural expression occasionally. More than a quarter of pre-service teachers of foreign languages (27%) pointed out that they rarely contemplated on their cultural expression, and 9% admitted never doing it. Consequently, it can be concluded that pre-service teachers evaluatef their own cultural intelligence quite critically. It is noteworthy that there was a considerable regression in their ranking of own capacities: the higher the rank (considering mechanical accomplishment of activities as the lowest, and reflection on cultural expression as the highest rank), the less frequent capacities of the respondents were indicated.

The last explored dimension of cultural competence of pre-service teachers of foreign languages was their *cultural experience*. It was explored as two-fold: firstly, it is assumed that experience acquired at different periods in life helps a person to develop and search for better decisions in other (similar or different) situations [37]. Secondly, linguistic experience, i.e. acquired foreign language skills, ensure success of a person's performance in different intercultural environments.

To explore the first aspect of cultural experience, the respondents were given an open-ended question to describe the factors of former cultural activity, which they considered as having affected their further development and ability to perform in different cultural situations. The replies were summarized and attributed to the following sections: involvement in arts activities (performing drama, painting, dancing, singing, writing literature, etc.); observation of arts (going to theatres, exhibitions, reading books, etc.); involvement in sports activities; travelling; meeting representatives of different cultures; art studies and other activities not related to culture. The obtained results arc presented in Picture 5.

Picture 5: Cultural experience of pre-service teachers of foreign languages in terms of involvement in cultural activities

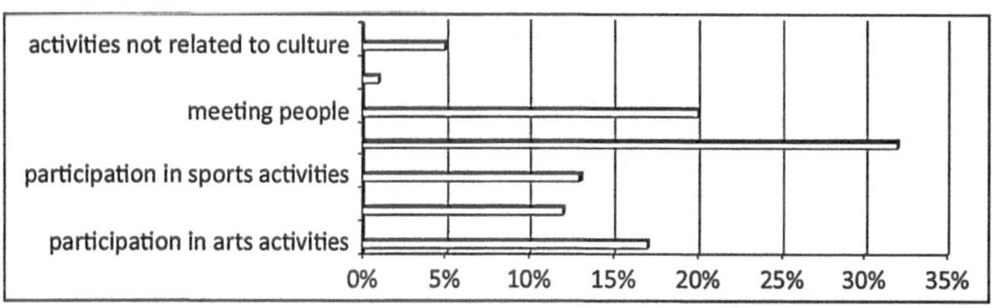

The obtained results revealed that the greatest importance pre-service teachers attached to travelling (32%) and meeting people of different cultures (20%). Involvement in different arts activities such as performing on stage, writing literature, etc. also received considerable percentage of the respondents' replies (17%). Less significance the future teachers of foreign languages attributed to observation of arts and involvement in sports activities (12% and 13% respectively). When asked to validate their choices, the greater part of students emphasized the impact of cultural experience on their future development and ability to cope with various cultural situations; however, only 13% were able to specify it by describing specific situations or providing specific examples. Thus, it leads to a conclusion that pre-service teachers of foreign languages understood the role of gained cultural experience in their life and were able to identify which factors most significantly affected their further development as personalities. However, they were not yet capable of deeper analysis or reflection.

The second factor of pre-service teachers' cultural experience rests on foreign language skills. The respondents were offered descriptions of different communicative activities (listening, reading, speaking and writing) at different levels and asked to mark the section which best of all reflected their current level of foreign language skills. The descriptor was designed in accordance with Common European Framework of Reference for Languages (2001). The distribution of the obtained results is presented in Picture 6.

Picture 6: Distribution of the level of pre-service teachers' communicative skills

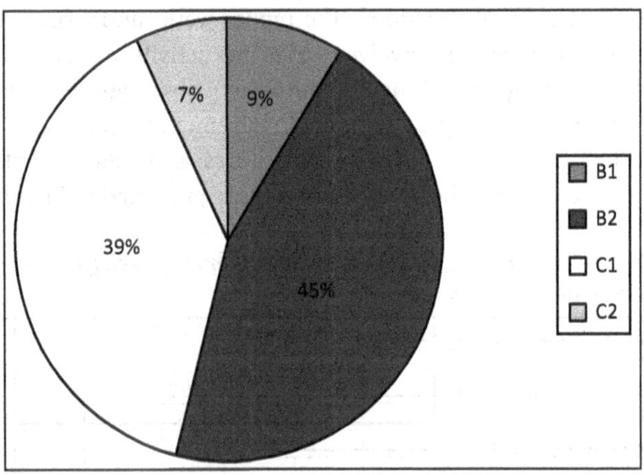

It is noteworthy that the greater part of respondents attributed the current level of their communicative language skills to Levels B2 (45%) or C1 (39%). And only a considerably smaller number of pre-service teachers considered the level of their communicative language skills as either lower (B1) or higher (C2). No respondents, event first-year students, considered attributed their skills to level A1/A2. Consequently, it is assumed that the pre-service teachers of foreign languages adequately assessed their foreign language skills.

Pre-service teachers of foreign languages were also asked whether their foreign language skills sufficed communication with representatives of foreign countries. They were asked to assess their skills of interacting in a foreign language as highly sufficient, sufficient, average and lower than average. Their answers are summarized in Picture 7.

Picture 7: Assessment of the sufficiency of foreign language skills of pre-service teachers of foreign languages

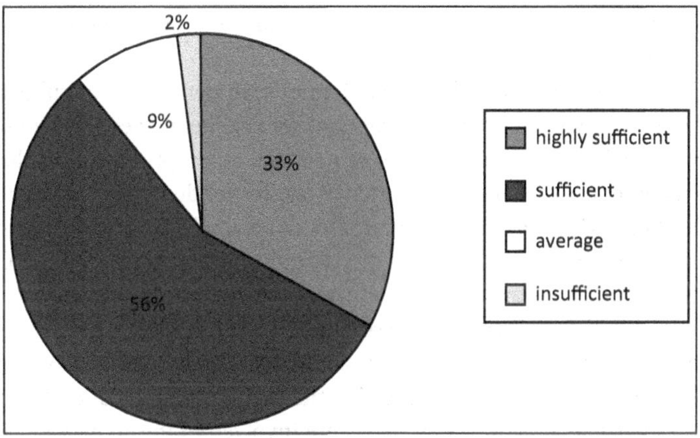

As it is seen from the picture, more than a half of the respondents considered that their foreign language skills were sufficient and a third more believed that they had highly sufficient skills. And only a tenth of pre-service teachers assumed that their foreign language skills were average or insufficient. Consequently, it can be concluded that pre-service teachers of foreign languages did not experience much difficulty while interacting with representatives of different cultures.

Conclusions

On the basis of scientific literature and research, a four dimensional model of the cultural competence of pre-service teachers of foreign languages has been developed, which comprises knowledge and awareness of cultural phenomena, emotional perception of cultural phenomena both in native and target countries, cultural behaviour or involvement in cultural activity, as well as cultural experience, which includes students' involvement in cultural activity at different periods of life and their foreign language skills. The expression of every dimension of cultural competence been explored individually.

The results of the research revealed that the first element of pre-service teachers' cultural competence, knowledge and awareness of the forms of culture is rather mediocre. The respondents managed to attribute only 50% of the outlined features to an appropriate culture form. An exception was ethnic (folk) culture, where they recognized 74% of the outlined properties. Only a quarter of future teachers of foreign languages could substantiate their understanding of culture forms by providing country-specific examples of cultural phenomena; whereas about a half of research participants showed the ability to provide examples compliant to individual culture forms, though could not relate them to a country-specific context.

The analysis of the research results of emotional perception of cultural phenomena shows that pre-service teachers of foreign languages largely attribute their emotional perceptions to different personal assumptions rather than cultural phenomena. Only about a quarter of the respondents were able to define their emotions with regard to cultural phenomena, which included artists and their works, preservation of objects of cultural inheritance, cultural education of children, etc.

As far as pre-service teachers' involvement in cultural activity is concerned, the obtained results show that more than a half (59%) of the respondents did not participate in the activities outlined in the questionnaire. Out of the remaining respondents, a half stated that they were involved in the activity of different art groups (music bands, folklore groups, drama circles, etc) or sports activities (approximately a quarter). Hence, it seems that pre-service teachers of foreign languages are not keen participants in cultural events.

The analysis of their capacity to perform culturally with regard to oneself and other people revealed that the respondents positioned own capacities rather high: around three quarters of the respondents attributed most frequent occurrences of their cultural behaviour to mechanical performance of activities and establishment of positive relationships with others. In terms of the ability to creatively interpret own cultural activity and defend own position interacting with other people, the respondents turn out to be more self-critical: only about a quarter of the respondents stated that they frequently interpreted their cultural activity, and about a half of them maintained being able to occasionally interpret their cultural behaviour and defend personal position when communicating with others. It seems that pre-service teachers of foreign languages tend to demonstrate a rather mechanical behaviour in different cultural contexts devoid of deeper analysis or reflection.

The last dimension of cultural competence, cultural experience, was revealed as two-fold: cultural experience acquired at different periods in life and acquired foreign language skills, which presumably affect the respondents' capacity to perform in different cultural and intercultural contexts. It appeared that about a half of the respondents attached greatest significance to such factors as travelling and meeting people of different cultures as the most decisive in developing their cultural experience and ability to cope with various (inter)cultural situations. However, only 13% were able to precisely specify the impact of the mentioned factors by giving concrete examples or situations.

Finally, more than a majority of respondents attached the current level of their foreign language skills to Levels B2 or C1 and considered them as highly sufficient or sufficient to interact with representatives of different cultures. It is assumed that the pre-service teachers of foreign languages adequately assessed their foreign language skills by stating that they did not experience much difficulty while interacting in different cultural contexts.

References

1. Andrijauskas A. 2006. Technogeninė civilizacija, medijos ir kultūros globalizacija. In: Kultūrologija 13. Kultūra globalizacijos sąlygomis. Vilnius: Kultūros, filosofijos ir meno institutas, p. 76-96.

2. Bauman Z. 2007. Globalizacija: pasekmės žmogui. Vilnius: Apostrofa.

3. Byram M. 2000. Social Identity and Foreign Language Teaching. In: Social Identity and the European Dimension: Intercultural Competence Through Foreign Language Learning. (ed. M. Byram. Tost Planet M.), p. 17–24.

4. Byram M., Gribkova B., Starkey H. 2002. Developing the Intercultural Dimension in Language Teaching: a Practical Introduction for Teachers. Strasbourg: the Council of Europe.

5. Bukraba-Rylska I. 2002. Kultura ludowa w społeczeństwie współczesnym. In: Kultura ludowa, wieś, mieszkańcy wsi : wybór artykułów z lat 1987-2002, Warszawa: Instytut Rozwoju Wsi i Rolnictwa Polskiej Akademii Nauk.

6. Chlopek Z. 2008. The Intercultural Approach to EFL Teachingand Learning. In: English Language Forum, No. 4.

7. Chodzkienė L. 2012. Europos Sąjungos pedagogo tarpkultūrinė komunikacinė kompetencija ir jos tobulinimo socioedukaciniai veiksniai: doctoral dissertation. Vilnius: Edukologija.

8. Common European Framework of Reference for Languages: Learning, Teaching, Assessment, 2001. Retrieved from: http://www.coe.int/t/dg4/linguistic/Source/ Framework_en.pdf (accessed on 25 November 2013).

9. Communication from the Commission Europe 2020: a Strategy for Smart, Sustainable and Inclusive Growth, 2010. Retrieved from: http://eur-lex.europa.eu/LexUriServ/LexUriServ.do?uri=COM:2010:2020:FIN:EN:PDF (accessed on 1 December 2013).

10. Coperias Aguilar M. J. 2009. Intercultural Communicative Competence in the Context of the European Higher Education Area. In: Language and Intercultural Communication, V. 9, No. 4, p. 242-255.

11. Deardorff D. K. 2004. The Identification and Assessment of Intercultural Competence as a Student Outcome of Internationalization at Institutions of Higher Education in the United States: doctoral dissertation. Raleigh: North Carolina.

12. DeCatanzano D. 2003. Motywacje i emocje. Poznan: Zysk I S-ka Wydawnictwo. EC Green Paper Unlocking the Potential of Cultural and Creative Industries, 2010. Retrieved from: http://ec.europa.eu/culture/documents/greenpaper_creative_industries_en.pdf (accessed on 1 December 2013).

13. EC White Paper On a European Communication Policy, 2006. Retrieved from: http://europa.eu/documents/comm/white_papers/pdf/com2006_35_en.pdf (accessed on 1 December 2013).

14. Fiske J. 2008. Populiariosios kultūros supratimas. Vilnius: Žara.

15. Gaižutis A. 2008. Daugiaveidis elitas. Vilnius: Vilniaus pedagoginio universiteto leidykla.

16. Giddens A. 2000. Modernybė ir asmens tapatumas. Vilnius: Pradai.

17. Gudykunst W. B. 2002. Intercultural Communication Theories. In: Cross-cultural and Intercultural Communication. Thousand Oaks: Sage Publishers, p. 167-189.

18. Holliday A. 2011. Intercultural Communication and Ideology. Sage Publications.

19. Holliday A. 2013. Understanding Intercultural Communication: Negotiating a Grammar of Culture. London and New Yourk: Routledge.

20. Kim Y. Y. 2005. Adapting to a New Culture: An Integrative Communication Theory. In: Gudykunst W.B. Theorizing about Intercultural Communication. Thousand Oaks, CA: Sage Publications Ltd., p. 375–400.

21. Kramsch C. 2000. Language and Culture. Oxford: Oxford University Press.

22. Kriaučiūnienė R. 2009. Būsimųjų užsienio kalbų mokytojų dorinės nuostatos ir jų ugdymas universitete: soctoral dissertation. Klaipėda: Klaipėdos universitetas, 2009.

23. Kundrotas M. 2009. Tauta amžių kelyje: tautinės pasaulėžiūros gairės ir tautinis judėjimas Lietuvoje. Vilnius: VPU leidykla.

24. Kuzmickas B. 2003. Globalizacija – priklausomybės ir tapatybės. In: Globalizacija: taikos kultūra, žinių visuomenė, tolerancija: collective monograph (ed. J. Morkūnienė). Vilnius: Lietuvos teisės universitetas, p. 30–42.

25. Lang K., Lang G. E. 2009. Mass Society, Mass Culture and Mass Communication: the Meaning of Mass. In: International Journal of Communication, No. 3 (2009), p. 998-1024.

26. Lietuvos kultūros politikos nuostatos. Pritarta LR Vyriausybės 2001 m. gegužės 14 d. nutarimu Nr. 542. Retrieved from: http://www.lrkm.lt/index.php (accessed on 1 December 2013).

27. Lietuvos Respublikos švietimo istatymas. Nauja įstatymo redakcija 2011 m. kovo 17 d. Nr. XI- 1281. Retrieved from: http://www3.lrs.lt (accessed on 1 December 2013).

28. LR Seimo nutarimas „Dėl Lietuvos Respublikos vyriausybės programos“, 2008. Retrieved from: http://www.lrv.lt/bylos/vyriausybes/14-vyr-dok/14-programa.pdf (accessed on 1 December 2013).

29. LR Seimo rezoliucija dėl Lietuvos Respublikos užsienio politikos krypčių, 2004. Retrieved from: http://www.urm.lt/popup2.php?item_id=156 (accessed on 1 December 2013).

30. Lukšienė M. 2000. Jungtys. (ed. R. Bruzgelevičienė). Vilnius: Alma Littera.

31. Mazlaveckienė G. 2012. The Cultural Dimension in Foreign Language Teaching and Learning. In: Language in Different Contexts, Vol. V (1), p. 302-309.

32. Mažeikienė N., Virgilaitė-Mečkauskaitė E. 2007. The Experience of Measurement and Assessment of Intercultural Competence in Education. In: Socialiniai mokslai, No. 4 (58), p. 70–82.

33. Mokytojų profesijos kompetencijos aprašas, 2007. Retrieved from: http://www3.lrs.lt/pls/inter3/dokpaieska.showdoc_p?p_id=291726 (accessed on 1 December 2013).

34. Pegrum M. 2008. Film, Culture and Identity: Critical Intercultural Literacies for the Language Classroom. In: Language and Intercultural Communication, V. 8, No. 2, p. 136-154.

35. Petkevičiūtė N., Būdaitė R. 2005. Vadovų tarpkultūrinė kompetencija Lietuvos organizacijose. In: Organizacijų vadyba: sisteminiai tyrimai, No. 36, p. 133-150.

36. Pukelis K. 2009. Ability, Competency, Learning/Study Outcome, Qualification and Competence: Theoretical Dimension. In: Aukštojo mokslo kokybė, No. 6, p. 12-35.

37. Qu, Y. 2010. Culture Understanding in Foreign Language Teaching. In: English Language Teaching, V. 3, No. 4, p. 58-61.

38. Ramanauskaitė E. 2002. Šiuolaikinės kultūros fenomenų tyrinėjimai. Kaunas: Vytauto Didžiojo universiteto leidykla.

39. Recommendation of the European Parliament and of the Council on Key Competences for Life-long Learning, 2006. Retrieved from: http://eur-lex.europa.eu/LexUriServ/LexUriServ.do?uri=OJ:L:2006:394:0010:0018:en:PDF (accessed on 25 November 2013).

40. Report of Directorate General for Foreign Policy REGIONS 2020: an Assessment of Future Challenges for EU Regions, 2008. Retrieved from: <http://www.eurada.org/site/files/Regional%20development/regions2020_en.pdf> (accessed on 1 December 2013).

41. Savignon, S. J. 2002. Interpreting Communicative Language Teaching: Contexts and Concerns in Teacher Education. New Haven and London: Yale University Press.

42. Sercu L. 2005. Foreign Language Teachers and the Implementation of Intercultural Education: a Comparative Investigation of the Professional Self-Concepts and Teaching Practices of Belgian Teachers of English, French and German. In: European Journal of Teacher Education, V. 28, No. 1, p. 87-105.

43. Stallman E. M. 2009. Intercultural Competence and Racial Awareness in Study Abroad: doctoral dissertation, 2009. Retrieved from: http://gateway.proquest.com/openurl%3furl_ver=Z39.88-2004%26res_dat=xri:pqdiss%26rft_val_fmt=info:ofi/fmt:kev:mtx:dissertation%26rft_dat=xri:pqdiss:3371916 (accessed on 1 December 2013).

44. Šernas V. Svarbi metodologinio kalbų mokymosi modelio dimensija. Acta paedagogica Vilnensia, 2006, Nr. 16, p. 118–126.

45. Thanasoulas, D. 2001. The Importance of Teaching Culture in the Foreign Language Classroom. In: Radical Pedagogy, V. 3/7.

46. The Europe of Elites: a Study into the Europeanness of Europe's Political and Economic Elites (ed. Best H., Lengyel G. and Verzichelli L.). 2012. Oxford, New York: Oxford University Press.

47. Tomlison J. 2002. Globalizacija ir kultūra. Vilnius: Mintis/ALK.

48. Vander Heijden V. L. 2010. Intercultural Communicative Competence: Assessing Outcomes of an Undergraduate German Language Program: doctoral dissertation. Retrieved from: http://repositories.lib.utexas.edu/bitstream/handle/2152/ETD-UT-2011-05-3414/VANDERHEIJDEN-DISSERTATION.pdf?sequence=1 (16 November 2012).

49. Virgailaitė-Mečkauskaitė E. 2011. Tarpkultūrinės kompetencijos ugdymas aukštojo mokslo internacionalizacijos kontekste (magistrantūros studijų aspektas): doctoral dissertation. Šiauliai: VšĮ Šiaulių universiteto leidykla.

50. Žemaitis V. 2005. Etikos žodynas. Vilnius: Rosma.

51. Ассаджиоли Р. 2002. Психосинтез: принципы и техника. Москва: ЭКСМО-Пресс.

52. Ильин Е. П. 2001. Эмоции и чувства. Санкт-Петербург.

53. Равен Дж. 1999. Педагогическое тестирование: проблемы, заблуждения, перспективы. Москва: Когито-центр.

The Peculiarities of Applying Information Communication Technologies in the Process of Musical Education

Jolanta Abramauskienė

Abstract
On the basis of research material analysis the paper aims to determine the changes happening in musical education, which were brought about by the application of information communication technologies (ICT). The possible trends of applying ICT in musical education are analyzed. The paper also presents the analysis of the trainee teachers' attitudes to the application of ICT in musical education and highlights the peculiarities of ICT use. The analysis of research results will help to improve the preparation of music teachers.

Keywords: information communication technologies, musical education, preparation of music teachers

The topicality and novelty of research

Nowadays information communication technologies (ICT) are no longer an innovation, but they give new opportunities for the process of education. Many modern researchers (Lawless, Pellegrino, 2007; Haris, Mishra, Koehler, 2009; Dagienė, Žilinskienė 2011; Bartaševičius, 2012 et al.) state that the use of ICT enhances the efficiency of the education process. R. Girdzijauskienė, P. Gudynas et al. (2010) believe that ICT is an integral part of modern life, and, if used appropriately, can improve the quality of teaching and learning. The application of ICT can enrich study programmes, develop innovative pedagogy, and create more efficient organizational structures and links with the external world (Paulionytė, Grabauskienė et al., 2010). Compared with the traditional means, ICT has obvious advantages in the visualization of teaching material, experimentation, cooperation, consolidation and evaluation of knowledge, search for information, encouragement of learners' independence, and enhancement of learning motivation etc. (Brazdeikis, 2009; Ignatova, Kurilovas, 2012; Kontvainas, Radčenko, 2013 et al.). If in traditional pedagogy only the teacher was active, now, when more and more different information communication technologies are integrated into the education process, both the teacher and the students are active participants (Čiužas, 2007). Previously the focus was on teaching, now the emphasis is laid on the interactive cooperation between the teacher and the student. Hence the traditional process of education is radically changing – there appear discussions whether the teacher will be replaced by technology in the future universities (Carrillo, 2012). All the aforementioned factors create the preconditions for implementing the changes of education paradigms. Simultaneously the introduction of ICT encourages the alteration of the education process.

The development of ICT has changed the content of all education, including the musical curriculum. In music, just like in other spheres, technological progress gives an opportunity to search for new ways of expression. In musical education ICT opens new possibilities: music and sound can be recorded, created, composed, edited, played on different instruments, sounds and music can be analyzed, new information can be searched on the internet etc. (Burneikienė, Jarienė et al., 2005). ICT can be a great assistant in a variety of musical activities. In one of his interviews Steve Jobs said: "We believe that when technology gets married with the humanities, we get results that make our hearts sing" (Isaacson, 2012, 533). It could mean that the content becomes very significant, therefore, is essential to know what possibilities ICT can have in musical education. Nevertheless, the integration of ICT in the process of musical education is not simple. In modern music pedagogy three spheres of musical activity are typically distinguished: *musical expression; listening to music, its description, evaluation and recognition of music in the socio-cultural environment* (Lithuanian Primary and Basic Education General Programmes, 2008). In each of these spheres ICT can play a significant role, yet it is not enough to be acquainted with the musical technological means. One has to be able to use them in a reflective way: evaluate, understand when the computer can be useful, and when it should not be used. ICT tools are not always favorably assessed. There still are skeptical and cautious attitudes to teaching by means of ICT. Juteikienė and Motekaitytė (2009) believe that the variety of technologies causes chaos and one's inability to correctly choose and use them can bring such negative results as the decline of literacy, absence of critical thinking, and distortion of reality. Fuchs and Woessmann (2004) conducted a research whose results show that a computer in children's room determines worse grades at school since pupils conceive the computer as a toy, a game, not as a source of knowledge. Thus the role of ICT in the process of education is controversial and a professional application of ICT is an aspiration that requires significant endeavor.

In Lithuania the application of ICT in musical education has been analyzed by the following researchers: Šakalys 1998; Vitkauskas, 2006; Vilutienė, 2007; Girdzijauskienė, Rimkutė-Jankuvienė, 2008 et al. Abroad this issue was considered by Rudolph, 2004; Brown, 2007; Burns, 2008; Watson, 2011 et al. The conducted studies theoretically and practically analyze the role of ICT in music, the integration of technologies in primary school musical education, the use of technologies for the development of creativity, etc. However, there has been no research that fully reveals the possibilities of applying ICT in the process of musical education.

These are the key factors aiming to determine the main peculiarities of using ICT in the process of musical education. Such research would help to reveal the ways how modern information communication technologies can be efficiently applied to teach music and improve the training of music teachers. Having this in mind, **the object of the research** is ICT in the process of musical education.

The goal of the research is to determine the peculiarities of applying ICT in the process of musical education.

Objectives

1. To characterize the role of ICT in the study process.
2. To survey the trainee teachers' opinion about the application of information technologies and their use in musical education.

Research methods: literature review, interview, methods of mathematical statistics for the quantitative processing of data. On the basis of literature review there was a questionnaire prepared in advance. The questionnaire consisted of closed- and open-ended questions.

The respondents' characteristics

Students were interviewed in the period of 2010-2013. 93 first year students of music pedagogy at Lithuanian University of Educational Sciences (LUES) took part in the survey.

ICT in the process of musical education

The teacher's profession is exceptional since it requires a variety of different skills and, most importantly, continuous development and search for innovations. The life of the modern person, including teachers of music, cannot be imagined without the computer and the Internet. Nowadays a music teacher who is not computer literate can hardly find a job at school. Hence, the trainee teachers of music have to be interested in, aware of new information technologies, they have to master them and be able to apply them in the process of musical education. At the Lithuanian University of Educational Sciences while training music teachers the following ICT issues are considered: how can technologies be efficiently used in musical education; which technologies appropriate in music pedagogy should be taught; how should students be trained to become technologically aware music teachers who are able to appreciate ICT. Answers to these questions could reveal the possible trends of changing the study programmes of music pedagogy. The study programmes of music pedagogy at LUES pay significant attention to ICT as can be seen from the following subjects: "Music Technologies", "Computer Technologies in Musical Education", "Musical ICT in Education". Today LUES possesses computerized classrooms and library, the Internet connection, interactive boards, there is a possibility to use the facilities of the Lithuanian network of distance education, the university funds the development of special infrastructures. All this improves the training of music teachers and helps to achieve most goals and objectives of musical education.

ICT can be used in music in a variety of ways and the possibilities are considerable. For example, information communication technologies can perform the function of musical instruments (sources of sound) or they can facilitate creating music, performing, recording and analyzing it. In most cases the computer helps to record music: in the form notes, a matrix, a list, various schemes or sound waves (The General Programme of Computerized Music Technologies for Secondary Education Students, 2007). In the process of musical education the functions of ICT can be classified. They can be used as teaching tools, learning tools, a learning environment, tools of cooperation, therapeutic aids, diagnostic tools, administrative tools, etc. (Dudzinskienė et al., 2010).

The ways of learning music based on ICT offer new possibilities. In the process of musical education it is possible to apply the following:

- demonstration, video and sound playback software (*Microsoft Power Point, Prezzi.com, Open Office Presentation, Windows Media Player, Karaoke Player,* etc.);
- educational music software (*Musition, Music Ace, Groovy Music,* etc.);
- software for modeling, composition and music creation (*Mažasis Mocartas, Magic Music Factory, Rythm Maker, Soundation, Dance E-Jay, Music Maker, Band in a Box, Cubase,* etc.)
- systems of testing music and simulators (*Auralia, Ear Master, EarPower, Earope,* etc.)
- software for rendering musical texts (*Sibelius, Finale, Maestro,* etc.);
- regular software (*Microsoft Word; Microsoft Excel; Windows Media Player, Winamp; Windows Movie Maker; e-mail,* etc.);
- interactive boards (*ActivInspire, SMART Notebook,* etc.);
- virtual learning environments (*Moodle, discussion forums, video and audio conferences,* etc.);
- online music resources (*Encyclopedia of Classical Music, New Groowe Dictionary of Music,* etc.);

- technical tools (digital Dictaphone, multimedia projector, headphones with a microphone, electronic keyboard attached to a computer, digital camera, mobile phone, recorder, etc.)

All the aforementioned ICT tools are used in the process of musical education at LUES, and the higher the module the more complicated, conceptual tasks are assigned. In the last module the elective musical education is based on creative activity using ICT. Hence, today the process of musical education at the university is inseparable from the use of ICT.

The results of the survey on the use of ICT in the process of musical education

With the view to analyze the aspects of using ICT in the process of musical education, there were two groups of questions distinguished: students' experience of using ICT and the application of ICT in musical education. The first group of questions aimed to reveal the use of ICT, the possibilities and aims. The respondents were asked how much time they spend working on the computer every week. 37.6 % responded that they spend more than 5 hours, 19.4 % said that they work on the computer from 3 to five hours a day, 35.5 % - from 1 to 3 hours, and 7.5 % - less than 1 hour.

The Standard of Teachers' Computer Literacy (2001) defines the term "computer literacy" broadly: it covers the knowledge and skills of using ICT in education and defines the general level of teachers' information culture. When the respondents were asked to assess their general level of computer literacy, their answers showed that more than two thirds (69.9 %) of the respondents are able to work on the computer quite well, 26.9 % of the interviewees said that their level of computer literacy is low, and only 3.2 % of the respondents said that the level of their computer literacy is very good and they possess the European Computer Driving License (ECDL). Hence, the computer literacy of the trainee teachers of music is sufficient, almost all the students have computers at home and the majority has the internet connection.

Table 1: The Students' Purposes of Using ICT

Purposes	Daily or almost every day	Once a week	Once a month	Never or hardly ever
Surfing the internet for study material	53.8	37.6	8.6	0
Surfing the internet for leisure	60.2	21.5	8.6	9.7

Purposes	Daily or almost every day	Once a week	Once a month	Never or hardly ever
Playing games	10.8	5.4	25.8	58.1
Personal interaction (Facebook, e-mail, Skype, MSN etc.)	76.3	16.1	6.5	1.1
Online services (shopping, paying the taxes, e-banking etc.)	62.3	18.8	15.4	3.5
Working with educational music software	15.1	24.7	34.4	25.8
Working with music creation software	11.8	8.6	18.3	61.3

The research data show (Table 1) that the most popular purpose of using the computer is personal interaction (76.3 %), more than two thirds (62.3 %) prefer online facilities, a similar number of the respondents (60.2 %) surf the internet for leisure and more than a half (53.8 %) of the interviewed students surf the internet to find information for their studies. The students are not well acquainted with educational and music creation software. 25.8 % of the respondents say that they have never used educational music software and 61.3 % have never used any software for music creation. It shows that the computer is rarely used as a teaching aid in musical education and the students are not aware of these technological possibilities. In comparison with the survey conducted in 2009 (Bijeikienė et al., 2009), the students at Vytautas Magnus University also use the computer as a daily or frequent form of leisure. As a result, it is possible to assume that the integration of ICT into the process of musical education can increase the students' interest since they find technologies as an attractive environment for communication, search for information and other interactive activities.

The use of innovative ICT is inseparable from the application of computer software. The survey aimed to determine what software is mostly used by the students. The results show (Figure 1) that the students mostly use the following software: *Internet Explorer, MS Power Point, Media Player, MS Word.* Less than a third of the respondents use the following musical software: *Music Maker, Karaoke Player, Frooty Loops, Cubase.* Consequently, it can be stated that the students use the widely spread, most popular types of software, but are poorly acquainted with the application of ICT and their possibilities in musical education. The students' inability to work with music software can be determined by the fact that in the majority of cases students become acquainted with those programmes informally, unconsciously, while studying individually. Therefore it is essential to plan and prepare methodological and technical aids that would help the students improve their skills necessary for work with musical ICT.

Figure 1: The Use of Computer Software

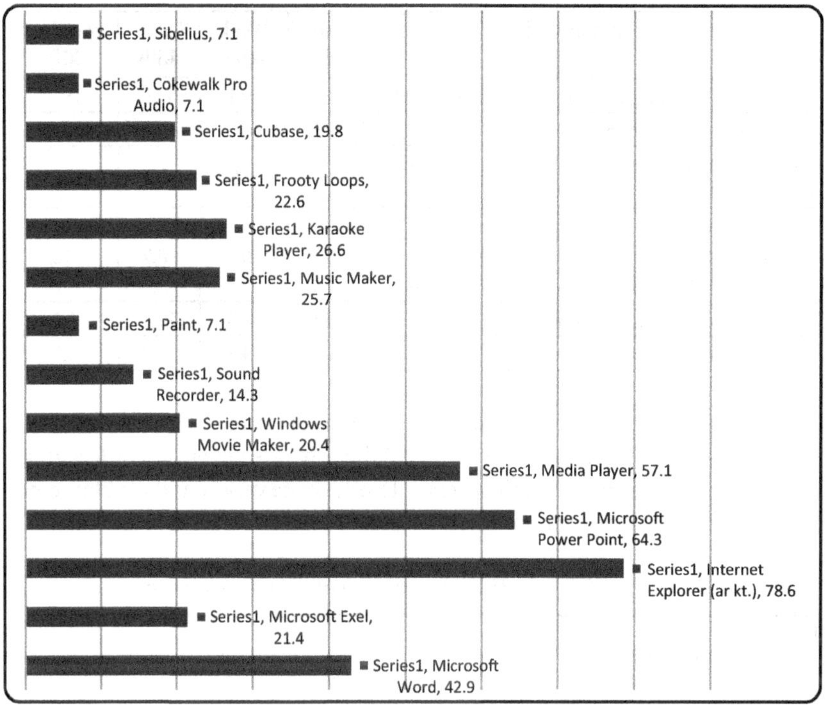

Note: some of the respondents indicated several variants.

The development ICT elicits the appearance of skills and abilities. The acquisition of computer skills is constantly changing; therefore, the attitudes towards computer skills should be dynamic and regularly adjusted. When the respondents were asked to assess their computer skills, their answers (Table 2) showed that two thirds (64.5 %) of the respondents evaluated their ability to prepare textual material well. 44.1 % of the interviewees said that they have good skills to prepare presentations. The smallest percentage of the respondents can work well with educational software (40.9 %) and musical software (39.8 %). The acquired data show that the students' computer skills in the process of musical education are insufficient; hence, there is a need to improve the students' computer literacy, skills and introduce them to the application of information technologies in musical education.

Table 2: The Students' Computer Skills

Computer Skills	No skills	Weak skills	Good skills	Excellent skills
Preparation of textual material on the computer	0	16.1	64.5	19.4
Preparation of presentations	6.5	36.6	44.1	12.9
Awareness of the main internet services (search, e-mail, etc.)	0	1.1	50.5	48.4
Transfer of one's activity to the electronic environment (publishing information on the internet, e-banking facilities, e-shopping, etc.)	3.2	26.9	38.7	31.2
Awareness of educational software	28.6	40.9	29.4	1.1
Awareness of musical software	29.0	39.8	26.9	4.3

The survey also aimed to determine what methods of using ICT are used in music lessons at schools of general education. All the ways are beneficial for teaching / learning as they encourage the learners' interest, create variety and facilitate comprehension and memorization.

Figure 2: Ways of Using ICT in Music Lessons at General Education Schools

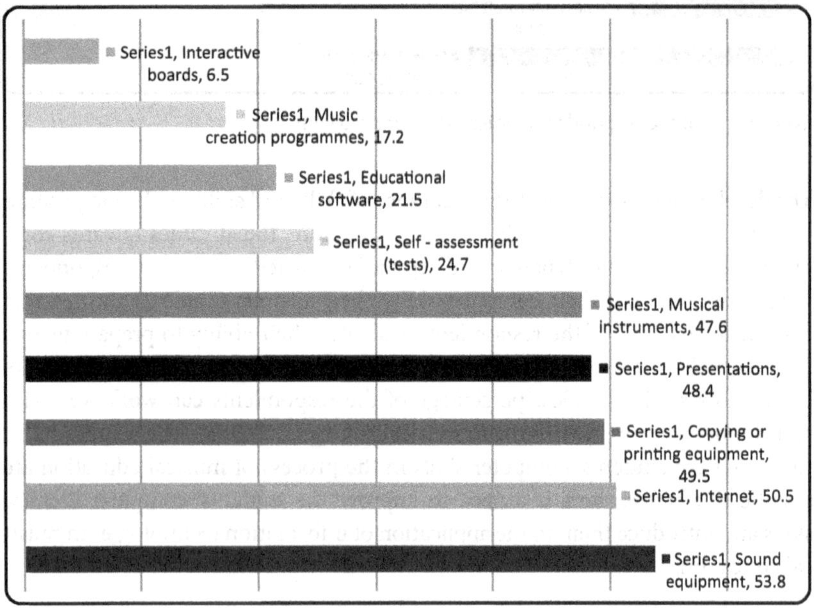

Note: some of the respondents indicated several variants.

The students' answers (Figure 2) reveal that the following methods of using ICT were most frequently applied in music lessons: sound equipment, the internet, copying equipment, presentations, musical instruments. The rarest way of using ICT at general education schools was the interactive boards (6.5 %). 5.4 % of the respondents also mentioned "other" ways: the telephone and virtual learning environments. The analysis of the results makes it possible to state that music lessons at schools of general education were not very interesting or untraditional because of the absence of ICT. Information technologies in music lessons can be using when speaking about music history, its theory, creation, when learning to sing, play an instrument, etc. The use of information technologies is beneficial since theoretical knowledge is consolidated in practical activity; moreover, such way of presentation is more appealing for young people. The results of an international survey on the application of IT in education (SITES, 2006) revealed that ICTs were used in the education process by 63 % of mathematics and science teachers. They mostly use ICT for projects and presentations (about 60 %), for testing (40 %) and least frequently – for research activity and different investigations. These results show that in Lithuanian schools in the process of musical education information communication technologies are used insufficiently. Therefore, the application of ICT has a significantly smaller influence on the efficiency of educational process.

The question if it is necessary to use information technologies in musical education was answered by the respondents in the following way: most of the interviewees (78.8 %) think that they are obligatory, 12.2 % believe that they are sometimes necessary, and 10.0 % said that they are necessary. ICT is most appreciated for the possibilities it gives: "*to record and sound*", "*create music*", "*analyze audio material*", "*control the learning process individually*", "*more vividness*", "*more variety*" etc. The interviewed students mostly appreciate mixed teaching, i.e. when technologies are used with traditional classes which emphasize "*live interaction with music*", "*live music*", "*live playing and singing*" etc.

The results of the survey demonstrate that there appear aspects of using ICT in the education process that need to be improved. Hence, it is important to look for ways to convey the knowledge of music and skills more efficiently. ICT can facilitate rendering the knowledge and designing the process of musical education.

Conclusions

1. The changes in the society and the development of ICT promote a new approach to the process of musical education. There appears new ways of thinking, a new environment of learning and interaction, which help to manage the process of education more efficiently and purposefully. The use of ICT in the

process of musical education presents new opportunities: music can be recorded, created, composed, it is possible to play different instruments, learn the theory of music, develop one's hearing, search for new information by means of the internet resources. ICT can be used in various spheres of musical activity: while learning to sing, play, conceive the rhythm or while listening to works of music. ICT is significant for the professional competences of the trainee teachers of music and can complement the training of teachers of music.

2. The results of the empirical research revealed that the computer literacy of the trainee teachers of music is sufficient, they use ICT quite frequently and spend quite a lot of time working on the computer. The most popular purposes of using the computer are: personal interaction, online services, leisure, search for information on the internet for studies; however, in musical education ICTs are rarely used and students are not aware of the technological possibilities. The respondents prefer using the usual software and are poorly acquainted with the possibilities of using ICT in musical education. The respondents assessed their abilities to work with the usual software well; however, the students know little about the educational and musical software. The trainee teachers of music support the application of ICT in musical education.

References

1. Bartaševičius R. Mokymo(si) aplinka XXI amžiuje. *Švietimo problemos analizė*. Vilnius: Švietimo aprūpinimo centras, 2012.

2. Bijeikienė V., Linkevičiūtė E., Pundziuvienė D. ICT in foreign language learning and teaching: attitudes and practices. *Kalba ir kultūra: nauji iššūkiai Europos mokytojui* Vilnius: Vilniaus universiteto leidykla, 2009, p. 252–262.

3. Brazdeikis V. Informacinėmis ir komunikacinėmis technologijomis papildytų edukacinių aplinkų kaita. *Informacijos mokslai*. 2009, Nr. 50, p. 57–63.

4. Brown R. A. *Computers in music education*. New York: Routledge, 2007.

5. Burneikienė N., Jarienė R., Jašinauskas L., Motiejūnienė E., Neseckienė I., Vingelienė S. *Informacinių komunikacinių technologijų taikymo ugdymo procese galimybės*. Vilnius: Švietimo plėtotės centras, 2005.

6. Burns A. M. *Technology Integration in the Elementary Music Classroom*. USA: Hal Leonard Corporation, 2008. http://www.amazon.com/Technology-

Integration-Elementary-Music-Classroom/dp/1423427572/ref=sr_1_12?s=books&ie=UTF8&qid=1356597480&sr=1-12&keywords=music+technology - #.

7. Carrillo F. A. G. Can Technology Completely Replace Human Interaction in Class? *World Journal on Educational Technology*. Volume 4, Issue 3, 2012. p. 153-164.

8. Čiužas R. Mokytojo ir mokinio vaidmenų kaita edukacinės paradigmos virsmo sąlygomis. *Pedagogika,* 2007, t. 87, p. 64–70.

9. Dagienė V., Žilinskienė I. Mokymosi veiklos samprata skaitmeninėje plotmėje. *Pedagogika,* t.102, 2011, p. 94–103.

10. Dudzinskienė, R., Kalesnikienė, D., Paurienė, L., Žilinskienė, I. *Inovatyvių mokymo(si) metodų ir IKT taikymas.* II knyga. Vilnius: Ugdymo plėtros centras, 2010.

11. Fuchs T., Woessmann L. Computers and student learning: Bivariate and multivariate evidence on the availability and use of computers at home and at school. *Brussels Economics Review,* 47, 2004, p. 359–385.

12. Girdzijauskienė R., Gudynas P., Jakavonytė D., Jevsikova T. *Inovatyvių mokymo (-si) metodų ir IKT taikymas.* Vilnius: Ugdymo plėtotės centras, 2010.

13. Girdzijauskienė R., Rimkutė-Jankuvienė S. Aukštesniųjų klasių mokinių kūrybinių gebėjimų muzikinėje veikloje ugdymas informacinėmis komunikacinėmis technologijomis (IKT). *Kūrybos erdvės,* 2008, t.8, p. 8–19.

14. Harris J., Mishra P., Koehler M. Teachers' technological pedagogical content knowledge and learning activity types: Curriculum-based technology integration reframed. *Journal of Research on Technology in Education.*41 (4), 2009, p. 393–416.

15. Ignatova N., Kurilovas E. Informacinėmis ir komunikacinėmis technologijomis grįsto mokymo ir mokymosi individualizavimo kryptys Lietuvos švietimo kontekste. *Pedagogika,* t. 106, 2012, p. 21–29.

16. Isaacson, W. *Steve Jobs. Oficiali biografija.* Kaunas: Obuolys, 2012.

17. Juteikienės V., Motekaitytė V. Efektyvus žiniasklaidos naudojimas vidurinio lavinimo švietimo įstaigose. *Informacinių technologijų taikymas švietimo*

sistemoje 2009: E-studijų patirtis, aktualijos ir perspektyvos. Kauno kolegija, 2009, p. 18–21.

18. The General Programme of Computerized Music Technologies for Learners of SecondaryEducation. Vilnius: Švietimo aprūpinimo centras, 2007.

19. Kontvainas R., Radčenko M. Geografijos mokytojų požiūris į virtualios mokymo(si) aplinkos taikymą geografijos ugdyme. *Pedagogika,* 2013, t. 109, p. 118–125.

20. Lawless K., Pellegrino J. Professional development in integrating technology into teaching and learning: Knowns, unknowns, and ways to pursue better questions and answers. *Review of Educational Research,* 77(4), 2007, p. 575–614.

21. The General Programmes for Lithuanian Schools of Primary and Basic Education. Vilnius: Švietimo aprūpinimo centras, 2008.

22. Merkys G., Urbonaitė-Šlyžiuvienė D., Balčiūnas S., Mikutavičienė I. *IKT taikymas ugdyme (SITES 2006 Lietuva).* Kaunas, Šiauliai: Švietimo informacinių technologijų centras, 2007.

23. Paulionytė J., Grabauskienė V., Žemgulienė A., Schoroškienė V., Makarskaitė-*Petkevičienė R. IKT ir inovatyvių mokymo (si) metodų taikymo pradiniame ir specialiajame ugdyme pasiūla, taikymo praktika ir perspektyvos Lietuvoje ir užsienyje.* Vilnius: Švietimo aprūpinimo centras, 2010.

24. The Standard of Teachers' Computer Literacy. Vilnius: Švietimo aprūpinimo centras, 2001.

25. Rudolph T. *Teaching Music with Technology.* Chicago: GIA Publications, 2004.

26. Šakalys A. Technikos pritaikymo galimybės mokykloje (muzikos pamokų metu). *Muzikinio ugdymo problemos: žvilgsnis į XXI a.* Vilnius: LMA, 1998.

27. Vilutienė A. Informacinių technologijų taikymas muzikiniame ugdyme ir projektinėje veikloje. *Informacinės technologijos mokykloje.* 2007, p. 124–127.

28. Vitkauskas R. Application of Digital Technologies in Music Education: objective and subjective dimensions. *ATEE Spring University.* 2006.

29. Watson S. *Using technology to unlock musical creativity.* USA: Oxford University Press, 2011.

Positive Experience in Teaching Practice of Pre-Service Teachers of Music: Analysis of Written Reflections

Jolanta Lasauskienė, Asta Rauduvaitė

Abstract
The paper highlights the characteristics of pre-service music teachers' professional experience of implementing the teaching practice goals. The qualitative analysis determined categories which revealed the teaching practice spheres (relations with pupils, mentors and other teachers, personal self-education and professional development) that conditioned the students' positive experience. The research revealed that positive experience during the teaching practice depends on the collegial interaction with pupils, mentors and other teachers, the ability to interest the pupils in the subject of music, efficient planning of a music lesson and feedback.

Keywords: teaching practice, positive experience, reflection

Introduction

In the present-day world education is perceived as one of the essential priorities of the domestic politics of the European Union, and people's education, their knowledge and competences are viewed as the key to the European future (*Leuven Communicate*, 2009). Modern education theory and practice are aiming at answering the challenges of the 21st century – implementation of lifelong learning, education quality assurance, people's preparation for a successful professional career in the context of the rapidly changing social and cultural conditions. In the present-day situation teacher training institutions acquire a responsible mission to prepare competent educators for the school of the future and constant changes. The works of scholars from different countries (Shaffer, 2004; Pollard, 2008; Black, Plowright, 2010, et al.) highlight the significance of the connection between the theory and practice of studies and emphasize the interrelation of professional experience. These opportunities are revealed in the teaching practice when the students' personal experience is reflected upon. Only practical activity shows if a trainee teacher has a sufficient knowledge and is able to act and create, communicate and cooperate responsibly.

According to Pečiuliauskienė (2011), teaching practice is the study period when the first steps of professional activity are made, when the students form their attitude to the teacher's job, rights, duties and responsibility. It is the educational sphere which reveals the quality of competences acquired during the studies. Hence, the purpose of teaching practice is to help the trainee teacher acquire the professional competences and experience necessary for the practical teaching job (*Regulations of Teacher Training*, 2012). The success and failures experienced at the beginning of one's professional career condition the further professional road of a pre-service teacher. Rodzevičiūtė (2008) believes that the students' teaching practice and the teacher's pedagogical competence should be linked by a cause and effect relationship: the quality of teaching practice determines the development of the trainee teachers' pedagogical competence.

In the rapidly changing world an educator has to be a continuously developing personality. Adomaitienė et al. (2008, p. 14) argue that every teacher has to become a "director" of planning; implementing and improving his/her own activity. Moreover, the processes of reflection during the teaching practice enable its participants acquire a competence of research activity (Pečiuliauskienė, 2009). Thus, pre-service teachers have to become reflecting practitioners who constantly investigate and reflect on their own activity. Reflection in the process of teaching practice facilitates self-observation, interaction, cooperation with others, correction of one's goals of activity, learning achievements and drawbacks (Bubnys, Žydžiūnaitė, 2010).

While investigating real problems of education and reflecting upon their personal experience, pre-service teachers aim at new knowledge and competences which are essential for the improvement of teaching practice and the enhancement of professional motivation. In the present-day school the teachers are not passive observers, they deal with practical issues: why, how and what has to be done in order to escape failures and attain better results in their educational practice. Personal professional experience is the sum of knowledge and skills which results from activity. Empirical knowledge is acquired while dealing with direct experience, observing activity and acting practically. The positive experience in teaching practice of the pre-service teachers of music can become a significant basis for the improvement of the study process, fostering of pedagogical vocation, professional purposefulness, development of the career vision, and promotion of their motivation to participate in pedagogical activity.

In the present-day context of teacher training development, there is qualitatively new project of the teaching practice model which emphasizes the trainee's experience (Martišauskienė, 2007; Barkauskaitė, Pečiuliauskienė, 2009). It has to be noted that teachers of music possess distinctive personal and professional (subject-related) qualities which to a great extent are formed at school. Hence, music education is very specific, and has distinctive methods of cognition of artistic musical world (in the aesthetic-spiritual sense) which are implemented in the process of musical education (Reimer, 2003; Rinkevičius, Rinkevičienė

2006). However, up to now there has not been any exhaustive research on the pre-service music teachers' learning from their experience in the teaching practice. Such research is necessary for the analysis of the possibilities of teacher training development and study quality enhancement at schools of higher education.

The works of music education specialists, which focus on the change of music teachers' practice, highlight only separate aspects of this problem (Girdzijauskiene, 2008; Lasauskienė, 2010; Tavoras, 2011, et al.). However, not enough attention is given to the subjective discourse of training students of music education, there is not enough consistent research focusing on the analysis of the characteristics of the students' personal development during their pedagogical studies. The practical aspect of developing the training of pre-service teachers of music has also been little investigated just like the implementation of competences acquired during their studies during the teaching practice, and the assessment of students' personal professional experience as a factor of improving teacher training and study quality.

The problem of the research is the development of pre-service music teacher training in the context of personal positive experience in the teaching practice.

The object of the research is the pre-service music teachers' personal professional experience in the teaching practice.

The goal of the research is to reveal the positive experience of pre-service teachers of music by analyzing their written reflections and determine the spheres where positive experience is felt during the teaching practice.

The Research Methods

Data acquisition. The data were collected by applying the method of written reflection (Yip, 2006), which prompts the research participant to recall, reflect, evaluate and write down a specific experienced case of teaching practice. The students were given an open-ended questionnaire. This paper presents one part of the research and analyzes the respondents' considerations in answer to the following questions: *Describe a situation when you experienced success and positive feelings during the teaching practice; what is the significance of the teaching practice personally to you? What did you learn about yourself as a future teacher of music?* The research participants could choose individually how many of such situations they want to describe and give specific examples.

Data analysis. The content of the written reflections was explored by means of a qualitative content analysis (Bitinas, Rupšienė, Žydžiūnaitė, 2008, p. 225). The qualitative content analysis revealed how successful the situation of teaching practice seems to the author of the reflection, what ways of thinking, conception and positive feelings are revealed, how deep the reflection is. The analysis aimed to distinguish the meaningful units (categories, subcategories) of the text. Their generalized formulations are presented in the form of tables. The research ethics was based on the principles of goodwill, voluntary participation and confidentiality.

The Research Sampling

The research involved 34 third and fourth year students of Music Education at Lithuanian University of Educational Sciences. The reflections were written when the students returned from their teaching practice.

The Research Results

The spheres of teaching practice where the students acquired positive experience

The analysis of the students' reflections on their experience during the teaching practice revealed 3 qualitative categories which can be assigned to the spheres of positive experience (Table 1).

Table 1: The Results of the Students' Written Reflections: Spheres of Positive Experience

Category	Subcategory
The students' positive experience of the relationship with pupils	Humane relationships which are based trust and respect Ability to interest the pupils in the subject of music Efficient planning of the lesson Class management Giving and receiving feedback
The students' positive experience of the relationship with the mentor and other teachers	Close cooperation with the mentor Collegial relationship with the other teachers of the school Encouragement of the students' self-confidence and creation of good-natured atmosphere of activity
Personal self-education and professional development	Self-assessment Self-development Acquisition of new experience Enhancement of professional motivation

The research revealed that the positive experience in the teaching practice depends on the positive relationships with the pupils, mentors and other teachers, the personal self-education and insights of professional development (the assessment of personal characteristics, their development, acquisition of new experience, and enhancement of professional motivation). The content of reflections demonstrated that the students' positive experience in the teaching practice is an important factor for professional motivation. The choice of profession is determined by a number of factors: certain needs, calls and interests, abilities and intellect, character and temperament, goals and value-related attitudes (Rodzevičiūtė, 2008). Hence, aiming to validate the strategies of maturing the pre-service teacher's professional attitudes, it is purposeful to investigate the positive professional experience more thoroughly and in greater detail.

The students' positive experience of the relationship with pupils:

The analysis of the students' positive experience of the relationship with pupils distinguished the subcategories of humane relationships, ability to interest pupils in the subject of music, efficient planning of the lesson, class management, and giving and receiving feedback (Table 2).

Table 2: The Expression of the Students' Positive Experience of Relationship with Pupils

Subcategory	Extracts from the reflections
Humane relationships based on trust and respect	*Children need love, understanding and support <...> it is important to make pupils remember their teachers with warm feelings*
	It is important to come to understanding with pupils <...> when you speak with them as if they were your equals, you can get really good results
	There was a situation when a student asked me for advice <...> if he should choose the road of music. I felt really good that the student confided in me
Ability to interest pupils in the subject of music	*I managed to involve the students, to offer something new, to conduct a creative lesson. I saw their eyes "burning"*
	I chose a topic that was interesting for the pupils. During the teaching practice I hope to find my own model of teaching

Subcategory	Extracts from the reflections
Efficient planning of the lesson	*I imagine success in the teaching practice as a successful musical activity in the lesson. Implementation of the lesson plan is my success of the teaching practice*
	<...>I was successful because I was ready, considered all the possibilities, planned everything and had some prior experience
Class management	*I experienced success when I had to conduct a lesson to a disobedient class*
	I managed to express my wishes and lesson goals perfectly well for the first time
Giving and receiving feedback	*<...> I managed to attain a good psychological contact which is really important for the interaction with pupils*
	<...> I felt necessary for the children, I knew how and managed to help them
	I was glad about the good results for the music test which included the topic I had taught.

The communication of the educator and pupils is the essential part of pedagogical interaction (Martišauskienė, 2007). According to Rinkevičius and Rinkevičienė (2006), in a music lesson the teacher should attain a spiritual contact with the pupils, awaken their musical creative self-expression, and be able to manage the arising conflicts. The research revealed that the students frequently managed to establish a contact with their pupils: they felt expected and accepted, the pupils communicated with them frankly. The reflections speak about the relationships with the pupils that are based on respect and trust: "*Children need love, understanding and support <...> it is important to make pupils remember their teachers with warm feelings.*" The respect and trust were showed b mutual attentiveness, friendly attitudes, and frank, willing cooperation. The students believe that "*it is important to come to understanding with pupils <...> when you speak with them as if they were your equals, you can get really good results.*" Collegial interaction with the pupils helped to experience positive feelings and helped to conceive the significance of the teacher's activity and personal recognition: "*There was a situation when a student asked me for advice <...> if he should choose the road of music. I felt really good that the student confided in me.*"

The reflections reveal the fact that the students aimed to hear out their pupils, understand their needs, sympathize with them, help and trust their abilities. On the other hand, it helped to successfully organize the musical activity in the lesson, improved the students' self-confidence and affected the students' good feelings during the teaching practice. These findings are supported by Martišauskienė's (2007, p. 16) study which proved that the students who had been firmly determined to become teachers before they entered the university, i.e. they had formed their preconceived attitudes, are more tolerant of the different opinions of the pupils and are better at socialization with their pupils. The presented data of the qualitative research make it possible to assume that a pedagogical calling is one of the key preconditions for one's suitability for the teacher's profession. It is the individual's self-awareness which manifests itself as the cognition of oneself as an educator, a positive relation with the teacher's profession, the experience of suitability for the teacher's profession and the goal to implement it (Kavaliauskienė, 2001, p. 165). Thus, the teaching practice should create favorable conditions for the operation and development of these attitudes, and the practical reflection should help the students deal with the professional objectives and motivate them to create a more efficient process of teaching / learning.

The students' positive experience of the relationships with the mentor and other teachers

The analysis of the expression of the students' positive experience of the relationships with mentors and other teachers revealed the subcategories of close cooperation, collegial interrelations, and creation of a good-natured atmosphere and promotion of the students' self-confidence (Table 3).

Table 3: The Expression of the Students' Positive Experience of Relationships with Mentors and Other Teachers

Subcategories	Extracts from the reflections
Close cooperation with the mentor	I felt that the agreement with the mentor was really important. The teacher <...> helped to prepare for the lesson
	The relationships with the teachers are really significant. It is important how the teacher presents the students to the pupils
Collegial interrelations with the other teachers at school	<...> the teachers of English trusted me and asked to help them to prepare the pupils for the events of "The Week of English" at school
	The other teachers called me their "colleague", and that gave me really positive emotions
Encouragement of the students' self-confidence and creation of a good-natured atmosphere for activity	After the lesson the teacher encouraged me <...> I noticed that at the end of the teaching practice I became more self-confident and braver
	<...> I could distinguish the teacher's confidence in me (she let me conduct the lessons, organize and implement different events), and her praise
	<...> the mentors gave me advice, demonstrated the possible solutions to problem situations

A mentor is typically defined as an experienced educator who is in charge of a student or a trainee teacher, promotes the development of their professional practical competences and guarantees a successful integration into the educational institution. The interrelation of the mentor, the teaching practice supervisor and the student is an important condition for the implementation of the teaching practice goals since frequently students look for an example teacher who could be followed. This is confirmed by the students' arguments: "*the relationships with the mentor are significant*", "*it is important to get on well with the practice supervisor*". The students' reflections reveal the significance of a competent mentor's help in the acquisition of practical experience – the mentor can encourage and promote reflection and self-assessment of activity ("*after the lesson the teacher encouraged me*", "<...> *she really helped to prepare for the lesson*", "<...> *gave advice, demonstrated the possible solutions to problem situations*"). Hence, the data enables us to argue that the pre-service teachers of music think it is important to have congenial people in the professional environment who could understand and support, provide help at the beginning of their professional road, promoted self-confidence, approve of their professional goals and guarantee their professional development since such people form the trainee teachers' authentic relation to the teacher's profession.

It was determined that the students' activity in the teaching practice is motivated by the creation of a favorable atmosphere: welcoming the students at school, creating favorable conditions for the implementation of teaching practice tasks, self-assessment in the context of new situations of pedagogical activity, partnership, cooperation and efficient feedback. It can be assumed that this conditions psychological security and the students' greater involvement in the educational activity at school. The experience of sincere communication with the mentors and other teachers enabled the students to become more self-confident and braver: "<...> *I could distinguish the teacher's confidence in me (she let me conduct the lessons, organize and implement different events)*"; <...> *I noticed that at the end of the teaching practice I became more self-confident and braver.*" Obviously, the teaching practice gives students exceptional experiences which make them braver and encourage the attainment of professional goals.

The expression of personal self-education and professional development

The third category of the students' positive experience reflects the pre-service music teachers' subcategories of personal self-education and professional development (Table 4).

Table 4: The Expression of the Students' Personal Self-Education and Professional Development

Subcategory	Extracts from the reflections
Self-assessment	The teaching practice provides with opportunities for self-assessment, self-cognition, you can feel if you are available for pupils and teachers
	<…> during the teaching practice I learned and experienced a lot. I saw my mistakes which can be corrected<…>
Self-development	Teaching practice for me was a challenge to overcome myself, to understand my calling, to acquire new knowledge and courage, pedagogical confidence in myself
	I understood that one of the most important things is that teachers have to work with themselves. The best teacher for that purpose is children
Acquisition of new experience	Practice is very important for the developing teacher of music
	Practice is perfect since I acquire a lot of experience, attitudes and knowledge. During the teaching practice I improve my skills and abilities
Enhancement of professional motivation	In the future I see myself as a teacher who works successfully
	<…>I want to get back to the school atmosphere as soon as possible, to meet professional teachers and learn something new that will be useful for me in the future

Pedagogical activity requires a deep self-analysis. Hence, the teaching practice has to develop the skill of reflecting upon one's experience (Davies, 2005). The content of the students' reflections revealed that the pupils, mentors and other teachers that the students interacted with during the teaching practice had a lot of influence on the students' attitude to themselves as future educators: "*Practice is perfect since I acquire a lot of experience, attitudes and knowledge.*", "*I saw that it is not easy to be a teacher. It requires a lot of preparation for lessons and good psychological knowledge <…> good eloquence and a sense of humour*", "*The teacher has to constantly develop, learn, adjust to the situation and be modern*". It can be assumed that this helped the pre-service teachers of music understand the significance and value of educational activity, promoted the development of professionalism and pride in the profession of a teacher of music.

As it is known, the relationships with pupils, mentors and other teachers are in a way the basis of forming professional self-identification. The students' reflections highlighted a fairly favorable attitude to the assessment of the teaching practice: "*Teaching practice for me was a challenge to overcome myself, to understand my calling, to acquire new knowledge and courage, pedagogical confidence in myself.*" They feel the significance of developing personal teaching / learning competences (*"I understood that one of the most important things is that teachers have to work with themselves."*). There is no doubt that this positive experience motivates the students' desire to become teachers. According to Navaitienė (2001), the main incentive to develop professional purposefulness is the success of professional activity and the smooth process of revealing professional skills. For a trainee teacher this process is the teaching practice, conducted at a general education school, which should promote the search for information, create, investigate and analyze one's own experience.

On the other hand, the content of some students' reflections reflected a negative experience in the teaching practice, namely: fear, excitement, the lack of self-confidence, pedagogical skills and experience, the pupils' destructive behavior in the classroom, and the lack of cooperation and feedback. This shows that the students have insufficient teaching practice, skills to communicate and manage the pupils' behavior and discipline, and ability to define their own role in the school community. To some extent, this represents the "emotionally complex" and "tiring" pedagogical work at school and reflects the real situation of the modern teacher at a general education school (Barkauskaitė, Pečiuliauskienė, 2007).

Discussion

Most foreign researchers focus a lot on teachers' practical activity (*National System Overview on Education systems in Europe and Ongoing Reforms. Eurodyce*, 2011). In foreign countries universities and other teacher training institutions emphasize the significance of the teaching practice, develop the network of schools where it can be accomplished and so guarantee a continuous support for the inexperienced teachers. Teaching practice is considered to be the main part of pedagogical studies since it helps trainee teachers to participate in the life of school community.

The comparison of the results of the survey with the data of some foreign studies (Baskan, Yildiz, Tok, 2013; Fook, 2012; Güven, 2010; Li, Lowe, 2006, et al.) make it possible to state that during the teaching practice in different countries students experience similar positive feelings, acquire positive experience, and face similar difficulties while implementing the goals of their teaching practice. The results reveal the trainee teachers' attitude to the teaching practice at school, the prestige of the teacher's profession and its social status in different countries. The data of the aforementioned studies show that practical teacher training has to be enhanced and constantly developed. Moreover, teaching practice cannot limit itself to education technologies; it has to teach the methodology of pedagogical and personal self-education.

Conclusions

1.1 The analysis of the students' written reflections determined the following spheres of the teaching practice which enable the students acquire positive experience: 1) the sphere of the students' relationships with pupils, 2) the sphere of the students' relationships with the mentor and other teachers, 3) personal self-education and professional development.

1.2 *The students' positive experience of relationship with pupils* depends on humane interrelations (based on respect and trust), ability to interest pupils in the subject of music, efficient planning of the lesson, class management, giving and receiving feedback.

1.3 *The students' positive experience of relationships with the mentor and other teachers* depends on the close cooperation with the mentor and collegial relations with the other teachers at school, as well as in the students' self-confidence and the creation of a good-natured atmosphere for activity.

1.4 *Personal self-education and professional development* during the teaching practice depend on the students' ability to assess and develop their personal characteristics, acquire new experience and enhance the motivation of choosing the teacher's profession.

2. The teaching practice forms the students' attitude to the future educational activity. Their attitudes are determined by a number of factors (their wish to become a teacher, their subjective attitude to studies, their goals, results, and the internal processes which affect and constitute the studies). The analysis of the written reflections highlighted the significance of the pre-service music teachers' positive experience in the teaching practice, the change of their professional attitudes concerning their role in the school community. The teacher's activity was viewed by the trainee teachers as complex and responsible, requiring a good professional and psychological preparation, independence, reflective skills, personal maturity and the ability to perceive the value-related significance of the teacher's personality.

3. Even though some students are still having difficulty with meta-cognitive evaluation, the students' overall attitude to their activity and the assessment of the teaching practice at school was fairly favorable. Hence, the development of the organization of trainee teachers' teaching practice requires an emphasis on applying reflection in the study process as the basis for matching and integrating theoretical and practical knowledge and learning from one's own experience. It is purposeful to enrich the syllabi of other academic subjects with the means developing these skills.

References

1. Adomaitienė J., Zubrickienė I., Teresevičienė M. (2008). Profesinis mokytojų ir dėstytojų tobulėjimas, atliekant veiklos tyrimus (Professional Teacher Development in Performing Action Research). *Profesinis rengimas: tyrimai ir realijos*, No. 15, p. 10–22 (In Lithuanian).

2. Barkauskaitė M., Pečiuliauskienė P. (2009). Universitetinės pedagoginės praktikos turinio ir struktūros projektavimo konstruktyvistinės įžvalgos (Constructivist Insights on Content and Structure Projection in University Students' Teaching Practice). *Pedagogika*, vol. 96, p. 43–48.

3. Baskan, G. A., Yildiz E. P., Tok G. (2013). Teacher training system in Finland and comparisans related to Turkey. *Procedia - Social and Behavioral Sciences*, vol. 83, p. 1073–1076.

4. Bitinas B., Rupšienė L., Žydžiūnaitė V. (2008). Kokybinių tyrimų metodologija. Klaipėda: S. Jokužio leidykla-spaustuvė (In Lithuanian).

5. Black P. E., Plowright D. (2010). A Multi-Dimensional Model of Reflective Learning for Professional Development. *Reflective Practice*, vol. 11, (2), p. 245–258.

6. Bubnys R., Žydžiūnaitė V. (2010). Reflective Learning Models in the Context of Higher Education: Concept Analysis. *Problems of Education in the 21st Century* (Issues in Educational Research-2010), vol. 9, p. 58–70.

7. Davies E. A. (2005). Characterizing Productive Reflection among Pre-Service Elementary Teachers: Seeing what Matters. *Teaching and Teacher Education*, vol. 22, p. 281–301.

8. Fook, Ch.Y. (2012). Best Practices in Teaching in Higher Education in the United States: a case study. *Procedia - Social and Behavioral Sciences*, vol. 46, p. 4817–4821.

9. Girdzijauskienė R. (2008). Muzikos mokytojo profesinės kompetencijos ypatumai. *Tiltai*, Nr. 37, p. 5–12.

10. Güven S. (2010). Vocational Education Faculty Students' Impressions of Practice Teaching Schools and Teachers. *Procedia - Social and Behavioral Sciences*, vol. 2, p. 2501–2505.

11. Yip K. (2006). Self-reflection in Reflective Practice: A Note of Caution. *British Journal of Social Work*, vol. 39, p. 777–788.

12. Lasauskienė J. (2010). Muzikos mokytojo veikla ir kompetencijos (Activity and Competencies of Music Teacher). Vilnius: VPU leidykla.

13. Li Z., Lowe J. (2006). Practice Training in Initial Teacher Education: Case study comparisons in England and China. *Preparing Teachers for a Changing Context' Institute of Education*, 3-6 May, London, 2006.

14. Martišauskienė E. (2007). Studentų asmenybinio ugdymo(si) pedagoginių kompetencijų sklaida per pedagoginę praktiką (Dissemination of Students' Pedagogical Competences of Personality (Self-) Development During Teaching Practice). *Pedagogika*, vol. 86, p. 14–22.

15. *National System Overwiew on Education systems in Europe and Ongoing Reforms. Eurodyce, 2011.*

16. Navaitienė J. (2001). Būsimųjų mokytojų profesinio kryptingumo lygiai. *Ugdymo psichologija*, vol. 4, No, 1 (8), p. 20–25.

17. Kavaliauskienė V. (2001). *Pedagoginis pašaukimas ir jo ugdymas*. Klaipėda: Klaipėdos universiteto leidykla.

18. Pečiuliauskienė P. (2011). *Pedagoginės praktikos mokykloje vadovas*. Vilnius: Edukologija.

19. Pečiuliauskienė P. (2009). Pre-Service Teachers' Research Activity during the Pedagogical Pratice: Situation and its Development. *Quality of education and culture: Theory and Practice*, Baltische Studien zur Erziehungs-und Sozi Sozialwissenschaft. Frankfurt am Main; Berlin; Bern; Bruxelles; New York; Oxford; Wien: Peter Lang.

20. *Regulation of Teacher Training*, confirmed by the Minister of Education and Science of the Republic of Lithuania on May 15, 2012, Decree No. V-827 (Žin., 2012, Nr. 149-7654). Available from: http://www.smm.lt/teisine_baze.

21. Pollard A. (2008). *Reflective Teaching:* Evidence- informed Profesional Practice (3rd ed.). Continuum International Publishing Group. Ltd.

22. Reimer B. (2003). *A Philosophy of Music Education: Advancing the Vision* (3rd ed). NJ: Prentice Hall.

23. Rinkevičius Z., Rinkevičienė R. (2006). Žmogaus ugdymas muzika. Klaipėda: S. Jokužio leidykla-spaustuvė (In Lithuanian).

24. Rodzevičiūtė E. (2008). Pedagogo profesijos pasirinkimo motyvacija ir jos kaita pedagoginių studijų metu (Motivation to Choose Teaching as Profession and its Change During Pedagogical Studies). *Profesinis rengimas: tyrimai ir realijos (Vocational Education: research and reality)* vol. 15, p. 74–86.

25. Shaffer D. W. (2004). Pedagogical Praxis: The Professions as Models for Postindustrial Education. *Teachers College Record,* vol. 106, No. 7, p. 1401–1421.

26. Tavoras V. (2011). Būsimų muzikos mokytojų meninės individualybės raiška (Expression of Artistic Individuality of Future Music Teachers). Vilnius: Vilniaus pedagoginio universiteto leidykla.

27. *The Bologna Process 2020 –The European Higher Education Area in the new decade.* Communicate of the Conference of European Ministers Responsible for Higher Education, Leuven and Louvain-la-Neuve, 28-29 April 2009.

The Characteristics of Developing the Communicative Competence of Primary School Children

Giedrė Strakšienė

Abstract
The paper analyzes the characteristics of developing the communicative competence of primary school children: it validates the relevance of developing the communicative competence, presents the key communicative skills and reveals the nature of the development of these skills. The paper presents the results of the research which surveyed 318 primary school teachers and revealed the situation of developing the communicative competence in Lithuania. The article also emphasizes the teachers' opinion on the factors influencing the learners' communicative competence, the applied measures and the problems that arise in the process.

Keywords: communicative competence, the development of communicative competence, primary school children

Introduction

One of the priority goals in the modern society is the development of personal competences. The relevance of their development is highlighted in most of the European Union and Lithuanian education documents: A Common European Framework Reference, (2013); the Project of the State Strategy for Education of 2013-2022 (2013); the General Programmes of Primary and Basic Education of Lithuania (2008); the Methodological Recommendations for Primary Education (2009) etc.

The significance of developing the communicative competence in the primary school age has been recently increasingly emphasized. In ordcr to solve the problem of its development, it is important to determine the types of activity and teaching methods that would be effective to develop the communicative knowledge and skills of learners of this age and to form their value-related attitudes.

The General Programmes of Primary and Basic Education (2008) state that the integral character of the curriculum enables students to develop their communicative competence in an integrated way, i.e. in all the teaching subjects, taking into account the subject matter, the specificity of the field, the character of a specific topic or task. Lessons of all subjects should include the major requirements for communication "literacy" in the widest sense: linguistic culture, the norms of etiquette, the structure of an oral text, style etc.

Taking into consideration the fact that education is an active creative process, the learners' communicative competence is developed actively, through creative activity. The active process of education is guaranteed by a variety of methods. The variety of teaching and learning methods was revealed in the review of scientific literature. Different authors (Teresevičienė, Gedvilienė, 1999; Kardelienė, Gudzinevičiūtė, Mykolaitienė, 2007; Jovaiša, 2007; Gedvilienė, Zuzevičiūtė, 2007; Šaučiukėnienė, Stankevičienė, Čiužas, 2011 et al.) describe a lot of different teaching / learning methods and classify them in a variety of ways. Despite the number and variety of methods, there are two major trends: traditional/classical and modern/active methods. The authors agree that the application of the classical/traditional teaching methods cause a typically low level of physical and mental activity, and conversely the application of active teaching methods involves learners into active physical and mental activities, there appears a possibility to develop their communicative skills: express their own opinions, find out the opinions of the other members of the learning process and discuss.

The relevance of the development of the communicative competence highlights the need for a greater variety of education technologies (including methods) the application of which enable teachers to develop the learners' communicative competence. However, it has been noted that in the depth of the educational process one can still find such methods as teachers' explanation, questions-answers (Kardelis, Kardelienė, 2006). This allows an observation that the teacher has the right of talking/communication in the lesson, he/she focuses on the process of education rather than on the learners' individual experience. Bukantienė (2010) believes that practice at school shows that in the classroom the teacher initiates speaking, and learners are left with the role of a passive listener.

The development of the communicative competence of primary school children at school is organized on the basis of meaningful, active practice and the teacher's appropriate preparation for this activity. It is important for the teacher not only to select teaching methods correctly, but also to apply them in a flexible and creative way. Gedvilienė, Zuzevičiūtė (2007) emphasize the role of the teacher as the creator of a situation and organizer of activity in class. The authors state that the teacher has to select appropriate methodology which would help children to learn freely, creatively, without limitations; he/she has to match and apply the active teaching methods which would help to develop the learners' value-related attitudes in communication, would promote independence, critical thinking and the skill to apply the acquired information flexibly and creatively in the communicative activity.

A significant factor for developing the communicative competence is the communicative situations and their participants in the school environment. Research shows that in Lithuania the primary school teachers' verbal communication with children is insufficient because of their inability to set up communicative situations in class (Kardelis, Kardelienė, 2006; Bukantienė, 2010). The authors draw a conclusion that pedagogues find it difficult to create an atmosphere which is favourable for discussions, to lead discussions, and the issue of greatest concern is the ways of initiating and maintaining a conversation with learners. In the process of communication the teacher not only conveys the information related to the teaching subjects, but also forms the learners' attitudes to the correct use of the language, polite behavior while communicating. Teachers' working style is also important for the development of learners' communicative competence as it makes the process of education more attractive and enhances the interaction of teachers and students.

The educational function in the society is performed not only by teachers, but also by parents. Researchers (Jovaiša, 1995; Bitinas, 2001; Juodaitytė, 2003; Kvieskienė, 2005 et al.) state that the family conditions the formation of communication stereotypes which influence children's communicative skills. Gedvilienė, Zuzevičiūtė (2007) state that the effect of education within the family is ambivalent: 1) formal – when the parents' actions and words consciously and purposefully aim to affect the child's development; 2) informal – i.e. the family members' interrelations which condition a certain psychological climate in the family (communicative experience, the amount and quality of children's interaction with their family members). Children learn to communicate when parents read books to them, answer their questions or play with them. Communication in the family develops children's phonemic hearing, teaches the correct pronunciation of sounds, the correct use of cases, and enriches their vocabulary (Crain-Thoreson, Dahlin, Powell, 2001). The researchers draw a conclusion that children's communicative skills are best developed by the situations when children discuss some events in their life with their parents, by the social context and the circumstances of the described event (Kardelienė, Kardelis, 2006). Moreover, children's communicative competence is affected by various random educational factors of the environment: leisure activities at home (reading books, watching TV, playing computer games, listening to music records, etc.), or socialization with friends in the yard etc.

In the context of developing the communicative competences of primary school children the aspects of integrity and consistency of the curriculum, the organization of education as an active and creative process (selection and application of the active teaching methods), education as a cooperative process based on value-related attitudes. However, at the same time it is acknowledged that the existing system of education does not guarantee a high level of primary school children's communicative competence. The practical experience shows that only some teachers reach the desired results. The reasons can be different: the polysemy of the conception of communicative competence, the complexity of its assessment, the lack of teaching methods and teaching material, the lack of primary school teachers' knowledge and experience.

Hence, there rightly appear some questions: why are teachers having difficulty in the development of learners' communicative competence? Why do teachers not get the desired results while developing their learners' communicative competence? If pedagogues have insufficient understanding of the characteristics of developing learners' communicative competence or of the factors and conditions affecting them, it is hard to expect success. Obviously, in order to efficiently develop learners' communicative competence,it is important to reconsider the search for the most favourable educational environment for the expression of this competence.

The ***goal*** of the paper is to reveal the characteristics of the development of primary school children's communicative competence.

The object of the research is the development of primary school children's communicative competence.

The research methods are review of pedagogical, psychological literature and education documents, an interview of primary school teachers.

The Research Methodology

In the scientific literature there are a lot of works on the structure of a person's communicative competence. However, the author of this paper did not determine any research surveying the expression of communicative skills in the primary school age. Therefore, the study first of all prepared a theoretical model of primary school children's communicative competence (Table 1). On the basis of research on the communicative competence by different authors (Nauckūnaitė, 1998, 2000; Chreptavičienė, 1999, 2003; Baršauskienė, 2002; Cesevičiūtė, 2003, 2004; Grebliauskienė 2004; Baršauskienė, Janulevičiūtė-Ivaškevičienė, 2005;Kardelienė, Kardelis, 2006;Šernas, 2006; Adler, Proctor, 2007; Bukantienė, 2010; Madaule, 2010;Karaliūnas, 2008; et al.) there were the key indicators of the communicative competence distinguished: the comprehension of information, a fluent, clear speech, enriched vocabulary, attentive listening, a tactful interruption, expressive speech, expressive non-verbal language and value-related attitudes: a belief in one's abilities and the benevolence of others, attempts to help the interlocutor understand the information, respect for oneself and the interlocutor, the control of one's emotions, actions and words during a conversation, a wish to share one's ideas, experiences, a bold expression and defense of one's opinion.

Table 1: A Theoretical Model of Primary School Children's Communicative Competence

Components of communicative competence	Indicators of communicative competence	Characteristics of Communicative Competence Indicators
Socio-linguistic	Comprehension of information (arrangement and interpretation)	• Conceives the content of the intended message • Selects the necessary information • Gives the message an authentic meaning
	Coherent speech	• Composes appropriate word combinations and sentences • Expresses his ideas logically and consistently • Speaks concisely and precisely
	Rich vocabulary	• Uses various common language words • While speaking, selects emotional and expressive linguistic means • While speaking, selects words on the basis of the interlocutor's speaking style
	Clear speech	• Pronounces clearly • Speaks rhythmically • Accentuates correctly
Value-related attitudes		• Believes in his/her abilities and benevolence of others • Tries to help the interlocutor understand the information
Social	Attentive listening	• Retains attention for a longer time • Does not interrupt the interlocutor • Hears out without preconceived judgment
	Tactful interruption	• Waits for the right moment to interrupt • Uses polite phrases when interrupting • Uses non-verbal expressive means when interrupting
Value-related attitudes		• Respects him/herself and the interlocutor • Controls emotions, actions and words

Components of communicative competence	Indicators of communicative competence	Characteristics of Communicative Competence Indicators
Emotive-expressive	Expressive speech	• Speaks melodiously • Speaks lively and non-monotonously • Adjusts the speaking tone and loudness according to the circumstances
	Expressive non-verbal language	• Uses the possibilities of mimicry and look to convey the information • Uses the possibilities of pose and gestures to convey the information • Shows involvement
Value-related attitudes		• Willingly shares ideas and experiences • Is not afraid to express his/her opinion, defends it

The Survey of the Primary School Teachers

On the basis of the theoretical model of primary school children's communicative competence, there was a questionnaire for primary school teachers prepared which aimed to select information about how the teachers organize the development of learners' communicative competence: how much attention is given for its development (how frequently is it developed), what methods and how frequently they are used, what factors, in the teachers' opinion, influence the development of communicative competence, what problems arise.

The questionnaire consists of the following: the instruction which presents the purpose of the research and guarantees anonymity on the basis of research ethics principles, and 6 sections of questions (closed- and open-ended questions). The questions in the first section (closed-ended type) aim to determine how frequently learners' communicative competence is developed in lessons of different subjects (Lithuanian, mathematics, natural science, art, technologies and physical education); the questions in the second section (closed-ended type) helped to determine how frequently separate indicators of the communicative competence are developed (Table 1); the questions in the third section (closed-ended type) helped to establish the methods applied for developing the learners' communicative competence and the frequency of their use. For the purpose of developing the communicative competence there was an inventory of methods prepared on the basis of research conducted by Teresevičienė, Gedvilienė (1999); Šaučiukėnienė, Stankevičienė (2002), Kardelis, Kardelienė (2006), Kardelienės,

Gudzinevičiūtė, Mykolaitienė (2003), Jovaiša, (2007), Gedvilienė, Zuzevičiūtė (2009) and other authors. The questions in the fourth section (closed-ended type) were given to investigate the teachers' opinion about the significance of factors affecting their learners' communicative competence. The questions in the fifth and sixth sections (open-ended type) helped to highlight the problems the teachers face when developing the learners' communicative competence.

All the answers in the first, second and third groups included a *Likert*scale with five options of frequency: very frequently (every lesson), frequently (every second or third class), sometimes (every fourth or fifth lesson, depending on the situation, or on the topic of the lesson), very rarely (once a month, when there appear significant communication or linguistic mistakes), never (not applied yet). The answers to the questions in section 4 were listed in the order of significance: very important, important, neither important nor unimportant, unimportant, totally unimportant. The group of social-demographic variables consisted of the social-demographic characteristics of the respondents: gender, age, teaching experience, qualification category, the form the teacher teaches, and the location of the school.

The method applied was an anonymous written survey (via the internet). According to Kardelis (2002), this is a promising way of collecting sociological information, when there is a possibility to engage the respondents in the research, a well-planned survey can bring more answers to the questions, and the anonymity of the survey increases the feedback. Moreover, the choice of such a survey is determined by its access online and the possibilities to survey a relatively big number of respondents in quite a short time.

The survey of primary school teachers was conducted by publishing the questionnaire online and sending the internet address by e-mail to different schools in Lithuania. The authors of the survey aimed to cover as many general education and primary schools in as many different locations in Lithuania as possible. For that purpose the lists of general education and primary school were consulted, their contacts were corrected: the e-mails and contact telephone numbers of different schools. After sending informative e-mails-invitations with the attached website address of the questionnaire, the organizers of the survey also telephoned the heads of schools and their assistant principals responsible for primary education and asked to distribute the questionnaire among the primary school teachers.

The internet survey had 318 respondents – Lithuanian primary school teachers working in different educational institutions of Lithuania. The sample of the respondents is random. The respondents possess different qualification categories, different working experience, all the interviewed teachers were female (Table 2).

Table 2: The Characteristics of the Sample of the Surveyed Teachers (no.,%)

Qualification Category							
Teacher		**Senior teacher**		**Teacher-supervisor**		**Teacher-expert**	
94	29.56 %	119	37.42 %	80	25.16 %	25	7.86 %
Teaching Experience							
5–10 years		**11–20 years**		**21–30 years**		**30 and more years**	
45	14.15 %	100	31.45 %	117	36.79 %	56	17.61 %
The Form the Teachers Teach							
Form 1		**Form 2**		**Form 3**		**Form 4**	
59	18.455%	81	25.47 %	98	30.82 %	80	25.16 %
Age							
23–30 years old		**31–40 years old**		**41–50 years old**		**51 and older**	
21	6.60 %	145	45.60%	125	39.31%	27	8.49%
The Location of School							
Teachers working at schools in towns (cities and towns)				**Teachers working at village schools**			
277	87.1 %			41	12.9%		

The research data were summarized by means of descriptive statistics. The research data were accumulated and processed using *Microsoft Office Excel*software, the data were presented by means of systemization method (the percentage of the empirical data, grouping the answers) (Kardelis, 2002).

The Research Results

The results of the primary school teachers' reveal the characteristics of the development of primary school children's communicative competence in Lithuania. The analysis of the teachers' answers about the frequency of developing the communicative competence in lessons of different subjects (Table 3) showed that most teachers pay attention to the development of children's communicative competence, i.e. develop it very frequently or frequently in the lessons of the Lithuanian language and natural science.

Table 3: The Frequency of Developing Primary School Children's Communicative Competencein lessons of Different Subjects (no.,%)

The School Subject	Very frequently		Frequently		Sometimes		Very rarely		Never	
	no.	%	*no.*	%	*no.*	%	*no.*	%	*no.*	%
Lithuanian	288	90.57	30	9.43	0	0	0	0	0	0
Mathematics	21	6.60	22	6.92	262	82.39	13	4.09	0	0
Natural science	191	60.06	99	31.13	28	8.81	0	0	0	0
Art and Technologies	22	6.92	23	7.23	231	72.65	10	3.04	32	10.06
Physical Education	105	33.02	78	24.53	95	29.87	40	12.58	0	0

The teachers said that they develop the communicative competence very frequently in lessons of the Lithuanian language: as many as 90.57 % of the respondents indicated that they develop it in every lesson and only 9.43 % focus on that every second or third lesson. The lessons of Lithuanian, whose curriculum consists of reading, retelling and comprehending short texts of prose and poetry, dialogues, answering the questions after texts and discussions on different topics, are most frequently related to the development of the communicative competence. A slightly smaller percentage (60.06 %) of the respondents said that the learners' communicative competence is very frequently developed in lessons of natural science. However, only 6.60 % of the respondents develop the competence in lessons of mathematics. Having in mind the relatively big number of mathematics lessons per week and the respondents' answers, it is possible to assume that there is no focus on the development of the communicative competence in mathematics classes (only one sixth of the respondents said they do that very frequently and frequently). A similar ratio of the frequency of development was observed in art and technology lessons: the learners' communicative competence was very frequently developed by 6.92 % and frequently by 7.23 % of the respondents. Moreover, it was determined that quite frequently the learners' communicative competence was developed in physical education classes (very frequently – 33.02 %, frequently – 24.53 %). The survey did not ask how frequently teachers develop the communicative competence in lessons of ethics, religion, foreign languages, music, dance or drama. Not all primary school teachers conduct these classes, thus, the obtained data would have been unreliable.

The survey of the teachers showed how frequently pedagogues develop separate skills of primary school children's communicative competence (Table 4).

Table 4: The Frequency of Developing Primary School Children's Communicative Competence Skills (no., %)

Indicators of communicative competence	Very frequently		Frequently		Sometimes		Very rarely		Never	
	no.	*%*	*no.*	*%*	*no.*	*%*	*no.*	*%*	*no.*	*%*
Comprehension and interpretation of information	150	47.14	168	52.83	0	0	0	0	0	0
Fluency	213	66.98	105	33.02	0	0	0	0	0	0
Rich vocabulary	86	27.04	127	39.94	105	33.02	0	0	0	0
Clarity of speech	127	39.94	159	50.00	32	10.06	0	0	0	0
Attentive listening	175	55.03	127	39.94	16	5.03	0	0	0	0
Tactful interruption	128	40.25	150	47.17	40	12.58	0	0	0	0
Expressive speech	67	21.07	124	38.99	119	37.42	8	2.52	0	0
Expressive body language	51	16.04	104	32.70	141	44.34	22	6.92	0	0

The teachers' survey revealed that most frequently they develop the learners' fluency: two thirds of the teachers (66.98 %) do that very frequently. Slightly fewer respondents (55.03 %) indicated that very frequently they develop attentive listening. More than half of the respondents frequently develop primary school children's ability to comprehend and interpret information, and to speak clearly. The expressive speech and body language are developed noticeably less rarely: only 21.07 % of the teachers very frequently develop their students' expressive speech and 16.04 % - their expressive body language. Theatre educators (Nosevičiūtė, 1987; Prėskėnienė, 2004; Kazragytė, 2011 et al.) believe that clear and correct speech is insufficient. They note the significant role of intonation, logical pauses, the speaker's mimicry and body language in a conversation. Researchers note that the intonation of children's reading or speaking often lacks vitality, naturalness, use of mimicry, gestures or body language. Quite frequently there appear students who can read very fluently, but their reading speed is too quick, they read in a monotonous way, the speaker is frightened, stressed and constrained. Therefore, the development of learners' communicative competence requiresmore frequent attention not only to the reading technique (correct pronunciation, joining of sounds), but also to the verbal and non-verbal expression of reading.

The examination of the frequency of developing primary school children's value-related attitudes revealed the following results (Table 5).

Table 5: The Frequency of Developing Primary School Children's Value-Related Attitudes of the Communicative Competence (no, %)

Value-related attitudes of the communicative competence	Very frequently		Frequently		Sometimes		Very rarely		Never	
	no.	%	*no.*	%	*no.*	%.	*no.*	%.	*no.*	%.
Belief in one's own abilities and good-will of others	48	15.09	133	41.82	82	25.79	55	17.30	0	0
Attempts to help the interlocutor to understand the information	77	24.21	107	33.65	100	31.45	34	10.69	0	0
Respect for one's self and the interlocutor	95	29.87	179	56.29	44	13.84	0	0	0	0
Control of emotions, actions, words	49	15.41	109	34.28	117	36.79	43	13.52	0	0
Willing interchange of ideas, experience	53	16.66	102	32.08	120	37.74	43	13.52	0	0
Boldness to express one's opinion and defend it	31	9.75	132	41.51	92	28.93	63	19.81	0	0

The results showed that the teachers give insufficient attention for the development of value-related attitudes which are essential in the process of communication.Practically all the indicators of value-related attitudes are very frequently developed only by one sixth of teachers. In the context of value-related attitudes only the development of one value stands out, i.e. respect for one's self and the interlocutor. Teachers focus on this value more than on others: 29.87 % of the respondents said they developed it very frequently, and 52.29 % - frequently.

The results of the other data of the survey (Table 6) revealed the methods used most frequently by teachers to develop their learners' communicative competence. The results show that they quite frequently apply information and practical-operational methods: as many as 60.06 % of the respondents use the method of a revision discussion very frequently and 33.02 % use it frequently; reading practice is very frequently used by more than a half (52.83 %) of the respondents and 40.25 % use it frequently. The method of explanation is still quite popular – 48. 74 % of the respondents use it very frequently and 31.76 % use it frequently. The method of practical activity in the development of learners' communicative competence was used by slightly more than half of the interviewed teachers (50.94 %). The method of narration was very frequently used by slightly less than a half of the respondents (39.31 %) and half of the teachers use it frequently (50.63 %). It has been determined that the greatest percentage of primary school teachers (73.27 %) develop their learners' communicative competence by only sometimes using the method of informative conversation, and 16.98 % of the respondents apply this method very rarely. Creative methods are also rare – only one third of the respondents use them frequently, and a similar part of the interviewees state that they rarely apply them.

The application of the active teaching methods in the lesson is significantly rarer. One fifth of the respondents (21.38 %) very frequently use the method of group work and only slightly more than one tenth of the teachers use the method of interview (12 %). The survey revealed that one tenth of the teachers never used the methods of pair reading and contemplation and *Echo*. But more frequently they used the methods of expressive narration and role-play (12 % of the respondents use them very frequently), role-playing dialogues with puppets (15.72 % of the respondents said that they use them very frequently). The results of the research demonstrated that teachers do not take on roles themselves (94.97 %), or perform roles very rarely – 5.03 % of the respondents. There is a possibility that primary school teachers are not well acquainted with these methods and the possibilities of their application in the development of the communicative competence.

The analysis of the frequency of applying different methods for the development of primary school children's communicative competence revealed that primary school teachers more frequently use classical (traditional) teaching methods (e.g. explanation, revision, conversation, narration), and the application of modern active methods (discussion, talking out, consult-consider-share, improvisation etc.) is much rarer.

Table 6: The Frequency of Using Teaching Methods for the Development of Primary School Children's Communicative Competence (%)

Teaching methods	Very frequently	Frequently	Sometimes	Very rarely	Never
Classical teaching methods					
Information methods					
Informative					
Narration	16.98	62.89	20.13	0	0
Explanation	48.74	31.76	19.20	0	0
Informative conversation	0	9.75	73.27	16.98	0
Demonstration	5.66	69.50	19.21	5.03	0
Use of literature	20.13	49.37	20.75	9.75	0
Recovery					
Retelling	39.31	50.63	10.06	0	0
Revision discussion	60.06	33.02	6.92	0	0
Illustration	30.19	40.25	24.21	5.35	0
Practical-operationalmethods					
Reading practice	52.83	40.25	6.92	0	0
Practical activity	50.94	39.31	9.76	0	0
Creativemethods					
Heuristic					
Heuristic conversation	10.06	34.91	35.22	19.81	
Logical proof	9.75	25.16	24.21	20.75	20.13
Investigation	15.72	34.59	20.13	19.50	10.06
Exploration					
Observation	0	9.02	72.96	17.92	0
Exploratory conversation	7.23	25.79	53.14	13.84	0
Modernactiveteaching/learningmethods					
Group work / project activity	11.95	21.38	47.80	18.87	0
Pair reading	9.75	30.50	29.25	30.50	0
Reading and contemplation in pairs	5.35	28.30	30.50		10.6
Unfinished sentences	10.38	30.50	28.30	30.82	0
Consult, consider, share	4.72	29.87	9.75	50.00	5.66
Echo	10.06	14.78	37.74	27.67	9.75
Interview	12.26	26.42	52.83	8.49	0
I and others	5.03	16.67	60.06	18.24	0
Talking out	4.72	15.09	47.17	33.02	0
Buzzing bees/groups	0	8.49	72.01	19.50	0

Teaching methods	Very frequently	Frequently	Sometimes	Very rarely	Never
Expressive narration	11.95	14.15	28.30	37.11	8.49
Role-play	11.95	25.79	52.83	9.43	0
Linguistic games	8.49	10.06	75.16	6.29	0
Role-play games	7.86	37.11	35.22	19.81	0
Role-playing dialogue with puppets	15.72	48.43	31.13	31	4.72
Role-playing teacher	0	0	0	5.03	94.97

The survey of the teachers revealed the significance of the factors which affect the learners' communicative competence (Table 7). In teachers' opinion, the communicative competence of primary school children is mostly affected by the teacher's personal example of communication. It is important to remember that children from 7 to 14 years old search for an authority, i.e. a mediator between themselves and the world, and when they find such a person, trust him/her unconditionally and follow his/her example. Thus, it has to be noted that the teacher's personal example is very important for learners. Another important factor, in teachers' opinion, is the application of the active teaching methods in the process of school education and reading books at home (respectively, 81.13 % and 79.87 % of the respondents marked this as a very important factor). The other significant factors are: the parents' education, the amount of parent-child communication at home, a systemic development of the communicative competence in lessons of different subjects and appropriate textbook tasks for the development of the communicative competence.

Table 7: The FactorsAffecting the Development of Primary School Children's Communicative Competence (%)

Factors Affecting the Development of Communicative Competence	Very important	Important	Neither important nor unimportant	Unimportant	Totally unimportant
School environment					
The guidelines for developing the communicative competence in the general programmes of primary education	27.04	46.86	20.13	5.97	0

Sufficient number of textbook tasks and their appropriateness for the development of the communicative competence	36.79	50.0	13.21	0	0
Systemic development of the communicative competence in all subjects (integration)	56.60	37.11	6.29	0	0
Purposeful application of the active teaching methods for the development of the communicative competence	81.13	18.86	0	0	0
Organization of the education process on the basis of communication and cooperation	37.73	47.17	15.09	0	0
Teacher's personal example of communication	80.19	19.81	0	0	0
Family environment and children's leisure at home					
Succession of children's births	0	27.04	40.26	19.81	12.89
Amount of parent-child communication	66.98	33.02	0	0	0
Parents' education	39.94	53.14	6.92	0	0
Number of family members	0	12.89	6.67	7.23	13.21

Factors Affecting the Development of Communicative Competence	Very important	Important	Neither important nor unimportant	Unimportant	Totally unimportant
Reading books at home	79.87	20.13	0	0	0
Communication of the phone	0	0.94	64.78	26.73	7.55
Playing computer games	7.23	6.92	65.41	13.21	7.23
Watching TV	7.55	12.89	65.41	7.23	6.92
Listening to the radio	6.92	13.21	64.15	9.12	6.60

53.45 % of the respondents answered the question about the problems they face while developing the learners' communicative competence. The teachers' responses revealed the following problems: the volume of the curriculum and the lack of time for the development of the communicative competence (this was mentioned by 88.63 % of the interviewees); the problems of controlling the class when involving students in the communicative situations (22.14 %). Some other problems were also singled out: the lack of methodological literature and teaching aids for the development of the communicative competence, lack of teachers' communicative competence. It was also determined that teachers poorly assess their own creative abilities, they lack knowledge and experience to organize the process of education in the lesson. Only a small part of the respondents said that they have difficulty with the big amount of additional responsibilities, the social activity at school, the intensity of work with parents, administration of abundant documentation.

Conclusions

The results of the research showed that the learners' communicative competence is most frequently developed in the lessons of the Lithuanian language and natural sciences, and most rarely – in mathematics classes. Most frequently teachers focus on the development of the following components of the communicative competence: fluency and attentive listening. More than half of the respondents stated that in every lesson they develop the learners' clear speech and the ability to conceive and interpret information. Value-related attitudes, which are essential in the process of communication, do not receive sufficient attention from teachers. Some more focus is given to the development of one value-related attitude – respect for one's self and the interlocutor.

For the development of the communicative competence primary school teachers more frequently choose the classical (explanation, revision discussion, narration, reading practice) rather than the modern active methods (group work, consult-consider-share, brainstorming etc.). While developing the learners' communicative competence, teachers quite often apply informative (narration, explanation) and practical-operational methods (reading practice, practical activity, observation). The application of modern active teaching methods in the lesson is noticeably rarer. Only one fifth of the interviewed teachers very frequently use the method of group work and only one tenth – pair reading and contemplation. Teachers do not involve their students in active and creative activity frequently enough, the process of education lacks communicative situations. There is a possibility that the educational power of artistic subjects has not been sufficiently evaluated yet in the promotion of pedagogical innovations. The cognitive-creative activity (e.g. comprehension, imagining what we speak about) is an obligatory precondition for the development of learners' communicative abilities and its significance has been emphasized by both researchers and teachers. Piaget (2002) believes that primary school children possess such qualities as the joy of discovery, creativity, volatility of ideas, imagination and modeling of their own world, self-expression. It can be assumed that there are not enough possibilities for the learners to develop their communicative competence, and the child's creative potential is poorly exploited.

As it was determined from the teachers' answers to the question about the problems they face when developing the learners' communicative competence, primary school teachers develop this competence rarely because of the big requirements of the educational programme and the lack of time which is essential for the application of the active, creative teaching methods.

The teachers' opinions helped to determine the factors which influence the development of primary school children's communicative competence. The respondents said that the following factors are very important: the teacher's personal example, the application of the active teaching methods in the lesson and reading books at home. The least significant factor for the development of the communicative competence, in primary school teachers' opinion, is the succession of children's birth and the number of family members.

References

1. *A Common European Framework Reference.* Available from: http://www.coe.int/t/dg4/linguistic/source/framework_en.pdf [accessed on 15-09-2013].

2. Adler, R. B., Proctor II, R. F. (2007). *Looking Out. Looking In, Twelft Edition.* Belmont: Thomson Learning Inc.

3. Baršauskienė, V. (2002). Dalykinėkomunikacija. *Kaunas: Technologija.*

4. Baršauskienė, V., Janulevičiūtė-Ivaškevičienė, B. (2005). *Komunikacija: teorijairpraktika.* Kaunas: Technologija.

5. Bitinas, B. (2000). *Ugdymofilosofija.* Vilnius: Enciklopedija.

6. Bukantienė, J. (2010). *Mokiniųirmokytojųkomunikavimaslietuviųbendrinetartimi.* Monografija. Klaipėda: Klaipėdosuniversitetoleidykla.

7. Canale, M., Swain, M. (1980). Theoretical bases of Communicative approaches to second language teaching and testing. *Applied Linguistics, Vol. 1, p. 1–47. Retrieved June 27, 2012.*

8. Cecevičiūtė, I. (2003). *Studentųkomunikacinėskompetencijosugdymastikslinė jeedukacinėjeaplinkoje. PhD thesis. VDU.*

9. Cesevičiūtė, I. (2004). Specific Aspects of a Targeted Educational Environment for the Development of Students Communicative Competence at University: their Manifestations in the Dimensions of Context and Background. *Socialiniaimokslai,* No. 2 (44), p. 82–94. Kaunas: Technologija.

10. Chreptavičienė, V. (1999). Komunikaciniųmokėjimųdidaktinėssistemosgal imybiųplėtrataikantšiuolaikinesedukacinesirinformacinestechnologijas. *Socialiniaimokslai,* No. 4 (21), p. 30–38. Kaunas: Technologija.

11. Chreptavičienė, V. (2003). *Dalykinėkomunikacijaorganizacijoje.* Kaunas: Technologija.

12. Crain-Thoreson, C., Dahlin, M. P., Powell, T. A. (2001). Parent-Child Interaction in three Conversational Contexts: Variations in style and strategy. *New Directions for Child and Adolescent Development,* No. 92, p. 23–37.

13. Eisner, E. W. (2002). What can education learn from the arts about the practice of education? *The encyclopedia of informal education. Available from:* www.infed.org/biblio/eisner_arts_and_the_practice_or_education. htm[accessed on 11-08-2013].

14. Gedvilienė, G., Zuzevičiūtė, V. (2007). *Edukologija.* Kaunas: VytautoDidžiojouniversitetas.

15. Grebliauskienė, B., Večkienė, N. (2004). *Komunikacinėkompetencija. Komunikabilumougdymas*. Vilnius: Žara.

16. Grebliauskienė, B. (2006). Komunikaciniųmokėjimųstruktūrairturinys. Available from: http://www.leidykla.vu.lt[accessed on 17-09-2013].

17. Jackūnas, Ž. (2000). Meniniougdymovietaintegruotomokymosistemoje. Kn.: V. Matonis (ed.). *Šiuolaikinėsmeniniougdymokoncepcijos*. Vilnius: Enciklopedija.

18. Jakavonytė-Staškuvienė, D. (2009). *Tekstoskaitymaskaippradiniųklasiųmokiniųk omunikavimokompetencijosugdymopagrindas*. Klaipėda: Klaipėdosuniversitetas.

19. Jovaiša, L. (1995). *Hodegetika. Auklėjimomokslas*. Agora.

20. Jovaiša, L. (2007). *Enciklopedinisedukologijosžodynas*. Vilnius: Gimtasisžodis.

21. Juodaitytė, A. (2003). *Vaikystėsfenomenas: socialinis-edukacinisaspektas*. Šiauliai: Šiauliųuniversitetoleidykla.

22. Karaliūnas, S. (2008). *Kalbosvartojimasirsocialiniskontekstas*. Vilnius: Lietuviųkalbosinstitutas.

23. Kardelienė, L., Gudzinevičiūtė, O. L., Mykolaitienė, L. (2007). *Pradiniųklasių mokytojųkompetencijųplėtotė*. Šiauliai: VšĮŠiauliųuniversitetoleidykla.

24. Kardelis, K. (2002). *Moksliniųtyrimųmetodologijairmetodai*. Kaunas: JUDEX.

25. Kardelis, K., Kardelienė, L. (2006). *Pedagoginėkomunikacijakūnokultūrosvei kloje*. Monograph. Kaunas: Lietuvoskūnokultūrosakademija.

26. Kazragytė, V. (2011). *Mažųjųvaidinimai. Literatūriniaitekstaiirkomentarai*. Vilnius: Tyto Alba.

27. Kievišas, J., Sližauskaitė, J., Aleksandravičiūtė, V. (1996). Žaidimųpobūdži oužduotyspamokoje. *Pradinisugdymas*. Šiauliai: ŠP institutoPedagogikosfakultetas.

28. Kvieskienė, G. (2005). *Pozityvioji socializacija*. Vilnius: VPU Leidykla, 2005, ISBN 9955-20-021-9, 184 p.

29. *The General Programmes of Lithuanian General Education Schools and the Standards of Education. Pres-school, primary and basic education.* (2008). Vilnius.

30. Madaule, P. (2010). *Kai klausymasatgyja. Veiksmingomokymosiirbendravimovadovas.* Vilnius: Vertėjųbiuras „Magistrai".

31. Nauckūnaitė, Z. (1998). Sakytinėskalbosugdymoproblemos. *Ugdymoproblemos.* Mokslodarbai, p. 26-39.

32. Nauckūnaitė, Z. (2000). Diskursokomponentaskomunikacinėskompetencij osstruktūroje. *Žmogusiržodis,* T. 2, Nr. 1, p. 79–84.

33. Nosevičiūtė, S. (1987). *Scenoskalba.* Kaunas: Šviesa.

34. Piaget, J. (2002). *Vaikokalbairmąstymas.* Vilnius: Aidai.

35. *The Methodological Recommendations of Primary Education* (2009). Available from: http://www.upc.smm.lt/ugdymas/pradinis/rekomendacijos/ [accessed on 17-09-2013].

36. Prėskėnienė, E. (2004). *Garso, žodžio, sakinio, tekstoraiška.* Šiauliai: ŠiaurėsLietuva.

37. Šaučiukėnienė, L., Stankevičienė, N. (2002). *Bendrosiosdidaktikospagrindai.* Kaunas: Technologija.

38. Šaučiukėnienė, L., Stankevičienė, N., Čiužas, R. (2011). *Didaktikosteorijairpraktika.* Kaunas: Technologija.

39. Šernas, V. (2006). Komunikacinėskompetencijos, jųugdymo(si) kaikuriosst rateginėsirtaktinėsproblemos. *Santalka. Filologija. Edukologija,* T. 14, No. 4, p. 52–53.

40. Teresevičienė, M., Gedvilienė, G. (1999). *Mokymasisbendradarbiaujant.* Vilnius: Garnelis.

41. The Provisions of the State Strategy of Education of 2003–2013. Available from: www.smm.lt/svietimo_bukle/docs/1562_Svietimo_strategija% 202012.pdf. [accessed on 15-09-2013].

The Vision of Pre-Service Music Teachers' Personal Career

Asta Rauduvaitė, Daiva Ramanauskienė

Abstract
The paper discusses the vision of pre-service music teachers' personal career in the context of vocational decision making and choices. On the basis of research results the paper analyzes different aspects of the process of students' choice of their career which help to consolidate their professional motivation and create the vision of their personal career.

***Keywords*:** personal career, vocational decision making

Introduction

The rapid development of social and cultural life alters school, the teacher's role, the general attitude to the teacher's profession. The modern creative society is complex, dynamic and technologically advanced. The challenge of the modern society is the constant change of knowledge. The dynamic development changes people's attitude to knowledge and skills. Hence, teacher training institutions acquire a responsible mission to prepare competent teachers able to work in the context of change and the school of the future (Teacher Training Regulation, 2012). The modern teacher cannot gain knowledge for the lifetime, it is important to have the ability to search for knowledge and the possibilities of its application in professional activity, to identify problems and solve them, to have a good command of technologies and the ability to use them in pedagogical practice, to know languages and other cultures, to be able to successfully work in a group, to work both individually and in teams, and to be able to be loyal.

The teacher's profession and professional career today relate to the global and local phenomena of the environment as well as to the internal changes of school as organization. The choice of profession and planning of one's personal career is a responsible process which has a great influence on a person's well-being and his/her future visions. As it is known, the choice of a profession and the realization of one's abilities is one of the main elements of personal self-realization. A successful career is only possible when an individual fulfills him/herself, when his/her profession answers the person's interests and likes (Kirkpatrick, Johnson, Mortimer, 2002).

Nowadays researchers increasingly focus on the period between leaving school and entering the labour market. This is the time when young people choose their future profession, begin their studies and accumulate experience. At the beginning of the 20th century there appeared the first theories on vocational decision making, the choice of profession and career planning which were developed by Super (1985), Levinson (1986), Holland (1996) et al. The authors acknowledge the fact that career planning is the ongoing personal attempts to envisage the tendencies of labour market changes and to plan and implement the processes of professional activity improvement in the context of those changes. Super (1985) was the first to use and justify the notion of the term *career* (lat. *carraria* – a person's road of life). He defined career as a whole of a person's roles in life, his positions at work and certain stages of activity. The modern personal, individual model of career has the following characteristics: an active personal position, a strategy of continuous improvement, professional flexibility, mobility.

Today a personal career is conceived as a personal achievement, a profession, a continuation of steady work, as a consistent follow-up of life experience of various roles. For trainee teachers of music, who aim to control the process of their careers, it is essential to take into consideration the psychological, social economic and cultural factors. They have to master modern spheres of competences and apply them to search for information, create, explore, and reflect on their professional experience etc. For a teacher of music this reveals new career possibilities, but at the same there appears a demand for more effort. According to Pukelis (2003), a personal career is the result of a rational choice of profession and career planning. Apart from the personal characteristics, the author emphasizes satisfaction (will the chosen profession bring joy, answer the interests etc.) and the labour market requests (the possible perspectives etc.). When an individual plans a personal and professional career, he/she participates not in a single, but rather in several processes at the same time – fulfillment, development and making decisions.

On the basis of the aforementioned theories it is possible to state that a young person, who is nurturing a dream to become a teacher of music, evaluates his/her music abilities early, plans his/her studies, and visualizes their results. Different foreign researchers (Mayrhofer, Meyer, Iellatchitch, Schiffinger, 2004; Cochran-Smith, 2005; Gelissen, Graaf, 2006; Blasio, 2013; etc.), who investigated the issues of vocational decision making and career, note that people constantly observe themselves and evaluate their own activity. Successfully accomplished tasks and higher achievements can encourage choosing a profession which requires similar skills and abilities. Past experience (successful musical education, active participation in musical cultural life, performing), character qualities necessary for pedagogical activity (self-discipline, diligence, sociability, creativity, flexibility etc.) and personal expectations (conception of the value-related significance of the teacher's personality). Hence, the process of a trainee music teacher's personal career is multifold and varied, and requires a deeper analysis and cognition.

It is assumed that a young person has to responsibly assess, plan and manage his/her personal career, aim for autonomy which is possible only if one is well aware of his/her personality, if one can embed his/her independence, take on responsibility and conceive the significance of continuous development.

Lithuanian researchers have analyzed the questions of teacher training, discussed the issues of pedagogical career planning and pedagogical motivation. A lot of attention has been given to teachers' professional development; there were studies on the expectations of pre-service teachers and teachers who have just started their professional career, on the tendencies of teachers' employment (Dzemyda, Šaltenis, 2001; Merkys, 2006), on the motives of beginning teachers' activity (Monkevičienė, Autukevičienė, 2008; Rodzevičiūtė, 2008; Pečiuliauskienė, 2011; et al.). There has been a discussion on the development of teachers' value-related world and the directions of improvement (Barkauskaitė, 2006; Aramavičiūtė, Martišauskienė, 2009; Martišauskienė, Tavoras, 2012; et al.), as well as on the issues of music teachers' professional training (Piličiauskas, 1998; Lasauskienė, 2010; Balčytis, 2012; et al.). However, the questions of planning the personal career of music teachers have not been analyzed in greater detail. Thus, it is necessary to investigate what incentives determine the personal career vision of trainee music teachers on their road to vocational decision making.

The object of the research is the vision of trainee music teachers' personal career.

The goal of the research is to determine the factors which affect the vision of trainee music teachers' personal career in the context of vocational decision making.

The research methods: scientific literature review, a thematic essay, a notional interpretative analysis of the research data.

The research methodology

With the aim to determine the factors affecting the vision of trainee music teachers' personal career, the methodology of qualitative research was applied. The study employed the *Life* method (Bitinas, 2002), *a thematic essay* "The Beginning of My Professional Road". The essays were analyzed using the method of content analysis. The analysis of the essays revealed respective categories and subcategories whose notional contexts were based on the authentic ideas of the research participants. The research group consisted of 17 students of the second and third years of Music Education from Panevėžys College and 27 students of the third and fourth years of Music Education from Lithuanian University of Educational Sciences. The choice was determined by the fact that these two schools of higher education have long traditions of training music teachers. The research did not show any significant differences between the two research groups, hence, the data were not compared.

The research results

The analysis of the essays written by the students of Music Education revealed several key concepts – planning a profession, the assessment of students' personalities, the factors of vocational decision making, a vision of a personal career.

Career planning is revealed in the overview of the period of the beginning of students' career which starts with dreaming about the future profession of a music teacher, creating the vision of their profession, imagining what should a music teacher be, and forming close professional relations (Table 1). This period is called by Levinson (1986) as the apprentice stage at the beginning of one's career. It is likely that a successful solution of problems in this stage can determine success in the further career of a trainee music teacher.

Table 1: Career Planning

Categories	No. of statements	Examples of notional contexts-statements
Creating a dream	28	"I think I will work with talented children who love music (…)" "I have a dream, but I can't exactly say what I really want yet" "I want to open a music studio, find similarly-minded people"
Creating a professional vision	15	"At school it is important to spread musical culture, and develop a child by means of music" "The purpose of our chosen profession is to give the art of music to people" "Nobody can do things which can be done by a teacher of music (…)"
Creating a teacher's professional image	19	"I want to be just like my teacher of music, she is an example of this profession." "A teacher has to be an exceptional personality, since she has to teach others how to love music." "A music teacher has to be able to play well, sing and understand children."

Forming professional relations	27	"It is important for me to have a possibility to reveal myself at work so that my colleagues accept me for who I am." "Trust among colleagues is an essential condition for mutual work (...)" "I know some music teachers who will always help me. I can trust them."

A purposeful activity is possible when the needs are conceived. Kuliutkinas and Suchobskaja (1972) note that in vocational decision making one of the most significant moments to envisage a certain desired state, orient oneself to the future and wait for it. Then it is possible to imagine, foresee and plan one's activity. The research revealed that students create professional dreams and implement them, project their future and create the vision of their profession, observe and master the teacher's roles, create and develop professional connections. In the students' dreams of their chosen profession, the category of *Creating dreams* reveals the students' wish to work at school, to open a music studio or a private school, work with talented children etc. The professional vision in the essays is not limited to the formulation of their activity goals, it is related to the teacher's mission, the perception of the value-related significance of his/her personality ("*to develop the traditions of musical culture*", "*to develop children by means of music*", "*concern about cultural education and rendering of values*", "*to give the art of music to people*" etc.). Hence, in the preparation for pedagogical activity a trainee teacher has to understand who he/she really is, perceive him/herself from the social, cultural, qualifying, interest- and value-related standpoints (Kavaliauskienė, 2001).

The essays revealed that the students' professional and personal vision is formed by their previous positive experience and impressions. The analysis showed that a teacher of music is frequently the greatest authority, advisor and guardian who inspired the students choose this profession. It confirms the idea that the teacher's system of values has a special social role since teachers affect others by their own example. In the majority of cases there was a teacher who was looked up to. Trainee music teachers identify themselves with the educators who impressed them most: "*I want to be just like my teacher (...)*"; "*A teacher has to be an exceptional personality (...)*" etc. The essay analysis showed that at the beginning of the professional road the teacher's role is idealized, perfection is the primary goal. Some statements present a doubt whether such an ideal teacher can really exist. The analysis of the essays reveals that this topic is interrelated with the issue of professional relations whose emphasis shows that trainee music teachers find it important to have similarly-minded people in their professional environment as they could understand and support, help at the beginning of their professional road, and would approve of their professional goals. Trainee music teachers also want their professional parts to believe in their abilities and support them. This prompts a young person's self-confidence, enhances his/her self-worth and helps to solve professional problems.

When students dream of their professional future, most of them note that it is important for them to have a possibility to reveal themselves and be their true selves. Students value the atmosphere of trust and cooperation, and believe that more experienced teachers can help the young specialists who are just starting their professional career. When choosing one or another profession, every person has to be aware of his/her character qualities which could help to reach for the target professional goals.

The assessment of students' personality is revealed by highlighting their key character qualities, achievements, aspirations, recognition or disappointment after confrontation with professional difficulties (Table 2).

Table 2: The Assessment of Students' Personality

Categories	Subcategories	No. of statements	Examples of notional contexts – statements
Self-assessment	Character qualities	23	"I am responsible, diligent, punctual, dutiful (...)" "I can state that I am sociable, I like interaction"
	Achievements	31	"I can sing, play and I have a musical feeling" "I have innate musicality (…)" "I can do a lot of things, especially, if they are related to creation (…)"
	Aspirations	11	"I firmly know what I seek and know that I can do that" "(…) I want to be unique, the best of all" "I always strive for perfection in everything I do (…)"
	Recognition	14	"My students love me and I love them" "I am good at socializing. Other people tell me that (…)" "I think that I have chosen what I know best – music"
	Disappointment	8	"I don't have big dreams, I don't expect a miracle (…)" "(…) I feel disappointed, I am afraid of wishing…" "Sometimes it feels that I can't do anything, I just no good at anything"

The modern school aims to educate a person, to develop his/her human feelings, aesthetic taste, and to instill a sense of duty and responsibility. This idea makes it obvious that the research participants' expectations of their personal career are directly related to their self-conception. This is categorized as *Self-assessment.* When students speak about their key character qualities, at the same time they show what is most important to them in this stage, what character qualities they value most. This fact supports the idea that the teacher's personal values merge with his/her professional needs and values, they become inseparable and have influence on pedagogical activity (Kavaliauskienė, 2001). A lot of the essays emphasized diligence, responsibility, ability to socialize and interact. It is likely that students consider these qualities necessary to establish themselves in professional activity and acquire the necessary competences.

In this category the major attention is given to personal abilities, creativity, musicality, ability to play or sing etc. The notional context of aspirations is related to desires, the inner wish to be exceptional, seek for professional heights, experience of completeness. Trainee teachers tend to emphasize the evaluative categories, have ambitions, and yet it is fully clear to what degree their desires and possibilities correspond. In some essays the dreams of a bright future are accompanied by fear of uncertainty or suspicion of the possible future difficulties. The following statements prove that: *"I don't have big dreams and I do not expect a miracle"; "(...) I am afraid of desires"; "Sometimes it seems that I can't do anything and know nothing"* etc. Hence, it is possible to assume that during their studies students confront ambiguous feelings – their ambitious desires are followed by the first disappointments, self-confidence – by doubts of their abilities which appear facing some difficulties. Sometimes students lack confidence in their abilities, and their motivation to reach for the target goals can start decreasing as soon as they familiarize with the peculiarities of the profession.

In summary it is possible to state that the students' personal qualities, which reveal themselves on their vocational decision making road, define their relation with the environment. For example, diligence reveals their relation with the professional activity, their need to work and express their true "self" (the motive of creativity), punctuality – their attitude to the norms of behavior established in the society etc. At the same time character qualities encourage trainee teachers to act respectively – to finish their job, aim for professional perfection, meet the determined requirements, socialize with people etc.

The factors of vocational decision making in the essays are closely related to the professional motivation. It is possible to see how students assess their personal professional possibilities, the peculiarities of the market, the significance of family beliefs, some social factors, as well as what professional consciousness the students possess (Table 3).

Table 3: The Factors of Vocational Decision Making

Categories	Subcategories	No. of statements	Examples of notional contexts - statements
Professional motivation	Professional possibilities	19	"I knew nothing about this job (…) I loved music" "I thought I knew everything, but now I see I was wrong" "I don't know if I will work at school, but I will have a profession"
	Perception of the market peculiarities	23	"Now it is difficult to establish oneself, especially at school (…)" "I don't know if I get a job, but I would like to be a teacher and will try to find a position"
	Family attitudes	17	"My mother is a music teacher, I had no doubts" "My parents were surprised that I entered here, but now they are calm"
	Social factors	15	"I love communication, and this gives a lot of possibilities for that" "The instructors are good, the fellow students are good, there is a lot of self-expression"
	Professional consciousness	18	"I have always known that this is my profession (…)" "I have chosen my vocation (…)" "I am best at what I am doing now"

The analysis of the factors of vocational decision making demonstrated that the research participants chose their studies because they loved music, wanted to connect their future activity with music or were encouraged by their teachers to acquire this profession. The family members, friends, their interests or vocations were also a big influence. When assessing the market peculiarities, the students noted that *"now it is difficult to establish oneself"*, yet they say that their decision to be a music teacher is firm as *"they chose what they wanted"* etc. Obviously, before choosing a profession, students do not consider where they will work after graduation; they focus on emotional experience rather than on conscious planning of their professional career. During their studies students start thinking about the future prospects. Hence, the world of music and the pedagogical process of music are very specific, they distinguish themselves by the cognition of the artistic-spiritual curriculum of music, by the focus of the basic values, personal sense and the variety of meanings (Rinkevičius, Rinkevičienė 2006).

Moreover, the research revealed that students do not know much about the market conditions and tendencies. They do not know how to look for work, or who could help in the sphere. Nevertheless, despite the aforementioned doubts, they are not disappointed about their studies, and positively assess the study process, instructors and fellow students.

A personal career vision is developed under the influence of a variety of incentives. A young person's relation to the desired profession is defined by the motives that prompt this decision, by their domination. In the essays the students' ***personal career vision*** revealed itself in the consideration of professional resources, the presupposed professional roles and emotional state.

Most students are convinced that the period of studying is the time when they gather experience – they develop the basic skills, their thinking patterns and improve in the chosen professional sphere. The essay analysis shows that during their studies students find it important *"determine their working style"*, *"creatively realize their ideas in practical activity"*, *"maintain sustainable relationships between teachers and learners"*. The essays also included such aspects of career vision as the multiple professional roles – those of an assistant, counselor, friend, teacher, a close person, etc. It is interesting to note that some students said that the profession of a music teacher is *"female, beautiful"*, that there are more women working as music teachers than men, however, men can have their advantages – they are *"stricter and more interesting for children"* (Table 4).

Table 4: The Factors Determining Personal Career Vision

Categories	Subcategories	No. of statements	Examples of notional contexts – statements
Professional resources	Basic skills	10	"One has to find his own working style and be creative (...)" "The cognition of culture, knowledge of music, skills and pedagogical abilities are essential for a teacher"
	Thinking abilities	13	"Creative thinking helps to find your way" "I've learned to improve and understood the limits of my possibilities"
	Professional improvement	29	"At school one can test himself and develop (...)" "I strive for perfection, cognize my profession and move forward!" "It is important to create and sustain harmonious relations between the teachers and learners"

Categories	Subcategories	No. of statements	Examples of notional contexts – statements
Professional roles	Multiple roles	16	"When you become want you want to be, you can help others" "I want to become children's friend (…) and to be there when they need me" "A teacher is like a mother who doesn't stop loving even if she is angry. I will always love them and will try to be a second mother to them"
	Professional segregation	5	"That's a very female and spiritual profession" "There are more female teachers, but a man can be stricter and more interesting for children (…)"
Psychological state	Satisfaction	29	"I am studying what I wanted and I will reach for my dream (…)" "My expectations were fulfilled 100 percent. I am happy" "I like my choice, in three years I see myself as a teachers"
	Dissatisfaction	10	"It is so irritating that no matter how you try there is always somebody better than you (…)" "Children are wild these days, they are not interested in what you say…" "Nobody is to blame. There too few possibilities in Lithuania, too little work"

The research revealed that a significant part of the research participants are glad about their personal decision, feel happy and believe that they will work in the chosen sphere, i.e. their personal vision reflects high professional aspirations. However, sometimes they feel dissatisfied when they understand that their attempts do not always guarantee the best results. This reason appears due to their inability to satisfy their needs, realize goals and overcome difficulties. Thus, during study years there appears a possibility to realistically assess the professional resources and the limits of their personal abilities. In other words, the success and failure experienced at the beginning of pedagogical activity determine the further professional road of a teacher (Pečiuliauskienė, 2011). The teaching practice is a period when students make their first steps on the professional road, when they form their attitude to the teacher's work.

The research results make it possible to state that during the study years the process of planning a personal career continues, students assess their individual abilities, form professional goals and create a personal vision of the future.

Discussion. Most foreign countries pay a lot of attention to a successful personal career planning and vocational decision making. The results of the conducted research show that in comparison with the data acquired by foreign researchers (Watt, Richardson, Wilkins, 2008; Vilhjálmsdóttir, Arnkelsson, 2013; Veldman, Tartwijk, Brekelmans, Wubbels, 2013; et al.) it is possible to state that different countries face practically identical peculiarities of defining a personal career at its beginning with a focus on the vision of a personal career. The aforementioned research data show that trainee teachers relate their vocational decision making with their prior positive experience, individual skills, the teacher's authority, encouragement and support of the correct decisions, and professional aspirations. Teaching practice is considered to be an integral and the most important part of pedagogical studies focusing on the teacher's personality from the standpoint of his/her maturity, value-related commitment and competent professionalism.

Conclusions

1. The research demonstrated that trainee music teachers' vision of personal career is directly related to the planning of professional activity, the assessment of students' personality, the factors determining their vocational decision and the creation of a personal career vision.

1.1 *Career projection* is connected with the creation of a dream, professional vision, the teacher's professional image, and the formation of professional connections.

1.2 *The students' personality assessment (self-assessment)* is inseparable from character qualities, achievements, aspirations, recognition, and disappointment.

1.3 *The factors of vocational decision making* are closely related to professional motivation; professional possibilities, perception of the market peculiarities, family attitudes, social factors, professional consciousness.

1.4 *The vision of a personal career* reveals itself with the emphasis of professional resources (basic skills, thinking abilities, professional improvement), professional roles (their multiplicity, professional segregation), and psychological state (satisfaction, dissatisfaction).

2. The research revealed that trainee music teachers believe that searching for the meaning of life is especially important and they relate that search with the creation of a professional vision, the goals of their professional activity, and the teacher's mission. The research participants positively assess the study process and are convinced that the years of studies are the period of

gathering experience – when basic skills are formed, thinking abilities are developed, and trainee teachers consistently improve in their professional sphere. During the study years the process of projecting the trainee music teachers' personal career continues, the expectations of the beginning of a personal career are matched with personal musical and pedagogical abilities, and the latter are adjusted to the professional requirements. If the process of musical studies is efficiently organized, it can help to meet the student' needs for self-expression and self-development, and promote the value-related meaningfulness of music teachers' activity.

3. The research results showed that at the beginning of their studies, most students have high aspirations, want to acquire the necessary professional skills, reach for professional perfection. However, their experience acquired during teaching practice and the more realistically assessed professional career possibilities help the students create a more realistic vision of their personal career.

References

1. Aramavičiūtė V., Martišauskienė E. Vertybės: švietimo strategija ir realybė// Pedagogika. – Vol. 95 (2009), p. 28-34.

2. Balčytis E. Muzikinio ugdymo labirintais. – Šiauliai: ŠU leidykla, 2012.

3. Barkauskaitė M. Suaugusiųjų pedagogikos gairės. – Vilnius: VPU leidykla, 2006.

4. Bitinas B. Pedagoginės diagnostikos pagrindai. – Vilnius: Parama, 2002.

5. Blasio B. On the Role of Life Stories in Vocational Choice // Procedia - Social and Behavioral Sciences. – Vol. 82, No. 3 July (2013), p. 487-491.

6. Cochran-Smith M. Studying Teacher Education. What we know and need to know // *Journal of teacher education. – Vol. 56, No. 4 (2005), p. 20-25.*

7. Gelissen J., Graaf P. M. Personality, social background, and occupational career success // Social Science Research. – Vol. 35, No. 3 (2006), p. 702-726.

8. Holland J. L. Exploring careers with a typology: What we have learned and some new directions // American Psychologist. – Vol. 51, No. 4 (1996), p. 397–406.

9. Kavaliauskienė V. *Pedagoginis pašaukimas ir jo ugdymas.* - Klaipėda: Klaipėdos universiteto leidykla, 2001.

10. Kirkpatrick, M., Johnson, J., & Mortimer, J. T. Sociological Perspective to Career choice and development. In D. Brown & Associates (Eds.), *Career choice and development* (4th ed.). - San Francisco: Jossey-Bass, 2002.

11. Lasauskienė J. Muzikos mokytojo veikla ir kompetencijos. - Vilnius: VPU leidykla, 2010.

12. Levinson D. J. A conception of adult development // American Psichologist. - Vol. 43, No.1 (1986), p. 3-13.

13. Lietuvos pedagogų duomenų bazės kokybinis laidavimas, tyrimas (regioninė specifika) ir pedagogų poreikio prognozė. Auth. G. Dzemyda, V. Šaltenis et al. - Vilnius, 2001. Available from: < http://science.mii.lt/pedagogai>.

14. Mayrhofer W., Meyer M., Iellatchitch A.,, Schiffinger M. Career and human resource management a European perspective // Human Resource Management Review. - Vol. 14, No. 4 (2004), p. 473-498.

15. Martišauskienė E., Tavoras V. Mokytojų vertybinės orientacijos kaip jų meninės individualybės formavimosi veiksnys. Acta Pedagogica Vilnensia. Vol. 28 (2012), p. 73-82.

16. Mokytojų poreikio prognozė iki 2015 m. Auth. G. Merkys. - Kaunas, 2006, Available form: <http://*smm.lt/svietimo_bukle/docs/tyrimai/ppp_2015_ataskaita*>.

17. Monkevičienė O., Autukevičienė B. Pradedančiojo pedagogo profesinės raidos etapai // Pedagogika. - Vol. 91 (2008), p. 27–34.

18. Pečiuliauskienė P. Pedagoginės praktikos mokykloje vadovas. - Vilnius: Edukologija, 2011.

19. *Teacher Training Regulations,* confirmed by the Minister of Science and Education of the Republic of Lithuania on May 15, 2012 by the decree No. V-827 (Žin., 2012, No. 149-7654). Available from: http://www.smm.lt/teisine_baze.

20. Piličiauskas A. Muzikos pažinimas. 2-oji knyga. Stadijos, rezultatai ir reikšmė. - Vilnius: LAMUC, 1998.

21. Pukelis K. Karjeros projektavimo gebėjimai žinių visuomenėje: nauji iššūkiai profesiniam konsultavimui ir karjeros planavimui // Profesinis rengimas: tyrimai ir realijos. – Vol. 6 (2003), p. 66-76.

22. Rinkevičius Z., Rinkevičienė R. Žmogaus ugdymas muzika. – Klaipėda: S. Jokužio leidykla, 2006.

23. Rodzevičiūtė E. Pedagogo profesijos pasirinkimo motyvacija ir jos kaita pedagoginių studijų metu // Profesinis rengimas: tyrimai ir realijos. – Vol. 15 (2008), p. 74–85.

24. Super D. E. Coming of age in Middletown: Careers in the making // American Psychologist. – Vol. 40, No. 4 (1985), p. 405-414.

25. Veldman I., Tartwijk J., Brekelmans M., Wubbels T. Job satisfaction and teacher –student relationships across the teaching career: Four case studies // Teaching and Teacher Education. – Vol. 32 (2013), p. 55-65.

26. Vilhjálmsdóttir G., Arnkelsson G. B.. Social aspects of career choice from the perspective of habitus theory // Journal of Vocational Behavior. – Vol. 83, Iss. 3 (2013), p. 581-590.

27. Watt H. M.G., Richardson P. W. Motivations, perceptions, and aspirations concerning teaching as a career for different types of beginning teachers // Learning and Instruction. – Vol. 18, Iss. 5 (2008), p. 408-428.

28. Кулюткин Ю. Н., Сухобская Г.С. Мотивация познавательной деятельности (сбоник научных трудов). – Ленинрад, 1972.

Social Aspects of Gender Representation in Efl Textbooks

Evelina Brusokaitė, Daiva Verikaitė

Abstract
The aim of the present research was to determine social gender representation patterns in the Lithuanian, British and Argentinian EFL textbooks. To achieve the aim, EFL textbooks circulated in Lithuania (*Super English 4*), the UK (*Speakout*) and Argentina (*My Life 2*) were analyzed with regard to social aspects of gender representation. The methods chosen for the study were content analysis and the calculation of the relative frequency of occurrence. The results revealed that the Lithuanian EFL textbook depicts females in a stereotyped way with regard to household duties. However, females are visible in the field of IT which is treated as a male specific attribute in the other two analysed textbooks. The content analysis of illustrations in the British EFL textbook revealed that males dominate the labour market whereas females are engaged only in the artistic occupations as dancers, models or singers. Furthermore, in the Argentinian EFL textbook females are excluded from the field of IT; however, they are not associated with the domestic sphere. In contrast to Lithuanian EFL textbook, females are presented lying on the beach, sunbathing, and watching TV. Further studies can examine other aspects of gender representation and cover wider range of EFL textbooks.

Keywords: gender representation, gender-awareness, gender-bias, visibility, gender roles and gender attributes

Introduction

Gender is the object of interdisciplinary investigation. It is an important area of study in sociology, sociolinguistics, psychology and linguistics. Sociologists claim that the perception of gender is developed through the observation and is predetermined by "gender agents" (Brym and Lie, 2007, 97). Teaching material presented in textbooks is one of the mostly influential "gender agents". Gender representation in teaching materials reflects the attitude of a particular society towards gender that is subconsciously acquired by learners and may cause long term drawbacks on their performances and social behaviour.

The EFL textbooks are the best source for investigation of gender representation as they are designed to reflect the most common and prevailing features of gender representation in a society. The studies made in the field of gender representation in EFL textbooks (Foroutan, 2012; Mustapha 2012; Kim 2012; Amini & Birjandi 2012; Jean & Yuit 2012; Nagatomo 2010; Hamdan 2010; Johansson & Malmsjo 2009; Kobia 2009; Skliar 2007; Lee & Collins 2006) revealed that females are underrepresented and trivialized in relation to males. The results demonstrated that regardless of culture and religion, social activities, family roles, occupations and attributes ascribed to women and men are stereotyped.

Taking into consideration the previous studies made in the field of gender representation, we can construct a ***hypothesis*** that EFL textbooks circulated in different regions of the world that have different social, cultural and economic background possess different patterns of gender representation.

Therefore, the ***aim*** of the present research was to analyze Lithuanian, British and Argentinian EFL textbooks with regard to gender representation in order to determine the recurrent patterns of gender representation from the social perspective. To achieve the aim of the research, the following ***objectives*** were set:

- To analyse and compare the ways of gender representation in different cultural settings.
- To analyze gender-biased use of visual items.
- To determine the relative frequency of the occurrence of gender-biased items in the analysed textbooks.

Quantitative and qualitative approaches were used for the analysis of the items under the investigation. The content analysis method was employed for the analysis. The major categories analysed in the textbooks were visual materials with respect to gender representation.

All the analysed instances were classified under the three main categories: gender-marked, gender-balanced and gender-neutral.

The relative frequency of the occurrence of gender-marked instances was calculated and the comparative analysis of the results with respect to gender-biased gender representation in Lithuanian, British and Argentinian EFL textbooks was performed.

To achieve the above set objectives, three EFL textbooks were subjected to analysis, i.e. *Super English 4*, *Speakout* and *My Life 2*. *Super English 4* is offered for the market in Lithuania. It was published by five Lithuanian and one foreign author in 2009 and is one of the newest publications of EFL textbooks in Lithuania. The textbook consists of two parts and is aimed at students studying in the eighth form. The textbook comprises twenty-eight units covering topics from festivals to special occasions and from free time interests to law and crime. The textbook consists of one hundred and fifty one pages. *Speakout pre-intermediate* is the British EFL textbook published by *Longman* in 2011. *Speakout* is a comprehensive six-level general English course for adults. The textbook consists of twelve units which include the most essential topics covering issues related with everyday life, nature and society. The textbook consists of one hundred and seventy five pages. *My Life 2* is the Argentinian EFL textbook published by two Argentinian authors in 2009. The textbook is originally based on the textbooks published by *Pearson*. The textbook consists of the student's book and a workbook. *My Life 2* comprises six units of the student's book and six coinciding units of the workbook. The units cover six main topics such as free time, meals, past events, travel, choices and big events. The textbook consists of ninety-five pages.

In the analysis the British EFL textbook is referred to as "Textbook A", the Argentinian EFL textbook is referred to as "Textbook B", the Lithuanian EFL textbook is referred to as "Textbook C".

The selection of the textbooks was made randomly; however, the textbooks from different regions were chosen intentionally in order to compare the treatment of gender in the textbooks involving different cultural perspectives. These textbooks are recent publications therefore they provide an up-to-date data with regard to gender representation. The overall analysis comprised four hundred and twenty-one pages of the text.

The analysis of the textbooks with regard to gender representation in EFL textbooks is one of the newest and widely examined issues. The results of the study will contribute to the development of gender-awareness of the teaching process participants, authors and publishers of EFL textbooks. Further studies can examine other aspects of gender representation and cover wider range of EFL textbooks.

Results

Visibility

As defined by Stockdale (2006, 1), visibility is the number of occurrences of male and female illustrations in textbooks. Amini and Birjandi (2012, 137) define visibility as one of the main gender bias manifestations in the textbooks which is also referred to as omission. To specify, visibility is the number of occurrences of persons representing genders in visual material of textbooks which may reveal the omission of one gender in preference to another.

Visibility may reflect the authors' ability to design gender fair visual content in order to create well balanced teaching materials. The number of pictures illustrating gender directly influences the overall results of gender representation. To be precise, if the number of pictures representing one gender is greater than that of representing another, naturally, the analysis of particular aspect of gender representation will provide results in favour of one gender and that gender will have an advantage over the other.

The analysis of visibility in the Lithuanian, British and Argentinian textbooks helped us measure the equality in gender representation. The analysis was based on the examination of gender-marked pictures or illustrations. The pictures including both genders were not taken into consideration. Analyzed pictures included authentic photographs and illustrations. Moreover, in order to investigate the relation between visibility in teaching materials' content and the possible influence made by the authors, the gender of authors was also taken into consideration.

The analysis was conducted regarding two aspects: whether pictures depict male or female clear and whether pictures or illustrations depicting male or female characters had equal frequency of the occurrence in Lithuanian, British and Argentinian textbooks. The analysis revealed that males outnumbered females in representation in the three analyzed textbooks (*see Figure 1*).

Figure 1: The gender visibility in Lithuanian, British and Argentinian EFL textbooks

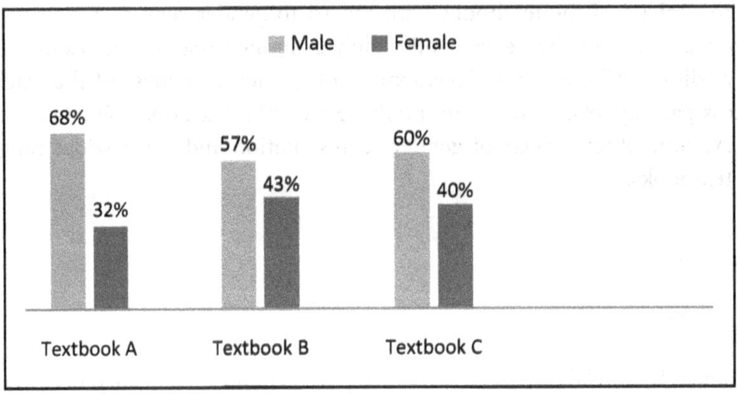

As seen in Figure 1, all three textbooks exhibited a prevailing number of male illustrations. The difference between female and male representation was significant, the majority of illustrations depicted males, therefore it could be concluded that one gender, viz. male gender, was given preference to female gender.

It might be surprising that the British EFL textbook possessed the most substantial difference in gender visibility. Similarly to the British EFL textbook, in the Lithuanian ELF textbook males outnumbered females in illustrations. However, the Argentinian EFL textbook was the best balanced with regard to gender visibility in comparison with other two textbooks.

Previous studies on gender representation in EFL textbooks demonstrated that the deviation in favour of one gender might be associated with the gender of authors as textbooks are designed through the feminine or masculine perspective. However, taking into consideration the gender of the authors (eight females and one male) of the analyzed textbooks, it can be stated that gender of the authors did not influence gender representation in textbooks, i.e. female authors did not promote feminine perspective.

There are two possible explanations for this. Firstly, the content of the textbooks is selected by the authors which means that the reality reflected in the textbooks is the reality perceived by the authors. As a result, if the authors themselves view gender in a stereotypical way, they are not able to promote gender equally in textbooks. Secondly, authors are more concerned about the subject matter rather than issues of "secondary importance" such as gender-awareness and gender-balanced presentation of gender in the teaching materials. However, the visual material makes the biggest influence on the subconscious learning which highly influences a learner's perception. So while designing a textbook it is of vital importance to take into account gender-awareness issues as it is an integral part of learning.

Evaluating the acquired results, we can conclude that males and females are not seen as equal members of the society; it is obviously reflected in the analyzed EFL textbooks which are designed in favour of males with regard to gender visibility. Even though textbooks are recently published, the content still needs to be improved and balanced. Moreover, it is important to raise authors' gender-awareness and to emphasize the importance of gender-balanced representation of gender in teaching materials.

Gender Attributes

Another aspect of social gender representation is gender attributes. As described by Mustapha (2012, 257), attributes ascribed to genders may reflect the social reality or can be very close to the real life of people at a given time. So the analysis of the attributes ascribed to genders in Lithuanian, British and Argentinian textbooks may reveal the attitude towards males and females in the present societies of different cultures.

Analysis of the quantitative results of character involvement in certain activities demonstrated that the difference between male and female involvement was not significant in all three analyzed textbooks; males were only slightly more actively involved in activities in comparison to females (57% and 43%, respectively). However, the type of activities females and males were engaged in was different. Males were the most visible in sports, computing and studying, however, they were absolutely not associated with shopping whereas females were the most visible in sports, computing and speaking on the phone (*see Table 1*).

Even though the relative frequency of occurrence of pictures representing male and female characters demonstrated that both genders are engaged in sports and computing, the analysis of separate instances revealed that males were twice as often depicted engaged in sports and studying as females whereas females were depicted six times more often engaged in shopping (*see Table 1*). Therefore, it can be stated that sports is regarded as a male specific attribute whereas shopping is regarded as a female specific attribute.

Table 1: The distribution between activities with regard to gender

	Males	Females
Computing	14	11
Sports	37	15
Shopping	0	6
Speaking on the phone	9	9
Crime	6	2
Studying	14	8
Travelling	9	6

The analysis of the textbooks published in three different regions i.e. Western Europe, Eastern Europe and Latin America revealed some similar attributes that are ascribed to men and women (*see Table 2*).

Table 2: The distribution between activities with regard to gender in different regions

	Textbook A		Textbook B		Textbook C	
	M	F	M	F	M	F
Computing	5	7	7	1	2	3
Sports	17	3	15	11	5	1
Shopping	0	4	0	2	0	0
Speaking on the phone	8	4	1	5	0	0
Crime	4	0	0	0	2	2
Studying	9	5	1	0	3	3
Travelling	4	2	5	2	0	2

The first feature that is evident in all three textbooks is that male and female characters are engaged in using a computer. In the Argentinian EFL textbook the field of computing is associated mainly with males. The number of instances showing a male character computing is the greatest among all three analyzed textbooks (*see Table 2*). The findings suggest that Argentinian EFL textbook predominantly associates the activity of computing with males. However, the British and Lithuanian EFL textbooks demonstrated that the activity of computing is gender-unmarked, i.e. the activity of using a computer is depicted by presenting both males and females. In addition, it is interesting to note that the British EFL textbook depicts an elderly woman using a computer which suggests that IT is not associated with the representation of age group and a particular gender. Moreover, in the Lithuanian EFL textbook there are pictures presenting situations when both genders are engaged in using a computer, nonetheless, females are those who are typing or in other active way using a computer whereas males are sitting nearby and watching.

Furthermore, the results of analysis concerning gender engagement in sports activities demonstrated that in all three textbooks sport is regarded as a male specific attribute. In all three analyzed textbooks males outnumbered females in the activity of doing some sport. For example, in the British EFL textbook males were engaged in sports activities five times more often than females. Similarly in the Lithuanian EFL textbook there is one case of a female character presented engaged in sport, i.e. cycling. In contrast, males are engaged in team sports such as basketball, cricket and football. Moreover, male characters are predominantly involved in other sports such as squash, badminton, tennis, rollerblading, windsurfing, surfing and jogging. Despite that, mental games such as chess are also presented as being engaged in by male characters. However, the findings are more favourable to females in the Argentinian textbook. Even though males outnumber females in the pictures representing certain sports (*see Table 2*), females are engaged in more various sports such as football, surfing, rollerblading, tennis and cycling. From the results of the three analyzed textbooks it can be stated that females are associated as being involved in yoga and gymnastics as no males were depicted in these sports, respectively, basketball is regarded as a male specific sport.

In addition, the analysis of the relation between gender and crime revealed some unfavourable results to males. Males are shown as criminals, thieves and burglars. Moreover, they are depicted stealing, breaking in, shooting and being arrested. In the British EFL textbook the crime vocabulary is presented portraying male characters whereas females are not presented involved in crime. Contrarily, in the Lithuanian EFL textbook females appear visible under the topic of "wrongdoing", however, males are involved in more serious crimes such as damaging road signs and stealing whereas females are plagiarizing material during the test. Even though males are presented as criminals, in the Lithuanian EFL textbook there is an example of male being a victim of bullying. In addition, among those who are bullying both genders are depicted.

Furthermore, the aspect of travelling was associated mainly with men during the previous studies made in the gender representation field; presenting males going abroad and being "outside" the homes. However, the results obtained during the present study of the Lithuanian, British and Argentinian textbooks demonstrated that both genders are mobile and able to travel on their own. The analyzed pictures presenting characters travelling mainly included both genders which suggest that the activity of travelling is gender-bias free.

Similarly, the relative frequency of the occurrence of a gender being involved in studying is well balanced in all the three textbooks (*see Table 2*). Both genders are engaged in reading books or doing other tasks for learning purposes. The Lithuanian EFL textbook stands out as having a considerable number of authentic pictures presenting both genders in classrooms having a lesson or doing group activities as well as oral presentations. Nevertheless, in the British EFL textbook the gender-marked attitude towards the subjects in relation to gender exits. Information technologies (IT), physics, mathematics and physical education (PE) are presented with pictures showing only males. In contrast, females are presented studying such subjects as biology, history and drama. The results do not reflect the reality where both genders are present in different fields of study. Naturally, it may create the stereotypical perception of gender abilities and capabilities.

Free time activities can also contribute to the overall analysis of social gender representation in the Lithuanian, British and Argentinian EFL textbooks. For example, the activity of barbequing is associated mainly with males. Even though the whole family or a group of friends are spending time together and having a barbeque, only men are shown grilling meat or sausages. The findings reveal the attitude that barbequing is males' responsibility.

Moreover, the activity of playing computer games was mainly associated with males in the previous studies of gender representation. The investigation of the Argentinian and British EFL textbooks confirmed the results of the other studies, i.e. males are those who are keen on playing computer games. The results of the present study demonstrated that males preferred to play such sport games as tennis and football. However, in the Lithuanian EFL textbooks there was one case showing female character engaged in computer games. The analysis of the Lithuanian EFL textbook demonstrated that the activity of playing computer games was gender-unmarked.

Furthermore, the most evident attribute ascribed to females in all three analyzed textbooks was the activity of shopping. Cases of males doing some shopping were not detected in the analyzed textbooks (*see Table 2*). Women were the ones that are presented doing shopping. It is worth mentioning that in the Argentinian EFL textbooks in the chapter where innovative shopping like, e-commerce was presented, a woman was the one who was engaged in such the new way of buying items. Even though computing in the Argentinian textbook was presented as a male specific field, in the case of shopping online a female was the one who used a computer.

Another significant aspect in the social gender representation was the interdependency between gender and the activity of speaking on the phone. Even though the results showed balanced presentation of gender with regard to mobile phone use (*see Table 2*), the content analysis of the situations revealed that males spoke on the mobile phone for business whereas females spoke on the mobile phone for leisure. In addition, in the Argentinian EFL textbooks females were portrayed multi-tasking, i.e. they were speaking on the phone and using a computer, speaking on the phone and watching TV or speaking on the phone and unpacking luggage. Moreover, women were depicted speaking on the phone as the way of spending their leisure time, doing it for relaxation or speaking loudly in the public places which annoyed other people.

Moreover, there were several cases of culture-specific attributes ascribed to genders. For example, collecting mushrooms was the activity that was depicted exclusively in the Lithuanian EFL textbook as a free time activity done by a male character. It is a culture-specific activity typical of Lithuanians. To add, in the Argentinian EFL textbook there were a considerable number of pictures representing female and male characters having a pet. The most frequently displayed pets were cats, dogs and rabbits. The importance of pets was expressed in the feature of presenting them together with people in the family trees.

To sum up, the following patterns of gender perception were reflected in the analyzed textbooks: the most frequent attribute associated with males was sports and the attribute associated with female was shopping. It is interesting that the activity of speaking on the mobile phone was performed by genders with different purposes. All other attributes with regard to gender were well balanced in the analyzed textbooks and could not be treated as biased.

Gender Roles and Occupations

For the analysis of gender role representation in the textbooks, family roles and other household duties were taken into consideration. The Lithuanian EFL textbook stood out from the other two in the high frequency of females engaged in household activities (*see Figure 2*): cleaning, ironing, cooking, serving a table, and washing dishes. Females were doing housework, looking after children, serving table, cooking and frequently wearing an apron. The results demonstrated that the British EFL textbook possessed well balanced attitude towards gender with regard to household duties. Moreover, there were no cases of male or female characters being involved in household duties in the Argentinian EFL textbook (*see Figure 2*).

Figure 2: Household activities with regard to gender

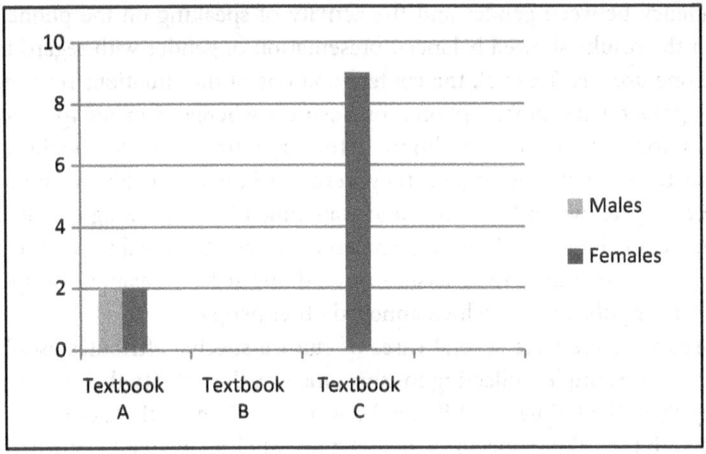

The Lithuanian EFL textbook possessed rather stereotypical and gender-marked representation of female roles. Females were attached to home and were involved in such household activities as ironing, cooking and washing dishes. In one of the illustrations a female character was presented sweating while washing dishes which reflected the attitude that doing housework is not an easy activity and that it takes all energy out of the doer. Despite the difficulty, these tasks were ascribed to and associated with women. In contrast, in the British EFL textbook a male character was presented multi-tasking and being involved in washing dishes, ironing and looking after a child which firmly denies the stereotype that only females are attached to homes or household chores. In addition, in the Argentinian EFL textbook females were not depicted in any household activities.

An apron is an attribute stereotypically ascribed to females in the Lithuanian EFL textbook. The total number of pictures displaying female characters was twenty-five; out of these twenty-five illustrations three presented women wearing an apron which made up 12% of all illustrations. It could be treated as a gender specific attribute reflecting a stereotypical view of females.

Furthermore, in the Lithuanian EFL textbook females were presented as responsible for children's nurture. Females were depicted reading books for a child or educating them. In addition, in the British EFL textbook a female character was shouting at her children and trying to control them as they were misbehaving in the shop. Interestingly, there were no cases of males being involved in the nurture of children in the three analyzed textbooks.

Contrarily, the analysis of the British and Argentinian EFL textbooks demonstrated that females were not attached to home, the household chores were not stereotypically ascribed to them and they were visible in public sphere. In the British EFL textbook in the unit called "changes" females were the leading characters. The stories told about women who changed their lives from ordinary and secure to challenging and requiring courage. For example, one female character changed her occupation from being a doctor to a professional musician, and the other left her boring job and decided to travel around the world. Female characters were presented as self-confident, self-assured and free to make their own choices.

Additionally, the analysis of the Argentinian EFL textbook revealed that females were frequently presented as having a rest. Females were lying on the coach, comfortably sitting and watching TV, sunbathing, lying on the beach or having dance lessons. These significant findings which were not identified in the other two textbooks, demonstrated the attitude that women in Argentina might not be associated with household duties. In contrast, they were depicted enjoying their free time and doing the activities they liked or were interested in.

The investigation of the distribution of occupations with regard to gender focused on two aspects, i.e. the relative frequency of illustrations depicting gender in certain occupations (which helped to measure the visibility of genders in the field of labour) and the analysis of occupations ascribed to males and females (which allowed to define the gendered occupations).

In the three analyzed textbooks there were eighty-five cases of illustrations depicting males and females engaged in certain occupations. The British EFL textbook possessed the greatest number of illustrations (57) in comparison with the Argentinian (19) and Lithuanian (10) EFL textbooks which showed that the labour market is an important topic in the British EFL textbook.

Moreover, the relative frequency of pictures depicting males and females involved in certain occupations revealed that males dominated the labour market. In the Lithuanian EFL textbook males were depicted in certain occupation twice as often as females. In addition, in the British EFL textbook males were visible in certain occupation almost three times more often in comparison with females. However, the most male-centered textbook with regard to the distribution between occupations with regard to gender was the Argentinian EFL textbook as males were presented almost four times more often than females.

Furthermore, males were involved in a greater variety of occupations than females. The occupations presented in the textbooks could be grouped under the headings: manual work, risky work, IT, travelling, the mass media, occupations requiring a degree, sport industry and artistic occupations. Males were engaged in the occupations under all headings whereas females were the most frequently engaged in the artistic occupations. For example, in the British EFL textbook female characters were depicted as *fashion designers*, *sculptors*, *singers* and *models*. Similarly in the Argentinian EFL textbook females were presented as *singers*, *actors* and *artists*.

Moreover, in the British EFL textbook there was the greatest number of illustrations depicting males engaged in various occupations. For instance, males were depicted doing manual work as *plumbers*, *electricians*, *cleaners* and *drivers*. Moreover, occupations requiring an advanced degree such as *doctors*, *lawyers*, *architects* and *vets* were also ascribed to males. It is worth mentioning that the presentation of males in risky occupations such as *rescues*, *firefighters* and *police officers* predominated in all the three textbooks. In addition, in the Argentinian EFL textbook males were associated with the catering industry. Males were depicted as *waiters*, *chefs* and *restaurant administrators*.

Among the three analyzed textbooks males were depicted as *police officers*, *football* or *basketball players*, *TV* or *radio presenters* and *actors*. From the acquired results it could be stated that the prevailing groups of occupations ascribed to males in the Lithuanian, British and Argentinian EFL textbooks were risky and artistic occupations and occupations related with the sports industry and mass media.

Furthermore, the rather positive finding was that an occupation of *a teacher* was depicted using a male character which was not a common feature in the previous studies. Additionally, in the Argentinian EFL textbook a female character was depicted as *a professional football player* which suggests that females are visible in the sport industry.

However, the stereotypically ascribed occupation of being *a housewife* which was very common in the previous studies still existed in the analyzed textbooks. In the British EFL textbook a word *housewife* was being taught as an occupational word presented using a female character for illustrations. A positive finding was that being a housewife was treated as a separate occupation which is not only an obligation of being a wife. However, the representation of the occupation revealed the existing stereotypical attitude that only females can be housewives and proves the attachment of females to the domestic sphere.

Conclusions

The perception of gender is developed through observation and is predetermined by "gender agents" among which the mostly influential are textbooks. Stereotypical gender representation and gender-biased use of language in textbooks may cause long term drawbacks on students' performances and social behaviour. The studies made in the field of gender representation in EFL textbooks revealed that females are underrepresented and trivialized in relation to males. The results of the present study confirmed the results of the previous studies and demonstrated that gender roles, attributes and occupations ascribed to male and female characters are stereotyped.

Lithuanian, British and Argentinian EFL textbooks are not designed to promote gender-bias, however, certain stereotypical attitudes towards gender that can be interpreted as being based on culture and traditions are evident. The values and norms of the particular society influence stereotypical representation of gender in EFL textbooks.

- The Lithuanian EFL textbook presents a rather stereotypical attitude towards females with regard to gender roles. Females are associated with household duties such as cooking, washing dishes, ironing, serving a table that are perceived as stereotypical female activities. In addition, an apron is treated as a female specific attribute since in a significant number of illustrations representing females they are wearing an apron.
- In the British EFL textbook females are underrepresented in the labour market since they are involved only in the artistic occupations such as *actors, models* and *designers* whereas males are engaged in a much wider range of occupations including manual occupations, risky occupations, IT, travelling, mass media, occupations requiring a degree and sport industry.
- The Argentinian EFL textbook excludes women from the field of IT. The use of the word "techies" referring to males prove that IT is treated as a male-specific attitude. Moreover, the results of quantitative data revealed that males dominate the labour market; they are depicted three times more often engaged in various occupations in comparison to females. However, females are not associated with the domestic sphere – they are not depicted engaged in any household activity; they are presented as *relaxing*.

- The comparative analysis of the results of the present study demonstrated that the analyzed textbooks exhibit the following similar patterns of gender representation:
  - Males outnumber females in illustrations.
  - *Sport* is regarded as a male specific attribute whereas *shopping* is regarded as a female specific attribute.
  - Males speak on the phone for business; females speak on the phone for leisure.
  - Males dominate the labour market.
- Attributes that are stereotypically ascribed to genders suggest that textbook authors lack gender-awareness and that they are more concerned about the subject rather than about issues of "secondary importance" such as gender-awareness and gender-balanced presentation of gender in the teaching materials.

References

Amini, M. & Birjandi, P. Gender Bias in the Iranian High School EFL Textbooks, *English Language Teaching*, 5/2: 134 – 147. 2012.

Brym, R.J. & Lie, J. *Sociology Your Compass for a New World*. The USA: Thomson Wadsworth. 2007.

Foroutan Y. Gender representation in school textbooks in Iran: The place of languages. *Current Sociology*, 60/6: 771 – 787. 2012.

Hamdan, S. English-language Textbooks Reflect Gender Bias: A Case Study in Jordan, *Advances in Gender and Education*, 2: 22 – 26. 2010.

Jean, L. H., Yuit, C.M. *Gender Representation in the Malaysian Fifth Form English Language Text Book*. 2012, accessed 7 March 2013, available from www.fbmk.upm.edu.my/micollac/proceedings

Johansson, S. & Malmsjo, K.B. *Gender Bias in ELF Textbook Dialogues*. BA paper. Malmö högskola. 2009.

Kim, H. Social and Cultural Issues in Some EFL Textbooks in Korea, *Hawaii Pacific University TESOL Working Paper Series*, 10: 30-39. 2012.

Kobia, J.M. Femininity and Masculinity in English Primary School Textbooks in Kenya, *The International Journal of Language Society and Culture*, 28: 57 – 71. 2009.

Lee, J.F.K. & Collins, P. *Gender Representation in Hong Kong English Textbooks.* 2006, accessed 18 March 2013, available from http://www.eoc.org.hk/eoc/upload/2006711112336211184.pdf

Mustapha, A.S. Dynamics of Gender Representations in Learning Materials, *Multidisciplinary Journal of Gender Studies*, 1/3: 243 – 270. 2012.

Mustapha, S. *Gender Equality in and through Education in Nigeria: Gender Representation in Learning Materials.* 2012, accessed 14 March 2013, available from http://shura.shu.ac.uk/5503/

Nagatomo, D.H. 2010. Critical Analysis of Gender Representation in an EFL Textbook, *Journal of the Ochanomizu University English Society*, 1: 53 - 61.

Skliar, O.S. *Gender Representations and Gender Bias in ELT Textbooks Published in the Middle East: A Case Study of ELT Textbooks Published in Turkey and Iran.* MA thesis. Middle East Technical University. 2007.

Stockdale, D.A. *Gender Representation in an EFL Textbook.* MA paper. The UK: The School of Humanities of the University of Birmingham. 2006.

Sources

Clare, A. & Wilson, J.J. *Speakout: Pre-intermediate.* Longman: Pearson Education Limited. 2011. 176 p. (Textbook A).

Rabinovich, P. & Dumas, A. *My Life 2.* Pearson Education S.A. 2009. 95 p. (Textbook B).

Rupainienė, V., Baskakovienė, B., Shaw, S., Liutkuvienė, R., & Šneiderienė, I. *Super English 4: Part 1.* Kaunas: Šviesa. 2009. 74 p. (Textbook C).

Rupainienė, V., Baskakovienė, B., Shaw, S., Liutkuvienė, R., & Šneiderienė, I. *Super English 4: Part 2.* Kaunas: Šviesa. 2009. 78 p. (Textbook C).

Bildungswissenschaftliche Aspekte Bei Der Motivationssteigerung Der Litauischen Hochleistungssportler Im Kontext Der Spe Zifischen Nationalidentitätsmerkmale

Sniegina Poteliūnienė, Stanislav Sabaliauskas

Abstract
The article provides analysis on educational aspects which make influence on athletes' motivation for elite sport mastership, together with the highlight on the role of a coach as on an educational factor. Impact of a coach's managed educational environment on the athlete's motivation at the stage of his elite sport mastership is also analyzed. Research instrument - haft structured interview. For analysis of the research data, method of grounded theory research was chosen. The research embraced 13 Lithuanian elite athletes - Lithuanian Olympic team members and candidates, as well as 9 coaches, training elite athletes - Lithuanian Olympic team members and reserve team. The research disclosed the interrelation context of Lithuanian elite athletes and their coaches with its characteristic features of particular factors under the influence of sociocultural context, which has both reinforcing and hardening influence on athletes' motivation. The authors of this article aim to reveal sport's role in disclosing person's identity development, which is evidenced and being formed in certain social and educational environment.

Einleitung

Einen wichtigen Bestandteil der sozialen Identität eines jeden Menschen stellt die nationale Identität dar, die sich im Prozess der Sozialisation einer Person im Kontakt mit anderen Menschen, Gruppen, Völkern entwickelt und formt. Der Sport als spezifischer von der Gesellschaft anerkannter und sichtbarer Bereich der Kultur übt den Einfluß auf die Entwicklung und Formung der Identität des Sportlers aus - das Selbstbewusstein, das Selbstwertgefühl. Der Sport bringt ideal in Einklang sowohl die persönliche, als auch die soziale Anerkennung des Sportlers. So kann die sportliche Tätigkeit als Kennzeichen der sozialen Identität, Ausdrucksmittel der Individualität bezeichnet werden, weil dem Sportler ausgezeichnete Möglichkeiten zum Beweisen der sozialen Identität zur Verfügung stehen, durch die Sporttätigkeit kann er seine Identität wahrnehmen und bestätigen.

Die Identität der Sportlers spiegelt den Grad des Sichidentifizierens mit seiner Rolle (Brewer u.a., 1993) wider und beginnt sich mit der Teilnahme des Sportlers am Training und an den Wettkämpfen zu formen. Horton, Mack (2000) behaupten, dass sich die Stärke der Identität des Sportlers in ständiger Entwicklung befindet und von der sportlichen Erfahrung einer Person und von dem Wahrnehmungsniveau von seinem „Erfolg" und „Misserfolg" abhängig ist. Die Resultate der Untersuchungen (Brewer u. a., 1999; Grove u. a., 2004) zeugen davon, dass die Sportler, die eine bestimmte Zeit in den Wettkämpfen den Misserfolg erlebt haben, sich durch den niedrigeren Grad der Identifikation gekennzeichnet haben, als diejenigen, deren Teilnahme an den Wettkämpfen erfolgreich war. Die Resultate der Forschungen von den obenerwähnten Wissenschaftlern bestätigen die Rolle und den Einfluss der situativen Faktoren auf die Bildung der persönlichen Identität.

Die Forscher (Skarbalius, 2003; Baker, Horton, 2004) weisen darauf hin, dass bei der Heranbildung der Hochleistungenssportler eine bedeutende Rolle solche Kontextfaktoren spielen, wie die Sportpolitik des Landes, die Kultur, die Naturbedingungen, die geographische Lage des Landes, die speziphischen Tendenzen in der Entwicklung des modernen Sports, die Unterstützung der Teilnehmer im Bildungsprozess. Die Bildungsumgebung und das Motivationsklima werden durch verschiedene soziale Faktoren bestimmt: physische Aktivität in der Familie, Unterstützung der Familienangehörigen, den Inhalt der Übungen, den Sportpädagogen, Teamfreunde, Sportinventar, Medien, Kulturwerte und soziale Einstellungen (Hassandra u. a., 2003; Lenzen u.a., 2004).

Eine wesentliche Ressource stellt das soziale Netz und dic daraus resultierenden sozialen Unterstützungsleistungen dar (Tietjens, 2001). Der Trainer sollte im sozialen Netz der jugendlichen Athleten eine besondere Rolle einnehmen (Schulze, Burrmann, Stucke (2007). Die wahrgenommene soziale Unterstützung hängt auch davon ab, in welcher Karrierephase (z.B. Beginn oder Aufrechterhaltung der Sportkarriere) sich die Person befindet oder in welchem Sportbereich (Leistungssport, Breitensport) die Person agiert (Tietjens, 2001).

Einer der wichtigsten Gründe, die die Entscheidung der Sportler Sport zu treiben und dabei nach einer hohen Leistung zu streben beeinflussen, ist die Motivation. Bei der Untersuchung der Motivation der Hochleistungssportler stoßen die Wissenschaftler auf mehrere Unklarheiten und Missverständnisse. Die Untersuchungen von Willimczik, Kronsbein (2005) haben festgestellt, dass sich sowohl die persönlichen Motive als auch die Innen-und Außenanlässe vom Anfang der sportlichen Karriere bis zum Erreichen der Hochleistung wesentlich unterscheiden.

Chantal u. a. (1996) behaupten, dass man alle Tendenzen bei der Motivationsäußerung unter der Berücksichtigung des soziokulturellen Bildungskontextes einschätzen und forschen sollte. Da die Heranbildung der Hochleistungsportler in der spezifischen soziokulturellen Umgebung unter dem individuellen Verhältnis zwischen dem Trainer und dem Sportler geschieht, können die Faktoren der soziokulturellen Umgebung von entscheidender Bedeutung für die Möglichkeiten des Sportlers beim Erzielen der Hochleistung sein. Sportliche Leistungen werden nicht nur durch persönliche Eigenschaften und Talent des Sportlers, sondern auch durch die Bildungsumgebung beeinflusst. Viele Kinder, die sich in ihrer frühen Kindheit durch besondere Begabungen auszeichnen, erbringen im späteren Alter keine hohe Leistungen, die die Faktoren im Entwicklungs-und Reifewerdenprozess (Malina, Bouchard, 1991) oder wegen der unpassenden Bedingungen der sozialen Umgebung nicht ausnutzen (Perleth u. a., 2000).

Jowett, Cockerill (2002) behaupten, dass die Verhältnisse, die sich im Kontext des Sports zwichen dem Trainer und dem Sportler entwickeln, die Hauptrolle im physischen und psychosozialen Entwicklungsgang des Sportlers spielen. Die gegenseitige Interaktion zwichen dem Trainer und dem Sportler wird von den Wissenschaftlern am meisten von der Perspektive der Leistung (Riemer, Chelladurai, 1998) und der Verhältnisse (Jowett, Meek, 2002) analysiert. Wenn man die Zusammenwirkung von der Position der gegenseitigen Verhältnisse betrachtet, ist zweckmäßig, sie als soziale Situation zu untersuchen, in der das Denken, die Emotionen, das Benehmen des Trainers und des Sportlers einander beeinflussen.

Da auf die Trainer-Sportler-Interaktion die Eigenschaften der Persönlichkeit einen Einfluss ausüben, hängt es in erster Linie von ihnen ab, was man unter der gegenseitigen Zusammenwirkung versteht. Jowett (2012) behauptet, dass die Sportler mit den positiven Persönlichkeitseigenschaften günstliger die Qualität der Beziehungen zwischen dem Trainer und dem Sportler einschätzen. Es ist auch festgestellt, dass die Vertreter der spielerischen Teamsportarten weniger zufrieden mit der Qualität der gegenseitigen Beziehungen mit dem Trainer sind als die Sportler der individuellen Sportarten (Saborowski, Würth, Alfermann, 2000; Jowett, 2012).

Pensgaard, Roberts (2002), Poczwardowski u.a. (2002) betonen die Bedeutung der Trainerpersönlichkeit bei der Schaffung einer günstigen Entwicklungsumgebung sowie des Klimas im Team; beides garantiert die Effektivität der Zusammenarbeit des Trainers und des Sportlers. Die Qualität dieser Zusammenarbeit ist sehr wichtig, weil das Motivationsklima dabei einen großen Einfluß auf die Leistungen sogar der Elite – Sportler ausübt (Pensgaard u.a., 2002). Der Trainer trägt die Verantwortung nicht nur für die Leistungen und die Entwicklung der sportlichen Karriere des Sportlers, sondern auch kann er Ursache der Beendigung dieser Karriere für den Sportler sein (Stirling, Kerr, 2009). Deswegen stellt das Problem der Verhältnisse und Zusammenarbeit zwischen dem Trainer und dem Sportler in den letzten Jahren ein wichtiges Forschungsobjekt vieler Wissenschaftler dar (Poczwardowski u.a., 2002; Jowett u.a., 2002; Pfeffer u.a., 2004; Jowett, 2006).

Das Ziel der Arbeit besteht in der Aufdeckung der bildungswissenschaftlichen Aspekte bei der Steigerung der Motivation der Hochleistungssportler aus der Sicht der Spezifik der nationalen Identität.

Methode

Für die Untersuchung wird die Methode der *grounded theory* verwendet. Die grounded theory-Forschungsmethode stützt sich auf die systematische Datenverarbeitung und geht bei der Analyse des Forschungsobjekts von realen und konkreten Daten aus, die sich auf das Zusammenwirken von sozialen Prozessen konzentrieren, darum entspricht die verwendete Theorie völlig den untersuchten Realien und damit hat sie eine feste empirische Grundlage (Strauss, Corbin, 1997). Es ist sehr wichtig, für eine hochqualitative Forschung, von ihren Zielen ausgehend, die informativen Respondenten auszuwählen, die eine langjährige persönliche Erfahrung gesammelt haben und dadurch eine zuverlässige Information über die geforschte Erscheinung liefern können, weil kaum ein anderer besser Bescheid über die Probleme der Untersuchung weiß. Eine solche Ausleseweise lässt am besten die spezifischen Bedürfnisse der Untersuchung befriedigen. Die Zusammensetzung der Respondenten waren 9 Trainer (TR) (8 Männer und 1 Frau), die die Hochleistungssportler heranbilden, und 13 Hochleistungssportler - potenzielle Mitglieder der litauischen Mannschaft für die Olympischen Spiele. Die Kriterien für die Wahl der Untersuchten - Sportler (SR) der internationalen Kategorie in verschiedenen Sportarten, die:

- an den Olympischen Spielen Plätze 1 bis 24;
- in den Weltmeisterschaften Plätze 1 bis 18;
- in den Europameisterschaften Plätze 1 bis 12;
- in den Jugendweltmeisterschaften Plätze 1 bis 8 belegt haben.

Zur Datenerhebung wird an Hand transkribierter Aufnahmen auf halb strukturierte Interviews zurückgegriffen. Es wurden typische für die Entwicklungsinteraktion zwischen Hochleistungssportlern und Trainern Faktoren, Voraussetzungen, Kontext, Strategien und Konsequenzen zugeordnet.

Die Untersuchungsdatenanalyse wurde auf Grund der Kodierung unternommen: induktiv wurde Subkategorienentwicklung festgestellt und in Kategorien verbunden. Die Datenanalyse der Forschung wurde von den Experten durchgeführt. Dabei wurden die Prinzipien der Anonymität und Vertraulichkeit berücksichtigt, die beschreibende, verallgemeinernde, interpretative und innere Validität und Zuverlässigkeit gesichert (Maxwell, 2002). Die Resultate der Untersuchung werden mit den Kategorien der hohen Abstraktionsstufe präsentiert. Bei der Analyse der Daten wurden die ausgesonderten Kategorien nicht an die schon existierenden theoretischen Modelle gebunden, die Ergebnisse der Untersuchung unterlagen nicht der Analyse und dem Vergleich nach dem Aspekt der Geschlechter.

Ergebnisse

Die Untersuchung läßt die sich in konkreter Umgebung entwickelnde Zusammenarbeit zwischen dem Trainer und dem Sportler ergründen und die darauf beruhende grounded theory der Bildungsinteraktion „Trainer-Sportler“ und „Sportler-Trainer“ für die Hochleistungssportler schaffen. In der Arbeit werden nur die Kontextfaktoren analysiert, die die Fakten, Geschehnisse und Erscheinungsumstände sowie Besonderheiten deren Interaktion zum Ausdruck bringen.

Obwohl der soziokulturelle Kontext nicht das Ziel der Forschung war, haben die Aussagen der Sportler während der Interviews gezeigt, dass alle Sportler diesen Aspekt als eine der wichtigsten Voraussetzungen beim Streben nach der Hochleistung erwähnt haben. Bei der Analyse der Kontextfaktoren (aus der Sicht der Sportler) im Umfeld, in dem die Interaktion zwischen den Hochleistungssportlern und den Trainern in Litauen stattfindet, wurden 21 Kategorien zusammengestellt. Diese lassen sich in Faktorengruppen zusammenfassen: *soziokultureller Kontext, Trainer als Entwicklungsfaktor, Zustand des Sportlers.* Die Kategorien, die den soziokulturellen Kontext charakterisieren, und die sie bestätigenden Aussagen werden in der Tabelle 1 präsentiert. Wie aus der Tabelle 1 ersichtlich, gelten als die wichtigsten Merkmale im sozialkulturellen Kontext Litauens für die Sportler beim Streben nach der Hochleistung folgende Merkmale: *die Anzahl hochqualifizierter Trainer ist in Litauen beschränkt; es gibt einen Mangel an adäquaten Bedingungen und finanziellen Ressourcen für die Heranbildung von Sportlern; es gibt eine geringe Konkurrenz unter den Sportlern im Land; der auf die Sportler ausgeübte Druck von der Seite der Gesellschaft.*

Tabelle 1: Der soziokulturelle Kontext bei der Heranbildung litauischer Hochleistungssportler (aus der Sicht der Sportler)

Kategorien	**Unterstützende Behauptungen**
Die Anzahl hochqualifizierter Trainer ist in Litauen beschränkt.	*„..mein ehemaliger Trainer erwies sich als aussgezeichnet nämlich im Jugendsport. Ob er so auch beim Training von Erwachsenen ist – weiß ich nicht, ob er mich zu solchen Leistungen führen würde, die ich jetzt besitze. Vielleicht war es zweckmäßig, den Trainer zu wechseln".*
Es gibt eine geringe Konkurrenz unter den Sportlern im Land.	*„Die Konkurrenz hilft mir bei meiner weiteren Vervollkommnung… Nachdem ich mich dem Hochleistungssport angeschlossen habe und mit ihm zu trainieren begann … Gerade das hat mir sehr geholfen."*
Der auf den Sportler ausgeübte gesellschaftliche Druck.	*"…du bekommst den Druck von der Seite der Föderation, des olympischen Zentrums, wo nur das entsprechende Resultat verlangt wird… Ein jedes Mißgeschlick in einem Wettkampf würde vielleicht den völligen Krach und das Ende der Karriere für den Sportler bedeuten".*
Die Klarheit in Bezug auf das System der materiellen Förderung bei knappen finanziellen Ressourcen.	*„Wenn du das eine Resultat erzielst, kommst du auf eine Liste, wenn das andere, wirst du in eine andere eingetragen".*
Es gibt einen Mangel an adäquaten Bedingungen und finanziellen Ressourcen für die Heranbildung der Sportler.	

Die Forschungsresultate haben gezeigt, dass die Sportler zugeben, dass für das Endresultat bei der Heranbildung des Sportlers die Effektivität der Zusammenarbeit des Sportlers, des Trainers und des Service-Personals, die klare Zuordnung der Verantwortlichkeit unter ihnen, ihre Wahrnehmung und Ausführung ausschlaggebend sind. Der Trainer, der die Hochleistungssportler heranbildet, trägt große Verantwortung – die Teamatmosphäre zu schaffen und nach der engen Zusammenarbeit aller am Prozess beteiligten Personen zu streben.

Die Untersuchung ließ sowohl die Aspekte des Trainers, des bildungswissenschaftlichen Faktors bei der Motivationssteigerung, als auch die die Zusammenarbeit zwischen dem Sportler und dem Trainer belastenden Umstände aufdecken (Tabelle 2):

Tabelle 2: Trainer als Entwicklungsfaktor (aus der Sicht der Sportler)

Kategorien	**Unterstützende Behauptungen**
Die Partnerschaft zwischen dem Trainer und dem Sportler beruht auf dem gegenseitigen Vertrauen.	*„…mit dem Trainer waren wir nicht im Verhältnis Schüler-Lehrer, sondern Kollektiv zweier Menschen. Die Entscheidungen wurden zusammen getroffen".*
Das Verständnis für die Bedeutung des Strebens nach dem gemeinsamen Ziel.	*„es ist nicht so, dass du zum Training erscheinst und nur daran denkst, wie du das Resultat erzielst. Der Trainer denkt auch in der Nacht darüber nach, wie du das oder jenes ausführen oder ausprobieren musst".*
Die Fürsorge des Trainers für den Sportler.	*„meine Trainerin gilt für mich als die zweite Mutter, die immer nur hilft"*
Die Trainingsphilosophie, die den Leitungsstil des Trainers bestimmt.	Die Konservativität des Trainers (Subkategorie): *„das sind Trainer des alten Schnitts, was die vor 30 Jahren gemacht haben… Für sie ist es schwer zu verstehen, dass alles im Leben vorangeht"* Keine Stabilität in der Strategie der Sportlerausbildung und die Unterwerfung des Trainers den Außenfaktoren (Subkategorie): *„wir machen auf unsere Art … dann sehen wir, wie das von Olympiasiegern und Weltmeistern ausgeführt wird, und es wird uns auf eine andere Art und Weise erklärt".*
Unwille des Trainers zur Weiterbildung.	*„Unsere Trainer naben keinen Wunsch zu lernen, sich zu vervollkommnen. Sie sind der Meinung, sie haben selbst Sport getrieben, deswegen kennen und können sie alles".*
Die Inkompetenz des Trainers, die sie sich in unprofessionellem Verhalten äußert.	Subkategorien: Negative persönliche Eigenschaften des Trainers: *„Ich kenne keinen einzigen Sportler bei uns, der wegen des Trainers nie mal eine Träne vergossen hätte.* Kein Vertrauen in die Leistungsfähigkeit des Sportlers (im Falle einer Niederlage im Wettkampf): *"Der Trainer hat gefragt: „Lohnt es sich, weiter Sport zu treiben?" Wenn der Trainer so was gefragt hat…"* Öffentliche Demonstration von schädlichen Gewohnheiten: *„Der Sportler kommt zum Training, während der Trainer in der Bar sitzt und Bier trinkt".* Das Ignorieren der Meinung des Sportlers: *„Unsere Meinung wird nicht akzeptiert … Der Trainer, der gewohnt war, andere Meinung zu ignorieren, hielt für richtig, allein Recht überall gehabt zu haben".* Die Nichtakzeptanz eigener Fehler: *„Wenn der Trainer seine eigenen Fehler nicht einsieht, die begangen worden sind, ist es einfach unmöglich, normal mit solch einem Trainer zu arbeiten".*

Kategorien	Unterstützende Behauptungen
	Das Verlieren der Beherrschung der Emotionen durch den Trainer während des Wettkampfes: *„Das stört sehr den Sportler im Prozess der Vorbereitung auf den Wettkampf, wenn du siehst, dass „der Trainer mehr nervös ist als der Sportler selbst...“* Keine Klarheit in Bezug auf die Finanzierung: *"Du kannst nicht mehr dem Trainer finanziel vertrauen. Wenn du ins Sportlager fährst, bekommst du kein Tagesgeld, was dir zugeteilt ist".* Die Erinnerung an die misslungene Erfahrung: *„Nach dem verlorenen Wettkampf sind vielleicht schon ein paar Jahre vergangen, und er erinnert dich immer noch daran. Das ist so eine peinliche Sache.“*

Für die andere ***Gruppe der Respondenten*** wurden als Forschungsobjekt zwei Kategorien verwendet: *das Verständnis der Teammitglieder untereinander, gute Beziehungen unter der Sportlern und dem Service-Personal.*

Die Sportler, die an der Untersuchumg teilgenommen haben, haben die Bedeutung der guten Beziehungen der Sportler zu dem Service-Personal betont. Besonders positiv wird von ihnen die informelle Verständigung mit den Ärzten und Masseuren des Teams eingeschätzt. Es ist sehr wichtig, dass die Sportler während des Trainings beim Streben nach den gemeinsamen Zielen die ständige Unterstützung der Teampartner fühlen. Das gemeinsame Training ruft eine gewisse Konkurrenz hervor: *„wenn du alleine bist, ist es schwer zu trainieren. Wenn du aber in einer Gruppe trainierst, entsteht gleich die Konkurrenz. Das hilft, selbstverständlig, bei der Vervollkommnung“* (SR-3), schafft zugleich eine entsprechende „Arbeitsatmosphäre“, lässt die Monotonie im Trainingprozess vermeiden, die Teamarbeit spornt den Sportler zur Aktivität an: *„du siehst, wie es ihnen schwer fällt, und es wird dir leichter sein, weil du nicht alleine dabei bist“* (SR-8), *„wenn wir zusammen anfangen – du beginnst mit dem Lächeln“* (SR-2), *„wenn im Training freudige Laune herrscht, vergeht die Zeit schneller. Nicht so, wie es ist, wenn du alleine kommst und nur daran denkst, schneller das Training zu enden...* „ (SR-12). Die moralische Unterstützung der Teammitglieder hilft den Sportlern leichter die großen Auslastungen im Training überwinden.

In der Tabelle 3 werden die Kategorien angegeben, die den ***Zustand der Sportlers*** charakterisieren, der einen großen Einfluss auf die Motivation der Hochleistungsportler ausübt:

Vertrauen des Sportlers in die eigene Stärke; Sicherheitsgefühl des Sportlers, das durch das Vertrauen in den Trainer entsteht; Identifizierung des Sportlers mit dem für den Sport typischen Umfeld; Hingabe und Begeisterung des Sportlers für den

Trainingsprozess; erhöhtes Verantwortungsbewusstsein des Sportlers; physische und psychische Ermüdung; psychische Anspannung, die durch die Qualifikationen für wichtige Wettkämpfe ausgelöst wird; Schuldgefühl des Sportlers, seiner Familie nicht genügend Aufmerksamkeit widmen zu können.

Tabelle 3: Zustand des Sportlers

Kategorien	**Unterstützende Behauptungen**
Sicherheitsgefühl des Sportlers, das durch das Vertrauen in den Trainer entsteht.	*„Du fühlst immer, dass der Trainer in seinem Element ist, er kontrolliert alles, Ruhe herrscht ringsum".*
Vertrauen des Sportlers in die eigene Stärke.	*„Du weisst, dass du das meistern kannst, dass das in deinen Kräften liegt, weil du das nicht einmal bewiesen hast. Es bleibt nur dasselbe noch einmal zu machen".*
Die Identifizierung des Sportlers mit der sporttypischer Umgebung.	*„Ich kann mir jetzt mein Leben ohne Trainingslager, meine Trainingsfreunde nicht vorstellen".*
Hingabe und Begeisterung des Sportlers für den Trainingsprozess.	*„Wenn während des Trainings fröhliche Stimmung herrscht, verläuft die Zeit schneller. Das bedeutet aber nicht, dass du alleine mit dem Gedanken kommst, schneller das Training zu beenden. Dort aber geht es launig zu. Das ist für mich wichtig".*
Erhöhtes Verantwortungsbewusstsein des Sportlers.	*„Wenn du für Litauen auftrittst, ist die Aufmerksamkeit aller auf dich gerichtet, alle Trainer schauen auf dich, du trägst Verantwortung für dich selbst, erst dann vor Öffentlichkeit. Je besser deine Leistungen sind, desto größer wird Verlangen danach und du wirst ernster beobachtet".*
Physische und psychische Ermüdung.	*„Wenn du dich in dem Zeitraum sowohl physisch als auch geistig nicht erholst, kannst du zum Schluß kommen, dass du Sport nicht mehr treiben willst oder sogar nicht mehr kannst. Vielleicht könntest du das, aber der Organismus ist nicht mehr im Stande, solche physischen und psychischen Anforderungen zu überwinden".*
Psychische Anspannung, die durch die Qualifikationen für wichtige Wettkämpfe ausgelöst wird.	*„Man muss eiserne Nerven haben, um nicht daran denken zu müssen, was daraus wird, sondern nur an das Resultat".*
Schuldgefühl des Sportlers, seiner Familie nicht genügend Aufmerksamkeit widmen zu können.	*„Die Familie leidet darunter, dass ich Leistungssportler bin. Wenn man den Hochleistungssport treiben will, muss man zehnmal pro Woche trainieren und für das Training entsprechende Bedingungen auch außerhab Litauens haben. Dann geht alles von sich selbst voran, du bist nicht zu Hause, kannst weniger Aufmerksamkeit der Familie, der Frau, dem Kind schenken".*

Bei der Analyse der Antworten der Trainer in den Interwievs wurden neun qualitative Ketegorien ausgesondert, die den Kontext bei der Interaktion: Trainer - Sportler charakterisieren:

- die geschlechtlichen Unterschiede zwischen dem Trainer und dem Sportler;
- die Verständigung zwischen dem Trainer und Sportler;
- unterschiedliche Erwartungen des Trainers und des Sportlers;
- das Begreifen durch den Trainer der Kompliziertheit des Hochleistungssports;
- die Erweiterung der Funktionen des Trainers;
- der Mangel an finanziellen Ressourcen für die entsprechenden Bedingungen bei der Heranbildung der Sportler;
- der von der Gesellschaft ausgeübte Druck auf den Trainer;
- die Konkurrenz unter den Sportlern;
- die extremen physischen Auslastungen, denen der Sportler ausgesetzt wird.

Die geschlechtlichen Unterschiede zwischen dem Trainer und dem Sportler

Die Forschungsergebnisse haben die Behauptungen der Wissenschaftler (Seller u.a., 1999) bestätigt, dass die Geschlechtsunterschiede des Trainers und des Sportlers die Spezifik in ihrer Zusammenwirkung entscheiden und bedingen. Die Trainer weisen darauf hin, dass die Geschlechtsunterschiede den Einfluss auf die Arbeitsspezifik des Trainers und der Verhältnisse zu den Sportlern ausüben: *„die Frau ist mit der Mutterschaft verbunden: sie kann manchmal mehr bemitleiden. Die Verständigung, die Vorahnung, das Verständnis fallen ihr leichter. Bei den Männenr ist das anders.“*

Die Verständigung zwischen dem Trainer und Sportler

Diese Kategorie spiegelt die Qualität der Kommunikation zwischen dem Trainer und dem Sportler wider. Die Trainer betonen die Bedeutung der Beziehungen untereinander. Die Untersuchung hat gezeigt, dass dabei gegenseitiges Kennenlernen geschieht und sich der eigene Kommunikationsstil entwickelt: *„mit den älteren Sportlern, mit denen man schon eine längere Zeit arbeitet, bilden sich automatisch irgendwelche Regeln. Unter den Hochleistungssportlern gibt es keine Regelverletzung, weil sie genau wissen, was sie machen; sie wissen, wozu sie das machen; sie wissen, was dabei schadet, was sie tun dürfen und was nicht“* (TR-2). Die Teilnehmer der Untersuchung haben betont, dass im Prozess der

Zusammenwirckung die nichtformalen Beziehungen zwischen dem Trainer und dem Sportler am typischsten sind: *„zwischen dem Sportler und dem Trainer muß eine Beziehung, eine Verständigung sein, ein jeder von ihnen hat eigene Sorgen, Probleme, das trifft sogar einen Hochleistungssportler“* (TR-9). Die Untersuchung hat aber auch Merkmale der formalen Verständigung festgestellt, wenn der Trainer nur seine direkten Pflichten ausführt: *„man will die Arbeit erfüllen und sie alle verlassen. Wir sind weder Freunde, noch Feinde“. „Ich bin für die direkten Forderungen und für die formalen Beziehungen, damit nach dem Training der Trainer mit seinen Sportlern zur Diskothek oder anderswohin nicht gehen muss“ (TR-8).* Wir sind der Meinung, dass der „notwendige“ Kontakt, der zwischen den Mitgliedern im Prozess der Heranbildung der Sportler entsteht, als das Merkmal der Professionalität betrachtet werden sollte.

Unterschiedliche Erwartungen des Trainers und des Sportlers

Die die berufliche Tätigkeit des Trainers und die Heranbildung der Hochleistungssportler reglementierenden Dokumente, Anforderungen, die ständige Aufmerksamkeit und der Druck von der Seite der Gesellschaft, das für den Hochleistungssport charakteristische hohe Konkurrenzniveau üben einen großen Einfluss auf die Orientierungsweise des Trainers hinsichtlich der hohen Leistungen und Resultate im Sport aus: *„das verlangt von uns der Arbeitgeber...“* (TR-5); *„ich möchte, dass der Sportler im Mittelpunkt stehen würde, aber das Leben und die Realität haben sich nicht verändert - ich muss nach dem Resultat streben, anders verliere ich meine Arbeit“* (TR-8); *„die Sportler, die kein Resultat erzielen, sind der Gesellschaft uninteressant. Die Leute mögen irgendwelches ästhetische Bild, den Konkurrenzkampf, das Dominieren der Persönlichkeiten im Sport“* (TR-8).

Die Untersuchung hat gezeigt, dass sogar bei der Heranbildung der Hochleistungssportler die berufliche Orientierung, die Motive und Erwartungen des Trainers nicht immer mit den Motiven und Erwartungen des Sportlers zusammenfallen. Die Trainer stoßen oft auf die unterschiedlichen Motive und Einstellungen der Sportler: *„die Motivation in der beruflichen Tätigkeit der Sportler unterscheidet sich manchmal sehr von deiner Motivation, von deinen Einstellungen, die Interessen und die Wünsche sind sehr verschieden“* (TR-8). Die Trainer bestätigen, dass nach den ersten guten Sportleistungen die Motivation der Sportler sehr stark ist, aber nach der Teilnahme an den internationalen Wettkämpfen werden bei den Sportlern bedeutende Motivationsveränderungen beobachtet: *„wenn die Sportler ein bestimmtes Niveau erreicht haben, verstehen sie, dass es nicht so ein-*

fach ist. Die Motivation wird zu einem komplizierten Ding... das Gewehr <der Sportler> muss manchmal den Schuss erzeugen, das geschieht aber nicht – das Gewehr selbst fehlt... Wenn der Sportler die Motivation in jüngerem Alter verliert, können wir etwas noch machen. Später, wenn die Reife kommt, können wir den Sportler nicht mehr behalten – er fährt fort" (TR-7).

Das Begreifen durch den Trainer der Kompliziertheit des Hochleistungssports

Diese Kategorie spiegeln die Behauptungen wider, die mit der Fähigkeit des Trainers, real und kritisch die Heranbildung der Sportler einzuschätzen und ihre Vielseitigkeit zu begreifen, verbunden sind: „*die Leistung besteht aus vielen Faktoren: es gibt in Litauen solch ein Maß, das „kapa" heißt und das aus 60 Stückeinheiten besteht. So kann man ähnlich auch im Sport 60 Faktoren zusammenzählen. Du wirst aus 60 alle nicht gewinnen können, aber es reicht in 6 Faktoren den Rivalen zu besiegen und ihm überlegen zu sein, und du bist Champion*" (TR-7). Die Trainer behaupten, dass entscheidend für die Konkurrenz unter den litauischen Sportlern, denen primitive Sporteinrichtungen zur Verfügung stehen, die Fähigkeit des Trainers ist, mit Hilfe von verschiedenen Trainingsverfahren maximal effektiv die vorhandenen Bedingungen auszunutzen: „ *es ist nicht so einfach, die Trainer Europas „zu hintergehen", sie verfügen über viel bessere Bedingungen, als wir: nur das Primitive: Metallhaufen, einfaches Trainingsgerät, altes Gebäude, darum bist du gezwungen, methodisch überlegener zu sein*" (TR-7).

Die Erweiterung der Funktionen des Trainers

Mit den zunehmenden Leistungen des Sportlers erweitert sich auch das Feld der Funktionen, die von dem Trainer erfüllt werden: „*ich persönlich bin sowohl Trainer, Arzt, Manager und Masseur, als auch Psychologe und sogar Koch im Trainingslager... in einer Person... man muss eigenes Geld investieren, nach irgendwelchen Sponsoren suchen. Es fällt manchmal schwer, all das zu bewältigen*" (TR-6). Viele Teilnehmer der Untersuchung haben eingestanden, dass sie eine große Verantwortung für die Gesundheit und die Resultate der Sportler tragen: „*und all das müssen die Trainer auf ihre Schultern nehmen. Die Verantwortung trage ich vor allem für die Gesundheit der Sportler, damit sie kein Trauma bekommen. Man muss zugeben, dass ich die Verantwortung fühle, weil alle die Resultate verlangen*" (TR-3).

Der Mangel an finanziellen Ressourcen für die entsprechenden Bedingungen bei der Heranbildung der Sportler

Eine der wichtigsten Ursachen, die die Erweiterung der Funktionen des Trainers entscheiden, sind, laut der Teilnehmer der Forschung, der Mangel an finanziellen Ressourcen und die ungeeigneten Trainingsbedingungen. Die Trainer haben anerkannt, dass die litauischen Sportler unter etwas schlimmeren Trainingsbedingungen heranbildet werden, als ihre Konkurrenten aus anderen Ländern: *„vieles hängt, wirklich, von den Bedingungen und besonders von der materiellen Ausstattung ab“* (TR-8); *„wenn man alle Bedingungen, Sporteinrichtungen, genügende Finanzierung hat, kann man sich völlig dieser Arbeit ergeben – dich stören keine organisatorischen Fragen, die dich von der direkten methodischen Arbeit ablenken“* (TR-3).

Der von der Gesellschaft ausgeübte Druck auf den Trainer

Alle an der Untersuchung beteiligten Trainer haben unterstrichen, dass bei der Heranbildung der Hochleistungssportler ein großer Druck von der Seite der Geselschaft ausgeübt wird, der indirekt höhere Forderungen sowohl an den Trainer, als auch an den Sportler stellt: *„die Aufmerksamkeit der Gesellschaft wirkt sehr negativ, das bedeutet eine bestimmte Verpflichtung, weil du immer komplizierter Aufgaben bewältigen musst“* (Tr-8). Die Teilnehmer der Untersuchung behaupten, dass *„der moderne Sport anders geworden ist. Alles wird unter dem Gesichtswinkel der Resultate betrachtet und geschätzt.“* Die ständige gesellschaftliche Aufmerksamkeit und „das Verlangen nach dem Resultat“ löst eine große innere Spannung aus.

Die Konkurrenz unter den Sportlern

Als einer der wichtigsten Faktoren im Prozess der Heranbildung der Hochleistungssportler gilt die Zunahme der Konkurrenz unter der Sportlern. Die strengen Auswahlkriterien für die Europa-, Weltmeisterschaften und für die Olympischen Spiele tragen zur Konkurrenzsteigerung unter den Sportlern verschiedener Länder bei, was sich positiv auf die Konkurrenz unter der Sportlern im eigenen Land auswirkt. Die Trainer haben bestätigt, dass beim hohen Konkurrenzniveau unter den Sportlern der Welt die Effektivität der Heranbildung und Resultate durch das niedrige Niveau der Konkurrenz unter den Sportlern in Litauen zu erklären sind: *„eine geringe Konkurrenz, eine geringe Auswahlmöglichkeit; wir haben keine solche Auswahl, woraus man wählen könnte“* (TR-3).

Die Trainer verstehen die Rolle der Konkurrenz für die Vervollkommnung der Sportler und betonen die fair-play-Prinzipien: *„die Konkurrenz muss auf der gegenseitigen Achtung beruhen. Es gibt Sportler die einander hassen“ (TR-2). Neben den konstruktiven Konkurrenzmerkmalen werden von den Trainern auch destruktive bemerkt: „die Konkurrenz kann die Sportler, deren Begabungen sich nicht völlig entfaltet haben, „isolieren“; unter der Konkurrenz leiden am meisten die Menschen, die keine Lebenserfahrung haben, die Person, die psychologisch schwächer, überhaupt schwächer sind“ (TR-8).*

Die extremen physischen Auslastungen, denen der Sportler ausgesetzt wird

Die Zunahme der Konkurrenz unter den Sportlern spornt zur Suche nach den mehr effektiven Verfahren bei der Heranbildung der Sportler an, verlangt die Trainingsauslastungen zu vergrößern. Die Trainer haben die extremen physischen Auslastungen, denen die Sportler ausgesetzt werden und die eine große physische Ermüdung hervorrufen, unterstrichen, was für den Trainingsprozess bei der Heranbildung des Sportlers von Bedeutung ist.

Diskussion

Heute ist das Problem der nationalen Identität im Kontext der Herausforderungen durch die Globalisierung sehr aktuell. Es wird heftig diskutiert, welcher Platz und welche Rolle der nationalen Identität in der heutigen multikulturellen Welt zugemessen werden können. Die nationale Identität wird von dem nationalen Selbstbewusstsein bestimmt, die nationale Identität, im Gegenteil, bildet den Inhalt des Selbstbewusstseins (Liubinienė, 2003). Der reife Sportler, der seine Zugehörigkeit zur einen oder anderen Nation bewusst wahrnimmt und begreift, übernimmt die von dem Volk im Laufe der Jahrhunderte gepflegten Werte, Traditionen, Sitten und Bräuche, das Symbolsystem, die geschichtliche Erfahrung, Ansichten, Normen u.a. Auf diese Weise formt sich sein Selbstbewusstsein. Unter dem System der nationalen Identität wird die Erscheinung verstanden, die sich im ständigen inhaltlichen Entwicklungsprozess befindet, der durch die zeitlichen und sozialen, wirtschaftlichen, politischen und kulturellen Veränderungen beeinflusst wird. Die Bildung der nationalen Identität beginnt mit der primären Sozialisation des Individuums und dauert bis zum

Lebensende, aber beim Wechsel der sozialen Umgebung durch das Individuum kann sich auch das Verstehen des Begriffs „nationale Identität" ändern. Laut Giddens (2000), wird die nationale Identität so verstanden und interpretiert, auf welche Weise sich selbst das Individuum durch seine biographische Perspektive indentifiziert.

Die Wissenschaftler werfen die Frage auf, welche Rolle der Sport für die Entwicklung der Identität einer Person spielt, welche Erfahrung und Faktoren die persönliche Identität und das nationale Selbstkonzept schützen (Dóczi, 2012). Der Sport stellt eine wichtige soziale Dimension dar, die ermöglicht, das Selbskonzept und die soziale Anerkennung einer Person vollkommmen in Einklag zu bringen (Weiss, 2001). Der Sport wirkt auf die Pflege der nationalen Identität, weil die Leistungen der Sportler auf dem internationalen Niveau auch den nationalen Stolz stärken. Wenn man die Rolle des Sports für die Bildung und Entwicklung der Identität einer Person verstehen will, sollte man auch die soziobildungswissenschaftliche Umgebung analysieren, in der die Ausbildung geschieht, um auch die Rolle des Trainers für die Entwicklung der persönlichen und der sozialen Identität des Sportlers aufzudecken und hervorzuheben.

Die Resultate der Untersuchungen von den Wissenschaftlern (Seller u.a., 1999; Pfeffer, Gallitschke, 2008; DeWeese, 2010) zeugen davon, dass auf die Zusammenwirkung Trainer-Sportler viele Faktoren den Einfluss ausüben: die Spezifik der Sportart, die persönlichen Eigenschaften des Trainers, seine Qualifikation und die Schaffenskraft, die Kommunikationsfähigkeiten, die Emotionskontrolle, der Leistungsstil, die Fähigkeit zum Motivieren des Sportlers und sein Leistungsniveau.

Die Naturbedingungen und das Interesse der Gesellschaft verschaffen günstige Bedingungen für die Pflege der einen oder anderen Sportarten im Lande oder in der Region. Zum Beispiel, in den skandinavischen Ländern und in Russland sind die Wintersportarten sehr populär, besonders das Skilaufen. In Österreich und in Norditalien sind ausgezeichnete Naturbedingungen für den Alpinskisport. Eine kleine Einwohnerzahl bedingt relativ kleine Anzahl von begabten und talentierten Sportlern (Skarbalius, 2003). Außerdem beeinflussen die Entwicklungs-und Ausweitungstendenzen der populären Sportarten in einem kleinen Land ihre Popularisierung unter den Kindern, ungeachtet ihres angeborenen Hangs zur Tätigkeit einer ganz anderen Art und Spezifik.

Die Untersuchung bestätigt die Notwendigkeit der Entwicklungsinteraktionsanalyse zwischen dem Trainer und dem Sportler im konkreten soziokulturellen Kontext. Ungünstige Kontext- soziokulturelle Aspekte erschweren die Entwicklungsinteraktion zwischen dem Trainer und dem Sportler und kann negativ auf die zwischenmenschliche Beziehungen auswirken. Es entsteht die Frage, wie viel das alles den Entschluss des Sportlers beeinflusst, die Karriere fortzusetzen oder zu beenden.

Die Familie, die Verwandten, die Freunde, die Lehrer, die sportliche Umgebung (die Teamfreunde, die Trainer, das Service-Personal, die Mediker, die Masseure, die Sportwissenschaftler), die Kirche, die Kulturtraditionen des Landes, die Sportpolitik bilden die Umgebung, in der der Sportler lebt, lernt, trainiert, sich vervollkommnet als Persönlichkeit (Morgan, Giacobbi, 2006). Mit den immer höheren Leistungen des Sportlers und seiner Aufnahme in die Landesmannschaft wird mehr Zeit für seine Heranbildung während der Trainingslager, Teilnahme an der Wettkämpfen gewidmet. Der Sportler verbringt immer mehr Zeit mit dem Trainer, die Rolle und der Einfluss des Trainers auf die sozialen und akademischen Interessen des Sportlers nehmen bedeutend zu.

Die Hochleistungssportler müssen oft wegen der Wettkämpfe und bestimmer Trainingsbedingungen in unterschiedlichen soziokulturellen Verhältnissen leben, was den zusätzlichen Stress verursacht. Als der wichtigste psychologische Faktor, der das Verhältnis der Persönlichkeit zur sozialen Umgebung gestaltet und reguliert (zusammen mit der sozialen Adaptation), gilt der Lebensstil, der die Persönlichkeit ermöglicht, maximal ihre Lebensprojekte zu verwirklichen (Giddens, 2005). Der Erfolg bei der Anpassung an die Umgebung, die sich ständig verändert, hängen, laut Juodraitis (2004), von vielen Persönlichkeiteigenschaften, Forderungen der sozialen Umgebung sowie der Unterstützung ab. Für einen Sportler, der hohe Leistungen im Sport erziehlt hat, entstehen viele Probleme bei der Suche nach der sozialen Identität, weil er oft gezwungen ist, nicht in einem sozialen Kontext, sondern parallel in zwei oder mehreren sozialen Umgebungen zu handeln: er muss im Stande sein, seine Tätigkeit integrieren, sie in Übereinstimmung in einigen sozialen Kontexten bringen und gleichzeitig sie voneinender unterscheiden. Darum kann hier der Trainer als einer der Faktoren auftreten, der die Schwierigkeiten bei der Adaptation mildern kann, dass die Persönlichkeitsentwicklung des Sportlers, der nach den Hochleistungen strebt, nicht nur mit den Leistungen im Sport verbunden wäre, sondern auch mit der Speicherung der sozialen Erfahrung, der Vorbereitung auf die Erfüllung bestimmter sozialer Rollen in der Gesellschaft.

Es sollte betont werden, das die Sozialisation der Sportler oft in einer sehr begrenzten sportlichen Umgebung geschieht, darum ist sie auf soziale Rolle der Sportler beschränkt, es werden oft die Grenzen ihrer Tätigkeit als Sportler nicht überschritten. Aus diesem Grund stoßen sie auf die Probleme beim Unterbrechen oder nach dem Abschluss der Sportkarriere.

Die Untersuchungen von Schulze, Burrmann, Stucke (2007) haben gezeigt, dass bei der Heranbildung der jungen Sportler eine sehr große Rolle der Trainer als Garant der sozialen Unterstützung spielt. Die Resultate unserer Forschung zeugen davon, dass in der Etappe der Heranbildung der Hochleistungssportler die Rolle der sozialen Unterstützung durch die Trainer neben der fachlichen immer noch wichtig bleibt.

Die Wissenschaftler betonen die Wichtigkeit des Einklangs der Persönlichkeitstypen des Trainers und des Sportlers und der Ähnlichkeit im Verstehen der Erscheinungen im Prozess der effektiven pädagogischen Zusammenarbeit. Für die erfolgreiche Kooperation ist es wichtig, dass der Trainer den Sportler als den gleichberechtigten Partner betrachtet. Es wird diskutiert, ob der Trainer in verschiedenen Etappen der Heranbildung der Hochleistungssportler den verschiedenen Stil der Leitung und Betreuung, unterschiedliche Einflussmethoden verwenden solle, um eine motivierende Trainingsatmosphäre zu schaffen. Würth, Saborowski, Alfermann (1999) haben festgestellt, dass die Zufriedenheit der angehenden Sportler mit der Kooperation bei ihrer Heranbildung mit dem positiv wahrgenommenen Feedback, dessen Grundlage die spürbare soziale Unterstützung ist, verbunden ist, während dessen im Hochleistungssport das positive Feedback der Sportler mehr mit der Aufgabenorientierung korreliert. Die Forschungsresultate der Wissenschaftler (Pfeffer, Würth, Alfermann, 2004) zeigen, dass die Korrelation Trainer-Sportler abhängig auch von der Sportart (individueller oder Mannschaftssport) analysiert werden soll.

Die Resultate unserer Untersuchung bestätigen, dass es in der Etappe der Heranbildung der Hochleistungssportler sehr wichtig ist, zu erforschen, wie der Trainer und der Sportler die Qualität der pädagogischen Interaktion und der Zusammenarbeit verstehen. Bei der Gestaltung der günstigen Bildungsumgebung muss der Trainer verstehen, dass die Motivation der Sportler, die nach den Hochleistungen streben, mit dem Wunsch, ihr eigenes Potenzial zu realisieren, wesentlich korreliert. Für die Vervollkommnung der Sportler werden originelle Herausforderungen verwendet, die „die schöpferische Spannung“ schaffen und unterstützen.

Die Forschungen zeigen (Duffy u.a., 2001), dass die Sportler mit den zunehmenden Leistungen immer positiver die Rolle der Spezialisten einzelner Berufe, der Mediziner, Wissenschaftler, die Bedeutung der kontruktiven Kritik einzuschätzen beginnen, die Verantwortung übernehmen können. Aber in kritischen Momenten, nach den Niederlagen und Enttäuschungen, mit dem Entstehen der Schwierigkeiten, den Sport mit den anderen Lebensbereichen zu vereinbaren, sind die Sportler auf die Hilfe des Trainers angewiesen.

Die Untersuchung hat aufgedeckt, dass die beim Motivieren Hochleistungsportlers auftretenden hemmenden Bildungsfaktoren sind: nichtprofessionelles

Verhalten des Trainers, geringe Kommunikationskompetenz, Anwendung nicht passender Strategien nach den Niederlagen, unterschiedliche Hoffnungen des Trainers und des Sportlers, keine Stabilität in der Strategie der sportlichen Ausbildung und die Unterwerfung des Trainers den Außenfaktoren.

Die Auswahl der Teilnehmer unserer Untersuchung hat uns nicht erlaubt, die typischen Erscheinungen bei der Korrelation Trainer-Sportler hinsichtlich der Geschlechter zu analysieren. Wir sind der Meinung, dasss sich ohne Zweifel die Erfahrungen der Hochleistungssportler von denen der Sportler, die das erste Mal hohe Resultate erzielt haben, unterscheiden sollen. Die Analyse solcher Erfahrungen könnte die wertvolle methodische Information sowohl den die Sportler heranbildenden Spezialisten, als auch den Sportlern selbst, die sich auf der Suche nach der effektiven Kooperation befinden, um hohe Sportleistungen zu erbringen. Alle von uns untersuchten Sportler und ihre Trainer waren Bürger eines und desselben Landes, deren Zusammenarbeit in der beiderseits bekannten soziokulturellen Umgebung verlief. Es bleibt nicht klar, wie sich der Interaktionskontext Trainer – Sportler verändern würde, wenn die Elite-Trainer aus dem Ausland herangezogen würden.

Sportliche Leistungen werden nicht nur durch persönliche Eigenschaften und Talent des Sportlers, sondern auch durch die Bildungsumgebung beeinflusst. Die Identität einer Person als multidimensionales Konstrukt, das psychologische, soziale und Aspekte des sozialen Verhaltens einschließt, kommt zum Ausdruck und entwickelt sich in konkreten sozialen und bildungswissenschaftlichen Umgebungen. Die Bildungskorrelation zwischen dem Trainer und dem Sportler wird als Konzept verstanden, das die bildungswissenschaftlichen und psychologischen Erscheinungen miteinschließt, die für die spezifische soziokulturelle Umgebung typisch sind, und sie kann als Widerspiegelung der nationalen Identität betrachtet und analysiert werden. Anders gesehen, kann diese Interaktion im Ausbildungsprozess auch als die Voraussetzung für die Formierung der sozialen Identität beider Teilnehmer analysiert werden.

Schlussfolgerungen

1. Es wurde festgestellt, das der bildungswissenschaftliche Kontext bei der Steigerung der Motivation von Hochleistungssportlern aufs engste mit dem soziokulturellen Kontext des Landes verbunden ist, die die nationale Identität widerspiegeln und beeinflussen. Der litauische soziokulturelle Kontext wird durch die spezifischen Aspekte, die negativ auf die Interaktion zwischen dem Trainer und dem Sportler wirken, gekennzeichnet: *die Anzahl hochqualifi-*

zierter Trainer ist in Litauen beschränkt; es gibt eine geringe Konkurrenz unter den Sportlern im Land; es gibt einen Mangel an adäquaten Bedingungen und finanziellen Ressourcen für bei der Heranbildung von Sportlern, was die Funktionen des Trainers einschränkt.

2. Die Motivation des Sportlers wird sowohl von den Trainern, als auch von den Sportlern selbst als die wichtigste Voraussetzung für das Erzielen von Hochleistungen betrachtet. Für die Sicherung der günstigen Bildungsumgebung, die die litauischen Sportler für die Hochleistungen motivieren, können folgende typische Merkmale erwähnt werden, die aufs engste verbunden sind mit: dem Trainer und dem Sportler *(das Streben nach den gemeinsamen Zielen; Partnerschaftsbeziehungen; gegenseitiges Verständnis und gegenseitige Achtung; das Vertrauen des Sportlers zur Kompetenz des Trainers; das Vertrauen des Trainers zu den Fähigkeiten des Sportlers; vertrauliches Nähegefühl; die Pflege der Beziehungen des Trainers zu den Familienangehörigen des Sportlers)*; den anderen Teilnehmern des Bildungsprozesses (*wechselseitiges Verständnis zwischen den Teammitgliedern; gegenseitige Unterstützung der Teammitglieder; gute Beziehungen zwischen den Sportlern und dem Service-Personal; die Partnerschaft der Sportler und des Service-Personals bei der verantwortlichen Erfüllung ihrer Pflichten*).
3. Die Faktoren, die die Motivation der Hochleistungssportler behindern und erschweren, sind mit folgenden Komponenten der pädagogischen Interaktion verbunden:
   a) emotionale Komponente *(große Konkurrenz unter den Sportlern, extrem hohe physische und psychische Belastung; der Druck von der Seite der Gesellschaft; Schuldgefühl des Sportlers, seiner Familie nicht genügend Aufmerksamkeit widmen zu können);*
   b) kognitive Komponente *(das Verstehen der Kompliziertheit des Hochleistung ssports;Unterschiede hinsichtlich des Geschlechts;keine Stabilität in der Strategie in der Ausbildung der Sportler, unterschiedliche Hoffnungen und Erwartungen des Trainers und des Sportlers; geringe Kommunikationskompetenz);*
   c) die das Verhalten steuernde Komponente *(nichtprofessionelles Verhalten des Trainers, Anwendung nicht passender Strategien nach den misslungenen Wettkämpfen und Niederlagen; die Unterwerfung des Trainers den Außenfaktoren).*

Literatur

1. Baker, J., Horton, S. (2004). A review of primary and secondary influences on sport expertise. *High Ability Studies*, 15, 211–228.

2. Brewer, B.W., Boin, P.D., Petitpas, A.J. (1993). *Dimensions of athletic identity.* Paper presented at the American Psychological Association annual conference, Toronto, ON.

3. Chantal, Y., Guay, F., Dobreva-Martinova, T., Vallerand, R. (1996). Motivation and Elite Performance: an Exploratory Investigation with Bulgarian Athletes. *International Journal of Sport Psychology*, 27 (2), 173–182.

4. DeWeese, B. (2010). Shedding Light On The Elite Coach-Athlete Dyad: Perspectives Of The Participants In The 2008 Men And Women's Olympic Trials Marathon. *Track Coach*, 192, 6143–6148.

5. Dózci, T. (2012) Gold fever: Sport and national identity – The Hungarian case. *International Review for the Sociology of Sport*, Vol. 47, no 2, p. 165-182.

6. Duffy, P., Lyons, D., Moran, A., Warrington, G., Macmanus, C. (2001). Factors Promoting And Inhibiting The Success Of High Performance Players And Athletes In Ireland. *Publications and Press Releases.* [žiūrėta 2010 m. kovo 3 d.]: <www.coachingireland.com/publication_and_press_release.htm>.

7. Giddens, A. (2000). *Modernybė ir asmens tapatumas: asmuo ir visuomenė vėlyvosios modernybės amžiuje.* Vilnius: Pradai, 2000.

8. Grove, J.R., Fish, M. & Eklund, R.C. (2004) Changes in athletic identity following team selection: Self-protection versus self-enhancement. *Journal of Applied Sport Psychology* 16: 75-81.

9. Hassandra, M., Goudas, M., Chroni, S. (2003). Examining Factors Associated with Intrinsic Motivation in Physical Education: a Qualitative Approach. *Psychology of Sport and Exercise*, 4, 211–223.

10. Horton, R.S., Mack, D.E. (2000) Athletic identity in marathon runners: Functional focus or dysfunctional commitment? *Journal of Sport Behavior* 23(2): 101-119.

11. Jowett, S. (2006). Interpersonal and Structural Features of Greek Coach-Athlete Dyads Performing in Individual Sports. *Journal of applied sport psychology*, 18, 69–81.

12. Jowett, S., Cockerill, I. M. (2002). Incompatibility in the coach–athlete relationship. In I. M.Cockerill (Ed.), Solutions in sport psychology (pp. 16–31). London: Thomson Learning.

13. Jowett, S., Meek, G. (2002). Closeness, Co-orientation and Complementarity in the family coach – athlete relationship: A case study. Manuscript under review.

14. Jowett, S. (2012). The coach-athlete relationship. *Sport & Exercise Scientist, 33, 16.*

15. Juodraitis, A. (2004). *Asmenybės adaptacija: kintamųjų sąveika.* Šiauliai: VšĮ Šiaulių universiteto leidykla.

16. Lenzen, B., Brouwers, M., Dejardin, R., Lachi, B., Cloes, M. (2004). Comparative Study of Coach-Athlete Interactions in Mixed Traditional Japanese Martial Art, Female Amateur Track and Field, and Male Professional Basketball. *International Journal of Sport Psychology*, 35 (1), 77–90.

17. Liubinienė,V. (2003). Value orientations in Estonian and Lithuanian society: sociological analysis from the gender and age perspective. *Socialiniai mokslai*, Nr. 3 (40), p. 48-55.

18. Malina, R. M. Bouchard, C. (1991) *Growth, Maturation and Physical Activity.* Champaign, IL: Human Kinetics. Prieiga per internetą: <http://books.google.lt> [žr. 2009 09 25].

19. Maxwell, J. A. (2002). Understanding and Validity in Qualitative Research. In Huberman, A. M., Miles, M. B. (Ed.), *The Qualitative Research Companion.* Thousand Oaks, CA, Sage, pp. 37–64.

20. Morgan, T., Giacobbi, J. (2006). Toward Two Grounded Theories of the Talent Development and Social Support Process of Highly Successful Collegiate Athletes. *Sport Psychologist, 20* (3), 295–313.

21. Pensgaard, A. M., Roberts, G. C. (2002). Elite Athletes‘ Experiences of the Motivational Climate: The Coach Matters. *Scandinavian Journal of Medicine & Science in Sports*, 12 (1), 54–59.

22. Perleth, C., Schatz, T., Mőnks, F. J. (2000). Early Identification of High Ability. in: Heller, K. A., Mőnks, F.J., Sternberg, R. J., Subotnik, R. F. (ed.), *International handbook of giftedness and talent (2nd edn).* Elsevier: Oxford., pp. 297–316.

23. Pfeffer, I., Würth, S., Alfermann, D. (2004). Die subjektive Wahrnehmung der Trainer-Athlet-Interaction in Individualsportarten und Manschaftspielen. *Zeitschrift für Sportpsychologie, Vol. 11, Issue 1,* p. 24-32.

24. Poczwardowski, A., Barott, J., Henschen, K. (2002). The Athlete and Coach: their Relationship and its Meaning, Results of an Interpretive Study. *International Journal of Sport Psychology*, 33 (1), 116–140.

25. Riemer, H.A., Chelladurai, P. (1998). Development of the Athlete satisfaction questionnaire (ASQ). *Journal of Sport and Exercise Psychology*, 20, 127-156.

26. Saborowski, C., Würth, S, Alfermann, D. (2000). Trainerinnen und Trainer im Leistungsorientierten Nachwuchtssport. *Leipziger Sportwissenschaftliche Beitrage*, vol. 41, Issue 2, p. 30-52.

27. Schulze, K., Burrmann U., Stucke Ch. (2007). Soziale Unterstützungsleistungen des Trainers. *Leistungssport, 37 (4), 4-10.*

28. Seller, R., Kevesligeti, C., Valley, E. (1999). Coaches and Female Athletes: the Basis of Interaction. *Motricidade Humana*, 12 (1), 67–76.

29. Skarbalius, A. (2003). *Didelio meistriškumo rankininkų rengimo optimizavimas: habilitacinis darbas.* Vilnius: VPU.

30. Stirling, A., Kerr, G. (2009). Abused Athletes' Perceptions of the Coach-Athlete Relationship. *Sport in Society, 12* (2), 227–239.

31. Strauss, A.L., Corbin, J. (1997). Grounded Theory in Practice, Sage.

32. Tietjens, M. (2001). *Sportliches Engagement und sozialer Rückhalt im Jugendalter. Eine repräsentative Surveystudie in Brandenburg und Nordrhein-Westfalen.* Lengerich: Pabst Science Publisher.

33. Weiss, O. (2001). Identity Reinforcement in Sport. Revisiting the Symbolic Interactionist Legacy. *International Review for the Sociology of Sport*, vol. 36, no. 4, 393-405.

34. Willimczik, K., Kronsbein, A. (2005). Leistungsmotivation in Verlauf von Spitzensportkarrieren. *Leistungssport*, 5 (35), 4–10.

35. Würth, S, Saborowski, C., Alfermann, D. (1999). Trainingsklima und Führungsverhalten aus sicht jugendlicher Athleten und deren Trainer. *Psychologie und Sport*, vol. 6, Issue 4, p. 146-157.

Sprache und Kultur im Wandel

Zu Transformationstendenzen der litauischen Grundwerte und Identität

Vida Rudaitienė, Ernesta Račienė

Abstract

The aim of the current paper is to analyze fundamental values transformation trends; to look at the English language as one of the most important factors of globalization and its influence on the Lithuanian language and in particular on its vocabulary; to highlight the importance of the Lithuanian language as a fundamental value to the Lithuanian nation and its traditional culture. The research of the Lithuanian nation's fundamental values and the Lithuanian language, as one of those values, under the social and cultural change is a relevant problem. The analysis of changes in fundamental values, especially the Lithuanian language, and Lithuanian culture can help predict the development of Lithuanian identity. The results of the research make it possible to blueprint some measures to foster the Lithuanian nation's fundamental values and to preserve the Lithuanian language and culture in the intensive globalization process. In this article, the manifestation of values transformation in the Lithuanian language is considered on the basis of the most recent observations data. The relationship between the society and the language it uses is analyzed from a new dimension. The main source of data is the public spoken language used on the radio and television.

Keywords: ethnic culture, phraseological unit, internationalization, zoomorphism

Santrauka

Straipsnyje analizuojamos pamatinių vertybių transformacijos tendencijos, nagrinėjama anglų kalbos, kaip vieno iš svarbiausių globalizacijos veiksnių, sustiprėjusi įtaką lietuvių kalbai, ypač jos leksikai. Straipsnyje iškeliama lietuvių kalbos, kaip pamatinės vertybės, svarba lietuvių tautai, jos tradicinei kultūrai.

Lietuvių tautos pamatinių vertybių, lietuvių kalbos, kaip vienos iš šių vertybių, tyrimas kintant socialinėms ir kultūrinėms aplinkybėms – aktuali problema. Pamatinių vertybių, ypač lietuvių kalbos, lietuvių kultūros pokyčių analizė gali padėti prognozuoti lietuvio tapatybės raidą. Atsižvelgiant į tyrimo rezultatus, galima numatyti ir kai kurių priemonių, padedančių puoselėti lietuvių tautos pamatines vertybes, išsaugoti lietuvių kalbos ir kultūros savitumą vykstant intensyviai globalizacijai.

Šiame straipsnyje vertybių transformacijos apraiškos lietuvių kalboje nagrinėjamos remiantis naujausiais stebėjimo duomenimis. Nauju aspektu analizuojama visuomenės ir jos vartojamos kalbos ryšys, t.y. kokia visuomenė, tokia ir jos kalba. Pagrindinis duomenų šaltinis – viešoji sakytinė kalba, paprastai per Lietuvos radiją ir televiziją ir rašytinė, t.y. periodinės spaudos kalba.

Reikšminiai žodžiai: etninė kultūra, frazeologizmas, internacionalizmas, zoomorfizmas.

1. Einleitung

Das Ziel des vorliegenden Beitrags ist die Analyse der Transformationstendenzen von Grundwerten und Identität des litauischen Volkes, die Untersuchung des Einflusses der englischen Sprache als eines der wichtigsten Faktoren der Globalisierung auf die litauische Sprache und besonderes auf ihre Lexik, die Hervorhebung der Bedeutung der litauischen Sprache als des Grundwertes für das litauische Volk und seine traditionelle Kultur.

Die Untersuchungsmethoden sind empirisch und analytisch. Die Grundlage für die Analyse bilden die neuesten Beobachtungsdaten aus den Jahren 2009-2013.

Die Untersuchung der Grundwerte des litauischen Volkes sowie der litauischen Sprache als eines von diesen Werten im Wandel von sozialen und kulturellen Kontexten stellt ein aktuelles Problem dar. Die Analyse der Transformationen von Grundwerten, besonders der litauischen Sprache und Kultur, kann bei der Prognose der Entwicklung von litauischer Identität helfen. Unter der Berücksichtigung der Untersuchungergebnisse könnten auch einige Maßnahmen vorausgesehen werden, um die grundlegenden Werte des litauischen Volkes bei der intensiven Globalisierung aufzubewahren.

Die Manifestation der Transformationen von Grundwerten in der litauischen Sprache ist ein wenig erforschtes Problem. Wechselbeziehungen von Gesellschaftsentwicklung, Sprache und kulturellem Wandel untersuchen unter verschiedenen Aspekten solche Forscher wie Aloyzas Aloyzas Gudavičius, Laima Kalėdienė, Simas Karaliūnas, Regina Koženiauskienė, Rūta Marcinkevičienė u.a.

Im vorliegenden Beitrag werden die Manifestationen der grundlegenden Werte in der litauischen Sprache aufgrund der neuesten Beobachtungsdaten analysiert. Unter neuem Aspekt wird die Beziehung der Gesellschaft und der Sprache untersucht, d.h. wie ist die Gesellschaft, so ist auch ihre Sprache.

Die Hauptquellen für Daten sind die öffentlich gesprochene Sprache im litauischen Rundfunk und Fernsehen (im Text werden nach Belegen die üblichen Abkürzungen der Radio-und TV-Sender angegeben; Zahlen bedeuten das Datum, wann der sprachliche Fakt vernommen wurde) und geschriebene Sprache der Periodika (siehe Quellenabkürzungen). Die eine oder andere angegebene Quelle bedeutet nicht, dass der sprachliche Fakt nur in dieser Quelle vorkommt. Die Illustrationsbeispiele werden gewöhnlich verkürzt.

2. Grundwerte und die Tendenzen ihrer Transformationen

2.1. Die Transformation der grundlegenden Werte wird von verschiedenen sozialen und kulturellen Faktoren beeinflusst. Der Globalisierungsprozess führte zu den Veränderungen im Leben der litauischen Nation, zum sozialen Wandel. Besonders großen Einfluss üben ständig entstehende neue Technologien, intensive Integration Litauens in die EU und in andere wirtschaftliche politische Strukturen, geopolitische Veränderungen, internationaler Informationsaustausch aus. Mit den Veränderungen im Leben der Nation ändern sich auch ihre Werte oder Einstellungen dazu. Manchmal verstehen Leute gar nicht mehr, was Werte sind. Typischerweise entsprechen die Werte dem Leben, dass heißt, wie das Leben ist, so sind auch die Werte. Werte und ihre Transformation manifestieren sich in der Sprache eines jeden Volkes, so auch in der litauischen Sprache. Die Sprache erfüllt ihre Funktion, indem sie verschiedene Seiten des menschlichen Lebens, alle Tätigkeitsbereiche widerspiegelt. Alle Formen des sozialen Lebens sind mit der Sprache verbunden. Die Entwicklung der litauischen Sprache hängt auch mit dem Wandel des Lebens der Nation und der grundlegenden Werte zusammen.

2.2. Sprache ist für jedes Volk ein grudlegender Wert. Laut Aloyzas Gudavičius, „ jede Sprache ist die Schatzkammer der Erfahrung des Volkes, die Widerspiegelung seines Lebens, seiner Geschichte, materieller und geistiger Werte, die Vorstellung der Nation von der Welt - ein besonderes, einzigartiges Phänomen“ (Gudavičius, 2000, 150). Sprache ist nicht nur der Ausdruck der nationalen Identität, die Grundlage für das Leben der Nation, sondern auch einer der wichtigsten Bestandteile der Kultur. Sprache als Teil der ethnischen Kultur zeichnet sich dadurch aus, dass sie die wesentlichen Merkmale der Gesamtheit der Kultur beinhaltet, durch sie wird das System der Werte, die gesellschaftlich verwurzelte Beziehung zu der Welt präsentiert. Jede Sprache erstellt ein anderes Weltbild, drückt die Weltsicht, die Vorstellung von der Welt aus. Wie Algis Mickūnas behauptet, „litauische Sprache drückt direkt und rein alles aus, worüber man spricht; man muss nur dieser Sprache *zuhören*, weil durch sie uns unsere Welt *gehört* und wir der Welt gehören“ (Mickūnas, 2011, 214). Die Sprache jedes Volkes widerspiegelt die Welt nicht direkt, sondern durch die von diesem Volk wahrgenommene Weltinterpretation. Für Litauer ist die betont emotionale

Beziehung zu der Welt charakteristisch. Der Ausdruck der emotionalen Wertung wird durch Reichtum litauischer Wortbildungsmittel, besonders Suffixe ermöglicht. Litauische Sprache, ihre Ausdrucksmöglichkeiten sind eng mit dem litauischen Geist, mit litauischer ethnischer Kultur verbunden. Sie bringt am besten die Besonderheiten der litauischen Kommunikation, der inneren Welt der Litauer zum Ausdruck, „sogar die Zugehörigkeit der menschlichen Seele zur Erde und Natur" (Mickūnas, 2011, 201). Es besteht der Zusammenhang von Sprache, menschlichem Weltbild und zwischenmenschlichen Beziehungen. Wichtig ist die Verbindung zwischen geistiger Welt des Menschen und seiner Sprache, dass heißt, wir sind so, wie wir sprechen.

Im Prozess der Globalisierung entstanden Veränderungen in den Beziehungen zwischen dem Litauischen und anderen Sprachen. Es werden die Bereiche übernommen, in denen allgemein die Landessprache der internationalen (englischen) Sprache weicht. Soziale Bedingungen sind für litauische Sprache ungünstig. Bei einer Anstellung sind fast nie gute litauische Sprachkenntnisse erforderlich. Gewöhnlich wird die Beherrschung der Fremdsprache, besonders des Englischen, betont. Wird eine andere Sprache bevorzugt, so werden eigene Sprache und eigene Kultur herabgesetzt. In Litauen gilt das Englische als Prestigesprache. Da das Ansehen des Englischen sehr hoch ist, so ist es schwer seinem Einfluss zu widerstehen. Zugleich sinkt das Ansehen des Litauischen. In der Zeit der Germanisierung, Polonisierierung, insbesondere der langen Periode der Russifizierung hat sich das litauische Volk den mit Gewalt aufgezwungenen Dingen widersetzt. Das Englische wird freiwillig übernommen. Verschwunden ist innere Stärke, diese soziale Absicherung, die geholfen hat, während der russischen Besatzung zu widerstehen.

Der Einfluss des Englischen auf die litauische Sprache ist groß. Die englische Sprache ist abstrakter als die litauische. Wegen des englischen Einflusses spricht man abstrakter. Litauische Wörter sind konkreter, und konkrete Wörter sind genauer. Viele Entlehnungen veringern die Genauigkeit des Ausdrucks. Im Litauischen steigt die Zahl der Entlehnungen, die anstatt von litauischen Wörtern verwendet werden. Die Bilanz des litauischen und fremden Wortguts wird gestört, die Eigenart des Litauischen wird beeinträchigt. Ein großer Teil von Entlehnungen kommt in die litauische Sprache über Übersetzungen. Besonders viel wird aus dem Englischen übersetzt. Durch die übersetzte Lexik wird auch das lexikalische System des Litauischen verändert. Das semantische System des Litauischen wird dem Englischen angenähert.

Im Litauischen steigt die Zahl der Wörter der englischen Sprache, die anstelle der litauischen Wörter oder der üblichen Internationalismen verwendet werden, z. B.: *Autsitas* (= Pailginta darbo savaitė) *yra tik keliose Europos šalyse (verlängerte Ar-*

beitswoche) LTV 2009-01-06; *Koalicijos sutarties pasirašymo dedlainas* (= galutinis terminas) *lapkričio 6 d. (Endtermin der Unterzeichnung von Koalitionsvertrag)* LRT 2012-11-02; *Kitais metais mes galime patirti defaultą* (= nemokumą), *valiutos devalvavimą (Zahlungsunfähig werden)* „Lietuvos rytas" tv 2010-01-24; *Dabar daug diskutuojama apie dženderinę* (= lyčių lygių galimybių) *ekonomiką* (*Genderwirtschaft*) LTV 2009-05-18; *Reikia susipažinti su opozicinių partijų siūlymų draftu* (= planu, projektu) *ekonominei krizei įveikti* (*Enfwurf der Vorschläge für Überwindung der Wirtschaftskrise*) „Lietuvos rytas" tv 2010-01-24; *Vyriausybės programa yra emeiluose* (= elektroniniuose paštuose), *internete* (*Regierungsprogramm ist in E-Mails*) LRT 2012-12-07; *Partijų imidžui* (= įvaizdžiui) *kenkia jų narių nesąžiningumas* (*Image der Parteien*) BTV 2009-01-24; *Reikia nutiesti šešis elektros linkus* (= šešias elektros jungtis) į Rytus (*elektrische Verbindungen nach Osten*) LRT 2012-11-22; *Kiek Lietuvos įmonių įregistruota į ofšorus* (= lengvatinių mokesčių, neapmokestinamas zonas) (*steuerfreie Zonen*)? LTV 2009-02-24; *Bankai nutraukė overdraftus* (= kredito perviršius) (Überziehungskredite) LRT 2012-11-09; *Siūloma vykdyti odos vėžio pilotinę* (= bandomąją) *programą* (*Pilotprogramm für Hautkrebs*) LRT 2012-12-08; *Būtinas įvykių politinis skryningas* (= vertinimas) (*politische Evaluierung*) LTV 2009-04-07; *Svarbu teikiamų finansuoti projektų objektyvus skryningas* (= objektyvi atranka) (*objektive Auswahl von Projekten*) LTV 2009-01-02. *Vykdoma moterų sveikatos skryningo* (= patikros) programa (*Untersuchung von Frauengesundheit*) LTV 2009-01-14; *Prezidentas padarė taimautą* (= pertraukėlę) (Präsident hat *Pause gemacht*) LTV 2009-06-04.

In der litauischen Sprache nimmt die Zahl der Internationalismen, die litauische Lexik aus dem Gebrauch verdrängen. Ständig werden internationale Verben verwendet, wie z.B.: *generuoti (generieren), identifikuoti (identifizieren), inspektuoti (inspizieren), komunikuoti (kommunizieren), koreliuoti (korrelieren), realizuoti (relisieren), subsidijuoti (subventionieren), traktuoti (betrachten) u.a.* Viele internationale Verben werden aus dem Englischen übernommen und der litauischen Sprache angepasst. Viele von diesenn Verben sind ungewöhnlich und verletzen auch die litauischen Sprachnormen. Es entstehen immer neue solche Verben im Litauischen, obwohl es litauische Entsprechungen oder mehr geläufige entsprechende Internationalismen gibt, z. B.: *Defliacijos augimas nėra tvarus, todėl neindikuoja* (= nerodo) *ekonomikos gerėjimo* (Deflation *zeigt* keine Wirtschaftsverbesserung) BTV 2009-12-13. *Tikimasi, kad atsiras su Rusija dialogas, kuris indikuos* (= reikš, rodys) *santykių pagerėjimą* (der Dialog mit Rusland *bedeutet/ zeigt* Verbesserung der Beziehungen) LTV 2010-02-22; *Nesutinku, kad rusofobija yra mums indoktrinuota* (= įskiepyta; įdiegta), *nes pažįstu ir tremtinių, norinčių gerų santykių su Rusija* (*Russophobie wird indoktriniert*) (LTV 2010-02-22. *Seimas indoktrinavo* (= įteisino; įdiegė) *Statybos įstatymo pataisas* (*Seimas indotri-*

nierte Baugesetzänderungen) LRT 2012-11-20. *Politikai neturėtų intervencijuotis* (= kištis) į STT darbą (*Politiker sollten sich nicht einmischen*) LTV 2009-05-04. *Teismų darbas neturėtų intervencijuojamas* (= trukdomas) (*Die Arbeit von Gerichten sollte nicht gestört werden*) LRT 2012-10-22; *Mes pozicionuojame* (= pristatome) *Kauno oro uostą kaip pigių skrydžių bendrovę* (*Wir positionieren den Flughafen von Kaunas als kostengünstige Fluggesellschaft)* LTV 2010-03-08. *Mes pozicionuojame* (= pristatome; teikiame) *Lietuvą kaip globalią valstybę* (*Wir positionieren Litauen als globalen Staat*) LTV 2010-04-19. *Kai kuriuose darbo kolektyvuose profesionalūs, principingi žmonės stigmatizuojami* (= niekinami) (*professionelle, prinzipielle Menschen werden verachtet*) (LRT 2012-10-08. *Kodėl egzistuoja principas stigmatizuoti* (= niekinti) *visuomenę?* (*Gesellschaft wird stigmatisiert*) LRT 2012-12-06; *Nėra būtinybės simplikuoti* (= supaprastinti) *požiūrį į lietuvių pamatines vertybes* (*Es besteht keine Notwendigkeit den Ansatz zu litauischen Grundwerten zu simplifizieren (= vereinfachen*) LNK 2012-09-06; *Nenorėčiau trivializuoti* (= sumenkinti; subanalinti) įmonėms keliamų reikalavimų dėl įtraukimo į sąrašą Lietuvos biržoje (*die Forderungen an die Unternehmen trivialisieren*) BTV 2011-06-12. *Pamatinės lietuvių tautos vertybės trivializuojamos* (= sumenkinamos; subanalinamos) (*die Grundwerte des litauischen Volkes werden trivialisiert*) LRT 2012-12-04.

Wenn es litauische Entsprechungen für Internationalismen gibt, wird es empfohlen sie zu verwenden. Im Artikel 12 des Gesetzes über das Verfahren der Vorbereitung von Gesetzen und anderen juristischen Vorschriften, das die Sprache von juristischen Akten reglementiert, wird festgelegt, dass „ internationale Wörter erst dann verwendet werden, wenn es keine litauischen Entsprechungen für diese Wörter gibt“ (*Valstybės žinios*, 1995, Nr. 41–991). Der Gebrauch von internationalen Wörtern in der litauischen Sprache hängt auch vom funktionalen Stil ab. Außerdem können auch die synonymisch verwendeten internationalen und litauischen Wörter nicht immer einander ersetzen, weil nur „ der Kontext bestimmt, welches von ihnen für die Genauigkeit des Ausdrucks passt„ (Koženiauskienė, 2005, 304). Manchmal kann der Gebrauch von internationalen Wörtern Mangel an Respekt nicht nur für die litauische Sprache, sondern auch für das litauische Volk, die Kultur zeigen. Für diejenigen, die das Gefühl der litauischen Sprache nicht verloren haben, ist es schwierig folgende Ausdrücke zu hören, wie z. B.: *Bibliotekoms trūksta kultūros žurnalų, tai trukdo cirkuliuoti kultūrai (Bibliotheken fehlt es an Kulturzeitschriften, das verhindert die Zirkulation der Kultur, wörtlich: verhindert Kultur zu zirkulieren); Kaimo verslo diversifikavimas kuria darbo vietas* (*Diversifizierung des ländlichen Geschäfts schafft Arbeitsplätze)* LTV 2010-08-04. *Lietuviškas kaimas diversifikuojasi (Litauisches Dorf diversifiziert sich)* LTV 2009-05-19; *Lietuvos žemdirbiai linkę inovuotis* (*Litauische Bauern neigen zu Inovationen, wörtlich: zu inovieren)* LRT 2013-03-08; *Pastebima tam tikrų Lietuvos kaimo raidos indikacijų (Es gibt einige Indikationen der litauischen ländlichen Entwicklung)* LTV 2009-01-02; *Lietuvos etninės kultūros indikatoriai prastėja (Indikatoren der litauischen ethnischen Kultur verschlechtern sich)* LTV 2009-02-13; *Susirūpinta moterų reprodukcine sveikata (Man ist für die reproduktive Gesundheit von Frauen besorgt)* LTV 2012-05-30.

Natürlich werden Internationalismen als passende Entlehnungen für das Litauische betrachtet. Ohne internationale Wörter wäre der Umgang mit anderen Nationen nicht möglich, doch ihre Verwendung „sollte nicht die Tür in die gemütliche Welt schließen, die das litauische Sprechen eines Menschen zeichnet„ (Mickūnas, 2011, 195).

Den Internationalismen werden auch neue Bedeutungen verliehen, die der litauischen Sprache untypisch sind. Diese Bedeutungen basieren auf den Bedeutungen der entsprechenden englischen Wörter. Einige Beispiele können diese Terndenz illustrieren. Nach den Normen des Litauischen kann man nur Personen *delegieren* und nicht die Funktionen, Pflichten, Kompetenzen usw. Also ist das internationale Wort *deleguoti (delegieren)* mit den Wörtern, die keine Person bedeuten, ungebräuchlich. Vieleicht auf Grund der Bedeutung des entsprechenden englischen Wortes *delegate* wird das oben genannte internationale Wort in der Bedeutung *suteikti (verleihen), perduoti (übergeben)* verwendet, z. B.: *Apskričių reforma vykdoma skubotai, neaišku, kokios funkcijos deleguotos* (= perduotos) *savivaldybėms* (Funktionen werden an Selbstverwaltung delegiert) „Lietuvos rytas" tv 2010-03-21. *Ar prezidentas turi naudoti jam deleguojamus* (= suteikiamus) *įgaliojimus?(die Vollmacht wird an Präsidenten delegiert)* LTV 2009-03-16. *Sveikatos apsaugos ministerijai deleguota* (= suteikta) *teisė parengti pinigų banką* (an das Gesundheitsministerium wurde das Recht delegiert…) LTV 2011-06-06.

Ungewöhnlich wird das Partizip *motyvuotas, -a verwendet.* Dieses Partizip wird verwendet, wenn man über eine Person spricht und sie dabei bewertet, meistens über einen Fachmann, Beamten, Politiker, Studenten usw. Besser passt die Wortverbindung *turintis, -i motyvaciją* (= wörtlich Motivation habende) oder *atsidavęs (-usi) specialistas (-ė)(Fachmann, Fachfrau), valdininkas (-ė) (Beamter, Beamtin), politikas (-ė)(Politiker, Politikerin), studentas (-ė)(Student, Studentin) usw., z.B.: Dauguma motyvuotų (= turinčių motyvaciją) žmonių gali rinktis geresnės kokybės aukštąsias mokyklas* (motivierte Leute) LTV 2012-01-11. *Mes galime pasirinkti motyvuotus (= turinčius motyvaciją; atsidavusius) mokytojus* (motivierte Lehrer) LTV 2012-02-23. Solche Verwendung des Partizps *motyvuotas, -a* ist eine Übersetzung aus dem Englischen. Den Normen der litauischen Sprache nach motiviert sein können nur das Benehmen des Menschen, seine Arbeit oder Tätigkeit, seine Handlungen, aber nicht der Mensch selbst.

Nach dem Beispiel der englischen Sprache wird Amnestie nicht nur Personen, sondern Gegenständen, Erscheinungen gewährt, z.B.: *nelegalaus verslo amnestija(Amnestie des illegelen Unternehmens)* BTV 2010-03-01, *skolų amnestija (Amnestie der Schulden)* LTV 2010-04-15, šešėlinių pinigų *amnestija* (*Amnestie des Schattengeldes*) TV3 2010-03-01. Besser wäre die üblichen Wörter *legalizuoti, legalizavimas; įteisinti, įteisinimas (legalisieren, Legalisierung)* zu benutzen.

Die Sprache jeder Nation ist einzigartig. Also kann man nicht alle Wörter oder ihre Bedeutungen von anderen Sprachen, z. B. dem Englichen, übernehmen und in der litauischen Sprache legitimieren. Man kann bemerken, dass verschiedene Völker bei der Übernahme derselben Fremdwörter ihnen unterschiedliche Bedeutungen verleihen. Zum Beispiel, das Adjektiv lateinischer Abstammung *originalus, -i* hat im Englischen nicht nur die im Litauischen verwendeten Bedeutungen 1. tikras, neperdirbtas (echt, unverarbeitet); 2. savitas, savotiškas (eigentümlich, eigenartig, unverwechselbar); sondern auch „pirmasis, pirminis, pradinis; autentiškas; naujas, naujoviškas“ (ursprünglich; authentisch; neu, innovativ). Das Adjektiv lateinischer Abstammung *aktualus, -i* enttspricht im Englischen der im Litauischen verwendeten Bedeutung „zu einem Zeitpunkt wichtig“ und auch der Bedeutung von Adjektiven *faktinis, -ė, faktiškas, -a* „basierend auf Fakten, sicher, unausgedacht“. Also is es sehr wichtig, ob das internationale Wort in dieser Bedeutung genutzt wird, in der es seit langem im Litauischen verwendet wird. Wenn die Grenzen der Wortbedeutung gebrochen werden, führt es unvermeidlich zu Zweideutigkeit, manchmal zu Mißverständnissen

Es besteht der Bedarf, litauische Sprache als nationalen Wert zu fördern, das Prestige der litauischen Sprache zu erhöhen. Das Vertrauen in litauische Sprache ist auch die Grundlage für ihr Überleben. Laut Romualdas Grigas, „ist Sprache, ihre Nutzung und ihre Perspektive die empfindlichste Stelle der litauischen Identität und daher auch der Raum für die Kontinuität des Volkes“ (Grigas, 2004, 31).

2.3. Pragmatismus, Konsum sind Begleiterscheinungen der Globalisierung. Dies wirkt sich auf die Werte-Bildung aus. Idealismus jener Zeiten verwandelt sich in eine Sache der Vergangenheit. Existenzbedürfnisse werden zu zentralem Wert. Gewinn, wirtschaftlicher Nutzen sind oft verwendete Begriffe. Betont wird z.B. *miško ekonominė nauda* (wirtschaftlicher Nutzen des Waldes) LTV 2011-06-03, *Baltijos jūros ekonominė nauda* (wirtschaftlicher Nutzen der Ostsee) LTV 2011-03-06. Zu oft spricht man von Geld, häufig verwendet man Diminutivform *pinigėliai (Geldchen)*, pvz.: *Trūkstamų pinigėlių turėtume daugiau surinkti čia, viduje, o ne iš tarptautinių fondų* (fehlendes Geldchen) LTV 2010-06-10. *Trūksta pinigėlių atlyginimams ir pensijoms didinti* (Es fehlt an Geldchen für Gehälter) LNK 2012-06-14.

Die Entfaltung des Pragmatismus, des Konsums der Gesellschaft zeigt der aktive Gebrauch der sogenannten Marktlexik in der litauischen Sprache. Als

Beispiel kann man das internationale Wort *produktas* (lot. *productus* – pagamintas) anführen. Nach dem Vorbild der englischen Sprache ersetzt dieses Wort das litauische Wort *kūrinys* (Werk, Schöpfung)), pvz.: *Atpigo meno produktai(Kunstprodukte sind billiger geworden)* LTV 2012-04-03. *Dabar dažnai nuvertinami kultūros produktai* (Kulturprodukte) LTV 2012-02-10. *Jie* [aktoriai] *nuoširdžiai ir sunkiai dirbdavo visą savaitę, todėl ir parengdavo gerą produktą(Schauspieler haben gutes Produkt vorbereitet)* LR (TVA) 2012 8 11. *Knygų mugėje unikalus produktas – knyga slaugantiems ligonius namuose (Buch als unikales Produkt)* LRT 2013-02-21. Sogar Menschen werden als Produkte bezeichnet, z.B: *Abi mūsų dukterys – angliškos kultūros ir anglų kalbos produktas* (Unsere Töchter sind das Produkt der englischen Kultur und Sprache) LR (TVA) 2011 43 53. *Baltarusijos prezidentas – unikalus politinis produktas* (Der Präsident Weißrusslands it ein unikales politisches Produkt) LTV 2011-05-02. *Britai plaukikę R.M. mato kaip savo produktą* (Briten betrachten die Schwimmerin R.M. als ihr Produkt) LRT 2012-08-06. Šalys gauna tikrą *produktą – medicinos specialistus* LR 2011 251. Unseres Erachtens sollte man zwischen *produktas* (Produkt) und *kūrinys* (Werk, Schöpfung) unterscheiden, das Schaffen von der Produktion trennen.

Ein weiteres Beispiel für den Pragmatismus der Gesellschaft, schnelles Lebenstempo ist auch die sinkende Verwendung des Verbs *kurti* (schaffen). Man ersetzt es durch *daryti (machen).* Radio- und Fernsehsendungen, verschiedene Strategien, Programme, Vorträge werden *gemacht.* Sogar Festspiele, Konzerte, Aufführungen, Musik, Lieder macht man oft. Natürlich ist es einfacher zu machen als zu schaffen, weil das Schaffen eine höhere Stufe der menschlichen Tätigkeit ist. Etwas zu tun ist manchmal schneller und einfacher als zu schaffen, doch gemachte Dinge sind auch einfacher. Es wird empfohlen, sogar Freude zu *machen*, z. B.: *Pasidaryk draugų, ir verslas klestės* (wörtlich: mache dir Freunde und das Unternehmen wird florieren) LTV 2010-11-06. In der englischen Sprache wird der übliche Ausdruck *to find a friend* auch durch den neuen Ausdruck *to make friends* verdrängt (vgl. Gudavičius, 2000, 157). Die ganze Zeit hat man das Geld verdient (*pinigus už(si)dirbti)*, jetzt hört man immer häufiger den Ausdruck *Geld machen* (*daryti pinigus)*, z.B.: *Aš organizuoju paramą arba darau pinigus (Ich organisiere Unterstützung oder mache Geld)* LTV 2011-03-15. Früher sollten Litauer *darbą, darbus (nu)dirbti (*Arbeit, Arbeiten erledigen, arbeiten), jetzt neigt man dazu *darbą, darbus daryti(*Arbeit, Arbeiten machen), z.B.: *Kai reikėjo sunkiausius darbus daryti, jūs bėgote į krūmus* (die schwersten Arbeiten machen) LNK 2011-03-14.

2.4. Der übertriebene Konsumismus, das Streben nach materiellem Gewinn fördert den moralischen Niedergang der Menschen, man beobachtet Manifestationen der moralischen Krise. Laut Stanislovas Juknevičius, gehört das Erosionstempo der moralischen Vergebung und Glauben der Litauer zu den schnellsten in Europa (Juknevičius, 2006, 131). Betroffen sind solche grundlegenden Werte des litauischen Volkes wie Ehrlichkeit, Respekt, Wahrheit, Fleiß. Bei der Beurteilung einer Person wird der Begriff *(k)ein anständiger Mensch* immer weniger verwendet. Solche Gefühle wie Scham, Anstand werden verloren. Die Verantwortung für seine Worte, Handlungen, Lebensstil ist geschwächt. Es beginnen Aggression, Respektlosigkeit, Schmälern des anderen Menschen zu dominieren. Ethische Normen werden lässsig betrachet oder ignoriert. Litauische Sprache, besonders ihre Lexik passt sich diesen Erscheinungen an. Beim Sprechen oder beim Schreiben bemüht man sich schärferes Wort zu verwendet, Emotionen zu beeinflussen, z.B.: *Ministrai guodžiasi, kad juos išdūrė arba pakabino* (wörtlich: Minister wurden *ausgestochen oder aufgehängt* = umg. betrogen) LTV 2011-02-11. *Vyriausybė mus išdūrė* (wörtlich: Die Regierung hat uns *ausgestochen* = umg. betrogen) LRT 2012-08-20; *Pasvarstykime, kas liktų iš A.K. vyriausybės iškirtus du ministrus* (wörtlich: was bleibt von der Regierung, wenn zwei Minister *gefällt werden* = entlassen werden) LTV 2012-04-14.

In der öffentlichen gesprochenen und geschriebenen Sprache nehmen herabwürdigende, sogar erniedrigende Worte zu, z.B.: *FNTT vadovai – puspročiai* (Leiter von FNTT sind *Halbverrückte)* LNK 2012-02-19. *Į valstybinį sektorių papuola bepročiai arba vagys* (Im öffentlichen Sektor sind *Verrückte oder Diebe*) LR 2010 254. *Laikas baigti su tuo subinlaižiavimu* (Schluss mit dem *Arschlecken*) init 2009-03-17. *Konstitucinis Teismas – tai ne bokštinis kranas, kuris vienu kabliu visą Seimą iš smirdančios mėšlo krūvos galėtų ištraukti* (*stinkender Haufen Scheiße*) LR 2010 260. Šitiems *politiniams banditams reikėtų baigti spekuliuoti mergaitės likimu* (*diese politischen Banditen*) LNK 2012-05-22. Es entsteht immer neue Lexik, die litauische Ethik und die Etikette der litauischen Sprache verletzt, z.B. *politikos kūdra* (Teich der Politik)LR 2012 175, *politinė pliurzė* (politisches Schlamassel) LTV 2012-02-26, *politinis turgus* (politischer Markt) LR 2011 82, *politinis tvaikas* (politischer Gestank)LR 2012 128, *valdžios lovys* (der Trog der Macht) LR 2011 18.

Abwertende Einstellung drückt auch die Verwendung der Wörter in übertragener Bedeutung. Um sich schärfer, beleidigender, verletzender, frecher auszudrücken, verwendet man medizinische Termini und andere Lexik, z. B.: *Ar miškininkai yra impotentai, kad negali priimti įstatymų?* (Sind Forstwirte *impotent*, dass sie keine Gesetze verabschieden können) init 2011-03-29. *Kai kurie Seimo nariai ir toliau demonstruoja savo politinę impotenciją* (einige Parlamentsmitglieder zeigen *ihre politische Impotenz*) LRT 2012-07-29; *Ar tai tikrai nuoširdi kova su alkoholizmu, o gal nuoširdus valdžios vyrų debilizmas?* (*Debilismus der Machthabenden*) LR 2011 47; *Ne visi politikai yra idiotai* (nicht alle Politiker sind *Idioten*) LR 2012 5; *Vyriausybės nariai elgiasi kaip politiniai autistai* (Regierungsmitglieder verhalten sich wie *politische Autisten*) init 2009-05-06; *Mes neturime skleisti įtarumo ir paranojos* (Wir sollen nicht *Misstrauen und Paranoia* verbreiten) LTV 2010-10-15; *Mūsų Vyriausybės nariai linkę į šizofreniją* (Mitglieder unserer Regierung sind neigen zu *Schizophrenie*) LTV 2010-02-28; *Negalima vieną dieną kažką rašytis su vienais, o jau tos pačios dienos vakare pasirašyti su kitais – tai politinė prostitucija* (*politische Prostitution*) LR 2010 51. Natürlich zeigt diese verletzende Lexik den kulturellen Niedergang der Personen, die sie verwenden.

Einigen gewöhnlichen Wörtern werden negative Konnotationen verliehen, z. B..: *gaidys (Hahn), (riebus)katinas (fetter Kater), kurmis (Maulwurf), palydovas, -ė (Begeleiter, Begleiterin), plaštakė (Schmetterling).* Berufsbezeichnungen *kasininkas (Kassierer), laborantas (Laborant), santechnikas (Klempner), sąskaitininkas (Buchhalter)* werden verwendet, um schlecht arbeitende Menschen zu benennen. Alle diese Wörter bekommen pejorative Bedeutungsschattierungen, zugleich werden auch die mit diesen Wörtern bezeichneten Menschen verschmäht, denn „konnotative Bedeutung ist die Information des Wortes über emotionale Beziehung des Sprachnutzers zu dem genannten Sachverhalt“ *(Lietuvių kalbos enciklopedija*, 1999, 327).

Auch andere grundlegende Werte des litauischen Volkes werden herabgesetzt und sogar verachtet. Einige Beispiele dazu: in pejorativer Bedeutung wird der Ausdruck *Lietuva – Marijos žemė (Litauen ist Mariens Land)* LTV 2010-08-14 verwendet. Mutter Gottes Maria wird zum Symbol des Banalen, der Verachtug, Geringschätzung. Wörter *patriotizmas, patriotas, -ė* (*Patriotismus, Patriot, Patriotin*) werden zum Objekt des Spottes, verspottet werden auch Menschen, die sich zu patriotischen Ideen bekennen. *Patriotas, -ė* (Patriot, Patriotin) wird auch mit dem in der negativen Bedeutung verwendeten Wort *nacionalistas, -ė (Nationalist, Nationalistin)* gleichgesetzt. Der Gebrauch von Begriffen *Lietuva, Lietuvos valstybė* (Litauen, litauischer Staat) tritt zurück. Anstatt dieser Begriffe werden immer häufiger Ausdrücke *mūsų kraštas, mūsų šalis (unser Land)*, sogar *mūsų regionas (usere Region)* verwendet. Auch der Name Litauens (Lietuva) wird oft entstellt, falsch ausgesprochen *Letuva*, man verwendet *letuvis, -ė, letuvys, -ė* (anstatt richtig lietuvis, lietuvė = Litauer, Litauerin) usw. Ständig betont man Begriffe *pilietis, -ė, (Staatsbürger, Staatsbürgerin), pilietinė visuomenė (Zivilgesellschaft), pilietiškumas (Staatsbürgerlichkeit)*, doch man vergißt *tauta (Volk), tautinė visuomenė (nationale Gesellschaft), tautiškumas (Nationalbewusstsein).* Die Bedeutung des Nationalbewusstseins ist sehr groß, weil „nur das Volk als Organisationsform am besten den Lebensbedürfnissen der Individuen entspricht. Mit der Vernichtung des Nationalbewustseins wurde auch die Entwicklungsmöglichkeit des Individuums vernichtet“ (Neimantas, 1999, 107).

2.5. Fleiß gehörte für Litauer immer zu Grundwerten. Litauer schätzten immer fleißige Menschen. Im *Wörterbuch der litauischen Sprache* finden wir viele

Bezeichnungen des fleißigen Menschen: *darbštuolis, -ė, darbininkas, -ė, darbuočius, darbuolis, -ė, darbštukas, -ė, darbius, -ė, darbišius, -ė* ir kt. In zahlreichen Phraseologismen Idiomen, bildhaften Vergleichen wird diese schöne Eigenschaft des Menschen gepriesen. Litauer verhönten Faulenzer und verspotteten nie tüchtige, pflichtbewusste Menschen. Diejenigen, die ein Gefühl für litauische Sprache haben, werden nie fleißige Menschen als *darbomanas, -ė (wörtlich: Arbeitssüchtige)* bezeichnen, was auch ironische Konnotation hat. Umso mehr *darboholikas (vgl. alkoholikas = Alkoholiker)*. Bei dieser Bezeichnung werden krankhafte Abhängigkeit, krankhafter Fleiß betont, aber Litauer haben nie Fleiß als krankhafte Eigenschaft betrachtet. Allerdings verändert sich das Konzept des Fleißes, mehr als fleißiger Mensch wird ein schlauer, starker unnachgiebiger, betrügerischer geschätzt.

Litauer schätzten und schätzen auch die Arbeit hoch. Arbeit ist in Litauen für viele unvermeidliche ökonomische Notwendigkeit, Existenzgrundlage und auch ein großer Wert. Nach den Angaben von Alfonsas Algimantas Mitrikas schätzen Arbeit 60% Respondenten der Befragten hoch (Mitrikas, 2006, 155). Doch der Begriff der Arbeit selbst wird erweitert, auch seine Bezeichnung ändert sich. Z. B.: synonymisch verwendet werden Ausdrücke *sekso paslaugos (Sexdienstleistungen, Sexdienste)* ir *sekso darbas (Sexarbeit)* LR (RV) 2011 1. Als Synonyme verwendt man auch Bezeichnungen *prostitutė (Prostituierte) und sekso darbininkė (Sexarbeiterin)* LR (RV) 2011 1. Immer häufiger verwendt man Begriffe *verslas (Unternehmen, Geschäft), verslininkas, -ė (Unternehmer, Unternehmerin).* Das Konzept des Unternehmens wird erweitert, es entstehen neue Arten der Unternehmen und ihre Bezeichnungen, z. B.: *smulkusis verslas (Kleinunternehmen), stambusis verslas (Großunternehmen), vidutinis verslas (mittelständisches Unternehmen), kaimo turizmo verslas (ländliches Tourismus-Geschäft)* LR (GB) 2012 28, *lošimo verslas (Glücksspiel-Geschäft)* LR 2010 12, *narkomanijos verslas (Drogen-Geschäft)* LR 2012 198, *pažinčių verslas (Bekanntschaften-Geschäft)* LR (TVA) 2012 29 9, *prostitucijos verslas* (*Prostitution- Geschäft* LR (RV) 2011 1. Ein Teil der Beschäftigungen, die als Unternehmen genannt werden, sind für litauische ethnische Kultur fremd, nur schwer zu akzeptieren, wie z.B. *laidojimo verslas* (Bestattungs-*Geschäft*) LR 2011 255. Besser passt die für Litauer übliche Bezeichnung *laidojimo paslaugos (Bestattungsdienstleistungen).* Auch solche Erscheinungen wie *spekuliacija (Spekulation), sukčiavimas (Betrug) u.a.* werden als *verslas (Unternehmen, Geschäft) bezeichnet.* Es entstehen neue Begriffe, die als Masskierung, Euphemismen für diese konkreten negativen Erscheinungen verwendet werden, z. B.: *nelegalus verslas (illegales Unternehmen)* LR 2012 163, šešėlinis verslas (Schattenunternehmen) LR 2010 4; *juodoji buhalterija (schwarze Buchhaltung)* LR 2009 14, *juodieji pinigai (schwarzes Geld)* LR 2009 10, *nešvarūs pinigai (schmutziges Geld)* LR 2011 6, šešėliniai pinigai (Schattengeld) LR 2012 7. Unehrliche, oft auch hinterhältige Taten nennt man *juodosios technologijos* (*schwarze Technologien*) LR 2011 58.

2.6. Ernest Hemingway hat gesagt, dass zwischenmenschliche Beziehungen der einzige Wert in der Welt sind. Beziehung des Menschen zu dem Menschen, Beziehungen in der Gemeinde waren immer sehr wichtig für das litauische Volk. Heutzutage sind Gemeinschaftsgefühle, Gefühle für aufrichtige Beziehungen geschwächt. Laut einer Umfrage, schätzen Freunde als eine sehr wichtige Sache litauische und pakistanische Befragte am geringsten (vgl. Mitrikas, 2006, 154–155). Tendenzen der Vereinsamung, Entfremdung werden deutlich sichtbar. Ohne direkten persönlichen Umgang entsteht laut Antanas Andrijauskas „der durch Computerbildschirm-Kultur geformte Kommunikationstyp" (Andrijauskas, 2006, 93). Die Normen der ethnischen zwischenmenschlichen Beziehungen verschwinden. Als die Menschen den Normen der ethnischen Beziehungen folgten, gab es mehr Ehrlichkeit, Respekt, Wahrheit. Neue Technologien verändern die Kultur des gesellschftlichen Verkehrs, ihre Ausdrucksformen, Arten der Informationsdarstellung. Das für Litauer übliche vertraute Wort *bendrauti* (*verkehren, umgehen*) wird durch *komunikuoti, kontaktuoti (kommuniziehren, kontaktieren)* ersetzt. Erwachsene, Kinder kommunizieren, kontaktieren (*komunikuoja, kontaktuoja)* miteinander, aber nicht verkehren, in Beziehung, in Verbindung treten, stehen (*bendrauja, susisiekia, palaiko ryšį).* Selbst *bendravimas, bendravimo būdai* (*Beziehungen, Verkehrsarten*) werden als *bendravimo technologijos* (*Kommunikationstechnologien*) bezeichnet. Vielleicht deswegen entfremdet man sich, verliert man Identität.

Die Einheit von einer kleinen Menschengruppe ist eine Familie. Die Familie stellt den Grundwert des litauischen Volkes dar. Es ist die Quelle des Kulturerbes und moralischen Ressourcen. Nach den Angaben der Untersuchung von A. A. Mitrikas, Familie ist sehr wichtig im Leben für 67 Prozent der Befragten in Litauen (Mitrikas, 2006, 155). Obwohl die Familie als Wert anerkannt wird, wird die Zahl der Eheschließungen in vielen Ländern deutlich reduziert. Immer mehr Menschen unterstützen die Ansicht, dass die Ehe eine veraltete Sache ist. 1999-2001 stimmten dieser Meinung in Litauen 21 Prozent der Befragten zu, 1990 nur 9 Prozent (Mitrikas, 2006, 161). Die Veränderungen des Konzepts von Familie widerspiegeln sich auch in der litauischen Sprache. In der sowjetischen Periode kamen in den Wortschatz der litauischen Sprache neue Begriffe *vienišа motina (alleinerziehende Mutter), sugyventinis, -ė (Lebensgefährte, Lebensgefährtin).* Der Begriff der Familie wird erweitert. Zu den litauischen Traditionen entsprechenden Begriffen *santuoka (Ehe), vyras (Gatte), žmona (Gattin), sutuoktinis, -ė (Ehegatte, Ehegattin)* kommen *partnerystė (Partnerschaft), (gyvenimo) partneris, -ė,*

(Lebenspartner, Lebenspartnerin, (gyvenimo) draugas, -ė (Lebensfreund, Lebensfreundin hinzu. Mann spricht auch über die Legalisierung der homosexuellen Familienbildung in Form der Ehe, es entstand ein neuer Begriff *homoseksualų santuoka* (*Homosexuellenehe*) LR 2012 110. Schon in der Sowjetzeit war der Begriff *nesantuokinis vaikas (uneheliches Kind*) geläufig. Jetzt wird dieser Begriff nicht betont, obwohl die Zahl der unehelichen Kinder zunimmt. Fakten zeigen, dass jedes dritte Kind in Litauen unehelich geboren wird.

2.7. Kultur, vor allem ihr ethnischer Teil, stellt den grundlegenden nationalen Wert dar, die Überlebensbasis eines jeden Volkes, auch des Litauischen, dar. Globalisierung ist einer der Faktoren, die die Grundlagen der traditionellen Kultur der Völker zerstören. Jedes Volk, das Identität schätzt, widersetzt sich der Globalisierung, der Nivellierung. Globalisierung in der heutigen Welt manifestiert sich in der Regel als eine weltweite Ausbreitung der westlichen Kultur. Auf der anderen Seite, laut St. Juknevičius, „ist es unmöglich, westliche wissenschaftliche und technische Errungenschaften, die Demokratie zu nutzen und ihre Werte oder Lebensweise nicht zu übernehmen" (Juknevičius, 2006, 119). Man kann sehen, dass territorialer, politischer, auch ökonomischer Imperialismus durch den kulturellen Imperialismus ersetzt wird.[1] Samuel P. Huntington hat noch im Jahr 1993 vorhergesagt, dass „die Hauptquellen von Konflikten in der neuen Welt nicht ideologisch oder wirtschaftlich sein werden. Zu der dominierenden Quelle von Fragmentierung der Menschheit und Konflikten werden kulturelle Unterschiede" (Huntington, 1998, 1).

Kultur, ihre Transformationsmanifestationen sind auch in der litauischen Sprache festzustellen (vgl. Rudaitienė, 2006, 363–367; Račienė, Rudaitienė, 2008, 209-224). Kulturelle Prozesse zeigen auch neue Begriffe in der litauischen Sprache, z.B.: *kultūriniai mainai* (*kultureller Austausch*) LR 2010 12, *kultūrų dialogas* (*Kulturendialog*) LR (RV) 2012 34, *kultūrų sąveika* (*Wechselwirkung der Kulturen*) LR (RV) 2009 20, *tarpkultūrinis būvis* (*interkultureller Zustand*) LR 2011 14. Fremde Lexik festigt sich in der litauischen Sprache in Form der Realienbezeichnungen, die aus anderen Kulturen, vor allem westlichen, übernommen werden. Die Einstellung zu der litauischen ethnischen Kultur ändert sich, eigene Kultur wird geschmälert. Vor allem junge Menschen nehmen Feste und Bräuche anderer Völker an. Nach Litauen kommen solche Feste /Feiertage, die den litauischen Traditionen nicht fremd sind. Man entfernt sich von der alten baltischen Kultur. In Litauen hat man begonnen, St. Patricks Tag zu feiern. St. Patrick

---

1 Dieses Problem wurde behandelt auch auf der internationalen Konferenz „Prozesse der Kulturtransformationen in den Ländern des Mittel-und Osteuropas" an der Universität Erfurt (Deutschland) am 29.09.-01.10. 2000.

bedeutet für Irland das gleiche, was für Litauen der heilige Kasimir, also haben wir, wen wir ehren können. Es gibt Litauer, die *Padėkos dieną* (*Thanksgiving day*) feiern. *Thanksgiving Day* (*Padėkos diena)* wird in Amerika seit 1621 gefeiert, als Emigranten für reiche Ernte dankten. Wenn es ein Wunsch besteht, kann man in Litauen *Derliaus dieną* (*Erntefest*) feiern. Siesta ist eine mediterrane Tradition. Wenn wir Mittagsrast als *siesta* bezeichnen, anstatt *perpietė, pogulis* zu gebrauchen, *priešpiečius (Mittagessen)* oder *pusryčius (Frühstück) lanču, lenču (lunch)* nennen, so drücken wir Traditionen und Bräuche ganz anderer Nationen aus. Es ist unakzeptabel, eigene Feste mit dem spanischen Wort *fiesta* zu nennen. in der neuen Lexik der litauischen Sprache widerspiegeln sich auch die für das litauische Volk fremde Lebensrealien, z. B.: *sekso turistas, -ė* (*Sextourist, Sextouristin*) LR (GB) 2010 47, *sekso turizmas* (*Sextourismus*) LR (GB) 2010 47; *svingerių vakarėlis* (*Swinger-Party*) LR 2011 64 u.a.

Es fällt schwer, folgende Begriffe mit der litauischen Kultur in Einklang zu bringen, z.B.: *kultūros fabrikas* (*Kulturfabrik*) LRT 2013-02-25, *kultūrų inkubatorius (Kultureninkubator*) LRT 2013-02-25, *muzikos dirbtuvės (Musikwerkstatt*) LRT 2012-12-03, šokių ferma (*Tanzfarm*) LR (TVA) 2012 27 4.

Der Ausdruck *blusų turgus* (*Flohmarkt)* ist auch ein Produkt der Globalisierung. Diese Bezeichnung wurde aus dem Westen übernommen. Zum ersten Mal wurde sie im Jahre 1885 in Frankreich erwähnt. Auf diesem Markt hat man alte Sachen verkauft, die auch Flöhe hatten. Übliche litauische Bezeichnungen waren *senų daiktų turgus, sendaikčių turgus (wörtlich der Mark, auf dem man alte Sachen verkauft).* Jetzt verwendet man immer häufiger Bezeichnungen *blusų turgus, blusturgis* LR (S) 2012 17.

Westliche Sprachen bildeten die Grundlage für die Entstehung der für litauische ethnische Kultur fremden Loblexik. Als ein expressives Lobwort wird häufig der aus dem Englischen übernommene Latinismus *super* verwendet, z.B.: *super atliktas darbas (super Arbeit)* LNK 2012-08-13, *super gražu stadione* (*super schön auf dem Stadion*) LRT 2012-08-12, *superidėja* (*Superidee*) LRT 2012-08-07, *superklausimas* (*Superfrage*) BTV 2011-09-02, *supernuolaida* (*Super-Rabatt*) LNK 2011-07-30, *superpograma (Superprogramm*) LRT 2012-08-16, *superprojektas* (*Super Projekt*) TV3 2011-08-07. Außerdem wird *super* in den westlichen Sprachen, wie z.B. im Englischen, als selbständiges Lexem und auch als präpositionale Wortkomponente gebraucht. In der litauischen Standardsprache ist es nur als präpositionale Wortkomponente akzeptabel und wird zusammen mit dem Wort, das es bestimmt, geschrieben. Dieses Element bekommt auch litauisches Suffix und Endung, was eine Art seiner Befestigung in der litauischen Sprache zeigt, z.B. *superinė apžvalgos aikštelė (super Aussichtsplattform)* LTV 2012-07-09, *superinis darbas* (*super Arbeit*) LNK 2010-07-18, *superinis gydytojas* (*super Artzt*)

BTV 2011-08-12, *superinis koncertas* (*super Konzert*) LRT 2012-08-20, *superinis verslas (super Geschäft)* TV3 2011-06-18. Nach dem Beispiel des Englischen hat der Latinismus *seksualus, -i (sexy)* im Litauischen auch neue Bedeutungen bekommen, solche wie *mielas (lieb), malonus (nett), gražus (schön), dailus (gut ausehend), žavus (charmant), šaunus (cool), puikus (großartig), nuostabus (fantastisch), patrauklus (attraktiv)* u.a. ir kt.

Es steigt die Zahl der für litauische Kultur fremden übersetzten Grußwörter, Glückwünsche und Ausdrücke, z.B.: *hello!* (= *sveikas*!), aus der englichen Umgangssprache *hai* (engl. *hi*), *geros sėkmės* (vgl. engl. *good luck*), *turėk gerą dieną* (vgl. engl. *have a good day*), *turėk gerą laiką* (vgl. engl. *have a good time)*, *turėk gerą savaitgalį* (vgl. engl. *have a good weekend*).

Die Identität und Verschiedenheit der ethnischen Kultur einer jeden Nation zeigen Zoomorphismen – Bezeichnungen des Menschen, seiner Eigenschaften und Tätigkeiten aus der Tierwelt. Sie lassen erkennen, dass man nicht alles aus anderen Kulturen übernehmen kann. Zum Beispiel, im Litauischen ist die Biene (*bitė*) ein Symbol für Fleiß und Sorgfalt/Gründlichkeit, im Kasachischen für Zorn und Unzufriedenheit. In Indien schmeichelt man einer Frau, wenn man sie mit einer Kuh (*karvė*) und ihren Gang mit dem eines Elefanten (*dramblys*) vergleicht. Für eine Litauerin wäre es eher Beleidigung als Kompliment. In der litauischen und russischen Kultur ist es beleidigend, eine Frau eine Gans (žąsis) zu nennen, aber in Ägypten ist das ein Liebeskompliment. Ein Schwein symbolisiert für Litauer und Russen Unsauberkeit, Unkultiviertheit, für Engländer Gefräßigkeit, in der vietnamesischen Kultur Dummheit und in der chinesischen Glück. Die größte Beleidigung für einen Chinesen ist ein Vergleich mit dem Hasen, in den Ländern des Budismus ist der Hase ein Symbol für Weisheit. *Labas, kalės vaike (Hey, Hurensohn*, wörtlich *Sohn einer Hündin*) – so eine Begrüßung würde einen Litauer beleidigen, doch Argentienier zeigen damit, dass sie beste Freunde sind. Viele Diferenzen solcher Art führt Vladimiras Želvys (1997, 229-236) an. In der litauischen Kultur haben Zoomorphismen meistens eine negative Bedeutung, vgl.: *avinas (Schaf), ožys (Ziegenbock), ožka (Ziege), varna(Krähe), veršis(Kalb), višta (Huhn)* u.a. Nach den Angaben von A. Gudavičius (2000, 93-94), Zoomorphismen mit positiver Kennzeichnung des Menschen machen nur 13 Prozent aus.

Phraseologismen/Idiome und phraseologische Einheiten bilden die Grundlage des Geistes eines Volkes, darin widerspiegelt sich auch das Weltbild der Nation. Außerdem zeigen sie die Lebendigkeit und Schönheit der Sprache, widerspiegeln ihre Eigenart, verhindern den Verlust des nationalen Klanges. Leider nimmt die Zahl der übersetzten Phraseologismen zu, die anstatt eigener, den litauischen Traditionen und Bräuchen entsprechenden verwendet werden,

z. B.: *ir mums moliūgas pavirs karieta* (wörtlich: *auch für uns verwandelt sich ein Kürbis in eine Kutsche*) LNK 2009-06-10 = ir mūsų namuose bus šventė[2], saulė nušvis, saulutė nusišypsos (wörtlich auch in unserem Haus wird ein Fest sein, die Sonne wird scheinen, lächeln); *kates karti* (wörtlich: *Katzen aufhängen*) LTV 2009-06-12 = šunis karti (karstyti) (wörtlich Hunde aufhängen); *kiaulei gali padažyti lūpas, bet ji vis tiek liks kiaulė* (wörtlich: *man kann einem Schwein Lippen schminken, aber es bleibt immer noch ein Schwein*) LTV 2010-08-10 = laukas gimė, laukas ir dvės, laukas gimęs, laukas ir iškarš (wörtlich: als Feld geboren wird auch als Feld sterben); *makaronus kabinti* (wörtlich: *Nudeln aufhängen*) LR 2011 221 = akis apdumti, akis dumti, akis apmauti (wörtlich: Augen betrügen, täuschen); *pakišti kiaulę* (wörtlich: *Schwein herunterschieben*) LR 2009 39 = iškrėsti kiaulystę, iškrėsti šunybę (wörtlich Scweinerei spielen, machen); *priskaldyti malkų* (wörtlich: *Holz hauen*) LNK 2010-06-10 = košės privirti, į košę įbristi (įsipainioti) (wörtlich: Brei kochen, im Brei waten, sich verwickeln); *stabdyti arklius* (wörtlich: *Pferde stoppen*) BTV 2009-04-06 = burną užčiaupti (uždaryti, užkimšti, užkišti), dantis užčiaupti (uždaryti) (wörtlich: Mund, Zähne schließen, verstopfen); *stogas važiuoja* (wörtlich: *Dach fährt*) LR (RV) 2011 6, *stogas nuvažiuoja* (wörtlich: *Dach fährt herunhter*) LR 2010 187 = nei kepęs, nei viręs, ne visi šulai galvoje, vieno šulo stinga (trūksta) galvoje, sraigtelių trūksta, kaip šlapiu maišu trenktas (pritrenktas) (wörtlich: weder gebraten noch gekocht, nicht alle Balken/ Schrauben im Kopf haben, wie mit nassem Sack geschlagen); šakėmis ant vandens rašyta (wörtlich: *mit Gabel auf dem Wasser geschrieben*) LR 2009-06-05 = muilo burbulas, vienas niekas (wörtlich: Seifenblase, Unsinn); *užmušinėti laiką* (wörtlich: *Zeit totschlagen*) LR (S) 2010 12 = stumti laiką (wörtlich: Zeit schieben), varles badinėti (wörtlich: Frösche stechen); *vaidenasi maurai su šakėmis* (wörtlich: *es geistern Mohren mit Forken*) TV3 2009-10-05 = kinkos dreba (virpa), širdis dreba, širdis nutirpo, širdis atsidūrė kulnyse, (wörtlich: Beine/ Herz zittert, Herz schmolz/ fliel in die Fersen); *vynioti į vatą* (wörtlich: *in Watte einwickeln*) LR 2011 58 = miglą pūsti į akis, akis dumti, jautį sukti už uodegos (wörtlich: Nebel in die Augen blasen, Augen täuschen, den Stier am Schwanz drehen). Einige von diesen Phraseologismen sind aus dem Russischen, andere aus den germanischen Sprachen übersetzt. Man sollte nicht übersetzte Idiome anderer Sprachen verwenden, weil so auch die litauische ethnische Kultur gefährdet wird. Es ist schlimm, dass fremde, für litauische ethnische Kultur ungewohnte Phraseologismen als eigene verstanden werden.

2 Die meisten litauischen Phraseologismen sind dem Phraseologiewörterbuch der litauischen Sprache entnommen *Lietuvių frazeologijos žodynas*. Vilnius: Lietuvių kalbos institutas, 2001.

Globalisierungsprozesse sind unaufhaltsam. Das litauische Volk ist auch für andere Kulturen offen. Aber Offenheit gegenüber anderen Kulturen sollte auf Ethnizität bezogen werden. Litauische Kultur, Sprache und andere Grundwerte sowie ihre Identität sollten aufbewahrt und gefördert werden. Jede Nation ist für die Welt so viel interessant wie sie einzigartig ist. Die Kultur jeder Nation ist ein Bestandteil der gesamten Weltkultur, aber man kann diese Welt sich nicht ohne bewusste Grenzen vorstellen. Unter den Bedingungen der Globalisierung spitzt sich das Problem der Universalität und Nationalität der Kultur generell zu. Das litauische Volk soll seine Grundwerte nicht verlieren, aber es soll auch sich nicht abgrenzen von allen menschlichen Werten, von den allgemeinen moralischen Grundlagen, um die Kultur anderer Völker zu verstehen. Laut Stasy Šalkauskis, „eigentümliche Form des allgemeinen Inhalts der gesamten Menschheit - so muss die Kultur sein, die allein Gewähr für das Leben der litauischen Nation sein kann" (Šalkauskis, 1995, 170). Man muss von anderen Kulturen das übernehmen, was litauische Kultur und Sprache bereichert und nicht ärmer macht. Auf solche Weise bereichern wir nicht nur eigene, sondern auch die Kultur der ganzen Welt.

Schlussfolgerungen

Nach der durchgeführten teoretischen und empirischen Datenanalyse von Ausdrucksmitteln der Grundwerte der litauischen Sprache und des litauischen Volkes kann man Folgendes schließen:

1) die litauische Sprache wandelt sich unter dem Einfluss des Englischen als eines der wichtigsten Globalisierungsfaktoren; das Prestige der litauischen Sprache sinkt; die Zahl der Entlehnungen, die aus dem Englischen oder über das Englische kommen, nimmt zu; das Gleichgewicht des Gebrauchs von litauischem und fremdem Wortgut wird gestört; das semantische System der litauischen Sprache wird dem Englischen angenähert; die Eigenart des Litauischen wird beeiträchtigt;
2) es is notwendig, die litauische Sprache als nationalen Wert zu pflegen, das Ansehen des Litauischen zu erhöhen;
3) Globalisierung mit den damit verbundenen sozialen und kulturellen Veränderungen beeinflusst litauische Identität und nationale Werte, darunter solche Grundwerte wie Ehrlichkeit, Respekt, Wahrheit, Fleiß; der Begriff der Arbeit wird erweitert; Patriotismus, Nationalbewusstsein werden herabgesetzt; Pragmatismus, Konsum, Grobheit, Agressivität, Verachtung des Menschen entfalten sich;

4) Gemeinschaftsgefühle sind geschwächt, Gemeinschaftsbeziehungen und das Konzept der Familie wandeln sich, es entstehen neue Begriffe, die diese Veränderungen manifestieren;
5) die Einstellung zur ethnischen litauischen Kultur ändert sich, sie wird auch geschmälert; Realien und Bräuche der westlichen Kultur werden angenommen; übersetzte Phraseologismen, die litauischen Traditionen und Bräuchen nicht entsprechen, nehmen zu;
6) die Offenheit gegenüber anderen Kulturen sollte mit Ethnizität, mit der Aufbewahrung und Förderung der litauischen Kultur, Sprache und anderer Grundwerte verbunden werden;
7) es ist notwendig, die kulturellen Unterschiede zwischen den Völkern zu erforschen und auf der Grundlage wissenschaftlicher Erkenntnisse bessere Voraussetzungen für interkulturelle Kommunikation unter den Bedingungen der untraditionellen Multikulturalität und Mehrsprachigkeit zu schaffen.

Quellenabkürzungen

LR – *Lietuvos rytas* (dienraštis).

LR (GB) – *Lietuvos ryto* priedas „Gyvenimo būdas".

LR (RV) - *Lietuvos ryto* priedas „Rytai – Vakarai".

LR (S) – *Lietuvos ryto* priedas „Sostinė".

LTR (TVA) – *Lietuvos ryto* priedas „TV antena".

Literatur

Andrijauskas, A. (2006), Technologinė civilizacija, medijos ir kultūros globalizacija, *Kultūrologija. Kultūra globalizacijos sąlygomis*, Vilnius: Kultūros, filosofijos ir meno institutas, T.13, pp. 76-96.

Grigas, R. (2004), Tautos valstybės sąmoningumas: lietuvių kalbos atvejis, *Lietuvių tautos – lietuvių kalbos likimas*, Vilnius: VĮ Mokslotyros institutas, II dalis, pp. 20-44.

Gudavičius, A. (2000), *Etnoligvistika*, Šiauliai: Šiaulių universitetas.

Huntington, S.P. (1998), *The Clash of Civilizations and the Remaking of World Order*, Touchstone Books.

Juknevičius, St. (2006), Vakarai ir islamas: konfrontacija ar tarpusavio supratimas? *Kultūrologija. Kultūra globalizacijos sąlygomis*, Vilnius: Kultūros, filosofijos ir meno institutas, T.13, pp. 116-152.

Koženiauskienė, R. (2005), *Juridinė retorika*, Vilnius: Teisės informacijos centras.

Lietuvių frazeologijos žodynas (2001), Vilnius: Lietuvių kalbos institutas.

Lietuvių kalbos enciklopedija (1999), Vilnius: Mokslo ir enciklopedijų leidybos institutas.

Lietuvių kalbos žodynas (1969), Vilnius: Mintis, T.2.

Mickūnas, A. (2011), *Estetika. Menas ir pasaulio patirtis*. Vilnius: Lietuvių literatūros ir tautosakos institutas.

Mitrikas, A.A. (2006), Šeimos gyvenimas: vertinimas ir tikroji padėtis. *Kultūrologija. Kultūra globalizacijos sąlygomis*, Vilnius: Kultūros, filosofijos ir meno institutas, T.13, pp. 154-182.

Neimantas, R. (1999), Lituanistika ir orientalistika: sąsajos ir sąveika, *Tarpdisciplininiai ryšiai lituanistikoje. Pasaulio lituanistų bendrijos darbai*, Vilnius: Homo liber, T.4, pp. 102-122.

Račienė, E., Rudaitienė, V. (2008), New lexis – an expression of transformation of Lithuanian culture, *Bildungs - und Kulturmanagement. The management of education and culture*. Frankfurt am Main: *Peter Lang*, Band 15, pp. 209-224.

Rudaitienė, V. (2006), Globalisation and transformations of Lithuanian culture, *Citizens and Governance for Sustainable Development*, Vilnius: Technika, pp. 363-367.

Šalkauskis, St. (1995), *Raštai*, Vilnius: Mintis, T.4.

Желвис, В.U. (1997), *Поле брани. Сквернословие как социальная проблема*. Москва: Ладомир.

Vergleichende Untersuchung der Redeetikette als Materialquelle für die Heranbildung der primären linguakulturellen Kompetenz (am Beispiel der russischen, litauischen und türkischen Abschiedsausdrücke)

Erika Andrijevskaja

Abstract
The subject of this article is speech etiquette as a fragment of concept "politeness". Competence in speech etiquette and knowledge of its national characteristics are defined as part of the primary linguo-cultural competence of linguistic personality. The article provides a comparative analysis of farewell etiquette formulas in the Russian, Lithuanian and Turkish language as well as reveals similar and specific means of expressing politeness in farewell situations.

Einleitung

Die Untersuchung der Redeetikette verlangt einen interdiziplinären Forschungsweg, auf den schon selbst die Kombination der Begriffe „die Rede" und „die Etikette" hinweisen. Laut N. Formanovskaja, „stellt die Redeetikette das vollständige sprachliche Weltbild dar, das mit einem würdigen, relevanten, höflichen Umgang der Menschen miteinander verbunden ist" [7: 5]. Unseres Erachtens, repräsentiert die Redeetikette nicht so das vollständige sprachliche Weltbild, sondern tritt als Bestandteil eines Fragments des sprachlichen Weltbildes (der sprachlichen Weltvorstellung), des axiologischen Konzepts „die Höflichkeit" auf, das mit einem würdigen und relevanten Umgang der Menschen miteinander verbunden ist.

Die vergleichenden Untersuchungen der Redeetikette lassen die Besonderheiten der erwähnten Fragmente des nationalen sprachlichen Weltbildes feststellen. Die Anwendung der vergleichenden Methode führt die Untersuchungen solcher Art zum Gebiet der vergleichenden Linqakulturologie.

Im Rahmen der Linqakulturologie besteht die theoretische Aktualität der vorgenommenen vergleichenden Analyse der Formen der Redeetikette darin, den linqakulturellen Kode zu finden und aufzuschließen, mit Hilfe dessen die Übertragung der Kultur auf das stereotypische Redeverhalten und ihre Widerspiegelung (die Fixation, „die Markierung") im nationalen sprachlichen Weltbild geschieht.

Das Sprachbild der Welt bildet eine Komponente des kollektiven Bewusstseins einer bestimmten linquakulturellen Gemeinschaft. Im Prozess der Aneignung der Muttersprache und der Sozialisation integriert das Bewusstsein des Individuums das Weltbild der muttersprachlichen Linquakultur. Es geschieht die Formierung der sogenanten linquakulturellen Identität, d. h. die Zugehörigkeit eines konkreten Menschen zu einer bestimmten Linquakultur, von der seine Werteeinstellungen determiniert und geformt werden.

Das Sprachbild der Welt als Kreuzungszone des Nationalen und des Individuellen soll unbedingt bei der Untersuchung des Prozesses der interkulturellen Kommunikation berücksichtigt werden, weil gerade diese Zone „als Zone der Nichtübereinstimmung" zwischen dem Inhalt und dem Ausdruck, den, scheinbar, universellen Konzepten, den Stereotypen im Denken und Verhalten gilt. Sie verursacht zugleich die sogenannten „semantischen Störungen" (Sinnentstellungen, Missverständnisse), die die Effektivität der interkulturellen Kommunikation senken.

Die Wirksamkeit und der Erfolg der interkulturellen Kommunikation hägen unmittelbar und direkt von dem Grad der Herausbildung der globalen Identität der Gesprächspartner, der die gegenseitige Toleranz der Zugehörigkeit zu verschiedenen linquakulturellen Gemeinschaften, zur Spezifik „der fremden" Weltanschauung zugrunde liegen, zugleich das Streben danach, die „fremde" Kultur zu verstehen.

Das Beherrschen des „fremden" linquakulturellen Kodes verrringert das Risiko des kommunikativen Misserfolgs, wird die adäquate Redekommunikation mit den Mitgliedern der „fremden" Linquakultur gewährleisten, was die angewandte Aktualität solcher vergleichenden linquakulturologischen Untersuchungen im Kontext der Verallgemeinerung der Probleme der interkulturellen Kommunikation darstellt.

Das Wissen über die Besonderheiten der Redeetikette bildet einen wichtigen Teil der linquakulturellen Kompetenz[1] in der gegenseitigen Wechselwirkung unter den Teilnehmern der Redekommunikation. In diesem Aspekt dient die vergleichende linquokulturologische Analyse als Instrument für die Feststellung der Kenntnisse, die für die Vorbereitung und Herausgebung der Lehrmaterialien für den Sprachunterricht unentbehrlich sind.

---

1 Linquakulturelle (oder linquakulturologische) Kompetenz stellt ein bestimmtes „Bild der Welt", das sich in einer jeden Person historisch gestaltet hat und zustande gekommen ist; das Verfügen der Kenntnisse durch den idealen Kommunikationspartner (den Sprechenden und den Zuhörenden) über das ganze System der kulturellen Werte, die in einer Sprache ausgedrückt werden [5: 251].

Das Ziel dieser Untersuchung besteht in der Feststellung der ähnlichen und spezifischen Rededarstellungseigenschaften bei der Realisation des sprachlichen Weltbildes, die am Material der russischen, litauischen und türkischen sprachlichen Redeformeln in einem höflichen Redeverhalten, der festen Redeausdrücke für die Verabschiedung (den Abschied) und präsentiert verwendet werden.

Als Hauptuntersuchungsmaterial für den Vergleich dienen normgemäße russische Etiketteausdrücke und situationsgebundene Parameter ihrer Untersuchung, die von N. Formanovskaja festgestellt und ausgewählt wurden [7].

Die litauischen und türkischen Entsprechungen wurden von uns auf der Grundlage ihrer Möglichkeit, denselben perlokutiven Effekt auszuüben, ausgewählt. Die Richtigkeit der litauischen Entsprechungen im Redeprozess und ihre Gebrauchlichkeit, die in der Presse fehlen, sind von der wissenschaftlichen Mitarbeiterin der Abteilung für die Standardlitauisch des Instituts für die litauische Sprache, der Chefredakteurin der periodischen Zeitschrift „Kalbos kultūra" („Sprachpflege") Rasuolė Vladarskienė bestätigt und gebilligt worden. Die anologischen türkischen sprachlichen Formeln, die auch in der Presseausgaben fehlen, wurden auf Basis der bestimmten kontextuell – pragmatischen Parameter während der Interviews der Türken, die die russische und / oder die litauische Sprache beherrschen oder nicht, ausgewählt.

Hauptteil der Untersuchung

Die konstruktive und effektive Kommunikation als verbindendes Mittel unter den Mitgliedern der Gesellschaft erfolgt nur unter der Bedingung der allgemein anerkannten Regelung der menschlichen Verhältnisse, ihrer konventionellen Orientierung. Als solcher Orientierungspunkt dienen die sittlichen (moralischen) **Werte**, die sich aus der Kultur der Gesellschaft ableiten lassen und die ihre Stabilität sichern. In der zwischenmenschlichen Kommunikation stellt einen besonderen Wert *die Höflichkeit* dar – „die Moraleigenschaft, die das Verhalten des Menschen charakterisiert, für den die Achtung den anderen Menschen gegenüber zur täglichen Handlungsnorm und zum gewöhnlichen Mittel im Umgang mit den Mitmenschen geworden ist" [4].

Wenn die Höflichkeit als moralische Norm anordnet, *was* die Leute machen und wie sie handeln sollen, d. h., in erster Linie, einander achten, so schließt die Kultur des Verhaltens (als Bestandteil der Kultur der Gesellschaft) auf, **auf welche** Weise sich die Achtung zu den Menschen im Verhalten der Menschen untereinander äußern sollte. Das Kulturniveau im Verhalten (wie äußerliche, als auch innere Kultur[2] des Menschen) wird danach eingeschätzt, in welchem Grad das Verhalten eines konkreten Menschen in Hinsicht auf den Ausdruck der Achtung im Umgang mit anderen den allgemein anerkannten Regeln entspricht.

Die Regeln des Verhaltens und der Kommunikation gehört zum Bereich der Etikette. „Die Etikette – (franz. Etiquette, russ. Ярлык, dt. Das Etikett) – das ist die Gesamtheit der herkömmlichen Umgangsregeln, die sich auf die Verhältnisse der Menschen untereinander beziehen (Umgang mit den Mitmenschen, Anrede- und Begrüßungsformeln, das Benehmen am öffentlichen Ort, Manieren und die Kleidung) [4]. Die in den Klammern erwähnten Anrede- und Begrüßungsformen gehören zu den Regeln der Redeetikette.

Die Redeetikette als System der Regeln und Mittel für den Ausdruck der Höflichkeit ist charakteristisch für alle Völker, d. h. stellt nach den Worten von N. Formanovskaja die *funktional – semantische und pragmatische Universalie* [6: 407], mit anderen Worten gesagt, die das Redeverhalten charakterisierende Universalie dar. Die Redeetikette ordnet die Forscherin dem Bereich des Redeverhaltens zu.

Als Grundlage des Redeverhaltens, genauso wie der kommunikativen Kultur und der Kultur des Verhaltens im Allgemeinen dient die Höflichkeit. T. Larina [2: 126 – 171] betracht sie als einen Regulator des kommunikativen Verhaltens. Ihrer Meinung nach, bedeutet die Höflichkeit vor allem „die Einhaltung der Kommunikationsnormen durch die Beachtung der kommunikativen Strategien, die die sozial – kulturellen Werte ausdrücken und den kommunikativen Erwartungen des Gesprächspartners entsprechen" [ebd: 166].

Die Höflichkeit, die allen Völkern typisch ist, kann man, unseres Erachtens, als moralisch – ethische Universalie betrachten. Die Höflichkeit als kultureller Wert eignet sich der Mensch an und macht davon Gebrauch während des sukzessiven Prozesses der Erziehung, Sozialisation, d. h., spiegelt sich in seinem Bewusstsein wider, wird zum Bestandteil seiner Weltvorstellung. Deshalb kann man sie als eine gewisse „Zelle" des Bewusstseins betrachten, die sich in engstem Zusammenhang mit der Kultur und der Sprache befindet, d. h., als linquakulturelles *Konzept* auftritt, dessen Inhalt und Ausdruck sich beim „Übergang" von einer Linquakultur in die andere ändern.

Das System der nationalspezifischen Redestrategien mit es begleitenden konventionellen sprachlichen Ausdrucksmitteln im Kommunikationsprozess nennt T. Larina die *sprachliche Höflichkeit* (engl. linguistic politeness) [ebd: 169].

---

2 Wenn wir über die Kultur des Menschen sprechen, wird der Bergriff „die Kultur" in der alltagsmäßigen Bedeutung „anständig, gebildet" erklärt.

Unseres Erachtens, gehört die sprachliche Höflichkeit zur Struktur des linquakulturellen Konzepts „Höflichkeit" in Form von einem System aus einzelnen Konzepten – Stereotypen des Redeverhaltens. Diese Konzepte befinden sich im linquakulturellen Bewusstsein und existieren, unserer Meinung nach, als Schemata, Szenarien. Die verschiedenartigen Strategien sind darauf gerichtet, dem Partner gegenüber ein ehrerbietiges und wohlwollendes Verhältnis zu demonstrieren, was das Hauptziel (die Intention) des höflichen Redeverhaltens und natürliche Erwartung des Gesprächspartners darstellt.

Die aktuelle Bedeutung (den Sinn) der Höflichkeit gewinnt die Formel der Redeetikette im Diskurs, die angebracht für eine bestimmte kommunikative Situation, für einen bestimmten kommunikativen *Kontext* erscheint. Die Angemessenheit / Nichtangemessenheit beim Gebrauch der Formeln der Redeetikette in einem bestimmten Kontext wird von den kulturspezifischen Normen des Verhaltens, der Kommunikation, d. h., von dem *kulturellen Kode* bestimmt. Folglich formt sich der Sinn der Einheit der Redeetikette und er wird als Resultat des Zusammenwirkens des linquistischen (sprachlichen) und kulturellen Kodes gefasst. Dieses bedeutungsbildende Zusammenwirken von Kodes kann man darum als *linquakulturellen Kode* bezeichnen.

Auf diese Weise scheint der Weg zur Sinnverlangung, der das *Verstehen* der etiketteadäquaten Äußerung darstellt, umständlich und erschwert nur auf der Basis der Sprachbeherrschung (des Sprachkodes), die die Komponente der sprachlichen (linquistischen) Kompetenz ist, zu sein. Für die Sinnaufschließung des Redehandelns ist auch das Verfügen über die extralinquistischen Kenntnisse (das Beherrschen des kulturellen Kodes) notwendig.

Die linquistischen und extralinquistischen Kenntnisse machen die kommunikative Kompetenz (darunter auch die linquakulturelle) unter den Kommunikationspartnern aus.

Der Fakt des Sinnverstehens der etikettemäßigen Äußerung – Handlung, d. h. das Erreichen vom *illokutiven Ziel*, entdeckt sich auf dem Niveau der Perlokution. Der positive *perlokutive Akt* der etiketteadäquaten Äußerung (beim übereinstimmenden Verstehen) bedeutet „das Einverständnis" des Adressaten zum Kontakt, zu seiner Aufnahme oder Vollendung mit der Beibehaltung / Voraussetzung der Möglichkeit der weiteren Kontakte. Der positive perlokutive Akt stellt ein gewisses Binde – und Übergangsglied, das von dem Sprechakt des Adressanten zu dem des Empfängers gerichtet ist, dar.

Die Sprechverhaltensszenen organisieren und strukturieren zusammen mit den Zeichen der Redeetikette, die an ihrer realen Gestaltung teilnehmen, den Diskurs. Die Phasen der Feststellung und der Auflösung des Kontakts unter den Kommunikationspartnern stellen „eine erhöhte Risikozone", das Risiko des

Entstehens von Kommunikationsmissverständnissen dar. Die unangebrachte (ungeschickte) Anwendung einer Formel der Redeetikette kann von dem Kommunikationspartner als Spott, Sarkasmus oder direkte Beleidigung verstanden werden. Deswegen können die Aufnahme oder die Wiederherstellung des Kontakts in der Zukunft nicht zustande kommen. Auf diese Weise kann behauptet werden, dass das Beherrschen der Redeetikette, das Verfügen der Kenntnisse über ihre kulturnationalen Besonderheiten einen bedeutenden Teil der primären linquakulturellen und sprachlichen Kompetenz der Persönlichkeit darstellen.

Der Abschied

- Der gebräuchlichste / neutrale:

1) До свидания!
   [L] Sudie(v)! [8: 124; 9: 21; 12: 40; 11: 283]
   [T] 1. Iyi günler / akşamlar / geceler! [10: 8] (Guten Tag / Abend / Nacht!)
   2. Hoşça kalın! (der Weggehende beim Abschied) [10:9] (≈ bleiben Sie mit Vergnügen)
   3. Allahaısmaradık! (der Weggehende) [10: 8] (≈ wir haben (Sie) dem Allah anvertraut)

Am häufigsten kommt in der russischen Sprache die Äußerung ‹ 1 › vor, die implizit den Abschied bis auf das nächste Wiedersehen ausdrückt. In der litauischen Sprache wird dafür am häufigsten der Ausdruck ‹ 14 L › gebraucht. Für die gekürzte neutrale Variante des traditionsgemäßen Ausdrucks Su Dievu! ‹ 1 L › gibt G. Čepaitienė folgende Erläuterung: „ … der Sender weiß, dass nur der Gott alleine den Adressenten vor Unglück schützen kann ‹ … › und wünscht ihm mit dem Gott zu bleiben, d. h., in seiner Obhut zu sein“ [8: 125].

Als russisches lexikalisch – formelles Äquivalent dieser Äußerung gilt die Formel С Богом! Sie wird als umgangssprachlich qualifiziert und als Abschiedswort verwendet [1: 65].

Die neutrale und am häufigsten gebräuchliche Abschiedsformel in der türkischen Sprache ist der Ausdruck ‹ 1 T 1 ›, der den Wunsch des guten Tags, Abends oder der guten Nacht ausdrückt. Es kommen auch die Ausdrucksformeln ‹ 1 T 2 › und ‹ 1 T 3 › vor, die aber nur beim Weg - / Fortgehen gebraucht werden. Die Äußerung Hoşça kalın! ist recht ähnlich der russischen Счастливо оставаться! (das Wünschen des Wohlergehens beim Abschied oder die selbständige Form eines freudschaftlichen Abschieds [1: 536]). Der Ausdruck Allahaısmarladık! wird auch beim Abschied von den Fortgehenden angewandt. Ihrer primären Bedeutung nach ist dieser Ausdruck der russischen gehobenen Form des Ausdrucks für das Wohlergehen Молю Бога! ähnlich [1: 296].

Die Ähnlichkeit unter den am meisten verbreiteten Abschiedsformeln lässt sich in den litauischen ‹1 L› und türkischen ‹1 T 3› Varianten mit der Semantik „unter Obhut Gottes stehen" beobachten.

- Der die Zeitspanne präzisierende bis zum neuen Wiedersehen:

2) До встречи! (Auf Wiedersehen!)
   [L] 1. Iki susitikimo! [9: 20]
   2. Iki / ligi/lig pasimatymo! [8: 130; 9: 20; 11: 283]
   3. Iki kito susitikimo! [8: 128]
   4. Iki malonaus pasimatymo! [8: 131]
   [T] (Tekrar) görüşmek üzere! [10:8] (= Bis zum (nächsten) Wiedersehen!)
3) До скорого свидания! (= Auf baldiges Wiedersehen!)
4) До скорой встречи! (= Auf schnelles Zusammentreffen!)
   [L] Iki greito pasimatymo! [8: 130; 9: 20] (= Auf baldiges Wiedersehen!)
5) До встречи + Ortsangabe (в театре, в университете)! (= Auf Wiedersehen (im Theater, an der Universität)!)
   [L] Iki pasimatymo + Ortsangabe (teatre, universitete)! [lebendige Rede]
   [T] Ortsangabe (tiyatroda / üniversitede) + görüşmek üzere! [lebendige Rede]
6) До встречи + Zeitangabe (в воскресенье / вечером) (= Auf Wiedersehen am Sonntag / am Abend)!)
   [L] Iki pasimatymo + Zeitangabe (Sekmadienį / vakare)! [lebendige Rede]
   [T] Die Zeitangabe (Pazar günü / aksam) + görüşmek üzere! [lebendige Rede]
7) До завтра! (= Bis morgen!)
   [L] 1. Iki rytojaus! [9: 20]
   2. Iki rytdienos! [8: 130]
8) До воскресенья! (= Bis Sonntag!)
   [L] Iki sekmadienio! [9:20]
9) До вечера! (= Bis zum Abend!)
   [L] Iki vakaro! [8:130] [9:20]
10) До праздника! (= Bis zum Feiertag!)
   [L] Iki švenčių! [9:20]
11) До лета! (= Bis zum / auf den Sommer!)
   [L] Iki vasaros! [9:20]

Die Konstruktionen ‹ 1 › - ‹ 11 › mit der Präposition до = iki (= bis) und dem Substantiv im Genitiv drücken den Abschied bis zum bestimmten Zeitpunkt aus. Die erwähnten Ausdruchsformeln enthalten die Semantik der zeitlichen Trennung und die Intention der Kontaktaufnahme in der Zukunft. Die Zeit des potenziellen Kontakts in der Zukunft wird in den äquivalenten russischen und litauischen Ausdruksformeln wie свидание / встреча = pasimatymas, susitikimas (= das Wiedersehen / das Treffen), manchmal mit der Begriffsbestimmung скорая(-ое) = greitas (= baldig) und / oder mit der Orts- / Zeitangabe bezeichnet. Es werden folgende Zeitangaben des nächsten Treffens angewandt: завтра = rytojus/ rytdiena (= Morgen), вечер = vakaras (= Abend), Benennung des Wochentages, der Jahreszeit u. s. w. In der litauischen Sprache wird die Variante mit dem Attribut malonus = приятный (= angenehm) (‹ 2 L 4 ›) verwendet. Im Wörterbuch von A. Balakaj ist ihr veraltetes russisches Äquivalent До приятного свидания (= bis zum angenehmen Treffen!) fixiert [1: 151].

In der türkischen Sprache werden die Ausdrucksformeln verwendet, die strukturell – semantisch äquivalent den russischen Varianten ‹ 2 ›, ‹ 5 › und ‹ 6 › mit den Komponenten = До встречи. + Ort/ + Zeit/ - Angabe sind. Im Türkischen fehlt die analogische Ausdrucksformel mit dem Attribut = скорая (-ое) (= baldig), die auf den Wunsch / die Erwartung der baldigen erneuten Kontaktaufnahme hinweisen würde. Die Semantik des Abschieds до следующей встречи (= bis zum nächsten Treffen) enthalten die Ausdrucksformeln ‹ 2 L 3 › im Litauischen und die ‹ 2 T› im Türkischen, implizit auch die russischen ‹ 1 › und ‹ 2 ›.

Die Ausdrucksformeln ‹ 7 › - ‹ 11 › stellen, vermutlich, die elliptische Variante des veralteten Ausdrucks Прощай(-те) до свидания (до вечера, до завтра ‹…›) [1: 451]. Die Umgangsformen solcher Art sind für die türkische Sprache nicht typisch. Es ist nicht klar, auf welche Weise sich ihre litauischen Äquivalente gebildet haben. Nach der Meinung von A. Kučinskaitė, konnten die Entstehung der Ausdrucksformeln des Abschieds solcher Art analogische Formen beeinflusst haben, die in den slawischen und der deutschen Sprache vorzutreffen sind [9: 22].

- <u>Der Abschied auf lange Zeit oder für immer</u>:

12) Прощай (-те)!
    [L] Sudie (-v)! [lebendige Rede]
    [L] Allahaısmarladık! [lebendige Rede]

In der litauischen und der türkischen Sprache werden in dem angegehenen situativen Kontext dieselben neutralen Ausdrucksformen: ‹ 1L › = ‹ 12 L ›, ‹ 1 T 3 › = ‹ 12 T › verwendet, die von uns davor analysiert wurden. Die Bedeutung der Ausdrucksform Прощай(-те)! kommt etymologisch auf den altertümlichen Brauch, beim Abschied gegenseitig um das Verzeihen zu bitten, zurück [1: 450, 447].

13) Не поминайте лихом![3] (≈ Gedenke meiner nicht im Bösen!)
[L] Atleiskit, jei kas buvo netaip! [lebendige Rede]
[T] Hakkına(-ıza) helal et(edin)! [lebendige Rede] (≈ Trete der Gott als Zeuge für deine (Ihre) Gerechtlichkeit auf!)

Die Redenausdrucksform ‹ 13 › lässt sich der etymologischen Bedeutung nach mit der der Formel ‹12 › vergleichen. Derselben Bedeutung nach kann sie den Ausdrucksformeln: *Не обижайтесь, не сердитесь, не судите слишком строго, если что не так*; *Прости(-те), если что не так* [1: 445] gegenübergestellt werden (≈ nehmen Sie das nicht übel; richten Sie uns nicht so streng, wenn etwas nicht in Ordnung war; Verzeihen Sie, wenn etwas nicht recht war). Die letzte Variante ist strukturell – lexikalisch als Äquivalent mit der litauischen Formel ‹ 13L › zu vergleichen, die häufig von den Gästen beim Abschied, besonders nach einem geselligen und lustigen Beisammensein, verwendet wird.

Den Ausdrucksformen der obenerwähnten Gruppe ist die türkische Variante ‹ 13T › ähnlich, in der nicht „die Schuld" des Senders / Adressanten, sondern „die Rechtlichkeit" des Empfängers / Adressaten betont und hervorgehoben wird.

- Der Abschied, der die versteckten / geheimen Wünsche ausdrückt:

14) Всего хорошего! (= Alles Gute!)
[L] Viso gero! [8: 126; 9: 19; 11: 283]
15) Всего доброго! (= Alles Gute!)
[L] Viso labo! [8: 126; 9: 19; 11: 283]
[T] Selametle! (dem Weggehenden gewidmet) [10:8] (≈ Wohlergehen Ihnen!)

Als lexikalisch – strukturelle Äquivalente können die russischen und litauischen impliziten Wünsche betrachtet werden. Eine ähnliche Konstruktion fehlt in der türkischen Sprache. Einen versteckten Wunsch enthält die Ausdrucksform, die an den Weggehenden (‹ 15 T ›) gerichtet ist und die semantisch und situationsgemäß der russischen Ausdrucksform С Богом! und der litauischen Sudie(v)! (Su dievu!) entspricht.

16) Будь(-те) здоров(-ы)! [8: 129] (= Sei(d) gesund!)
[L] 1. Būk sveikas/būkite sveiki! [8: 129]
2. Pasilik sveikas/likite sveiki! [8: 129; 9: 20; 11: 283]
[T] Sağlıcakla kal/-ın! (der Weggehende) [10: 9]

3 *Лихо* – зло. Ursprünglich hinsichtlich des Verstorbenen oder als der letzte Wunsch des Sterbenden; dem Volksglauben nach, wird sich der Hingeschiedene im Himmelreich unwohl fühlen, wenn man seiner auf Erde im Bösen gedenkt [1: 396].

17) Счастливо! (≈Viel Glück!)
[L] Laimingai! [9: 20; 11: 283]
[T] Güle güle! (dem Weggehenden gegenüber) (≈ Im Lachen!) [lebendige Rede]

In allen drei untersuchten Sprachen enthalten die Ausdrucksformen für den Abschied das Sem „die Gesundheit" (die Gruppe ‹ 16 ›). Es ist bemerkenswert, dass in allen drei Sprachen die Form des Imperativs = *будь(-те) / оставайся(-тесь)* ohne irgendwelche zusätzlichen Komponenten zur Milderung der Imperativität verwendet wird, was in manchen Formen der Bitte der Fall ist. Es scheint das mit der Semantik des Wünschens des Wohlergehens für den anderen, wie in diesem Fall, und für sich selbst, im Fall der Bitte, verbunden zu sein.

Die litauische Ausdrucksformel ‹ 17 L › wird der Form nach und semantisch als Äquivalent der russischen ‹ 17 › bezeichnet. In der türkischen Sprache ist sie semantisch der assoziativen Kette „Glück – Freude – Lachen" ähnlich.

- Der stilistisch gehobene (offizielle) Abschied:

18) Разрешите / позвольте попрощаться! (= Erlauben Sie mich zu verabschieden!)
[L] Leiskite atsisveikinti! [8: 131]
[T] Müsadenizle vedalaşmak / ayrılmak istiyorum [lebendige Rede]

Die Ausdrucksformen für den stilistisch gehobenen Abschied in allen drei Sprachen präsentieren die Formeln der Gruppe ‹ 18 › mit dem Sem „Erlaubnis / Genehmigung", das in den litauischen und russischen Entsprechungen im Unterschied zu den türkischen (≈ mit Erlaubnis / Genehmigung) mit dem Imperativ ausgedrückt wird = *разрешите / позвольте* (= erlauben / gestatten Sie).

- Der stilistisch ermäßigte / verringerte (in einer ungezwungenen, familiär – vertraulichen Situation) Abschied:

19) Всего! (= Alles (Gute)!)
[L] Viso! [8: 127]

20) Пока! (= Bis (später/ nachher)!)
[L] 1. Iki! [8: 131]
2. Iki malonaus! [8: 131]
[T] 1. Güle güle! (dem Weggehenden gegenüber) [lebendige Rede]
2. (Hadi) Eyvallah! (der Weggehende) [10: 9] (syn. Allahaısmarladık)

In der türkischen Sprache fehlt das elliptische Äquivalent dem russischen *Желаю всего хорошего/доброго* (= Wünsche alles Gute!) und den litauischen Entsprechungen ‹ 19 › und ‹ 19 L ›. Dem russischen *Пока!* entsprechen pragmatisch die litauischen Konstruktionen ‹ 20 › / ‹ 20 L 2 › (≈ *До! / До приятного!*). Die erwähnten elliptischen Konstruktionen haben im Türkischen auch keine strukturellen Äquivalente. In einer ungezwungenen und vertraulichen Situation werden die Konstruktionen ‹ 20 T › / ‹ 20 T2 › verwendet: die erste von ihnen ist an den Weggehenden gerichtet, die zweite wird nur von dem Weggehenden selbst hervorgebracht.

- Der Abschied am Abend, für die Nacht, unmittelbar vor dem Schlafengehen:

21) Спокойной ночи! (≈ Gute Nacht!).
22) Доброй ночи! (Gute Nacht!)
    [L] 1. Labos nakties! [8: 131; 9: 22; 11: 283]
        2. Labanaktis! [8: 131; 9: 21; 11: 283]
    [T] İyi geceler [10: 8] / hayırlı geceler! [lebendige Rede]

Standardisierte (genormte) Ausdrucksformen für den Abschied vor dem Schlafengehen in allen drei Sprachen beinhalten den allgemeinen Sinnkern für den Ausdruck des Wunsches „gute" Nacht. Die russische Ausdrucksform ‹ 21 › enthält einen mehr konkretisierten Wunsch „спокойной ночи" (ruhige Nacht).

Schluss

Die Resultate der vergleichenden Analyse lassen schlussfolgern, dass unter den gebräuchlichsten Formen für den Ausdruck des Abschieds die inhaltliche Ähnlichkeit zwischen der litauischen grundlegenden und der türkischen variativen Lakuneformeln mit der Semantik des Wunsches unter „Obhut, Schutz Gottes stehen" festzustellen ist.

Zu der populärsten russischen Ausdrucksform, die auch für den Abschied im Litauischen und Türkischen charakteristisch ist, zählt man die Konstruktion, die implizit die Erwartung eines neuen Wiedersehens ausdrückt. Auf diese Weise beinhalten die Ausdrücke beim Abschied den Wunsch der Kontaktwiederaufnahme, der folglich, einen hohen linquakulturell-kommunikativen Wert in den drei untersuchten Sprachen darstellt.

Spezifisch scheint der Ausdruck Прощай(-те)! zu sein, der beim Abschied für längere Zeit oder für immer verwendet wird und der den alten russischen Brauch, vor dem Abschied um die Entschuldigung zu bitten, widerspiegelt. Typisch für das Russische gilt der Ausdruck beim Abschied „Не поминай(-те) лихом!", d. h. недобрым словом (→ gedenke meiner nicht im Bösen).

In den analogen Lakunekonstruktionen ist die langjährige Tradition und die Sorge für die Beibehaltung wohlwollender Beziehungen einander gegenüber von der Seite des Gesprächspartners zu verfolgen, was eine unentbehrliche Voraussetzung für die weitere Kommunikation darstellt. In allen drei Sprachen sind die Ausdrücke für den Abschied, die genauso oder ähnlicherweise die entsprechende Intension ausdrücken, aber auch Ausdrucksformen mit der Semantik des konkreten Wunsches (des Guten, des Wohlergehens, der Gesundheit, des Glücks, der Freude), vorhanden.

Als allgemeingültig können die Sinneskerne der Ausdruckskonstruktionen beim Abschied für die Abreisenden / Weggehenden bezeichnet werden, die die Seme „хороший / добрый“ (→ gut, gutherzig), „счастливый“ (→ glücklich) beinhalten. Eine interessante Besonderheit im Gebrauch einiger türkischer Ausdrücke beim Abschied stellt ihre Anwendung nur von den Weggehenden, die sich verabschieden und die nur an die Weggehenden gerichtet sind, dar.

Im Prozess der vergleichenden Analyse von kommunikativ – pragmatischen Entsprechungen wurde festgestellt, dass einige Ausdrücke in den untersuchten Sprachen den verschiedenen Grad ihrer Äquivalenz darstellen, die in verschiedenen situativen Kontexten vorkommen, d. h. sie sind nicht situationsadäquat.

Deswegen dient schon selbst der Prozess bei der Auswahl der Übersetzungsentsprechungen, der von den Situationsparametern bestimmt und begründet wird, als Wissensquelle, die die interkulturelle Kompetenz einschließt.

Im Prozess des Vergleichs mit anderen Sprachen wird klar, dass als Folge der stereotypen Denkweise die grundlegende aktuelle Bedeutung der etiketteäquivalenten Form des Ausdrucks verlorengeht. Das erlaubt in die Vorstellungswelt des kollektiven Unterbewusstseins, in die noch tiefere Schicht des linquakulturellen Bewussten „hineinzuschauen“ und damit auf das axiologische System der linquakulturellen Gemeinschaft (sowohl der „eigenen“ als auch der „fremden“) hinauszugehen. Als Resultat der vergleichenden Analyse also gilt das Wissen, das zur kulturellen Kompetenz gehört, d. h. zur Gesamtheit von Kenntnissen und Fertigkeiten, die das Verstehen von Werteinstellungen, psychologischen Besonderheiten und soziokultureller Identität, die charakteristisch für eine bestimmte Kultur sind, gewährleisten [3: 262 - 263].

Resümee

Der Artikel behandelt das Problem der Redeetikette als vielfältiges linquakulturelles Phänomen. Es werden die theoretischen Aspekte der Redeetikette im Rahmen der Linquakulturologie und der interkulturellen Kommunikation untersucht. Die Redeetikette wird als Teil des sprachlichen Weltbildes betrachtet, das mit der höflichen Verständigung verbunden ist, d. h. als Fragment des axiologischen linquakulturellen Konzepts „Höflichkeit“. Das Beherrschen der Redeetikette, das Wissen über ihre nationalen Besonderheiten werden als Bestandteil der primären linquakulturellen Kompetenz der Persönlichkeit hervorgehoben, was die Aktualität vergleichender Untersuchungen auf diesem Gebiet darstellt, zugleich die Bedeutung der Anwendung ihrer Resultate im Mutter- und Fremdsprachenunterricht. Es wird die vorrangige Bedeutung bei der Bildung dieser Kompetenz betont, die notwendig für die Gewährleistung der adäquaten zwischenmenschlichen Kommunikation ist.

Es wird die vergleichende Analyse der Umgangsformen des Abschieds im Russischen, Litauischen und Türkischen unternommen. Auf dieser Basis werden die ähnlichen und spezifischen Ausdrucksmittel der Höflichkeit festgestellt, die auf die Besonderheiten der Fragmente des russischen, litauischen und türkischen sprachlichen Weltbildes hinweisen, das mit dem höflichen Umgang in der kommunikativen Situation des Abschieds verbunden ist.

Literatur

1. Балакай А. Г. Словарь русского речевого этикета. Москва, 2007.

2. Ларина Т. В. Категория вежливости и стиль коммуникации: сопоставление английских и русских лингвокультурных традиций. Москва, 2009.

3. Садохин А. П. Теория и практика межкультурной коммуникации. Москва, 2004.

4. Словарь по этике. (http://www.moralphilosophy.ru/index.htm).

5. Тхорик В. И., Фанян Н. Ю. Лингвокультурология и межкультурная коммуникация. Учебное пособие. Москва, 2006.

6. Формановская Н. И. Речевое взаимодействие: коммуникация и прагматика. Москва, 2007.

7. Формановская Н. И. Речевой этикет в русском общении. Теория и практика. Москва, 2009.

8. Čepaitienė G. Lietuvių kalbos etiketas: semantika ir pragmatika. Šiauliai, 2007.

9. Kučinskaitė A. Lietuvių kalbos etiketas. Vilnius, 1985.

10. Selçuk A. Kültürlerarası İletişim Açısından Gündelik İletişim Davranışları// S. Ü. Sosyal Bilimler Enstitüsü Dergisi, 13/2005 (S. 1-17), Konya.

11. Sisteminis lietuvių kalbos žodynas. Vilnius, 1987.

12. Slenkstis. Council of Europe Publishing, 1997.

Autoren

Abramauskienė, Jolanta, Assoc. Prof. Dr., Lithuanian University of Educational Sciences, Faculty of Education, Department of Music, Studentų g. 39, LT-08106 Vilnius, e-mail: jolanta.abramauskiene@leu.lt

Andrijevskaja, Erika, Ph.D. student, Lithuanian University of Educational Sciences, Faculty of Philology, Department of Russian Philology and Intercultural Communication, Studentų g. 39, LT-08106 Vilnius, e-mail: erikaa@yandex.com

Autukevičienė, Birutė, Dr., Lithuanian University of Educational Sciences, Department of Childhood Studies, Studentų g. 39, LT-08106 Vilnius, e-mail: birute.autukeviciene@gmail.com

Barkauskaitė, Marijona, Prof. Dr., Lithuanian University of Educational Sciences, Faculty of Education, Department of Education, Studentų g. 39, LT-08106 Vilnius, e-mail: marijona.barkauskaite@leu.lt

Beecroft, Raphaëlle, M.A., Research Assistant and Ph.D. candidate, Pädagogische Hochschule Heidelberg – University of Education, Modern Languages Institute, English Department. Im Neuenheimer Feld 561, 69120 Heidelberg, e-mail: beecroft@ph-heidelberg.de

Brusokaitė, Evelina, Lithuanian University of Educational Sciences, Studentų g. 39, LT-08106 Vilnius.

Bukantienė, Janina, Assoc. Prof. Dr., Klaipeda University, S. Nėries g. 5, LT-92228 Klaipėda, e-mail: bukanticnc@gmail.com

Bulajeva, Tatjana, Assoc. Prof. Dr., Vilnius University, Faculty of Philosophy, Department of Educology, Universiteto g. 9/1, LT-01513 Vilnius, e-mail: tatjana.bulajeva@fsf.vu.lt

Carlsburg, Gerd-Bodo von, Prof. Dr. Dr. h.c. mult., Pädagogische Hochschule Heidelberg – University of Education, Erziehungs- und Sozialwissenschaftliche Fakultät, Institut für Erziehungswissenschaft, Präsident des Weltbund für Erneuerung der Erziehung – Deutschsprachige Sektion (www: wef-wee.net), seit 1994 Gastprofessor an der Litauischen Universität für

Bildungswissenschaften in Vilnius, Ehrenmitglied des Europäischen Instituts für postgraduale Bildung an der Technischen Universität Dresden (EIPOS), Keplerstr. 87, D-69120 Heidelberg, e-mail: b.v.carlsburg @gmail.com

Celiešienė, Eglė, Dr., Lecturer, Lithuanian University of Educational Sciences, Faculty of Social Education, Department of Social Work and Sociology, Studentų g. 39, LT-08106 Vilnius, e-mail: egle.celiesiene@leu.lt

Dammer, Karl-Heinz, Prof. Dr., Pädagogische Hochschule Heidelberg – University of Education, Erziehungs- und Sozialwissenschaftliche Fakultät, Direktor des Instituts für Erziehungswissenschaft, Keplerstraße 87, D-69120 Heidelberg, e-mail: dammer@ph-heidelberg.de

Dudinskienė, Ligita, Karalius Mindaugas Vocational Training Center, Preschool Education Division, Karaliaus Mindaugo alėja 11, LT-44287 Kaunas, e-mail: ligita.dudinskiene@kaupa.lt

Gaižutis, Algirdas, Prof. Dr. Dr. h.c. mult., Rector, Lithuanian University of Educational Sciences, Academician of the Lithuanian Academy of Sciences, Studentų g. 39, LT-08106 Vilnius, e-mail: rekt.leu@leu.lt

Galkienė, Alvyra, Assoc. Prof. Dr., Lithuanian University of Educational Sciences, Faculty of Education, Department of Education, Studentų g. 39, LT-08106 Vilnius, e-mail: alvyra.galkiene@leu.lt

Grincevičienė, Vilija, Prof. Dr., Lithuanian University of Educational Sciences, Faculty of Social Education, Department of Social Work and Sociology, T. Ševčenkos g. 31, LT-03111 Vilnius, e-mail: vilija.grinceviciene@leu.lt

Jahn, Markus, Dipl.-Geogr., Pädagogische Hochschule Heidelberg – University of Education, Faculty of Natural, Human and Social Sciences, Department of Geography (Research Group for Earth Observation – rgeo), Czernyring 22/11-12, D-69115 Heidelberg, e-mail: jahn@ph-heidelberg.de

Klimka, Libertas, Prof. Dr., Lithuanian University of Educational Sciences, Faculty of History, Department of Baltic Prehistoric Studies, T. Ševčenkos g. 31, LT-03111 Vilnius, e-mail: libertas.klimka@leu.lt

Lange, Bernward, Dr. Dr., Prorektor für Forschung und Internationalität, Pädagogische Hochschule Heidelberg – University of Education, Erziehungs- und

Sozialwissenschaftliche Fakultät, Institut für Erziehungswissenschaft, Keplerstraße 87, D-69120 Heidelberg, e-mail: lange@ph-heidelberg.de

Lasauskienė, Jolanta, Assoc. Prof. Dr., Lithuanian University of Educational Sciences, Faculty of Education, Department of Music, Studentų g. 39, LT-08106 Vilnius, e-mail: jolanta.lasauskiene@leu.lt

Liimets, Airi, Prof. Dr., Universität Tallinn, Institut für Erziehungswissenschaften, Lehrstuhl für Pädagogische Philosophie und Soziologie, Narva mnt 25, EE-10120 Tallinn, Phone: +372-6199 760, e-mail: airi.liimets @ tlu.ee

Liimets, Reet, Assoc. Prof. Dr., Leiterin der Abteilung für Germanistik, Lehrstuhl für Fremdsprachendidaktik und Literaturanthropologie an der Universität Tartu, Tartu Ülikool, Ülikooli 18, EE-50090 Tartu, e-mail: reet.liimets @ ut.ee

Martišauskienė, Elvyda, Prof. Dr., Lithuanian University of Educational Sciences, Faculty of Education, Department of Education, Studentų g. 39, LT-08106 Vilnius, e-mail: martisauskiene@gmail.com

Mazlaveckienė, Gerda, Lecturer, Ph.D. student of Social Sciences (Education), Lithuanian University of Educational Sciences, Faculty of Philology, Department of English Didactics; Studentų g. 39, LT-08106 Vilnius, e-mail: gerda.mazlaveckiene@leu.lt

Möller, Martina, Dr., Akademische Mitarbeiterin, Pädagogische Hochschule Karlsruhe – University of Education, Institut für Erziehungswissenschaft, Abt. Allgemeine Pädagogik, Bismarckstr. 10, D-76133 Karlsruhe, e-mail: moeller@ph-karlsruhe.de

Monkevičienė, Ona, Prof. Dr., Lithuanian University of Educational Sciences, Dean of the Faculty of Education, Department of Education, Studentų g. 39, LT-08106 Vilnius, e-mail: ona.monkeviciene@leu.lt

Navickienė, Vida, Dr., Lecturer, Lithuanian University of Educational Sciences, Faculty of Education, Department of Education, Studentų g. 39, LT-08106 Vilnius, e-mail: vida.navickiene@leu.lt

Nagrockaitė, Šarūnė, Ph.D. student, Vilnius University, Faculty of Philosophy, Department of Educology, Universiteto g. 9/1, LT-01513 Vilnius, e-mail: sarune.nagrockaite@fsf.vu.lt

Pečiuliauskienė, Palmira, Prof. Dr., Lithuanian University of Educational Sciences, Faculty of Education, Department of Education, Studentų g. 39, LT-08106 Vilnius, e-mail: palmira.p@post.skynet.lt

Poteliūnienė, Sniegina, Prof. Dr., Lithuanian University of Educational Sciences, Vice-Dean of the Faculty of Sports and Health Education, Studentų g. 39, LT-08106 Vilnius, e-mail: sniegina.poteliuniene@leu.lt

Rabe, Dirk, Prof. Dr.-Ing., Hochschule Emden-Leer, Fachbereich Technik, Projektleiter des vom Bundesministerium für Bildung und Forschung geförderten Projekts BEST4HEL (01PL11058), e-mail: rabe@technik-emden.de

Račienė, Ernesta, Prof. Dr., Gediminas Technische Universität Vilnius, Lehrstuhl für deutsche Philologie und Didaktik, Leiterin des Lehrstuhls für Fremdsprachen, Professorin an der Litauische Universität für Bildungswissenschaften, Saulėtekio alėja 11, LT-10223 Vilnius 10223, e-mail: ernesta.raciene@vgtu.lt

Ragauskas, Aivas, Prof. Dr., Vice-Rector, Lithuanian University of Educational Sciences, Faculty of History, Department of Lithuanian History, Studentų g. 39, LT-08106 Vilnius, e-mail: aivas.ragauskas@leu.lt

Ragauskienė, Raimonda, Prof. Dr., Lithuanian University of Educational Sciences, Faculty of History, Department of Lithuanian History, Studentų g. 39, LT-08106 Vilnius, e-mail: raimonda.ragauskiene@leu.lt

Ralys Kęstutis, Assoc. Prof. Dr., Lithuanian University of Educational Sciences; Social Communication Institute, Studentų g. 39, LT-08106 Vilnius, e-mail: kestutis.ralys@gmail.com

Ramanauskienė, Daiva, Master of Educational Sciences, Lecturer, Panevėžys College, Faculty of Medical and Social Sciences, Department of Pedagogy, Laisvės alėja 23, LT-35200 Panevėžys, e-mail: kolegija@panko.lt

Rass, Eva, Prof. Dr., Analytische Kinder- und Jugendlichen-Psychotherapeutin, Einzel- und Gruppentherapie, kindzentrierte Familiengespräche, Säugling/Klein-kind-Mutter/Eltern-Psychotherapie, psycho-edukative Beratung, Supervision, Lehrtätigkeiten: Institut für Analytische Kinder- und Jugenlichenpsychotherapie Heidelberg, Ärztliche Akademie für Psychotherapie von Kindern und Jugendlichen e.V. München, Institut für Psychoanalyse und Analytische Psychotherapie Würzburg e.V., Hochschule Mannheim (Fakultät

Sozialwesen), Lehrbeauftragte an der Pädagogischen Hochschule Heidelberg – University of Education, Institut für Erziehungswissenschaft, Praxis als Therapeutin: Hoch-stadtstr. 36, D-74722 Buchen, e-mail: drevarass@aol.com

Rauduvaitė, Asta, Assoc. Prof. Dr., Lithuanian University of Educational Sciences, Faculty of Education, Head of the Department of Music, Studentų g. 39, LT-08106 Vilnius, e-mail: asta.rauduvaite@leu.lt

Rohlfs, Carsten, Prof. Dr., Pädagogische Hochschule Heidelberg – University of Education, Erziehungs- und Sozialwissenschaftliche Fakultät, Institut für Erziehungswissenschaft, Abt. Schulpädagogik – Methoden der Bildungsforschung, Keplerstraße 87, D-69120 Heidelberg, e-mail: crohlfs@icloud.com

Rudaitienė, Vida, Prof. Dr., Mykolas-Romeris-Universität Vilnius, Fakultät für Politik und Management, Institut für Philosophie und Humanistik, Ateities g. 20, LT-08303 Vilnius, e-mail: rudaitiene@mruni.eu

Sabaliauskas, Stanislav, Dr., Lithuanian University of Educational Sciences, Faculty of Sports and Health Education, Institute of Sports Science, Studentų g. 39, LT-08106 Vilnius, e-mail: stanislav.sabaliauskas@leu.lt

Šetkus, Benediktas, Prof. Dr., Lithuanian University of Educational Sciences, Faculty of History, Director of Centre of History Didactics, T. Ševčenkos g. 31, LT-03111, Vilnius, e-mail: benediktas.setkus@leu.lt

Schwarz, Joachim, Prof. Dr., FOM – Hochschule für Oekonomie & Management, Abt. Betriebswirtschaftslehre, Lehrgebiete Wirtschaftsmathematik und Statistik; Annaberger Straße 291a, 53175 Bonn, e-mail: jschwarz1@gmx.net

Schön, Henrike, Akademische Oberrätin, Leiterin des Akademischen Auslandsamts, Pädagogische Hochschule Heidelberg – University of Education, Keplerstraße 87, D-69120 Heidelberg, e-mail: henrike.schoen@vw.ph-heidelberg.de

Sederevičiūtė-Pačiauskienė, Živilė, Assoc. Prof. Dr., Lithuanian University of Educational Sciences, Faculty of Science and Technology, Department of Technology and Technological Education, Director of Academic Affairs, Studentų g. 39, LT-08106 Vilnius, e-mail: zivile.sedereviciute@leu.lt

Siegmund, Alexander, Prof. Dr., Pädagogische Hochschule Heidelberg – University of Education, Faculty of Natural, Human and Social Sciences, Department of Geography (Research Group for Earth Observation – rgeo), Czernyring 22/11-12, D-69115 Heidelberg, e-mail: siegmund@ph-heidelberg.de

Siemsen, Hayo, Ph.D., Ernst Mach Institut für Erkenntnistheorie; Institut für Nachrichtentechnik und Kommunikationssysteme (INK) der Hochschule Emden-Leer; Lehrbeauftragter an der Pädagogischen Hochschule Heidelberg – University of Education, Institut für Erziehungswissenschaft, Fasanensteg 18, D 66787 Wadgassen, e-mail: hayo.siemsen@gmail.com

Siemsen, Karl Hayo, Prof. Prof. h.c. Dr. phil. Dipl.-Ing., Hochschule Emden-Leer, Gründungsdekan des Fachbereichs Elektrotechnik und Informatik 1981, Constantiaplatz 4, D-26723 Emden, e-mail: siemsenKH@gmail.com

Sliwka, Anne, Prof. Dr., Universität Heidelberg, Fakultät für Verhaltens- und Empirische Kulturwissenschaften, Institut für Bildungswissenschaften, Gastprofessorin an der Pädagogischen Hochschule Heidelberg – University of Education, Akademiestr. 3, D-69117 Heidelberg, e-mail: sliwka@ibw.uni-heidelberg.de

Strakšienė, Giedrė, Dr., Klaipėda University, Herkaus Manto g. 84, LT-92294 Klaipėda, e-mail: giedre@ik.ku.lt

Tamulienė, Rasa, Ph.D. student, Vytautas Magnus University, Faculty of Social Sciences, Department of Education Sciences, Donelaičio g. 52, LT-44248 Kaunas, e-mail: rasa.tamuliene@go.kauko.lt

Targamadzė, Vilija, Prof. Dr., Lithuanian University of Educational Sciences, Faculty of Social Education, Department of Social Work and Sociology, Studentų g. 39, LT-08106 Vilnius, e-mail: vilija.targamadze@leu.lt

Tijūnėlienė, Ona, Prof. Dr., Klaipėda University, Faculty of Pedagogy, Department of Education Sciences, S. Nėries g. 5, LT-92228 Klaipėda, e-mail: onatij@gmail.com

Urnėžienė, Emilija, Assoc. Prof. Dr., Lithuanian University of Educational Sciences, Faculty of Education, Department of Education, Studentų g. 39, LT-08106 Vilnius, e-mail: emilija.urneziene@leu.lt

Vaičiulienė, Aldona, Assoc. Prof. Dr., Lithuanian University of Educational Sciences, Faculty of Education, Department of Psychology, Studentų g. 39, LT-08106 Vilnius, e-mail: aldona.vaiciuliene@leu.lt

Valantinaitė, Ilona, Assoc. Prof. Dr., Lithuanian University of Educational Sciences, Faculty of Education, Department of Education, Studentų g. 39, LT-08106 Vilnius, e-mail: ilona.valantinaite@leu.lt

Verikaitė, Daiva, Assoc. Prof. Dr., Lithuanian University of Educational Sciences, Faculty of Philology, Department of English Philology, Studentų g. 39, LT-08106, Vilnius, e-mail: daiva.verikaite@leu.lt

Vilkelienė, Aldona, Prof. Dr., Lithuanian University of Educational Sciences, Department of Arts Education, Studentų g. 39, LT-08106 Vilnius, e-mail: aldona.vilkeliene@leu.lt

Vitytė, Birutė, Assistant, Ph.D. student, Lithuanian University of Educational Sciences, Faculty of Education, Department of Arts Education, Studentų g. 39, LT-08106 Vilnius, e-mail: birute.vityte@leu.lt

Vogel, Thomas, Prof. Dr., Pädagogische Hochschule Heidelberg – University of Education, Erziehungs- und Sozialwissenschaftliche Fakultät, Institut für Erziehungswissenschaft, Abt. Schulpädagogik – Berufspädagogik, Keplerstraße 87, D-69120 Heidelberg, e-mail: t.vogel@ph-heidelberg.de

Wehr, Helmut, Dr., M.A., Akademischer Oberrat, Pädagogische Hochschule Heidelberg – University of Education, Erziehungs- und Sozialwissenschaftliche Fakultät, Institut für Erziehungswissenschaft, Lehrbeauftragter an der Pädagogischen Hochschule Ludwigsburg – University of Education, Geschäftsführer des Weltbund für Erneuerung der Erziehung – Deutschsprachige Sektion (www: wef-wee.net), Mitglied des Redaktionsteams der Zeitschrift „Lehren & Lernen", Keplerstr. 87, D-69120 Heidelberg, e-mail: dr.helmut.wehr @gmail.com

Wiebe, Joachim, Prof. Dr.-Ing., Fachbereich Technik: Nachrichtentechnik, Elektrotechnik, Institut für Nachrichtentechnik und Kommunikationssysteme (INK), Hochschule Emden-Leer, Constantiaplatz 4, D 26724 Emden, e-mail: wiebe@technik-emden.de

Wilke, Maren-Kristina, M.Ed., Akademische Mitarbeiterin, *Pädagogische Hochschule He*idelberg – University of Education, Projekt Experts in Education, Keplerstraße 87, D-69120 Heidelberg, e-mail: wilke@ph-heidelberg.de

Želvys, Rimantas, Prof. Dr., Vilnius University, Faculty of Philosophy, Department of Educology, Universiteto g. 9/1, LT-01513 Vilnius, e-mail: rimantas.

zelvys@fsf.vu.lt, Lithuanian University of Educational Sciences, Faculty of Education, Department of Education, Studentų g. 39, LT-08106 Vilnius, e-mail: rimantas.zelvys@leu.lt

Žibėnienė, Gintautė, Assoc. Prof. Dr., Mykolas Romeris University, Faculty of Social Technologies, Institute of Educational Sciences and Social Work, Ateities g. 20, LT-08303 Vilnius, e-mail: zibeniene@gmail.com

Žydžiūnaitė, Vilma, Prof. Dr., Vytautas Magnus University, Faculty of Social sciences, Department of Education, Head of the Center for Quality and Innovations, S. Daukanto g. 27-314, LT-44249 Kaunas, e-mail: v.zydziunaite@kic.vdu.lt

Baltische Studien zur Erziehungs- und Sozialwissenschaft

Reihenherausgeber:
Gerd-Bodo von Carlsburg, Algirdas Gaižutis und Airi Liimets

Band 1 Gerd-Bodo Reinert / Irena Musteikienė (Hrsg.): Litauische Gespräche zur Pädagogik. Humanismus – Demokratie – Erziehung. 1999.

Band 2 Gerd-Bodo Reinert / Irena Musteikienė (Hrsg.): Litauische Gespräche zur Pädagogik II. Staat und Schule. 2000.

Band 3 Jaan Mikk: Textbook: Research and Writing. 2000.

Band 4 Gerd-Bodo Reinert / Irena Musteikienė (Hrsg.): Erziehungswandel und moderne pädagogische Verfahren. 2001.

Band 5 Airi Liimets (Hrsg.): Integration als Problem in der Erziehungswissenschaft. 2001.

Band 6 Gerd-Bodo Reinert / Irena Musteikienė (Hrsg.): Wissenschaft – Studium – Schule auf neuen Wegen. 2002.

Band 7 Palmira Jucevičienė / Gediminas Merkys / Gerd-Bodo Reinert: Towards the Learning Society. Educational Issues. 2002.

Band 8 Gerd-Bodo Reinert / Irena Musteikienė (Hrsg.): Bildung im Zeitalter der Informationsgesellschaft. 2003.

Band 9 Dalia Ohlinger: Argumentation in der Erst- und Fremdsprache. Pragmalinguistische und grammatikalische Aspekte anhand von Argumentationen deutscher und litauischer Studierender. 2003.

Band 10 Inga Šukevičiūtė: Interkulturelle Interferenzen im Bereich Wirtschaftskommunikation. Zur Analyse von Geschäftsbriefen im Vergleich Deutschland, Litauen und Russland. 2004.

Band 11 Gerd-Bodo Reinert von Carlsburg / Irena Musteikienė (Hrsg./eds.): Innovation durch Bildung. Innovation by Education. 2004.

Band 12 Gerd-Bodo von Carlsburg / Palmira Jucevičienė / Gediminas Merkys (eds.): Learning and Development for Innovation, Networking and Cohesion. 2005.

Band 13 Gerd-Bodo von Carlsburg / Irena Musteikienė (Hrsg./eds.): Bildungsreform als Lebensreform. Educational Systems Development as Development of Human Being. 2005.

Band 14 Gerd-Bodo von Carlsburg (Hrsg./ed.): Entwicklung erziehungswissenschaftlicher Paradigmen: Theorie und Praxis. Development of Educational Paradigms: Theory and Practice. 2007.

Band 15 Gerd-Bodo von Carlsburg (Hrsg./ed.): Bildungs- und Kulturmanagement. The Management of Education and Culture. 2008.

Band 16 Jaan Mikk / Marika Veisson / Piret Luik (eds.): Reforms and Innovations in Estonian Education. 2008.

Band 17 Gerd-Bodo von Carlsburg (Hrsg./ed.): Qualität von Bildung und Kultur. Theorie und Praxis. The Quality of Education and Culture. Theoretical and Practical Dimensions. 2009.

Band 18 Denkkulturen. Selbstwerdung des Menschen. Erziehungskulturen. Festschrift für Professor Dr. Dr. h.c. Dr. h.c. Heino Liimets. Herausgegeben von Airi Liimets. 2010.

Band 19 Ellu Saar (ed.): Towards a Normal Stratification Order. Actual and Perceived Social Stratification in Post-Socialist Estonia. 2011.

Band 20 Marika Veisson / Eeva Hujala / Peter K. Smith/ Manjula Waniganayake / Eve Kikas (eds.): Global Perspectives in Early Childhood Education. Diversity, Challenges and Possibilities. 2011.

Band 21 Airi Liimets / Marit Mäesalu (eds.): Music Inside and Outside the School. 2011.

Band 22 Gerd-Bodo von Carlsburg (Hrsg./ed.): Enkulturation durch sozialen Kompetenzerwerb. Enculturation by Acquiring of Social Competences. 2011.

Band 23 Reet Liimets: Ich als raumzeitliches Konstrukt. Die Fiktionen vom Leben der estnischen und deutschen Jugendlichen. 2012.

Band 24 Raivo Vetik (ed.): Nation-Building in the Context of Post-Communist Transformation and Globalization. The Case of Estonia. 2012.

Band 25 Airi-Alina Allaste (ed.): ‚Back in the West'. Changing Lifestyles in Transforming Societies. 2013.

Band 26 Gerd-Bodo von Carlsburg (Hrsg./ed.): Bildungswissenschaft auf der Suche nach globaler Identität. Educational Sciences in Search of Global Identity. 2013.

Band 27 Airi Liimets / Marika Veisson (eds): Teachers and Youth in Educational Reality. 2014.

Band 28 Gerd-Bodo von Carlsburg / Thomas Vogel (Hrsg./eds.): Bildungswissenschaften und akademisches Selbstverständnis in einer globalisierten Welt. Education and Academic Self-Concept in the Globalized World. 2014.

www.peterlang.com